Second Edition

Sarcocystosis of Animals and Humans

Second Edition

Sarcocystosis of Animals and Humans

J. P. Dubey • R. Calero-Bernal • B. M. Rosenthal

Animal Parasitic Diseases Laboratory
Beltsville Agricultural Research Center
Agricultural Research Service
U.S. Department of Agriculture
Beltsville, Maryland

C. A. Speer

Healthwide Solutions
Vail, Colorado

R. Fayer

Environmental Microbial and Food Safety Laboratory
Beltsville Agricultural Research Center
Agricultural Research Service
U.S. Department of Agriculture
Beltsville, Maryland

CRC Press
Taylor & Francis Group
Boca Raton London New York

CRC Press is an imprint of the
Taylor & Francis Group, an **informa** business

CRC Press
Taylor & Francis Group
6000 Broken Sound Parkway NW, Suite 300
Boca Raton, FL 33487-2742

First issued in paperback 2020

ISBN 13: 978-0-367-57545-8 (pbk)
ISBN 13: 978-1-4987-1012-1 (hbk)

Library of Congress Cataloging-in-Publication Data

Dubey, J. P., author.
  [Sarcocystosis of animals and man]
  Sarcocystosis of animals and humans / J.P. Dubey, R. Calero-Bernal, B. Rosenthal, C.A. Speer, and R. Fayer. -- Second edition.
    p. ; cm.
  Preceded by Sarcocystosis of animals and man / authors, J.P. Dubey, C.A. Speer, R. Fayer. c1989.
  Includes bibliographical references and index.
  ISBN 978-1-4987-1012-1 (alk. paper)
  1. Sarcocystosis--Diagnosis. 2. Sarcocystosis in animals--Diagnosis. 3. Sarcocystis. 4. Sarcocystosis in animals. I. Calero-Bernal, R. (Rafael), 1983- , author. II. Rosenthal, B. (Benjamin), 1968- , author. III. Speer, C. A., author. IV. Fayer, R., author. V. Title.
  [DNLM: 1. Sarcocystis. 2. Sarcocystosis. SF 780.6]

QR201.S27D83 2016
614.5′53--dc23   2015009622

Visit the Taylor & Francis Web site at
http://www.taylorandfrancis.com

and the CRC Press Web site at
http://www.crcpress.com

Contents

Preface

Since the publication of the first edition of the *Sarcocystosis of Animals and Man* in 1989, considerable progress has been made in understanding the biology of the genus *Sarcocystis*. Sarcocystosis is one of the most frequent infections of warm-blooded and poikilothermic animals worldwide. Completion of the life cycle requires two host species: an intermediate (or prey) host and a definitive (or predator) host. Hosts can harbor more than one species of *Sarcocystis*. In intermediate hosts, whether normally involved in the life cycle or aberrantly involved, some species of *Sarcocystis* cause reduced weight gain, poor feed efficiency, anorexia, fever, anemia, muscle pain, and weakness, reduced milk yield, abortion, neurologic impairment, and death. Such infections are of economic importance in intermediate hosts such as cattle, sheep, goats, and pigs. Disease has also been observed in avian species. Humans occasionally have been involved as aberrant intermediate hosts. In definitive hosts, some species of *Sarcocystis* can cause digestive disturbances, including nausea, vomiting, and diarrhea. Humans serve as a host for at least two such species. The underlying mechanisms giving rise to disease in both muscular and intestinal sarcocystosis are reviewed.

Outbreaks of a mysterious disease with severe symptoms have been reported in vacationers returning from the Malaysian islands. Symptoms included fever, myalgia, headache, and cough, associated with elevated levels of enzymes indicative of muscle degeneration and inflammation. Muscle biopsies in a few patients revealed intramuscular sarcocysts. In the absence of other etiologies, these infections have been identified as muscular sarcocystosis, presumably acquired from ingestion of water and food contaminated with *Sarcocystis* sporocysts of reptile origin. The evidence linking the infectious agent to *Sarcocystis nesbitti* of nonhuman primates is herein reviewed. Methods of diagnosis and detection are discussed with regard to future needs. The status of chemoprophylaxis, chemotherapy, immunity, and vaccination is reviewed and future needs discussed.

After the publication of the first edition, the causative agent of a fatal disease of horses, equine protozoal myeloencephalitis (EPM), was found to be a new species of *Sarcocystis*—*S. neurona* was described in 1991, and its life cycle was completed in 2000. Unlike most other species of the genus, *S. neurona* was found to have a wide host range of intermediate hosts but only opossum as a definitive host. Additionally, an EPM-like fatal disease was recognized in many other hosts, especially marine mammals. Furthermore, *S. neurona* has been used as a model by biologists to study organelle synthesis, because it is easily cultivated in cell culture and because the nuclear division is prolonged. We have summarized available literature on *S. neurona* biology, clinical diagnosis, and prevention in a separate chapter in this second edition.

There has been considerable progress in molecular aspects of *Sarcocystis*, including characterization of recombinant antigens for diagnosis and the development of methodologies and markers that make possible studies on molecular epidemiology. We have added a separate chapter on the molecular biology of *Sarcocystis*.

We have attempted to preserve the format used in the first edition to add recent developments, and exclude information that has become redundant. Our objective has been to include all literature, and provide a comprehensive review of biology, clinical disease, economic losses, public health concerns, diagnosis, treatment, and prevention. We have tabulated information on all *Sarcocystis* species by host and listed species that should be considered species enquirende/invalid.

It is hoped that this book will be useful to biologists, veterinarians, physicians, and researchers.

We would like to acknowledge those who made this book possible; we feel we cannot possibly list all of them. Ana Beatriz Cassinelli, Camila K. C. Cézar, Oliver Kwok, Yuging Ying, and Francisco J. López Acevedo helped with the bibliography, Petras Prakas and Liuda Kutkiené translated Russian articles, and Yurong Yang and Junjie Hu translated Chinese papers. Mike Grigg, Mosaad Hilali, Dan Howe, David Lindsay, Gaston Moré, and Arvid Uggla, among others, made many helpful suggestions. Many scientists contributed to illustrations in the present edition and the first edition, and Shiv Kumar Verma helped with figure composites.

J. P. Dubey
Rafael Calero-Bernal
Benjamin M. Rosenthal
C. A. Speer
Ronald Fayer

Authors

J. P. Dubey, MVSc, PhD, was born in India. He earned his veterinary degree in 1960 and master's in veterinary parasitology in 1963, from India. He earned a PhD in medical microbiology in 1966 from the University of Sheffield, England. Dr. Dubey received postdoctoral training from 1968 to 1973 with Dr. J. K. Frenkel, Department of Pathology and Oncology, University of Kansas Medical Center, Kansas City. From 1973 to 1978, he was associate professor of veterinary parasitology, Department of Pathobiology, Ohio State University, Columbus, and professor of veterinary parasitology, Department of Veterinary Science, Montana State University, Bozeman, from 1978 to 1982. He is presently a senior scientist, Animal Parasitic Diseases Laboratory, Beltsville Agricultural Research Institute, Agricultural Research Service, U.S. Department of Agriculture, Beltsville, Maryland.

Dr. Dubey has spent over 50 years researching protozoa, including *Toxoplasma*, *Neospora*, *Sarcocystis*, and related cyst-forming coccidian parasites of humans and animals. He has published over 1400 research papers in international journals, more than 250 of which are on sarcocystosis. In 1985, he was chosen to be the first recipient of the "Distinguished Veterinary Parasitologist Award" by the American Association of Veterinary Parasitologists. Dr. Dubey is a recipient of the 1995 WAAVP Pfizer Award for outstanding contributions to research in veterinary parasitology. He also received the 2005 Eminent Parasitologists Award from the American Society of Parasitologists. The Thomas/Institute for Scientific Information identified him as one of the world's most cited authors in plant and animal sciences over the last decade. In 2003, he was selected for the newly created Senior Scientific Research Service (SSRS) and is one of the few scientists and executives within the USDA's Agricultural Research Service; selection for this position is by invitation only, on approval by the Secretary of Agriculture. In 2010, Dr. Dubey was elected to the U.S. National Academy of Sciences, Washington, DC and inducted in the USDA-ARS Hall of Fame.

Rafael Calero-Bernal, DVM, MSc, PhD, was born in Badajoz (Spain). He earned his degree in veterinary medicine at the University of Extremadura, Spain in 2006. One year later, he attended the Official Master's program in meat science and technology at the same institution. In 2011, Dr. Calero-Bernal earned a PhD in European framework in veterinary medicine. Since 2008, he has been a professor in the Animal Health Department at the University of Extremadura, developing teaching curricula at the Faculty of Veterinary Medicine of the University of Lisbon (Portugal). He has been a researcher in the Spanish National Microbiology Centre and the Tropical Medicine National Centre, both part of the Instituto de Salud Carlos III (Spain). Dr. Calero-Bernal developed several research protocols at Istituto Superiore di Sanità (Italy), Instituto Nacional de Saúde (Portugal), Fundação Oswaldo Cruz (Brazil), and Centro de Referencia para el Control de Endemias (Equatorial Guinea). Currently he is a postdoctoral researcher in the Agricultural Research Service, United States Department of Agriculture. He has authored more than 50 articles and 4 books related to veterinary sciences. His research interests are wildlife parasites and zoonoses, especially meat-borne pathogens and tissue cysts forming coccidia.

Benjamin M. Rosenthal is a parasitologist with primary interests in the population genetics, phylogenetics, genomics, and epidemiology of zoonotic parasites. He is a graduate of Oberlin College and the Harvard School of Public Health. Since 1999, he has led a research program in veterinary and zoonotic parasitic diseases for the USDA's Agricultural Research Service, where his studies have encompassed toxoplasmosis, sarcocystosis, trichinellosis, and related diseases. His work uses genetic variation within and among parasitic microorganisms to (1) develop improved methods for their detection and diagnosis, (2) clarify their routes of transmission and define what risk they may pose to public health, and (3) establish an accurate understanding of their evolutionary relationships and origins. He mentors secondary, undergraduate, graduate, and postdoctoral students. His course work at the University of

Maryland, College Park focuses on the ecology and evolutionary biology of infectious disease. Dr. Rosenthal developed microstatellite markers to characterize strains of *Sarcocystis neurona*, employed molecular diagnostics to differentiate among various suspected species of *Sarcocystis*, and provided the inference that *Sarcocystis nesbitti*, a newly recognized zoonotic species implicated in human tissue infections, might have its definitive host in a snake or related reptile.

C. A. Speer, PhD, earned a BS from Colorado State University in 1967 and MS and PhD degrees from Utah State University in 1969 and 1972, respectively. During 1972 and 1973, Dr. Speer was assistant professor of histology at the University of Texas Medical Center in Houston. From 1973 to 1975, he was a postdoctoral fellow of the U.S.-AID (Agency for International Development) Malaria Immunity and Vaccination Project at the University of New Mexico, Albuquerque. From 1975 to 1983, Dr. Speer was assistant and then associate professor of microbiology at the University of Montana, Missoula. From 1983 to 2000, Dr. Speer held several positions at Montana State University, Bozeman—as associate professor of veterinary science and director of the Electron Microscope Facility in the Department of Veterinary Science from 1983 to 2000 and as professor and head of Veterinary Molecular Biology from 1986 to 1994. In 1994, he founded and served as director of the Center for Bison and Wildlife Health until 2000. From 2000 to 2008, he served as Dean of the College of Agricultural Sciences and Natural Resources, director of the Agricultural Experiment Station, founder and director of the Center for Wildlife Health and the B. Ray Thompson Distinguished Professor of Cell and Molecular Immunology at the University of Tennessee, Knoxville. Currently, Dr. Speer is professor emeritus at Montana State University and the University of Tennessee and vice president of Strategic Planning for Healthwide Solutions located in Vail, Colorado.

Dr. Speer's scholarly activities have contributed to the education of numerous undergraduate and graduate students and postdoctoral fellows. He has published more than 220 research papers in refereed journals on malaria, Chagas' disease and Crohn's disease of humans and various infectious diseases of wildlife and livestock, including malaria, coccidiosis, trichomoniasis, equine protozoal myeloencephalitis (EPM), cryptosporidiosis, toxoplasmosis, whirling disease, and Johne's disease (paratuberculosis). He has also published on epitope analysis in leukocyte–endothelial cell interactions, genetic disruption of cell junctions, and expression of L-selectin by bovine T cells. He conducted research on human malaria and other protozoan diseases in the Brazilian Amazon, in India, and in Africa. He holds two international and two U.S. patents on the biochemical extraction and use of surface antigens to diagnose infectious diseases. In 1986, he received the Mershon Award for Excellence in Research from the Montana Academy of Sciences, the Honorary Alumnus Award in 1987 from the College of Natural Sciences, Colorado State University, Fort Collins and in 2002, the Gubernatorial Award for Biotechnology in Tennessee.

Ronald Fayer earned his master's and PhD degrees in 1964 and in 1968, respectively, from Utah State University. He joined the Agricultural Research Service of the U.S. Department of Agriculture in 1968 where he has worked on the life cycles, *in vitro* cultivation, pathological effects, epidemiology, immunology, and molecular biology of protist parasites of veterinary and public health importance, including *Sarcocystis*, *Cryptosporidium*, *Giardia*, *Microsporidia*, and *Blastocystis*. He has published over 400 peer-reviewed scientific articles, review chapters, and books, and shares four patents. He has served as president of the American Society of Parasitologists (ASP) and of the American Association of Veterinary Parasitologists (AAVP). For his scientific contributions he has received numerous awards, including the H.B. Ward medal from the ASP, the Distinguished Veterinary Parasitologist Award from the AAVP, USDA Superior Service Awards from two Secretaries of Agriculture, the rank of Distinguished Senior Professional conferred by the President of the United States, and other awards. He received a Fulbright Senior Scholar Award for research in Spain. He is currently working on molecular epidemiology of zoonotic protists at the Environmental Microbial and Food Safety Laboratory.

Abbreviations

ALT	alanine aminotransferase
AAT	aspartate aminotransferase
BUN	blood urea nitrogen
CPK	creatinine phosphase kinase
CNS	central nervous system
h	hours
DPI	day postinoculation
H and E	hematoxylin and eosin
ELISA	enzyme-linked immunoabsorbent assay
IgG	immunoglobulin G
IgM	immunoglobulin M
IHA	indirect hemagglutination
IHC	immunohistochemical staining
IFA	indirect fluorescent antibody
ITS	internal transcriber space
LDH	lactic dehydrogenase
KO mice	interferon gamma gene knockout mice
LM	light microscopy
min	minutes
PAS	periodic acid Schiff reaction
PCR	polymerase chain reaction
PCV	packed cell volume
PI	postinoculation
PV	parasitophorous vacuole
PVM	parasitophorous vacuolar membrane
RAPD	random amplified polymorphic DNA
RBC	red blood cells
RFLP	restriction fragment length polymorphism
SAT	*Sarcocystis neurona* direct agglutination test
SBDH	sorbitol dehydrogenase
SCID	severe combined immune deficiency syndrome
SEM	scanning electron microscopy
TEM	transmission electron microscopy
WB	Western blot

CHAPTER 1

General Biology

1.1 INTRODUCTION AND HISTORY

Sarcocystis was first reported by Miescher in 1843 as "milky white threads" in the skeletal muscle of a deer mouse caught in his house in Switzerland.[1191] Miescher did not specify the zoological name of the mouse. Subsequently, the parasite found in the house mouse (*Mus musculus*) was named *Sarcocystis muris* (Figure 1.1) and is believed to be the same as that discovered by Miescher. The life cycle of *Sarcocystis* remained unknown until 1972 (Table 1.1).

Historical accounts[803,1317,2020,2110,2111,2169,2170,2144,2167,2289] of *Sarcocystis* are given in the first edition of this book[393] and are not repeated here.

Sarcocystis species are coccidian parasites and belong to

Phylum—Apicomplexa; Levine, 1979
Class—Sporozoasida; Leuckart, 1879
Subclass—Coccidiasina; Leuckart, 1879
Order—Eucoccidiorida; Leger and Duboseq, 1910
Suborder—Eimeriorina; Leger, 1911
Family—Sarcocystidae; Poche, 1913
Subfamily—Sarcocystinae; Poche, 1913
Genus—*Sarcocystis*; Lankester, 1882
Type species—*Sarcocystis miescheriana* (Kühn, 1865) Labbe, 1899

1.1.1 Generic Diagnosis

Sarcocysts are found in muscles and in the central nervous system of homeothermic and poikilothermic animals. The life cycle includes: obligatorily 2 hosts, asexual multiplication in the intermediate host with "sarcocysts" in muscles and CNS, gamonts, fertilization, and endogenous sporulation of oocysts in the intestine of definitive hosts. Oocysts are not infective for the definitive host.

There is considerable debate concerning the number of species in the genus *Sarcocystis*. In the last review of the taxonomy of the genus, Odening[1317] listed 189 species; many of these are probably not valid. In the last chapter of this book (Chapter 24), we have summarized the current status. It appears that there are 196 valid *Sarcocystis* species; however, full life cycles are known only for 26 of them.

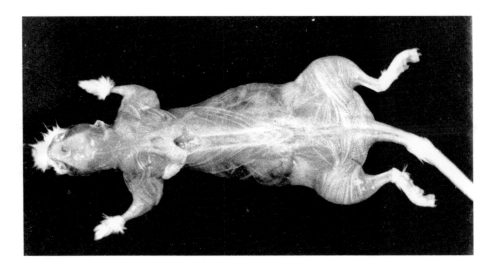

Figure 1.1 Macroscopic, criss-crossing, thread-like *S. muris* sarcocysts in skeletal muscles of a mouse (*Mus musculus*). (From Springer Science+Business Media: *Cardiovascular and Musculoskeletal Systems,* Sarcocystosis of the skeletal and cardiac muscle, mouse, 1991, pp. 165–169, Dubey, J.P.)

Table 1.1   Historical Landmarks Concerning Genus *Sarcocystis*

Year	Findings	References
1843	Sarcocysts found in the muscles of a house mouse	1191
1882	Genus *Sarcocystis* named	1022
1972	Sexual phase cultured *in vitro*	571
1972	Two-host life cycle found	1478, 1479
1973	Vascular phase recognized and pathogenicity demonstrated	572
1975	Multiple *Sarcocystis* species within a given host recognized	803
1975	Chemotherapy demonstrated	576
1976	Abortion due to sarcocystosis recognized	577
1981	Protective immunity demonstrated	358
1986	Vascular phase cultured *in vitro*	1655
1989	*Sarcocystis* species classified based on morphology of cyst wall	393
1991	Multiple host *Sarcocystis* species, *S. neurona* discovered as the etiological agent of Equine Protozoan Myeloencephalitis (EPM)	404
2015	First *Sarcocystis* genome annotated	109

1.2 STRUCTURE AND LIFE CYCLE

1.2.1 Structure

Sarcocysts (in Greek, Sarkos = flesh, kystis = bladder) are the terminal asexual stage found encysted, primarily in striated muscles of mammals, birds, and poikilothermic animals (intermediate hosts). The number and distribution of sarcocysts throughout the body vary greatly from host to host. Factors affecting the number and distribution of sarcocysts include the number of organisms (sporocysts) ingested, the species of *Sarcocystis*, the species of host, and the immunological state of the host. Although most sarcocysts develop in striated muscles of the heart, tongue, esophagus, diaphragm, and skeletal muscles, some sarcocysts have been found in the smooth muscles of the intestine. Sarcocysts of *S. mucosa* have been found in the gut of marsupials, apparently in smooth muscles.[108,1293] Infrequently, immature sarcocysts of some species (*S. cruzi*, *S. tenella*) also have been found in the muscularis layer of the gut.[360,367] Sarcocysts also have been found in the CNS and in the Purkinje fibers of the heart and muscle bundles, but always in relatively low numbers. They are found in different types of myofibers.[1412]

Sarcocysts vary in size and shape, depending on the species of the parasite (Figures 1.1 through 1.5). Some always remain microscopic (e.g., *S. cruzi*), whereas others become macroscopic (e.g., *S. gigantea*, *S. muris*). Microscopic sarcocysts vary from very long and narrow to short and wide. Macroscopic sarcocysts, which are nearly always in skeletal muscles or esophageal muscles, appear filamentous (e.g., *S. muris*), like rice grains (e.g., *S. rileyi*), fusiform (e.g., *S. fusiformis*) or globular (e.g., *S. gigantea*).

Sarcocysts are always located within a PV in the host-cell cytoplasm, often not visible under the LM (Figure 1.6). More than 1 sarcocyst may be found in 1 host cell. The sarcocyst consists of a cyst wall that surrounds the asexual metrocyte or zoite stages (Figures 1.7 and 1.8). The structure and thickness of the cyst wall varies among species of *Sarcocystis*, and within each species, as the sarcocyst matures. A connective tissue wall (secondary sarcocyst wall) surrounds the *S. gigantea*, *S. mucosa*, *S. hardangeri*, and *S. rangiferi* sarcocysts[691,692,893,1141,1164] (Figure 1.9). Histologically, the sarcocyst wall may be smooth, striated, or hirsute, or may possess complex branched protrusions (Figure 1.10). Internally, groups of zoites may be divided into compartments by septa that originate from the sarcocyst wall or may not be compartmentalized. When present, septa are usually less than 2 μm thick, but sometimes they are indistinct or not found (Figure 1.11).[392,935] The structure of the parasites within the sarcocysts varies with the maturation of the sarcocyst. Immature sarcocysts contain globular parasites called metrocytes (mother cells). Each metrocyte produces 2 progeny by an internal form of multiplication called endodyogeny (described in Section 1.3.1.1.1). After what appears to be several such generations, some of the metrocytes, through the process of

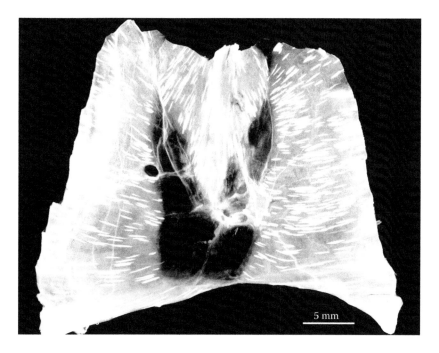

5 mm

Figure 1.2 Macroscopic *S. campestris* sarcocysts in the diaphragm of a Richardson's ground squirrel (*Spermophilus richardsonii*). (From Cawthorn, R.J., Wobeser, G.A., Gajadhar, A.A., 1983. *Can. J. Zool.* 61, 370–377. With permission.)

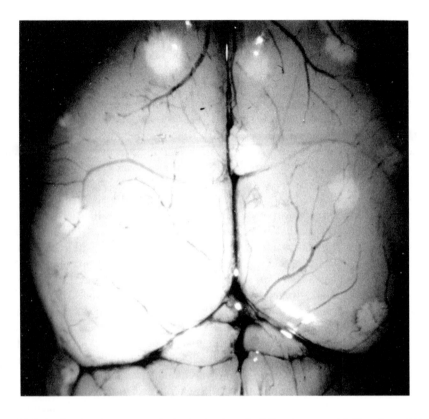

Figure 1.3   Globular *S. microti* sarcocysts in the brain of a vole (*Microtus agrestis*). (Courtesy of M. Rommel.)

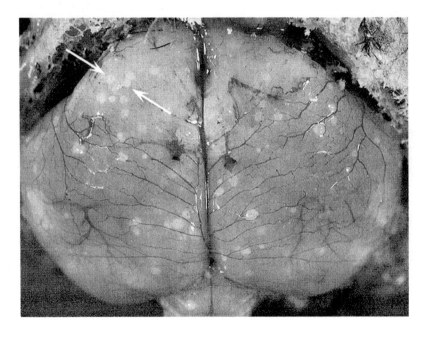

Figure 1.4   Globular *S. ochotonae* sarcocysts in the brain of a pika (*Ochotona* sp.). (From Odening, K. et al., 1998. *Zool. Garten* 68, 80–94. With permission.)

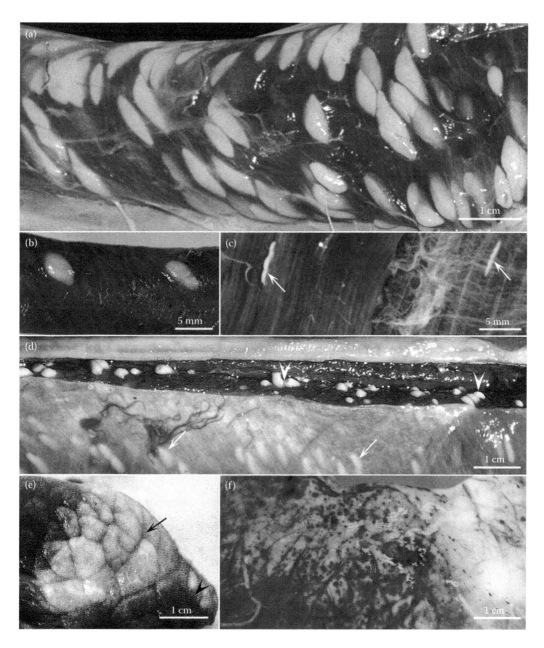

Figure 1.5 *Sarcocystis*-associated gross lesions in animals. Unstained. (a) Sarcocysts of *S. fusiformis* on the surface of the esophagus of a naturally infected water buffalo in Vietnam. (Courtesy of A. Uggla and L.T.T. Huong.) (b) Sarcocysts of *S. gigantea* on the surface of the esophagus of a naturally infected sheep in Sweden. (Courtesy of A. Uggla.) (c) Sarcocysts (arrows) of *S. hirsuta* in the flank muscle of a naturally infected cow in the United States. (Adapted from Dubey, J.P. et al. 1990. *J. Am. Vet. Med. Assoc.* 196, 1095–1096.) (d) Sarcocysts of *S. rileyi* in the pectoral muscle of a naturally infected mallard (*Anas platyrhynchos*) in Sweden. The sternum is on the top. Arrowheads point to sarcocysts in cut muscle. Arrows point to sarcocysts visible through the connective tissue. (Courtesy of A. Uggla.) (e) Eosinophilic myositis in the skeletal muscle of a naturally infected cattle in Belgium. Arrow points to a large greenish area and arrowhead points to smaller lesions. (Courtesy of the Laboratory of Veterinary Pathology, Ghent University, Belgium.) (f) Hemorrhages in the epicardium of the heart of a calf experimentally infected with *S. cruzi*. (From Dubey, J.P., Speer, C.A., Epling, G.P., 1982. *Am. J. Vet. Res.* 43, 2147–2164. With permission.)

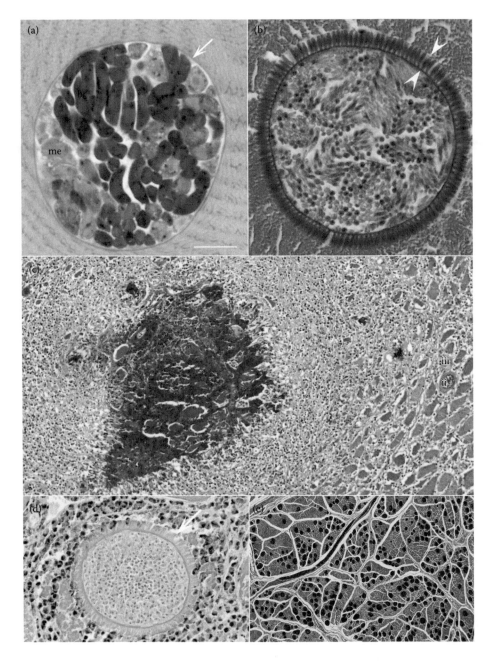

Figure 1.6 Sarcocysts and associated lesions. (a), toluidine blue stain. (b–e), H and E stain. Bar applies to all parts; bar in (a) and (b) = 10 μm, in (c) = 200 μm, in (d) = 25 μm, and in (e) = 350 μm. (a) *S. muris* in an experimentally infected *M. musculus* muscle. Note the thin sarcocyst wall (arrow), faintly stained metrocytes (me), and intensely stained bradyzoites (br). (b) *Sarcocystis* sp. in the skeletal muscle of a naturally infected turkey. (Adapted from Teglas, M.B. et al., 1998. *J. Parasitol.* 84, 661–663). (c) Eosinophilic myositis in a naturally infected cattle. Note an area of central necrosis (i), a peripheral rim of inflammatory cells (ii), and an intact sarcocyst (iii). (d) A degenerating *S. hominis*-like sarcocyst in an area of eosinophilic myositis. The bradyzoites have degenerated but the sarcocyst wall is mostly intact. (Adapted from Wouda, W., Snoep, J.J., Dubey, J.P., 2006. *J. Comp. Pathol.* 135, 249–253.) (e) Numerous *S. cruzi* sarcocysts in the tongue of an experimentally infected asymptomatic calf. (From Dubey, J.P., Speer, C.A., Epling, G.P., 1982. *Am. J. Vet. Res.* 43, 2147–2164. With permission.)

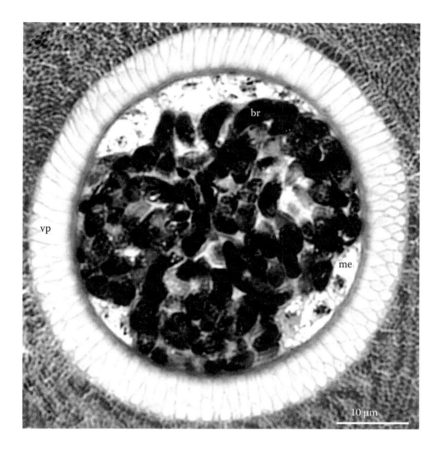

Figure 1.7 Cross section of a sarcocyst of *S. hominis* in the skeletal muscle of a cow. Villar protrusions (vp) on the sarcocyst wall enclose intensely stained bradyzoites (br) and pale staining metrocytes (me) arranged peripherally. 1 µm section, toluidine blue stain. (Adapted from Wouda, W., Snoep, J.J., Dubey, J.P., 2006. *J. Comp. Pathol.* 135, 249–253.)

endodyogeny, produce banana-shaped zoites called bradyzoites (also called cystozoites). Within the sarcocysts, metrocytes are generally located in the cortex, whereas bradyzoites are located in the medulla. In old, large sarcocysts, bradyzoites near the center of the sarcocyst are sometimes degenerate and are replaced by granules or globules. Live sarcocysts are probably fluid filled with metrocytes and bradyzoites, moving within the fluid. In histologic sections stained with H and E, metrocytes are paler than bradyzoites. The bradyzoites contain prominent amylopectin granules that stain bright red with PAS. In smears stained with Giemsa, metrocytes are oval to round in shape, and pink/magenta in color (see Figure 1.18).

1.2.2 Life Cycle

Sarcocystis has an obligatory prey–predator 2-host life cycle (Figure 1.12). Asexual stages develop only in the intermediate host, which, in nature, is often a prey animal. Sexual stages develop only in the definitive host, which is carnivorous. Intermediate and definitive hosts vary for each species of *Sarcocystis*. For example, there are 4 named species of *Sarcocystis* in cattle: *S. cruzi, S. hirsuta, S. hominis,* and *S. rommeli.* The definitive hosts for these species are canids, felids, primates, and unknown, respectively. Some species of *Sarcocystis* can complete the life cycle in the same host species, but not in the same animal[66] (see Chapter 22).

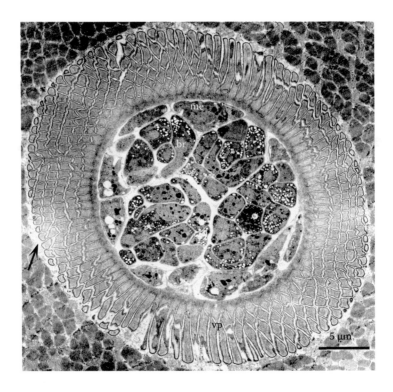

Figure 1.8 TEM cross section of a *S. hemioni* sarcocyst in the skeletal muscles of a mule deer. The host-cell cytoplasm is separated from the sarcocyst by a thin parasitophorous vacuole (arrow). Note villar protrusions (pv), metrocytes (me), and bradyzoites (br).

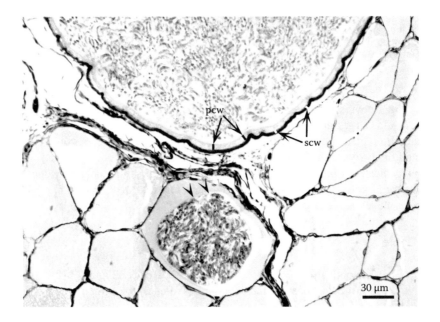

Figure 1.9 Sarcocysts of *S. gigantea* (top) and *S. arieticanis* (bottom) in the skeletal muscle of a sheep. The primary cyst wall (pcw) in *S. gigantea* is thick and surrounded by a connective tissue capsule (secondary sarcocyst wall, scw), while *S. arieticanis* (arrowheads) has no secondary sarcocyst wall. Gomori's methanine silver reaction, a connective tissue stain.

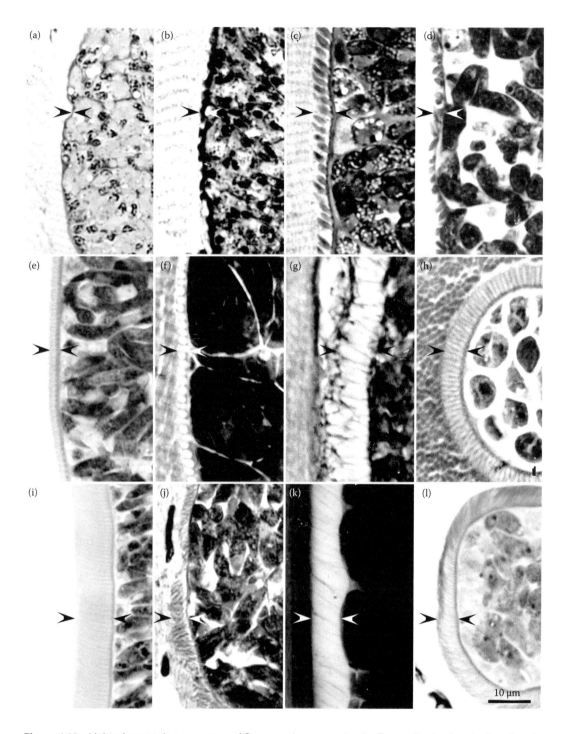

Figure 1.10 Light microscopic appearance of *Sarcocystis* sarcocyst walls. Bar applies to all parts. Arrowheads indicate the villar protrusions (vp) or lack of it on cyst walls. (a and b) Thin-walled *S. idahoensis*. The wall is smooth in (a) and irregular in (b). (c) *S. campestris* with vp. (d) *S. microti* with small vp of uneven length. (e) *S. odocoileocanis* with small stubby vp. (f) *S. hemionilatrantis* with inverted T-like vp. (g) *S. youngi* with vp of uneven width. (h and i) *S. hemioni* with thick wall with long vp. (j) *S. peromysci* with fine vp. (k) *S. arieticanis* with long, hirsute vp. (l) *S. sybillensis* with thick tufts of fine vp.

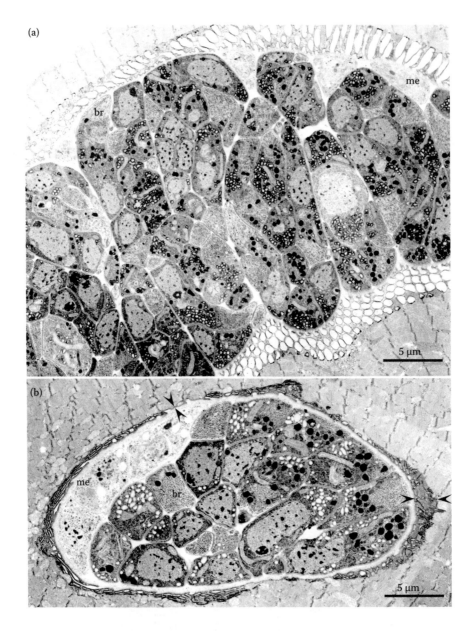

Figure 1.11   TEM of 2 mature sarcocysts in sheep at 103 DPI. Both sarcocysts contain numerous bradyzoites
(br) and a few metrocytes (me). (a) *S. tenella*. The sarcocyst wall is uniformly thick. The bradyzoites are arranged in packets and separated by septa. (b) *S. arieticanis*. Septa were not found. The
sarcocyst wall is thin at some places and thick at others (arrowheads).

For the following description of the life cycle and structure of *Sarcocystis*, *S. cruzi* will serve as
the example. Examples from other species are discussed when information on *S. cruzi* is deficient
or at variance.

Dogs, coyotes, red foxes, and possibly, wolves, jackals, and raccoons, are the definitive hosts,
whereas bison (*Bison bison*) and cattle (*Bos taurus*) are the intermediate hosts for *S. cruzi*.[361,586]
The definitive host becomes infected by ingesting muscular or neural tissue containing mature sarcocysts. Bradyzoites are liberated from the sarcocyst by digestion in the stomach and intestine.

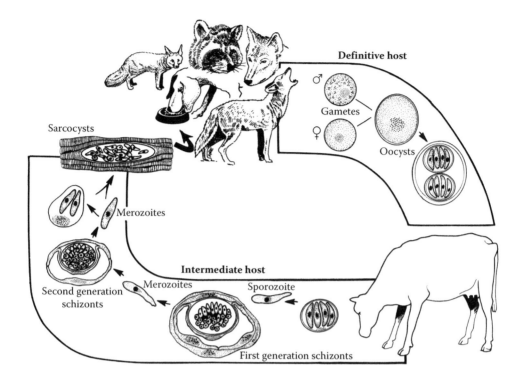

Figure 1.12 Life cycle of *S. cruzi*.

Bradyzoites move actively, penetrate the mucosa of the small intestine, and transform into male (micro) and female (macro) gamonts. Within 6 h of ingesting infected tissue, gamonts were found within a PV in goblet cells near the tips of villi or the lamina propria (Figure 1.13a). In certain species of *Sarcocystis* (e.g., *S. idahoensis*), gametogony is delayed for several days after ingestion of sarcocysts.[111,112] The ratio of macrogamonts to microgamonts is approximately 95:5. Macrogamonts are ovoid to round, 10–20 μm in diameter, and contain a single nucleus with compact chromatin. Microgamonts are ovoid to elongated, and contain 1 to several nuclei. The microgamont vesicular nucleus divides into several nuclei (usually up to 15), and, as the microgamont matures, the nuclei move towards the periphery of the gamont (Figure 1.13b–e). In *S. cruzi*, mature microgamonts, which are 7 × 5 μm, contain 3–11 slender gametes (Figure 1.13e). The microgametes, which are 4 × 0.5 μm in size, consist of a compact nucleus and 2 flagella. Microgametes liberated (Figure 1.13f) from the microgamont actively move to the periphery of the macrogamont, where their membranes fuse and only the nucleus of the microgamont passes into the macrogamont. After fertilization,[1598] a wall develops around the zygote and the oocyst is formed. The entire process of gametogony and fertilization can be completed within 24 h, but it is asynchronous. Thus, gamonts and oocysts can be found at the same time. The location of gametogony, and the type of cell parasitized, varies with species of *Sarcocystis* and stage of gametogenesis. For example, gamonts of *S. muris*, *S. hirsuta*, and *S. cruzi* initially develop in goblet cells at or near the surface of the intestine, whereas those of *S. idahoensis* develop in enterocytes next to the basement membrane adjoining the lamina propria. Occasionally, oocysts have been found in the mesenteric lymph nodes.[125,1604]

It is assumed that the infected enterocyte or goblet cell moves to the lamina propria, where sporulation occurs (Figures 1.14 and 1.15). Initially, the sporont was granular, eosinophilic, and filled the young oocysts. During the initial stages of sporont condensation, a clear cap-like structure was found in *S. cruzi*[360] and *S. idahoensis*.[111] When the sporont condensed further, a similar clear

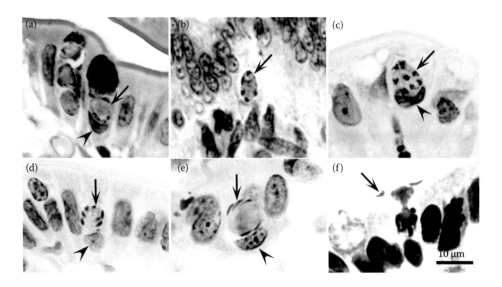

Figure 1.13   Gamonts of *S. hirsuta* (a, c, d, and f) and *S. cruzi* (b and e) in the small intestines of cats (*S. hir-suta*) and coyotes (*S. cruzi*), respectively. Each arrowhead points to a host-cell nucleus, and arrows point to parasites. (a) Four oval early gamonts in 2 adjacent goblet cells. (b) Microgamont (arrow) with 5 nuclei. (c) Microgamont with 8 nuclei. (d) Microgamont (arrow) with 7 peripheral nuclei. (e) Mature microgamont with gametes. (f) Ruptured microgamont liberating microgametes (arrow). (From Dubey, J.P., 1983. *Parasitology* 86, 7–9; Dubey, J.P., 1982. *J. Protozool.* 29, 591–601. With permission.)

area appeared at the opposite end. The oocyst contained one large nucleus with 1 or 2 prominent nucleoli, and several PAS-positive granules. As sporulation progressed, the nucleus elongated and became parallel to the longitudinal axis of the sporont.[111,174,1648] The elongated nucleus divided into 2 nuclei, 1 at each pole of the sporont. These nuclei underwent a second transverse division, and the sporont cytoplasm divided transversely into 2 sporoblasts (Figure 1.14). Each sporoblast contained 2 nuclei. The nuclei moved towards the opposite poles of each sporoblast, and the sporoblasts became surrounded by an eosinophilic wall; sporoblasts became sporocysts. Four sporozoites formed in each sporocyst, apparently by a third nuclear division. Because sporulation is asynchronous, unsporulated and sporulated oocysts are found simultaneously (Figure 1.14).

Sporulated oocysts are generally colorless, thin-walled (<1 μm), and contain 2 elongated sporocysts. An oocyst residuum and a micropyle are absent. Each sporocyst contains 4 slender sporozoites and a granular sporocyst residuum, which may be compact or dispersed, but a Stieda body is absent. Because sporozoites are flexed in the sporocyst, all 4 are often not seen in a single plane of focus (Figures 1.14 and 1.15). Each sporozoite has a central to terminal nucleus, and several cytoplasmic granules, but there is no refractile body.

The oocyst wall is thin and often ruptures (Figure 1.16). Free sporocysts, released into the intestinal lumen, are passed in feces (Figure 1.16). Occasionally, unsporulated and partially sporulated oocysts are excreted in feces. The prepatent and patent periods vary, but for most *Sarcocystis* species, oocysts are first excreted in the feces between 7 and 14 days after ingesting sarcocysts.

The intermediate host becomes infected by ingesting sporocysts in food or water. Sporozoites excyst from sporocysts in the small intestine. The fate of the sporozoite, from the time of ingestion of the sporocyst until initial development in mesenteric lymph node arteries, is not known. Sporozoites of *S. cruzi* were first found in the lumen and in the endothelium of arteries 4–7 DPI. At such time, free zoites have been seen in arteries in mesenteric lymph nodes. First-generation schizogony begins in endothelial cells as early as 7 DPI and may be completed as early as 15 DPI (Figure 1.17). Second-generation schizonts have been seen in endothelium from 19 to 46 DPI, predominantly in capillaries,

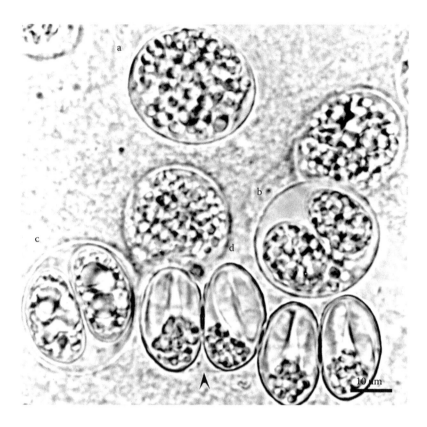

Figure 1.14   Sporulating *S. cruzi* oocysts in the lamina propria of the small intestine of an experimentally infected coyote (a) unsporulated oocyst, (b) oocyst with 2 sporoblasts, (c) oocysts with 2 sporocysts, (d) sporulated oocyst with 2 sporocysts and a thin oocyst wall (arrowhead). Unstained.

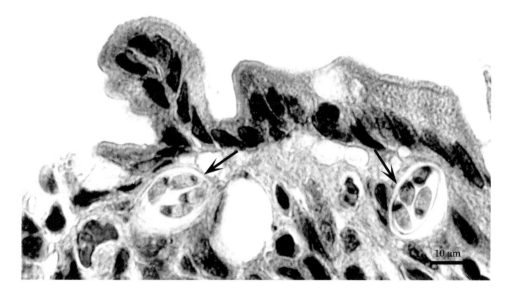

Figure 1.15   Two *S. cruzi* sporocysts (arrows), each with 4 sporozoites in the lamina propria of the small intestine of a dog. H and E stain. (From Dubey, J.P., 1976. *J. Am. Vet. Med. Assoc.* 169, 1061–1078. With permission.)

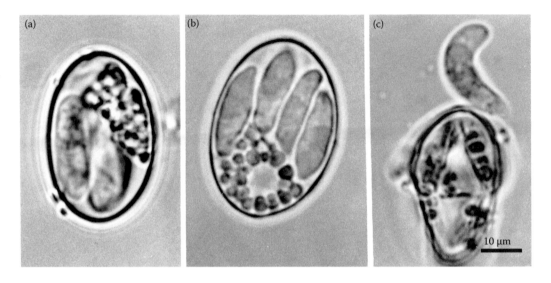

Figure 1.16 *In vitro* excystation of sporozoites of *S. cruzi* in trypsin-bile excystation fluid. (a) Untreated sporocyst. (b) Treated sporocyst with 4 sporozoites and sporocyst residuum. (c) One excysted sporozoite and collapsed sporocyst wall. (Adapted from Fayer, R., Leek, R.G., 1973. *Proc. Helminthol. Soc. Wash.* 40, 294–296.)

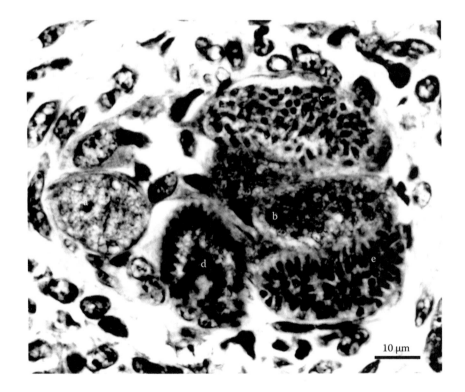

Figure 1.17 Development of first-generation schizonts of *S. cruzi* in a mesenteric lymph-node artery of an experimentally infected calf. Progressive developing stages are (a) lobulated nucleus, (b) undifferentiated nuclei, (c) differentiated nuclei, (d) developing merozoites, and (e) mature with merozoites. Iron hematoxylin plus Giemsa stain. (From Dubey, J.P., 1982. *J. Protozool.* 29, 591–601. With permission.)

but also in small arteries, virtually throughout the body. These schizonts were most numerous in the glomeruli of the kidney (Figure 1.18). Immature schizonts stained with H and E are basophilic and stain lighter than the host tissue. The nucleoplasm of the schizont is diffuse and difficult to recognize in 5-µm thick-histological sections, but it appears clear in 1- to 3-µm sections (Figure 1.19a). The nucleus becomes lobulated (Figure 1.19b) and divides into several lobes. Merozoites form at the

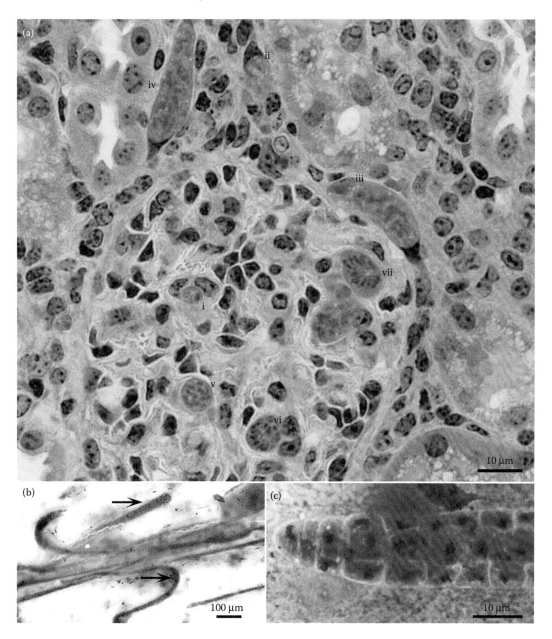

Figure 1.18 *Sarcocystis* schizonts and sarcocysts. (a) Second-generation schizonts within and outside the renal glomerulus of a calf experimentally infected with *S. cruzi*, in presumed order of development from uninuclate schizont (i) to mature (vii). Two schizonts (ii, iv) are located in capillaries in renal parenchyma. H and E stain. (b) Immature thread-like *S. speeri* sarcocysts (arrows) in a skeletal muscle smear of an experimentally infected KO mouse. 22 DPI. Giemsa stain. (c) Higher magnification of a terminal end of a sarcocyst in Figure 1.18b. Note rows of magenta-colored metrocytes.

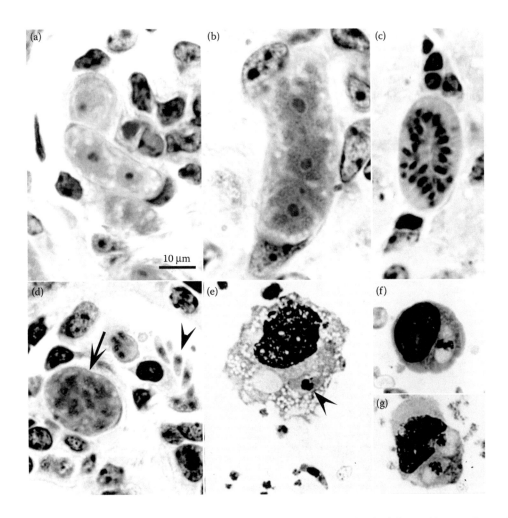

Figure 1.19 Second-generation schizonts (a–d) and intravascular zoites (e–g) of *S. cruzi* in experimentally infected calves. (a) Immature schizont with 1 or 2 nuclei. (b) Lobed nucleus with 3 prominent nucleoli. (c) Peripherally arranged merozoites. (d) Individual merozoites (arrowhead) and a multinucleated schizont (arrow). (e–g) Merozoites in peripheral blood. (e) Intracellular merozoite (arrowhead) in a monocyte and an extracellular merozoite. (f) Binucleated merozoite in a mononuclear cell. (g) Mononuclear cell with 2 merozoites. (From Dubey, J.P., 1982. *J. Protozool.* 29, 591–601. With permission.)

periphery (Figure 1.19c). The shape and size of schizonts vary considerably. Schizonts observed in skeletal muscle are longer than those seen in other tissues. Both first- and second-generation schizonts are located within the host cytoplasm, and are not surrounded by a PV.

Merozoites have been found in peripheral blood smears 24–46 DPI coincident with the maturation of second-generation schizonts (Figure 1.19). Merozoites in blood are extracellular or located within unidentified mononuclear cells (Figure 1.19e). Intracellular merozoites contain 1 or 2 nuclei, and some divide to form 2 merozoites, apparently by endodyogeny[582] (Figure 1.19f and g). Extracellular merozoites often appear degenerate. Division in blood cells has not been seen in some species of *Sarcocystis* (e.g., *S. hirsuta*). Individual merozoites have been seen in macrophage-like cells in tissues of animals infected with some species in cattle, deer, pig, and sheep.[75,362,379,1296]

The number of generations of schizogony and the type of host cell may vary with each species of *Sarcocystis*, but trends are apparent. For example, all species of *Sarcocystis* of large domestic animals (sheep, goat, cattle, pigs) form first- and second-generation schizonts in vascular endothelium, whereas only a single precystic generation of schizogony, has been found in the *Sarcocystis* species of small mammals. The species found in mice generally develop in hepatocytes, whereas *Sarcocystis* species in rats develop both in the endothelial cells and in the connective tissue cells of the spleen, kidneys and lungs.[1363] Schizonts of *S. neurona* in the horse are found only in the neural cells (see Chapter 3). *S. hemionilatrantis* represents a variation, with the first 2 generations of schizogony in vascular endothelium and an additional third or fourth generation in macrophages in muscles (Figure 1.20). Still other variations are seen in cattle, in which the second-generation schizont of *S. cruzi* develops in virtually every organ, whereas those of *S. hirsuta* are restricted to the muscles.

Sarcocystis schizonts divide by a special form of endopolygeny (Section 1.3.1.1.1) (Figures 1.21 through 1.23). Usually, schizonts are time-limited stages that disappear before sarcocysts form. However, schizonts of *S. falcatula* develop in blood vessels of birds (budgerigar) from day 2 to 5.5 months PI,[1626] with 2 distinct peaks of schizogony at 7–8 DPI and 28 DPI. All schizonts are structurally similar. The site of schizogony shifts progressively from capillaries to venules to veins.[1626] Merozoites liberated from the terminal generation of schizogony initiate the sarcocyst formation.

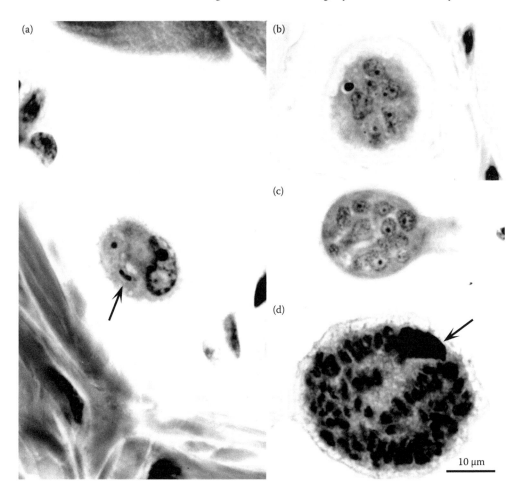

Figure 1.20 Third-generation *S. hemionilatrantis* schizonts in macrophages in histological sections of the skeletal muscles of a mule deer. H and E stain. (a) Binucleate schizont with 1 dividing nucleus (arrow). Myocytes are on the top right and connective tissue cells are at the bottom left, and the infected macrophage is in the edematous perimysial space. (b and c) Multinucleated schizonts. (d) Nearly mature schizont. The host-cell nucleus (arrow) is indented. (From Dubey, J.P., Kistner, T.P., Callis, G., 1983. *Can. J. Zool.* 61, 2904–2912. With permission.)

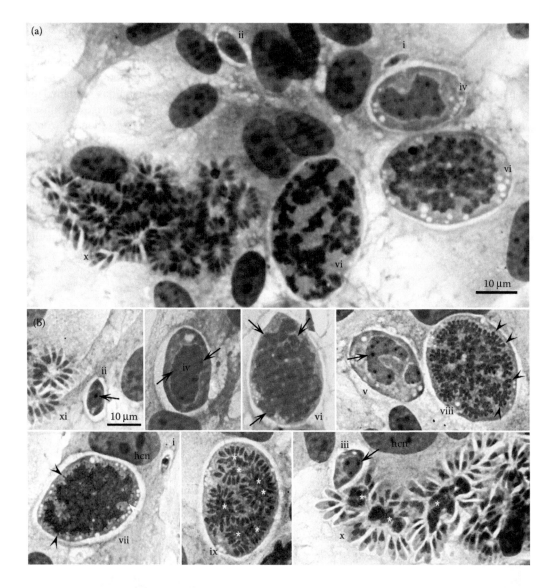

Figure 1.21   Development of *S. falcatula* schizonts in bovine turbinates cells. Illustrations are from 1 slide, 5
days after seeding with merozoites. Giemsa stain (Slide courtesy of D. S. Lindsay). hcn = host-
cell nucleus. There is considerable variability in staining of different stages of the parasite (a)
Different developmental stages in one field. (b) Stages selected from different fields to show
details. Stages (i–xi), in presumed order of development. Intracellular merozoite (i) with a single nucleus, earliest schizont (ii) with a prominent nucleus and 1 nucleolus (arrow), an immature schizont with 3 nucleoli (iii), and a schizont with 7 nucleoli (iv). Thereafter, the nucleus has
become lobulated (v, vi). In (vii, viii) the ends of merozoites are forming. In (ix) the merozoites are
arranged in different groups. In (x) the merozoites have formed around residual bodies (*). Note
different shapes and sizes of merozoites, some are more slender than others. Mature schizonts
without residual bodies are shown in (xi).

The intracellular merozoite surrounded by a PV becomes round to ovoid (metrocyte) (Figure 1.24).
After repeated divisions, the sarcocyst is filled with bradyzoites, the presence of which is the infective stage for the predator. Sarcocysts generally become infectious around 75 DPI, but there is
considerable variation among species of *Sarcocystis*. Immature sarcocysts and schizonts are not
infectious for the definitive host.

Figure 1.22 Immunofluorescence of *S. neurona* schizonts in BT cells. Note lobulated nucleus in (i), 24 h, merozoite formation in (ii), 48 h, and a mature schizont with merozoites in (iii), 72 h post infection. (a) Transgenic clone of *S. neurona* expressing yellow fluorescent protein (YFP). Differential interference contrast image with epifluorescence image overlay, showing a multilobed schizont and a mature schizont of a clone of *S. neurona* that stably expresses YFP. The host cell and parasite nuclei were stained with 4′,6-diamino-2-phenylindole (DAPI) (blue). (b) Genesis of apicoplast during schizogony of *S. neurona*. The apicoplast is stained green, micronemes are red, and nucleus/ nuclei are blue. The apicoplast follows the nucleus and divides/fragments during the nuclear division. Micronemes (stained red) are absent in the stage expressed in (i). The nucleus is stained blue. (From Vaishnava, S. et al., 2005. *J. Cell Sci.* 118, 3397–3407. With permission.)

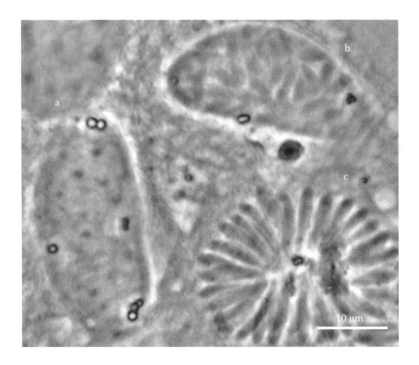

Figure 1.23   Same image as in Figure 1.22b. Phase contrast. (From Vaishnava, et al. 2005. *J. Cell Sci*. 118, 3397–3407. With permission.)

Sarcocysts may persist for the life of the host, but many begin to disappear after 3 months (Figure 1.25).

Rupture of sarcocysts may initiate host response, but the released bradyzoites do not initiate new development. In other words, there is no reactivation of chronic infection, irrespective of the immune status of the host. Experimentally, merozoites (from cell cultures) of certain species (e.g., *S. neurona*) are infectious to immunodeficient mice, but bradyzoites do not infect mice.[470]

1.3 ULTRASTRUCTURE

Sarcocystis spp. are single-cell, eukaryotic organisms that contain a nucleus, nucleolus, endoplasmic reticulum, ribosomes, a plastid, a Golgi complex, and a mitochondrion. They also have organelles that are characteristic of the phylum Apicomplexa, such as apical rings (also called conoidal or preconoidal rings), polar rings, a conoid (which may be absent in some apicomplexans, i.e., *Plasmodium*), a pellicle, subpellicular microtubules, micropores, rhoptries, and micronemes.

There are numerous reports on the ultrastructure of *Sarcocystis* spp. sarcocysts (see References **5, 6, 8–10, 23, 39, 82, 91, 96, 97, 99, 126, 127, 137, 172, 173, 190, 201, 220, 222, 224–226, 240, 244, 245, 253, 273, 289, 326, 327, 360, 367–370, 378, 380, 383, 385, 393, 394, 396, 446, 452, 453, 457, 461, 466, 467, 474, 478, 479, 525, 530, 533, 535, 536, 538, 540, 542, 543, 636, 673, 677, 691– 696, 698, 707, 708, 787, 797, 802, 819, 824, 833, 862, 863, 865, 895, 926, 936, 938, 961, 963, 964, 1112, 1132, 1135, 1137, 1142, 1164, 1167–1176, 1178, 1227, 1273, 1274, 1290–1293, 1296, 1299, 1303, 1309, 1310, 1313, 1316, 1318, 1326, 1348, 1367, 1369, 1404, 1406–1408, 1504, 1518, 1519, 1552, 1561, 1564, 1565, 1567–1569, 1576, 1583–1585, 1597, 1609, 1610, 1633, 1651, 1654, 1681, 1684, 1708, 1752, 1793, 1794, 1796, 1803, 1825, 1845, 1867, 1868, 1871, 1875, 2227**), schizonts (see References **39, 75, 184, 356, 359, 366, 436, 534, 708, 808, 1142, 1178, 1306, 1329, 1347, 1363,**

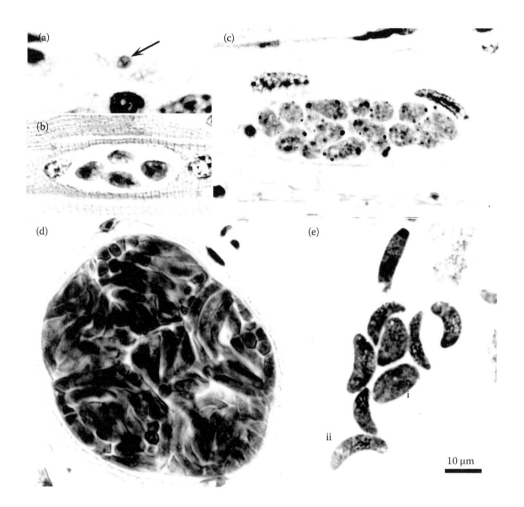

Figure 1.24   *S. cruzi* sarcocysts in myocytes of experimentally infected calves. (a–d), histological sections. (a and d), H and E stain, (c), PAS reaction, G, smear. Giemsa stain. (a) Merozoite in a parasitophorous vacuole (arrow), probably a unizoite sarcocyst. (b) Sarcocyst with 4 metrocytes. (c) Sarcocyst with numerous metrocytes. (d) Mature sarcocyst with banana-shaped bradyzoites. (e) Bradyzoites (i) and metrocyte (ii) with 2 organisms. Giemsa stain. (From Dubey, J.P., 1982. *J. Protozool.* 29, 591–601. With permission.)

1368, 1561, 1585, 1620, 1626, 1627, 1649, 1650, 1652, 1660, 1752, 1798), and gamonts, oocysts, and sporozoites (see References **86, 531, 537, 539–541, 543, 887, 888, 1018, 1142, 1165, 1177, 1178, 1364–1366, 1566, 1598, 1648, 1790, 1868**). We have summarized available information.

1.3.1 Sarcocysts

The ultrastructure section begins with sarcocysts because of the importance of this stage in transmission, and as a taxonomic aid. Each *Sarcocystis* species produces a sarcocyst wall that often has unique ultrastructural characteristics that can be used to distinguish it from other species within the same intermediate host.

Sarcocyst development begins when a merozoite enters a myocyte (Figure 1.26) or neural cell. The merozoite resides in a PV and is surrounded by a PVM which appears to develop immediately into a primary sarcocyst wall (PSW). The PSW consists of a PVM plus an underlying electron-dense layer. During transformation of a merozoite into a metrocyte, many of the organelles of the apical

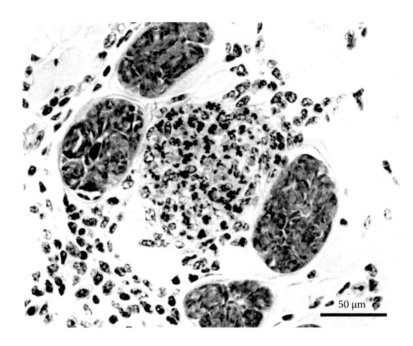

Figure 1.25   A degenerate sarcocyst in the center, surrounded by mononuclear cells and infiltrated by granu-
locytes, and 4 intact sarcocysts. H and E stain. (From reference Dubey, J.P., Speer, C.A., Epling,
G.P., 1982. *Am. J. Vet. Res.* 43, 2147–2164. With permission.)

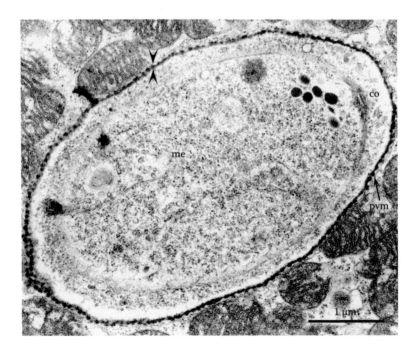

Figure 1.26   TEM of unizoite sarcocyst of *S. raushorum* at 9 DPI. The merozoite is in a late stage of transformation to a metrocyte (me). The infected myocyte has already formed a primary sarcocyst
wall (opposing arrowheads), and remnants of the conoidal (co) complex are visible. (Courtesy of
R. J. Cawthorn.)

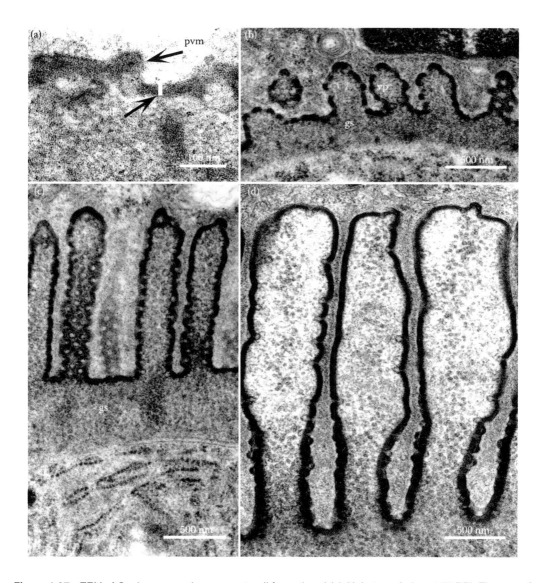

Figure 1.27 TEM of *S. singaporensis* sarcocyst wall formation. (a) Initial stage during at 25 DPI. The pvm of the sarcocyst wall has formed a bleb and is lined with an electron-dense layer (edl). (b–d), 30 DPI. (b) The primary sarcocyst wall has folded to form short villar protrusions (vp), (c). The vp are elongate with parallel sides and the granular layer is about twice as thick as in (b). (d) The vp are longer and club-shaped.

complex, such as micronemes, conoid, polar and apical rings, disappear (Figure 1.26), while ribosomes, endoplasmic reticulum, and mitochondrion become more abundant, and the nucleus becomes larger. A granular layer located immediately beneath the PSW develops early with the acquisition of minute granules and vesicles, which pinch off from invaginations in the PSW (Figure 1.27).

Mature sarcocysts of all *Sarcocystis* spp. have minute undulations in the PSW. In some species, the PSW may remain relatively simple (e.g., *S. rauschorum*), whereas other species may have highly complex (e.g., *S. fusiformis*) walls, in which the primary sarcocyst wall is folded or branched to form protrusions that project outwardly from the sarcocyst. The protrusions may contain minute granules, electron-dense bodies, microfilaments, microtubules, and small vesicles. *S. singaporensis* has a complex wall, which requires nearly 200 days to become fully formed. Figures 1.27 through

1.30 show a series of TEMs in the development of the primary sarcocyst wall of *S. singaporensis*. At 25–30 DPI, the PSW initially consists of minute undulations only (Figure 1.27a), which soon becomes folded to form short villar protrusions (vp) (Figure 1.27b), elongate protrusions with parallel sides (Figure 1.27c), and then much larger club-shaped protrusions that are approximately 2.3 μm long (Figure 1.27d). At 176 DPI, the vp are sausage-shaped with a narrow stalk (Figure 1.28) and are 10 μm long. At 184 DPI, most sarcocysts have protrusions that are shorter and usually irregular, measuring approximately 3 μm long (Figure 1.30).

As the sarcocyst develops, the contents of the vp also change, which possibly reflects a change in the metabolic activity within the sarcocysts. The vp of early and intermediate sarcocysts are filled with small vesicles and membranous whorls, which pinch off from the PSW at indentations in the PSW (Figures 1.27b–d and 1.28a and b). In mature sarcocysts, the vp contain granules and clusters of vesicles (Figures 1.28b through 1.30).

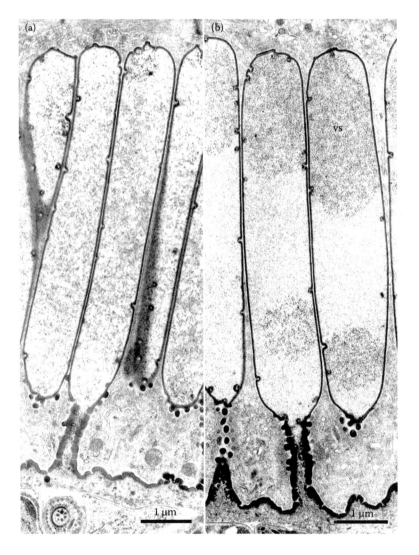

Figure 1.28 TEM of *S. singaporensis* sarcocyst at 176 DPI. (a, b) The vp have become sausage-shaped with a narrow stalk, and contain small vesicles (vs). The vp in (b) are slightly longer and wider, and contain more granules and aggregates of vesicles than in (a).

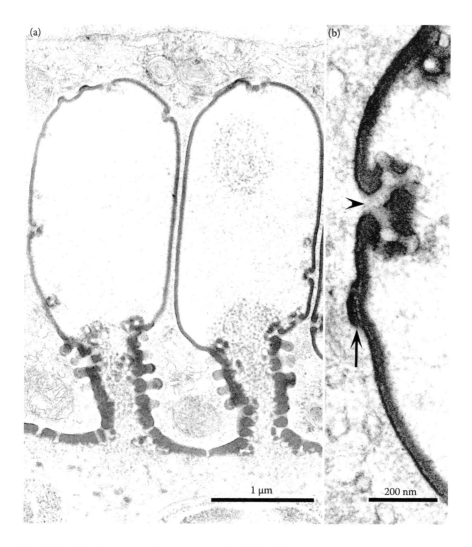

Figure 1.29 TEM of *S. singaporensis* sarcocyst at 184 DPI. (a) The vp have decreased in size to about one-fourth the length of those in Figure 1.28. (b) High magnification of the apex of a vp showing disk-shaped plaques (arrow) and indentations in the primary sarcocyst wall (arrowhead) that gives rise to small vesicles.

1.3.1.1 Metrocytes (Figures 1.31 through 1.35)

Metrocytes are ovoid, rapidly multiplying forms, which have an electron–lucent cytoplasmic matrix, numerous ribosomes, mitochondrion, 1 or more micropores, a pellicle consisting of 2 or 3 membranes, subpellicular microtubules, endoplasmic reticulum, fissures, 1 or 2 Golgi complexes, a few amylopectin granules, centrioles, electron-dense bodies, an occasional lipid body, and a diffuse nucleus (Figures 1.26, 1.31, 1.32, and 1.35). Metrocytes may also contain anlagen of 2 progeny formed by a process called endodyogeny. Metrocytes may arise from metrocytes or from dedifferentiated merozoites. Metrocytes in early stages of endodyogeny may also contain remnants of the conoid, micronemes, and rhoptries. Early sarcocysts contain only metrocytes (Figure 1.32), whereas relatively few metrocytes are contained in intermediate and mature sarcocysts (Figure 1.33).

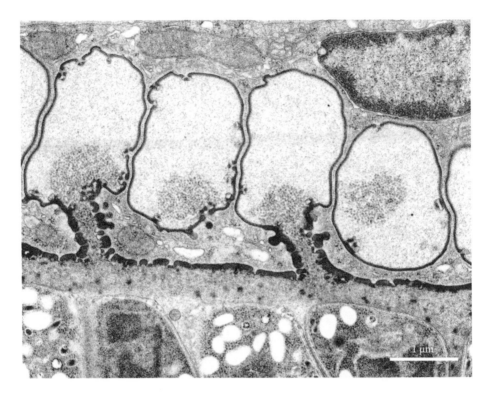

Figure 1.30　TEM of *S. singaporensis* sarcocyst at 184 DPI. Aged sarcocyst wall. The vp have become even shorter, with the distal portion irregular in shape.

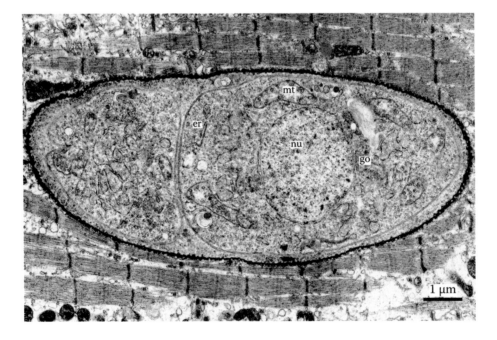

Figure 1.31　Young sarcocyst of *S. raushorum* containing 2 metrocytes with a large nucleus (nu), mitochondrion (mt), Golgi complex (go), ribosomes and endoplasmic reticulum (er). (Courtesy of R. J. Cawthorn.)

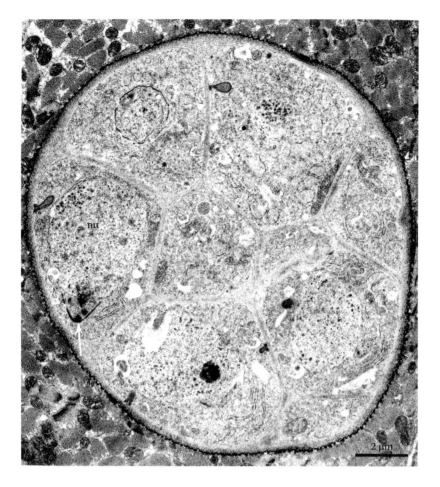

Figure 1.32   Young sarcocyst of *S. raushorum* containing 8 metrocytes; a developing apical complex (arrow) is visible in 1 metrocyte. 15 DPI. (Courtesy of R. J. Cawthorn.)

1.3.1.1.1 Endodyogeny

Endodyogeny is a form of asexual multiplication in which 2 progeny develop within a preceding stage that is ultimately consumed in the process. A series of TEMs shows endodyogeny in metrocytes of *S. singaporensis* (Figures 1.33 and 1.34). The process usually begins with the transformation of the merozoite, metrocyte, or bradyzoite, from a fusiform to an ovoid shape. The parasite cytoplasm becomes electron lucent, and micronemes, rhoptries, amylopectin, and electron-dense bodies become localized or scattered, and some disappear. The Golgi complex divides into 2 parts, and 2 centrioles appear at the anterior margin of the large vesicular nucleus. The nucleus becomes bean-shaped, with each of the 2 lobes containing a spindle apparatus consisting of nuclear spindle microtubules and a centrocone. The centrocone is a sharply conical structure from which the spindle microtubules originate. The spindle microtubules pass from the centrocone into the nucleus via perforations (not nuclear pores) in the nuclear envelope, which remains intact during nuclear division. Each of 2 centrioles is situated immediately above and slightly lateral to each centrocone. The inner membrane complex and subpellicular microtubules of each progeny form a conical-shaped structure above each spindle apparatus in the anterior part of the metrocyte. The apical rings and polar rings appear early, along with the inner membrane complex and subpellicular microtubules.

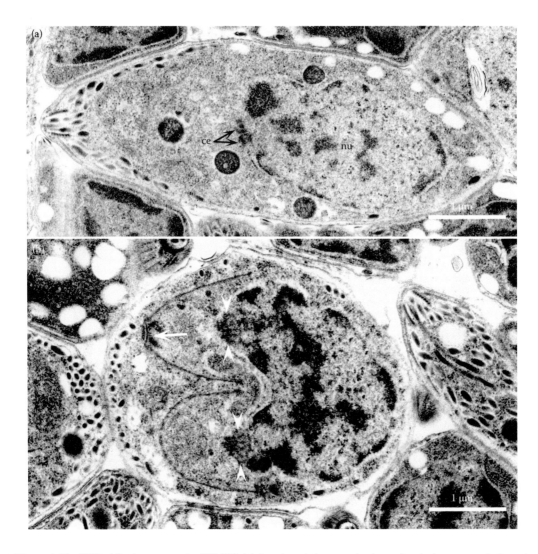

Figure 1.33   TEM of *S. singaporensis* at 60 DPI. (a) A metrocyte in an early stage of endodyogeny; note the pair
of centrioles (ce) at the anterior margin of the nucleus and micronemes scattered around the margin of the metrocyte. 176 DPI. (b) A metrocyte containing 2 developing zoites. The apical complex
→ of each zoite has begun to form immediately above each lobe of the U-shaped nucleus. Note
spindle apparatus (arrowheads).

During this early stage of endodyogeny, a conoid is gradually formed at the tip of the developing
complex. A centriole is often seen in close proximity to the developing conoid and may actually
be responsible for its formation in much the same manner that centrioles give rise to a subsequent
generation of centrioles.

   As division progresses, the inner membrane complexes and subpellicular microtubules grow
posteriorly, incorporating increasingly more of the dividing nucleus. The electron-dense areas, present at the edges of the inner membrane and at the ends of the subpellicular microtubules, probably
represent areas in which components of these structures are assembled. As the inner membrane
complex nears the posterior end of the metrocyte, the nucleus pinches into 2 nuclei, with each portion
incorporated into a progeny. At this point, each progeny usually has a completely formed conoid,
some micronemes and rhoptries, and various inclusion bodies. The progenies increase in volume,

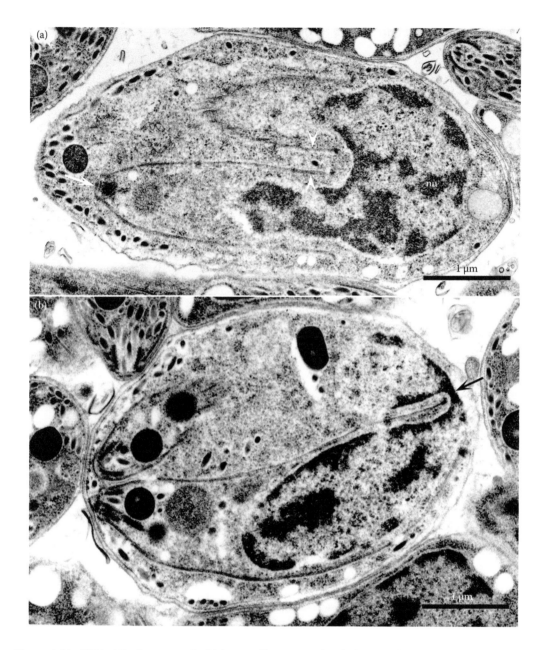

Figure 1.34 TEM of *S. singaporensis*. (a) Intermediate stage of endodyogeny showing a completely formed conoid (arrow), the inner membrane complex and subpellicular microtubules of each daughter zoite have extended further posteriorly (arrowheads), incorporating more of the parasite nucleus, 176 DPI. (b) TEM of advanced stage in endodyogeny; note that the 2 lobes of the nucleus are connected by a narrow isthmus (arrow) at 60 DPI.

acquiring more of the metrocyte cytoplasm and organelles, until they occupy most of the metrocyte. As the progenies approach the surface, the inner membrane complex of the metrocyte disappears, and the outer membrane becomes the outer membrane of each progeny. At this point, they separate into 2 progenies. During endodyogeny, various spheroidal structures appear in the anterior end, which probably represent anlagen of rhoptries and micronemes of the developing progeny.

1.3.1.2 Bradyzoites (Figures 1.35 through 1.40)

1.3.1.2.1 Pellicle

The pellicle consists of 3 membranes, an outer plasmalemma and an inner double membrane complex (Figures 1.35, 1.39, and 1.40). The plasmalemma is a continuous unit that completely encloses the whole parasite, whereas the inner pellicular membranes are interrupted at the anterior and posterior ends and at micropores. The plasmalemma is separated from the inner membranes by an electron–lucent space, 15–20 nm thick. In *S. tenella*, the intramembranous particles are randomly arranged in the plasmalemma, but have an orderly arrangement in the inner pellicular membranes.[1406] The inner pellicular membrane complex is similar to tight intercellular junctions

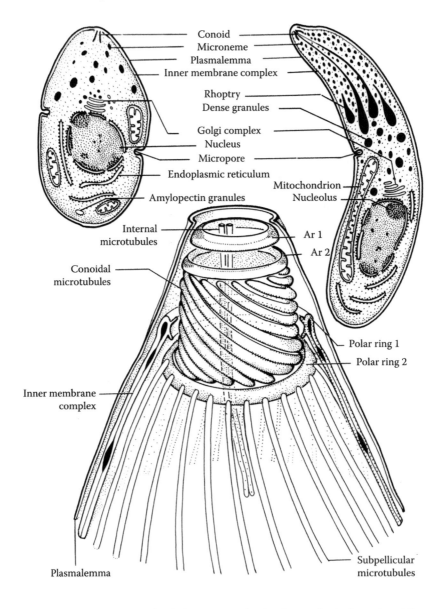

Figure 1.35   Schematic representation of typical metrocytes, bradyzoites, and apical complex of *Sarcocystis* spp.

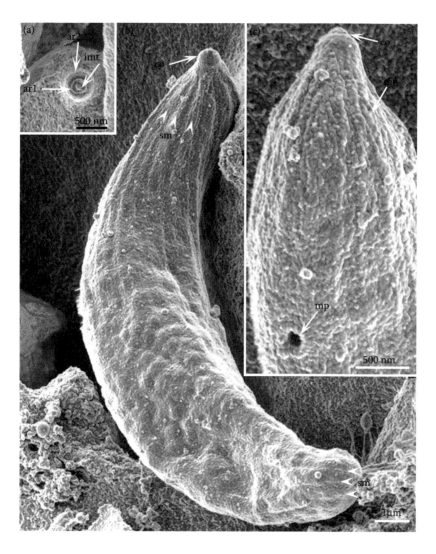

Figure 1.36   SEM of *S. fusiformis* bradyzoites. (a) Enface view of conoid (co). Note 2 inner microtubules (imt), and 2 apical rings (ar1, ar2). (b) Surface view of bradyzoite (br). Note subpellicular tubules (sm) arising from the polar ring of the co and extending the entire length of the br. The posterior half of the bradyzoite has irregular bumps. (c) Conoidal part of a bradyzoite with a conoid and micropore (mp); it is granular because of the numerous micronemes that are partly visible through the pellicle. (Adapted from Dubey, J.P. et al., 2015. *Parasitology*. 142, 385–394.)

of higher organisms and is composed of 11 rows of rectangular contiguous plaques (0.5 × 2 μm in size) that are arranged in a loosely coiled helical pattern. The plaques converge at the posterior end to form the posterior pore (posterior ring) which represents an interruption in the inner membrane complex. At 1.5–2 μm from the anterior tip, the rectangular plaques abut tightly to a conical inner membrane complex called the anterior cape. The anterior cape has an interruption at its anterior tip, which appears electron-dense in TEM, and forms the polar ring. At the anterior tip, the plasmalemma contains a rosette of intramembranous particles (1 central and 8 peripheral particles). The rosette of intramembranous particles may be involved in rhoptry secretion during host-cell penetration or serve as a receptor–processor system, which enables the parasite to recognize and penetrate host cells.

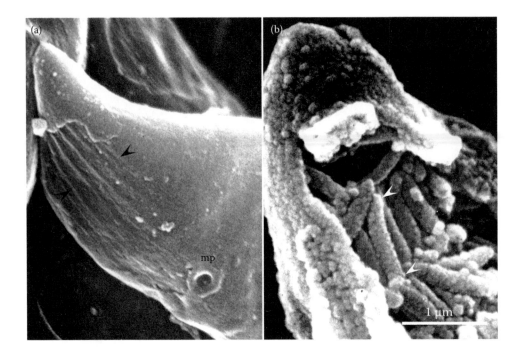

Figure 1.37 SEMs of the apical ends of *S. cruzi* bradyzoites. (a) Note opening of micropore (mp) and ridges (arrowheads) on surface created by underlying subpellicular microtubules. (b) Portion of pellicle has been removed exposing the micronemes (arrowheads). (From Dubey, J.P., Speer, C.A., Epling, G.P., 1982. *Am. J. Vet. Res.* 43, 2147–2164. With permission.)

1.3.1.2.2 Apical Rings

Two apical rings (also called preconoidal or conoid rings) are situated at the anterior tip, immediately beneath the plasmalemma and above the papillary conoid (Figures 1.36, 1.39, and 1.40). Bradyzoites of *S. tenella* may have 3 apical rings.[1406] The function of these apical rings is not known.

1.3.1.2.3 Polar Rings (Figures 1.35 through 1.40)

The polar rings of *Sarcocystis* species have been studied less extensively than those of other coccidia. A single polar ring appears to be present in *S. tenella* bradyzoites,[1408] whereas other species have 2 polar rings. Polar ring 1 appears as an electron-dense thickening at the anterior termination of the anterior cape of the inner membrane complex, and serves to anchor the subpellicular microtubules in those bradyzoites with a single polar ring.[1408] In those species with 2 polar rings, polar ring 2 is located immediately interior to and slightly posterior to polar ring 1, and serves an an anchoring point for the subpellicular microtubules (Figure 1.40). Two electron-dense projections are attached to the inner surface of polar ring 1 (Figures 1.35 and 1.40).

1.3.1.2.4 Microtubules (Figures 1.35 through 1.39)

Microtubules are found in virtually every stage of *Sarcocystis*. A few are cytoplasmic or nuclear microtubules, whereas most other microtubules are components of specialized structures, such as subpellicular microtubules, conoid, centrioles, flagellar axoneme, and the mitotic spindle apparatus. Bradyzoites of *Sarcocystis* species have 22 subpellicular microtubules that originate at evenly

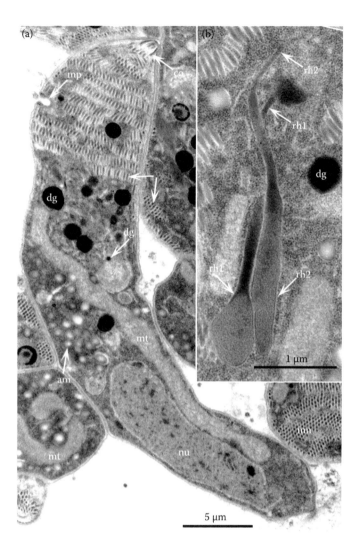

Figure 1.38   TEM of *S. fusiformis* bradyzoites. (a) Longitudinally cut bradyzoite (br) and parts of other br. Note conoid (co), numerous micronemes (mn), a micropore (mp), at least 8 dense granules (dg) of different sizes, numerous amylopectin granules (am), a convoluted mitochondrion (mt), and a large nulcleus (nu). (b) Part of a bradyzoite showing 2 rhoptries (rh1, rh2) that are twisted at their necks. Note variability in density of the rhoptry contents. (Adapted from Dubey, J.P. et al., 2015. *Parasitology*. 142, 385–394.)

spaced intervals around the polar ring, and extend posteriorly immediately beneath the inner membrane complex for about 1/2 to 2/3 the length of the zoite (Figures 1.35 and 1.39).

The subpellicular microtubules are similar to those found in a wide variety of plant and animal cells. They are cylinders approximately 25 nm in diameter. They usually contain 12 or 13 protofilament strands running longitudinally. They appear striated because the protofilaments appear to be composed of repeating subunits, probably α- and β-tubulin, arranged in transverse or helical rows. Subpellicular microtubules probably provide structural integrity to the overall shape of zoites, and may be involved in motility (gliding and flexing), transport of cytoplasmic components near the margin of the zoite, or serve to anchor protein molecules in the inner membrane complex. Treatment with the antimicrotubular agents—colcemid, colchicine, or vinblastine—stops motility of bradyzoites of *Sarcocystis*.[273]

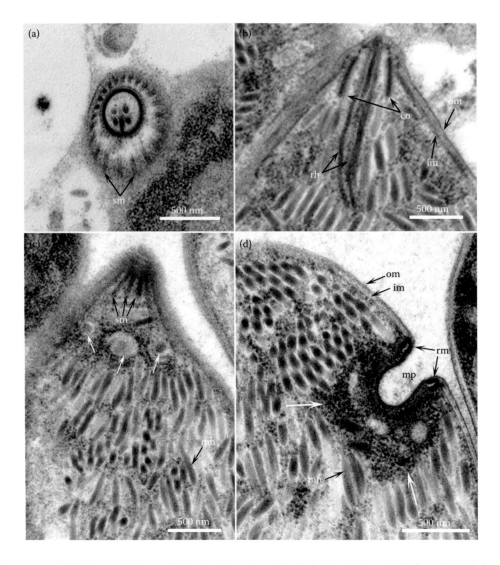

Figure 1.39 TEM of conoidal part of 4 bradyzoites. (a and b), *S. fusiformis*, (c and d), *S. cafferi*. (a) Cross/
oblique section through the conoid. Note 22 subpellicular tubules (sm) originating from the polar
ring. (b) Conoid with 2 rhoptries (rh) opening in the conoid. (c) Details of conoidal end with annu-
lar and polar rings, subpellicular tubules (sm), 3 ring-like structures at the conoidal end (white
arrows). Note haphazard arrangement of micronemes (mn). (d) Details of pellicle with outer (om)
and inner (im) membrane at the micropore (mp) junction. The im is interrupted at the micropore
opening and a rim-like (rm) structure is present at the opening. Electron-dense secretory material (white arrows) surround the micropore. Note numerous micronemes (mn). ((a and b) Adapted
from Dubey, J.P. et al., 2015. *Parasitology.* 142, 385–394; (c and d) Adapted from Dubey, J.P. et
al., 2014. *J. Parasitol.* 100, 817–827.)

1.3.1.2.5 Micropore

Micropores are present in all stages of *Sarcocystis* species, except microgametes. They consist
of an invagination of the parasite plasmalemma into the parasite cytoplasm (Figures 1.36 through
1.39). The invaginated membrane is encircled on the cytoplasmic side by an electron-dense membranous collar, which the inner membrane interrupts, and is juxtaposed at a 90°-angle to it. Although
most micropores appear to be inactive, some have appeared to be portals for ingestion of particulate

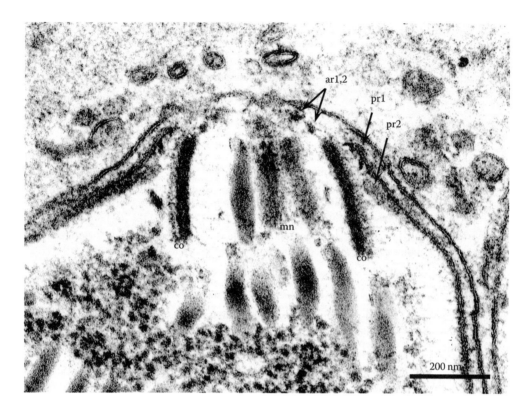

Figure 1.40 TEM of apical complex of *S. campestris* bradyzoite at 62 DPI. Note: apical (ar1,2) and polar (pr1,2) rings, conoid (co), and micronemes (mn).

matter in developing or multiplying stages such as metrocytes. Occasionally, electron-dense material surrounds the cytoplasmic side of the micropore (Figure 1.39d).

1.3.1.2.6 Rhoptries and Micronemes

Rhoptries and micronemes are electron-dense, membrane-bound, elongate structures occupying the anterior of the bradyzoite (Figures 1.36 and 1.38). These organelles are believed to constitute a single functional unit involved in secreting substances to aid penetration of cells and may secrete proteins that become antigenic or are inserted into the bradyzoite plasmalemma just anterior to the conoid.

Rhoptries consist of a narrow, duct-like anterior region found within the core of the conoid and a posterior club- or sack-like region. Most bradyzoites have only 2 rhoptries; in early studies, dense granules may have been mistaken as rhoptries.[474] To avoid misinterpretation, it is necessary to identify the elongated neck region, or a very clear bulbous end, because the rhoptry contents can appear differently within the same rhoptry (Figure 1.38). The contents of rhoptries are often electron-dense. Bradyzoites of all species of *Sarcocystis* contain numerous micronemes. Micronemes are elongate, rod-like structures that usually appear ovoid or round in cross section. In some Apicomplexa, narrow interconnections have been found between micronemes and rhoptries, but interconnections between the rhoptries and micronemes in bradyzoites of *S. tenella* are controversial. Micronemes appear to form within single membrane-bound vesicles in the apical region of bradyzoites, and apparently give rise to rhoptries. The arrangement of micronemes can vary within different regions of the bradyzoite. Their location in the bradyzoite is fluid, and they are considered to move towards the conoid.

1.3.1.2.7 Dense Granules

Dense granules are electron-dense and are scattered throughout the bradyzoite; but most are located in the middle of the parasite (Figure 1.38). They are secretory vesicles, and their contents play a major role in the modification of the PV and regulation of immune mediation.

1.3.1.2.8 Conoid

The conoid is a hollow, papillary organelle located at the anterior tip of sporozoites, bradyzoites, and merozoites. The conoid of bradyzoites of *S. tenella* is made up of 20 microtubules arranged in a helical pattern. Some investigators believe the conoid may be involved in penetration of host cells, because, when it is protruded, the anterior tip becomes shaped like a long, narrow stylet that could facilitate active penetration into the cell. Within the core of the conoid are the ducts of rhoptries and 1 or 2 eccentrically located microtubules that extend for a short distance into the zoite cytoplasm (Figure 1.35). These microtubules may be attached to the conoid, and serve to protrude or retract the conoid.

1.3.1.2.9 Golgi Complex

The Golgi complex is well-developed in the stages that are most metabolically active, such as metrocytes and schizonts, but are reduced in size and complexity in bradyzoites of most *Sarcocystis* spp. The functions of the Golgi complex are probably similar to those described for other cells.

1.3.1.2.10 Endoplasmic Reticulum

Both smooth and rough endoplasmic reticula are present but small in bradyzoites, especially in the area just posterior to the nucleus.

1.3.1.2.11 Ribosomes

Bradyzoites have an abundant amount of free ribosomes, found from the conoid to the posterior tip, filling the bradyzoite cytoplasm not occupied by organelles or inclusion bodies.

1.3.1.2.12 Mitochondrion

A mitochondrion is found in bradyzoites, as well as in all other stages of *Sarcocystis* spp. Bradyzoites appear to have 1 long twisted mitochondrion (Figure 1.38). This organelle is double-membrane-bound, with tubular cristae, which are characteristic of most protozoans.

1.3.1.2.13 Apicoplast

The apicoplast is a membrane-bound, alga-derived, obligatory endosymbiotic structure. Considerable biologic curiosity has been focused on this nonphotosynthetic organelle. It is essential for the survival of the parasite, and has its own genome. In the past, it has been called a multivesicular body, multimembrane-bound vesicle, or Golgi adjunct. It is surrounded by 4 membranes.[1754] Its synthesis was studied in detail in *S. neurona* schizonts (Figure 1.22)[1770] (see Chapter 3).

1.3.1.2.14 Inclusion Bodies

Bodies that appear free in the cytoplasm, such as amylopectin, lipid bodies, and electron-dense bodies, are generally referred to as inclusion bodies (Figure 1.35). Although electron-dense bodies are seen in nearly all stages, their function is still not known. Some inclusion bodies probably serve as a site for storage of energy reserves, such as amylopectin, and others may store proteins, lipids, and, perhaps, enzymes, whose functions are unknown.

1.3.1.2.15 Size

Bradyzoites vary in size. Bradyzoites of species of *Sarcocystis* in livestock are approximately 3 times larger than those species parasitizing reptiles. They are often banana-shaped; in few species they are lanceolate (e.g., *S. wenzeli* of chicken).

1.3.1.3 Types of Sarcocyst Walls (Figures 1.21 through 1.30, 1.41 through 1.61)

Numerous ultrastructural reports have shown that the sarcocyst walls of *Sarcocystis* species vary—from being relatively simple to highly complex. After examining published reports, there appear to be at least 82 distinct types of sarcocyst walls (Table 1.2), which herein are referred to as Types 1–42. It is likely that additional types will be discovered. To avoid

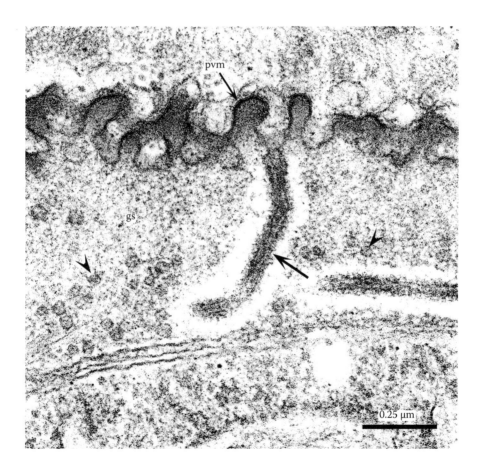

Figure 1.41 TEM of *S. rauschorum* sarcocyst wall at 60 DPI. The primary sarcocyst wall consists of an undulating pvm and an electron-dense layer immediately beneath the pvm. At certain points, the pvm invaginates (arrow) into the granular layer; the latter contains fine granules and small vesicles (arrowheads) that arise from the pvm.

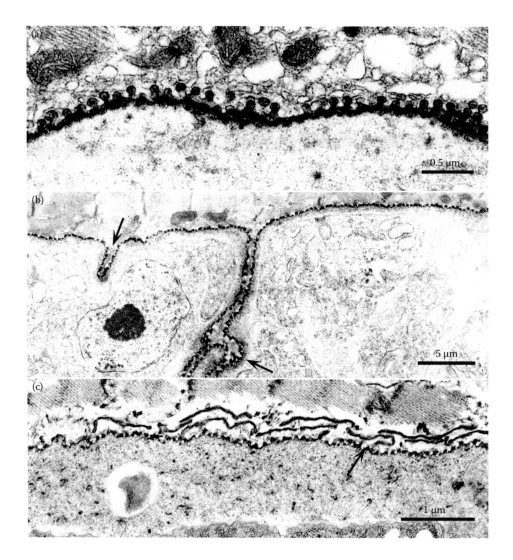

Figure 1.42 Types 1 and 2 sarcocyst walls. (a) *S. muris*, type 1a with the simplest structures with blebs on pvm. (b) *S. rauschorum*, type 1h. The sarcocyst wall invaginates (arrows) deeper into the sarcocyst at 60 DPI. (Courtesy of R. J. Cawthorn.) (c) Type 2 *S. wapiti* showing thin hirsute structures arising from the apex of the undulating primary sarcocyst wall (arrow). (From Speer, C.A., Dubey, J.P., 1982. *Can. J. Zool.* 60, 881–888. With permission.)

misinterpretations, the types reported earlier[393,432] were retained, and new types were added as subtypes (Figures 1.58 through 1.61). The wall types presented herein do not correspond in number with those reported by Beaver et al.[84] which were based on LM observations, and, therefore, were limited in the detail that could be observed.

The primary sarcocyst wall consists of a PVM and an electron-dense layer (edl) immediately beneath the PVM. A granular layer is immediately beneath the primary sarcocyst wall. Septa, which arise from the granular layer, traverse the sarcocyst, separating it into compartments (Figure 1.11a), which contain bradyzoites and metrocytes.

There are only a few reports that used SEM to study the ultrastructure of sarcocysts.[126,127,281–284,360,382,473,474,698,700,702,1176,1309,1316,1684] Compared to a TEM, an SEM provides little information because it is limited to examining surfaces (Figures 1.53 through 1.57).

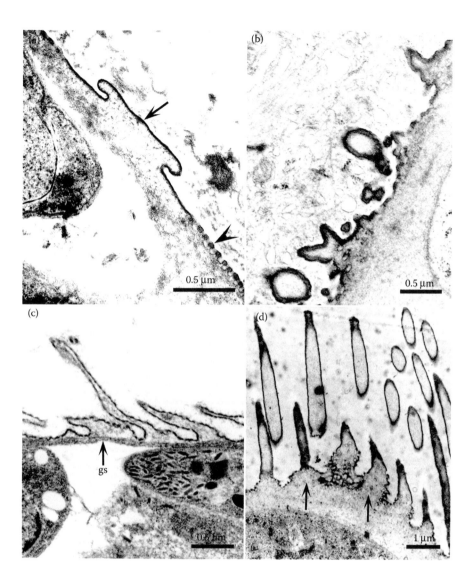

Figure 1.43 Types 3–6 sarcocyst walls. (a) Type 3 *S. ferovis* sarcocyst wall, showing undulating pvm with blebs (arrowhead) and a flattened mushroom-like vp (arrow). (From Dubey, J.P., 1983. *Proc. Helminthol. Soc. Wash.* 50, 153–158. With permission.) (b) Type 4 *S. sigmodontis* wall, showing short, irregularly shaped vp. (c) Type 5 *S. sulawesiensis* wall with hirsute vp and a relatively thin granular layer (gs). (From O'Donoghue, P.J., Watts, C.H.S., Dixon, B.R., 1987. *J. Wildl. Dis.* 23, 225–232. With permission.) (d) Type 6 *S. capreolicanis* wall, showing finger-like vp arising from a dome-shaped base (arrows). The pvm has minute undulations in the dome-shaped base, but is relatively smooth in the finger-like vp that are oriented perpendicularly to the sarcocyst wall. (Courtesy of R. Entzeroth.)

The sarcocyst surface of type 14 sarcocysts of *S. capracanis* has a honeycomb-like appearance, and the protrusions appear villus-like (Figure 1.54a and b).[382]

1.3.1.4 Gametogenesis

After ingestion by the appropriate definitive host, bradyzoites escape from sarcocysts, and penetrate enterocytes, usually goblet cells, in the small intestine (Figure 1.62a). Development

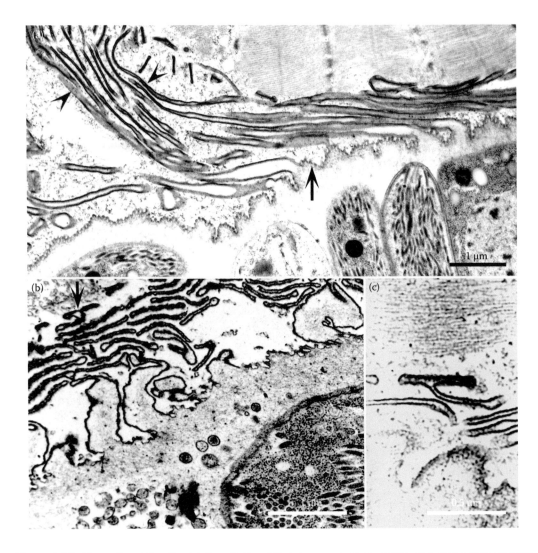

Figure 1.44 Types 7 and 8 walls. (a) Type 7 *S. arieticanis* sarcocyst consisting of a dome-shaped base (arrow) with an undulating pvm, an intermediate finger-like segment, and a thin, thread-like distal segment (arrowhead). The intermediate and distal segments are often bent 90° and run parallel to the sarcocyst surface, whereas in other areas, the protrusions are arranged collectively to form conical tufts. (b) Type 8a *S. gruneri*. This sarcocyst is similar to Type 7, except that the distal segment of the protrusion is branched (arrow). (From Entzeroth, et al., 1985. *Z. Parasitenkd*. 71, 33–39. With permission.) (c) Higher magnification of type 8a showing branched distal segment.

from bradyzoite to microgamonts or macrogamonts is rapid, being completed in 6–18 h or less. Bradyzoites are situated within a PV and begin immediately to develop into either a micro- or macrogamont. How the sex is determined or regulated is not known.[255] Bradyzoites quickly transform from a fusiform to an ovoid or spheroid form, and most of the organelles of the apical complex undergo dissolution. Micronemes become scattered in the gamont cytoplasm, and soon disappear. Simultaneously, the early gamont produces much endoplasmic reticulum, numerous ribosomes, mitochondrion, and a vesicular nucleus, with scattered heterochromatin and a nucleolus. Macrogamonts have 2 ultrastructural features that appear to be lacking in microgamonts. These are an exocytosis pore and electron-dense granules which are transported to a location near the surface

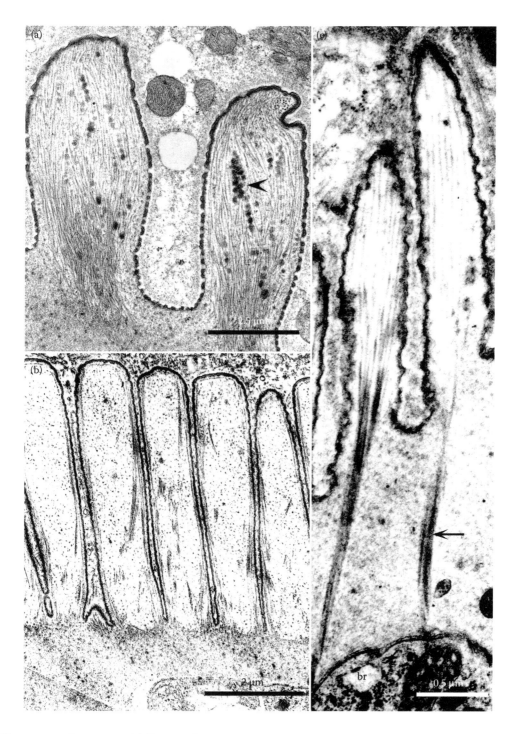

Figure 1.45   Types 9–11. (a) Type 9a *S. campestris* sarcocyst showing widely spaced vp. The pvm has minute undulations over the entire sarcocyst surface and the vp contain prominent electron-dense granules (arrowhead) and microtubules that extend from the villar tips into the granular layer. (b) Type 10a *S. odoi* sarcocyst wall, similar to Type 9a but with tightly packed vp. (c) Type 11a *S. fayeri* sarcocyst wall.The microtubules extend from the villar tips to the plasma-lemma of bradyzoites (arrow). (From Tinling, S. et al., 1980. *J. Parasitol.* 66, 458–465. With permission.)

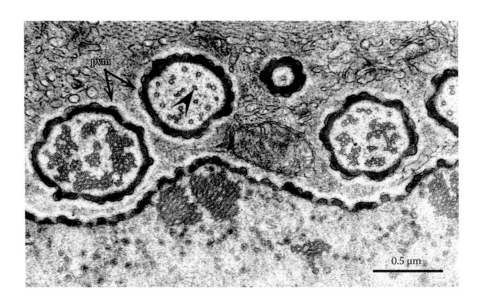

Figure 1.46 Type 11b *S. falcatula* wall, showing vp and microtubules (arrowhead) in cross section. The pvm is wavy and lined underneath with an electron-dense layer. Also see Figure 17.1, showing that microtubules do not extend to the pellicle of bradyzoites. (From Box, E.D., Meier, J.L., Smith, J.H., 1984. *J. Protozool.* 31, 521–524. With permission.)

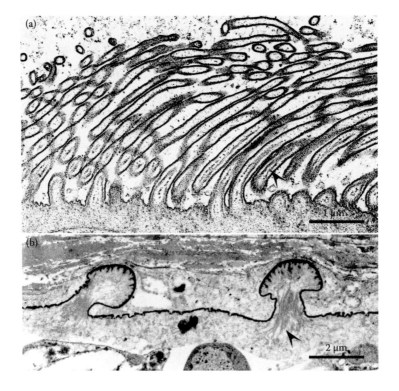

Figure 1.47 Types 12 and 13. (a) Type 12 *S. sybillensis* sarcocyst wall consisting of tightly packed, thin, finger-like vp with microtubules in the basal one-third of the protrusion. (b) Type 13 *S. mucosa* sarcocyst with widely spaced, mushroom-like vp with microtubules that extend from the core of the protrusion into ground substance (gs) (arrowhead). (From O'Donoghue, P.J. et al., 1987. *Parasitol. Res.* 73, 113–120. With permission.)

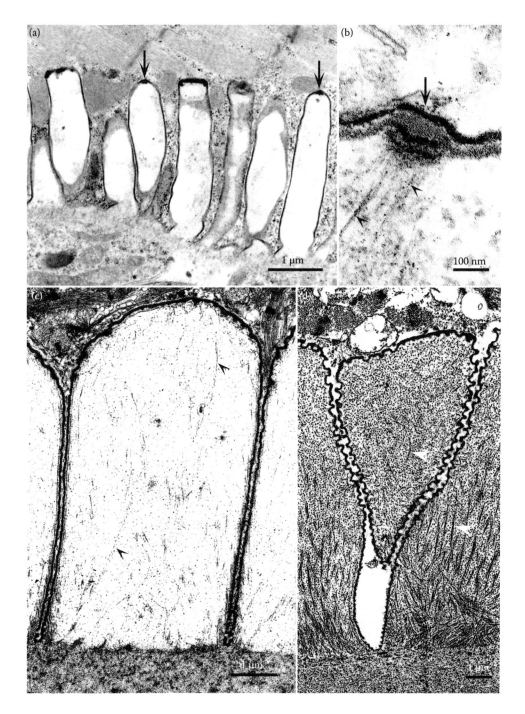

Figure 1.48   Types 14–16. (a) Type 14 *S. tenella* sarcocyst with tightly packed vp, the tips of which contain disc-shaped plaques (arrows). Microtubules are absent. (b) High magnification of a disc-shaped plaque (arrow) in *S. hemionilatrantis*. (From Speer, C.A., Dubey, J.P., 1986. *J. Protozool.* 33, 130–132. With permission.) (c) Type 15 *S. rangiferi* with tightly packed upright tombstone-like vp that contain widely scattered microtubules (arrowheads). In some species with type 15, the villar protrusions have interdigitating lateral margins. (d) Type 16 *S. youngi* sarcocyst wall with alternating conical and club-shaped vp with criss-crossing microtubules. (From Dubey, J.P., Speer, C.A., 1985. *J. Wildl. Dis.* 21, 219–228. With permission.)

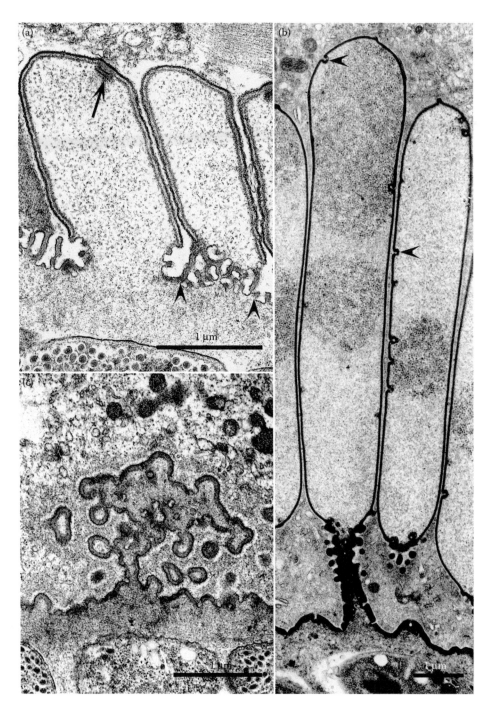

Figure 1.49 Types 17–19. (a) Type 17 *S. odocoileocanis* with disk-shaped plaques (arrow) at the apex and lateral margins of the vp and a highly branched primary sarcocyst wall at the base of the protrusions (arrowheads). (b) Type 19 *S. singaporensis* with club-shaped vp consisting of a cylindrical stalk and a sausage-shaped distal segment. In older sarcocysts, the distal segment may be shorter and irregular in shape (see Figure 1.30). Invaginations of the primary sarcocyst wall are numerous in the stalk (arrowheads) and widely scattered in the distal segment. (c) Type 18 *S. dispersa* with an irregularly shaped vp which may appear T-shaped. (From Sénaud, J., Cerná, Z., 1978. *Protistologica*. 14, 155–176. With permission.)

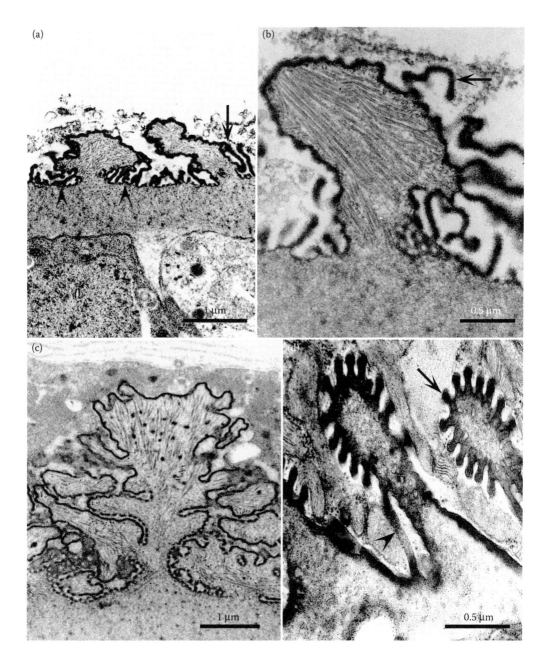

Figure 1.50 Types 20–22. Type 20 (a, b). *S. medusiformis* sarcocyst wall with angular vp that vary from pyramidal, trapezoidal, or mushroom-like. Serpentine projections arise from the surface of the vp (arrow) as well as from the base (arrowheads) and some connect the surface to protrusions. (a) Fine bristles (arrow) arising from the surface of the protrusions as well as from the surface of the serpentine projections. (b) The villar core contains tightly packed microtubules. (c) Type 21 *S. gigantea* wall that has cauliflower-like vp containing microtubules or microfilaments and fine or coarse granules. Also see Figures 1.53 and 1.57. (d) Type 22 *S. villivillosi* has protrusions with a proximal, sharply conical base and a distal segment that is cocklebur-like with short, radiating projections. The conical base is in oblique section. Cross-sections of the distal segment appear as cogwheels (arrow). Some of the projections from the conical base connect with the surface of the sarcocyst wall (arrowhead). Vesicles are located within the protrusions, especially in the distal half. (From Beaver, P.C., Maleckar, J.R., 1981. *J. Parasitol.* 67, 241–256. With permission.)

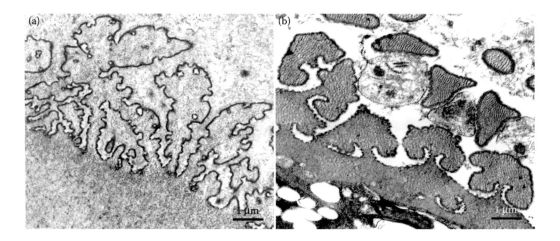

Figure 1.51   Types 23 and 24. (a) Type 23 *S. rileyi* consisting of anastomosing, cauliflower-like vp that contains fine granules and microfilaments. (Adapted from Dubey, J.P. et al., 2003. *J. Eukaryot. Microbiol.* 50, 476–482.) (b) Type 24 *S. cornagliai* sarcocyst type with mushroom-like projections arising from mushroom-like vp. (*Sarcocystis* sp. from mountain goat, courtesy of W. Foreyt, see Foreyt, W.J., 1989. *J. Wildl. Dis.* 25, 619–622.)

Figure 1.52   Two images (a, b) of type 38 *S. speeri* with steeple-like vp (arrow). The pvm is lined by an electron-dense layer, absent in invaginated areas (arrowheads). (Adapted from Dubey, J.P., Speer, C.A., Lindsay, D.S., 1998. *J. Parasitol.* 84, 1158–1164.)

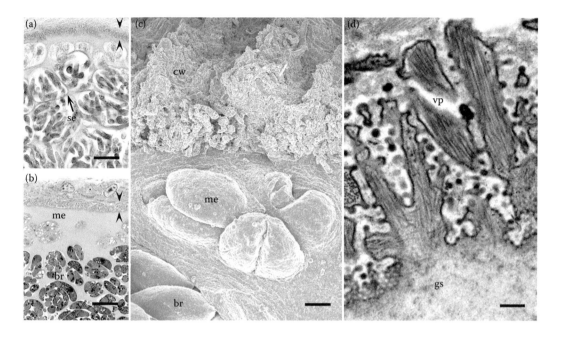

Figure 1.53 Structure of sarcocyst wall of *S. cafferi* observed with different techniques. (a and b), light microscopy; (c), SEM; and (d), TEM. Bar in (a) and (b) = 10 μm, in (c) = 2 μm, and in (d) = 300 nm. Note thin cyst wall (opposing arrowheads) in (a) and (b). (a) H and E stain, 5 μm section. (b) Toluidine blue stain, 1 μm section. (c) Surface and side views of the sarcocyst wall. (d) Branching of vp. Also note metrocytes (me), bradyzoites (br), and septa (se). (Adapted from Dubey, J.P. et al., 2014. *J. Parasitol.* 100, 817–827.)

of the gamont and discharged through the exocytosis pore (Figure 1.62b). In *S. cruzi*, the exocytosis pore is similar in size to micropores, but differs ultrastructurally (Figure 1.62b). The exocytosis pore protrudes slightly to produce a flattened dome on the surface of the gamont. The plasmalemma is invaginated at the apex of the dome, and the plasmalemma continues uninterrupted through the invagination. The single inner membrane is interrupted at the site of the micropore, allowing the invaginated plasmalemma to extend into the gamont cytoplasm. The margin of the interruption in the inner membrane is electron-dense, thick, and projects outward at approximately a 45° angle, instead of inwardly, as in micropores. The membranes surrounding the electron-dense bodies fuse with the invaginated membrane of the exocytosis pore, which causes the electron-dense bodies to be discharged into the PV, where it fuses to form a thin electron-dense layer (Figure 1.62b). This layer eventually becomes continuous to form the oocyst wall. Oocyst wall-forming bodies (WB), typical of *Eimeria* spp., and certain other coccidia, are lacking in macrogamonts of *Sarcocystis* spp. However, 2 types of electron-dense bodies in macrogamonts of *S. singaporensis* were interpreted as WB.[1364,1366] In *S. singaporensis*, the oocyst wall appears to be formed before fertilization and the WB disappear with formation of the oocyst wall.[1364]

Bradyzoites of *S. muris* have an exocytosis pore, through which dense granules are discharged into the PV.[529] The contents of the dense granules may stimulate the adhesion of the host-cell endoplasmic reticulum to the PV, to form a 3-membrane complex, or cause antigenic mimicry, preventing a host-cell immune response or inflammation.[529]

In most *Sarcocystis* species, macrogamonts begin development in the epithelium, but as development proceeds to an intermediate stage, the host cell degenerates and lyses (Figure 1.63), and the macrogamonts enter the lamina propria, where they undergo oogony and sporogony. In addition to forming the oocyst wall, the contents of the dense granules may also contain enzymes that cause

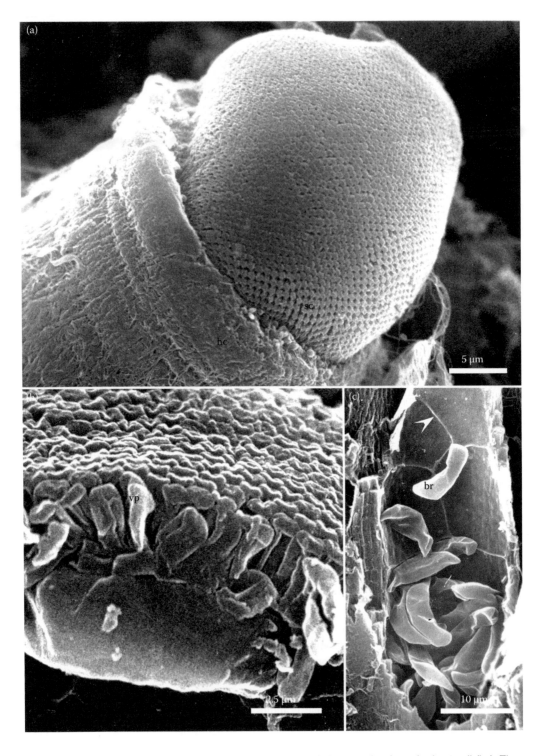

Figure 1.54   SEMs of sarcocysts. (a) *S. capracanis* sarcocyst (sc) protruding from the host cell (hc). The vp
give a honeycomb appearance to the surface of the sarcocyst. (b) Cross-sectional view of the
margin of a *S. capracanis* sarcocyst showing vp. (From Dubey, J.P. et al., 1984. *Int. Goat Sheep
Res.* 2, 252–265. With permission.) (c) Ruptured *S. cruzi* sarcocyst showing bradyzoites (br) and
septa (arrowhead). (From Dubey, J.P., 1982. *J. Protozool.* 29, 591–601. With permission.)

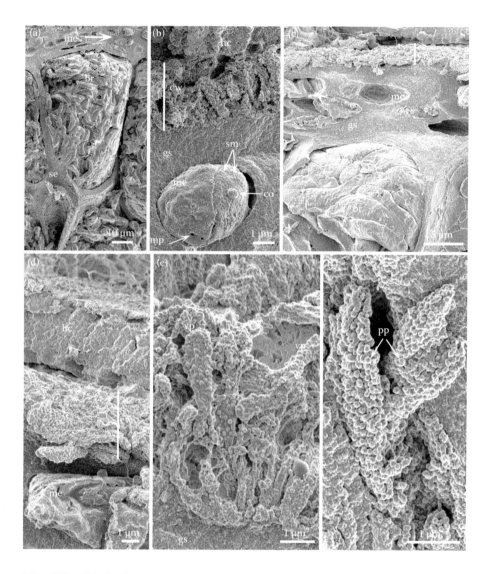

Figure 1.55 SEM of *S. fusiformis* sarcocysts. (a) Sectioned part of a sarcocyst. Note bradyzoites (br) are arranged in sacks enclosed by thick septa (se). Metrocytes are located in the gs. Empty spaces represent location of fallen metrocytes (me) during processing. (b) Sectioned sarcocyst. Note host cell (hc) enclosing the cyst wall (cw), thick, finely granular gs, and a single or group of metrocytes embedded in the ground substance. Part of outer covering of the metrocytes has ripped off showing a conoid (co), subpellicular tubules (sm), and micropore (mp). (c) Partly exposed sarcocyst. Note that vp are folded over the cyst wall (cw). The gs has several groups of metrocytes, 1 of which has 2 zoites (arrowhead). Also note a group of bradyzoites (br) enclosed in se. (d) Note combed appearance of the cw. (e) Higher magnification of the cw with irregularly shaped villar protrusions. (f) Higher magnification of the vp to show nearly identical papillomatous structures (pp) on the vp. (Adapted from Dubey, J.P. et al., 2015. *Parasitology*. 142, 385–394.)

lysis of the enterocyte, enabling the parasites to enter the lamina propria. Because the oocyst wall is being formed during this time, it seems likely that the electron-dense granules may contain cytolytic substances that cause lysis of the host cells. Neighboring cells are apparently not harmed.

In contrast to macrogamonts, microgamonts of some species of *Sarcocystis* can complete their development within the intestinal epithelium enterocyte or goblet cell (Figure 1.64). Several

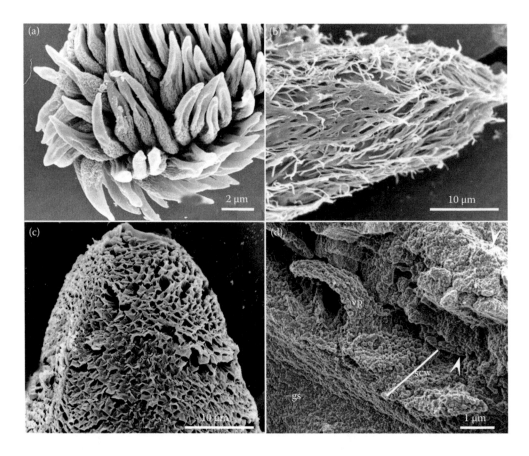

Figure 1.56   SEM of sarcocyst walls of 4 *Sarcocystis* species. (a) *S. mihoensis* with vp like a sea anemone; the vp are tapered towards the free end. (b) *S. arieticanis* with thread-like vp that are folded over the sarcocyst wall. (c) *S. tenella* with scale-like vp. (d) *S. fusiformis* shown at the surface view of the sarcocyst wall (scw). Note papillomatous growth on villar protrusions (vp), and host cell (hc). (Figure 1.56a–c courtesy of M. Saito, see Saito, M., Shibata, Y., Kubo, M., Itagaki, H., 1997. *Sarcocystis mihoensis* n. sp. from sheep in Japan. *J. Vet. Med. Sci.* 59, 103–106.)

ultrastructural features distinguish microgamonts from macrogamonts. Early microgamonts increase in size and become ovoid (Figure 1.65), have a well-developed endoplasmic reticulum, mitochondrion, numerous ribosomes, and a vesicular nucleus; but in contrast to macrogamonts, they lack electron-dense bodies and an exocytosis pore but do contain a mitotic spindle apparatus (Figure 1.64). The microgamont increases in size to about 10 µm in diameter, the inner membrane disappears, and the nucleus becomes lobulated, with each lobe situated near the gamont surface. A single microgamete develops immediately above each nuclear lobe. Two centrioles become basal bodies that form 2 flagella, each of which projects outward from the gamont. The chromatin becomes more compact and dense and, eventually, is situated in that portion of the nucleus nearest the gamont plasmalemma. The microgamete buds at the surface of the gamont (Figure 1.66), incorporating the dense portion of the nucleus and a mitochondrion and, eventually, the microgamete plasmalemma, separates from that of the gamont. Fully formed microgametes have a plasmalemma, 2 basal bodies associated with the base of each of 2 flagella, a dense nucleus, a mitochondrion, and several longitudinally oriented microtubules that extend from near the basal bodies to a point about midway along the nucleus (Figure 1.66). In *S. singaporensis*, the final stage of maturation on microgametes coincided with the degeneration of the parasitized enterocytes and loss of their brush border.[1365]

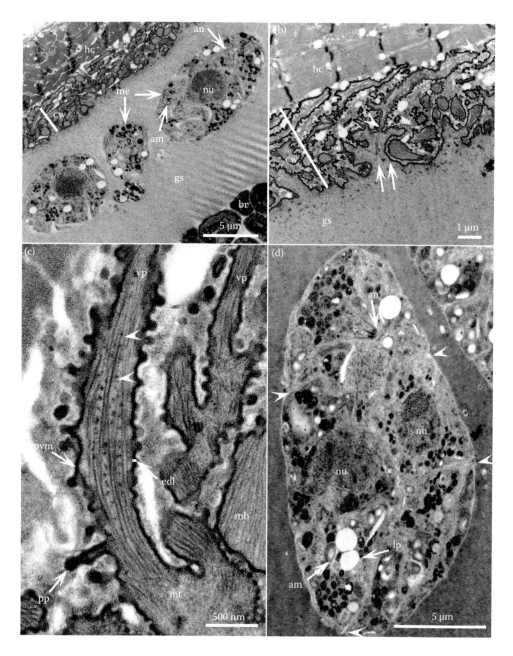

Figure 1.57 TEM of *S. fusiformis* sarcocyst wall. (a) The cyst wall (cw) is poorly demarcated from the host cell (hc). Three metrocytes (me) are located in gs. The me on the far right has a daughter anlagen (an), a few amylopectin (am) granules, a large nucleus (nu), and numerous electron-dense granular structures. (b) Highly branched vp. Double arrows point to branching of vp, 1 of the vp is very long (arrowheads) and juxtaposed against the host cell. The gs has a few granules at the junction of vp and the gs. (c) Longitudinal view of a vp with longitudinally oriented microtubules (mt). Arrowheads point to few indistinct structures, probably microtubules cut in cross section. Note dumbbell-like endings of pappilomatous structures (pp). (d) A metrocyte with 2 nuclei (nu), several am, lipid bodies (lp), fissure-like structures (arrowheads), and absence of micronemes and rhoptries. Note zoite anlagen (an) with its conoid. Numerous unidentified electron-dense structures of varying sizes are present. (Adapted from Dubey, J.P. et al., 2015. *Parasitology.* 142, 385–394.)

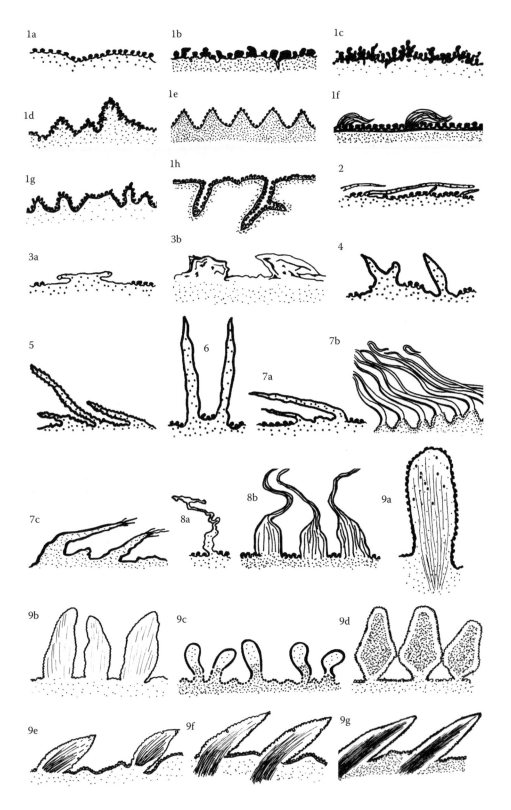

Figure 1.58 Line drawings of different sarcocyst wall types, 1a–9g (see Table 1.2 for description).

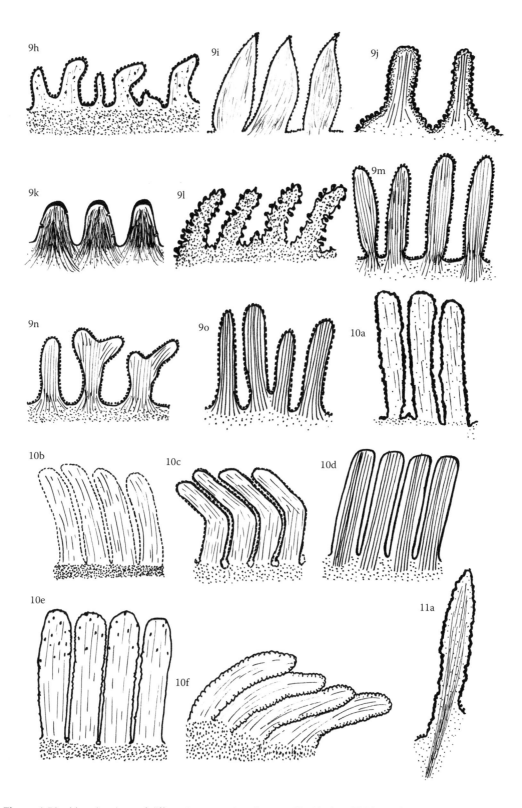

Figure 1.59   Line drawings of different sarcocyst wall types, 9h–11a (see Table 1.2 for description).

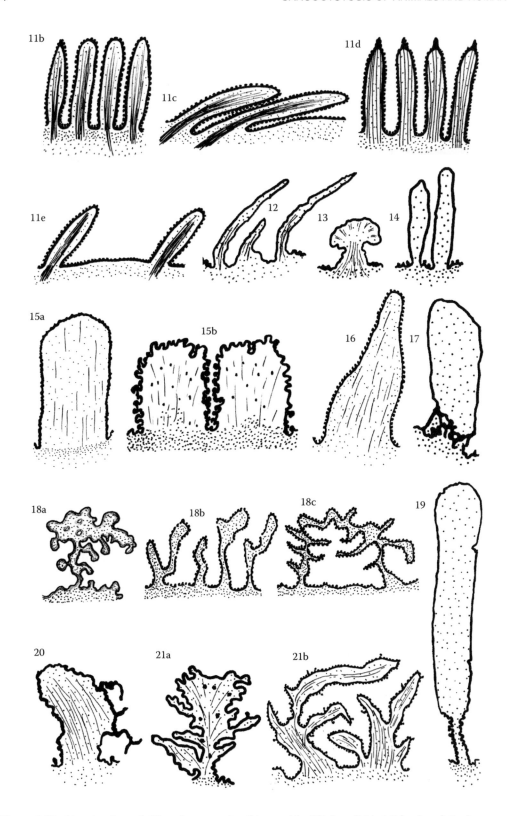

Figure 1.60 Line drawings of different sarcocyst wall types, 11b–21b (see Table 1.2 for description).

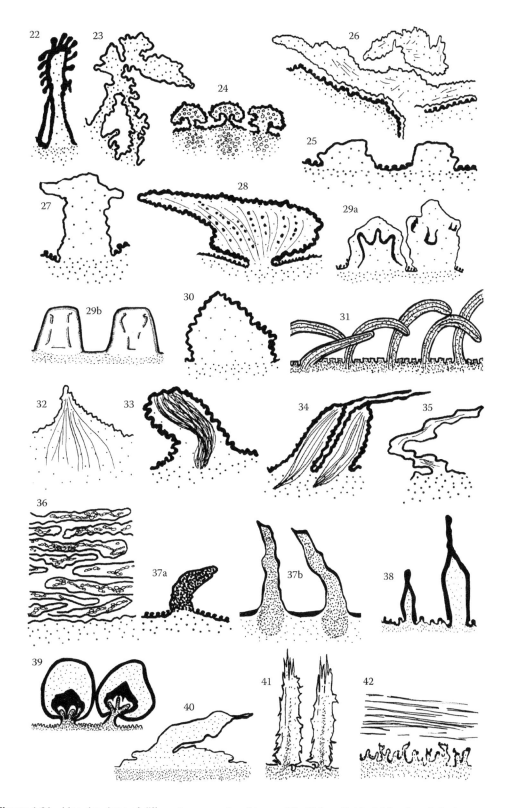

Figure 1.61 Line drawings of different sarcocyst wall types, 22–42 (see Table 1.2 for description).

Table 1.2 Characteristics of *Sarcocystis* Sarcocyst Wall Types

Type	Subtype	Sarcocyst Wall	Representative Species	Location in the Book
1	a	Pvm with minute undulations or knob-like blebs with rounded ends and spherical vesicles with short stalk inpocketing into host cell, gs smooth	*S. muris*	Figure 1.42b
	b	Pvm with pleomorphic blebs, conical, irregular	*S. schneideri*	Chapter 22
	c	Similar to type 1a but pvm with hyphenated blebs	*S. cymruensis*	Table 21.2
	d	Similar to type 1a but blebs interspaced with conical protrusions	*S. wobeseri*	Table 17.3
	e	Similar to type 1d but pvm with conical vp at uneven distances and blebs with conical to round tops; the sarcocyst wall appears spiny by LM	*S. sehi*[a]	Table 21.5
	f	Similar to type 1a but blebs interspersed with filliform vp lined with smooth mt	*S. melis*[a]	Table 19.1
	g	Pvm with stump-like protrusions at irregular distances	*S. cornixi*	Table 17.3
	h	Similar to type 1a, but invaginations into the interior of the sarcocyst	*S. rauschorum*	Figure 1.42a
2		Similar to type 1a, except that thin, hirsute structures arise from some of the apices of the undulating pvm	*S. wapiti*[a]	Figure 1.42c
3	a	Pvm undulated, vp flattened mushroom-like	*S. ferovis*[a]	Figure 1.43a
	b	Like type 3a but irregular mushroom-like vp not flattened	*S. sp. from black jackal*[a]	Table 19.1
4		Irregularly shaped vp with a finely granular core, pvm undulating at the base of the vp and smooth or undulating over the surface of the vp	*S. sigmodontis*	Figure 1.43b
5		Pvm highly folded to form hirsute vp with granular cores, gs very thin	*S. sulawesiensis*[a]	Figure 1.43c
6		Vp finger-like, arising from a dome-shaped base, pvm with minute undulations in the dome-shaped base, but relatively smooth over the surface of the vp. The vp with fine granular core and oriented perpendicularly to the sarcocyst surface	*S. capreolicanis*	Figure 1.43d
7	a	Elongated vp arising from a dome-shaped base, without fibrilar elements, and folded over the cyst wall	*S. cruzi*	Figure 7.2a
	b	Vp with 3 distinct regions: a dome-shaped base, an intermediate finger-like segment, and a thin thread-like distal segment. The intermediate and distal segments are often bent 90° and run parallel to the sarcocyst surface. In certain areas, however, the vp are arranged collectively in conical tufts. The core contains coarse granules in the base segment and fine granules in the intermediate and distal segments	*S. arieticanis*[a]	Figure 1.44a
	c	Vp very long, slender, folded over the cyst wall. Villar tips with a sharp spike-like structures	*S. stehlinii*[a]	Table 22.1
8	a	Similar to type 7, except that the basal and intermediate segments appear as 1, vp narrow abruptly to form the branched distal segment	*S. gruneri*[a]	Figure 1.44b and c
	b	Similar to type 8a, but vp have mt	*S. sp. from Sika deer*[a]	Table 18.1
9	a	Pvm with minute undulations, lined by thick edl, vp conical or tongue-shaped, close together, mt extending from tip to the gs, mt with edg	*S. campestris*	Figure 1.45a
	b	Same as type 9a but vp with indistinct mt without edg	*S. greineri*	Table 20.2
	c	Same as type 9a but pv wide apart at uneven distances, vp of variable sizes and without mt and edg	*S. felis*	Figure 15.1

(Continued)

Table 1.2 (*Continued*)   Characteristics of *Sarcocystis* Sarcocyst Wall Types

Type	Subtype	Sarcocyst Wall	Representative Species	Location in the Book
	d	Same as type 9a but vp leaf-like with a narrow stalk, wide in the middle, and rounded tapered end. The interior of the vp filled with vesicular electron-dense material	*S. gongyli*	Table 22.1
	e	Same as type 9a, except vp at irregular distances, folded over the cyst wall, vp with invaginations of the pvm, mt in vp denser at the base and without edg	*S. ramphastosi*	Table 17.3
	f	Same as type 9e but mt of vp extending deeper in gs and the mt much denser in gs versus in vp	*S. sulfuratusi*	Table 17.3
	g	Same as type 9f but mt in vp very dense both in the gs and in vp	*S. ursusi*	Table 19.1
	h	Same as type 9a, but with straight to sloping vp, with indistinct mt and dark granules	*S. kitikmeotensis*	Table 19.1
	i	Vp elongated and spear-like, mt without edg, numerous indentations of the pvm on the entire villar surface	*S. markusi*	Chapter 5
	j	Vp upright finger-like with knob-like structures arising from the pvm	*S. cameli* [a]	Figure 12.2a
	k	Vp short, stubby with criss-crossing mt, extending to gs. Villar tips with thick edl	*S. wenzeli* [a]	Figure 16.1
	l	Vp vertical to sloping with fine microtubules not extending into the gs	*S. phoeniconaii*	Table 17.3
	m	Vp similar to type 9d, but with profuse mt without edg extending into the gs	*S. mephitisi*	Table 19.1
	n	Vp similar to type 9m, but some villar tips bifurcated	*S. inghami*	Table 19.1
	o	Vp sloping to straight, mt without edg and extending from villar tips into the gs	*S. jaypeedubeyi*	Table 21.3
10	a	Similar to type 9a, but vp tightly packed, upright, few mt, not extending into gs	*S. odoi*	Figure 1.45c
	b	Similar to type 10a, but vp with profuse fine mt throughout the villar core	*S. hominis*	Figure 7.2
	c	Similar to type 10a, but vp elongated and bent at an angle and vesicles on pvm at the base of villi	*S. dubeyi*	Figure 10.1f
	d	Similar to 10b, but mt extending into the gs	*S. hemioni*	Table 18.1
	e	Similar to 10a, but vp with patchy electron-dense material on pvm and the core at the villar tips	*S. tarandi*	Table 18.1
	f	Vp folded over the cyst wall	*S.* sp. from black jackal [a]	Table 19.1
11	a	Vp similar to types 9e, but mt extend from the villar tips to the plasmalemma of metrocytes/bradyzoites	*S. fayeri* [a]	Figure 1.45b
	b	Similar to type 11a, but vp microtubules not extending deeper in gs, never reaching bradyzoites/metrocytes	*S. falcatula*	Figures 1.46, 17.1
	c	Similar to type 11b, but vp folded over the cyst wall giving it a thin wall appearance	*S. bertrami*	Chapter 11
	d	Vp similar to 11b, but with stylet-shaped villar tips	*S. lindsayi*	Table 17.2
	e	Vp similar to type 11b, but wide apart	*S.* sp. from sea lion [a]	Table 17.3
12		Tightly packed, thin, finger-like vp with mt in the basal one-third of the vp	*S. sybillensis*	Figure 1.47a
13		Vp widely spaced, mushroom-like with mt extending from the core into the gs. The pvm has irregularly spaced indentations	*S. mucosa* [a]	Figure 1.47b
14		Vp tightly packed, cylindrical-shaped, the tips of which contain disk-shaped plaques	*S. tenella*	Figure 1.48a and b

(Continued)

Table 1.2 (*Continued*)   Characteristics of *Sarcocystis* Sarcocyst Wall Types

Type	Subtype	Sarcocyst Wall	Representative Species	Location in the Book
15	a	Upright, closely packed "tombstone-like" vp with round tops, and a core containing scattered granules, scattered thin mt	*S. rangiferi*	Figure 1.48c
	b	Like type 15a but vp with deeply indented margins	*S. africana*[a]	Table 18.2
16		Similar to type 15, except that the sarcocyst wall consists of alternating conical and club-shaped vp with criss-crossing mt	*S. youngi*	Figure 1.48d
17		Rectangular vp with disc-shaped plaque at the apex and lateral margins and the pvm highly branched at the base of vp	*S. odocoileocanis*	Figure 1.49a
18	a	Vp branched and irregularly shaped, appearing T-shaped in appropriately cut sections, with a granular core	*S. dispersa*	Figure 1.49b
	b	Like type 18a, but vp elongated and U-like branching, and indentations on pvm	*S. muriviperae*	Table 21.1
	c	Vp highly branched anastomosing	*S. zamani*[a]	Table 21.2
19		Vp club-shaped with a cylindrical stalk and a sausage-shaped distal segment. In older sarcocysts, the distal segment may be shorter and irregular in shape. Invaginations of the pvm are numerous in the stalk and widely scattered in the distal segment. In young and intermediate sarcocysts, the core is filled with numerous small vesicles and membranous whorls, whereas in mature sarcocysts the core contains fine granules with localized areas of small vesicles.	*S. singaporensis*	Figures 1.28–1.30, 1.49c
20		Vp angular that may be pyramidal, trapezoidal, or mushroom-like. Serpentine projections arise from the surface of the vp as well as from the surface of the sarcocyst. The entire sarcocyst surface is covered by tiny bristles that project outward at 90° from the pvm. Vp core contains tightly packed mt	*S. medusiformis*[a]	Figure 1.50a and b
21	a	Vp with cauliflower-like branches containing microtubules with edg	*S. gigantea*	Figure 1.50c
	b	Vp highly branched, appear mesh-like by SEM, and like a dead tree by TEM, uniform papillomatous projection on vp with door knob tops	*S. fusiformis*	Figures 1.53, 1.55, 1.57
22		Vp with a proximal, sharply conical base and a distal segment that is cocklebur-like with short, radiating projections. Cross sections of the distal segment appear as cogwheels. Some of the projections from the distal segment connect with the margin of the sarcocyst wall. The core contains numerous vesicles especially in the distal half	*S. villivillosi*[a]	Figure 1.50d
23		Vp with anastomosing branches that contain fine granules and fine mt	*S. rileyi*	Figure 1.51a
24		Vp mushroom-like with a core of tightly packed mt, and mushroom-like protrusions also arise from the surface of other protrusions	*S. cornagliai*	Figure 1.51b
25		Similar to type 1a but blebs interspersed with dome-shaped small elevations at irregular distances and containing few mt	*S.* sp. from badger[a]	Table 19.1
26		Vp flattened and very irregular linguiform containing small invaginations on pvm. Pvm at villar tips lined by thick electron-dense material	*S. hardangeri*	Table 18.1
27		Vp palisade-like and thick T-shaped smooth surface	*S. sibirica*	Table 18.2

(Continued)

Table 1.2 (*Continued*)   Characteristics of *Sarcocystis* Sarcocyst Wall Types

Type	Subtype	Sarcocyst Wall	Representative Species	Location in the Book
28		Vp undulating, palisade-like or rhombic-like with a stalk at the base, expanded laterally, mt with edg	*S. hirsuta*	Figures 7.4, 10.1d
29	a	Vp molar tooth-like by SEM. By TEM interior of vp with variable-shaped condensations of electron-dense structures	*S. danzani*[a]	Table 18.2
	b	Similar to type 29a but vp with smooth margins and hollow on one side	*S. phacochoeri*	Table 18.2
30		Parasite induced host-cell encapsulation, vp irregularly triangular with indented margins looking like shark teeth	*S. dubeyella*	Table 18.2
31		Vp up to 14 µm long, folded over the sarcocysts surface, vp with mt in the core without edg. Battlement-like invaginations of pvm	*S. suihominis*[a]	Chapter 6
32		Vp thorn-like with mt radiating into the gs	*S. ippeni*[a]	Figure 12.2b
33		Vp thumb-like with a compact central bundle of mt in the core and with small invaginations on the surface	*S. hippopotami*[a]	Table 18.2
34		Vp finger-like with mt penetrating the ground substance, small invaginations on the surface and hirsute at the tip	*S. giraffae*[a]	Table 18.2
35		Vp kinked finger-like with scattered mt in the interior	*S. klaseriensis*[a]	Table 18.2
36		Vp in rows parallel to the cyst wall surface and strap-like with chain-like osmiophilic structures in the interior	*S. camelopardalis*[a]	Table 18.2
37	a	Pvm with invaginations deep in to the gs. Vp spine-like, arising at irregular distances. Interior of vp edg expanding into the middle of gs with irregularly indented margins and condensed matrix	*S. simonyi*	Table 22.1
	b	Vp slender, up to 7 µm long and lined by electron-dense material from the tip to knob-like ending inside the gs	*S. gallotiae*	Table 22.1
38		Vp steeple-shaped surmounted by a spire	*S. speeri*[a]	Figure 1.52
39		By SEM, vp pallisading, like a sea anemone. By TEM, vp elongated with tapered end. Condensation of electron-dense material at the base of vp giving mushroom-like appearance	*S. mihoensis*[a]	Figure 1.56a
40		Vp with flat wart-like elevations. Short pit-like parts on pvm between vp. Vp without mt and tipped ends	*S. melampi*[a]	Table 18.2
41		Vp upright, finger-like corrugated pvm. Villar tips with deep invaginations giving serrated appearance. Villar cores with electron-dense material that appears thick in the middle	*S.* sp. from Hispanic ibex[a]	Table 18.2
42		Macroscopic with secondary cyst wall. Pvm lined with thick edl, numerous irregular vp	*S. hoarensis*[a]	Table 22.1

Pvm = parasitophorous vacuolar membrane, gs = ground substance, vp = villar protrusions, mt = microtubules, edl = electron dense layer, edg = electron dense granules.

[a] Unique types reported only for 1 *Sarcocystis* species.

1.3.1.5 *Fertilization*

Fertilization usually occurs within 18 h PI.[819,1364,1598] Microgametes penetrate the macrogamont by an unknown process (Figure 1.67). The plasmalemma of the microgamete fuses with that of the macrogamont to create a cytoplasmic bridge through which the microgamete nucleus enters the macrogamont (Figure 1.67). Evidently, soon thereafter, the cytoplasmic bridge closes and the gamete nuclei fuse to form a zygote. The remnant of the microgamete becomes vesicular and eventually disappears *in situ* (Figure 1.68). What proportion of microgametes succeeds in fertilizing the macrogamete is unknown.

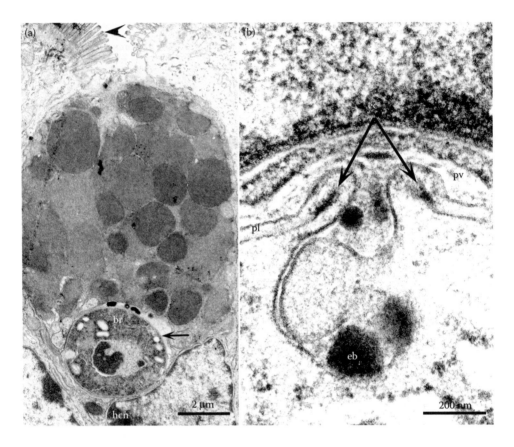

Figure 1.62 TEM of *S. cruzi* bradyzoites transforming to gamonts. (a) Bradyzoite (br) located between mucin droplets and nucleus of intestinal goblet cell of an experimentally infected coyote. Arrowhead points to microvillus border. (b) Exocytosis pore discharging electron-dense bodies into parasitophorous vacuole.

1.3.1.6 Oocyst

Unsporulated oocysts consist of a sporont surrounded by an oocyst wall. The oocyst wall has an electron-dense, finely granular outer layer, and an inner layer of 1–4 membranes (Figure 1.68).[803] Electron-dense projections arise from the outer surface of the oocyst wall (Figure 1.68). The sporont is limited by a plasmalemma and a single inner membrane and contains a nucleus, endoplasmic reticulum, mitochondrion, ribosomes, Golgi complex, amylopectin bodies, micropores, and an occasional electron-dense body (Figure 1.68).

1.3.1.7 Sporocyst

There are no reports on the ultrastructural changes during sporulation of oocysts of *Sarcocystis* spp. The sporocyst walls are composed of a thin continuous outer layer, and a thick inner layer consisting of 4 plates joined at sutures similar to those in related coccidia. The sporocyst wall of *S. cruzi* has a lamellar outer layer, approximately 50 nm thick, consisting of alternating electron-dense and electron-lucent layers and an electron-dense inner layer, approximately 130 nm thick, that is separated from the outer layer by a single membrane (Figure 1.69a). The inner layer has alternating bands (≈5 nm thick) of electron-dense and electron-lucent material perpendicular to the surface

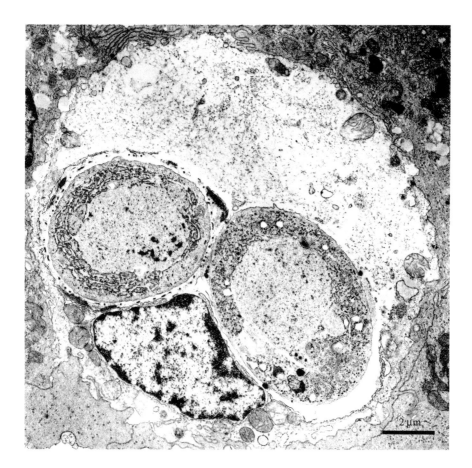

Figure 1.63 TEM of 2 *S. cruzi* macrogamonts in a degenerate goblet cell in the small intestine of an experimentally infected coyote.

of the sporocyst. A thickening (150–180 nm thick) is seen at the margin of each plate. A thin strip of electron-dense material (≈20–30 nm thick) is interposed between the thickenings of 2 apposing plates forming a suture (Figure 1.69b). In SEMs, the sutures appear as ridges on the surface of the sporocyst (Figure 1.69a). Treatment of sporocysts with NaOCl (commercial laundry bleach) usually results in the removal of the outer layer of the sporocyst wall (compare Figure 1.70a and b). Sporulated oocysts and sporocysts are located in the lamina propria. The mechanism of migration and release of sporocysts in the intestinal lumen is unknown.

During excystation, bile salts and/or trypsin act upon the sutures, causing the plates to separate from the interposed strip. The edges of the plates curl inward and openings appear where plates had been adhered to one another, enabling the release of sporozoites (Figures 1.69 and 1.70). However, *S. muris* sporocysts can excyst in mouse peritoneum, in the apparent absence of trypsin or bile.[683,1503] *S. neurona* sporocysts are equally infective to mice by the oral and the subcutaneous routes, indicating that trypsin is not necessary for excystation of some *Sarcocystis* species.[470]

1.3.1.8 Sporozoites

Sporozoites of *S. cruzi* are banana-shaped and have all the ultrastructural features described earlier for bradyzoites, such as conoid, apical rings 1 and 2, polar rings 1 and 2, micronemes,

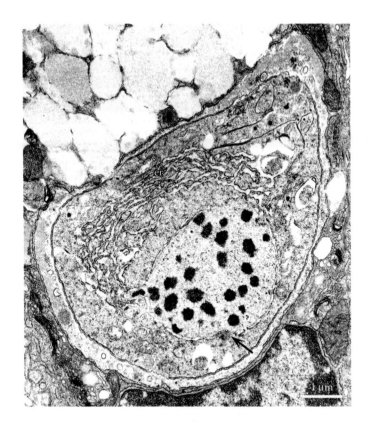

Figure 1.64 TEM of an early *S. cruzi* microgamont in the small intestine of an experimentally infected coyote showing a portion of the spindle apparatus (arrow). Note profuse endoplasmic reticulum (er).

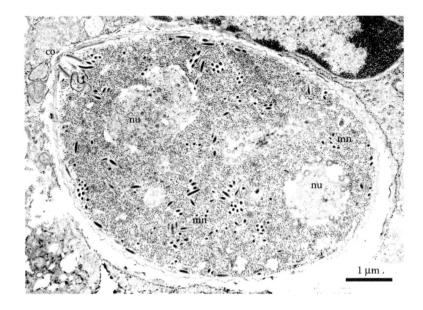

Figure 1.65 TEM of an early *S. cruzi* microgamont in the intestine of an experimentally infected coyote showing 2 lobes of a nucleus. Note the nuclear pores and scattered micronemes (mn) and conoid (co) of the original bradyzoite.

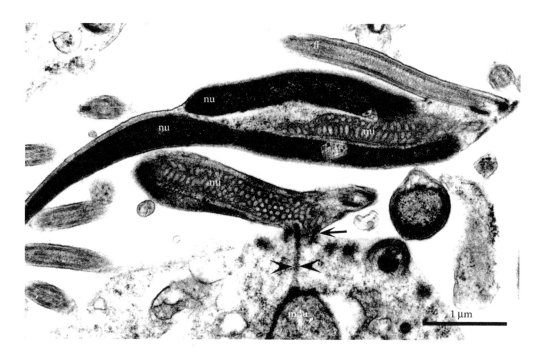

Figure 1.66   TEM of 2 *S. suihominis* microgametes in longitudinal section, 1 of which is still connected (arrow) to the surface of the microgamont in which a nuclear bridge (double arrowheads) extends from the residual nucleus into the microgamete. Note the flagellum (fl), nucleus (nu), and mitochondrion (mi) of the microgametes. (From Mehlhorn, H., Heydorn, A.O., 1979. *Z. Parasitenkd.* 58, 97–113. With permission.)

rhoptries, micropore, nucleus, mitochondrion, Golgi complex, endoplasmic reticulum, ribosomes, amylopectin granules, pellicle, and subpellicular microtubules. Sporozoites differ, however, from bradyzoites and merozoites by possessing a virus-like crystalloid body, and structures similar to rhoptries and micronemes in the posterior half of the body (Figure 1.71). The crystalloid body consists of electron-dense and electron-lucent granules (≈38 nm in diameter), and is usually located in the posterior half of the sporozoite. Some sporozoites have several crystalloid bodies randomly scattered throughout the sporozoite, with some bodies anterior to the sporozoite nucleus. Also, various other organelles and inclusion bodies, such as rhoptries, micronemes, and amylopectin granules, may be dispersed among the granules of the crystalloid body. This body is probably analogous to the homogenous refractile bodies of *Eimeria* spp., in which the body is believed to represent an energy or amino acid reserve.[1991] The crystalloid body was not found in *S. neurona* sporozoites.[1079]

1.3.1.9 Schizogony

Sarcocystis schizonts multiply by endopolygeny, in which the nucleous becomes multi-lobbed, and numerous merozoites bud simultaneously at the surface of the schizont (Figures 1.21 and 1.22).

Schizonts differ from other intracellular stages in the life cycle of *Sarcocystis* spp. in that they are located free within the cytoplasm of the host cell and are not surrounded by a PV (Figures 1.72 through 1.82). Schizonts also lack subpellicular microtubules. Early phases of sporozoite entry and development of *S. singaporensis* sporozoites into schizonts were reported.[887,891] *S. singaporensis* sporozoites entered several types of rat cells but developed only in rat endothelial and pneumocytes,

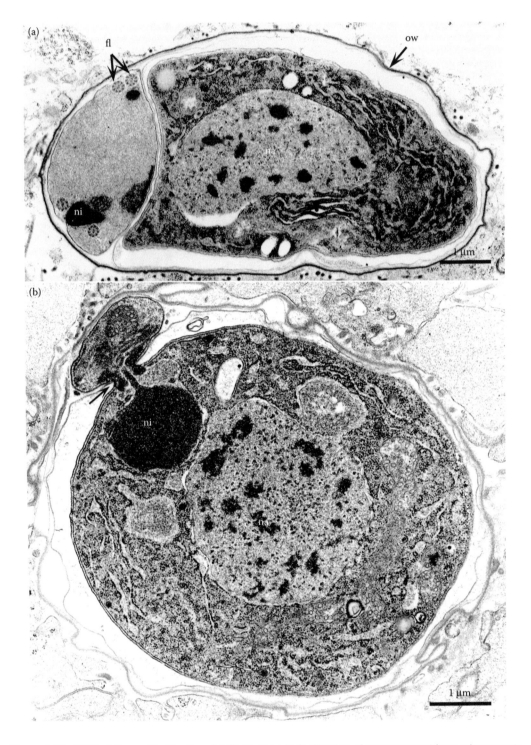

Figure 1.67   TEM of *S. cruzi* fertilization. (a) Early stage in fertilization of macrogamont by a microgamete. Note the oocyst wall (ow), a large nucleus (na) of the macrogamete, and small nucleus (ni), and flagella (fl) of microgamete. (b) Intermediate stage in fertilization in which the microgamete nucleus (ni) is entering the macrogamont through a cytoplasmic bridge (arrow) created by fusion of the plasmalemma of the microgamete and macrogamont. (Adapted from Sheffield, H.G., Fayer, R., 1980. *Proc. Helminthol. Soc. Wash.* 47, 118–121.)

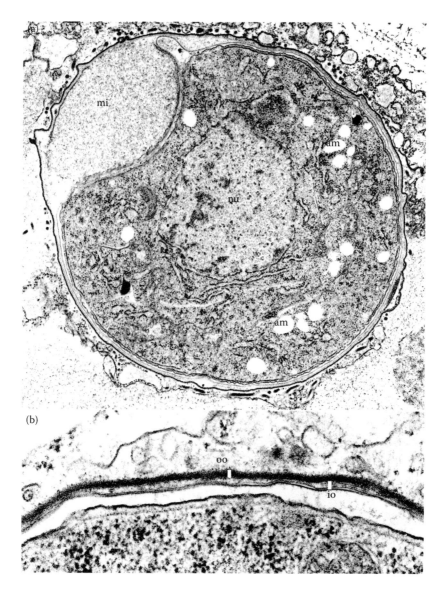

Figure 1.68   TEM of *S. cruzi* zygote. (a) A zygote, soon after fertilization, showing a remnant of the microgamete. Note the large nucleus (na) and amylopectin granules (am) of the macrogamete. (b) The oocyst wall and pellicle of the zygote. Note that the oocyst wall consists of an inner layer (io) of 2 membranes and an electron-dense outer layer (oo).

indicating that specific receptors or physiological conditions may be needed for development of schizonts. Most sporozoites entered cells within 2 h of inoculation but schizont development was delayed until the third day. At 1 h PI, some sporozoites were located within PVs, and the findings were confirmed ultrastructurally, while some were located free in the cytoplasm; these intravacuolar sporozoites had disappeared by 2 DPI. Even young schizonts, with an enlarged nucleus and condensed crystalloid body, were seen free in the host cytoplasm. The formation of a PV and subsequent escape may be an adaptation from the free-living state to the intracellular state.

Figure 1.69 SEMs showing sporocysts and excystation of *S. miescheriana* sporozoites. (a) Two sporocysts held together by a thin oocyst wall. Junctions between apposing plates appear as ridges (arrow) in the sporocyst wall. (b) Intermediate stage in excystation in which the apposing plates in the sporocyst wall have partially separated from each other in response to excysting fluid. (c) A single sporozoite (sp) is visible within a sporocyst in which the walls have almost completely separated. (d) Final stage in excystation in which the plates that formed the sporocysts wall have separated and released sporozoites. (From Strohlein, D.A., Prestwood, A.K., 1986. *J. Parasitol.* 72, 711–715. With permission.)

Soon after entering the appropriate host cell, the sporozoite transforms into an ovoid schizont ($\approx 10 \times 8$ μm), which contains a large nucleus with a single nucleolus, a double pellicular membrane, mitochondrion, endoplasmic reticulum, ribosomes, electron-dense bodies, Golgi complex, micropores, autophagic vacuoles, scattered micronemes and rhoptries, and lipid, amylopectin, electron-dense, and multivesicular bodies (Figures 1.72 through 1.75). Autophagic vacuoles contain parasite cytoplasm, ribosomes, and micronemes, indicating that this may be 1 of the sites for recycling the micronemes and other cytoplasmic components. Subsequently, the crystalloid body and

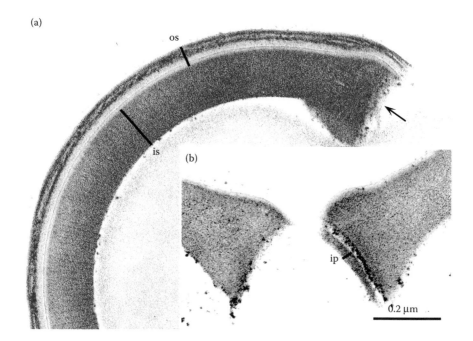

Figure 1.70 TEM of *S. cruzi* sporocyst wall. (a) The outer layer (os) is lamellated with alternating electron-lucent and electron-dense layers. The inner layer (is) is cross-striated and has a thick end at its margin (arrow). (b) Site of apposition between 2 plates of the sporocyst wall after exposure to excysting fluid. Note that the plates have separated and the interposed strip remains attached to the fish lip-like thickening of 1 plate; also note that exposure of the sporocyst to sodium hypochloride prior to incubation in excysting fluid has removed the outer layer of the sporocyst wall.

most organelles disappear. In more advanced stages of the parasite, the nucleus becomes irregularly shaped with several nucleoli, and the micronemes are completely absent (Figure 1.75).

Intermediate schizonts contain anlagen of developing merozoites and a highly lobulated nucleus (Figure 1.76). A spindle apparatus appears in association with each lobe of the nucleus (Figure 1.77). The spindle consists of several microtubules that extend completely across the nucleus from 1 centrocone to the other, and of shorter diverging microtubules that cover an arc of 35°. Each of the 2 centrioles is located immediately above and slightly lateral to the apex of each centrocone. In some species of *Sarcocystis*, 1 or 2 multivesicular bodies may be situated laterally along each centrocone, just outside the nuclear envelope. Coincidental with the appearance of centrocones, the conoid, apical rings, a single-unit membrane, and subpellicular microtubules of a merozoite anlagen form immediately above the centrioles (Figure 1.77). As each merozoite anlagen elongates, the single-unit membrane, and its associated subpellicular microtubules, extend posteriorly, incorporating a part of the nucleus, which eventually pinches off from the rest of the nucleus. Cytoplasm, ribosomes, endoplasmic reticulum, and mitochondrion, are also incorporated into each merozoite anlagen. Micronemes also appear at this stage in association with the Golgi complex or a multivesicular body, between the apical end and the nucleus (Figures 1.78 through 1.80), and continue to be formed until the merozoite is complete. The double-membrane pellicle folds in around the developing merozoites until they appear to bud at the surface of the schizont (Figures 1.78 through 1.80). The double-membrane pellicle becomes closely associated with the single-unit membrane of each merozoite anlage to form the merozoite pellicle. Usually, 1 of the 3 membranes disappears so that the fully formed merozoite has a double membrane pellicle.

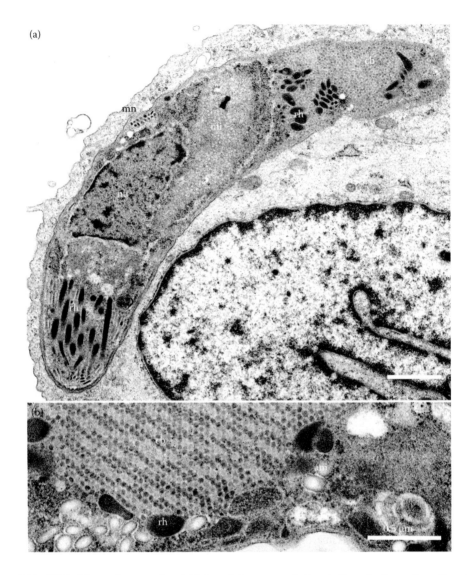

Figure 1.71   TEM of *S. cruzi* sporozoites. (a) Intracellular sporozoite free in the cytoplasm of a cultured cardio
pulmonary arterial endotelial (CAPE) cell. Note the absence of a parasitophorous vacuole, and
the presence of a large mitochondrion (mi), crystalloid body (cb), and numerous rhoptries (rh).
(b) High magnification of a portion of a sporozoite showing the crystalloid body, amylopectin
granules (am) and rhoptries.

Merozoites of all *Sarcocystis* species contain virtually all the organelles observed in bradyzoites, except rhoptries (Figure 1.82). The merozoite is twisted slightly to form a helical pattern on
its surface (Figure 1.83a). This helical pattern is probably due to a series of rectangular, contiguous strips (plaques) of the inner membrane, aligned in longitudinal rows, as has been described
for other coccidia. At the anterior tip, concentric ridges are visible by SEM, which are formed
by the apical rings and polar ring 1, which are located just beneath the merozoite plasmalemma
(Figure 1.83b and c).

In those species of *Sarcocystis* that undergo schizogony in blood vessels, the host cell appears
to be an endothelial cell. In renal glomeruli, the host endothelial cell is completely surrounded by

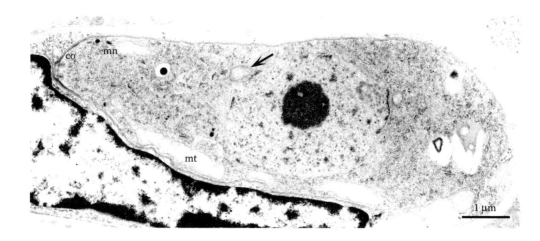

Figure 1.72   TEM of a *S. falcatula* merozoite transforming into an early schizont in a pulmonary capillary cell of an experimentally infected budgerigar. Note a central, spheroidal nucleus with a large nucleolus, conoid (co) pushed against the host-cell nucleus, a few micronemes (mn), a mitochondrion (mt), and apicoplast (arrow). (Adapted from Speer, C.A. Dubey, J.P., 1999. *J. Parasitol.* 85, 630–637.)

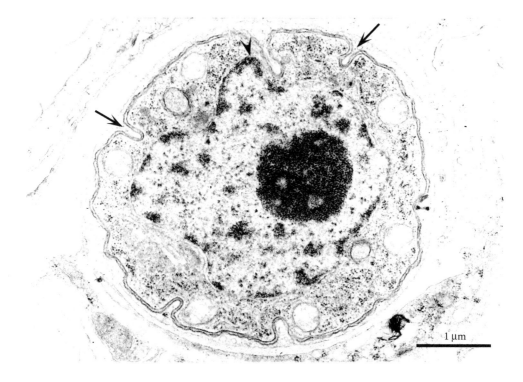

Figure 1.73   Cross section of early schizont of *S. falcatula* in a pulmonary capillary cell of the experimentally infected budgerigar in Figure 1.72. Note the infolded pellicle (arrows) of the schizont. The nucleus has begun to lobulate (arrowhead). (Adapted from Speer, C.A., Dubey, J.P., 1999. *J. Parasitol.* 85, 630–637.)

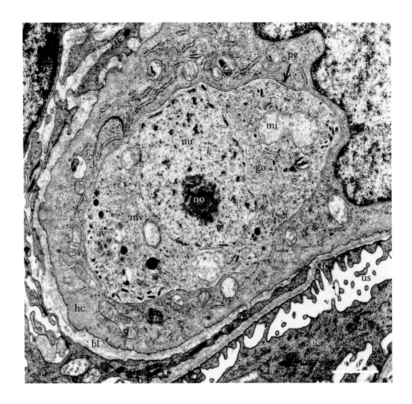

Figure 1.74   TEM of a uninucleate *S. tenella* schizont lying free in the cytoplasm of a glomerular capillary endothelial cell in the kidney of an experimentally infected sheep. (From Speer, C.A., Dubey, J.P., 1981. *J. Protozool.* 28, 424–431. With permission.)

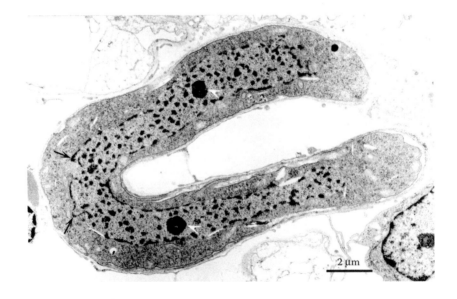

Figure 1.75   TEM of an elongate *S. falcatula* schizont conforming to the shape of the pulmonary capillary in a pulmonary capillary cell of the experimentally infected budgerigar in Figure 1.73. Note the elongated nucleus with 2 nucleoli (arrowheads) and nuclear spindles (arrows). (Adapted from Speer, C.A., Dubey, J.P., 1999. *J. Parasitol.* 85, 630–637.)

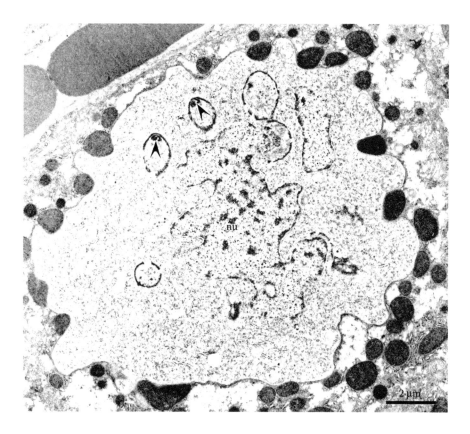

Figure 1.76   TEM of *S. rauschorum* schizont with irregular nucleus and developing apical complexes (arrowheads) in merozoite anlagen.

a basal lamina, on one side of which is a typical capillary endothelial cell, and on the other side are foot processes of podocytes and the urinary space (Figure 1.81). Merozoites may invade cells, normally not parasitized by schizonts. For example, *S. singaporensis* merozoites were shown inside red blood cells of rats.[1368]

Merozoites population differs structurally and biologically. Jäkel et al.[888] reported 2 types of merozoites in schizonts that developed 11–13 DPI of rats with *S. singaporensis* sporocysts. Type 1 merozoites were PAS-negative and these produced schizonts in capillary endothelial cell cultures. Type 2 merozoites were PAS-positive and these induced unizoite sarcocyst-like structures when inoculated in myoblast cultures.[888]

1.4 TAXONOMIC CRITERIA

Before the discovery of the life cycle of *Sarcocystis* in 1972, the 2 major criteria for naming a new species were the structure of sarcocysts and the species of the host. Because the age of sarcocysts and the method of fixation greatly influenced the sarcocyst structure, some species are not clearly described. Life cycle studies have indicated that some structurally similar sarcocysts (e.g., *S. tenella* of sheep and *S. capracanis* of goats) are actually different species, based on host specificity for the intermediate host. Furthermore, some species of *Sarcocystis* (e.g., *S. falcatula* and *S. neurona*) can infect numerous species of hosts belonging to different orders (discussed later). Because it is often very difficult to complete the life cycle of the *Sarcocystis* species of large animals

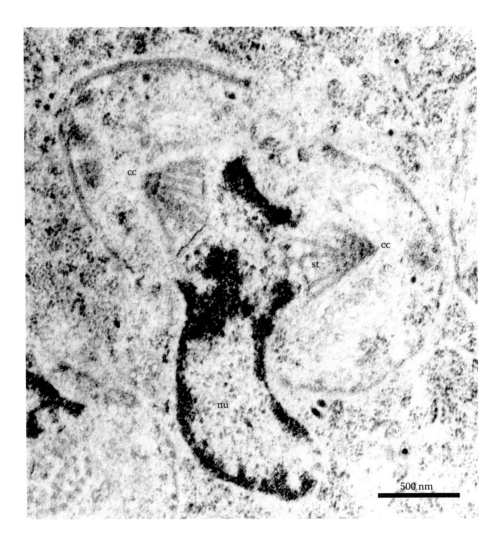

Figure 1.77   TEM of spindle apparatus and developing apical complexes (cc), and spindle microtubules (st) of 2 *S. tenella* merozoites. (From Speer, C.A., Dubey, J.P., 1981. *J. Protozool.* 28, 424–431. With permission.)

under laboratory conditions, the relative value of different criteria that might be used to validate the *Sarcocystis* species will be discussed. Consequently, multiple factors contribute to the naming of species, relying on what information is actually available, including known host species from natural and/or experimental infections, sarcocyst morphology, and, most recently, on DNA data. As new data become available, the taxonomic status of some species has changed and the status of others is expected to change.

1.4.1 Sarcocysts

The shape and size of the sarcocyst vary with the age of the sarcocyst, the type of host cell parasitized, and the techniques used for study. For example, sarcocysts of the same species in cardiac muscles and in the CNS are always smaller than those in skeletal muscles.[360,367,382] The size and shape of sarcocysts will also vary, depending on fixation (they are smaller in fixed specimens than

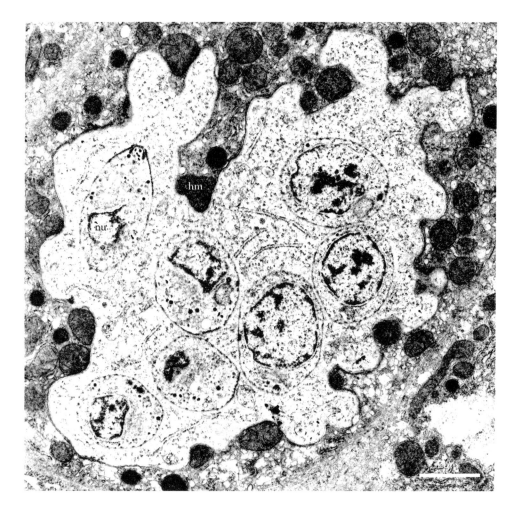

Figure 1.78   TEM of advanced *S. rauschorum* intermediate stage schizont in which each developing merozoite contains a nucleus but the merozoites are still located within the original parasite membrane.
Note numerous host-cell mitochondria (hm) surrounding the schizont.

in live specimens), and possibly might vary with the type of fixative. Because they are often located
in contractile muscles, the sarcocyst size will vary, depending on whether the host cell was relaxed
or contracted at the time of fixation.

Some *Sarcocystis* species continue to grow in size for several years after they have reached
infectivity. For example, *S. bertrami* of the horse and *S. medusiformis* of sheep do not attain their
maximum length (15 mm) in their hosts until they are approximately 4 years old, whereas *S. cruzi*
attains a maximum size of 0.5 mm at approximately 4 months PI. The shape of sarcocysts also varies with relation to the location. For example, *S. gigantea* sarcocysts in the esophagus are globular
to pear-shaped, whereas sarcocysts of the same species in the diaphragm are elongate and slender
(see Chapter 8). Sarcocysts may continue to grow in size, but the wall structure stabilizes once
bradyzoites have formed.[1597]

Metrocyte morphology is not a useful criterion for speciation because metrocytes are often
irregularly shaped and their size is highly variable, depending on the stage of division. The structure of bradyzoites also varies. Bradyzoites of some *Sarcocystis* species are densely packed within

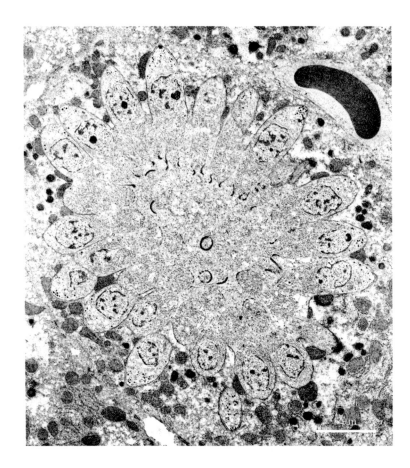

Figure 1.79   TEM of nearly mature *S. rauschorum* schizont in which the merozoites are still attached but are protruding at the margin of the meront.

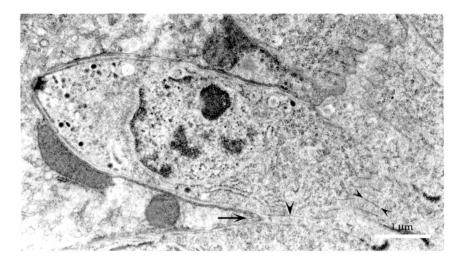

Figure 1.80   Higher magnification of a portion of Figure 1.79 showing a budding *S. rauschorum* merozoite; note that the merozoite plasmalemma (arrow) has invaginated just posterior to the merozoite nucleus. Single arrowhead points to the inner membrane complex of the merozoite and opposing arrowheads point to subpellicular microtubules.

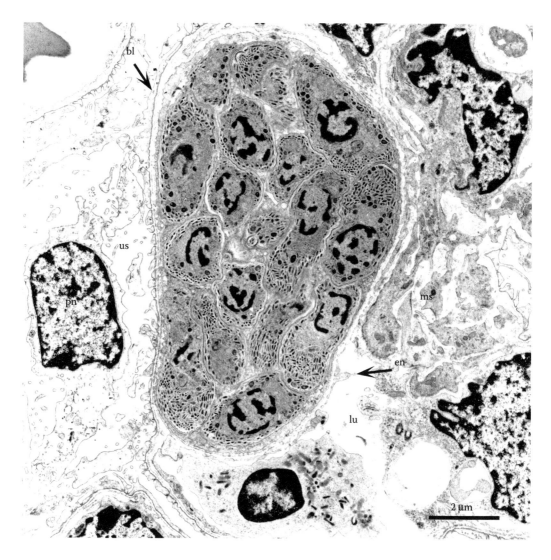

Figure 1.81 TEM of a mature *S. tenella* schizont in a glomerular capillary endothelial cell. Note the endotelial cell (en), basal lamina (bl), urinary space (us), podocyte nucleus (pn), mesenchymal area in glomerulus (ms), and lumen (lu) of the blood vessel. (From Speer, C.A., Dubey, J.P., 1981. *J. Protozool.* 28, 424–431. With permission.)

sarcocysts, whereas those of other species are sparsely found. Such conditions may affect size or shape. Because bradyzoites in most *Sarcocystis* species are banana-shaped, with great variation in curvature, it is difficult to measure them accurately. Measurements obtained from live preparations are unreliable, because the size varies considerably depending on the pH and osmolarity of the medium used, the weight of the cover glass, the volume of fluid under the coverslip, movement during measurement, measured when fixed or live, and the effect of birefringence. The banana-shaped bradyzoites of some *S. cruzi* become globular in acid–pepsin solution used to digest away surrounding host tissue. Therefore, measurements of zoites fixed *in situ* within sarcocysts are more reliable than measurements of zoites liberated by digestion. The width of the bradyzoites at the level of the nucleus may be more reliable than the length of bradyzoites, because in histologic sections it is difficult to measure bradyzoites cut longitudinally along their entire length. Overall, unless the

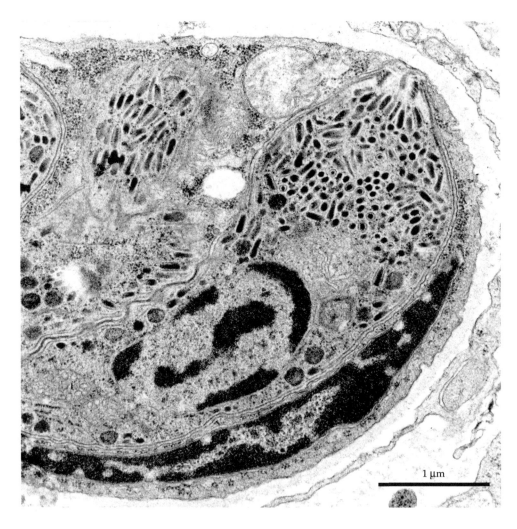

1 µm

Figure 1.82   TEM of *S. tenella* merozoite. Note the absence of rhoptries. (From Speer, C.A., Dubey, J.P., 1981. *J. Protozool.* 28, 424–431. With permission.)

size differences of bradyzoites are obvious, this criterion should not be used for identification. The structure of organelles, such as mitochondrion, rhoptries, and micronemes, is not a good taxonomic criterion.

The structure of the sarcocyst wall can sometimes be a useful criterion for speciation. For example, all 4 species of ovine *Sarcocystis* (*S. gigantea*, *S. tenella*, *S. medusiformis*, and *S. arieticanis*) have sarcocyst walls characteristically different from one another, even though the sarcocyst wall of *S. tenella* is similar to that of *S. capracanis* of the domestic goat. The sarcocyst walls of various species of *Sarcocystis* are illustrated in Figures 1.42 through 1.51.

1.4.2 Schizonts

The structure of the schizont is of limited taxonomic value because it varies greatly, depending on the developmental cycle (e.g., first versus second generation) and the host-cell parasitized. For example, second-generation schizonts of *S. cruzi* and *S. tenella* are slender and elongated in skeletal muscles but are globular in renal glomeruli.[360,382]

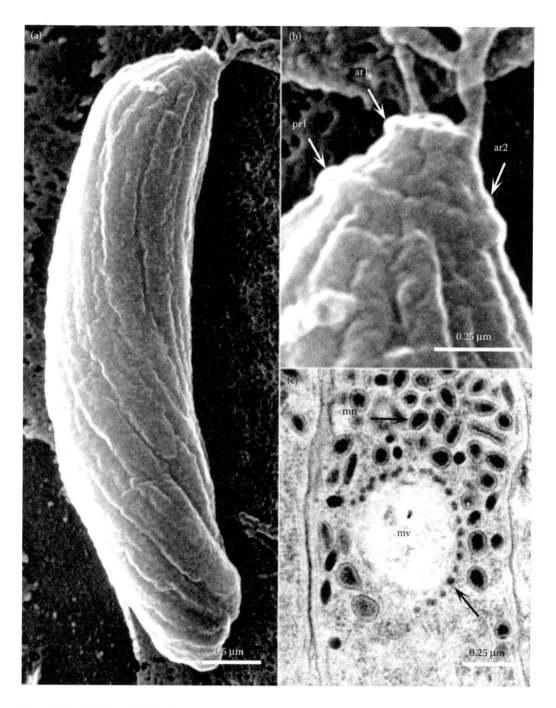

Figure 1.83 SEMs and TEM of merozoites. (a) A helical pattern is seen on the surface of a *S. cruzi* merozoite obtained at 57 days after sporozoites were inoculated into cultured CPAE cells. (b) High magnification of the anterior tip of the merozoite is seen in Figure 1.83a. Note the ridges created by underlying apical rings 1 and 2, and polar ring 1. (c) Anterior region of *S. tenella* merozoite shows multivesicular body (mv) which appears to give rise to micronemes (mn) at its margin (arrow). (From Speer, C.A., Dubey, J.P., 1981. *J. Protozool.* 28, 424–431. With permission.)

1.4.3 Oocysts and Sporocysts

The structure of the oocyst and sporocysts, traditionally used to determine species of other coccidia, is of little or no taxonomic value in *Sarcocystis* because

a. Except for minor variations in size, all *Sarcocystis* sporocysts and oocysts are structurally similar.
b. Sporocysts of numerous species of *Sarcocystis* in a given host overlap in dimensions. For example, dogs are definitive hosts for 22 species and cats are host for 12 species, and sporocysts of most canine species are about 15×10 μm and those of feline species are about 12×10 μm (see Chapters 14 and 15).

The prepatent and patent periods vary tremendously and are of no systematic value in *Sarcocystis*. For example, the prepatent period for *S. cruzi* is 7–33 days, and sporocysts can be excreted for many months.

1.4.4 Host Specificity

The species of *Sarcocystis* are generally more host-specific for their intermediate hosts than for their definitive hosts (Table 1.3 through Table 1.5). For example, for *S. cruzi*, the ox and bison are the only intermediate hosts, whereas the dog, wolf, coyote, raccoon, jackal, and fox, can act as definitive hosts. Coyotes and foxes also serve as efficient definitive hosts for other dog-transmitted species (e.g., *S. tenella* and *S. capracanis*). None of the species transmissible via dogs are proven to be transmitted by cats and vice versa (Table 1.4). However, *S. muris* transmitted via domestic cats can be transmitted experimentally by ferrets,[1484] and a species of *Sarcocystis* in chickens is reported to be transmitted by dogs and cats, although details require confirmation (see Chapters 14 through 16).

Table 1.3   Attempted Transmission of *Sarcocystis* Species of Large Animals to Other Animals

Sarcocystis Species	Intermediate Host	Attempted Transmission to	Results[a]	References
S. arieticanis	Sheep	Goat	Negative	816
S. tenella	Sheep	Goat	Negative	354
S. tenella	Sheep	Ox	Negative	1468
S. capracanis	Goat	Sheep	Negative	1341,1392
S. hircicanis	Goat	Sheep	Negative	1768
S. cruzi	Ox	Sheep, monkey, rabbit, pig, rat	Negative	578
S. cruzi	Ox	Bison	Positive	361
S. cruzi	Bison	Ox	Positive	586
S. cruzi	Ox	Buffalo	Negative	882
S. bertrami	Donkey	Horse	Positive	1131
S. ferovis	Bighorn sheep	Domestic sheep	Negative	368
S. wapiti	Wapiti	Ox	Negative	359
S. sybillensis	Wapiti	Ox	Negative	359
S. gracilis	Roe deer	Sheep, mice	Negative	538
S. odocoileocanis	White-tailed deer	Goat	Negative	1064
S. odocoileocanis	White-tailed deer	Sheep, ox	Positive	266

[a] Sarcocysts in the muscles of recipient animals.

Table 1.4 Excretion of Sporocysts by Carnivores Fed *Sarcocystis*-Infected Meat

Intermediate Host	*Sarcocystis* Species	Sporocysts Excreted[a]	Sporocysts Not Excreted[a]	Prepatent Period (days)	Year	Reference
Cattle (*Bos taurus*)	*S. cruzi*	11 dogs, 2 foxes, 2 raccoons	16 cats, 3 Macaca mulatta, 4 skunks, 2 ferrets (*Mustela putorius*), 6 rats, 6 guinea pigs, 6 rabbits	8–15	1976	**578**
	S. hominis	4 Rhesus monkeys, 4 baboons, 1 human	Cats, dogs	12	1976	**805**
	S. hirsuta	Cats	Rhesus monkey, baboon, human, dog	9		
	S. cruzi	Dogs	Cats	11–12	1985	**1223**
	S. hirsuta	3 cats	1 human	12	1990	**397**
	S. cruzi	8 dogs	8 cats	10–12	1992	**654**
	S. cruzi	4 dogs	2 cats	12–16	1994	**1372**
	S. hominis	2 *Macaca fascicularis*	2 dogs, 2 cats	10	1998	**1521**
Horse (*Equus caballus*)	*S. equicanis*	3 dogs	6 cats	8	1975	**1481**
	S. fayeri	10 dogs	10 cats	10–15	1977	**351**
	S. bertrami	1 dog	1 cat, 1 raccoon	9–10	1983	**1131**
	S. bertrami	1 dog	2 humans, 3 cats	13	1982	**833**
Donkey (*Equus africanus asinus*)	*S. bertrami*	1 dog	1 cat, 1 raccoon	9–10	1983	**1131**
Water buffalo (*Bubalus bubalis*)	*S. fusiformis*	4 cats	1 dog, 2 monkeys	8–14	1978	**327**
	S. fusiformis	4 cats	4 dogs	14–16	1995	**861**
	S. buffalonis	7 cats	4 dogs	10	1997	**863**
	S. levinei	2 dogs	2 cats	16–18	1997	**862**
Camel (*Camelus dromedarius*)	*S. cameli*	3 dogs	3 cats	10–14	1980	**822**
	S. cameli	3 dogs	5 cats	9–10	1995	**826**
	S. sp.	2 dogs	4 cats	9–10	1996	**568**
	S. camelicanis	13 dogs	13 cats	11	2009	**11**
Llama (*Llama* spp.)	*S. aucheniae*	1 dog	1 cat	11	1984	**1561**
	S. sp.	4 dogs	4 cats	9–16	1984	**722**
Chickens (*Gallus domesticus*)	*S. horvathi*	NS	Dog, cat, polecat, martin, goshawk	6–9	1982	**1822**
	S. wenzeli	Dog, cat	NS	15		
	S. sp.	3 dogs, 3 cats	NS	10–11	1974	**714**
	S. wenzeli	4 dogs, 4 cats	NS	8–13	1994	**1109**
Pig (*Sus scrofa*)	*S. suihominis*	2 humans	Dogs, cats	9–10	1976	**806**
	S. miescheriana	6 dogs, 2 raccoons	4 cats, 2 opossums	12–14	1980	**1424**
	S. sp.	4 dogs	2 cats	12	1984	**762**
	S. miescheriana (wild pig)	2 dogs, 2 red foxes, 1 wolf, 2 arctic foxes	2 tigers (*Panthera tigris*), 2 wild cats (*Felis chaus*), 2 bears (*Ursus arctos*)	10–12	1976	**546**
	S. miescheriana (wild pig)	Dogs	Cats	13–18	1981	**73**

(Continued)

Table 1.4 (*Continued*)   Excretion of Sporocysts by Carnivores Fed *Sarcocystis*-Infected Meat

Intermediate Host	*Sarcocystis* Species	Sporocysts Excreted[a]	Sporocysts Not Excreted[a]	Prepatent Period (days)	Year	Reference
	S. miescheriana (European wild boar)	Dogs	Cats and rats	9–10	1989	**33**
	S. suihominis	NS	2 dogs, 2 cats	–	1998	**1520**
Sheep (*Ovis aries*)	*S.* sp.	4 dogs	4 cats	12–13	1976	**350**
	S. medusiformis	5 cats	12 dogs	15–30	1979	**240**
	S. tenella	2 dogs	4 cats	12–13	1982	**553**
	S. arieticanis	12 dogs	1 cat	12	1985	**817**
	S. gigantea, S. medusiformis	4 cats	4 dogs, 4 ferrets, 2 humans	11–14	1986	**613**
	S. tenella	4 dogs, 4 foxes	1 cat	9–11	1987	**615**
	S. microps	4 dogs	4 cats	13–14	1988	**1807**
	S. arieticanis	2 dogs	2 cats	12	1996	**1517**
	S. mihoensis	6 dogs	2 cats	11	1997	**1519**
	S. tenella	2 dogs	2 cats	12	2008	**1753**
Rabbit (*Oryctolagus cuniculus*)	*S. leporum*	13 cats	4 dogs	11–20	1977	**581**
	S. cuniculi	2 cats	2 dogs, 1 fox, 1 weasel (*Mustela nivalis*), 1 krestel (*Falco tinnunculus*)	10	1977	**1706**
	S. sp.	Cats	Dogs	NS	1979	**238**
	S. cuniculi	3 cats	NS dogs	12–13	1980	**1253**
Goat (*Capra hircus*)	*S.* sp.	Fox, wolf, dog	Cat, lion, ferret, krestel (*Falco tinnunculus*)	NS	1975	**103**
	S. sp.	Dogs	Cats	NS	1979	**238**
	S. capracanis	5 dogs	4 cats	10–11	1983	**68**
	S. capracanis	30 dogs	30 cats	12–15	2011	**1226**
Richardson's ground squirrel (*Spermophilus richardsoni*)	*S. campestris*	2 badgers	1 cat, 1 skunk	9	1983	**172**
Rat (*Rattus norvegicus*)	*S. cymruensis*	Cats	Dogs, ferrets	4	1978	**41**
	S. sp.	Cats	Dogs	NS	1979	**238**
Moose (*Alces alces*)	*S. alceslatrans*	3 dogs	2 cats, 1 coyote	10–14	1983	**245**
Roe deer (*Capreolus capreolus*)	*S.* sp. type 1 and 2	10 dogs, 2 foxes	16 domestic cats, 2 wild cats, several birds	10–14	1978	**548**
Alpine ibex (*Capra ibex*)	*S.* sp.	1 dog, 1 fox, 1 wolf	1 cat, 1 ferret	12–21	1975	**103**
Mouflon (*Ovis musimon*)	*S.* sp.	4 dogs	2 cats	12–13	1991	**1277**
Fallow deer (*Dama dama*)	*S.* sp.	5 dogs	2 cats	10–11	1988	**1404**
Mule deer (*Odocoileus hemionus*)	*S.* sp.	1 dog	1 cat	9	1983	**379**
Red deer (*Cervus elaphus*)	*S.* sp.	2 dogs, 1 fox	2 cats	18	1985	**61**

(*Continued*)

Table 1.4 (*Continued*)   Excretion of Sporocysts by Carnivores Fed *Sarcocystis*-Infected Meat

Intermediate Host	*Sarcocystis* Species	Sporocysts Excreted[a]	Sporocysts Not Excreted[a]	Prepatent Period (days)	Year	Reference
	S. cervicanis	3 dogs	2 cats, 1 man	11–12	1981	**796**
Wapiti (*Cervus elaphus*)	*S. wapiti*	1 coyote, 1 dog	1 cat	10	1982	**1651**
Yak (*Poephagus grunniens*)	*S. poephagi*	Cat	Dog	NS	1990	**1815**
	S. poephagicanis	Dog	Cat	NS		

NS, not stated.

Additional information: There are other studies (see References **266, 267**) in which authors detected at least 2 *Sarcocystis* species in venison (white-tailed deer, *Odocoileus virginianus*) and which resulted in yield of sporocyst by several dogs, cats and 1 red fox. Subsequently, after isolation in dog and not in cat, the description of a new species, *S. odocoileocanis,* was revealed.

[a] In several cases the number of animals fed was not reported or not clearly specified.

Some species of *Sarcocystis* transmitted by opossums have a wide range of intermediate hosts, but opossums are the only known definitive hosts (see Chapters 3 and 20).

The species of *Sarcocystis* vary considerably in their biology; therefore, caution is needed before making generalized statements concerning the genus.

1.4.5 Molecular

Molecular characterization can aid speciation. However, the specificity/accuracy depends on the genes analyzed (see Chapter 23). The 18S rRNA gene used in earlier investigations is now considered of limited taxonomic help. For example, structurally and biologically different *S. neurona* (Chapter 3) and *S. falcatula* (Chapter 17) were found to be almost identical in their 18S rRNA sequences.

1.5 PATHOGENICITY

1.5.1 Intermediate Hosts

Not all species of *Sarcocystis* are pathogenic for intermediate hosts. Generally, species transmissible via canids are more pathogenic than those transmissible via other definitive hosts. For example, sheep have 4 universally recognized species of *Sarcocystis*: *S. tenella*, *S. arieticanis*, *S. medusiformis*, and *S. gigantea*. Of these, *S. tenella* and *S. arieticanis*, transmitted via canids, are pathogenic, whereas *S. gigantea* and *S. medusiformis*, transmitted via cats, are nonpathogenic.

The severity of clinical sarcocystosis is dependent on dose, as shown in Table 1.6. The size or weight of the host does not appear to be relevant to resistance or susceptibility of clinical disease. Mature cattle are as susceptible to a specific dose of sporocysts as calves. However, stress can play an important role in the severity of illness and the susceptibility to infection. Pregnancy, lactation, poor nutrition, weather, or other stresses, can influence the severity of clinical sarcocystosis.

1.5.1.1 Clinicopathological Findings

Cattle usually do not develop acute sarcocystosis under experimental conditions unless 200,000 or more *S. cruzi* sporocysts have been ingested at a given time. Except for fever ($\geq 40°C$) between 15

Table 1.5 Attempted Transmission of Sarcocystis Species of Small Mammals to Other Small Mammals

Sarcocystis Species	Intermediate Host	Definitive Host	Transmission Attempted to	Results[a]	References
S. atheridis	Laboratory mouse (Mus musculus), Barbary striped grass mouse (Lemniscomys barbarus)	Nitsche's bush viper (Atheris nitschei)	Norway rat (Rattus norvegicus)	Negative	1620
S. cernae	Common vole (Microtus arvalis)	Krestel (Falco tinnunculus)	House mouse (Mus musculus)	Negative	185
S. clethrionomy-elaphis	Voles (Microtus spp., Clethrionomys glareolus)	Aesculapian snake (Zamenis longissimus)	Tundra vole (Microtus oeconomus)	Positive	1135
			Günther's vole (Microtus guentheri)	Positive	
			Bank vole (Clethrionomys glareolus)	Positive	
			Common vole (Microtus arvalis)	Positive	
			Wood mouse (Apodemus sylvaticus)	Negative	
			Sand lizard (Lacerta agilis)	Negative	
			Common wall lizard (Podarcis muralis)	Negative	
S. cymruensis	Norway rat (Rattus norvegicus), deer mouse (Peromyscus maniculatus), greater bandicoot rat (Bandicota indica)	Cat (Felis catus)	Dog, ferret, house mouse (Mus musculus), white-footed mouse (Peromyscus leucopus)	Negative	41
			Bank vole (Clethrionomys glareolus)	Positive	740
S. dispersa	House mouse (Mus musculus)	Barn owl (Tyto alba)	Voles (Microtus arvalis)	Negative	183
S. gerbilliechis	Subfamily Gerbillinae (Gerbillus spp., Psammonys spp., Pachyuromys spp.)	Arabian saw-scaled viper (Echis coloratus)	Cairo spiny mouse (Acomys cachirinus)	Negative	885
			House mouse (Mus musculus)	Negative	
			Norway rat (Rattus norvegicus)	Negative	
			Multimammate mouse (Mastomys natalensis)	Negative	
S. idahoensis	Deer mouse (Peromyscus maniculatus)	Gopher snake (Pituophis melanoleucus)	House mouse (Mus musculus)	Negative	113
			White-footed mouse (Peromyscus leucopus)	Negative	
S. muris	House mouse (Mus musculus)	Cat (Felis catus), ferret (Mustela putorius)	Norway rat (Rattus norvegicus)	Negative	1155,1503, 1839
			Golden hamster (Mesocricetus auratus)	Negative	
			Guinea pig (Cavia porcellus)	Negative	
			Meadow vole (Microtus pennsylvanicus)	Negative	
			Mongolian jird (Meriones unguiculatus)	Negative	
S. muriviperae	House mouse (Mus musculus)	Palestinian viper (Vipera palaestinae)	House mouse (Mus musculus)	Positive	1142
			Günthers vole (Microtus guentheri)	Negative	
			Jirds (Meriones unguiculatus,M. tristrami)	Negative	
			Multimammate mouse (Mastomys natalensis)	Negative	
			European rabbit (Oryctolagus cuniculus)	Negative	

(Continued)

Table 1.5 (Continued) Attempted Transmission of Sarcocystis Species of Small Mammals to Other Small Mammals

Sarcocystis Species	Intermediate Host	Definitive Host	Transmission Attempted to	Results[a]	References
S. rauschorum	Varying lemming (Dicrostonyx richardsoni)	Snowy owl (Nyctea scandiaca)	Norway rat (Rattus norvegicus)	Negative	173
			House mouse (Mus musculus)	Negative	
			Red-neck vole (Clethrionomys gapperi)	Negative	
			White-footed mouse (Peromyscus leucopus)	Negative	
			Brown lemming (Lemmus sibiricus)	Negative	
S. singaporensis	Norway rat (Rattus norvegicus)	Python (Morelia reticulatus)	Golden hamster (Mesocricetus auratus)	Negative	140,1869
			Multimammate mouse (Mastomys natalensis)	Negative	140
			Vole (Microtus arvalis)	Negative	
			Mice, amphibians, rodents[b]	Negative	886
			Rats[c]	Positive	
			Other snakes[d]	Negative	
			Rhesus monkey (Maccaca mulatta)	Negative	85
			House mouse (Mus musculus)	Negative	
			8 Rattus spp. and 3 Bandicota spp.[e]	Positive	767
			15 species of Muridae[f]	Negative	
S. villivillosi	Norway rat (Rattus norvegicus)	Python (Morelia reticulatus)	Rhesus monkey (Macaca mulatta)	Negative	85
			8 Rattus spp. and 3 Bandicota spp.[e]	Positive	767
			15 species of Muridae[f]	Negative	
S. zamani	Norway rat (Rattus norvegicus)	Python (Morelia reticulatus)	House mouse (Mus musculus)	Negative	85
			Rhesus monkey (Macaca mulatta)	Negative	
S. zuoi	Norway rat (Rattus norvegicus)	King rat snake (Elaphe carinata)	Mouse (Mus musculus)	Negative	855

[a] Sarcocysts in the muscles of recipient animals.
[b] Bufo regularis, Rana mascareniensis, Chalcides ocellatus, Mabuya quinquetaeniata, Acanthodactylus boskianus asper, Ptychodactylus hasselquisti, Gerbillus gerbillus, Meriones shawi isis, Jaculus jaculus, Acomys cahirinus, Mus musculus praetextus.
[c] Rattus rattus frugivorous, Rattus norvegicus, Arvicanthis niloticus, Nesokia indica.
[d] Snakes of Families Elapidae, Viperidae, Colubridae, Boide.
[e] Sigmodon hispidus, Cricetulus griseus, Mesocricetus auratus, Clethryonomys glareolus, Arvicola terrestris, Microtus arvalis, Gerbillus perpallidus, Meriones unguiculatus, Saccostomus campestris, Apodemus agrarius, Apodemus flavicollis, Mastomys natalensis, Acomys dimidiatus, Mus musculus, Rattus surifer.
[f] Rattus novegicus, Rattus rattus, Rattus argentiventer, Rattus tiomanicus, Rattus exulans, Rattus losea, Rattus villosissimus, Rattus colletti, Bandicota indica, Bandicota bengalensis, Bandicota savilei.

Table 1.6　Abbreviations for Figures 1.26 through 1.81

am—amylopectin granule	mp—micropore
ap—apical complex	mt—microtubule
ar1,2—apical rings 1 and 2	na—nucleus of macrogamont
br—bradyzoite	ni—nucleus of microgamete
cc—centrocone	no—nucleolus
ce—centriole	nu—nucleus of parasite
co—conoid	oo—outer layer of oocyst wall
eb—electron-dense body	os—outer layer of sporocyst wall
edg—electron-dense granule	ow—oocyst wall
edl—electron-dense layer of primary sarcocyst wall	pe—pellicle
gs—ground substance	pi—plasmalemma of parasite
go—Golgi complex	pvm—parasitophorous vacuolar membrane
hc—host-cell cytoplasm	pr—polar rings
hm—host-cell mitochondrion	pv—parasitophorus vacuole
hn—host-cell nucleus	pcw—primary sarcocyst wall
im—inner membrane complex	rh—rhoptry
io—inner layer of oocyst wall	se—septa
ip—interposed strip in sporocyst wall	sm—subpellicular microtubule
is—inner layer of sporocyst wall	sp—sporozoite
me—metrocyte	sw—sporocyst wall
mi—mitochondrion	vp—villar protrusion
mn—microneme	

and 19 DPI, clinical signs have not been observed until 24 DPI.[366,572,591] Beginning the fourth week after inoculation, cattle can develop anorexia, diarrhea, weight loss, weakness, muscle twitching, prostration, and, sometimes, death. Pregnant animals can undergo a premature parturition, abortion, or produce a stillborn fetus.[577] Some or all of these clinical signs can last from a few days to several weeks.

Clinical laboratory findings indicate anemia, tissue damage, and clotting dysfunction. Anemia with packed cell volumes (PCV) below 20% has been found in animals with moderate-to-severe infections. The mechanism of anemia is not fully understood.[234–236,584,629,1097,1098,1422] Elevation of clotting factors during acute sarcocystosis can affect anemia.[299]

Serum bilirubin, LDH, AAT, SDH, and CPK are generally elevated for brief periods during the acute phase.[366] BUN becomes elevated approaching terminal sarcocystosis. Prothrombin time is longer, whereas platelet counts, clotting time, activated partial thromboplastin time, and thrombin time, are generally not altered.[629,1097,1422] Acquired Factor VII deficiency and platelet dysfunction are found in some acutely infected cattle.[1422]

As infections become chronic, other signs become apparent. Growth is adversely affected. Animals can become hyperexcitable; they might hypersalivate and they might lose hair, especially on the neck, rump, and tail switch[591] (Figure 1.84). Some become emaciated. Some eventually develop CNS signs, including recumbency, opisthotonous, nystagmus, cycling gait while laterally recumbent, and, occasionally, death. Clinical disease in relation to phases of the life cycle of the parasite is summarized in Figure 1.85.

With few exceptions, clinical signs similar to those seen in *S. cruzi*-infected cattle have been seen in goats, sheep, and pigs infected with large doses of *S. capracanis*, *S. tenella*, and *S. miescheriana*, respectively (see Chapters 6 through 9). In sheep and goats, neural signs were more prominent than in cattle, but hypersalivation was not seen in sheep or goats.

Figure 1.84 An emaciated steer naturally infected with *S. cruzi*. Note the loss of hair at the tip of the tail. (Adapted from Giles, R.C. et al., 1980. *J. Am. Vet. Med. Assoc.* 176, 543–548.)

1.5.1.2 Gross Lesions

Edema and focal necrosis in gut-associated lymph nodes are seen about 15 DPI.[366] The next gross lesions are seen at about 26 DPI. The most striking lesion seen at this time is hemorrhage[366,911] (Figure 1.5f). Hemorrhages are most evident on the serosal surface of viscera, in cardiac and skeletal muscles, and in the sclera of the eyes. On necropsy, characteristics of acute sarcocystosis include skeletal muscles that appear mottled or striped with pale areas interspersed with areas of dark hemorrhage. Hemorrhages vary from petechiae to ecchymoses several centimeters in diameter. Following acute infection, body fat becomes scanty, yellow-hued, and gelatinous. Body cavities contain straw-colored fluid, and organs become icteric. In chronically affected animals, the most notable lesion is serous atrophy of fat, especially pericardial and perirenal fat, with white flecks of mineralization.

1.5.1.3 Microscopic Lesions

The earliest lesion observed is hypertrophy of endothelial cells, associated with the development of first-generation schizonts in arteries and arterioles[366,371] (Figure 1.86). This is followed by perivascular and interstitial infiltration of many organs with mononuclear cells (Figure 1.87). The mononuclear cells are seen as early as 11 DPI and can be present for several months. The predominant inflammatory cells are lymphocytes and macrophages, with a few plasma cells and eosinophils. Hemorrhage can be generalized and often not associated with inflammation. Necrosis may be found in many organs, especially in skeletal muscles, heart, and kidneys, probably associated with

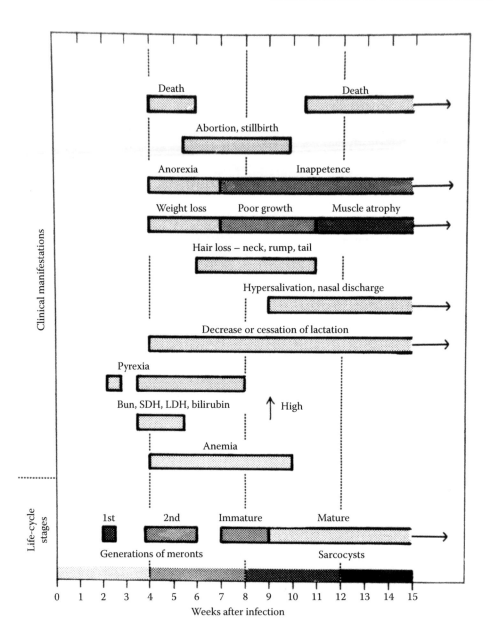

Figure 1.85 The chart compares the life-cycle stages of *S. cruzi* with the weeks after infection and the clinical signs of sarcocystosis in cattle. (Adapted from Fayer, R., Dubey, J.P., 1988. *Transplacental Effects on Fetal Health*. Alan R. Liss, Inc., New York, NY, pp. 153–164.)

vasculitis (Figures 1.87 through 1.92). The overall predominant lesion in sarcocystosis is inflammatory rather than degenerative.

Degenerating sarcocysts might be surrounded by mononuclear cells, neutrophils (Figure 1.93), eosinophils, giant cells, or a combination of these cells. In most livestock species, the cellular response is mainly mononuclear. How sarcocysts are removed from host tissue is not known. It is likely that some sarcocysts rupture spontaneously and inflammation follows. However, macrophages and plasma cells can be seen around seemingly intact live sarcocysts. However, many sarcocysts can be seen with no adjacent inflammatory reaction. There is no reactivation of chronic infection.

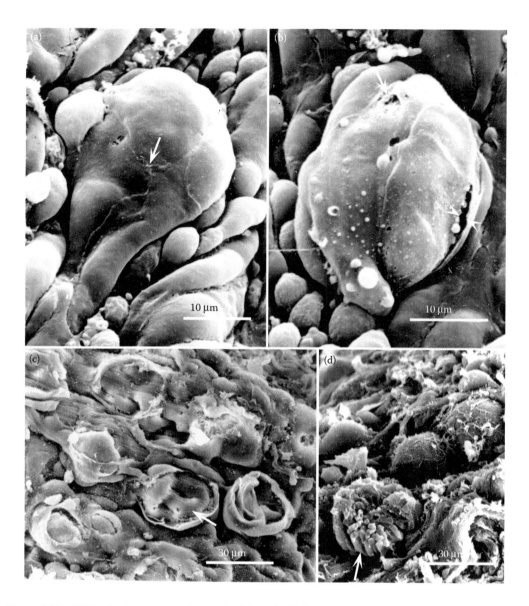

Figure 1.86   SEMs of ovine mesenteric arteries infected with first-generation schizonts of *S. tenella*. (a) The
protuberance in the center includes 4 endothelial cells demarcated by intercellular junctions
(arrows) covering a host cell containing a schizont (not visible). (b) Perforation (arrow) and separation of 2 endothelial cells (arrowheads) covering a schizont. (c) Sloughing of endothelial cells
exposing the internal elastic membrane and host cells harboring schizonts (arrow). (d) Most of
the endothelium has sloughed, exposing a schizont and merozoites (arrowheads). (From Speer,
C.A., Dubey, J.P., 1982. *Can. J. Zool.* 60, 203–209. With permission.)

1.5.1.4 Definitive Hosts

Sarcocystis generally does not cause illness in definitive hosts. Dogs, cats, coyotes, foxes, and
raccoons, fed tissues infected with numerous species of *Sarcocystis* excreted sporocysts, but were
otherwise normal. A few dogs and coyotes vomited or were anorexic for 1–2 days following ingestion of meat, but such signs might have resulted from the change in diet from laboratory chow
to raw meat. However, human volunteers who ingested beef and pork infected with *S. hominis*

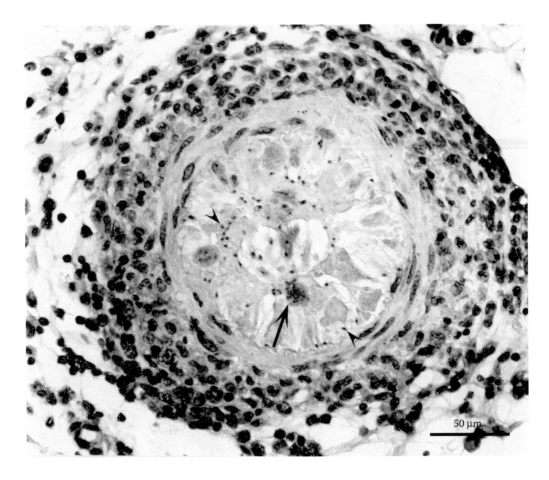

Figure 1.87   Necrosis and vasculitis in a mesenteric lymph node artery of a sheep infected with *S. tenella*. The endothelial cells have been destroyed by rupture of second-generation schizonts (arrow), releasing merozoites (arrowheads) into the lumen. Mononuclear cells are in and around arterial walls. H and E stain. (From Dubey, J.P., 1988. *Vet. Parasitol.* 26, 237–252. With permission.)

or *S. suihominis*, respectively, developed symptoms, including vomiting, diarrhea, and respiratory distress. These symptoms were more pronounced in volunteers who ate infected pork than in those who ate infected beef (see Chapters 4, 6, and 7).

1.6 PATHOGENESIS

1.6.1 Tissue Necrosis

Schizonts cause necrosis of cells and tissues, depending on the species of *Sarcocystis*, location, and the multiplication potential.[366] For example, *S. falcatula* of birds multiplies extensively in endothelial cells and produces several generations of schizogony.[1626] The physical damage alone as a result of vasculitis might result in the death of birds heavily infected with *S. falcatula*. The same hypothesis might apply to *S. idahoensis* infection in deer mice, in which hepatocytes are destroyed as a result of schizont multiplication.[112] However, localized tissue necrosis does not appear extensive enough to cause the severe illness or death seen in large animals (cattle, sheep, goats, and pigs).

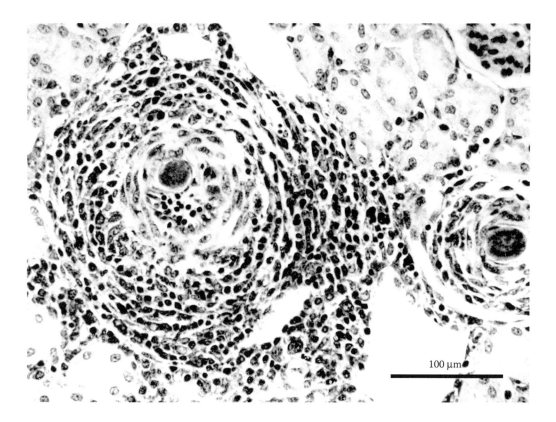

Figure 1.88 Severe mononuclear cell infiltrations in tunica adventia and tunica media in arteries in the renal cortex of a calf infected with *S. cruzi*. The vascular lumen is almost occluded by hypertrophied endothelial cells and first generation schizonts. Giemsa stain. (From Dubey, J.P., Speer, C.A., Epling, G.P., 1982. *Am. J. Vet. Res.* 43, 2147–2164. With permission.)

1.6.2 Inflammation

The perivascular mononuclear cell infiltration seen around *S. cruzi*-infected arteries at 7–11 DPI[366] indicates a host reaction to antigens liberated from sporozoites or immature schizonts, or the expression of parasite antigens by host cells, because merozoites are not liberated from these first-generation schizonts until 14 DPI or later. An intense inflammatory reaction is usually observed about the time when the second-generation schizonts mature and rupture, during the fourth to sixth week PI. The myositis, during the penetration of myocytes by merozoites, might be related to products liberated from merozoites or myocytes.[366] The intense mononuclear cell infiltrations in the kidneys, liver, lungs, and other organs, are probably stimulated by similar parasite antigens.

1.6.3 Immune Regulation of Pathogenesis

Vascular endothelial cells have MHC II antigens[1634] and can process antigen for presentation to lymphocytes.[1151] Adherence of thymus-derived (T) lymphocytes to endothelial cells via the lymphocyte function-associated molecule, LFA-1, or other mechanisms[785], is the first step toward emigration of these cells from the blood into the tissues. Lymphocytes also recognize specific MHC II antigen on vascular endothelial cells.[1402] In a rabbit model, a strong correlation was found

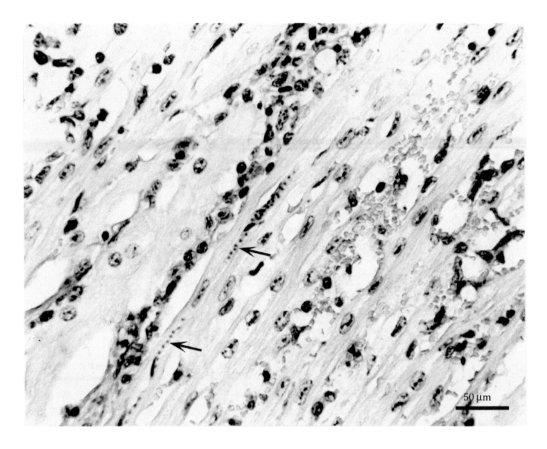

Figure 1.89 Hemorrhage and mild mononuclear cell infiltration in the myocardium of a calf experimentally infected with *S. cruzi*. Arrow points to merozoites and schizonts. H and E stain. (From Dubey, J.P., Speer, C.A., Epling, G.P., 1982. *Am. J. Vet. Res.* 43, 2147–2164. With permission.)

between the magnitude of the *in vivo* immune response and the vascular endothelial expression of MHC II antigens.[1635]

Thus, during acute *S. cruzi* infections, endothelial cells might participate in eliciting an immune response, leading to pathogenesis by processing and presenting antigens immunogenically, which results in sensitization of T lymphocytes.[1659] These antigen-sensitive T lymphocytes might then recognize antigen plus MHC II[1402] on vascular endothelium and give rise to destructive immune responses, such as local delayed hypersensitivity reactions involving mononuclear cell infiltration of vascular tissue.[366,1634,1635] Cytokines induced by *S. cruzi* infection can adversly affect growth regulatory hormones and interfere with weight gain.[594]

1.6.4 Edema

Ascites and edema in tissues are probably related to hypoproteinemia and vasculitis.[366,1626]

1.6.5 Fever

The peaks of fever coincide with maturation of schizonts and release of merozoites into the bloodstream.[591] Fever is probably related to the release of pyrogen from mature rupturing schizonts directly on the hypothalamus, or indirectly by stimulating the release of prostaglandins.

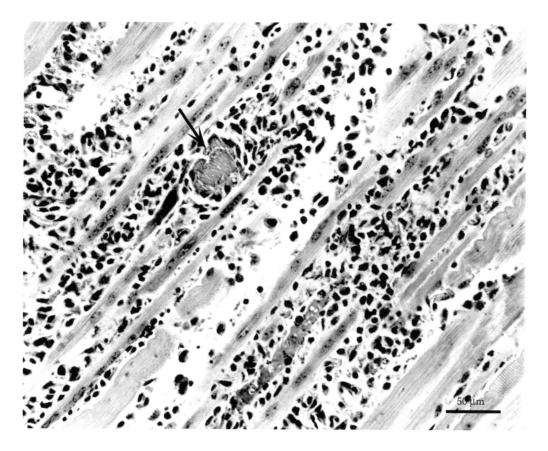

Figure 1.90 Severe necrosis, infiltration of mononuclear cells, and mineralization (arrow) in semitendinosus muscle of a calf infected with *S. cruzi*. H and E stain. (From Dubey, J.P., Speer, C.A., Epling, G.P., 1982. *Am. J. Vet. Res*. 43, 2147–2164. With permission.)

1.6.6 Anemia

Anemia is the most evident clinical finding of acute sarcocystosis in cattle, sheep, goats, and pigs, but the underlying mechanism is unknown.[359,366,584,1036,1097,1098,1422] Although the anemia is regenerative, few or no reticulocytes are found. The anemia is normocytic, normochromic, and primarily hemolytic. Hemorrhage can account for part of the loss of blood cells. In addition, many red-blood cells are removed from circulation and sequestered in the spleen, probably via immunologic mechanisms. It is possible that some unknown toxic factors, or metabolite released from schizonts, or infected host cells, contribute further to the anemia. The increased prothrombin time and increased fibrin degradation product concentration in infected calves indicate intravascular coagulopathy, which can result in capillary extravasation.[304]

1.6.7 Abortion

Abortion and fetal death can result when animals become infected with pathogenic species of *Sarcocystis* during pregnancy. Most cattle, sheep, swine, goats, and roe deer that developed clinical sarcocystosis from experimental infections induced in mid to late gestation aborted,[70,357,548,549,577,1037,1672] whereas most infected animals without signs of infection carried their fetus to term.[357,1254] Lesions were not found in fetal tissues from any of these experimentally

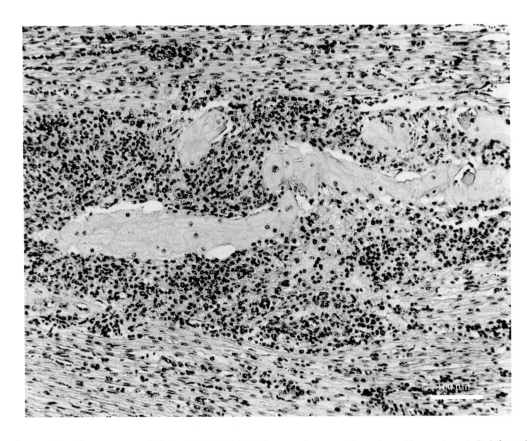

Figure 1.91 Severe myocarditis involving Purkinje fibers and myocardium of a calf experimentally infected with *S. cruzi*. H and E stain. (From Dubey, J.P., Speer, C.A., Epling, G.P., 1982. *Am. J. Vet. Res.* 43, 2147–2164. With permission.)

infected host species, with the exception of 1 lamb. Parasites were found in only 2 of 18 bovine fetuses and only 4 schizonts were found in fetal membranes of 3 sheep. In contrast, parasites and lesions were found in maternal placentomes of cattle, sheep, and goats. Thus, results from these experimentally infected animals indicate that although *Sarcocystis* is present in the maternal placenta, it rarely infects the fetus or fetal membranes. Unlike experimentally infected cattle, those with natural infections have had parasites, lesions, or both, in the fetuses (Figure 1.94).[30,844,908,913,988,1152,1246,1558,1741] Unlike experimental infections, parasites, as well as lesions, were found in the placentas in several natural infections.[256,365,1246] Furthermore, some naturally infected cows were asymptomatic, although organisms were detected in the fetus or placenta.[365] It is not possible to confirm that the natural infections and the experimental infections were due to the same species of *Sarcocystis*, although the parasites were morphologically similar, parasitized the same type of host cells, and, in some cases, responded serologically to the same antigen. Thus, the question remains as to whether differences observed between experimentally infected animals and naturally infected animals were due to different species or strains of *Sarcocystis*.

There are many unanswered questions and observations that appear contradictory within the subject of how sarcocystosis affects fetal health. Some fetuses are infected and have lesions, others are unaffected. Some fetal placentas are infected and have lesions, others are unaffected. Some maternal placentomes are infected and have lesions, others are unaffected. Most pregnant animals have overt clinical illness at the time the fetus is affected, others do not. Several possible mechanisms have been hypothesized by which sarcocystosis can directly or indirectly affect

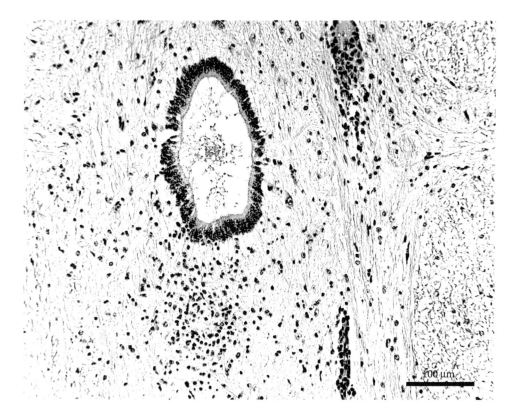

Figure 1.92 A glial nodule and perivascular infiltration of mononuclear cells in the spinal cord of a calf infected with *S. cruzi*. H and E stain. (From Dubey, J.P., Speer, C.A., Epling, G.P., 1982. *Am. J. Vet. Res.* 43, 2147–2164. With permission.)

fetal health.[592] In naturally infected bovine fetuses, parasites were found in virtually all organs, most often in the brain. Immature and mature schizonts and free merozoites were found, usually within endothelial cells of capillaries, but occasionally free in the lumen of a vessel or in neural tissue.[30,71,365,844,908,913,988] Neural lesions include nonsuppurative encephalitis or meningitis, with small foci of glial cells surrounding a central necrotic area throughout the gray and white matter of the cerebrum, cerebellum, or brainstem, with some perivascular mononuclear cell infiltration and occasional microthrombi in vessels in reaction foci. Other affected organs are similar to those of infected postnatal animals, including nonsuppurative myocarditis, pneumonitis, hepatitis, and renal glomerulitis accompanied by focal necrosis and hemorrhage.

One premature lamb born to an experimentally infected ewe had cerebral congestion and edema, as well as areas of leukoencephalomalacia in the cerebrum and midbrain.[1254] Another lamb, from a naturally infected ewe, was found with encephalitis characterized by mononuclear cell infiltration and gliosis.[386] Abortion and stillbirths were reported in sheep from Brazil[1390] and Denmark[17] (see Chapter 8), and in yak from China.[1816]

1.6.7.1 Placental Infection and Lesions

Free merozoites and schizonts, usually associated with lesions, have been observed in maternal or fetal placentas of some cattle, sheep, and goats. Immunofluorescent staining of placentomes from 6 experimentally infected cows revealed numerous merozoites and schizonts in the maternal, but

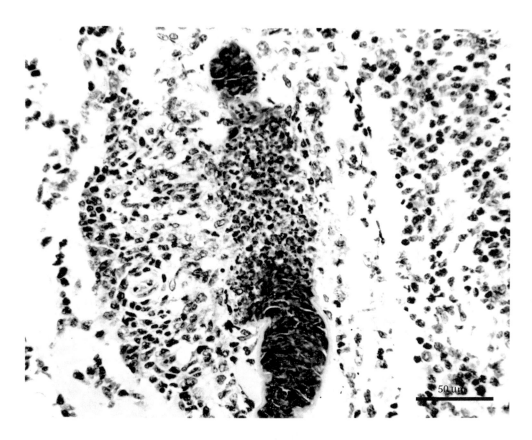

Figure 1.93   Part of a degenerating *S. capracanis* sarcocyst being replaced by neutrophils in the diaphragm of a vaccinated-challenged goat. Mononuclear cells surround the sarcocyst. H and E stain. (From Dubey, J.P., 1983. *Vet. Parasitol.* 13, 23–34. With permission.)

not the fetal, placentome.[70] Placentomes of all 6 cows were atrophied. Maternal placentomes from naturally infected aborting cows were also found to contain numerous parasites, but were hemorrhagic and necrotic with calcification of placental villar epithelial cells and of the fibrous connective tissue.[256] In other natural infections resulting in abortion, fetal placental lesions were mild to severe, with accompanying mononuclear cell infiltration.[365,1246] Of 11 experimentally infected sheep, only 4 schizonts were found in fetal membranes from 3 ewes.[1037] In experimentally infected goats, schizonts were found in the maternal placentome and endometrium, but not in fetal membranes.[357] Lesions were restricted to the maternal placentome and endometrium with heavy mononuclear cell infiltration and some focal necrosis.

1.6.7.2 Possible Mechanisms Causing Abortion or Fetal Death

Based on the forgoing descriptions of parasite locations and lesions, there appear to be several possible ways in which sarcocystosis might initiate abortion or fetal death (Figure 1.95).

One hypothetical sequence of events leading to labor might begin in the hypothalamus of the fetal brain, with transmission via the pituitary to the fetal adrenal, where cortisol provides the fetal trigger to maternal endocrine changes.[195] In turn, there might follow a release of steroids and prostaglandins from the placenta, the placentome, and the fetal membranes, which stimulate the myometrium to contract and the cervix to dilate. Under normal conditions, it is thought that there is a rise in concentration of plasma cortisol secreted from the fetal adrenal as it grows and

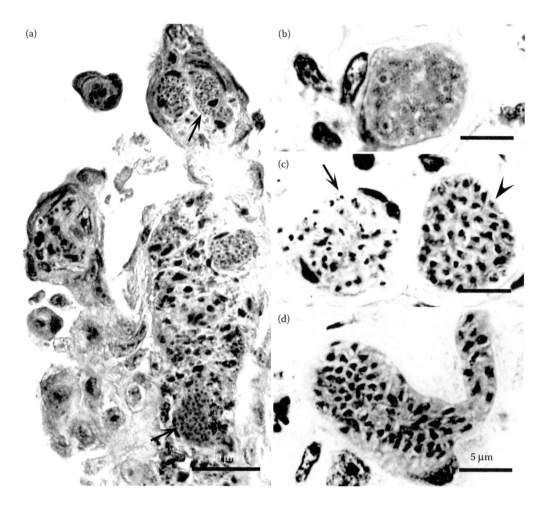

Figure 1.94 Necrosis and schizonts of *Sarcocystis* in a fetal placental cotyledon of a naturally infected cow. (a) Necrosis, schizonts (arrows), and numerous merozoites. (b) Early schizont with lobulated nucleus. (c) A degenerating (arrow) and normal (arrowhead) schizont. (d) An irregular-shaped schizont. H and E stain. (From Dubey, J.P., Bergeron, J.A., 1982. *Vet. Pathol.* 19, 315–318. With permission.)

develops, with the highest concentrations during the last few days of fetal life.[195] Mechanisms that result in high adrenocorticotrophic hormone (ACTH) or cortisol concentrations in the fetus leading to premature parturition or abortion are not known. One sequence of events could begin with anemia in the dam resulting in hypoxia in the fetus. Hypoxia could also result from insufficient blood-flow through the maternal placentome due to occlusion of capillaries by schizonts, localized hemorrhage, or intravascular coagulation. Evidence of hypoxia is the finding of leukoencephalomalacia in some fetuses. Vasopressin, known to stimulate ACTH release, rises severalfold with the approach of spontaneous parturition and in response to hypoxia.[195] Hypoxia itself has been shown to raise the concentration of immunoreactive ACTH in fetal plasma at all times during gestation, when it was studied. Isolated fetal adrenal cells appear to be responsive and secrete large amounts of cortisol in response to ACTH, a form of GTP, or dibutyrylcyclic AMP at various times during gestation.

Vascular damage associated with sarcocystosis in the dam might lead to changes in the concentration of hormones associated with the maintenance of gestation. The plasma concentration

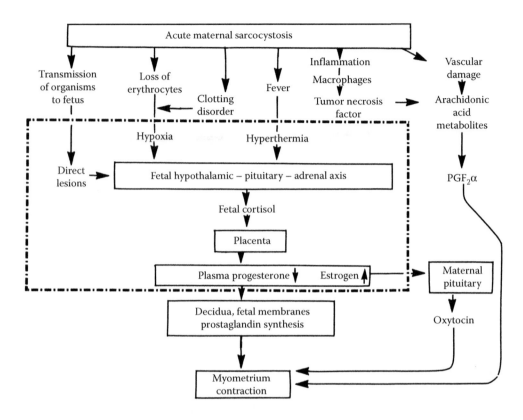

Figure 1.95   A hypothetical diagrammatic presentation of events leading to *Sarcocystis*-induced abortion. Fetal hormonal events are shown within the area marked by dotted lines. (Adapted from *Transplacental Effects on Fetal Health*. Alan R. Liss, Inc., New York, NY, pp. 153–164.)

of progesterone, E1, E2, PGFM, and alpha-fetoprotein, was monitored in pregnant cows experimentally infected with *S. cruzi*.[55,56]

Progesterone production after day 50 of gestation in sheep is primarily in the placenta. It acts by maintaining the myometrium refractory to stimulation by oxytocin or PFG2α. Decrease in placental progesterone at term follows and is caused by a rise in fetal cortisol levels. Progesterone withdrawal initiates increased PGF2α secretion and parturition.

In late pregnancy, estrogens usually promote uterine contractility, increase responsiveness of the uterus to agonists such as oxytocin and PGF2α, inhibit progesterone production, and stimulate prostaglandin synthesis and release. Maternal estrogen increases in response to fetal adrenal activity.

Unfortunately, changes in progesterone, E1 and E2 concentration in the experimentally infected cows could not be interpreted as cause or effect. Although the changes preceded or were concurrent with abortion, similar changes were documented at the time of normal parturition. Prostaglandins initiate labor by stimulating myometrial contractions and by softening the cervix. Two specific lines of observation suggest that events in acute sarcocystosis might result in increased concentration of prostaglandins in the peripheral circulation of the dam or in the maternal placentome and fetal placenta. First, release of second-generation merozoites is associated with damage of capillary endothelium, an event known to trigger the arachidonic acid cascade leading to PGF2α release. Parasitic disruption of cell membranes and the action of inflammatory cells could also stimulate platelet activity, including the release of PGF2α by platelets, and by damaged endothelium. Second, when second-generation merozoites mature, numerous mononuclear inflammatory cells are found perivascularly throughout the body and in the placenta. Many of these cells are macrophages, and *in vitro* studies indicate that macrophages can be stimulated by *Sarcocystis* lysates to release the tumor necrosis factor (TNF).[592,2009] One activity of TNF is the triggering of an arachidonic acid cascade leading to PGF2α release. Because the findings in identifying those events leading to high maternal plasma PGF2α concentration preceding bovine abortion were inconclusive, studies need to be conducted to clarify the relationship between sarcocystosis and prostaglandins.

Fever has been associated with premature parturition and sometimes regarded as the cause. However, it is not clear if fever itself is a cause or an effect. Pyrogens from infectious agents such as protozoa, or from degenerating body tissues, can cause fever. Pyrogens can act directly on the hypothalamus to raise the thermostatic control or can act indirectly by stimulating leukocytes to produce endogenous pyrogen. Endogenous pyrogen is thought to cause fever by inducing prostaglandin E1 formation, which elicits the fever reaction. Thus, it is not clear if fever directly affects the fetus, or if the initiators of fever—in this case, *Sarcocystis* pyrogens or prostaglandins—induce lesions. Prostaglandins were elevated during acute sarcocystosis also in pigs.[300]

In addition to the forgoing hypotheses leading indirectly to death or premature parturition of the fetus, invasion of fetal tissues by the parasite might result in specific lesions with similar end results. Distribution of parasites throughout the brain with associated lesions might stimulate the fetal hypothalamic–pituitary–adrenal axis or destroy significant brain tissue leading to death. Evidence of fetal myocarditis, pneumonitis, and hepatitis, indicate other possible avenues leading directly to fetal death.

1.6.8 Eosinophilic Myositis and Sarcocystosis

Eosinophilic myositis (EM) is a specific inflammatory condition of striated muscles, principally due to accumulations of eosinophils. It has been found mainly in cattle,[12,151,309,526,654,724,725,871,907,1046,1460,1472,1511,1778,1943] occasionally in sheep,[907] and rarely in pigs and horses. The affected animals might appear clinically normal. Most of findings are at the abbatoir level; thus, EM is generally detected at the time the surface of the carcass is inspected or when it is cut into prime cuts or into quarters.[907] Sex and breed have no influence on the prevalence. The condition was reported with equal frequency in steers, cows, and heifers. Virtually all striated muscles, including skeletal muscles, the muscles of the eye, larynx, and heart, can be affected.

In the United States, federal meat inspection regulations require that parts, or, in severe cases, the entire carcass, affected with EM be condemned. Generally, EM has been found in 1 of 100,000 slaughtered cattle, but in some feedlots up to 5% of the cattle were condemned.[907]

Two types of lesions are found associated with EM.[907] The most common lesion found on the surface of muscles is multifocal, spindle-shaped to round, and 5 to 15 × 1 to 3 mm in size. The color of the lesion can vary. Green lesions result from accumulations of eosinophils, principally between myocytes (Figure 1.1e). With the progression of EM, eosinophils and myocytes degenerate, resulting in granulomas with a central area of necrosis. Later, the central necrotic tissue becomes surrounded by zones of giant cells, epithelial cells, eosinophils, lymphocytes, and fibrocytes.[907]

A second type of lesion is less prevalent but more conspicuous. It can be up to 15 cm long, bright green to pale yellow, and firm in consistency.

The etiology of EM remains uncertain. Because sarcocysts are found in the same location as EM, *Sarcocystis* has been traditionally considered to be the cause of EM. The *Sarcocystis*-induced etiology of EM was supported by the finding of *Sarcocystis*-specific IgE antibody in lesions.[526,724,725] Because of the small number of cases of EM compared with the high prevalence of intramuscular cysts, the confirmation of *Sarcocystis* as the cause, and EM as the effect, requires more substantial evidence. It is necessary to note that EM has never been seen in any experimentally infected ox, sheep, or other animals. In experimentally infected animals, degenerating sarcocysts are often surrounded by mononuclear cells and neutrophils. Because most of these observations were made in animals infected less than a year of age and with only 1 dose of sporocysts, repeated natural infections can result in a different reaction, perhaps involving other cell types such as eosinophils.

Recently, EM-like lesions were induced in 2 calves by repeated injections of *Sarcocystis* antigen with adjuvants.[1777] EM has been found associated with all 3 species of *Sarcocystis* in cattle, *S. cruzi*, *S. hirsuta*, and *S. hominis*.[654,1778] In the study from Belgium, of 97 beef samples condemned because of EM, sarcocysts were found in 27; most (82%) of the sarcocysts were *S. hominis*.[1778] In the study from Colorado, United States, sarcocysts were found in 32 of 363 EM lesions and all of them were *S. cruzi*.[654] *S. hirsuta* sarcocysts were found in all 18 cattle that were condemned because of grossly visible sarcocysts.[397]

1.6.9 Chronic Sarcocystosis and Toxins

Little is known of the pathogenesis of chronic sarcocystosis. Anatomic lesions are insufficient, especially in cattle, sheep, and pigeons,[1099] to explain the etiology of CNS signs. Although sarcocysts in muscles or in the CNS are well adapted, usually without any host reaction, some sarcocysts probably rupture from time to time and might release toxic products. It has been known for about a century that an aqueous extract of bradyzoites is toxic (sarcotoxin) when inoculated into rabbits. Of mice, guinea pigs, chickens, and rabbits injected with *S. cruzi*, extract toxic reaction was observed only in rabbits.[1061,1513,1516,1527,2271] The crude *S. cruzi* extract was water soluble, acid–alkali stable, and thermolabile. Harada et al.[781] commented on 27 outbreaks of food poisoning in humans after eating uncooked horse meat in Japan. They proposed that the cause was a toxin associated with a 15 kDa protein from *S. fayeri*; freezing the meat for 2 days at −20°C abolished the toxic property of the infected meat. It is not clear how, or if, sarcotoxin might be released from intact sarcocysts and, if released, what role it might play in chronic sarcocystosis. Although lectins are also associated with bradyzoites, their role, if any, in causing disease is unknown.

It is possible that substances released from *Sarcocystis* stimulate production of TNF.[584] TNF is known to be associated with wasting disease. Poor weight gain and low feed efficiency in sarcocystosis are regulated through growth-regulating hormones.[505,506] Any relationship between TNF and the growth-regulating hormones has not been established.

Attempts have been made to characterize antigenic/biochemical properties of extracts prepared from sarcocysts, particularly *S. gigantea* of sheep, because of the availability of macroscopic sarcocysts from sheep esophagus and their easy separation from the host tissue.[1140,1527,1734,1746–1748,1945,2133,2176] Lectins and species-specific polypeptides were demonstrated. However, the role of these substances in the pathogenesis of sarcocystosis is not clear.

1.7 IMMUNITY

Cellular and humoral immune responses in infected animals indicate that *Sarcocystis* species are immunogenic in intermediate hosts. Unfortunately, most information is available only from responses directed against antigens derived from bradyzoites. Because bradyzoites are obtained from the terminal stage (sarcocysts), only cross-reactive antigens are actually tested.

1.7.1 Antigenic Structure

Relatively large numbers of highly purified parasites are needed to study the proteins and antigens of *Sarcocystis* spp. Parasites can be obtained from experimentally infected animals or from *in vitro* cultivation. Parasites such as bradyzoites should be obtained from experimentally infected animals, and not from naturally infected ones, to insure accurate species identification. Whether parasites are obtained from *in vivo* or *in vitro* sources, biochemical analyses should also include controls of solubilized noninfected host cells or tissues.

There is little information concerning the proteins and antigens of *Sarcocystis* spp., and most information is on *S. cruzi*, *S. neurona*, and *S. muris*. Proteins and antigens of *S. cruzi* sporozoites, merozoites, and bradyzoites were studied with sodium dodecyl sulfate–polyacrylamide gel electrophoresis (SDS–PAGE), and Western blotting with immune serum or monoclonal antibodies, in the hope of developing a commercial diagnostic test for bovine sarcocystosis that did not materialize because of lack of interest by industry. Bradyzoites and merozoites had similar proteins with Mr 15.7 kDa, several between Mr 18–21.4,29, and 110 kDa; proteins with Mr 15.7,29,49.9 and 54K kDa were common to sporozoites and merozoites. Only 2 proteins (Mr 15.7 and 16.5 kDa) were present in all 3 parasite stages.

Based on SDS–PAGE analysis of *S. cruzi* merozoite proteins, some appeared, disappeared, and then reappeared during *in vitro* development of merozoites. Merozoites harvested at 36 and 48 DPI (called 36D and 48D merozoites) from bovine pulmonary artery endothelial cells each had a unique protein of Mr 50 and 70 kDa, respectively. Merozoites harvested at 36 and 60 DPI had many proteins with similar molecular weights as well as several unique proteins.[393] For example, 36D merozoites had a 43-kDa protein that was not present in 60D merozoites, and 60D merozoites had a prominent 63.1-kDa protein not present in 36D merozoites. A 63.1-kDa protein first appeared in 37 or 38D merozoites and was also present in 38, 40, 48, 52, and 60D merozoites but was absent in 41 and 58D merozoites. Two other proteins also showed temporal variation in appearance. A 43-kDa protein was present in 31, 36, and 52D merozoites, whereas a 42.3-kDa protein was present in 29, 40, and 48D merozoites, and both proteins were present in 33, 38, 41, 58, and 60D merozoites.

In Western blot analyses, antibodies in bovine serum reacted with increasing numbers of antigen bands as the animals progressed through the course of infection. Proteins with Mr 21,40, and 50 kDa might be potent antigens because serum reacted strongly with them beginning early in infection.

Monoclonal antibodies (MAbs) have been generated against *S. cruzi*[154,1659] by immunizing female BALB/cBY mice 8–10 weeks of age with 10^6 merozoites or sporozoites per mouse. Merozoites were mixed 1:1 (v/v) in complete Freund's adjuvant and given by intraperitoneal injection. Mice were intravenously injected 1 month later with 8×10^5 merozoites or 5×10^5 sporozoites, respectively; 3 days later the spleens were removed and fused to the myeloma cell line P3/X63-Ag8.6.5.3 and selected as previously described. Hybridoma culture supernatants were screened on acetone-fixed merozoites or sporozoites by an IFA, cloned by limiting dilution, re-tested, and expanded in culture. MAbs were tested for their ability to react with whole parasites in a live IFA or with parasite proteins in Western blots.[393]

A variety of reactivities occurred when either antisporozoite or antimerozoite MAbs were used to detect antigens of *S. cruzi* on Western blots of whole parasite (sporozoite, merozoite, or bradyzoite) antigens separated by SDS–PAGE. Several MAbs reacted in IFA with acetone-fixed sporozoites and merozoites, and 3 MAbs reacted with the surface of live sporozoites or merozoites. In Western blots, the surface-reactive MAbs reacted with polypeptides ranging from Mr 20 to 74 kDa. Interestingly, only 1 of the MAbs analyzed reacted with blots of antigens prepared from homologous (eliciting) stages, although several did react with the surface of homologous stages in the live IFA. Thus, certain epitopes on the surface of both merozoites and sporozoites might have been irreversibly denatured in the blotting procedure, labile to certain proteases or nonprotein in composition (e.g., lipid, glycolipid). In addition, several epitopes on the surface of live merozoites appeared to be constituents of molecules within sporozoites. Such molecules can be precursors of surface components of merozoites or might simply share these epitopes on unrelated molecules. Antimerozoite MAbs also identified shared epitopes in sporozoites and bradyzoites that appeared to be located on molecules of both similar molecular weight (i.e., 40 kDa) and distinct molecular weight (i.e., 40 vs.

20 and 60 kDa). An MAb reactive with the surface of *S. cruzi* merozoites also identified epitopes on sporozoite and bradyzoite antigens, indicating that some of the 20- to 60-kDa sporozoite molecules may be precursors to internal or surface antigens of bradyzoites. Further investigations will be needed to elucidate the molecular relationships and functions of these molecules in the development of the major stages in the life cycle of *S. cruzi*.

Several MAbs generated against *S. cruzi* also revealed several patterns of fluorescence on infected bovine cardial pulmonary artery endothelial (CPA) cells.[393] For example, all 3 surface-reactive MAbs reacted with single, intracellular or extracellular merozoites and with merozoites in mature schizonts but did not react with schizonts in intermediate stages of development. Thus, the antigens identified by these 3 MAbs were expressed only by mature merozoites and not by intermediate schizonts. Results from the IFA indicated that intracellular merozoites shed the antigens into the host-cell cytoplasm. These antigens were then expressed on the surfaces of the host cell, as well as on adjacent noninfected CPA cells. Such observations indicate that active processing of antigen by neighboring CPA cells might occur after indirect acquisition of parasite antigens from adjacent infected CPA cells. Similar antigen processing and presentation might also occur in endothelial cells in infected animals.

For characterization of antigens/antibodies of *S. neurona*, see Chapter 3.

Three surface antigens of Mr 27, 43 and 90 kDa have been identified by ^{125}I labeling and immunoprecipitation of *S. muris* bradyzoites with specified rabbit or mouse antiserum.[1] Surface antigens of *S. muris* and *S. miesheriana* bradyzoites and sporozoites have been reported.[1642,1643,2336] Two major proteins were found in micronemes of *S. muris*.[1403]

Monoclonal antibodies have also been generated against some other *Sarcocystis* species for species differentiation and diagnostics.[1285] All 6 monoclonal antibodies against *S. gigantea* and *S. tenella* of sheep were species-specific.[2335,2337] One of these monoclonal antibodies was directed against micronemes of *S. tenella*, whereas 2 antibodies against *S. gigantea* reacted with micronemes and amylopectin granules.[1285] Species-specific and cross-reactive epitopes have been characterized in *S. muris*.[545,1183,1733,1998] Jäkel et al.[891] characterized monoclonal antibodies against *S. singaporensis* and their role in invasion and escape of the parasites *in vitro*. Tenter et al.[1731] used isoelectric separation of species-specific proteins to differentiate *S. gigantea*, *S. tenella*, and *S. arieticanis* of sheep.

1.7.2 Humoral Responses

Cattle,[577,580,589,626,681,1087] pigs,[1286,1287,1289,1461,1813,1826,1827] sheep,[666,1250,1294] mice,[1,186,193,666,839,1044,1729,1730] and rats[890] inoculated with *Sarcocystis* sporocysts developed IgG antibodies starting 3–5 weeks PI with respective species of *Sarcocystis*. These antibodies were detectable by IHA,[193,1087] ELISA or dot-ELISA,[666,1287,1729,1730] IFA,[186,1730] or complement fixation[1250] tests. The IgM antibodies appeared earlier than IgG antibodies, but were short-lived, usually declining to low levels by the time sarcocysts matured. The IgG antibody concentration in serum peaked during the early period of sarcocyst formation and persisted at a relatively high concentration during the chronic infection. There was no anamnestic antibody response in calves orally challenged with sporocysts 3–5 months after primary infection.[666] IgA and IgG2 antibodies were not found.[666]

The onset and persistence of *Sarcocystis* antibodies varied with the species of the host, species of parasite, source of antigen, and serological test. Although *Sarcocystis* species share antigens, the antibody titers were higher using antigen from homologous species of *Sarcocystis* than from heterologous species.[666] For example, antibody titers in pigs inoculated with *S. miescheriana* were higher when the antigen was derived from *S. miescheriana* than from *S. muris* of the mouse or *S. gigantea* of sheep.[1286]

Both bradyzoites and merozoites can be used in ELISA, but titers were higher by using the merozoite antigens.[1545,1549]

1.7.3 Cellular Responses and Immunosuppression

As might be expected of an intracellular parasite, immune cells are mobilized during the *Sarcocystis* infection. The predominant cells infiltrating visceral and muscular tissue during *Sarcocystis* infection are lymphocytes and macrophages.[366,911] This mononuclear cell infiltration begins during the third week of infection and can last for several months, long after the parasite is no longer demonstrable in visceral tissues. Not only visceral tissues are affected, but lymphocytes from peripheral circulation also show a blastogenic response when stimulated with antigen-specific *Sarcocystis*. In calves and sheep inoculated with *Sarcocystis*, this blastogenic response was evident 2 weeks after infection, coincident with the release of first-generation merozoites in cattle and sheep.[666]

Whether these cellular events participate in the recovery of the host from sarcocystosis has not been established, and passive transfer of resistance via cells or antibodies has not been reported. In certain animals, sarcocystosis can depress immunity.[628] Goats with subclinical infection with 1000 sporocysts were more susceptible to intestinal coccidial infections than controls.[389]

The intense cellular response seen in immune animals that survive lethal challenge indicates cell-mediated immunity against *Sarcocystis* (Figure 1.93). Cytotoxic antibodies or metabolites are known to destroy second-generation extracellular merozoites[1650] (Figure 1.96).

Immune responses might differ with the strain of *Sarcocystis* and the strain of the host, as demonstrated with *S. singaporensis* infection in rats.[890] The strain S5, originally obtained from feces of a wild python in Thailand, was highly pathogenic for laboratory rats and less pathogenic for wild rats. After 10 serial passages between pythons and Fischer's rats, the *S. singaporensis* strain (designated S1) became completely nonpathogenic for wild rats, irrespective of the dose.[890] Different classes of immunoglobulins were characterized in brown rats during acute and chronic infection with the modified S1 strain of *S. singaporensis*. Some strains became more virulent after serial passage in the laboratory.[892] Immune responses were higher for hypervirulent strains versus avirulent strains.[892] These findings should be considered when evaluating results of *Sarcocystis* infection in different hosts. Little is known of the comparative virulence of strains of *Sarcocystis* circulating in nature.

1.7.4 Protective Immunity and Vaccination

Cattle, sheep, goats, and pigs inoculated orally with a dose of sporocysts that resulted in subclinical infection were protected against a challenge dose that normally would have been lethal. This protective immunity persisted for 80 days, but not 120 days, in pigs,[1827] at least 252 days in cattle,[590] 274 days in goats,[372] and at least 90 days in sheep.[612] The size of the immunizing dose is important. Goats and pigs dosed with 100 or more sporocysts of *S. capracanis* and *S. miescheriana*, respectively, were protected; those dosed with 10 sporocysts were not protected. The degree of protection was better with 1000 sporocysts than with 100 sporocysts.[358,1880] Immunized animals had only mild illness after a putative lethal challenge, even though parasites from the challenge inocula underwent some development in the host.[358,1880] The protective immunity was induced only by homologous *Sarcocystis* species because cattle experimentally infected with *S. hirsuta* were not protected against challenge infection with *S. cruzi*.[590] The interruption of the second-generation schizogony by chemoprophylaxis did not affect the outcome of protective immunity. Cattle inoculated with lethal doses of sporocysts and treated with amprolium from days 21 to 35, survived acute sarcocystosis and developed protection against lethal challenge.[590]

The stage of the parasite responsible for protective immunity is not known but is likely to be either the sporozoite or first-generation schizonts. Pigs were vaccinated with live, killed, or fractions of *S. miescheriana*-bradyzoites-induced antibodies but provided no protection.[1289] Similar results were obtained in mice immunized with killed *S. dispersa* bradyzoites. Inoculation with *S. muris* sporocysts irradiated with 5 or 10 krad induced protective immunity in mice.[1044] Irradiated vaccines have not been tried in large animals.

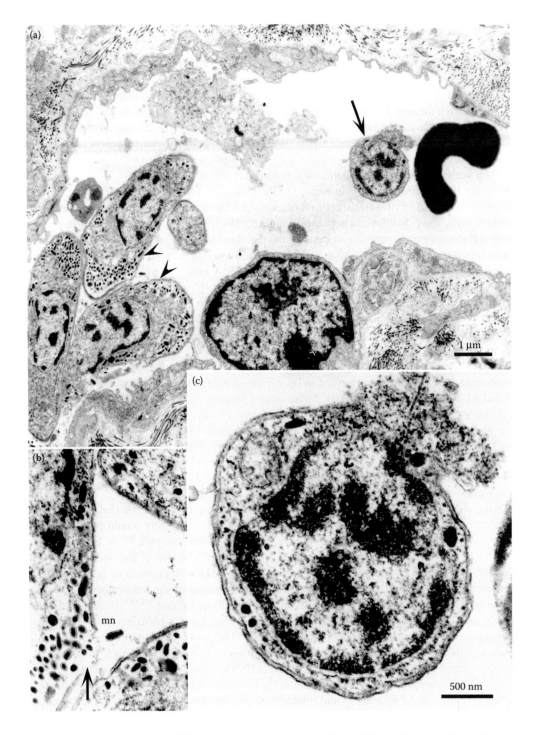

Figure 1.96   TEM of *in vivo* lysis of *S. cruzi* merozoites in a venule in the kidney of an experimentally infected calf. (a) Five merozoites are present in the vascular lumen. Cytoplasm is extruding from 1 merozoite (arrow) adjacent to a red-blood cell (dark, U-shaped). (b) Higher magnification of the merozoites (2 arrowheads) showing partial lysis of double pellicular membranes (arrow). Micronemes (mn) are being released from a pellicular lesion. (c) Higher magnification of a merozoite in (a), showing cytoplasm escaping (top right side) from a lysed merozoite. (From Speer, C.A., Dubey, J.P., 1981. *J. Parasitol.* 67, 961–963. With permission.)

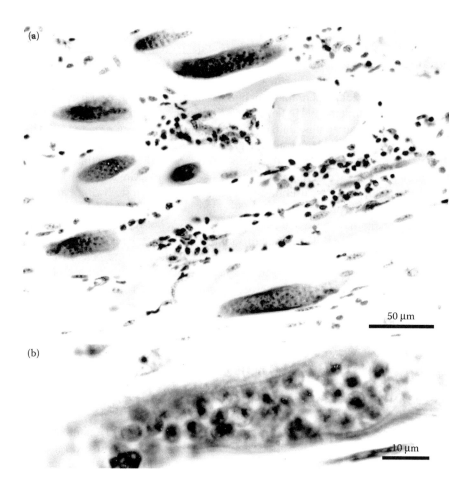

Figure 1.97 Myocarditis in a steer naturally infected with *S. cruzi*. H and E stain. (a) Myocardial cell degeneration (arrow), infiltration by mononuclear cells, and 7 immature sarcocysts. (b) Higher magnification of a sarcocyst to show spheroid metrocytes. (Adapted from Giles, R.C. et al., 1980. *J. Am.
Vet. Med. Assoc.* 176, 543–548.)

The administration of colostrum has no protective response in cattle and sheep, as these animals
remain susceptible to clinical sarcocystosis from birth and throughout their lives despite repeated
natural infection.[577,1250]

The persistence of live sarcocysts is probably not essential for the maintenance of protective
immunity. Goats were protected from challenge infection even when sarcocysts were no longer
demonstrable or were markedly reduced in number.[372]

1.8 DIAGNOSIS

Diagnosis of acute sarcocystosis is difficult. The disease is generalized in nature, with no specific signs; finding parasites in tissues of acutely infected cattle is unlikely. There is no commercially
available or standardized serologic test, and virtually all cattle, regardless of clinical involvement,
have been found with some sarcocysts in muscles. A diagnosis of bovine sarcocystosis is based on
the elimination of other causative agents, a good epidemiologic evaluation of the herd and its relationship to other animals (especially dogs), and clinical findings.

A presumptive diagnosis of acute sarcocystosis can be made if anemia, anorexia, fever, excessive salivation, abortion, loss of body hair (especially at the tip of the tail), increased levels of LDH, SBDH, CPK, BUN, and bilirubin, or lowered PCV are present. Finding *Sarcocystis* antibody or antigen in serum can aid diagnosis of acute sarcocystosis. There is a parasitemia during acute sarcocystosis. Although merozoites were found in buffy coat smears of cattle with acute sarcocystosis from 16 to 46 DPI[363]; this method is tedious, time consuming, and not practical for routine diagnosis. An ELISA to detect circulating antigen in mice and pigs has been reported.[1286] This method might eventually prove useful in diagnosis of acute sarcocytosis in naturally infected animals. The detection of humoral antibodies can aid diagnosis, even though titers sometimes do not correlate with the clinical state. Comparison of antibody titers in acutely ill animals—with those in animals not showing clinical signs—can be useful.[626,681] The more animals tested within an affected herd, the greater the ability to interpret trends in serologic data.

Sarcocystis antibodies have been detected by IHA, ELISA, dot-ELISA, and IFA tests.[393,488,1213,1215,1216,1700] At present, there is a lack of standardization of these tests. Because antigens are obtained from sarcocysts in the muscles of experimentally infected animals and consist of lysate of the bradyzoites,[1087] variations in preparative methods yield antigens varying greatly from one batch to another. Although antigens obtained from merozoites might be more suitable for serologic diagnosis of acute sarcocystosis than antigens from bradyzoites, merozoite antigens have not been utilized for diagnostic purposes. Serologic cross reactivity among *Sarcocystis* species is inconsistent. Although *S. muris* bradyzoite antigens from mice are more easily obtained than *S. cruzi* bradyzoite antigens from cattle, *S. muris* antigens are not suitable for the diagnosis of acute sarcocystosis in cattle. Antigens from *S. cruzi*, however, have been found suitable to detect antibodies against *S. hirsuta* in cattle and *S. tenella* in sheep.[666]

Finding schizonts in biopsies of muscles and lymph nodes, or in additional tissues post mortem, can aid diagnosis. However, schizonts are often too few to be found in histologic sections, because they might disappear by the time clinical disease is obvious. Finding large numbers of immature or mature sarcocysts at the same stage of development can also aid diagnosis (Figure 1.97). Most clinical outbreaks are likely to occur between 5 and 11 weeks following the initial ingestion of sporocysts. Coupled with the history of feeding raw beef to local dogs (or the consumption of a carcass by dogs or wild carnivores), this time period might be a useful guide. Factored into this guide is the consideration that it might take as long as 2–3 weeks for a carnivore to excrete sporocysts after ingesting beef.

In histologic sections, *Sarcocystis* must be differentiated from *Toxoplasma gondii* and other closely related coccidians. *Sarcocystis* schizonts develop in endothelium of blood vessels. The immature schizonts are basophilic structures with or without differentiated nuclei (Figure 1.98). Ultrastructurally, *Sarcocystis* schizonts are free in the host-cell cytoplasm, whereas all stages of *T. gondii* are separated from the host-cell cytoplasm by a PV and can develop in virtually any cell in the body. Because each *T. gondii* organism divides into 2, there is no immature stage in the intermediate host.

The diagnosis of *Sarcocystis*-induced abortion presents additional problems because the parasite is not consistently found in fetal tissues. Sarcocystosis produces gliosis and placental necrosis; therefore, brain and placenta should be examined (Figure 1.98). A detailed description of a case of congenital infection was described in a stillborn lamb from Denmark.[17] Within the placental cotyledons, schizonts were in both the lamina propria and submucosa. The chances of diagnosis are improved if numerous fetal tissues are examined.

IHC with specific antibodies is very useful in diagnosis of parasites in histological tissue sections. However, the specificity depends on the purity of reagents, antigenic variability among the strains of *Sarcocystis* used, and hosts used to produce antibodies.[436] IHC is only an aid, and other factors should be considered for final diagnosis. The IHC procedure is detailed in Chapter 2, and has been commonly employed for the diagnosis of *S. neurona* (Chapter 3). Cross reactivity among

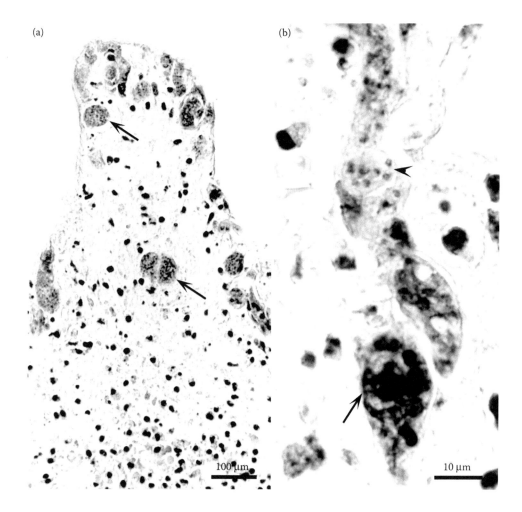

Figure 1.98 Placenta of an aborted cow naturally infected with *S. cruzi*. (a) Several schizonts (arrows) are located in the lamina propria of a villus exhibiting necrosis and infiltration of macrophages. (b) Higher magnification showing deeply stained immature schizont (arrow) and mature schizonts with merozoites (arrowheads). H and E stain.

Sarcocystis species using polyclonal antibodies can be problematic. For example, anti- *S. cruzi* bradyzoite polyclonal rabbit antibodies stains not only *Sarcocystis* species, including *S. neurona* and *S. canis* (see Figure 3.11), but also *T. gondii*.[724,726] Additionally, bradyzoites or bradyzoite extracts are often lethal for rabbits. However, species-specific *S. neurona* antibodies were prepared in rabbits using cell cultured merozoites.[420] Although specific monoclonal antibodies have been characterized against some *Sarcocystis* species, these are not available commercially for routine diagnostic use.

1.9 ECONOMIC LOSSES

Millions of dollars have been lost each year from condemnation or downgrading of meat containing grossly visible sarcocysts[241,591] but economic losses from clinical and subclinical infections are difficult to calculate because: (1) nearly 100% of cattle and sheep are infected, thus it is difficult to differentiate affected and nonaffected animals that are infected with microscopic sarcocysts; (2)

no dollar values are available for the losses due to poor feed efficiency, failure to grow, reduced milk and wool production, reproductive problems, and obvious clinical disease; and (3) clinical disease is difficult to diagnose and has been recognized in relatively few outbreaks involving cattle in natural conditions. The wide prevalence of pathogenic species of *Sarcocystis* in domestic animals leaves no doubt that clinical disease could be recognized when better diagnostic tests are available. Economic losses due to *S. neurona* are discussed in Chapter 3.

1.10 TRANSMISSION

Ingestion of sporocysts and oocysts in food or water is the only major mode of transmission to the intermediate host. Although transplacental infection has been documented in cattle and sheep in nature by finding schizonts in some aborted fetuses, it is rare. It is possible that all transplacentally infected cattle and sheep die because *Sarcocystis* sarcocysts have never been found in newborn calves or in lambs. Sarcocysts has been found only once in a 3-day-old foal.[269] The possibility of lactogenic transmission via milk or colostrum was tested experimentally, but evidence of transmission was not found.[587]

The ingestion of mature sarcocysts (i.e., those containing bradyzoites) is the only known means by which the definitive host acquires *Sarcocystis* infection.

1.11 EPIDEMIOLOGY

Sarcocystis infection is common in many species of animals worldwide. Virtually 100% of adult cattle in the United States are infected with this parasite. Several conditions exist that permit such an unusually high prevalence of *Sarcocystis* in animals:

1. A host may harbor any of several species of *Sarcocystis*. For example, sheep may become infected with as many as 5 species, and cattle may have as many as 4 species of *Sarcocystis*.
2. Many definitive hosts are involved in transmission. For example, cattle sarcocystosis is transmitted via felids, canids, and primates. *S. cruzi*, the most common species of *Sarcocystis* of cattle worldwide, is transmissible via dogs, coyotes, foxes, wolves, and raccoons. Wild carnivores, such as coyotes, may travel many miles for food each day and spread the parasite over long distances.
3. *Sarcocystis* oocysts and sporocysts develop in the lamina propria and are excreted over a period of many months.
4. Sporocysts or oocysts remain viable for many months in the environment.[1160,1547] They can be further spread or protected by invertebrates.[1111,1625]
5. Large numbers of sporocysts can be excreted. For example, dogs ingesting a relatively small amount of beef (250 g) excreted 100–6000 sporocysts per gram of feces.[579] With an average fecal output of 250–350 g/day, a dog may excrete from 250,000 to over 2 million sporocysts per day. Coyotes and foxes are even more efficient producers of *S. cruzi* sporocysts. As many as 200 million sporocysts were recovered from the intestinal scrapings of a coyote that was fed 1 kg of naturally infected beef.[354] *S. cruzi* is not the only species that produces large numbers of sporocysts. As many as 90 million *S. tenella* sporocysts were recovered from the feces of a dog fed naturally infected mutton.[611]
6. There is little or no immunity to reexcreting of sporocysts.[579,801,1479] Therefore, each meal of infected meat can initiate a new round of production of sporocysts.
7. Oocysts or sporocysts are resistant to freezing and, thus, can withstand winter on the pasture.[95,173,352,575,811,1042,1918] They are also resistant to disinfectants.[79,1159] Apparently sporocysts can be killed by drying and by 10 min exposure at 56°C. Sporocysts can survive for months even at low humidity.[1547]
8. Unlike many other species of coccidia, *Sarcocystis* is passed in feces in the infective form and is not dependent on weather conditions for maturation and infectivity.

Not all species of *Sarcocystis* are highly prevalent. Most species of *Sarcocystis* transmissible via cats have been found less frequently than those transmissible via canids. There may be several reasons for this. One reason may be that cats are very poor producers of *Sarcocystis* sporocysts. Cats fed several grams of macroscopic sarcocysts of *S. muris* and *S. gigantea* containing millions of bradyzoites produced relatively few sporocysts. Another reason may be that sarcocysts of feline-transmitted species, such as *S. gigantea* and *S. medusiformis*, require several months or years to become infective. Such species are not able to complete the cycle in young animals. Still another reason may be that some host species are inherently more susceptible to infection with some agents than are others. For example, the pronghorn (*Antilocapra americana*) is rarely found infected with *Sarcocystis*. Whether breeds of livestock species vary in susceptibility to *Sarcocystis* is not known, but certain strains of mice appear more susceptible to infection than others.[1486]

1.12 CONTROL

There is no vaccine to protect livestock against clinical sarcocystosis. However, experimental studies indicate that cattle, sheep, goats, and pigs can be immunized by small numbers of live sporocysts. Thus, there is hope of developing a vaccine for sarcocystosis in the future. For the present, prevention is the only practical method of control.

Excretion of *Sarcocystis* in the feces of definitive hosts is the key factor in the spread of *Sarcocystis* infection. Therefore, to interrupt this cycle:

1. Carnivores should be excluded from animal houses and from feed, water, and bedding for livestock.
2. Uncooked meat or offal should never be fed to domesticated carnivores. Freezing can drastically reduce or eliminate infectious sarcocysts in meat. Overnight freezing in a household freezer is effective in killing sarcocysts, based on the thickness of the meat.[672,722,1280,1486,1666] Exposure to heat at 55°C or higher temperatures for 20 min kills Sarcocystis.[242,574,972]
3. Dead livestock should be buried or incinerated. This is particularly important where dead animals are left in the field where carnivores can have access to the carcasses.
4. The prophylactic use of anticoccidials may be another practical method of controlling sarcocystosis in livestock.
5. Rodent bait containing lethal doses of *S. singaporensis* sporocysts have been suggested as a method to control rat populations[886,889] but this has not been approved for use in some countries; *S. singaporensis* is thought to be host-specific for rats.

1.13 CHEMOPROPHYLAXIS AND CHEMOTHERAPY

Several drugs routinely used as anticoccidials in poultry are effective against sarcocystosis in cattle, sheep, and goats, when administered continuously for a month, starting at or before inoculation with *Sarcocystis*.[576,813,1040,1041] Amprolium (100 mg/kg body weight) administered from 0 to 30 DPI, prevented acute disease or death in cattle infected with *S. cruzi* and in sheep infected with *S. tenella*; 50 mg/kg body weight of amprolium was less effective in sheep than the larger dose.

Calves treated with 100 mg/kg of body weight of amprolium from 21 to 35 DPI with 100,000 sporocysts had only mild signs of acute sarcocystosis, indicating that second-generation schizonts and any subsequent asexual stages preceding sarcocyst development were affected.[590]

Salinomycin (1 or 2 mg/kg body weight), administered from 1 to 30 DPI also prevented deaths in sheep with *S. tenella*. Halofuginone (0.22 mg/kg body weight) similarly reduced or prevented acute sarcocystosis in goats infected with *S. capracanis*, and in sheep infected with *S. tenella*; the drug was administered in feed from day 5 before inoculation and continued up to 36 DPI.[813,1797]

Lasalocid, decoquinate, and monensin, each incorporated in feed at 33 mg/kg of feed, had minimal or no effect on the course of acute sarcocystosis in cattle inoculated with 250,000 *S. cruzi* sporocysts; the medicated food was given from day 7 before inoculation through 80 DPI.[618] For chemotherapy, only halofuginone has been found effective.[813,1797] Halofuginone (0.67 mg/kg body weight) given once or 4 times, prevented death in goats and sheep acutely ill with sarcocystosis. Once the clinical signs of acute sarcocystosis appeared, the administration of salinomycin, sulfadoxin and trimethoprim, lasalocid, robenidin, or spiramycin, did not prevent death in sheep and goats.[813,1041,1797]

Although oxytetracycline (30 mg/kg body weight), given intravenously, prevented death in 2 sheep with acute sarcocystosis, this drug may be too expensive.[813]

Most anticoccidials affect *Sarcocystis* schizonts. Results with *S. muris* infections in mice provide further insight. Of the 4 phases of infections examined, sporozoite migration (2–10 DPI), schizogony (11–17 DPI), merozoite migration (18–27 DPI), and sarcocyst formation (28–50 DPI), anticoccidials were most effective against schizogony.[1486] Of the 12 anticoccidials tested in the *S. muris* model, including all 4 phases of *S. muris* infection, amprolium, monensin, arprinocid, and sulfaquinoxaline plus diaveridine had no antisarcocysticidal activity, and lasalocid, halofuginone, sulfadoxine plus trimethoprim, and sulfadimethoxine had only marginal anti-*S. muris* activity, whereas Zoalene®, primaquine diphosphate, sulfaquinoxalin plus pyrimethamine, and Bay® G 7183 had excellent anti-*S. muris* activity. Of particular interest is the activity of sulfaquinoxalin plus pyrimethamine against immature and mature sarcocysts, because no other drugs are known to be effective against sarcocysts.[1486,1487] Results with the murine *S. muris* model also indicate that results with 1 species of *Sarcocystis* may not be directly applicable to other species of *Sarcocystis*. For example, amprolium was effective against *S. cruzi*, but not against *S. muris*.

Chemotherapy of the pathogenic species *S. neurona* is discussed in Chapter 3.

CHAPTER 2

Techniques

2.1 EXPERIMENTAL INFECTION OF INTERMEDIATE AND DEFINITIVE HOSTS

Intermediate hosts are infected by feeding sporocysts that can be mixed in grain or suspended in a small volume at the back of the tongue to avoid spillage. The age of the animal does not appear to affect infectivity; 1-day-old calves, goats, or lambs, as well as older animals are easily infected with *Sarcocystis*. Inoculated animals should be kept in isolation for 10 days, and all the waste and bedding should be incinerated to kill sporocysts that might pass through the gut unexcysted and be excreted in the feces[136,381,1257] where they can be infectious to other animals.

Definitive hosts are infected by feeding intermediate host muscle tissue with mature sarcocysts containing bradyzoites; immature sarcocysts containing only metrocytes are not infectious. Time to maturity varies with each species of *Sarcocystis*. For example, in sheep, *S. tenella* and *S. arieticanis* become infectious within 3 months, whereas *S. gigantea* sarcocysts do not mature for more than 1 year or longer. The numbers of sarcocysts of some species (e.g., *S. miescheriana*, *S. capracanis*, *S. muris*) are reduced dramatically after 3 months.

2.2 ISOLATION, PURIFICATION, AND PRESERVATION

2.2.1 Bradyzoites

Bradyzoites can be released from sarcocysts by incubation in digestive fluid (trypsin or acid pepsin). The sarcocyst wall is dissolved by digestive fluid, but the released bradyzoites survive in it.[878] The following method[567,1087] is useful for obtaining *S. cruzi* bradyzoites:

a. Select heavily infected tissue by microscopic examination. The number of sarcocysts in different tissues may vary among animals; heart and tongue are the most heavily infected organs.
b. Remove as much connective tissue and fat as possible, and grind muscle in a meat grinder or mince into small pieces.
c. Suspend about 50 g of ground meat in 100 mL of warm (37°C) digestion fluid (pepsin, 2.5 g; concentrated HCl, 10 mL; water to make 1000 mL; dissolve pepsin in water, centrifuge to remove undissolved pepsin), and incubate at 37°C on a magnetic stirrer for approximately 10 min or longer. The objective is to release bradyzoites from sarcocysts that were broken during grinding, and this method is best suited for heavily infected tissues. Lightly infected tissues may have to be digested longer in acid–pepsin solution, and, in that case, the digestion fluid should be isotonic.[567]
d. Pour the homogenate through cheesecloth to remove large particles and centrifuge at 400*g* for 5–10 min using a floating-head centrifuge and 50-mL tubes.
e. Discard the supernatant fluid and suspend the sediment in Hanks' balanced salt solution (HBSS) at pH 7.4, preferably at 5–10°C.

f. Recentrifuge at 400g for 5–10 min and discard the supernatant. The pellet has 2 layers, the whitish bradyzoite layer at the bottom and the brownish host tissue layer on the top. Remove host tissue by gently aspirating it from the whitish layer; this separation is not always possible.

g. To further clean zoites of host tissue, mix concentrated zoite suspension with isotonic Percoll® solution, one part of zoite suspension with two parts of Percoll. To prepare isotonic Percoll, mix Percoll with 9% NaCl solution (9 parts Percoll + 1 part NaCl).

h. Centrifuge Percoll-zoite suspension at 400g or more for 10 min. The bradyzoites settle at the bottom and host tissue floats in Percoll.

i. Pour off Percoll and wash bradyzoite pellet three times in HBSS by centrifugation.

j. Bradyzoites recovered by the above procedure remain viable.

2.2.2 Sporocysts

Sarcocystis sporocysts, after sporulating in the intestinal lamina propria, are normally released into the intestinal lumen over several months. To optimize the recovery of sporocysts:

a. Euthanize the definitive host 3–7 days after oocysts or sporocysts are first passed in the feces.

b. Remove the small intestine, cut it lengthwise, and spread over paper towels with the luminal side up.

c. Lightly scrape the intestinal epithelium with a glass slide, removing only the tips of the villi, where most sporocysts are concentrated.

d. Suspend scrapings in water and then homogenize in a blender for about 2 min at top speed.

e. Centrifuge the homogenate at 400g for 10 min.

f. Discard supernatant and suspend the sediment in water. Repeat the above process until most sporocysts are released from the host tissue.

g. Concentrate sporocysts by centrifugation, filter through a 25-μm stainless-steel sieve, suspend in HBSS, and centrifuge.

h. Suspend the pellets in HBSS–antibiotic mixture (given below).

i. Although most sporocysts are concentrated toward the tips of the villus, some are located deep in the villus. To obtain these sporocysts, deeply scrape the intestine and homogenize the intestinal scrapings in water.

j. Emulsify homogenate in 5.25% sodium hypochlorite solution (1:1 ratio) in a cold bath for 30 min.

k. Centrifuge at 400g for 10 min and discard supernatant.

l. Suspend the sediment in water and wash by centrifugation until the smell of chlorine is gone.

m. Suspend sporocysts in HBSS–antibiotic mixture (penicillin, 10,000 U; streptomycin, 10 mg; Fungizone®, 0.05 mg; and Mycostatin®, 500 U/mL [mixture = PSFM] of HBSS–PSFM mixture).

n. Sporocysts stored in PSFM solution at 4°C remain viable for 12 months or more (1 L of antibiotic solution can be prepared and stored at −20°C in 100-mL aliquots. Each aliquot is thawed as needed).

o. Sporocysts can also be obtained from feces of definitive hosts by flotation methods used for other coccidia.

2.3 DIAGNOSTIC TECHNIQUES

2.3.1 Examination of Feces for Sporocysts

The number of sporocysts in feces is usually low; therefore, concentration methods are often necessary to detect *Sarcocystis* in feces. Either sugar or salt (NaCl or $ZnSO_4$) solutions of specific gravity 1.15 or more can be used to float sporocysts free of fecal debris. Sugar solution is less deleterious to sporocysts than salt solutions, and has been used with the following method:

a. Mix 5–10 g of feces thoroughly in 50–100 mL of Sheather's sugar solution (sugar, 500 g; water, 320 mL; add liquid phenol, 6.5 mL if the solution needs to be stored for more than few days, to prevent fungal growth). To thoroughly dissolve sugar, heat water to 70°C and stir solution continuously, then cool before use. To obtain a homogenous suspension, the feces should be mixed well in about 5–10 mL of water before adding sugar solution.

b. Filter through cheesecloth.

c. Centrifuge fecal suspension in 50-mL centrifuge tubes with a cap at about 400g for 10 min.

d. With a pipette, remove several small drops from the very top, put drops on a glass slide and cover the drop with a coverslip.

e. Let the slide coverslip lie flat for about 5–10 min so that the fecal particles can settle and the sporocysts can rise to the top before the slide is examined.

f. Examine at magnification of ×100 or more. *Sarcocystis* sporocysts are small (usually 15 × 10 μm), thin-walled, and of lower density than most other parasite eggs or oocysts; therefore, they lie just beneath the coverslip and are often missed (Figure 2.1).

g. To collect sporocysts from feces, aspirate the top 5 mL of the solution from the 50-mL centrifuge tube. Mix the aspirate with 45 mL of water and centrifuge at 400g for 10 min. Discard the supernatant fluid and resuspend the sediment in water, centrifuge, and repeat the process. Suspend the final sediment in PSFM solution.

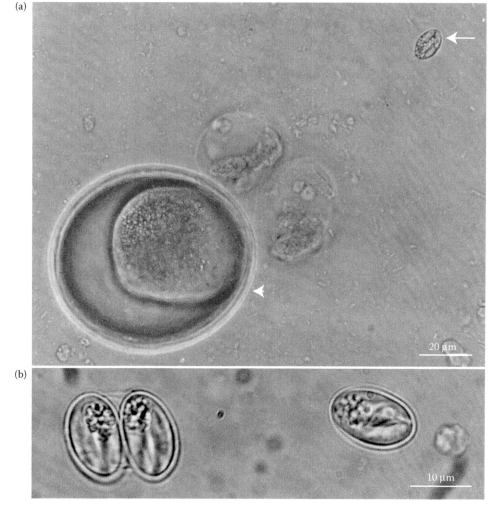

(a)

(b)

Figure 2.1 *Sarcocystis* sporocysts in sugar solution float of dog feces. (a) Sporocyst (arrow) and a roundworm egg, *Toxascaris leonina* (arrowhead). (b) Sporocyst on the right and an oocyst on the left. (From Streitel, R.H., Dubey, J.P., 1976. *J. Am. Vet. Med. Assoc.* 168, 423–424. With permission.)

2.3.2 Examination of Muscles for Sarcocysts

Sarcocysts can be detected in muscles by gross inspection, examination of unstained squash preparations, histologic examination, or by a digestion method. Each of these methods has its advantages and disadvantages. Only a few species of *Sarcocystis* form macroscopic sarcocysts. Examination of unstained squashes is useful to study sarcocyst structure (Figure 2.2); by this method, one can examine relatively more tissue than by histologic examination. However, this method is unsuitable for fixed tissue. Histologic examination is necessary to study sarcocyst morphology, but this method is the least sensitive because only a small amount of tissue can be sampled.

Digestion of host tissue is the most sensitive method to detect light *Sarcocystis* infection. In the digestion procedure, host tissue (50 g) is incubated in 10 volumes of either 1% trypsin solution at pH 7.4 or HCl–pepsin solution for 1–4 h.[132,243,392,878,1775] The digest is examined for bradyzoites. By this method, one may detect even a few sarcocysts in 50 g of tissue because several hundreds or thousands of bradyzoites are released from sarcocysts as the host tissue and sarcocysts are digested. The bradyzoites can be used for antigen, and the species can be identified by molecular methods. Disadvantages of this method are: (1) it is not possible to distinguish species of *Sarcocystis* because the sarcocyst wall is dissolved; (2) metrocytes are not resistant to digestion, therefore, immature sarcocysts may not be detected; and (3) it may not be possible to detect sarcocysts with relatively small bradyzoites (5–6 μm). To obtain intact sarcocysts, we suggest the digestion of small amount of tissue for 8–10 min.[547,1896,2072]

The following method is used to isolate individual sarcocysts from infected pigs (Heydorn, A. O., personal communication):

a. Remove fat and connective tissue from 100 g of heavily infected swine muscles (102–110 DPI after experimental infection).
b. Grind it in a meat grinder and suspend ground meat in 1 L of digestion fluid (0.3% trypsin [Merck], 0.05% ethylenediaminetetraacetic acid [EDTA] in phosphate buffer saline [PBS], pH 7.4–7.6).

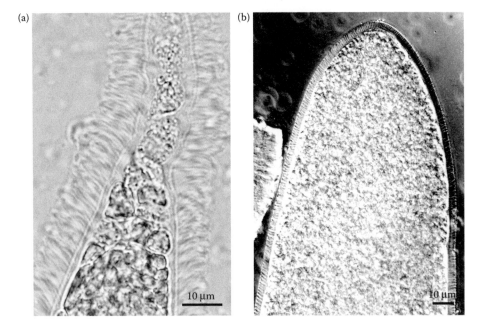

Figure 2.2 Sarcocyst walls in unstained wet preparations. (a) *S. cruzi*. Although the sarcocysts appear thin in H and E-stained sections, numerous elongated, hirsute protrusions are visible in this preparation. (b) *S. tenella* with striated wall.

c. Stir gently at room temperature with magnetic stirrer for 45–60 min.
d. Sieve through a coarse metallic sieve (or tea strainer) and save the filtrate.
e. To stop digestion, add 1 L of PBS with 15 mL of fetal calf serum.
f. Sediment sample in a tall glass cylinder for 15–20 min. Decant and save 100 mL of the sediment. Refill with PBS, repeating sedimentation and decanting 3–4 times, using the PBS as the diluent.
g. Examine the sediment for intact and broken sarcocysts.

2.3.3 Serological Techniques (for *S. neurona* Diagnosis, See Chapter 3)

2.3.3.1 Preparation of Soluble Antigen from Bradyzoites

In IHA and ELISA tests, soluble antigen is prepared by the following method.[666]

a. Obtain bradyzoites as described and then freeze and thaw the suspension 3 times or more.
b. Suspend the bradyzoite suspension in Dulbecco's phosphate-buffered saline (DPBS).
c. Homogenize the bradyzoite suspension in a homogenizer (Polytron®, Brinkmann Instruments, Westbury, NY) at low speed until the zoites are disrupted. Let the bradyzoites stand at 4°C for 1–2 h to leach out soluble material.
d. Centrifuge at 16,000–20,000*g* for 30 min. Collect the supernatant fluid.
e. Dialyze the supernatant fluid overnight at 4°C against DPBS without added calcium and magnesium, using a dialysis membrane with a 6000- to 8000-M_r cut-off.
f. Filter and sterilize the dialysate through a 0.2-µm filter unit.
g. Estimate the protein concentration using a Bio-Rad® Coomassie blue dye binding assay or an alternative method.
h. Store the antigen in aliquots at −70°C.

2.3.3.2 Method for ELISA Test

Method initially used for *S. cruzi* with crude bradyzoite lysate antigen was reported earlier.[393] Newer methods of ELISA, using cell culture-derived merozoites antigens (whole or recombinant), are discussed in Chapter 3.

2.3.3.3 Immunohistochemical Staining

Formalin-fixed paraffin-embedded tissues can be used in this technique. Although parasite antigens can be detected even 1 year after fixation in 10% formalin, fixation for short periods (24 h) is recommended. Sections are cut 3–5 µm thick. It is necessary to use grease-free clean slides, otherwise the sections detach from the slide; also, an addition of chrome-gelatin in the water bath helps adherence of sections to the slide. To avoid mortality due to toxicity of bradyzoites, rabbits can be infected with merozoites or oocysts; the latter are preferred because oocysts can be cleaned of fecal matter and treated with 5.25% sodium hypochlorite solution (Clorox®) to remove host material.

We carry out the following procedure:

a. De-paraffinize slides in old xylene I for 30 min. Discard xylene I. Pour xylene II into xylene I container. Soak 5 min.
b. Pour fresh xylene into xylene II. Soak slides in fresh xylene for 5 min.
c. Soak in 100% ethanol I for 5 min. Discard ethanol I. Pour ethanol II into ethanol I container. Soak 5 min. Pour fresh 100% ethanol into ethanol II container. Soak another 5 min.
d. Soak in 95% ethanol for 5 min.
e. Quench endogenous peroxidase in 3% H_2O_2 in methanol for 15 min. (20 mL 30% hydrogen peroxide and 180 mL methanol).

f. Soak in 80% ethanol for 5 min.
g. Soak in 70% ethanol for 5 min.
h. Soak in 50% ethanol for 5 min.
i. Soak in saline for 5 min.
j. Soak in warm saline (37°C) for 5 min.
k. Pepsin digestion: incubate slides in 0.4% pepsin in 0.01 N HCl for 15 min at 37°C. (0.8 g pepsin, 0.333 mL 6 N HCl, 200 mL saline, prewarmed to 37°C for 30 min).
l. Soak in 0.75% Brij-PBS for 5 min twice (2.5 mL 30% Brij-35 per liter PBS).
m. Block nonspecific binding with 0.5% Na caseinate in Brij-PBS for 10 min.
n. Drain and, without rinsing, apply the properly diluted primary antibody to the tissues for 30 min at 37°C.
o. Soak slides in Brij-PBS for 5 min.
p. Treat tissues with Dako EnVision Rabbit peroxidase solution® at 37°C for 30 min.
q. Soak slides in Brij-PBS for 5 min twice.
r. Apply Dako AEC substrate chromogen solution® to slides at 37°C for 15 min.
s. Collect chromogen in hazard waste bottle, and soak slides in Brij-PBS for 5 min.
t. Rinse slides in running tapwater for 2 min. Rinse in deionized water for 2 min. Twice.
u. Counterstain with Mayer's hematoxylin for 1 min.
v. Wash slides in running tapwater for 1 min.
w. Soak slides in Scott's tapwater substitute (MgSO$_4$7H$_2$O 20 g, NaHCO$_3$ 2 g, water 1 L) for 1 min.
x. Soak in running tapwater for 1 min. Soak in deionized water for 2 min. Twice.
y. Apply Crystal Mount® and incubate slides overnight at room temperature.
z. Apply Permount® and coverslip, if desired.

2.4 *IN VITRO* CULTIVATION

The first stages to be grown *in vitro* were sexual stages[86,569,571,1177] (Table 2.1). Only a few species of *Sarcocystis* have been grown *in vitro*. After that, an *in vitro* system was developed to grow large numbers of merozoites of several species of *Sarcocystis* that infect livestock.[1655,1656] The development of merozoites to sarcocysts in cultured cells has not yet been reported.

Two factors determine the success of *in vitro* growth experiments. First, a large supply of sterile stage-specific parasites (i.e., sporozoites, merozoites, or bradyzoites) must be readily available. Second, the selection of the cell line to be tested for its ability to support growth of the parasite should be based on the cell or tissue in which the parasite develops *in vivo*. For example, if schizonts develop *in vivo* in mouse hepatocytes, then a mouse epithelioid liver cell line should be tested. Because other coccidia have been shown to readily develop in some specific cell types, such as kidney cells, even though they do not normally develop in such cells *in vivo*,[1655] the literature on *in vitro* cultivation of coccidia should serve as a guide for possible cell types to test for *in vitro* growth of *Sarcocystis* spp.

Table 2.1   *In Vitro* Culture of *Sarcocystis* spp. Gamonts

Species	Host	Type of Cells Successful	Cells Unsuccessful	Reference
S. muris	Mouse	MDCK, FE, MF	Not stated	**539, 544**
S. muris	Mouse	CL, DK	HF, SK	**86**
S. suihominis	Pig	HB	CL, DK, SK	**86**
S. tenella	Sheep	DK	HF, SK, CL	**86**
S. capracanis	Goat	DK	HF, SK, CL	**86**
S. sp.	Grackle	MDCK, EBK, EBTr, ETK, ECM	ECK	**569, 571**
S. suihominis	Pig	HEI, HSF	–	**1177**

Abbreviations: MDCK = Madin–Darby canine kidney, FE = feline embryo, CL = cat lung, DK = dog kidney, SK = swine kidney, MF = mouse fibroblasts, HF = human fibroblasts, EBK = embryonic bovine kidney, EBTr = embryonic bovine trachea, ECK = embryonic chicken kidney, ECM = embryonic chicken muscle, ETK = embryonic turkey kidney, HEI = human embryonic intestine, HSF = human skin fibroblasts.

2.4.1 Excystation and Cultivation of Schizonts (for *S. neurona* Cultivation, See Chapter 3)

One of the most important preliminary steps in achieving *in vitro* development of sporozoites to merozoites is to obtain a relatively high rate of sporozoite excystation from sporocysts. After the sporocysts have been isolated from the definitive host, purified, and concentrated, they can be stored until used at 4°C in HBSS with antibiotics.[1039] Several factors are important in storing and successful excystation of sporocysts, and these conditions may differ with different species of *Sarcocystis*. For example, *S. cruzi* sporocysts should not be stored in aqueous $K_2Cr_2O_7$, which is routinely used to store coccidian oocysts, because it has a deleterious effect on excystation.[176,1039] McKenna and Charleston[1159] reported no excystation of *S. gigantea* sporocysts after 5 day exposure in 2% sulfuric acid, compared with 85% excystation by storage in tapwater and 15% excystation after storage in 2.5% $K_2Cr_2O_7$.

Pretreatment of sporocysts with NaOCl helps remove contaminating microorganisms from sporocysts and in subsequent excystation. Traditionally, 5.25% NaOCl (available commercially as undiluted laundry bleach) has been used for this purpose, and the treatment was done in an ice bath to avoid the deleterious effect of heat generated during the procedure. Horn et al.[846] reported higher efficiency of excystation of *S. gigantea*, *S. tenella*, *S. arieticanis*, and *S. hircicanis* by using 6%–8% NaOCl at room temperature for 20 min instead of an ice bath. The NaOCl treatment removes the outer lipid layer of the *Sarcocystis* sporocysts. Incubation in 1%–3% formic acid for 24–72 h at 4°C was reported to enhance excystation.[2282] The mechanism of excystation of *Sarcocystis* species is not fully understood. Trypsin, required for excystation of the *Eimeria* species, is not necessary for *Sarcocystis*, whereas bile is important.[573] Also, different concentrations (5%–15%) of bovine bile have been used for excystation of sporocysts.

The following method has proven useful for excysting sporozoites from sporocysts of *S. cruzi*, *S. capracanis*, and *S. tenella*:

a. Treat the sporocysts with 2.6% (v/v) NaOCl at 4°C for 30 min (i.e., Clorox, Purex®) to kill other microorganisms that would be certain to contaminate the *in vitro* cultures. A significantly greater excystation rate can be obtained by treating the sporocysts of *S. capracanis*, *S. cruzi*, and *S. tenella* with 2.6% NaOCl than with 1.3% or 5.3% NaOCl.[1039] Treatment of the sporocysts with cysteine HCl for 18 h may also improve the rate of excystation.[1039] Cysteine HCl and NaOCl evidently have similar beneficial effects on excystation, but NaOCl acts more rapidly (30 min vs. 18 h), and it must be used anyway to disinfect the sporocysts.

b. Wash the sporocysts several times (usually at least four times) by centrifugation and resuspension in cold saline or culture medium (i.e., RPMI 1640 without serum) until the odor of NaOCl can no longer be detected or until the pH color indicator in the culture medium remains unchanged in the presence of sporocysts.

c. Suspend the sporocysts in excysting fluid and incubate at 37°C for 1.5–4 h. The excysting fluid should consist of a balanced salt solution (i.e., HBSS, pH 7.2–7.4), containing 2% (w/v) trypsin or chymotrypsin (activity of 1:250 or 1:300) plus 5%–10% bile or 2.5%–5% (w/v) bile salt (i.e., sodium taurocholate obtained from the appropriate intermediate host such as bovine, ovine, etc.).

d. Remove the excysting fluid by washing the excysted sporozoites at least twice in the culture medium without serum, and resuspend in 5–10 mL culture medium.

e. To remove debris and sporocyst walls, the sporozoites may be passed through a nylon-wool column.[1026] Too much debris and sporocyst walls in the inoculum may have a deleterious effect on parasite development.[1655]

f. Count the number of sporozoites with a hemacytometer and resuspend them at the appropriate concentration in the culture medium containing 5% (v/v) fetal bovine serum, 2 mM *l*-glutamine, 50 μg/mL dihydrostreptomycin, and 50 U/mL penicillin G.

g. Inoculate 2 to 5×10^4 sporozoites per square centimeter of the monolayer of the cell line to be tested and incubate at 37–39°C in 5% CO_2. Sporozoites of *S. cruzi*, *S. capracanis*, and *S. tenella* will develop in bovine pulmonary artery endothelial cells (CPA; American Type Culture Collection,

Rockville, MD) or bovine monocytes (BM; see Reference **1653** for details concerning the BM). The CPA cell line will produce approximately seven times more merozoites of *S. cruzi* than the BM line,[1657] but the BM line is especially useful for investigating immunological parameters such as opsonizing antibodies and lymphokines.[1653]

h. Remove the culture medium on day 1, and then at 2- to 7-day intervals after sporozoite inoculation, replace it with a culture medium containing 1% or 2% fetal bovine serum (FBS).

i. The infected cell cultures might require maintenance for long periods to allow sufficient time for the parasites to undergo one or more generations of schizogony. Long cultivation periods can be achieved by varying the FBS concentration in the culture medium. As the cell cultures become sparsely populated due to parasite development (usually first observed at 24–28 DPI), then treat them with cell culture medium (CM) containing 10%–20% FBS, which can then be reduced to 1%–2% FBS once the monolayer has reestablished itself.

j. Parasite development can be monitored and photographed at various intervals after sporozoite inoculation by direct microscopic observation with an inverted phase-contrast microscope or by examining double-coverslip preparations[1373] with a compound phase-contrast microscope.

Schizonts of only a few species of *Sarcocystis* have been grown in cell culture, and the selection of cell type for culture is an important consideration (Table 2.2). Schizonts of *S. capracanis*, *S. cruzi*, and *S. tenella* develop similarly *in vitro*, except that *S. cruzi* produces considerably more merozoites per culture vessel than do *S. capracanis* or *S. tenella*.[1656,1657] *S. cruzi* and *S. tenella* develop to schizonts in BM and CPA cells, whereas *S. capracanis* develops in BM only. Soon after inoculation, sporozoites of all three species penetrate cells (Figure 2.3a) and gradually transform in approximately 2–3 weeks into ovoid or spheroidal schizonts with irregularly shaped nuclei, with small nucleoli, and small cytoplasmic granules (Figure 2.3b and c). More advanced stages appear to have numerous nuclei and nucleoli (ultrastructural studies

Table 2.2 Species of *Sarcocystis* Schizonts Cultivated *In Vitro*

Host	Species	Type of Cells Successful	Cells Unsuccessful	Schizonts Days	Reference
Ox	*S. cruzi*	BM, CPA	Mouse macrophages	18–1,320	**31, 1657, 1659**
Ox	*S. hirsuta*	BPE	BM	14–62	**177**
Rat	*S. singaporensis*	Rat brain endothelial cells, rat pneumocytes	Hepatoma, fibroblasts, myoblasts	3–18	**887, 891**
Sheep	*S. tenella*	BM, CPA	MDBK, OM	14–50	**1656**
Goat	*S. capracanis*	BM	MDBK, OM, CPA	60–100	**1656**
Budgerigar-*Didelphis virginiana*	*S. falcatula*	BT	–	3–4	**1070**
Budgerigar-*Didelphis-albiventris*	*S. falcatula*	BM	–	7	**418**
Budgerigar-*Didelphis-marsupialis*	*S. falcatula*	CV-1, BT	EK	4	**441**
Budgerigar-*Didelphis-albiventris*	*S. falcatula*	BT, Vero	ED, Hep-2	4–6	**510**
Budgerigar-*Didelphis-albiventris*	*S. lindsayi*	EK, BT, CV-1	–	2–3	**425**
Mouse-*Didelphis virginiana*	*S. speeri*	BM, EK	CV-1	3	**423**
Mouse-*Didelphis marsupialis*	*S. speeri*	BT	–	15	**427**

Abbreviations: BM = bovine monocytes, BPE = bovine pulmonary endothelial, BT = bovine turbinates, CPA = cardiopulmonary endothelial cells, CV-1 = African green monkey kidney, ED = equine dermal, EK = equine kidney, Hep-2 = human hepatocytes, MDBK = Madin–Darby bovine kidney.

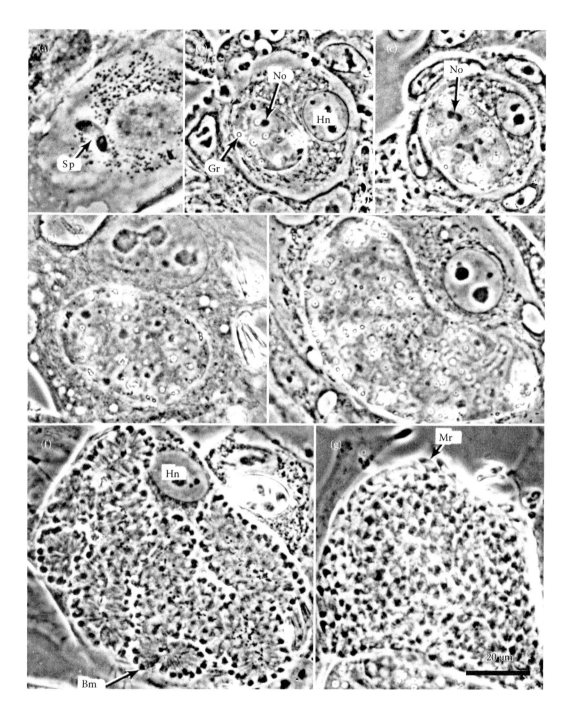

Figure 2.3 Phase-contrast photomicrographs of the development of sporozoites to large schizonts of *S. cruzi* in cultured bovine monocytes (a, f, g) and CPA cells (b–e). Day after sporozoite inoculation is given in parentheses. (a) Intracellular sporozoite (24). (b) Young schizont (18). (c) Young schizont with irregular nucleus and nucleoli (18). (d) Small intermediate schizont (23). (e) Large intermediate schizont (42). (f) Large schizont with budding merozoites (35). (g) Mature large schizont (35). (From Speer, C.A. et al., 1986. *J. Parasitol.* 72, 677–683. With permission.)

have shown that the nucleus is highly lobulated, but by light microscopy the lobes appear as separate nuclei) (Figure 2.3d and e). Eventually, these schizonts form large structures in which merozoites bud radially from 8 to 12 residual bodies (Figure 2.3f). Large mature schizonts contain 150–350 merozoites (Figure 2.3g) and may be present at 2.5–10 weeks or more after sporozoite inoculation. Merozoites frequently glide rapidly through the host-cell cytoplasm, escape through the host-cell plasmalemma, and some of these immediately penetrate and glide through the cytoplasm of cells in the vicinity of the original host cell. Within a few minutes after being penetrated by 1 or more merozoites, these previously uninfected cells retract their cellular processes, become spheroidal, and some detach to float freely in the culture medium. Spheroidal host cells that remain attached transform within a few hours to flattened, epithelioid cells. Some of the merozoites that enter other cells or remain within the cytoplasm of the original host cell develop into small schizonts (Figure 2.4a–d).

Small schizonts contain 36–100 merozoites and may be present at 16–138 days or more after sporozoite inoculation, but are most numerous at 30–52 days. Several generations of small schizonts

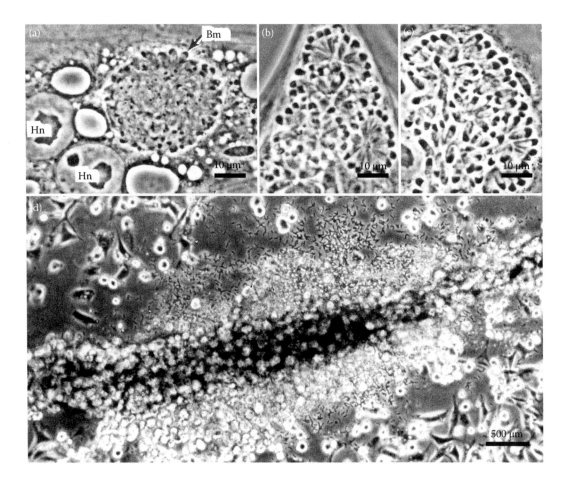

Figure 2.4 Phase-contrast photomicrographs of schizonts and merozoites of *S. cruzi* in cultured CPA cells. Day after inoculation of sporozoites given in parentheses. (a) Merozoites in early stage of budding at surface of schizont (42). (b) Schizont with merozoites budding from four residual bodies (48). (c) Mature schizont (42). (d) Large area of CPA culture showing numerous infected cells and extracellular merozoites (76). ([a–c] From Speer, C.A. et al., 1986. *J. Parasitol.* 72, 677–683; [d] Speer, C.A., Burgess, D.E., 1987. *Parasitol. Today* 3, 2–3. With permission.)

occur during this time, but the precise number of generations cannot be determined because schizogony is asynchronous. Merozoites bud radially from 4 to 8 residual bodies (Figure 2.3a and b), are highly motile, easily exit the host cell, exhibit pivoting and gliding movements, and readily penetrate other cultured cells in the immediate vicinity. Noninfected CPA cells adhere to CPA cells that are infected with small schizonts to form aggregates of cells, and some of these aggregates reorganize into branching and, occasionally, anastomosing capillary-like structures, which become centers of intense schizont activity and merozoite release. Cells that adhere to infected cells serve as host cells for additional schizogonous generations. Eventually, relatively large, elongated aggregates of cells (Figure 2.4d) are formed, within which nearly all cells are infected with schizonts and merozoites (Figure 2.5). Production of *S. cruzi* merozoites cycles during the cultivation period; peak merozoite numbers are present at 36 and 44 DPI and at 44 and 50 DPI for CPA and BM, respectively.[1657]

The host-cell type *in vivo* for large and small schizonts of *S. cruzi*, *S. capracanis*, and *S. tenella* has been considered, but not proven, to be vascular endothelial cells. The recent success in culturing merozoites *in vitro* indicates that the preferred host cell *in vivo* may indeed be an endothelial cell, because approximately 7–12 times more merozoites of *S. cruzi* and *S. tenella* were harvested from CPA than from BM.[1656,1657] In contrast, *S. capracanis* underwent schizogony in BM, but not CPA cells.[1656]

Ultrastructural studies of infected BM and CPA cells revealed that sporozoites, schizonts, and merozoites of *S. cruzi* were located free in the host cell cytoplasm (i.e., parasites were not surrounded by a parasitophorous vacuole)[1655,1658] (Figures 2.5 and 2.6), which substantiates this finding in earlier studies of *in vivo* infections with *S. cruzi* as well as other species of *Sarcocystis*.[356,1649] Certain endothelial cells are known to be part of the reticuloendothelial system and are capable of phagocytosing and digesting ingested matter and microorganisms. If such reticuloendothelial cells serve as host cells for schizonts of certain *Sarcocystis* species, being located free in the host cell cytoplasm may represent a means of escaping a potentially hostile host-cell environment.

This system for culturing merozoites of *Sarcocystis* species[356,1649] can be particularly useful in elucidating the biochemistry, as well as the mechanisms, of immunity and pathogenesis, especially because the parasite develops *in vitro*, as well as *in vivo*, in cell types (i.e., endothelial and mononuclear cells) that have been shown to participate in other immunologic and pathologic mechanisms.[1402,1634,1635] With minor modifications (i.e., selecting the appropriate cell line), this system may prove useful for culturing other species of *Sarcocystis* that are known to develop slowly within endothelial or endothelial-like cells in the intermediate host.

2.4.2 Cultivation of Gamonts

Fayer[569,571,2005] was the first to describe the *in vitro* development of bradyzoites of *Sarcocystis* to microgamonts, macrogamonts, and oocysts (Figure 2.7c and d), which was also the first description of such stages in the life cycle of *Sarcocystis*. Becker et al.[86] were able to achieve even greater *in vitro* development of this phase of the *Sarcocystis* life cycle, in which the oocysts of four species of *Sarcocystis* underwent sporulation in various mammalian cell lines. Gamonts of several other species have been grown in cell culture.[537,539,541,898,1177]

The ultrastructure of gametogony, oogony, and sporogony has been described for several species of *Sarcocystis* developing in cell cultures.[86,1177,1790] The ultrastructural features of gamonts and oocysts of five different species of *Sarcocystis* formed in cultured cells were essentially identical to those formed in the intestines of the natural hosts. These studies showed that bradyzoites developed directly into microgamonts and macrogamonts, usually within 18 h after inoculation. Each microgamont had a large, irregularly shaped nucleus, which gave rise simultaneously to 20–30 microgametes. At 22 h after inoculation, oocysts began to undergo sporulation to form sporocysts.

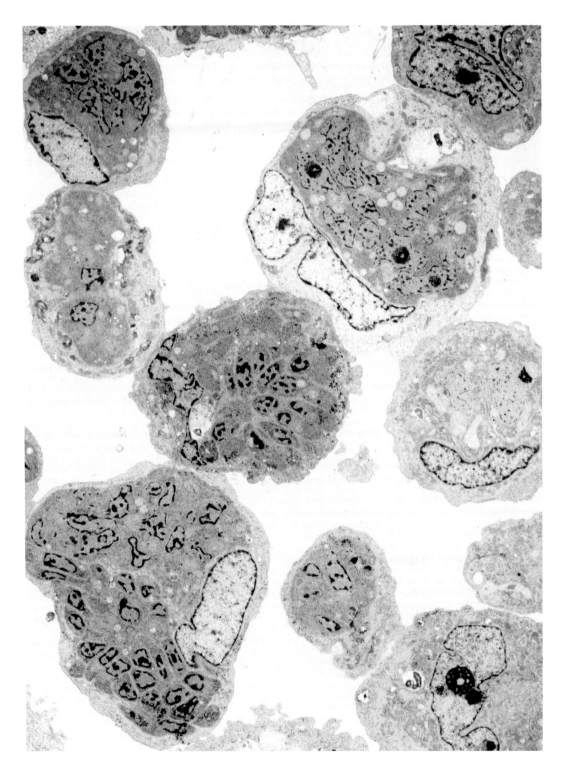

Figure 2.5   TEM of several CPA cells infected with schizonts of *S. cruzi* in various stages of development. 78 DPI.

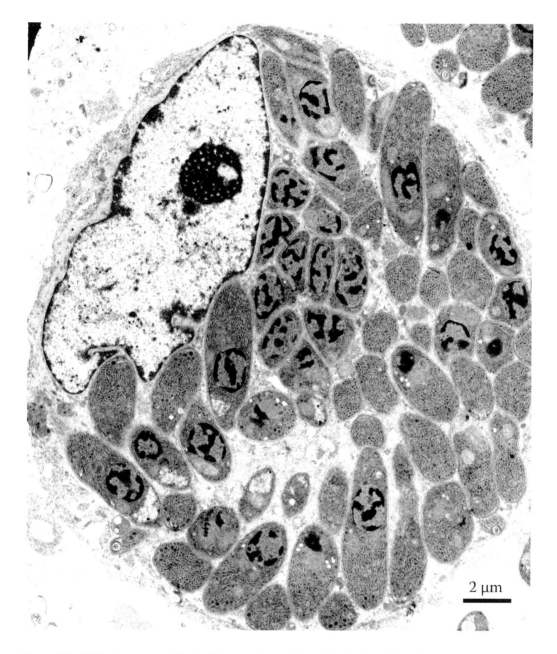

Figure 2.6   TEM of a mature schizont with merozoites of *S. cruzi* in a bovine CPA cell 35 days after sporozoite inoculation. (From Speer, C.A., Burgess, D.E., 1987. *Parasitol. Today* 3, 2–3. With permission.)

2.5 MOLECULAR METHODS

2.5.1 PCR from Tissues and Purified Tissue Cysts

Parasite DNA can often be extracted from infected host tissues, in quantities sufficient for successful PCR, without first isolating tissue cysts. This direct method will be most successful when infections are heavy, and when the tissue is fresh, or frozen before extraction.

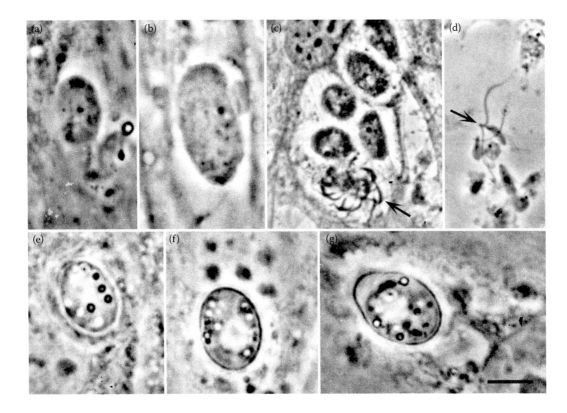

Figure 2.7   Phase-contrast photomicrographs of gamonts and oocysts in cultured cells. (a, b, e, f, g) Gamonts
and oocysts of *S. muris* in mouse kidney cells. (c and d) Gamonts of *Sarcocystis* sp. from grackles
in embryonic bovine trachea cells. Arrowheads point to host-cell nuclei. (a) Early microgamont. (b)
Advanced microgamont. (c) Mature microgamont (arrow) and five macrogamonts. (d) Microgamete
(arrow) with two flagella. (e) Macrogamont with four refractile granules (arrow). (f) Unsporulated
oocyst with oocyst wall (arrow). (g) Unsporulated oocyst with slightly condensed sporont. (From
Fayer, R., 1972. *Science* 175, 65–67. With permission.)

Purer preparations, which may be necessary in the case of light infections or when downstream
applications may be impaired by contamination with host DNA, can be obtained by first isolating
sarcocysts from the surrounding host tissue. Proteins and nucleic acids can be effectively extracted
from mechanically isolated sarcocysts or by digestion in solutions of either trypsin or pepsin, as
indicated earlier.

2.5.2 Preservation of Tissues That Cannot be Immediately Processed

When DNA cannot be extracted immediately, freezing at −20°C can safeguard the tissue.
Preservation in 70%–95% ethanol can be used, but this diminishes subsequent success, and even
residual ethanol has a low flashpoint, rendering such tissues hazardous for transport. Immersion
in RNALater or Trizol reagent can help maintain nucleic acids until they can be extracted in the
laboratory.

2.5.3 Options for Transporting DNA Prior to Analysis

Aqueous extracts of DNA should be kept cold before being analyzed, including during ship-
ping. Drying a DNA pellet or applying an aqueous DNA solution to FTA (Whatman®) cards

(impregnated with DNAse inhibitors) enable DNA to be transported at ambient temperatures (at significant cost savings). In any event, templates should not be subjected to x-ray or direct sunlight.

2.5.4 Extraction of DNA from Sporocysts

DNA can be successfully extracted from sporocysts.[1844] Clean sporocysts can be stored on filter paper in Eppendorf tubes submerged in 500 μL distilled water at −20°C before DNA extraction.

The quality of DNA obtained depends upon the purity of sporocysts and the method used to extract. It is often difficult to obtain DNA from intact sporocysts. Excystation of sporozoites facilitates DNA extraction because sporozoites, as opposed to intact sporocysts, are easily solubilized. Several methods have been used to excyst *Sarcocystis* sporozoites as explained earlier in Section 2.4.1.

Sporocysts can be crushed between the coverslip and glass slide to release sporozoites from sporocysts. DNA can then be extracted via a variety of methods; phenol–chloroform extraction methods, and the Invitrogen Easy-DNA kit® (using the tissue and blood protocol, as specified by the manufacturer) routinely yield success.

2.5.5 Freeze–Thaw Methods to Free Sporocysts from the Oocyst

Liquid nitrogen may be employed to accomplish cycles of freezing and thawing, and some success has been reported by using a "bead beater" to break the oocyst wall (although damage to DNA can also be expected when doing so). The methods used to extract DNA from other coccidian oocysts/sporocysts are applicable to *Sarcocystis*. The method for *Toxoplasma gondii* oocyst DNA extraction was published previously.[464]

Two rounds of PCR were needed to obtain sufficient product to visualize restriction fragment-length polymorphisms (RFLP) in ethidium bromide-stained agarose gels.[1844,1858] Less product might suffice for other downstream purposes, such as sequencing or visualizing by more sensitive means. In each round of PCR, primer 2L was employed: 5′-GGA TAA ACC GTG GTA ATT CTA TG 3′. The reverse primer in the first round (2H) is 5′-ACC TGT TAT TGC CTC AAA CTT C, and the reverse primer in the second round (3H) is 5′ GGC AAA TGC TTT CGC AGT AG.

Genes used to characterize *Sarcocystis* species are given in Tables 23.1 and 23.2. Additionally, genes/methods used to distinguish *Sarcocystis* species for each host or group of hosts are indicated in Chapters 3 through 22.

CHAPTER 3

Unique Multiple-Host *Sarcocystis* Species

3.1 INTRODUCTION

Most *Sarcocystis* species have a 2-host life cycle as explained in Chapter 1. Therefore, *Sarcocystis* infections are discussed by hosts or group of hosts. However, 2 species, *S. neurona* and *S. canis* have a wide host range and their biology is slightly different than other *Sarcocystis* species. *S. neurona* was discovered in 1991 after the publication of the first edition of this book. Since then, more than 300 papers have been published on this subject. The following information is drawn largely from 2 reviews.[438,475]

3.2 *SARCOCYSTIS NEURONA* INFECTIONS IN ANIMALS

Sarcocystis neurona Dubey, Davis, Speer, Bowman, de Lahunta, Granstrom, Topper, Hamir, Cummings, and Suter 1991.

3.2.1 Introduction and History

An equine clinical syndrome was referred by 3 different names in the 1970s—"segmental myelitis"[1491,2123] "Equine Protozoan Encephalomyelitis,"[87] and "Equine Protozoal Myeloencephalitis (EPM)"[1146]; the term EPM has been used by most authors. The etiology of EPM was not ascertained until 1991, when the protozoan was cultivated and named *Sarcocystis neurona*[404]; historical developments are summarized in Table 3.1. An EPM-like disease has also been recognized in several other hosts.

3.2.2 Biology

3.2.2.1 Hosts

S. neurona has a wide host range relative to other species in the genus *Sarcocystis* (Figure 3.1, Table 3.2). The North American opossum (*Didelphis virginiana*) and the South American opossum (*D. albiventris*) are its known definitive hosts; whether *S. neurona* can infect other species of South American opossums continues to be unknown. Several other animal species are its intermediate or aberrant hosts (Figures 3.2 and 3.3, Table 3.2). In some hosts, only schizonts have been identified with certainty, and these are considered aberrant hosts. Intermediate hosts are those in which sarcocysts mature. Laboratory-raised opossums excreted sporocysts after feeding naturally infected skunk, armadillo, raccoon, and sea otter muscles,[207,440,445,1723,2390] confirming that these are proven intermediate hosts. One study[1108] reported brown-headed cowbirds as

Table 3.1 History of *S. neurona* and Equine Protozoal Myeloencephalitis (EPM)

Contribution	Reference
1. Clinical syndrome first recognized without a defined etiology	1491
2. Protozoa first identified in lesions	270, 88, 348
3. EPM defined as a disease	87, 1146
4. Antiprotozoal chemotherapy introduced	87, 1146
5. Protozoa causing EPM considered to be a *Sarcocystis* species	349, 1611
6. Protozoa first isolated in cell cultures and named *Sarcocystis neurona*	404
7. *In vitro* development described	307, 308
8. *Sarcocystis neurona* Western blot developed for diagnosis	728
9. PCR primers introduced for diagnosis and molecular characterizations	600, 729
10. Opossum (*Didelphis virginiana*) proposed as the definitive host for *S. neurona*	601
11. Clinical syndrome simulating EPM induced in horses by feeding sporocysts from opossums	603
12. *Sarcocystis neurona* separated from *S. falcatula* based on susceptibility of immunodeficient mice to *S. neurona*	412, 1114, 1115
13. Opossum proven to be definitive host by inducing EPM-like disease in immunodeficient mice fed sporocysts	412
14. *In vitro* and *in vivo* testing of chemotherapeutic agents begun	439, 1068
15. *Sarcocystis neurona* isolated from CNS of a non-equine host	1072
16. South American opossum (*Didelphis albiventris*) from Brazil found to be another definitive host for *S. neurona*	442
17. Genome of *S. neurona* annotated	109

Source: Modified from Dubey, J.P. et al., 2001. *Vet. Parasitol.* 95, 89–131.

intermediate hosts for *S. neurona*, but this finding needs confirmation. *S. neurona*-like sarcocysts were reported in muscles of 1 horse,[1239] 2 dogs,[472,1780] 1 mink,[1445] and 1 bobcat.[1788]

3.2.3 Structure and Life Cycle

3.2.3.1 Structure

Schizonts and sarcocysts are tissue stages in intermediate hosts (Figures 3.4 through 3.10).

Schizonts divide by endopolygeny where the nucleus becomes multilobed before merozoites are formed (Figure 1.22). The schizogonic cycle may be asynchronous; schizonts of different maturity can be found in a single cell (Figure 3.6). Only 1 morphologic type of schizonts is known.

3.2.3.2 Mouse Modeling Schizogony

Only schizonts are produced in the KO mouse. After oral inoculation of *S. neurona* sporocysts, sporozoites excyst in the small intestine, and parasitemia has been detected 1–8 DPI.[430] By 1 DPI, sporozoites can be detected histologically in mesenteric lymph nodes. Mature schizonts with merozoites were first detected histologically starting 8 DPI.[431] Beginning 13 DPI, schizonts were seen consistently in the brain. The other organs parasitized were the lung, heart, liver, intestine, retina, and kidney. Although all regions of brain and spinal cords were parasitized, over 90% of organisms (Figure 3.3) were seen in the cerebellum of the KO mice.[637] The infectivity of sporocysts in the KO

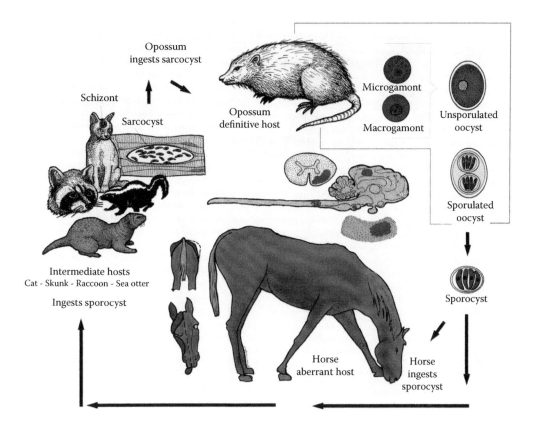

Figure 3.1   Life cycle of *S. neurona*. Opossums are the definitive host, and other animals are aberrant/intermediate hosts. *S. neurona* parasitizes and causes lesions (in red) in the brain and spinal cord of horses. Affected horses can have neurological signs, including abnormal gait, dysphagia, and muscle atrophy.

Table 3.2   Intermediate/Aberrant Hosts of *S. neurona*

Proven Intermediate Hosts[a] (See Table 3.5)	*S. neurona* Isolated from Host Tissue by Bioassay in Mice or Cell Culture (See Table 3.5)	*S. neurona*-Like Sarcocysts in Sections	Clinical Disease Reported (See Tables 3.9 and 3.10)	Only Antibodies Found (See Table 3.7)	Only *S. neurona* DNA Detected in Sections
Cat	Horse	Dog	Horse	Ring tailed lemur	Fur seal
Skunk	Sea otter	Horse	Pony	Blue eyed black lemur	Elephant seal
Raccoon	Pacific harbor seal	Mink	Zebra	Black and white ruffed lemur	Dolphin
Armadillo	Harbor porpoise	Bobcat	Raccoon	Beaver	Sperm whale
Sea otter	Raccoon	Fisher	Cat		Killer whale
			Ferret		
			Dog		
			Skunk		
			Lynx		
			Sea otter		
			Pacific harbor seal		
			Sea lion		
			Red panda		
			Fisher		

Source: Adapted from Dubey, J.P. et al., 2015. *Vet. Parasitol.* 209, 1–42.
[a] Laboratory-raised opossums shed sporocysts after consuming infected muscle; sporocysts were infective to KO mice.

Figure 3.2   An 18-year-old quarter horse gelding with ataxia and atrophy of hindquarter muscles. This horse improved after antiprotozoal therapy for EPM, and then relapsed 1 year later to the present state. (Adapted from Dubey, J.P. et al., 2001. *Vet. Parasitol.* 95, 89–131.)

mice by the oral route paralleled the subcutaneous route, irrespective of the strain of *S. neurona*.[434] All infected KO mice died, and the dose did not affect clinical signs.[203,207] The strain of the KO mouse (C57BL/6-Black or BALB/c White-derived) did not affect the outcome of the disease, and all affected mice died by 70 DPI; clinical signs were more severe in C57Bl/6-genetic background as compared to the BALB/c line.[470] SCID mice were not susceptible to *S. neurona* infection.[1114,1580] Thus, the genetic background and specific cellular immune components of mice play a role in their susceptibility, and the parasite's ability to replicate and cause disease. Cell-cultured merozoites were infective to KO mice, but bradyzoites from muscles of raccoon were not.[470]

3.2.3.3 Raccoon Modeling of Sarcocyst and Bradyzoite Development

Development of *S. neurona* was studied in 10 raccoons euthanized 1–77 DPI with SN37R strain *S. neurona* sporocysts derived from experimentally infected opossums[1673]; the SN37R strain had been passaged 10 cycles through laboratory-raised raccoons and opossums, ensuring the absence of extraneous species. Parasitemia was detected 3 and 5 DPI. Individual zoites, interpreted as sporozoites, were seen in histological sections of intestines at 1 and 3 DPI. Schizonts with merozoites were consistently detected in 5 raccoons euthanized 7–22 DPI. Schizonts were not detected in 2 raccoons euthanized 37 and 77 DPI. At 22 and 37 DPI, the sarcocysts were immature; they were up to 125 μm long at 22 DPI and up to 270 μm long at 37 DPI. Sarcocysts at 77 DPI were mature. The sarcocyst wall was up to 2.5 μm thick and the slender villar protrusions were 2.5 μm long.[1673] None of these raccoons had clinical signs. In another study, 2 raccoons were each fed 5 million sporocysts; 1 raccoon became lame and was euthanized 14 DPI.[445] The second raccoon developed neurological signs and was euthanized 22 DPI. Histologically, both raccoons had encephalomyelitis associated with schizonts. Additionally, sarcocysts were seen in the muscles of the raccoon euthanized 22 DPI.[445]

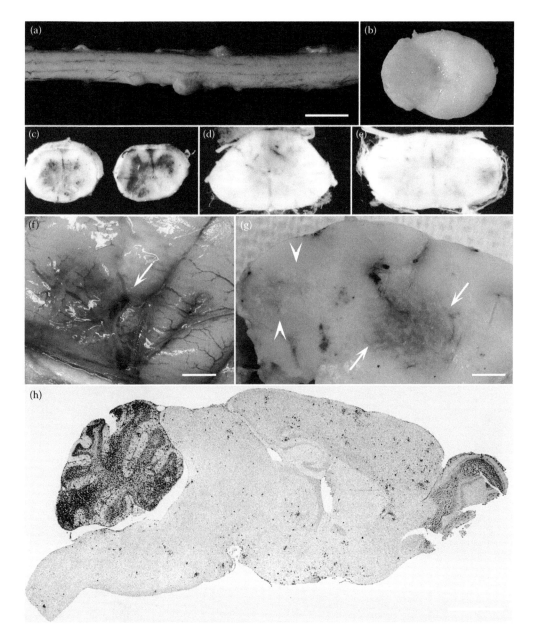

Figure 3.3 *S. neurona*-associated lesions. (a) Nodular growths between spinal nerve roots of the spinal cord of a dog in Reference **670**. (Specimens courtesy of Shelley J. Newman and Amanda Crews.) Unstained. Bar = 1 cm. (b) Cut section of the spinal cord in Figure 3.3a. The protruding lesion is extended into the central canal. Unstained. (c–e) Cross-sections of spinal cords of horses with histologically confirmed cases of EPM. Note the varying degree of hemorrhage and discolored areas, indicative of malacia. Unstained. (Figure (c) from Reference **438**, (e) from Reference **307**.) (f and g) Surface and cut view of cerebrum of a 20-year-old Paint horse with EPM confirmed histologically and by PCR. The horse had a 6-day history of muscle fasciculations, bruxism, difficulty in eating and drinking, and circling to the left with head pressing. Note hemorrhagic and yellow discolored areas, indicative of necrosis. Bar = 5 mm. (Courtesy of Dr. Uneeda Bryant.) (h) Longitudinal section of brain of an interferon gamma gene KO mouse experimentally infected with *S. neurona*. All red dots are merozoites and schizonts. Although the entire brain, from olfactory bulbs to the medulla oblongata, is parasitized, most parasites are concentrated in the cerebellum. Immunolabeling with *S. neurona* antibody. (Adapted from Fritz, D.L., Dubey, J.P., 2002. *Vet. Pathol.* 39, 137–140.)

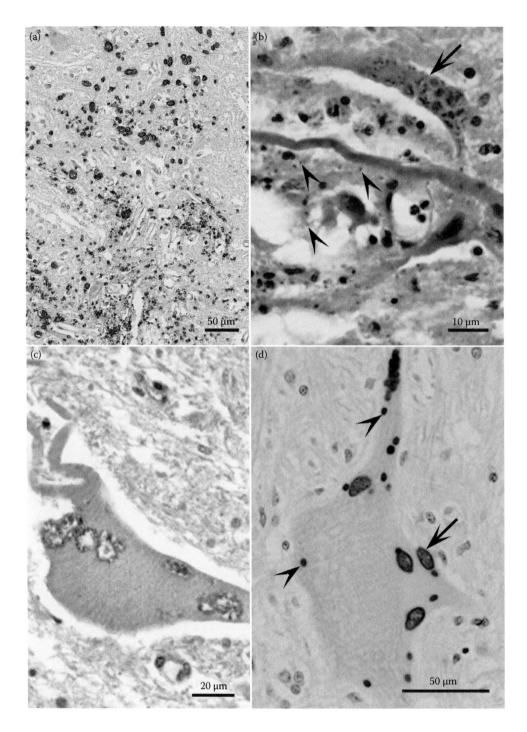

Figure 3.4 Histological sections of spinal cord of a horse with EPM. *S. neurona* was first named from the parasite observed in this horse with massive infection of the spinal cord. (a,c), immunostaining with *S. neurona* antibody, (b,c), H and E stain. (a) Severe myelitis and numerous parasites (all red stained structures). (b) Myelin degeneration, and numerous parasites in an axon (arrow). Merozoites (arrowheads) are 2–3 µm long (arrowheads) and difficult to see. The empty spaces are due to myelin degeneration. (c) Neuron with several schizonts. (d) Intraneuronal merozoites (arrowheads) and schizonts (arrow) in an intact neuron.

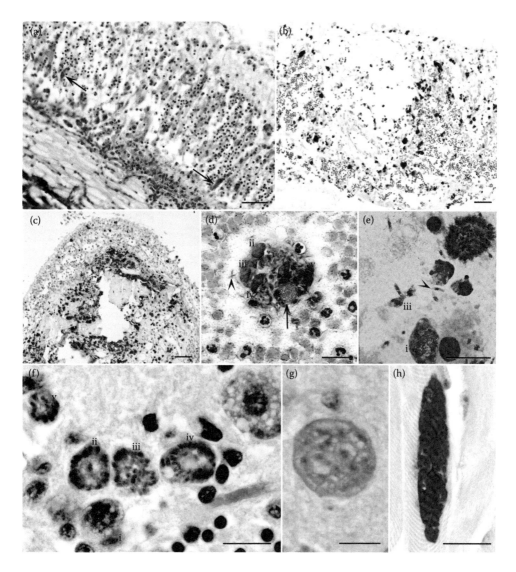

Figure 3.5 *S. neurona*-associated lesions in different hosts. (a, b, c, f, g, h), histological sections, (d and e) touch smears. Bar in (a,b and c) = 50 μm, (d, e, f, and h) = 20 μm, (g) = 10 μm. (a) Eye of a sea otter with 2 foci of retinitis (arrows). Numerous *S. neurona* schizonts and merozoites are present but not visible at this magnification. The retinal architecture is preserved. H and E stain. (From Reference **468**.) (b) Eye of a dog with retinitis and necrosis. The entire retina was riddled with numerous schizonts and merozoites (all red stained structures). Immunostaining with *S. neurona* antibody. (From dog reported in Reference **472**.) (c) Nasal septum from a ferret reported in Reference **145**. The lamina propria is severely parasitized by schizonts and merozoites. The epithelium is intact and mildly parasitized. Immunostaining with *S. neurona* antibody. (d) Smear from pus oozing from a dermal ulcer in a dog reported in Reference **401**. Note the 4 immature schizonts (i–iv), a mature schizont (v), several free merozoites (arrowhead). Arrow points to host cell nucleus. Compare size of merozoites with host red blood cells and neutrophils. Giemsa stain. (e) Touch smear from a grossly visible encephalitic lesion in a red panda. Note immature schizont (i), schizont with budding merozoites (ii), mature schizont with merozoites radiating from a residual body (iii), and free merozoites (arrowhead). Diff–Quick stain. (Courtesy of Timothy Walsch.) (f) Immature and mature schizonts (i–v) in the cerebrum of a sea otter. The section was stained deeply with H and E to reveal the structures. (g) Immature sarcocyst in the brain of a sea otter. Note the eosinophilic globular metrocytes and sarcocyst wall for differentiation from schizonts in Figure 3.5f. (h) Immature sarcocyst in the skeletal muscle of a sea otter. Immunostaining with *S. neurona* antibody. (Figure f–h, adapted from Thomas, N.J. et al., 2007. *J. Comp. Pathol.* 137, 102–121.)

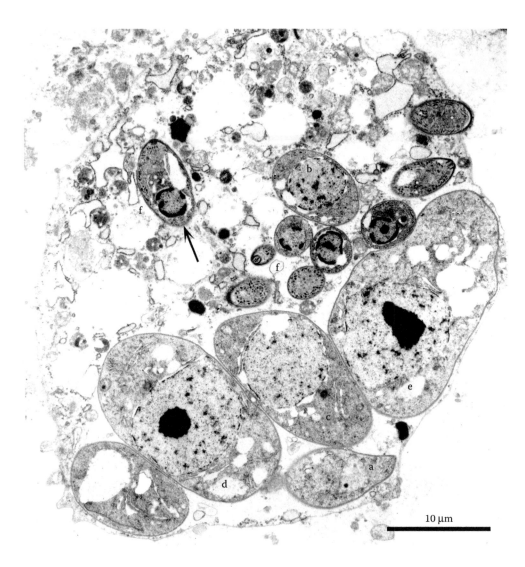

Figure 3.6 TEM of an infected neural cell in the brain of a raccoon naturally infected with *S. neurona*. Note asynchronous schizogony with 6 developing schizonts, in presumed order of development (a–e), and 9 merozoites (f). Arrow points to a longitudinally merozoite with a conoid at 1 end and a subterminal nucleus. The host cell is deteriorated but parasite structures are fairly well preserved. (Adapted from Dubey, J.P. et al., 1991. *J. Helminthol. Soc. Wash.* 58, 250–255.)

3.2.3.4 Feline Modeling of Sarcocysts and Bradyzoite Development

Sarcocysts also developed in cats after oral inoculation with *S. neurona* sporocysts.[429] At 36, 45, and 57 DPI, sarcocysts were immature. At 144 DPI, sarcocysts were up to 700 μm long, and opossums that were fed these sarcocysts excreted *S. neurona* sporocysts. Sarcocysts were also found in the muscles of 1 cat 43 days after the cat was inoculated parenterally with 10 million *S. neurona* merozoites; an opossum fed the sarcocysts from cat muscle did not excrete sporocysts.[157] Sarcocysts found in another cat euthanized 50 DPI were infective to an opossum, as evidenced by sporocyst excretion. Thus, bona fide *S. neurona* sarcocysts have been produced in cats following inoculation with culture-derived merozoites.

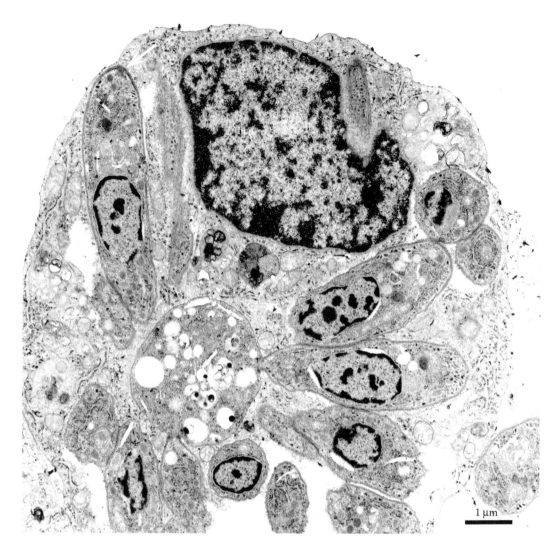

Figure 3.7 TEM of a BT cell culture infected with *S. neurona*. Mature schizont with non conoidal end of merozoites still attached to a residual body (rb). Note the 1 longitudinally cut merozoite (arrow) that has separated from the schizont, and the host cell nucleus (hcn). (Adapted from Dubey, J.P., 2004. *Infectious Diseases of Livestock with Special Reference to Southern Africa*. Oxford University Press, Ni City, South Africa, pp. 394–403.)

3.2.3.5 *Equine Modeling of Early Parasite Infection*

Attempts were made to study migration and development of *S. neurona* in 6 ponies after oral inoculation with 250,000,000 sporocysts[493]; the ponies were serologically negative to *S. neurona* and were euthanized 1, 3, 5, 7, and 9 DPI. Viable *S. neurona* were isolated by bioassay in KO mice, and in cell culture inoculated with tissue homogenates of pony mesenteric lymph nodes at 1, 2, and 7 DPI, liver at 2, 5, and 7 DPI, and from lungs at 5, 7, and 9 DPI. The parasite was not detected histologically in tissues of any pony. However, encephalitic lesions were detected in sections of brain and spinal cords of 2 ponies euthanized 7 and 9 DPI. One pony euthanized 9 DPI had IgM antibodies to *S. neurona*.[493]

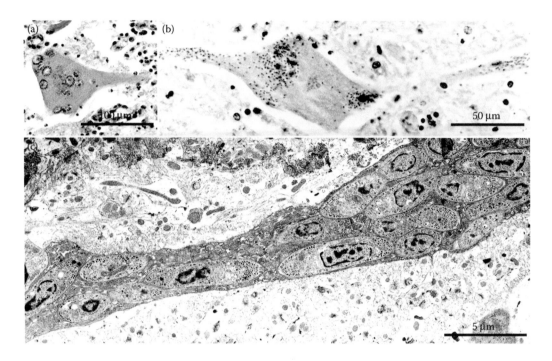

Figure 3.8 (a) A heavily infected neuron in the spinal cord of a horse with numerous schizonts. (b) Neuron with numerous merozoites. (From Dubey, J.P. et al., 1974. *J. Am. Vet. Med. Assoc.* 165, 249–255. With permission.) (c) *S. neurona* merozoites free in the cytoplasm of an unmyelinated axon in the cerebellum of an experimentally infected KO mouse. (Adapted from Fritz, D.L., Dubey, J.P., 2002. *Vet. Pathol.* 39, 137–140.)

3.2.4 Naturally Occurring Stages and Their Structure

In naturally infected horses with neurological signs, *S. neurona* schizonts have been found only in the CNS. Both neural and inflammatory cells in the CNS can be parasitized (Figure 3.8). As many as 13 schizonts and several hundred merozoites were present in 1 neuron.[348] As with other *Sarcocystis* species, the schizogonic cycle can be asynchronous; schizonts of different maturity can be found in a single cell (Figure 3.6). Mature schizonts in the CNS are up to 30 μm long and they can be oval, round, elongated, or irregular in shape. Ultrastructurally, merozoites were either stumpy (7.7 × 3.1 μm) or slender (7.3 × 1.7 μm), and they contained the same organelles as described in the merozoites of other *Sarcocystis* species, including the absence of rhoptries.[1661]

Sarcocysts have been demonstrated in skeletal and cardiac muscles and in the brain (Figure 3.5). Sarcocysts in the brain are often round and smaller in size than in skeletal muscles.[1196,1739] Additionally, sarcocysts in myocardium are smaller than those in skeletal muscles. Ultrastructurally, the mature *S. neurona* sarcocyst wall has vp that are up to 2.8 μm long and 0.4 μm wide.[435] The vp have microtubules that extend from tip to the base but have no granules (Figure 3.10). Few of these microtubules extend deeper in the ground substance, but they are electron-lucent and not prominent compared with the other *Sarcocystis* species of the horse in North America, *S. fayeri* (Figure 3.10). Bradyzoites in sections are 4.8–6.5 × 1.0–1.3 μm in size, and contain only 2 rhoptries.

S. neurona sporocysts are 11.3 × 8.2 μm in size. Sporozoites are slender, and have 2–4 rhoptries but no crystalloid body.[1079] The absence of a crystalloid body in sporozoites is notable, because the crystalloid body was present in *S. cruzi* and other species of *Sarcocystis* sporozoites examined ultrastructurally.[393]

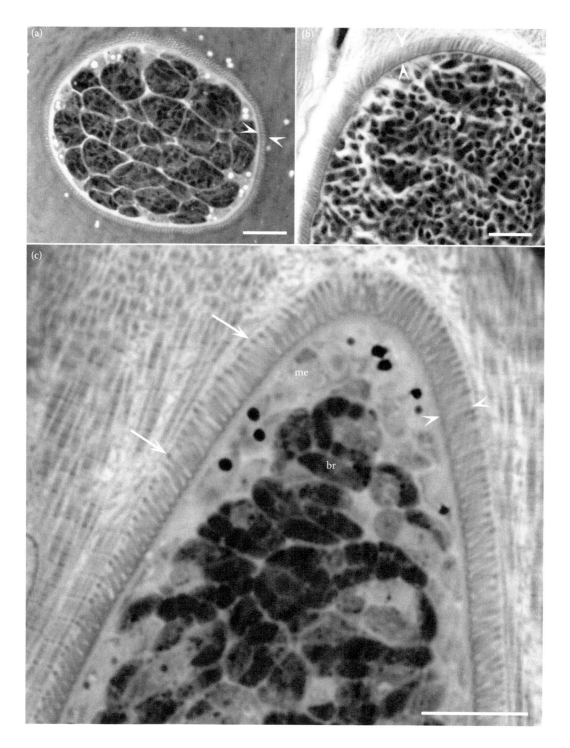

Figure 3.9 *S. neurona* sarcocysts in histological sections of skeletal muscle. Arrowheads point to striated cyst wall. Arrows point to thickening of the villar tips. (a) Cat, 144 DPI. (Adapted from Dubey, J.P. et al., 2001. *J. Parasitol.* 87, 1323–1327.) Toluidine blue stain. (b) Raccoon, 77 DPI. H and E stain. (c) Toluidine blue stain. (b and c, adapted from Stanek, J.F. et al., 2002. *Procyon lotor. J. Parasitol.* 88, 1151–1158.) Bars = 10 μm.

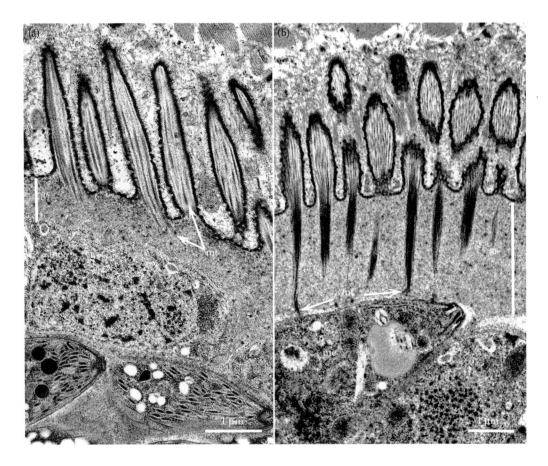

Figure 3.10   Comparison of the cyst walls of *S. neurona* (a) and *S. fayeri* (b) sarcocysts by TEM. The cyst walls, including the ground substance layer (gs) of *S. fayeri*, are thick, the microtubules (mt) are more electron-dense and extend up to the pellicle of the zoites, whereas the cyst walls of *S. neurona* are comparatively thin, the microtubules are few, and never extend deep in the gs. (Adapted from Stanek, J.F. et al., 2002. *J. Parasitol.* 88, 1151–1158; Saville, W.J.A. et al., 2004. *J. Parasitol.* 90, 1487–1491.)

3.2.5 Excretion of Sporocysts by Definitive Hosts

Large numbers of sporocysts can be excreted in feces of opossums (Table 3.3). Opossums are the definitive hosts for *S. neurona* and 3 other named species, *S. speeri*, *S. falcatula*, and *S. lindsayi*.[137,416,433,1722] Cheadle et al.[202] provided morphologic measurements of possibly 5 species of *Sarcocystis*-like sporocysts in the feces of 17 naturally infected opossums. Sporocysts of *S. neurona* were 10.7 × 7.0 μm, *S. speeri* were 12.2 × 8.8 μm, strain-1085 of *Sarcocystis* sp. were 10.9 × 6.8 μm, and *S. falcatula* were 11.0 × 7.1 μm; thus, the differences were within 2-μm range. Additionally, the identity of sporocysts of each species was from naturally infected opossums, and the species identification was not definitive.

Sporocysts of another organism were reported in feces of 10 opossums; sporocysts were 19.4 × 10.5 μm in size and had Stieda body-like structures at both poles.[202] Based on the large size, these are unlikely to be *Sarcocystis* sporocysts. Bioassay and molecular methods have been used to distinguish sporocysts of different species in opossum feces. The KO mice and budgerigars have been used to differentiate viable *S. neurona* and *S. falcatula* in opossum feces; *S. neurona* and *S. speeri* are not infective to budgerigars, and *S. falcatula* is not infective to KO mice.[412] *S. speeri*

Table 3.3 Prevalence of *Sarcocystis* Sporocysts in Feces/Intestines of Opossum (*Didelphis virginiana*) in the United States

Location	Year	No. Tested	With Sporocysts (%)	*S. neurona*	Other Species	*S. neurona* Isolates	Reference
LA, MD, VA, GA, FL, PA	1998–99	44	24 (54.4)	14 (31.8%)-bioassay in KOM	19 (43.1%) *S. falcatula*-bioassay in budgerigars, 8 (18.8%) *S. speeri*-bioassay in KO mice	SN8-15OP	**421**
MS	1999–00	72	24 (33.3)	19 (26.3%)	Not investigated, 1 opossum infected with *S. speeri*	SN16 to 34OP	**443**
MI	1996–02	206	37 (17.9)	23 (11.1%) PCR–RFLP	4 (1.94%) *S. falcatula* based on PCR–RFLP	Not stated	**513**
CA	2005–08	288	53 (18.4)	17–PCR–ITS-1	Not stated	OP-1-13	**1463**

Source: Adapted from Dubey, J.P. et al., 2015. *Vet. Parasitol.* 209, 1–42.
CA = California, FL = Florida, GA = Georgia, LA = Louisiana, MD = Maryland, MI = Michigan, MS = Mississippi, PA = Pennsylvania, VA = Virginia.

and *S. neurona* in KO mice can be distinguished immunohistochemically.[421] Molecular methods have been developed to distinguish *S. falcatula* and *S. neurona*.[1722] Few *S. neurona* sporocysts are lethal for KO mice,[470] and 1000 *S. speeri* sporocysts are pathogenic for KO mice.[416]

S. neurona sporocysts have been found in 6%–31% of opossums tested in the United States (Table 3.3). Three of the surveys listed in Table 3.3 examined factors associated with *S. neurona* positivity (Table 3.4). Season, body condition, and presence of young in pouch were associated with the presence of sporocysts. How these factors affect the presence of sporocysts in opossums is uncertain, because *Sarcocystis* does not multiply in definitive hosts, and little is known of immunity to reinfection. The availability of infected intermediate hosts during spring might account for the results shown in Table 3.4.

The South American opossum, *D. albiventris*, is also a host for *S. neurona*.[442]

Table 3.4 Factors Associated with *Sarcocystis* Infections in Opossums in the United States

Factor		Percentage of Positive (n)		
		Mississippi[1467]	Michigan[513,514]	California[1464]
Age	Adult	28 (14/50)	14.5 (30/206)	6.7 (14/206)
	Juvenile	22.7 (5/22)	3.3 (7/206)	3.7 (3/81)
Gender	Female	29.4 (10/34)	9.7 (20/206)	4.4 (6/136)
	Male	23.6 (9/38)	8.25 (17/206)	7.5 (11/146)
Female with pouch	Present	33.3 (6/18)	No data	8.3 (2/24)
	Absent	25 (4/16)	No data	5.2 (4/76)
Season	Spring	34.2 (13/38)	3.3 (7/206)	9.1 (14/154)
	Summer	No data	9.2 (19/206)	
	Autumn	No data	3.3 (7/206)	2.2 (3/133)
	Winter	17.6 (6/34)	1.9 (4/206)	
Total number tested		72	206	288
Total positive (%)		26.3 (19)	17.9 (37)	5.9 (17)

Source: Adapted from Dubey, J.P. et al., 2015. *Vet. Parasitol.* 209, 1–42.

3.2.6 Transplacental Transmission

Currently, there is no documented case of congenital *S. neurona* infection in land mammals. *S. neurona* DNA was recently found in tissues of 8 of 13 marine mammal neonates, supporting congenital infection.[210] These eight animals included three fetuses removed from the uteri of a naturally infected (with demonstrable DNA in tissues of mother) pygmy sperm whale (*Kogia breviceps*), a harbor seal (*Phoca vitulina*), and a Steller sea lion (*Eumatopias jubatus*). Additionally, five fetuses (four harbor seals and one harbor porpoise) were beach-cast carcasses, too young to have nursed.[210]

3.2.7 *In Vitro* Cultivation, Cell and Molecular Biology

3.2.7.1 In Vitro *Cultivation*

Viable *S. neurona* has been isolated from many hosts (Table 3.5). Numerous cell lines can support the growth of *S. neurona* (Table 3.5). In most EPM horses, the number of *S. neurona* in CNS tissue is low. Therefore, culture flasks seeded with CNS homogenates should be incubated for at least 2 months, because some strains are slow to adapt in cell cultures. Once established, *S. neurona* can complete schizogonic development in 3 days.[393,1770,2321] *S. neurona* has also been isolated from sporocysts purified from the intestines of opossums, either directly in cell cultures[511,1262] or by feeding sporocysts to KO mice and then recovering *S. neurona* in cell cultures from mouse brains. The isolation of *S. neurona* from sporocysts, via KO mice, is advantageous, because it removes *S. falcatula*, a common parasite in opossum feces; *S. falcatula* is not infective to KO mice.[412,1115] Cell cultures infected with *S. neurona* are useful for *in vitro* screening of anti-*S. neurona* compounds and other aspects of biology.[130,661,1118,1119,1993,1994]

3.2.7.2 *Cell Biology*

Because the parasite is easy to cultivate in many cell lines and can be genetically manipulated, *S. neurona* has been used as a model to study several aspects of cell biology. Biologically, *S. neurona* has many characteristics in common with *Toxoplasma gondii*, and multiple reagents and techniques developed for *T. gondii* have been applied in studying the biology of *S. neurona*. In turn, the *S. neurona* culture system has contributed to the understanding of organelle development, which is difficult to resolve with the *T. gondii* cultures. One example is the information on plastid replication. Compared with *T. gondii* (<6 h), the nuclear division in *S. neurona* merozoites is prolonged (3 days), and, thus, events during parasite development can be easily followed. Nuclear and plastid replication were followed[1770] in BT cells infected with the SN3 isolate of *S. neurona* (Figure 1.22). In this culture system, 64 merozoites were formed in 3 days. During schizogonic development, the nucleus continued to grow and became lobulated; 5 cycles of chromosome replication occurred without nuclear division.[1770] After the sixth and the final division, the nuclear lobes bifurcated into 2, giving rise to 64 merozoites. The apicoplast in *S. neurona* was a 4-layered, tubular structure without microcristae, and was closely associated with the nucleus at all stages of division. During nuclear division, the apicoplast stayed associated with the nuclear lobes, like a flexible hose. When the nucleus divided, the apicoplast fragmented, and 1 lobe was incorporated into the budding merozoite.

3.2.7.3 *Gene Discovery and Characterization*

There is limited information concerning the molecular composition of *S. neurona*. The initial efforts to generate nucleic acid sequence data for *S. neurona* were conducted in the context of phylogenetic analyses and included the use of RAPD assays[517,518,522–524,729,1722] and sequence analysis of

Table 3.5 *In Vitro* Cultivation of *S. neurona*

Host	Year	Tissue[a,b]	Cell Type Initial Culture[c,d]	Strain Designation	Reference
Horse (*Equus caballus*)	1990	Spinal cord	M617	SN1	404
	1990	Spinal cord	M617	SN2	307
	1991	Spinal cord	M617	SN3	727
	1991	Spinal cord	M617	SN4	307
	1992	Spinal cord	M617	SN5	729
	1998	Spinal cord	M617, Esp	SN6	420
	1997	Spinal cord	M617, EK, Esp	SN7	444
	1999	Spinal cord	ED (not BT, DT)	SN-MU1,2	1119
	1994–95	Spinal cord	M617	UCD1,2,3	1113, 1114
	2009	Brain	Not stated	H1756, H1801	1464
	1997–98	Spinal cord or brain	ED	MIH 1-8	1107
Sea otter (*Enhydra lutris*)	NS	Brain	CV-1, BT	SN-OT1	1072
	1999	Brain	MA104	SO SN$_1$	1195
	2009	Brain	MA104	SO1-3	1196
	2004	Brain	MA104	8 isolates	1197
	2005–08	Brain	MA104	23 isolates[e]	1464
Pacific harbor seal (*Phoca vitulina*)	1998	Brain, CSF	MA104	HS1	1194
	2011	Brain	MA104	PV16-19; PV22	210
Harbor porpoise (*Phocoena phocoena*)	2006	Brain	Not stated	HP060325	1464
	2011	Brain	MA104	PP11	210
Raccoon (*Procyon lotor*)	2001	Muscle[b]	Not applicable	SN37R	445, 1636
Cat (*Felis catus*)	2000	Muscle[b]	Not applicable	Sn-Mucat-2	157, 1765
Armadillo (*Dasypus novemcinctus*)	2000	Muscle[b]	Not applicable	Not stated	205
Sea otter (*Enhydra lutris*)	2001	Muscle[b]	Not applicable	SN-OT-2[f]	440
Brown-headed cowbird (*Molothrus ater*)	1999	Muscle[b]	Not applicable	MICB1	83
South American opossum (*Didelphis albiventris*)	1999	Intestine[a]	EK,BT	SN35-OP, SN36-OP	442
North American opossum (*Didelphis virginiana*)	1999	Intestine[a]	ED, M617	SN8 to 15OP	421
	1999	Intestine[a]	ED, BT, EK, M617	SN16 to 34OP	442
	2005–08	Intestine[a]	MA104	OP 26, 48, 68, 134, 166, 187, 201, 212, 226, L49, GA3, GA7, OPBF2	1464

Source: Adapted from Dubey, J.P. et al., 2015. *Vet. Parasitol.* 209, 1–42.

NS = not stated.

[a] By feeding sporocysts to KO mice, and *in vitro* cultivation from infected tissues of KO mice.

[b] By feeding naturally infected muscle to laboratory-raised opossums and the bioassay of sporocysts shed into KO mice.

[c] *In vitro* cultivation directly from sporozoites released from sporocysts.

[d] BT = bovine turbinate, CV1 = African green monkey, DT = deer testes, ED = equine dermal, EK = equine kidney, ESp = equine spleen, M617 = bovine monocytes, MA104 = monkey kidney.

[e] SO 4387, 4413, 4530, 4653, 4697, 4711, 4725, 4755, 4786, 4834, 4928, 4970, 4972, 5002, 5073, 5110, 5226, 5259, 5263, 5274, 5278, 5283, 5296.

[f] Designated here.

the 18S rRNA locus[290,600,1116] to identify species-specific genetic markers. These investigations led to the development of PCR-based tests to detect *S. neurona* and/or distinguish it from other closely related species.[601,1722]

Early studies to identify and characterize protein-encoding genes of *S. neurona* utilized traditional molecular biology approaches. Sequence analysis of random clones from a cDNA library produced from the UCD-1 strain, and standard immunoscreening of a cDNA library produced from the SN3 strain, independently identified the *S. neurona* major surface antigen SnSAG1,[499,849] which is homologous to the gene family of SAG/SRS surface proteins that have been investigated extensively in *T. gondii*.[917,1045,1812] Polyclonal antibodies against *T. gondii* enolase 2 (ENO2), followed by mass spectrometry, were used successfully to identify the *S. neurona* ENO2 homolog.[123,1829] Biochemical characterization of *S. neurona* merozoites demonstrated the presence of serine protease activity at defined molecular weights, but the encoding gene(s) was (or were) not identified.[72]

To better facilitate gene discovery in *S. neurona*, an expressed sequence tag (EST) sequencing project was performed using cDNA libraries constructed from the merozoite stage of the EPM isolates SN3 and SN4.[848,1054,1055] This project generated[848] partial gene sequences from *S. neurona* with another 6332 ESTs produced from the closely related *S. falcatula* for comparative purposes. The resulting database of sequences represented only a portion of the genes present in *S. neurona*, but it permitted more efficient approaches to identify and select parasite sequences for further investigation. While a majority of *S. neurona* genes were chosen for study based on their homology to *T. gondii* molecules, predicted structural features of the encoded protein and cDNA abundance (i.e., number of ESTs) were also useful criteria for selection of interesting genes.

Along with the major surface antigen, SnSAG1 that was identified by traditional approaches, 3 additional SnSAG paralogues—designated SnSAG2, SnSAG3, and SnSAG4—were revealed in the ESTs generated from the SN3 strain of *S. neurona*.[849] The SnSAGs are located on the surface of the extracellular merozoite stage of *S. neurona* and are present throughout intracellular development of the schizont as well. Additionally, analyses of the bradyzoite and sporozoite life cycle stages demonstrated that expression of the SnSAGs is differentially regulated during the life cycle,[668] similar to the SAG/SRS surface antigens in *T. gondii* and *N. caninum*. The specific function(s) of SnSAGs is unknown. However, evidence from *T. gondii* has implicated several of the TgSAGs as cell adhesins and modulators of host immunity,[483,958,1437,1812] so it seems reasonable to hypothesize that SnSAGs perform a similar role. Because the SnSAGs elicit strong humoral responses in infected animals,[498,499,849,1062] they have been used as target molecules to develop serologic tests for detection of antibodies against *S. neurona*.[500,837,1862] Interestingly, it is apparent that all strains of *S. neurona* do not possess the same repertoire of SnSAG genes. This antigenic diversity was initially revealed in an *S. neurona* isolate from a horse in Missouri.[1119] Further examination of this strain, designated Sn-MU1, revealed that it lacked the SnSAG1 gene that is transcribed abundantly in the UCD-1 and SN3 strains.[870] Subsequent immunologic analysis of a collection of 14 *S. neurona* strains, using antiserum against SnSAG1, demonstrated that this surface antigen is not expressed by multiple strains, and the lack of expression was shown to be due to the absence of the SnSAG1 gene.[850] Those parasite strains that lack SnSAG1 were found to express 1 of 2 alternative major surface antigens that were called SnSAG5 and SnSAG6.[263,1821] While it is conceivable that additional alternative major SnSAG paralogues exist, analyses of much more extensive collections of parasite strains suggest that SnSAG1, SnSAG5, or SnSAG6 will be predominant in the *S. neurona* strains circulating in nature.[1464,1820] The genes for these 3 SnSAG paralogues seem to be mutually exclusive to one another, because all strains of *S. neurona*, that have been analyzed, possess sequence for only SnSAG1 or SnSAG5 or SnSAG6; no strain has been found that possesses more than 1 of these genes. Analyses of synonymous versus nonsynonymous mutations in the SnSAG1 and SnSAG5 gene sequences suggest that there may be evolutionary advantages to altering some regions of these surface proteins.[512]

The *S. neurona* surface protein 1 (SnSPR1) gene is a seemingly novel sequence that is not paralogous to the SnSAG surface antigens but does encode a merozoite surface protein.[1877] Instead of relying on homology to select the sequence for investigation, this gene was chosen for study based on its abundance in the *S. neurona* EST collection, and the prediction that the encoded protein had an amino-terminal signal peptide and a carboxyl-terminal glycolipid anchor addition consistent with a cell-surface protein. In addition to localizing to the merozoite surface, SnSPR1 was shown to be a low molecular weight protein (approximately 14 kDa) that is present throughout the intracellular development of the *S. neurona* schizont. Despite being an abundant protein based on the number of ESTs that match the SnSPR1 sequence, this protein does not appear to elicit the robust immune response seen for the SnSAG surface antigens.

Genes encoding secretory proteins of *S. neurona* have been identified based on homology. A putative SnMIC10 sequence exhibited approximately 30% identity to the TgMIC10 and NcMIC10 orthologues of *T. gondii* and *N. caninum*, respectively, and examination of the native protein in merozoites using antiserum, raised against recombinant SnMIC10, revealed characteristics consistent with it being a microneme protein of *S. neurona*.[835] A dithiol-dependent nucleoside triphosphate hydrolase (SnNTP1) was similarly identified, based on sequence similarity to the 2 TgNTPase isoforms of *T. gondii*.[1876] Although localization to the apical end of the merozoite was unexpected for SnNTP1, it was found to be part of the secreted fraction of *S. neurona* consistent with it being a dense granule protein. Interestingly, both SnMIC10 and SnNTP1 are not expressed for much of *S. neurona* intracellular development, with these 2 proteins seen only in late schizonts containing newly forming daughter merozoites.[835,1876] While this was expected for SnMIC10, since microneme proteins participate in host-cell invasion and should not be needed during endopolygeny, it was unanticipated for SnNTP1 because the TgNTPases are important for intracellular growth of *T. gondii*.

3.2.7.4 Molecular Genetic Tools

To enhance the study of *S. neurona*, basic methods have been established for DNA transfection, transient expression of transgenes, and selection of stably transformed clones of this parasite. Luciferase, β-galactosidase (β-gal), yellow fluorescent protein (YFP), and red fluorescent protein (RFP) have been used successfully as reporter molecules in *S. neurona*.[656,657,1770] Mutant dihydrofolate reductase–thymidylate synthase (DHFR–TS) that confers resistance to pyrimethamine can be used to achieve stable transformation of *S. neurona*. This selection system was used to produce transgenic *S. neurona* clones that stably express either β-gal or YFP, which are useful for monitoring parasite growth or invasion rates *in vitro*.[656] Additionally, fluorescence-activated cell sorting (FACS) of YFP expression can be used to select stable clones of *S. neurona* that do not contain a drug resistance gene. Recently, the hypoxanthine–xanthine–guanine phosphoribosyltransferase gene of *S. neurona* (*SnHXGPRT*) was successfully disrupted by double crossover homologous recombination using a gene-targeting plasmid.[293] The SnHXGPRT-deficient mutant clones (SnΔHXG) were selected by their resistance 6-thioxanthine (6-TX), a toxic analog of xanthine, and could be complemented with the *T. gondii HXGPRT* gene, rendering the SnΔHXG parasites resistant to mycophenolic acid (MPA). Thus, the SnΔHXG clone provides an efficient system for both positive and negative selection of stable transgenic *S. neurona*. Collectively, the molecular genetic capabilities that have been developed for *S. neurona* make a variety of new experimental approaches possible, including gene knockouts, complementation studies, and gene regulation assays. Indeed, the value of these molecular tools has been demonstrated in studies examining the development and segregation of the *S. neurona* apicoplast during endopolygeny[1770] and identification of sequence elements involved in promoter function.[657]

3.2.7.5 Population Genetics

Compared to other prominent parasites within the Apicomplexa, there have been only limited analyses performed to examine the population genetic structure of *S. neurona* strains circulating throughout the Americas. As stated previously, initial efforts focused on developing markers capable of distinguishing between the different *Sarcocystis* species that infect the opossum definitive host, including *S. neurona*, *S. falcatula*, *S. lindsayi*, and *S. speeri*. The first molecular characterization was performed using 18S rRNA marker on *Sarcocystis* strains isolated from horses and established that *S. neurona* resolved phylogenetically within the family Sarcocystidae, suggesting a close relationship to *S. muris*.[600] Follow-up work at the 18S locus indicated that *S. neurona* was synonymous with *S. falcatula*, the parasite that cycles between opossums and birds.[290] Phylogenetic resolution between *S. neurona* and *S. falcatula* was finally achieved upon DNA sequencing the ITS-1 locus within the 18S rRNA gene array, which identified 12 nucleotide differences between the 2 parasites.[1116] In the absence of gene-specific molecular markers, panels of random amplified polymorphic DNA (RAPD) markers were next developed to differentiate *S. neurona* isolates from other related coccidia[729] and within the *S. neurona* species.[523,1722] These investigations led to the development of PCR-based tests used to detect *S. neurona*, and/or distinguish it from other closely-related coccidia.[601,1722]

During the last 15 years, a large collection of *S. neurona* strains have been isolated from a variety of geographic regions and host species (Table 3.5). To ascertain the true genetic diversity among circulating strains, and to address whether specific *S. neurona* strains are associated with increased disease risk, a serious effort has been pursued to develop a panel of genetic markers capable of resolving the parasite's population genetic structure, its evolutionary biology, and the extent to which it expands in nature.

To answer these important questions, a series of genetic markers of varying resolution have been developed. The first set of widely applied sequence-specific markers for population genetic analyses were derived from 2 RAPD markers and were described.[1722] An RFLP at the 33/54 locus was used to distinguish *S. neurona* from *S. falcatula*, and nucleotide sequence polymorphisms at the 25/396 locus identified 2 major alleles that resolved North American from South American *S. neurona* strains.[1492] To estimate whether the *S. neurona* population genetic structure was genetically diverse or clonal, microsatellites markers were developed[44,45] and applied 12 highly polymorphic microsatellite markers against 34 predominantly North American *Sarcocystis* samples collected from both definitive and intermediate hosts isolated from diverse geographical origins.[1693] This important study[1693] established that substantial allelic and genotypic diversity exists among circulating *S. neurona* strains and showed that 1 genotype is more prevalent than expected for a strictly outbred population, indicating that some degree of clonal expansion has occurred within the species.[45]

The identification of widespread *S. neurona* infections in marine pinnipeds and the discovery of a large epizootic that resulted in the death of approximately 2% of the federally listed threatened southern sea otter population, during a 3-week period in 2004, fostered the necessity to develop a more comprehensive genotyping scheme to determine the transmission dynamics of the parasite and whether specific genotypes of *S. neurona* circulating in the marine ecosystem were more pathogenic.[440,1072,1077,1194,1195,1197,1494] In 2010, a panel of gene-specific molecular markers of varying phylogenetic resolution were identified to increase the discriminatory power of the molecular markers used for *Sarcocystis* population genetic analyses. This work established a comprehensive multilocus sequence typing (MLST) approach capable of resolving strains at the genus, species, and intraspecies levels.[1464,1821] The markers consisted of a plastid-encoded RNA polymerase b gene (*RPOb*) and a cytochrome c oxidase 1 (*cox1*) gene encoded, respectively, within the apicoplast and mitochondrial organellar genomes, both of which are maternally inherited and exist as useful markers to detect genetic exchange (or hybridization) between strains based on incongruity between nuclear and organellar genome phylogenies. Hence, when 2 nearly identical MLST genotypes possess different organellar genomes, outcrossing is supported. In addition, the discovery of a family of polymorphic SnSAG surface antigen genes, orthologous to the highly informative SRS genotyping markers encoded by *T. gondii*,[849,850] identified a series of intraspecific genetic markers (annotated SnSAG1, 3, 4, 5, and 6) that possessed sufficient allelic diversity to chart the parasite's population genetic structure and produce the first genetic history model for *S. neurona*.[1820] The MLST sequence-level analysis identified 12 genetic types, 2 of which were predominant (56/87; 64%) and accounted for the majority of infections in the United States, based on the inheritance of different allele combinations encoded by the SnSAG genes among the 87 *S. neurona*-infected samples collected from varying hosts and geographic locations.[1820] This data supported a population structure that is both clonal and punctuated by a series of genetic types that evolved by sexual recombination through the definitive opossum host. Such an intermediate population structure is similar to that described by *T. gondii*.[737] To explain the existence of dominant clonotypes within the population structure for *S. neurona*, it was shown that *S. neurona*, like *T. gondii*, is capable of same-sex mating, and established selfing (or sexual amplification of a single clone) as a genetic mechanism for the clonal propagation of the species. Unlike *T. gondii*, the tissue stages of *S. neurona* are not infectious between intermediate hosts, and the sexual life cycle is obligatory. The extent to which uniparental mating versus outcrossing is impacting the population genetic structure of *S. neurona*, or its capacity to generate and expand specific strains capable of causing disease epidemics, is not currently known.[1820] This lack of knowledge certainly underscores the necessity to expand sample collection among intermediate and definitive hosts, to increase the number of informative polymorphic phylogenetic markers to improve the resolution of the current MLST typing scheme, and to pursue whole genome comparative studies in order to produce an accurate genetic history model for the species and assess the extent to which genetic recombination is impacting the parasite's population genetics in this genome-sequencing era.

Recently, a novel Type XIII *S. neurona* genotype was discovered; it was associated with severe disease in marine mammals in the north eastern Pacific Ocean.[210]

3.2.7.6 S. neurona *Genome*

As a genus, *Sarcocystis* arguably exists as the most successful protozoan parasite in nature, largely because all vertebrates, including birds, reptiles, fish and mammals can be infected by at least 1 *Sarcocystis* species. Despite its widespread prevalence and the relative ease of constructing drug-resistant, transgenic and gene-specific mutants in *S. neurona* that can be propagated *in vitro* in a variety of vertebrate cell lines, development of this parasite as a model system for genetic analyses has been hampered by the lack of a physical or genetic map, because *S. neurona* chromosomes do not condense and cannot be resolved by pulse–field gel electrophoresis. The recent whole genome shotgun sequencing of the *S. neurona* SO SN$_1$ genome, has rectified this knowledge gap and produced the first molecular karyotype for the genus, which should greatly facilitate future genetic and comparative genomic studies on this important pathogen.[109]

Combining Roche 454 and Illumina Hi-Scq reads at ~375X coverage, the *S. neurona* genome assembled into a molecular karyotype of 116 genomic scaffolds with a combined size of 127 Mbp, which was over twice the size of the *N. caninum*, *T. gondii*, and *H. hammondi* genomes.[1456,1806] The existence of a high proportion of repetitive Type II transposons, DNA- and LINE-element sequences, totaling 31 Mbp, accounted for nearly half of the increased genome size, with the remaining due to increased average intron length and intergenic region sizes, that were >3X that of *T. gondii*. The largest scaffold was 9.2 Mbp in length, and an additional 3.1 Mbp of sequence was encoded within 2950 unscaffolded contigs (that were each >500 bp in size). When RNA–Seq data was combined with the genomic data into an annotation pipeline, a complement of 7093 genes were identified, 5853 of which were expressed during merozoite growth. The predicted gene complement was similar to that found in *N. caninum*, *T. gondii*, and *E. tenella*; however, only limited chromosome-wide synteny was observed for homologous genes between *S. neurona* and *T. gondii*, with the largest block comprised of only 43 genes. This established that significant genome rearrangement between *Sarcocystis* spp. and that of *T. gondii* and *N. caninum* has occurred, which are largely syntenic across all chromosomes. In addition to its nuclear genome, the *S. neurona* apicoplast genome was largely conserved across the coccidia, with a few key differences. *S. neurona* has lost its *rpl36* gene, it had only one copy of the *tRNA-Met* ORF, and has two distinct RNA polymerase C2 genes. In common with *T. gondii*, it was missing *ORF A*. Furthermore, genome comparisons showed that *S. neurona* shared more orthologs with *T. gondii* (3169 genes) than with *Eimeria* spp. (1759 genes), indicating that *Sarcocystis* spp. is more closely related to *T. gondii*, than to *Eimeria* spp. Additionally, *Sarcocystis* spp. encoded 1285 (18%) genes that showed no detectable homology with any other species, underscoring the opportunity for investigators to now identify new gene families within the *Sarcocystis* genus, that promote their success in nature.

The tissue-encysting coccidia have evolved many families of dense granule (GRA), rhoptry kinase (ROPK), microneme (MIC), and surface protein adhesins, known collectively as the SRS, to promote their ability to disseminate infection among their intermediate hosts in order to establish long-term, chronic infections bearing infectious, transmissible sarcocysts. Comparative genomic, transcriptomic and metabolic data analyses between *S. neurona* and other coccidian genomes, has established that the *S. neurona* invasion machinery is largely conserved, but that it only has a limited set of ROPK and GRA proteins. Fifteen ROPK orthologs have been identified, which is significantly smaller than *E. tenella* (n = 27), *N. caninum* (n = 44), and *T. gondii* (n = 55). Moreover, none of the ROPK proteins previously identified as important murine virulence genes (ROP5, ROP18), or those that have the ability to alter host immune effector function by altering STAT3/6 (ROP16) or MAP kinase signaling (ROP38), are encoded within the *S. neurona* genome, indicating that *S. neurona* pathogenesis and infectivity is not dependent on the inactivation of these pathways, and may also explain why this parasite is not infectious to rodents in nature. In addition, *S. neurona* did not encode orthologs of the majority of GRA proteins expressed by *T. gondii* and *N. caninum*, including GRA6, GRA15, GRA24, and GRA25, which are known to regulate NFAT4, NF-kB, p38a MAP kinase and CXCL1/CCL2 levels, respectively, during acute infection. In fact, only 2 GRA protein orthologs have been identified in *S. neurona*, GRA10, and GRA12. This data underscores the differences between *S. neurona* and its related coccidia (*T. gondii* and *N. caninum*), and may indicate that *Sarcocystis* does not require an expanded repertoire of GRA and ROPK genes to promote its ability to recrudesce infection simply because, once encysted, it appears to undergo a terminal commitment to its gamont stage, thus requiring access to its definitive host to complete its life cycle.

3.2.8 Serologic Prevalence

3.2.8.1 Equids

Prevalence of antibodies is dependent on the distribution of opossums in the area, type, and age of horses sampled, geographical location, and the serological test used (Table 3.6). Sera in several studies were tested in 1 laboratory, by using immunoblots against low molecular proteins (marked WB[b] in Table 3.6). In these studies, prevalence was around 50% of horses tested. Seroprevalence increased with age.[93,114,419,1496,1535,1750,1863] In these surveys, the horses tested were mostly older than 1 year of age. It is likely that some of the foals tested had maternal-acquired antibodies.

Table 3.6 Seroprevalence of *S. neurona* Antibodies in Equids

Location	Year Sampled	Type, Source	No. Tested	Test, Cut-Off	Positive No	Positive %	Reference
Argentina	1996	Chaco Province	76	WB[a]	27	35.5	**419**
Argentina	2006–10	9 provinces	640	WB[a]	167	26.1	**1220**
Brazil	1998	Thoroughbreds	101	WB[a]	36	36.0	**417**
	NS	10 states	961	ELISA SAG4[b]	669	69.6	**838**
	NS	Rio Grande do Sul	181	ELISA-crude extract	61	33.7	**1397**
Canada	2001	Western Canada	239	WB[a]	0	0	**454**
Costa Rica	NS	7 provinces	315	SnSAG2 ELISA[b]	133	42.2	**292**
France	NS	2 farms in Normandy	50	SAT	18	36.0	**1396**
India	2003	3 states	123	WB[a]	1	0.8	**146**
Korea	NS	Thoroughbred Jeju island	191	WB	0	0	**752**
Mexico	NS	Durango state	495	rSnSAG2/4/3 ELISA	240	48.5	**1863**
Spain	NS	Galicia	384	SnSAG2 ELISA[c]	9	2.3	**32**
USA PA	1993–94	Thoroughbred farms[d]	117	WB[a]	53	45.3	**92**
OR	NS	Horses from 4 regions, >3 months old	334	WB[a]	149	45.0	**114**
OH	1993–94	37 breeds, statewide[e]	1056	WB[a]	560	53	**1535**
CO	1995–96	593 horses and 15 ponies	608	WB[a]	204	33.6	**1750**
OK	2001	Banked sera	798	WB[a]	712	89.2	**93**
WY	NS	Wild horses	276	WB[a]	18	6.5	**454**
				SAT, 1:50	39	14.1	
CA, FL, MO, MT	NS	Sera, diagnostic labs	208	IFA, 1:100	49	26.0	**1779**
MI	1997	1056 horses, 63 ponies, 3 donkeys, 1 mule from 98 herds	1121	WB[f]	627	56.0	**1496, 1499**
40 states	2010–11	40 states, sera submitted to diagnostics laboratory	3123	IFAT, 1:80	865	27.6	**1433**

Source: Adapted from Dubey, J.P. et al., 2015. *Vet. Parasitol.* 209, 1–42.
CA = California, CO = Colorado, FL = Florida, MI = Michigan, MO = Missouri, MS = Mississippi, MT = Montana, OH = Ohio, OK = Oklahoma OR = Oregon, PA = Pennsylvania, WY = Wyoming.
NS = not stated.
[a] Western blot, Granstrom et al.[728]
[b] Hoane et al.[836]
[c] Reactive to 30–35 kDA protein. Of these, 26.1% reactive to 16–17 kDA protein, 5 (3.62%) reactive to rSnSAG2 ELISA, 4 (3%) reactive in SnSAG4/3 ELISA.
[d] 20% of horses from each farm with a population of 580 horses; 4-week-old to 26-year-old, one 4-week-old might have colostrally acquired antibodies.
[e] Every 36th sample submitted to diagnostic laboratory for equine infectious anemia.
[f] WB, blot treated with bovine antibodies against *S. cruzi* antigen.[1495]

In 1 study, foals born to 33 seropositive mares were bled before suckling, 1 day after colostrum ingestion, and again at monthly intervals.[247] All foals were seronegative before suckling, all became seropositive 1 day after suckling, and 31 of 33 became seronegative by 9 months of age. The decay of antibody was probably related to the concentration of IgG in the colostrums. In another study, the median time of decay of maternal-acquired *S. neurona* antibodies in foals was 96 days, and these antibodies disappeared by 230 day of birth.[344] These results are in contrast to another study[1397] that reported antibodies in 61 (37.3%) of 181 mares at parturition, and in presuckle blood samples of 6.6% foals born to these mares. These observations need confirmation, because current evidence indicates that transplacental or lactogenic transmission of *S. neurona* in horses is very uncommon or absent. In an epidemiologic investigation of *S. neurona* seropositivity of horses in California tested by IFAT, there was no evidence for the transplacental transfer of *S. neurona* antibodies or parasite; all 366 presuckling foal sera were seronegative.[345] Another study of 174 foals born to mares on a farm with very high seroprevalence to *S. neurona* (90%) found only a single foal with a presuckle antibody titer.[1432]

Opossums are the only known definitive hosts for *S. neurona*, and they are found only in the Americas. Theoretically, therefore, horses in other countries should have no exposure to *S. neurona*. Consistent with this, *S. neurona* antibodies were not detected by immunoblotting in any of 191 horses from South Korea,[752] and in only 1 of 123 horses born in India.[146] However, seropositivity in horses from France[1395,1396] and Spain[32] is puzzling. In 28 French horses with EPM-like clinical signs, *S. neurona* antibodies were found by WB.[1395] In a followup study, they found seropositivity in 18 (36%) of 50 healthy horses tested by the SAT (*S. neurona* direct agglutination test). Even more intriguing are results from horses tested from Spain. Of 138 horses tested by immunoblotting, 26.1%–82.6% of horses were seropositive, based on the antibodies directed against immunodominant molecules at low (16–17 kDa) or high (30–35 kDa) molecular weights (MW); notably, Granstrom et al.[728] used antibodies directed against proteins lower than 16 kDA for interpretation of the WB. Retesting of these 138 samples with the rSnSAG2 ELISA and rSnSAG4/3 ELISAs, revealed that only 5 sera (3.6%) were reactive in rSnSAG2 ELISA, and only 4 (3%) were reactive in rSnSAG4/3 ELISAs; only 1 serum was strongly reactive in both types of ELISA tests.[32] Further, sera from 246 horses were tested by rSnSAG2 ELISA; 9 (2.34%) of 384 were seropositive. None of these 384 horses travelled out of Spain. Consequently, it was suggested that the immune-reactivity observed in the immunoblots was due to cross reactivity with another species of *Sarcocystis* that infects horses in Spain.

It has been suggested that there might be another definitive host for *S. neurona* in Europe, or there might be cross reactivity among *Sarcocystis* species in horses to account for the seropositivity in European horses.[1396] In this respect, another species of *Sarcocystis*, *S. bertrami* (with thin-walled sarcocysts—see Chapter 11) is found in horses in Europe.[393] Whether there is cross reactivity between *S. bertrami* and *S. neurona*, and if the type of antibody testing, or antigen preparations, account for this reactivity, has not been completely investigated. In the United States, only 1 non-*S. neurona* species of *Sarcocystis*, *S. fayeri*, has been found in horses. To resolve the question of cross reactivity between *S. fayeri* and *S. neurona*, 3 seronegative ponies were orally inoculated with 10^5, 10^6, or 10^7 *S. fayeri* sporocysts collected from dogs fed infected muscle obtained from horses in Texas.[1539] The fourth pony was an uninoculated control. Sera from ponies were tested by employing 3 tests (WB, IFA, and SAT) before dosing sporocysts, and then weekly, thereafter. With the WB using the strict interpretation criteria that were previously established,[728] antibodies specific to *S. neurona* were not found in any pony at 0, 2, 37, and 79 DPI. This negative reactivity was further confirmed using a recombinant SnSAG1 protein.[752] By SAT, only 1 sample collected 37 DPI from the pony dosed with 10^7 *S. fayeri* sporocysts was positive at 1:50 dilution. By contrast, all 3 inoculated ponies were seropositive by IFA, upto 1:400 dilution, whereas the control pony was seronegative.[1539] This experiment provides conclusive evidence that *S. neurona* infection can be distinguished from *S. fayeri* infection. It is important that the WB results are interpreted correctly, because it was apparent that these *S. fayeri*-infected ponies had antibodies that recognized numerous *S. neurona* antigens.[836]

3.2.8.2 Cats and Other Animals

Antibodies were found in 1%–40% of cats in the United States (Table 3.7). As a contrast, *S. neurona* antibodies were not found by SAT in any of the 502 domestic cats from Brazil.[448]

Antibodies (IFAT, 1:25) were found in 2 of 63 capybaras (*Hydrochoerus hydrochaeris*) and 8 of 11 free-living jaguars (*Panthera onca*) from Brazil.[1771,2397]

Antibodies to *S. neurona* were not found in experimentally[208] or naturally exposed[847] opossums. Antibodies were detected in skunks following oral inoculation of sporocysts.[204]

Table 3.7 Serological Prevalence of *S. neurona* in Nonequid, Nonmarine Mammals in the United States

Host	Location	Year	Type	No.	Test, Cutoff	No. Positive (%)	Reference
Cat (*Felis catus*)	VA			232	IFAT, 40	22 (9.0)	852
	PA			209		10 (5.0)	
	MS	NS	Stray	9	WB	1 (11.1)	1765
	MI	1999–01	Pets	196	IFAT, 20	10 (27)	1497
	OH	2001	Horse farms	35	SAT, 25	14 (40.0)	1674
		2001	Spay/neuter	275	SAT, 25	27 (10.0)	
	FL	NS	Humane shelter, feral	100	WB	5 (5.0)	684
Skunk (*Mephitis mephitis*)	CT	NS		24	SAT, 50	11 (46)	1201
Raccoon (*Procyon lotor*)	CT			12		100	1201
	PA, MA, FL, NJ			99	SAT, 50	58 (59.6)	1078
	CT			12		12 (100)	1201
	VA	2001–02	Feral	469	SAT, 50	433 (92.3)	779
Opossum (*Didelphis virginiana*)	CT			7		0	1201
	FL	NS	Feral	20	SAT, filter paper	0	208
Beaver (*Castor canadensis*)	MA			62	SAT, 25	4 (6.0)	915
Ring-tailed lemur (*Lemur catta*)				52	SAT, 50	2 (1.9)	1850
Black-and-white ruffed lemur (*Varecia variegata*)				4		1 (25.0)	
Black-eyed lemur (*Eulemur macaco flavifrons*)				6		0	
Armadillo (*Dasypus novemcinctus*)	FL	NS		2	SAT, 1:50	2 (100)	208

Source: Adapted from Dubey, J.P. et al., 2015. *Vet. Parasitol.* 209, 1–42.
CT = Connecticut, FL = Florida, NJ = New Jersey, MA = Massachusetts, MS = Mississippi, OH = Ohio, PA = Pennsylvania, VA = Virginia.
SAT = *S. neurona* agglutination test, IFAT = indirect fluorescent antibody test, WB = Western blot.

3.2.9 Clinical Infections

3.2.9.1 Horses, Ponies, and Zebras

EPM is often a progressively debilitating disease affecting the CNS of horses.[87,88,138,141,217,
270,328,333,438,602,777,1082,1090,1091,1096,1128,1147,1148,1210,1225,1489,1554,1555,1578,1755,1761,1992,2085,2115,2121,2122,2153–
2155,2177,2178,2246,2262,2285,2297,2377–2388] The clinical signs can vary from acute to insidious onset of focal
or multifocal signs of neurologic disease involving the brain, brainstem, spinal cord, or any combination of the areas of the CNS. Some horses affected with EPM have abnormal upper airway function, unusual or atypical lameness, or even seizures. In severe cases, the horse can have difficulty
with standing, walking, or swallowing, and the disease can progress very rapidly. In some horses,
the disease appears to stabilize or remain static for a time period.

The early clinical signs, of stumbling and frequent limb interference, are often easily confused
with a lameness of either the thoracic and/or the pelvic limbs. In many horses, the disease tends to
have a gradual progression of clinical signs, including ataxia, but, in some horses, mild clinical signs
followed by a rapidly progressive course have been observed. Blood monitoring of these horses,
using either Western blot, IFAT or ELISA testing, often results in dramatic changes in antibody
concentration. On physical examination, the vital signs are usually normal, although variations in
body condition, from very thin to obese, along with clinical depression, might also be observed.
Neurological examination often reveals an asymmetric weakness, ataxia and spasticity, involving
all 4 limbs. Frequently, areas of hypoalgesia or complete sensory loss may be noted. The most
frequent brain or cranial nerve deficits observed in horses appear to be head-tilt, depression, facial
nerve paralysis, difficulty in swallowing, upper airway dysfunction, such as dorsal displacement of
the soft palate and laryngeal hemiplegia, although signs are not limited to these areas (Figure 3.2).
Gait abnormalities are often a result of damage to the brainstem or spinal cord, and can be quite
variable, depending on the location and severity of the lesion.

Most horses affected with EPM are bright and alert at presentation; however, any horse with signs
of neurologic disease is a candidate to have EPM. At the time of initial examination, most horses
have normal blood values. One of the most helpful clinical signs is that horses with EPM often have
asymmetric gait deficits, with focal muscle atrophy. This can be a useful differentiating feature, and
may help direct one towards a clinical diagnosis of EPM, rather than another neurological disease.

The pathogenesis of EPM is not clear. Clinical signs of EPM are dependent on the area of
the CNS parasitized. For example, involvement of the cerebrum can cause depression, behavioral
changes, or seizures. Lesions in the brainstem and spinal cord often cause gait abnormalities, incoordination caused by involvement of ascending and descending tracts, and any of a variety of signs
attributable to damaged cranial nerves. Severe damage in the gray matter that innervates muscles of
the limbs can produce weakness, followed by atrophy of innervated muscles. The quadriceps and
gluteal, as well as the temporalis muscles, are often atrophied; *S. neurona* has not been found in
affected muscles.

Factors governing severity of EPM are unknown.[438,475] Clinical EPM is often reported in well-
cared race horses 3–6 years of age. Clinical EPM does not seem to be associated with poor nutrition
or known concurrent infections. There are no confirmed reports of clinical EPM in horses younger
than 6 months of age. EPM-like disease in a 2-month Appaloosa colt that had facial paralysis beginning 2 days after birth was reported.[731] *S. neurona* antibodies were found in its CSF by immunoblot,
and it responded favorably to pyrimethamine and trimethoprim–sulfadiazine medication.

EPM has been confirmed histologically only in horses born and raised in the Americas, coincident with the geographical range of opossums. However, EPM has been diagnosed in horses
exported from the Americas. In Japan, *S. neurona* was confirmed by IHC in a thoroughbred 15
months after importation from Kentucky.[945] From the published reports, it appears that EPM is rare

in equids other than horses. One case was diagnosed in a 10-year-old pony from Maryland.[387,422] EPM was also reported in an 8-year-old Grant's zebra (*Equus burchelli bohmi*) in California.[1117]

Prevalence of EPM in horses was estimated at 0.5%–1% of the horse population.[438] Clinical disease is sporadic, and more than 1 case is seldom seen at a particular farm. However, clusters of cases have occurred in a few instances, which would suggest that all of the risk factors necessary for disease were at those facilities. EPM has been reported in siblings.[1761]

The pathogenesis of *S. neurona* in horses is unclear, because it has been difficult to reliably induce disease experimentally in horses. Attempts at inducing EPM in horses are summarized in Table 3.8. Although horses developed clinical disease, attempts to demonstrate *S. neurona* in histological sections were essentially unsuccessful. Several factors, including the stage of the parasite inoculated, dose, age of the horse, immunological status, endogenous and exogenous stress during experiment, were examined by different investigators. Clinical outcome was not affected when sporocysts were derived from naturally infected versus experimentally infected opossums. The dose did not significantly affect outcome, because feeding as few as 100 sporocysts induced clinical disease.[1636] Administration of corticosteroids to horses also did not affect the outcome of the experiments.[272,1538] A transport stress model proved valuable in helping to induce clinical EPM. In a series of experiments, young horses selected from northwest Canada, where opossums are not present, were transported by road for 55 h and then dosed with *S. neurona* sporocysts as the horses were unloaded from trucks. In the transport-stressed horses, seroconversion to *S. neurona* was sooner and clinical signs were more severe than in horses not stressed. A second transport-related stress after a 2-week rest did not affect the severity of EPM.[1538,1540] Arabian foals with severe SCID, lacking T and B cells, were successfully infected with *S. neurona*, with demonstrable live parasites in blood and tissues; however, these foals did not develop clinical signs of EPM.[1083,1579] Parasitemia was demonstrable in an immunocompetent horse dosed daily with *S. neurona* for 112 days.[1500] All 5 horses inoculated with cultured *S. neurona* merozoites directly in the subarachnoid space seroconverted, but only 1 developed clinical signs.[1074] *S. neurona* can circulate in equine tissues within lymphocytes.[1080] To facilitate dissemination of *S. neurona* merozoites, and their entry into the CNS, equine peripheral blood mononuclear cells (PBMCs) were first infected with merozoites *in vitro*, and, after 5 h incubation, were inoculated intravenously back into the horses.[501,502] Five of the 6 inoculated horses developed signs of clinical EPM starting the first week PI. This experiment was repeated with 8 additional horses, with similar results. [1833]

Immunity is considered to play an important role in the clinical outcome of EPM, but data from naturally and experimentally infected horses is not conclusive.[642,644,1121,1279,1430,1574, 1662,1663,1756,1833,1855,2267] Interferon gamma is essential for controlling the development of EPM, as evidenced by studies in mice. Immunocompetent outbred and inbred BALB/c or C57Bl/6 mice are not susceptible to *S. neurona* infection, whereas the parasite is fatal to either BALB/c or C57Bl/6-derived KO mice.[412,470,1502,1830,1831,2303] Clinical disease develops in KO mice. Both KO and nude mice are susceptible to infection and disease development, whereas SCID mice do not show signs of parasite survival or disease development.[1114,1580] The humoral response is less critical in protection, as suggested by the fact that B cell-deficient mice are not susceptible to infection, and mice that die of severe disease have antibodies to *S. neurona*.[1832]

3.2.9.2 Marine Mammals (Sea Otter, Seal, Porpoise, Sea Lion)

3.2.9.2.1 Sea Otter (Enhydra lutris)

Reports of *S. neurona* infection in sea otters are summarized in Table 3.9. Among marine mammals, causes of mortality are best known for sea otters because, during long periods of time, all dead sea otters recovered by wildlife services were examined at necropsy. Sea otters are resident on

Table 3.8   Attempts to Induce Clinical Disease in Horses after Inoculation with *S. neurona*

No.	Type	*S. neurona*	Clinical Disease	Duration (days)	Diagnosis		Reference
					Bioassay	Histology	
5	3–6-month-old foals	Naturally infected opossums (NIO)	Clinical signs in all, 28–42 DPI, neurological, lesions in spinal cord of 4	40–110	Negative	Negative	**603**
8[a]	3–4 years	NIO	Spinal cord lesions in 7 of 8 horses	90	Negative	Negative	**272**
9[a]	Foals	NIO	Clinical signs more severe in stressed[b] horses	44–63	Negative	Negative	**1538**
20[a]	4–5 months	Experimentally infected opossums (EIO)[c]	Dose response 10^2, 10^3, 10^4, 10^5, 10^6, 4 horses per group Clinical signs in all, histologic lesions inconsistent	30	Negative	Negative	**1636**
24[a]	4–5 months	EIO[c]	Stressed horses[b], second time[d]	40	Negative	Negative	**1540**
6		EIO[c]	100–1000 sporocysts daily for 112 days		Positive, blood of 1 horse	No data	**1500**
1	Arabian 5-month-old SCID foal	NIO	Knuckling 39 DPI, euthanized 53 DPI. *S. neurona* isolated from brain and blood	53	Positive		**1083**
2	Arabian	NIO	Neurological signs in 1	90	Negative	Negative	**1579**
4	Arabian	500 million merozoites from culture, inoculated intravenously	3 horses neurological	41, 73, 80, 471	Positive, brain of 1 horse	Negative	**1579**
1	Arabian SCID foals	500 million merozoites from culture, inoculated intravenously into 1 foal	No clinical signs	21,32	Li, Sp, Sk of 1 foal, H and Sk of second foal	Negative	**1579**
2	Arabian SCID foals	NIO	No clinical signs	96	Li, Lu, Sp, H, Sk	Negative	**1579**
5	2–17 years, mixed breed	5×10^5 to 5×10^6 merozoites inoculated intrathecally	No clinical signs	132	Not done	Negative	**1074**
3	1–10 years	100, 1,000 and 10,000 cultured merozoites daily for 15 days	All horses developed clinical signs	90	Not done	Positive	**502**

(Continued)

Table 3.8 (*Continued*)   Attempts to Induce Clinical Disease in Horses after Inoculation with *S. neurona*

No.	Type	*S. neurona*	Clinical Disease	Duration (days)	Diagnosis Bioassay	Histology	Reference
8	1–19 years, several breeds	6000 merozoites from culture inoculated intravenously, daily for 14 days	All horses developed clinical signs	55	Not done	No data	**1833**
15[a]	2–15 years, standardbred and mixed breeds	612,500 sporocysts NIO	Mild ataxia in 3	84		Negative	**1092**

Source: Adapted from Dubey, J.P. et al., 2015. *Vet. Parasitol.* 209, 1–42.
SCID = severe combined immunodeficiency disease.
H = heart, Li = liver, Lu = lung, Sk = skeletal muscle, Sp = spleen.
[a] Horses obtained from Northwest Canada outside the range of opossums.
[b] Transport stress model. Horses transported for 55 h and dosed soon after arrival.
[c] Isolate SN-37R-obtained by serial passage between laboratory raised opossums and raccoons.
[d] Horses transported after rest period.

the coasts of California, Washington, and Alaska in the United States, British Columbia in Canada, and in Russia, but *S. neurona* has been definitively identified only in sea otters from California and Washington, which is consistent with the distribution of opossums. Data in Table 3.9 are based on dead or sick animals. *S. neurona* high titer antibodies were found in 2 of 74 sea otters from Kodiak, United States, but not in 89 otters from Bering Island, Russia.[713]

Most diagnosed cases were from southern sea otters (*Enhydra lutris neresis*) from California, compared with northern sea otters (*Enhydra lutris kenyoni*) from Washington (Table 3.9). Since 1992, more than 1000 sea otters were examined at necropsy, and protozoal encephalitis was identified as an important cause of their mortality.[978,1197,1739,2068,2375] Among numerous causes of mortality in 105 sea otters, *S. neurona* encephalitis was diagnosed in 7 individuals, which was an underestimate because diagnosis was made only when protozoa were identified in lesions based on conventional histological examination; immunohistochemistry was not performed.[978] In a detailed study of encephalitic lesions, *S. neurona* was identified alone in 22 of 39 cases of encephalitis, and dual infection, with *T. gondii*, in another 12 otters among 344 sea otters. Most of these deaths were sporadic.[1739]

An epizootic of protozoal encephalitis in sea otters has been recorded.[1196,1197] Within 1 month, 63 sick and dead otters were recovered near Morro Bay, California. This outbreak provided the opportunity for epidemiologic and etiological investigations. Seizures, muscle tremors, and paresis were noted in live stranded otters. A few animals had tachycardia and respiratory distress.[1197] Of these otters, 15 were studied in detail. Antibodies (IFAT >2560–81,920) to *S. neurona* were present in all 14 adult otters; 1 pup was seronegative (<1:40). Viable *S. neurona* were isolated from the brains of 8 adults. Predominant gross findings were enlarged heart with mottling, lymphadenopathy, and adipose petechiation. Microscopically, meningoencephalitis associated with schizonts and myositis associated with sarcocysts predominated. Immature and mature sarcocysts were seen in muscles, heart, and brain.[1196,1197]

In an epidemiological investigation, land-based runoff was associated with *S. neurona* mortality in sea otters.[1590] *S. neurona* was identified in 33 of 205 sea otters; however, it is not certain if these were new cases or previously included in the reports listed in Table 3.7. Increased runoff due to rain in the prior 1–2 months was 12 times more likely to be associated with *S. neurona* mortality.[1590]

Table 3.9   Reports of Clinical *S. neurona* Infections in Sea Otters in the United States

Location	Year	No. of Sea Otters Examined	No. Positive	Observations, Tissues Parasitized	Bioassay	Reference
Zoo, OR	1997	1	1	Antemortem diagnosis, biopsy muscle, CSF–WB positive, medicated with pyrimethamine, euthanized. Schizonts in brain, sarcocysts in muscle	Not done	**1494**
Monterey Bay, CA	2000	1	1	Meningoencephalitis, schizonts in brain, sarcocysts in muscle and heart	Positive	**1072**
Grays Harbor County, WA	2000	1	1	Convulsions, died. Meningoencephalitis, schizonts in brain. Coinfected with *T. gondii*	Positive	**1077**
Monterey Bay, CA	1999	1	1	Pup estimated 16-weeks old. Meningoencephalitis, schizonts in brain	Positive	**1195**
CA	1998–01	105	7	Only conventional histology. Encephalitis, schizonts in brain. Causes of mortality detailed. *T. gondii* in 17 otters	Not done	**978**
CA, WA	1985–04	344	39	Encephalitis in 17 otters from CA and in 5 from WA. *S. neurona* alone in 22 otters, dual *S. neurona* and *T. gondii* in 12, *T. gondii* in 7. Sarcocysts in muscles of all of 18 otters examined	Not done	**1739**
CA	2004	3	3	Lymphadenopathy, enlarged heart. Encephalitis, schizonts in brain. Sarcocysts in brain and muscle, proven by TEM	Positive	**1196**
Morro Bay, CA	2004	63 otters dead in an epidemic	8	16 necropsied and tested for *S. neurona. S. neurona* antibodies in 15, 1 pup seronegative. *S. neurona* isolated from 8, *T. gondii* isolated from 6. Encephalitis and myositis in 15	Positive	**1197, 1820, 1821**
Copalis Beach, WA	2010	1	1	Retinitis and encephalitis with schizonts	Not done	**468**

Source: Adapted from Dubey, J.P. et al., 2015. *Vet. Parasitol.* 209, 1–42.
CA = California, OR = Oregon, WA = Washington.

Encephalitis is the predominant lesion in sea otters, and both *T. gondii* and *S. neurona* can be found together, sometimes in the same histological section. In general, the *S. neurona*-associated lesions are more severe than those associated with *T. gondii*.[1739]

3.2.9.2.2 Pacific Harbor Seals (Phoca vitulina richardsi)

There are few reports of *S. neurona* mortality in seals. Encephalitis was found in 7 seals; 6 of them were stranded on the California coast, and the seventh was captive.[1023] Among these seals was a 1-week-old captured juvenile with tremors that lasted 17 days. Schizont-associated encephalitis was found in all 7 animals, and the diagnosis was confirmed immunohistochemically. *S. neurona* antibodies were found by immunoblot in 4 of 5 seals tested. Four seals had myocarditis, and sarcocysts were found in the heart of 1 seal.

Subsequently, viable *S. neurona* was isolated from an adult seal that was stranded near Monterey, California.[1194] The seal had encephalitis in association with *S. neurona* and *T. gondii* infection. Barbosa et al.[210] likewise isolated viable *S. neurona* from the brains of 4 harbor seals in 2011 and one in 2012 that stranded with severe encephalitis along the Washington state coastline.

Antemortem diagnosis and treatment were reported in a 20-year-old captive seal that developed tremors and dysphagia.[1267] Findings of *S. neurona* antibodies in serum and CSF and negative tests for other infections suggested the diagnosis of *S. neurona*. The animal improved clinically after treatment with ponazuril at 10 mg/kg/oral for 3 months.[1267]

3.2.9.2.3 *Pacific Harbor Porpoise (Phocoena phocoena)*

Barbosa et al.[210] identified *S. neurona* infection in 32 of 46 (70%) harbor porpoises that stranded along the Washington state coastline between 2006–2012; the majority of these infected animals had mild to severe forms of encephalitis upon histological examination. Notably, however, several animals that were PCR positive had no inflammatory lesions. One isolate recovered in 2011 was genotyped at multiple loci and identified to be a common Type VI strain that is widely distributed throughout the United States.

3.2.9.2.4 *California Sea Lion (Zalophus californianus)*

S. neurona infection was diagnosed antemortem in a California sea lion stranded ashore in Marin County, California.[165] The animal was lethargic. Acute myositis was diagnosed based on histological examination of biopsy, serum biochemical tests, and finding of elevated antibodies to *S. neurona* (IFAT >1:5120–1:20,480). The ALT and CK values were elevated. Treatment with ponazuril (10 mg/kg/oral for 5 days) resulted in clinical improvement. The animal was returned to its habitat 3 months after it was found stranded.

The finding of *S. neurona* in 9 of 10 asymptomatic sea lions is most interesting. Of the 10 adult healthy California sea lions euthanized to protect fish stocks, 9 were infected with *S. neurona*, and 1 had concurrent *T. gondii* infection.[680] These results were based on the detection of the parasite DNA in samples of heart, brain, muscle, and lymph nodes.

3.2.10 Other Marine Mammals

S. neurona infection has also been documented by molecular PCR and histology in a wide range of other marine mammals, including 11/25 (44%) Guadalupe fur seals (*Arctocephalus townsendi*), 3/8 (38%) Steller sea lions (*Eumatopias jubatus*), 1/5 (20%) Northern fur seals (*Callorhinus ursinus*), 3/3 (100%) Northern elephant seals (*Mirounga angustirostris*), a Pacific white-sided dolphin (*Langenorhynchus obliquidens*), a killer whale (*Orcinus orca*), and a pygmy sperm whale (*Kogia breviceps*).[210,680] These results indicate that *S. neurona* infects a vast array of marine mammal orders, including pinnipeds, cetaceans, and mustelids.

3.2.11 Miscellaneous Animals

Results are summarized in Table 3.10. Unusual findings are commented upon.

3.2.11.1 *Raccoons (Procyon lotor)*

There are several reports of *S. neurona* encephalitis in raccoons (Table 3.10). These raccoons were examined at necropsy because of suspicion of rabies. A concurrent morbillivirus infection (MVI) is considered to aggravate severity of disease. However, in 2 raccoons, evidence for MVI

Table 3.10   Reports of Clinical *S. neurona* Infections in Nonequid and Nonmarine Animals in the United States and Canada

Host	Location	No. Age	Clinical Signs	Diagnostic Methods	Organs Parasitized	Reference
Cat (*Felis catus*)	California	1, 13-week-old	Lame after fall, euthanized	IHC	B, Sc	**411, 422**
	Illinois	1, 4-month-old	Neurologic, post surgery	IHC	B, Sc	**450**
	Indiana	1, 5-month	Neurologic, antemortem dignosis	PCR	CSF	**104**
Canada lynx (*Felis canadensis*)	New York Zoo	1, 13-year	Encephalitis	IHC	B	**617**
Dog (*Canis familiaris*)	Illinois	1, 6-year-old	Myositis	IHC, PCR	Sk	**1780, 1987**
	Maryland, Illinois, Arkansas, Oklahoma, Louisiana	7, retrospective	Neurologic, dermal, pneumonia	IHC, PCR	B, K, Lu, skin,Sc	**460, 1988**
	Mississippi	1, 1.5-year-old	Ataxia, neurological	IHC, PCR	B	**249**
	Mississippi	1, 2-month-old	Neurologic	IHC	B, E, L, H, I, Nt, T,	**472**
	Tennessee	1, 2-year-old	Acute onset of parapresis	IHC, PCR	Sc	**670**
Raccoon (*Procyon lotor*)	Ohio	1	Neurological, euthanized	IHC	B	**398, 402, 422**
	Oregon	2	Encephalitis and myocarditis, no evidence of MVI	IHC	B, H	**778**
	New York	1	Neurologic, MVI	IHC	B	**1677, 422**
	Illinois	1	Neurologic, MVI	Hist.	B	**1742**
Mink (*Mustela vison*)	Oregon	2	Neurologic	IHC	B	**408**
Ferret (*Mustela putorius furo*)	British Columbia, Canada	1	Neurologic, respiratory, MVI	IHC, PCR	A, B, H, K, Li, Lu, Nt, Ln, Mo, Sk, Sp,	**145**
Fisher (*Martes pennanti*)	Maryland	1	Neurologic	IHC, PCR	B	**671**
Striped skunk (*Mephitis mephitis*)	Indiana	1	Neurologic	IHC, PCR	B, Li, Lu, Nt	**153**
Bald eagle (*Haliaeetus leucocephalus*)	Missouri	1	Neurologic signs	IHC	B	**1333**

Source: Adapted from Dubey, J.P. et al., 2015. *Vet. Parasitol.* 209, 1–42.
A = adrenal, B = brain, H = heart, I = intestine, Li = liver, Lu = lung, Mo = blood monocytes, Nt = nasal turbinates, Sk = skeletal muscle, Sc = spinal cord, Sp = spleen, T = tongue.
MVI = concurrent Morbillivirus Infection.

was not found; in these animals, the encephalitis was granulomatous rather than pyonecrotic or pyo-granulomatous, as seen in animals with concurrent MVI.[778] The most severe disease diagnosed was in a juvenile raccoon.[398] The raccoon had neurological signs, was unsteady, and could not maintain its balance. Within a week of onset of clinical signs, the raccoon's condition deteriorated, it was euthanized, and a complete necropsy was performed. Protozoa-associated lesions with numerous *S. neurona* schizonts and free merozoites were confined to cerebrum.

3.2.11.2 Mink (Mustela vison)

On a farm in Oregon, United States, 50 of 15,500 minks died within 2 weeks.[408] The affected minks were weak, ataxic, and anorectic. Complete necropsy was performed on two 1-month-old female minks. Grossly, a discolored area was found in the cerebrum of 1 of the 2 minks. Microscopically, both minks had severe meningoencephalitis associated with numerous schizonts of *S. neurona*. Minks with progressive neurologic disease can also have a variable number of sarcocysts present.[1445]

3.2.11.3 Domestic Cat (Felis catus)

Three reports of *S. neurona*-like infection in cats need explanation because of unusual presentations. The first cat diagnosed was a 3-month-old that became ill 2 weeks after a fall, and developed hemiparesis.[411] The second cat developed neurological signs 3 days after routine surgery for castration.[450] In both cats, lesions and *S. neurona* schizonts were confined to CNS. Additionally, *S. neurona*-like sarcocysts were seen in the brain of the second cat. In the third cat, the diagnosis was made antemortem, and, thus, a detailed clinicopathologic could not be documented.[104]

Efforts to induce clinical disease in cats were unsuccessful. Four inoculated with 10 million culture-derived *S. neurona* merozoites[157] and 5 cats orally inoculated with >1000 *S. neurona* sporocysts[429] became infected but remained subclinical; 3 of these cats had been administered rather high doses of methylprednisone acetate.[429]

3.2.11.4 Dog (Canis familiaris)

Of the 10 dogs diagnosed with *S. neurona*-like infections (Table 3.10), 3 reports need an explanation because of unusual presentations. The first case was a dog from Illinois.[401] Protozoa were seen in the skin of this dog. The dog was misdiagnosed as having an *S. canis* infection.[400] Subsequently, it was diagnosed as *S. neurona*, based on immunohistochemistry.[460] This dog had pustular dermatitis with many merozoites and schizonts found in a smear made from the oozing fluid (Figure 3.5d).

A 2-month-old dog was thought to be congenitally infected with *S. neurona*-like organisms; it had focal severe encephalitis, retinitis (Figure 3.5a and b), and myositis associated with schizonts and merozoites.[472] Mature sarcocysts were found in its skeletal muscle. Schizonts were also present in sections of nasal turbinates, but without inflammation.

Also, a 2-year-old dog had acute onset of paraparesis.[670] At necropsy, large nodular masses were found between nerve roots of the thoracic spinal cord (Figure 3.3). Microscopically, the nodular growth had mixed cellular infiltrates and necrosis associated with protozoa, thought to be *T. gondii*. However, these structures were *S. neurona* schizonts (not *T. gondii*) in different stages of maturation; they reacted strongly to *S. neurona* antibody.

Attempts to induce clinical disease in dogs were unsuccessful; 2 dogs orally inoculated with 1,000,000 sporocysts remained subclinical, despite administration of corticosteroids (J.P.D, unpublished).

3.2.11.5 Ferret (Mustela putorius furo)

The case of an infected ferret is unusual for 2 reasons. First, the animal had massive infection of *S. neurona* affecting the nasal turbinates.[145] Second, the protozoal infection might have been triggered by vaccination with live modified MVI vaccination. The ferret had been administered 1 dose of live canine distemper virus vaccine (CDV) 1 week before onset of rhinitis. In dogs, vaccination with CDV can trigger *T. gondii* infection.[464]

3.2.11.6 Canada Lynx (Felis lynx canadensis)

A 13-year-old lynx from a zoo had granulomatous encephalitis, both in grey and white matter throughout the cerebral cortex.[617] Schizonts were seen in neuropil as well as in vascular endothelium.

3.2.11.7 Fisher (Martes pennanti)

A wild-caught, ataxic fisher was found to have no fear of humans and was attacked by a dog.[671] After euthanasia, a complete necropsy was performed to exclude rabies. Numerous *S. neurona* schizonts were identified in encephalitic lesions, and the diagnosis was confirmed by PCR. Additionally, mature *S. neurona*-like sarcocysts were identified in the skeletal muscle.

3.2.11.8 Striped Skunk (Mephitis mephitis)

Disseminated *S. neurona* infection was diagnosed in a juvenile skunk with concurrent MVI.[153] Schizonts were seen in sections of brain, lung, liver, and the nasal epithelium.

3.2.11.9 Bald Eagle (Haliaeetus leucocephalus)

Sarcocystis-associated encephalitis was diagnosed in a captive bald eagle.[1333] Confirmation of *S. neurona* infection was needed because this was a unique report of this parasite in an avian species, and because another *Sarcocystis* species, *S. calchasi*, was diagnosed to cause encephalitis in birds[1329,1843] (see Chapter 17).

3.2.12 Diagnosis

3.2.12.1 Horses

3.2.12.1.1 General Observations

Although many neurologic disorders affect the horse, EPM remains the most commonly diagnosed infectious equine neurologic disease in the Americas.[438,1224,1225] A complete neurologic examination and implementation of a thorough diagnostic plan to rule out differential diagnoses are essential prerequisites to laboratory testing and appropriate interpretation of clinical signs and death. Many ancillary diagnostic procedures may be required to differentiate primary musculoskeletal disorders and other neurologic diseases from EPM.

3.2.12.1.2 Blood Chemistry

EPM does not produce consistent detectable changes in complete and differential cell counts or serum chemistry values. General analysis of CSF is also not informative. *S. neurona* merozoites have not been confirmed in CSF.

3.2.12.1.3 PCR

PCR testing of equine CSF may provide information regarding the presence of *S. neurona* DNA in the CNS. However, the sensitivity of the PCR test appears to be much lower than initially estimated, possibly because intact merozoites rarely enter CSF. Nonetheless, the PCR test may be a useful adjunct for the postmortem diagnosis of EPM in selected cases, when it is used on tissues collected at necropsy.

3.2.12.1.4 Serological Tests

Several serological tests have been used to detect antibodies to *S. neurona* in animals. The WB test, also called the immunoblot test, was the first assay developed for detection of antibodies against *S. neurona*, and diagnosis of EPM.[728] This assay has been available for over 2 decades and continues to be offered by several diagnostic testing laboratories.[230,1091,1198] Methods and data used to validate WB were reviewed.[438] Subsequent to its development, the *S. neurona* WB test has been modified to try to improve the diagnostic accuracy and/or to make the assay semiquantitative. Use of bovine serum against *S. cruzi* to "block" cross-reacting antigens was reported to yield sensitivity and specificity approaching 100%.[1495] However, evaluation of this modified WB protocol by other investigators found a lower sensitivity/specificity of 89% and 69%, respectively, when applied to a more extensive sample set.[279,343] A WB method used by a commercial testing laboratory (Neogen, Inc.) involved measuring the intensity of the antibody/antigen response to yield a unitless number referred to as the "relative quotient" (RQ). This test is no longer offered by the company.

Development of the WB test provided a tremendous boost to EPM diagnosis, but this technique is primarily a research tool that is fairly laborious and requires significant expertise to interpret accurately. Consequently, multiple "second generation" serologic assays, that provide greater throughput and are more informative, have been developed. Two assays developed, using whole *S. neurona* merozoites as antigen, were the SAT[1076] and the IFAT.[343,346] The usefulness of the SAT was further evaluated in experimentally infected cats, raccoons, and ponies.[447,1539] Both the SAT and the IFAT can be used to obtain an end-point antibody titer, which is a large improvement over the nonquantitative result provided by the WB test. The SAT is advantageous because it can be used for detection of antibodies in multiple different animal species, and the assay has been employed in a variety of research studies (Table 3.5). However, the SAT has not been routinely employed for EPM diagnosis and is not presently offered as a commercial test. The IFAT was optimized and validated at the University of California, Davis,[343,346] and testing is currently available from the University of California Veterinary Diagnostic Laboratory. Although specialized instrumentation is needed (i.e., a fluorescence microscope), interpretation of parasite fluorescence requires expertise, and concerns have been raised about cross reactivity with the nonpathogenic species *S. fayeri*,[1540] serologic results from the IFAT will be generally accurate when the assay is performed by individuals with proper training.

ELISAs are easy to perform, allow for relatively high-throughput testing, and provide a more objective interpretation of results relative to the other existing assays, particularly the WB. Several ELISAs have been developed using *S. neurona* antigens expressed as recombinant proteins in *E. coli*[500,836,1862] or in Baculovirus.[753] Thus far, these ELISAs are all based on the family of related *S. neurona* merozoite surface antigens (SnSAGs),[499,849] which are good serologic targets, because they are abundant and immunogenic. Importantly, antigenic diversity has been found among different strains of *S. neurona*,[263,850,870,1119,1821] so all of the SnSAGs are not equally appropriate for use in serologic assays.

ELISAs based on the SnSAG2, SnSAG3, and SnSAG4 surface antigens, have been shown to be accurate for detecting and quantifying antibodies against *S. neurona* in equine serum and CSF samples.[836,1862] To help overcome the negative impact of antigenic diversity in the *S. neurona* population and the varied immune responses that occur in different horses, fusion of SnSAG3 and SnSAG4 into a single chimeric protein (rSnSAG4/3) and concurrent analysis with 2 ELISAs (rSnSAG2 ELISA and rSnSAG4/3 ELISA) have been employed for commercial testing of equine samples (Equine Diagnostic Solutions, LLC). Although homologues of the SnSAG antigens are present in the related parasite *S. falcatula*,[1821] this does not hinder the specificity of the SnSAG2 and SnSAG4/3 ELISA results, since *S. falcatula* is incapable of infecting and causing seroconversion in horses.[271] Furthermore, extensive validation studies have shown that the SnSAG ELISAs are specific and do not cross-react with serum from horses infected with other species of *Sarcocystis*.[32,836] Hence, these surface proteins are valuable immunologic markers that accurately detect infection with *S. neurona*. Recently, a trivalent ELISA rSnSAG2/4/3 with highly comparable results to independent rSnSAG2 and rSnSAG4/3 tests have been developed.[2403]

An ELISA based on the SnSAG1 surface antigen was described[500] and has been offered for EPM diagnosis by Antech, Inc. This test provides an end-point titer and values greater than 1:100 in serum are reported to indicate active infection. This test and cut-off point have not been rigorously evaluated; however, the cut-off values used for a positive diagnosis appear to be arbitrary. More critically, the SnSAG1 protein has been shown to be absent from multiple *S. neurona* isolates,[850,870,1119] which explains the relatively low accuracy observed with this antigen in independent studies.[836,912] An assay that combines SnSAG1 with the two alternative major SnSAGs, SnSAG5[263] and SnSAG6,[1821] is presently offered as a stall-side test (Pathogenes, Fairfield, FL, USA), but no published reports have validated this assay so it is unclear whether the test accurately detects antibodies to *S. neurona*.

Because EPM occurs only in a small proportion of horses infected with *S. neurona*, the simple detection of serum antibodies against this parasite has minimal diagnostic value. Detection of antibodies in CSF is more informative, but this analysis is potentially confounded by blood contamination of the sample during collection as well as the passive transfer of antibody from the plasma across an intact and healthy blood–CNS barrier.[646,648] The blood–CNS barrier (BCB) can also be compromised in neonatal foals. Of 13 foals born to seropositive mares, antibodies were detected in CSF of 12 of 13 tested in *S. neurona* WB test 2–7 days after birth.[248] The CSF antibodies declined over time, but were detectable in 3 foals between 62 and 90 days; the foals were asymptomatic.[248] These obstacles to accurate immunodiagnosis of EPM have been reduced in recent years through the development of semiquantitative assays (e.g., IFAT and ELISA) and use of diagnostic methods that reveal intrathecal antibody production, which indicates active *S. neurona* infection in the CNS. The blood contamination of CSF had no effect on the diagnosis of EPM by using the IFAT.[606]

The Goldman–Witmer coefficient (*C*-value) and the antigen-specific antibody index (AI) are algorithms that use antibody endpoint titers in paired serum and CSF samples to assess whether the amount of pathogen-specific antibody in the CSF is greater than would be present from normal passive transfer across the BCB. Difficulties in the interpretation and value of these tests to diagnose EPM were reviewed previously.[438] Use of the SnSAG2 ELISA, and the *C*-value and AI algorithms, to examine a sample set of 29 clinical cases, demonstrated that these methods provide accurate diagnosis of EPM.[648] This study also confirmed prior findings that modest blood contamination of the CSF, up to 10,000 red blood cells/μL, will have minimal effect on test results.[606]

Many horses with EPM have CSF antibody titers that are profoundly higher than what should be present due to normal passive transfer across the BCB. Studies using the SnSAG2 and SnSAG4/3 ELISAs to examine a large collection of horses with neurologic disease showed that a simple serum:CSF ratio, calculated from the endpoint titers, can serve as a proxy for the more laborious and expensive *C*-value or AI.[912,1453] These studies found that the optimal serum:CSF titer ratio was 100:1, which approximates the normal partitioning of proteins (e.g., albumin or total IgG) between the blood and the CSF.[648,1498] Serum:CSF ratios equal to or lower than 100:1 were indicative of intrathecal antibody production against *S. neurona*, and yielded EPM diagnostic accuracy of approximately 93% sensitivity and 83% specificity for a collection of 128 cases examined.[1453] Of 44 EPM cases, 33 had serum:CSF ratios of 25:1 or less, with 9 horses exhibiting CSF antibody titers that approached or exceeded their serum titers (i.e., ratio of 1.6:1 or lower). Although IgM antibodies can be detected in CSF or the blood of horses experimentally infected with *S. neurona*,[1263] this test has not been used in the diagnosis of EPM in horses.

Based on statistical modeling of IFAT results from a sample set of naturally and experimentally infected horses, it has been proposed that EPM diagnosis can be achieved from serum antibodies alone, with minimal additional information provided by CSF testing.[344,347] This idea was not supported by the previously described studies using ELISAs[912,1453] which showed that serum antibody titers are not good indicators of EPM. This discrepancy is not likely due to differences in the serologic assays, but rather can be attributed to differences in the authenticity of the sample sets used for the studies. Specifically, the sample set used by Duarte et al.[344,347] had only 12 seropositive horses in the non-EPM population (N = 97), which does not represent the seroprevalence of *S. neurona* that has been documented in many parts of the United States and Central and South America (Table 3.6). In most regions where *S. neurona* is endemic, there will be numerous seropositive horses that exhibit signs of neurologic disease not due to EPM, and some of these horses will have serum antibody titers that are quite high. When sample sets that better represent the normal equine population are tested, it becomes apparent that serum antibody titers against *S. neurona* do not provide accurate EPM diagnosis.[912,1453]

3.2.12.2 Necropsy Examination

3.2.12.2.1 Gross Lesions

When present, the gross lesions of EPM in the horse are confined to the CNS. Acute lesions consist of multifocal, randomly distributed foci of hemorrhages, whereas subacute and chronic lesions show areas of discoloration ranging from pale to dark tan areas, and foci of malacia, respectively.[438] Although the brainstem is more often involved than other areas of the brain (Figure 3.3), the lesions are more frequently seen in the spinal cord. In rare cases, lesions can be present in both the brain and the spinal cord of a horse.

3.2.12.2.2 Microscopic Lesions

Microscopically, the predominant lesions range from multifocal to coalescing areas of hemorrhage, nonsuppurative inflammation, and small foci of necrosis. Perivascular cuffing by mononuclear cells is evident in some of the affected areas, particularly in the meninges. The numbers of *S. neurona* stages present are often few and difficult to locate in routine histological sections stained with H and E. Developmental stages of *S. neurona* are more easily seen if organisms are present in neurons rather than in inflammatory cells. The types of host cells infected are not definitively known, except that schizonts and merozoites are found in neurons and in macrophages.

3.2.12.2.3 Immunohistochemical Staining

S. neurona schizonts and merozoites are stained specifically with polyclonal rabbit antibodies raised against culture-derived *S. neurona* merozoites.[420,422,2044] The reactivity is less intense using polyclonal rabbit antibodies against bradyzoites and sporozoites (Dubey, J.P.-own observations). The reactivity of *S. neurona* sarcocysts to *S. neurona* antibodies is highly variable; it was found that immature sarcocysts reacted brilliantly, whereas mature sarcocysts were stained irregularly or not at all within the same histological section.[472,1673,1739] Monoclonal antibodies against *S. neurona* can also be used for immunostaining.[1120] In most EPM cases organisms are sparse or undetectable even when diagnosis is confirmed by bioassay in cell culture or *S. neurona* DNA is detected. Not all *S. neurona* organisms in EPM cases stain by immunohistochemistry, possibly due to antibody recognition of specific parasite antigens or lack of antigens in the tissue sections. For example, the Sn-Mu1 isolate (SnSAG1 minus) did not stain using polyclonal antibodies prepared against the Sn-UCD1 isolate (SnSAG1 positive).[1119] However, multiple isolates and histopathology sections from a variety of hosts containing *S. neurona* merozoites or schizonts are recognized using Mab 2G5-2, which targets a conserved *S. neurona*.[1196]

3.2.13 Treatment

Treatment of horses suspected to have EPM should be done as quickly as possible after clinical signs of the disease are recognized. Treatment appears to result in successful recovery in 70%–75% of the affected horses, although these estimates are somewhat suspect without postmortem confirmation.

Antiprotozoal drugs, such as sulfonamides and pyrimethamine, have been used to treat EPM since 1974, when the protozoal etiology of the disease was recognized. After the first cultivation of *S. neurona* in 1991, it became possible to test antiprotozoal drugs *in vitro*. Subsequently, when the KO mouse model was established in 2000, it became possible to test drugs against *S. neurona* *in vivo*.[438] Since then, various compounds have been developed and licensed for treatment of EPM in the United States. In addition to ponazuril (Marquis, Bayer), which was the first FDA-approved EPM medication, diclazuril (Protazil, Merck) and a sulfadiazine/pyrimethamine combination drug (ReBalance, PRN Pharmacal) are currently approved for treatment of EPM.

3.2.13.1 Ponazuril

Ponazuril (Marquis®, Bayer Animal Health) is widely used to treat EPM. The dosage commonly used is 5 mg/kg per day for a minimum of 28 days. Ponazuril is a benzeneacetonitrile compound that is related to the herbicide atrazine and may act by inhibiting apicoplast and/or mitochondrial function in the parasite.[1202] Ponazuril is technically considered to be a coccidiostat.[1092,1431] The significance of this mode of action *in vivo*, however, is totally unknown and is probably unimportant because efficacy studies between coccidiocidal and coccidiostatic drugs provide similar outcomes.

Ponazuril is well absorbed orally and achieves a steady state concentration of 0.16 ± 0.06 mg/L in the CSF of horses treated with 5 mg/kg body-weight.[643] *In vitro* studies have documented anti-*S. neurona* activity of ponazuril.[1073,1202] *In vivo* studies in KO mice inoculated with *S. neurona* sporocysts indicated reduction in the severity of clinical signs if the drug was administered 4 DPI.[624] Prophylactic administration of Ponazuril at 5 mg/kg also reduced severity of clinical signs in experimentally infected horses.[647] The time to reach steady-state concentrations of Ponazuril in the CSF of horses is approximately 1 week. For this reason, an initial loading dose is recommended in an effort to achieve therapeutic concentrations more quickly. While the clinical value of this approach has not been demonstrated, maximum CSF concentrations of Ponazuril were achieved within 28 h when horses were given a loading dose of 15 mg/kg body weight (3 times the normal dosage). Additionally, feeding 60 mL of corn oil immediately before administering Ponazuril results in blood concentrations that are 25% higher than if no corn oil is given. A field efficacy study of 101 horses demonstrated approximately 60% efficacy, with 8% of cases relapsing within 90 days if treatment is stopped after 28 days.[645] It is expected, although not proven, that the relapse rate (already low) will be even less if treatment is continued for longer than 28 days. Animals typically responded within 10 days of beginning treatment, although some horses may take up to 3 weeks, following the initiation of treatment, before tangible evidence of response is noted. Therefore, a refractory patient often continued to improve even after treatment stopped at 28 days. The baseline neurologic score did not influence outcome in that study. However, success was defined as improvement by 1 clinical grade, which may be considered unacceptable in severe cases.[645] In clinical cases, the animal should be reevaluated at the end of the treatment period, and then a determination made whether further treatment is needed. In general, if there has been a clinical response, yet the horse remains abnormal, a second month of Ponazuril is recommended. If finances are limited, treatment can be stopped after 28 days, but the horse should be reexamined after 1 month to ensure that there is no deterioration.

Toxicity studies have found Ponazuril to be very safe, with no systemic toxicity, even at high doses (30 mg/kg body-weight) for up to 56 days.[950] Uterine edema was noted in mares given 30 mg/kg body-weight each day for 30 days[950]; however, treatment of breeding stallions with 10 mg/kg body-weight did not affect androgenic hormone production nor spermatogenesis.[1819] Ponazuril has been used without obvious problems in pregnant mares, but the use of Ponazuril in pregnant animals is off-label, and owners should be made aware of this fact. As stated earlier, Ponazuril was used successfully to treat a sea lion with *S. neurona*-associated clinical disease.[1267]

3.2.13.2 Diclazuril

Diclazuril (Protazil®, Merck Animal Health) is also a benzeneacitonitrile compound that is chemically similar to Ponazuril. Diclazuril is administered at 1 mg/kg body-weight as alfalfa-based pellets that are top dressed in the daily grain ration. In a study of 49 horses, Diclazuril had a success rate of 58%, when success was defined as improvement by a minimum of 1 clinical grade.[1092] Recent investigation of Diclazuril pharmacokinetics has shown that a dose of 0.5 mg/kg body-weight (50% of the recommended dose) is sufficient to attain concentrations in plasma and CSF that will inhibit *S. neurona* growth in culture.[860] Further study is needed, but these findings suggest that effective treatment of EPM might be possible with lower doses of Diclazuril.

Diclazuril is absorbed quickly, especially the sodium salt, and has been found in serum 1 h after feeding to horses[324] Diclazuril has anti-*S. neurona* activity *in vitro*[1071] and *in vivo*.[439] Diclazuril in pelleted feed (50 parts per million), starting 5 days before or 7 days after feeding *S. neurona* sporocysts to mice, and continuing therapy for a month, resulted in the absence of *S. neurona* stages in the mice. Therapy was less effective when Diclazuril was given 12 days or more after feeding sporocysts. These results indicate that Diclazuril can inhibit the early stages of *S. neurona* and might be useful as a prophylactic against *S. neurona* infections in horses.[439]

3.2.13.3 Nitazoxanide

The broad-spectrum antimicrobial nitazoxanide (NTZ) and its other derivatives have anti-*S. neurona* activity *in vitro*.[662] NTZ was approved for the treatment of EPM and marketed as Navigator® (Idexx Pharmaceuticals, Inc.). Although treatment with NTZ was effective against EPM, there were prominent concerns about adverse side-effects (e.g., colic). Subsequently, NTZ was removed from the market and is no longer available.

3.2.13.4 Decoquinate

Decoquinate has been used to treat coccidiosis in poultry, cattle, sheep, and goats. *In vitro* testing has determined that it is effective against *S. neurona* at low concentrations.[1081] Decoquinate, in combination with the immunomodulator levamisole, has been investigated for treatment of EPM in 1 published study, with a very high rate of clinical improvement reported after only 10 days of treatment.[503] There are substantial concerns with regard to these findings, however, including case selection and assessment, and the diagnostic standard applied. Substantial additional research using confirmed cases needs to be performed before the use of this drug combination can be considered.

3.2.13.5 Sulfonamide and Pyrimethamine

The sulfonamide and pyrimethamine (S/P) combination has been used widely for the last 4 decades, but a premixed version of this combination has recently achieved FDA approval and can be purchased commercially (ReBalance®, PRN Pharmacal). The sulfonamide component of this compound competes with para-aminobenzoic acid to inhibit dihydropteroate synthetase activity, while pyrimethamine targets dihydrofolate reductase, which collectively inhibit folate metabolism. The synergistic effects of these compounds block synthesis of nucleic acids and amino acids, ultimately leading to parasite death. A field efficacy study reported a success rate of 57% after several months of treatment.[1092] The weaknesses of S/P suspension in the treatment of EPM include the prolonged duration of treatment required to affect a positive response, and the toxicity of the compound. Toxicity includes anemia, leukopenia, fetal loss, and fetal abnormalities. A benefit of S/P combinations is the lower cost, yet this may be offset by the extended dosing interval required.

Pyrimethamine has historically been given in combination with sulfa drugs in the treatment of EPM. There is, however, some evidence to suggest a synergistic effect of pyrimethamine when used with the benzeneacitonitrile-group antiprotozoals (Ponazuril, Diclazuril). While not empirically evaluated in horses with EPM, so that its effectiveness is not known at this time, it might be beneficial to add this drug to the treatment regimen in refractory cases.

3.2.13.6 Pyrantel Tartrate

Based on *in vitro* activity of pyrantel tartrate against *S. neurona* merozoites,[980] administration of pyrantel tartrate was considered for prophylactic use against EPM. However, this compound had no anti-*S. neurona* activity in experimentally infected KO mice.[1075] Similarly, daily administration of pyrantel tartrate for 134 days had no anti-*S. neurona* activity in experimentally infected horses.[1501]

3.2.13.7 Supportive/Ancillary Therapy

A variety of ancillary and supportive therapies may be indicated for the treatment of horses with EPM.[1581] Ancillary treatments may include various anti-inflammatory drugs such as phenylbutazone, flunixen meglumine, dimethylsulfoxide (DMSO) or steroids. Corticosteroids can be used to help stabilize horses with serious neurologic abnormalities during the early period of treatment. Long-term steroid treatment should be avoided due to their unknown effects upon immune clearance of *S. neurona*. However, it has been observed that some horses initially worsen slightly and transiently with treatment; hence, prophylactic use of nonsteroidal anti-inflammatory drugs to ameliorate this "treatment crisis" is sometimes advised. In seriously affected horses, prophylactic anti-inflammatory treatment is probably advisable for the first week. Additional ancillary treatments, such as Vitamin E, homeopathic medications, etc., have not been demonstrated to have any value, and use of these compounds or approaches is not recommended.

Immunostimulants have been recommended by some on the presumption that immunosuppression is a component of the pathophysiology of EPM. Levamisole (1 mg/kg orally daily), EqStim (5 mL IM on day 1, 3 and 7, then monthly), or Equimune IV, (1.5 mL intravenously weekly for 3 weeks) have all been advocated; however, there is no specific information to suggest that these have any positive effect. Another potential beneficial immune modulator is Parapox Ovis Virus Immunomodulator (Zylexis, Pfizer Animal Health). While licensed as an aid in the treatment of horses with EHV-1 and EHV-4, Parapox Ovis has been demonstrated to increase interferon gamma secretion in treated horses. As interferon gamma is considered a key cytokine in protection against *S. neurona* infection, this compound might be beneficial in selected animals.

3.2.13.8 Duration of Treatment

The duration of treatment for EPM is difficult to determine, and when to terminate treatment in a particular horse remains problematic. Duration of treatment appears to be more important than peak concentrations (as long as they exceed minimum inhibitory concentrations). Therefore, horses should be evaluated after 1 month of treatment with Ponazuril or Diclazuril. If improvement has been noted, but clinical signs remain, then a further month of treatment is recommended. If finances are limiting, or the horse appears clinically normal, then treatment can be discontinued. However, the horse should be examined one month after treatment is completed to ensure that there has been no relapse. Alternatively, a 1–2-month course of S/P can be given to help minimize the chance of relapse.

Treatment until the CSF–WB becomes negative has been advocated by some clinicians in the past. Experience has dictated that this is an unachievable goal for most cases. Antibodies can have a lengthy half-life, so horses will exhibit positive titers for long periods. The effects of treatment upon CSF or serum antibody concentrations is unclear, and, at the present time, repeating these tests appears to have little value in determining treatment success or duration.

3.2.14 Relapse of Clinical EPM in Horses

Horses can relapse with clinical EPM, even after prolonged therapy, remarkably with clinical signs similar to first episode. The cause of relapse is unknown because the encysted stage, sarcocyst, has not been confirmed. Whether schizonts can remain dormant in equine tissues is unknown.

3.2.15 Epidemiology of EPM in Horses

Few epidemiologic studies have assessed the risk factors associated with clinical EPM in horses in the United States.[1224,1536,1537,2283,2284] In 1998, a National Animal Health Monitoring System (NAHMS) study examined data collected from 1178 U.S. equine industry operators representing 83.9% of horses from 28 U.S. states and 51.6% of operations with 3 or more horses. A total of 164 operators reported information concerning the most recent EPM Case and a Control Horse, and 974 operators completed survey questions regarding a Control Horse from NonCase Premises. An extensive statically valid risk assessment was performed, and results were reported.[1224] The following conclusions were drawn:

a. EPM risk on operations is closely tied to environmental and management factors that impact exposures to opossums, their feces, and their environment. This finding suggests opossums, or contamination from these animals, may be more commonly found on premises where EPM has occurred. Similarly, climate and terrain factors impact suitability of opossum habitat as well as the likelihood that *S. neurona* sporocysts will survive in the environment.

b. Housing of horses, choice of bedding material, methods of storing concentrate feeds, and equine stocking density, all could impact the likelihood of horses being exposed to *S. neurona*, either through direct exposure in the environment or indirectly through exposure to contaminated feed and bedding.

c. Horses that were used primarily for racing had a greater risk compared with other horses, as did thoroughbred and warmblood horses.

d. When comparing horses diagnosed with EPM to non-neurologic controls, young horses (1–5 years old) and older horses (>13 years) had a higher risk of developing EPM than horses <1 year and 6–13 years of age; the youngest horses had nearly 5 times greater odds of developing EPM, and horses 18 months to 5 years old had twice the odds of developing EPM, when compared with horses >5 years old. Horses that were transported <1 month before disease onset or the date of interview did not have a detectable difference in disease risk compared with horses that had been transported at >6 months before these dates, but there was about 2.6 times greater odds of disease if owners reported transporting horses 1–6 months previously. There was also a 4-times greater odds of disease if owners could not remember when horses were transported, in comparison with horses transported >6 months previously.

Finding age-related differences in the likelihood of disease is suggestive of repeated exposure to *S. neurona* over time, and the subsequent development of specific immunity. Several signalment variables were associated with EPM occurrence in individual horses (e.g., age, sex, breed, and primary use), but only 1 of the variables marking hypothetical stressful events (e.g., participating in competition) was considered most important.

e. The variables most strongly associated with EPM occurrence were maximum stocking density of turnout areas, number of resident horses, primary use of resident horses, finding evidence of raccoons or skunks on premises, and being located near a marsh. It was expected that EPM cases were more likely to be identified on premises with more horses, because the likelihood of identifying a case increased with the horse population size. Accordingly, operations with >20 horses were 2–3 times more likely to identify an EPM case, when compared with smaller operations. Operations that had 2:1.4 horses/acre in turnout areas, were 5–10 times more likely to report an EPM case than were operations where turnout areas were less densely aggregated, or operations that did not have turnout available. Operations where the primary use of resident horses was for show, modest mean temperatures had the lowest odds of disease, and those with moderate temperatures (>9.4°C and <12.0°C) had the highest odds of disease, followed by those with the lowest mean annual temperatures. Operations that housed horses in stalls or paddocks during the day had small increases in the odds of EPM occurrence, when compared with operations that housed horses in pastures. Operations that stored concentrate feeds in rodent-proof containers had about 3 times lower odds of detecting EPM than those that did not. Operations that used wood products (e.g., shavings, chips, or sawdust) as the predominant bedding material had about 1.4 times lower odds of detecting EPM than did those that used other bedding materials (e.g., straw, corn stalks).

f. Horses used for breeding had about half the odds of disease compared with those used for pleasure, but there were no detectable differences in the odds of disease among horses used for pleasure, show or competition, farm or ranch work, or other uses. Thoroughbred and warmblood breeds had about 4 times greater odds of disease, compared with quarter horses; there was no detectable difference in disease risk when comparing other breeds. As expected, regional ecology was also associated with the likelihood of EPM occurrence in this final model; horses were twice as likely to have EPM if they resided in states from region A, than if they lived in the other 3 regions.

3.2.16 Prevention and Control

Preventing contamination of feed and water with opossum feces is essential to prevent EPM in animals. Opossums can excrete millions of sporocysts in their feces for months. Sporocysts are resistant to environmental influences, and most commonly used disinfectants do not kill *S. neurona* sporocysts (Table 3.11). Heating to 60°C for 1 min will kill sporocysts, but exposure to 55°C for 5 min will not. Although survival of sporocysts in different environmental conditions outdoors has not been tested, sporocysts remained viable at 4°C for months (Table 3.11).

Currently, there is no vaccine for EPM. A killed whole *S. neurona* merozoite vaccine was conditionally marketed by Fort Dodge in the United States, but the product is now withdrawn because of the difficulty of obtaining efficacy data.

Table 3.11 Resistance of S. *neurona* Sporocysts

Reagent/Condition	Duration	Viability[a]	Reference
5.25% NAOCl (bleach)	1 h	Killed	**449**
12.56% phenol (Wex-cide)	6 h	Live	
2% chlorhexide (Nolvasan)	6 h	Live	
1% iodine (Betadine)	6 h	Live	
5% O-benzyl-p-chlorphenol (TB-plus)	6 h	Live	
Formalin 10%	6 h	Live	
6% benzyl ammonium chloride (NPD)	6 h	Live	
Ammonium hydroxide			
100%	1 h	Killed	
10%	6 h	Live	
Thermal exposure			
60°C	1 min	Killed	
55°C	5 min	Live	
55°C	15 min	Killed	
50°C	60 min	Killed	
4°C	34 months	Live	**515**
4°C	44 months	Killed	

Source: Adapted from Dubey, J.P. et al., 2015. *Vet. Parasitol.* 209, 1–42.
[a] Determined by bioassay in KO mice.

3.3 *SARCOCYSTIS CANIS* DUBEY AND SPEER 1991 INFECTIONS IN ANIMALS

3.3.1 Introduction

Historically, a sarcocystosis-like illness was diagnosed in a dog from Maryland, United States.[405] In October 1989, 5 of the 8 pups born to a Rottweiler bitch became ill. Pup #1 had surgery for entropion (inverted eyelid) at 7 weeks of age and developed diarrhea and became anorectic. It was anemic and had slightly elevated liver enzymes. The pup was euthanized. At necropsy, only the liver and small intestine were collected for histological examination. The liver was enlarged grossly and had multiple foci of discoloration. Histologically, there was severe hepatic necrosis, associated with a *Sarcocystis*-like organism. Pup #2 died of massive toxoplasmic lympadentitis. The cause of illness in pups #3–5 was undetermined.

Dubey and Speer[400] named the parasite associated with hepatitis S. *canis* to differentiate it from S. *neurona* found in the neural cells of horses.[404] Subsequent investigations indicated that S. *neurona* can cause illness in several hosts other than horses (see previous sections in 3.2), and the diagnosis could be confirmed by PCR and an IHC test using S. *neurona*-specific antibody. The S. *neurona* antibody does not react immunohistochemically with S. *canis* in the liver of dogs.[460]

Since the initial report of S. *canis*-associated hepatitis in the Maryland dog, similar cases were diagnosed in a chinchilla, black bear, polar bear, sea lion, horse, Steller sea lion, and a Hawaiian monk seal from the United States (Table 3.12). Additionally, a case was diagnosed in a striped dolphin from Spain and in a dog from Costa Rica (Table 3.12). In all of the reports listed in Table 3.12, hepatitis was the only condition diagnosed and the parasite was confined to hepatocytes. Because the parasite has not been cultivated in cell culture or in laboratory animals, there are no specific

Table 3.12   Reports of *S. canis*-Like Infections in Animals

Host	Country	Signalment	Clinical	Diagnosis	Reference
Dog (*Canis familiaris*)	Maryland, USA	10-month-old Rottweiler	Diarrhea, listless, anorexic	Histology[a] (*S. neurona* antibody negative)	**400, 405, 460**
	Kansas, USA	3-month-old mixed breed, four dogs from 3 months old to 17 weeks old	Suspected toxoplasmosis	Histology[a] (*S. neurona* antibody negative but 3 more reacted positively)	**460, 1759**
	Colorado, USA	12-week-old Golden Retriever	Fever, lethargy, elevated enzymes, eosinophilia	Histology[a]	**28**
	Heredia, Costa Rica	10-week-old Rottweiler	Diarrhea, excessive salivation, convulsions, concurrent canine distemper	Histology	**98**
Polar bear (*Ursus maritimus*)	Anchorage, Alaska, USA	Two adult bears	Lethargy, anorexia	Histology, TEM	**663**
Black bear (*Ursus americanus*)	South Dakota, USA	2-year-old	Found dead	Histology	**1874**
	British Columbia, Canada	13-year-old, captive	Acute onset of vomiting and hepatic signs	Histology, PCR	**305**
Sea lion (*Zalophus californianus*)	Florida, USA	10-year-old	Found dead	Histology, TEM, also sarcocyst	**1184**
Chinchilla (*Chinchilla* sp.)	Georgia, USA	7-month-old	Found dead	Histology	**1443**
Horse (*Equus caballus*)	California, USA	4-year-old Trakehner	Fever, abdominal discomfort, enzymes elevated	Histology[a], TEM	**306**
Striped dolphin (*Stenella coeruleoalba*)	Mediterranean coast, Spain	80 kg adult female	Found dead	Histology[a]	**1465**
Hawaiian monk seal (*Monachus schauinslandi*)	Hawaii, USA	21-month-old	Gastrointestinal	Histology[a], TEM	**1860**
Steller sea lion (*Eumetopias jubatus*)	Alaska, USA	1100–1500 kg	Found dead	Histology, PCR	**1818**

[a]   Immunohistochemistry negative with *S. neurona* antibody.

diagnostic tests. Partial DNA characterization of the parasite from the liver of a polar bear that died of massive protozoal hepatitis revealed that the 18S and ITS-1 were distinct from *S. neurona* and other *Sarcocystis* species.[460] Based on IHC staining and DNA characterization, it was revealed that out of the 10 cases of *S. canis*-like infections in dogs, only the 2 dogs listed in Table 3.12 were *S. canis*.

   T. gondii, *N. caninum*, and *S. neurona* infections are frequent in marine mammals: That is why a differential diagnosis with respect to *S. canis* is a major concern.[455]

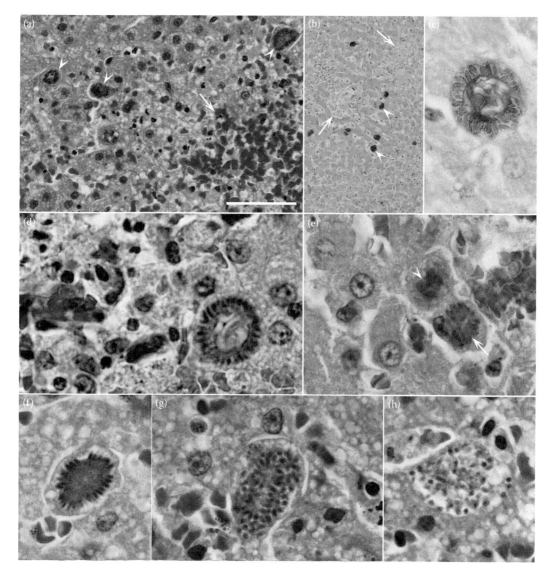

Figure 3.11 *Sarcocystis canis*-associated lesions in a sea lion (a,f,g,h described in Reference **1184**) and black bear (b,c,e described in Reference **1874**). (a,d–h). H and E stain. (b and c immunostaining with anti-*S. cruzi* rabbit polyclonal antibody. Bar in (a) = 100 µm, bar in (b) = 200 µm, and bar In (c–h) = 20 µm. (a) Hemorrhage (arrow) and several schizonts (arrowheads) at the periphery of the lesion. (b) Necrosis (arrows) and several schizonts (arrowheads). (c) Schizont with staining of both the residual body and merozoites reacting to *S. cruzi* antibody. (d) Inflammatory focus on left and a schizont on the right. (e) Two immature schizonts. Note lobulation of the nucleus (arrowhead) and budding of merozoites from the nucleus (arrow). (f) Merozoites budding from a large residual body. (g) Mature schizont without a residual body. (h) Ruptured schizont with cross sections of merozoites. Merozoites in (g) and (h) are indistinguishable from tachyzoites of *Toxoplasma gondii*.

3.3.2 Structure and Life Cycle

Only the schizont stages are known and the parasite multiplies in hepatocytes (Figure 3.11). The schizogonic development is typical of other *Sarcocystis* species. After merozoite development, a large eosinophilic residual body is left. By TEM, merozoites lack rhoptries. As indicated earlier, *S. canis* does not react with antibodies to *S. neurona*, *T. gondii*, and *Neospora caninum* antibodies, but react to *S. cruzi* antibodies (Figure 3.11c).

BIBLIOGRAPHY

Information pertinent to the subject matter of this chapter may be found in the following references:

28, 32, 44, 45, 72, 83, 87, 88, 92, 93, 98, 104, 114, 123, 130, 137, 138, 141, 145, 146, 153, 157, 165, 202–205, 207, 208, 210, 217, 230, 247–249, 263, 270–272, 279, 292, 293, 305–308, 324, 328, 333, 343–349, 387, 393, 398, 400, 402, 404, 405, 408, 411, 412, 416, 417, 419–422, 429–431, 433–435, 438–440, 442–450, 454–456, 460, 464, 468, 470, 472, 475, 493, 498–503, 511–515, 517, 518, 520, 522–524, 600–603, 606, 617, 624, 637, 642–648, 656, 657, 661–663, 668, 670, 671, 680, 684, 726–729, 731, 752, 753, 777–779, 800, 835–838, 847–850, 852, 870, 912, 915, 945, 950, 978, 980, 1023, 1026, 1054, 1055, 1062, 1068, 1070–1083, 1090, 1091, 1092, 1096, 1107, 1108, 1113–1121, 1128, 1146–1148, 1184, 1194–1198, 1201, 1202, 1210, 1220, 1224, 1225, 1239, 1262, 1263, 1267, 1279, 1329, 1333, 1395–1397, 1430–1443, 1445, 1453, 1463–1465, 1467, 1489, 1491, 1492, 1494–1502, 1535–1540, 1554, 1555, 1574, 1578–1581, 1590, 1611, 1636, 1661–1663, 1673, 1674, 1677, 1693, 1722, 1723, 1739, 1742, 1750, 1755, 1756, 1759, 1761, 1765, 1770, 1771, 1779, 1780, 1818–1821, 1829–1833, 1843, 1850, 1855, 1860, 1862, 1863, 1874, 1876, 1987, 1988, 1992–1994, 2044, 2085, 2115, 2121, 2122, 2123, 2153–2155, 2177, 2178, 2246, 2262, 2267, 2283–2285, 2297, 2303, 2321, 2377–2388, 2390, 2397, 2403.

CHAPTER 4

Sarcocystosis in Humans (*Homo sapiens*)

4.1 INTRODUCTION

There are 2 known species of *Sarcocystis, S. hominis,* and *S. suihominis*, for which humans serve as the definitive hosts. Humans also serve as the accidental intermediate or aberrant host for several unidentified species of *Sarcocystis.*

4.2 INTESTINAL SARCOCYSTOSIS

4.2.1 *Sarcocystis hominis* (Railliet and Lucet 1891) Dubey 1976

This species is acquired by ingesting uncooked beef containing *S. hominis* sarcocysts. The structure and life cycle are described in Chapter 7.

S. hominis is only mildly pathogenic for humans. A volunteer who ate raw beef from an experimentally infected calf developed nausea, stomachache, and diarrhea 3–6 h after ingesting the beef; these symptoms lasted 24–36 h. The volunteer excreted *S. hominis* sporocysts between 14 and 18 days after ingestion. During the period of patency, he had diarrhea and stomachache.[807] Somewhat similar but milder symptoms were experienced by other volunteers who ate uncooked, naturally infected beef.[40,821,1478]

Similar symptoms were reported in people from Brazil[1382] who ingested raw beef or raw kibbe made from uncooked beef. Of 7 volunteers in Brazil, who ate raw kibbe (beef), 6 began to excrete *S. hominis* oocysts/sporocysts 10–14 days later, and continued to excrete them for 5–12 days.[1382]

4.2.2 *Sarcocystis suihominis* (Tadros and Laarman 1976) Heydorn 1977

This species, acquired by eating undercooked pork, is more pathogenic than *S. hominis.* Its structure and life cycle are described in Chapter 6.

Human volunteers developed hypersensitivity-like symptoms—nausea, vomiting, stomachache, diarrhea, and dyspnea—within 24 h of ingestion of uncooked pork from naturally or experimentally infected pigs. Sporocysts were excreted 11–13 days after ingesting pork.[40,821,1394,1478] Similar symptoms were reported in people from China[1053,1057,1884] who ate *S. suihominis*-infected raw pork.

In India, the prevalence of *S. suihominis* in pigs and humans is high in an economically deprived sect,[64,1638] probably linked to slaughter practices. Street sweepers raise pigs that clean the streets and eat everything, including human waste. When the pigs are slaughtered, usually in the backyards of the owners, who live in slum conditions, the owners' children often gather around the slaughter to help with chores. In turn, the children are invariably rewarded with the tail of the slaughtered pig, some of which were found to be infected with *S. suihominis* sarcocysts and which the children consumed on the spot of slaughter.

4.2.3 Other Zoonotic *Sarcocystis* Species

In China, 2 volunteers who ate raw water buffalo meat excreted sporocysts; the buffalo had been fed sporocysts from a human who had ingested raw meat from a naturally infected bovine, but it is not clear if this buffalo was free from environmentally derived infection of a possible human *Sarcocystis* specific for the water buffalo. Two human volunteers who ingested raw *Sarcocystis*-infected buffalo meat became ill but did not excrete sporocysts[211]; the illness could be due to other foodborne microbes.

4.2.4 Natural Prevalence of Sporocysts in Human Feces

Reports on this topic are summarized in Table 4.1.

4.3 MUSCULAR SARCOCYSTOSIS

Until 1999, sarcocysts found in striated muscles of human beings were incidental findings, mostly reports of individual cases. Reports until 1978 were reviewed critically[84] and are included here for the sake of completeness (Table 4.2). Subsequent reports were summarized[595,596] (Table 4.3). Additionally, serologic findings were summarized[1410] for people from Malaysia,[1740] the former Yugoslavia,[1607] and Australia.[1264] All infections in humans until 2013 were reported as intramuscular sarcocysts of unknown species. The morphologies of 7 distinct types of sarcocysts were described from light microscope observations of histological sections from human tissues.[84]

A series of 3 outbreaks of acute sarcocystosis was first reported in 1999. In the first outbreak, 10 members of a United States Air Force team deployed for 1 week to a remote jungle village about 80 km northeast of Kuala Lumpur[36] became ill within 3 weeks after returning from deployment. They had fever, myositis, and bronchospasm, with elevated liver enzyme levels and eosinophilia. Seven members, 1 of whom was asymptomatic, from the jungle cohort were tested for eosinophilia, sedimentation rate, and serum enzyme levels, including ALT, CK, LDH, and AST, and virtually all had at least some levels elevated above the normal. Extensive laboratory investigation did not reveal a specific diagnosis. Myositis and immature sarcocysts were observed histologically in a muscle biopsy 3 months after the onset of the illness (Figure 4.1).

Another outbreak of acute muscular sarcocystosis involved college students and teachers who visited Pangkor Island off the west coast of peninsular Malaysia.[14,877] Of the 92 persons who attended a retreat on the island, 89 became symptomatic, with onset of fever (94%; 57% relapsing), myalgia (91%), headache (87%), and cough (40%), within 26 days of their return to the mainland. Eight persons, who had visible facial swelling for 4–6 weeks, showed changes, consistent with inflammatory edema, in their mastication muscles by "whole body" magnetic resonance imaging (MRI). Similar findings were observed in the back muscles of 4 persons and the calf muscles of 2 persons. A sarcocyst ~190 μm long was observed in cell culture inoculated with muscle homogenate from a patient, and a sarcocyst was observed in a histologic section of muscle from another patient. Neither had elevated serum CK levels, but both had eosinophilia.

Outbreaks in 2011 and 2012 involved at least 99 persons who vacationed on Tioman Island off the east coast of peninsular Malaysia; all were Europeans except for 2 Canadians and a British resident of Singapore.[559,560] The strict case definition of myositis, eosinophilia, and negative trichinellosis serology most likely underestimated the number of positive patients at 68. Myositis required at least 1 of the following: a complaint of muscle pain with a serum CK level greater than 200 international units per liter, muscle tenderness documented on physical examination, or histologic evidence of myositis in a muscle, confirmed by biopsy; 62 patients were considered probable, and 6

Table 4.1   Intestinal Sarcocystosis in Humans

Country	Human Subjects	Species	Source of Sarcocysts	Clinical Signs	Year of Diagnosis	Reference
Netherlands	12 of 72 persons 5 of 17 autopsies	*I. hominis*[a]	NI	NI	1962	1011
Germany	2 volunteers 4 volunteers	*S. hominis* *S. suihominis*	Beef Pork	1 of 6 persons had severe influenza-like symptoms and mild diarrhea	1972	1478
Poland	29-year-old F 7 (assumed children) 92 unidentified	*Sarcocystis* sp. *Sarcocystis* sp. *Sarcocystis* sp.	NI	Symptoms very scanty	1973	1398
Poland	200 persons	*I. hominis*[a]	NI	NI	1973	1399
Germany	1 volunteer	*S. tenella*	Sheep	None. No infection resulted from ingestion of macroscopic cysts from sheep	1974	1480
Germany	22 of 300 persons	*I. hominis*[a]	NI	10 of 22 persons had gastric and intestinal symptoms; 12, no clinical signs	1975	897
Germany	8 of 506 persons	*Sarcocystis* sp.	NI	Intermittent enteritis with diarrhea, rheumatic phenomena, or asymptomatic	1975	610
Poland	13 of 125 children 7 to 18 years old	*Sarcocystis* sp.	NI	NI	1976	1401
Germany	1 volunteer 3 volunteers	*S. hominis* *S. suihominis*	Beef Pork	Diarrhea and stomachache, bloating, inappetence, nausea, vomiting, diarrhea	1977	807
Slovakia	A 12-year-old female	*I. hominis*[a]	NI	Asymptomatic	1978	678
United States	2	*S. cruzi*	Beef	Not infectious, asymptomatic	1978	1038
Netherlands	5	*S. hominis*	Beef	Natural infection	1979	1714
	1	*S. hominis*	Beef	Experimentally infected with 15,000 sarcocysts		
France via tropical areas	2% of 3500 samples	*S. hominis*	NI	NI	1980	314
Thailand	30-year-old M 3-year-old F 60-year-old F 9-year-old M 19-year-old M 70-year-old M	*S. hominis* *S. hominis* *S. hominis* *S. hominis* *S. hominis* *S. hominis*	Beef Beef Beef Beef Beef Beef	All 6 had acute fever and acute abdominal pain, five had leukocytosis. Bacterial infection may have contributed to severity of illnesses	1982	152
China	48-year-old M	*Sarcocystis* sp.	Pork	Abdominal distension, pain, diarrhea, constipation, stomachache, dyspnea	1982	1884
	Adult M	*Sarcocystis* sp.	Unknown	No information		

(Continued)

Table 4.1 (Continued) Intestinal Sarcocystosis in Humans

Country	Human Subjects	Species	Source of Sarcocysts	Clinical Signs	Year of Diagnosis	Reference
Germany	8 of 403 persons	Sarcocystis sp.	NI	Intestinal complains	1983	667
China, Yunnan Province	2 of 12 persons	Sarcocystis sp.	Pork	Suspected schistosomiasis, abdominal distension, fatigue, diarrhea	1983	1885
China, Yunnan Province	Adult M volunteer	S. suihominis	Pork	No symptoms but excreted sporocysts infectious for pigs	1986	1057
China	1 volunteer, 2 monkeys	S. hominis	Beef	NI	1990	1059
Slovakia via North Vietnam	14 of 1228 workers	Sarcocystis sp.	Unknown	None	1991	1685
Tibet	20.5%–22.9% of 926	S. hominis	Beef	Asymptomatic	1991	1865
	0.6%–7% of 926	S. suihominis	Pork	Asymptomatic		
Laos	>10% of 1008	S. hominis	NI	NI	1991	679
China	123 of 414	S. suihominis	Pork	Asymptomatic	1991	1888
Australia	2 of 385 persons	Sarcocystis sp.	NI	NI	1993	1181
India	14 of 20 children, 3–12 years old	S. suihominis	Pork	Asymptomatic	1994	64
Thailand	23.2% of 362 Thai laborers	Sarcocystis sp.	NI, usual consumption of raw beef and pork	Asymptomatic	1996	1828
China	3 volunteers	Sarcocystis sp.	Cattle	All had clinical symptoms, including abdominal pain, distension, watery diarrhea, and eosinophilia	1999	210
	2 volunteers	Sarcocystis sp.	Water buffalo			
Brazil	6 of 7 volunteers	S. hominis	Kibbe (beef)	Diarrhea	2001	1382
Spain	1	S. hominis	Raw beef	Abdominal discomfort, loose stools	2001	218
China, Guangxi	27 (22 M, 5 F) of 501 persons older than 2 years	S. suihominis	Pork	8 of 27 had diarrhea and abdominal pain	2004	1051
China, Guangxi	32 of 489	S. suihominis	Pork	Asymptomatic	2005	1720
China	1 volunteer	S. suihominis	Pork	Distension, diarrhea, fever, pain	2007	1053

(Continued)

Table 4.1 (*Continued*) Intestinal Sarcocystosis in Humans

Country	Human Subjects	Species	Source of Sarcocysts	Clinical Signs	Year of Diagnosis	Reference
Thailand:						
Ubon Ratchathani	4.6% of 479	*Sarcocystis* sp	NI	NI	2007	**1764**
Khon Kaen	8% of 1124	*Sarcocystis* sp.	NI	NI		
Argentina	31-year-old HIV patient	*Sarcocystis* sp.	NI	Diarrhea, systemic muscular sarcocystosis	2008	**1782**
Argentina	0 of 69 children	NI	NI	Asymptomatic	2008	**1645**
China	2 volunteers	*Sarcocystis* sp.	Water buffalo	Became ill but did not excrete sporocysts	2011	**211**
NE Thailand:						
Khon Kaen	1 of 253 persons	*Sarcocystis* sp.	NI	NI	2013	**124**
Jordan	19-year-old M	*S. hominis*	Shawarma (beef)	Abdominal discomfort, diarrhea for 3 weeks	2014	**1278**
Malaysia	1 of 269 from indigenous subtribes	*Sarcocystis* sp.	NI	Asymptomatic	2014	**1035**

Source: Modified from Fayer, R., Esposito, D.H., Dubey, J.P., 2015. *Clin. Microbiol. Rev.* 28, 295–311.
NI = Not indicated.
[a] *Isospora hominis* was an early name used to identify sporocysts in feces and stages in lamina propria before the *Sarcocystis* species were known.
M = male, F = female

Table 4.2   Summarized Early Reports of Humans with Sarcocysts

Country	Patient Age, Gender	Sarcocyst Location	Biopsy or Autopsy	Sarcocyst Wall Type[a]	Original Author and Year[b]
Sudan	36-year-old, M	Abdomen	A	2	Kartulis (1893)
France	Adult, M	Larynx	A	1	Barbaran and St. Remy (1894)
France	Adult, unknown	Skeletal muscle	A	1	Vuillemin (1902)
Malaysia	30-year-old, M	Tongue	A	4	Darling (1919)
UK	Unknown	Heart	A	6	Manifold (1924)
India	55-year-old, M	Chest	B	2	Vasudevan (1927)
USA via West Indies	32-year-old, F	Heart	A	7	Lambert (1927)
Indonesia	20-year-old, M	Cheek	B	4	Bonne and Soewandi (1929)
China	Adult, M	Leg	B	1	Feng (1932)
UK	Adult, F	Heart	A	5	Hewett (1933)
Panama	11-year-old, F	Heart	A	7	Gilmore et al. (1942)
Panama	48-year-old, F	Heart	A	7	Kean and Grocott (1943)
Brazil	32-year-old, F	Heart	A	7	DeFreitas (1946)
USA via Germany	31-year-old, M	Heart	A	7	Arai (1949)
USA via Puerto Rico	34-year-old, M	Heart	A	7	
India	37-year-old, M	Leg	B	2	Dastur and Iyer (1955)
	20-year-old, M	Leg	B		
Sudan	45-year-old, M	Foot	B	3	McKinnon and Abbott (1955)
UK via India	60-year-old, M	Pectoral	B	2	McGill and Goodbody (1957)
Brazil	40-year-old, F	Heart	A	6	Koberle (1958)
Italy	17-year-old, F	Heart	A	5	D'Arrigo and Squillaci (1962)
Indonesia	51-year-old, M	Pectoral	B	4	Van Thiel and Van den Berg (1964)
UK via SE Asia	51-year-old, M	Leg	B	4	Mandour (1965)
Angola	30-year-old, F	Chest	B	2	Liu and Roberts (1965)
India	22-year-old, F	Leg, arm	B	2	Gupta et al. (1973)
SE Asia	21-year-old, M	Leg	B	3	Jeffrey (1974)
UK via Malaysia	34-year-old, M	Pectoral	B	4	
Malaysia	34-year-old, M	Larynx	B	4	Kutty and Dissanaike (1975)
Malaysia	12-year-old, F	Pharynx	A	4	Kutty et al. (1975)
India	56-year-old, M	Arm	B	3	Agarwal et al. (1976)
India	54-year-old, F	Leg	B	3	Thomas (1976)
Malaysia	20-year-old, M	Foot	B	4	Prathap and Dissanaike (1976)
Malaysia	23-year-old, M	Neck	A	4	Prathap and Dissanaike (1978)
Unknown	Unknown	Skeletal muscle	A	4	Frenkel (1976)
USA via Asia	40-year-old, M	Arm	B	4	McLeod et al. (1979)
Uganda	50-yr-old, M	Leg	B	1	Beaver (1979)
India	50-yr-old, M	Arm	B	3	
Singapore	Unknown	Skeletal muscle	B	4	
Singapore	Unknown	Skeletal muscle	B	4	
Costa Rica	9-yr-old, M	Heart	A	6	

Source: Complied from Beaver, P.C., Gadgil, R.K., Morera, P., 1979. *Am. J. Trop. Med. Hyg.* 28, 819–844.

A = Autopsy, B = Biopsy, F = Female, M = Male.

[a] Beaver et al.[84] reported 7 sarcocyst wall types that are not comparable with nomenclature stated by Dubey et al.[393] that classified the cysts wall by using TEM. Types reported by Beaver are: Type 1, thick, radially striated wall; large zoites often sparse in center; metrocytes. Type 2, sarcocysts generally large; thin, smooth wall; medium size zoites; septa. Type 3, sarcocysts small and medium size; thin wall, medium size zoites; septa. Type 4, sarcocysts long, medium size; thin wall; small zoites; septa evident in center. Type 5, sarcocysts small; thin wall; medium size zoites; in myocardium. Type 6, sarcocysts small to medium; thin wall; large zoites; in myocardium. Type 7, sarcocysts small; thin wall; small zoites; in myocardium.

[b] Original author and year can be found in Reference **84** and not all are included in the present Bibliography section.

Table 4.3 Summary of Reports of Muscular Sarcocystosis in Humans not Included in the review by Beaver[84] and Presented in Chronological Order of Publication

Country	No. of Patients, Age, Gender	Sarcocyst Location	Biopsy or Autopsy	Year of Diagnosis	Reference
USA	0 of 297 diaphragm specimens	Diaphragm	A	1978	**1738**
Malaysia	58-year-old, M[a]	Tongue	B	1981	**1374**
India	2 persons	Leg Gluteus	B B	1983	**20**
Denmark	4 of 112 specimens[b]	Muscle	A (not clear)	1985	**736**
Malaysia	45-year-old, M[a] 53-year-old, M[a]	Nasopharynx Nasopharynx	B B	1987	**1375**
Malaysia	1 person	Skeletal muscle	B	1988	**1376**
Australia via Thailand	31-year-old M	Skeletal muscle	B	1990	**1361**
Egypt	1 person	Skeletal muscle	B	1990	**2**
Malaysia	45-year-old, M[a] 67-year-old, F[a] 32-year-old, F[a]	Tongue Tongue Pectoral muscle	B B B	1992	**1377**
Malaysia	21 16–57 years	Tongue	A	1992	**1838**
Australia	0 of 50 specimens	NI	A	1992	**1762**
Belgium via Brazil, Kenya and Tanzania	31-year-old, M	Thigh	B	1995	**1773**
India	40-year-old, M 14-year-old, F 52-year-old, F 23-year-old, F	Thigh Arm Thigh Arm	B B B B	1996	**1179**
Malaysia	44-year-old, F[a] 19-year-old, F	Thigh Calf	B B	1998	**1602**
Malaysia	7 Adults, M	1 muscle	B	1999	**36**
Argentina	31-year-old HIV patient	Liver, blood smear	B	2008	**1782**
Thailand	66-year-old, M[a]	Larynx	B	2011	**1024**
India	20-year-old, M 50-year-old, M	Arm Neck	B B	2012	**1100**
India	47-year-old, M	Leg	B	2013	**60**
Malaysia, Pangkor Island	89 people	Leg	B	2013, 2014, 2014	**14, 877, 1034**
Malaysia, Tioman Island	39 F 29 M 4–72 years	Skeletal muscle from 15 patients; 1 PCR positive	B	2014	**560, 1624**
Malaysia, Tioman Island	3 F 3 M	Febrile myositis syndrome	NTE	2014	**1726**

Source: Modified from Fayer, R., Esposito, D.H., Dubey, J.P., 2015. *Clin. Microbiol. Rev.* 28, 295–311.
NTE = No Tissue Examined, A = Autopsy, B = Biopsy, F = Female, M = Male.
[a] Cancer patients.
[b] Tissues examined by trichinoscopy.

were confirmed by histological observation of intramuscular cysts compatible with sarcocysts. The onset of symptoms that clustered during the second week and the sixth week after returning from the island included fever (82%), myalgia (100%), and headache (59%), similar to the symptoms of visitors to Pangkor Island, as well as fatigue (91%) and arthralgia (29%). Blood eosinophilia and elevated serum CK levels were first observed during the fifth week postdeparture.

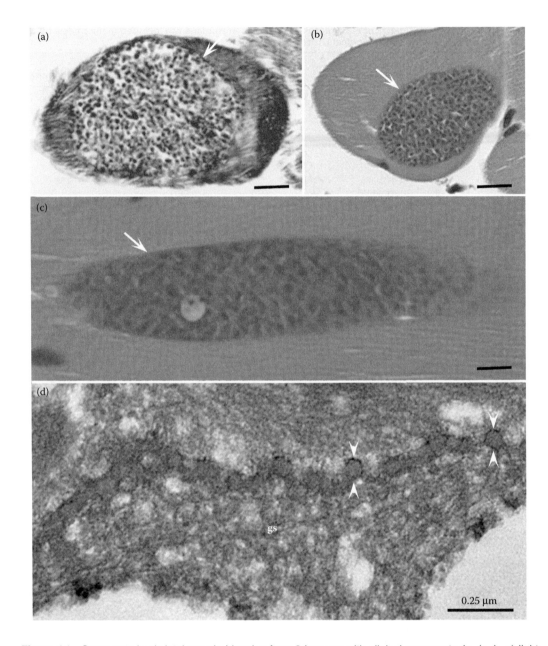

Figure 4.1 Sarcocysts in skeletal muscle biopsies from 3 humans with clinical sarcocystosis. (a, b, c) light microscopy, H and E stain. Bar = 10 μm. The sarcocysts are thin-walled (arrows). (d) TEM. The parasitophorous vacuolar membrane has tiny blebs (opposing arrowheads), and gs is relatively thin. ((a) Adapted from Arness, M.K. et al., 1999. *Am. J. Trop. Med. Hyg.* 61, 548–553. With permission; (b, c, d) from 2 patients reported in Esposito, D.H. et al., 2014. *Clin. Infect. Dis.* 59, 1401–1410; (b and c) Courtesy of Clifton P. Drew; (d) Courtesy of Cynthia S. Goldsmith.)

Additional details from 39 German patients with a history of travel to Tioman Island during 2011–2014, some of whom were included in the overall series during 2011–2012 (28 patients), were later reported.[1624,1891] These patients had a median duration of illness of 2.2 months (range 0–23 months), and 17% of patients had symptoms lasting >6 months; 2 patients had unresolved but diminished symptoms at 13 and 23 months. Severity of pain on a scale of 0–10 was a median of 6 (range 0–10). This represents the first data from multiple patients that indicate the duration of muscular sarcocystosis illness. Recently, an induction of Th2 cytokine polarization and biphase cytokine changes have been reported for muscular sarcocystiosis.[2402]

4.3.1 Evidence Linking Acute Sarcocystosis to *S. nesbitti*

S. nesbitti was originally described in muscles from a rhesus monkey from India, based on LM (see Chapter 5), but its taxonomic validity is questionable due to the lack of a good morphologic description. The first report of *Sarcocystis* in monkeys (*Macaca fascicularis*) in China was based on LM and TEM studies.[1859] Based on a perceived resemblance to *S. nesbitti*, the authors named the organisms they found in monkey muscles *S. nesbitti*. Morphological similarities suggested that a single species of *Sarcocystis* might infect *Macaca mulatta, M. fascicularis, Cercocebus atys,* and *Papio papionis,* as well as humans.[1743] When muscles containing sarcocysts identified as *S. nesbitti*[1859] were examined by molecular methods,[1743] DNA sequence data from 2 sarcocysts clustered with that of *S. atheridis* and *S. singaporensis* suggested that a snake might be a definitive host for *S. nesbitti*. The DNA sequence of *Sarcocystis* in feces from a cobra[1034] clustered with *Sarcocystis* from other snakes and with the *S. nesbitti* sequences obtained from monkey muscle.[1743] Sarcocysts were found in muscle biopsies from the swollen jaw of 1 patient and from the leg of another associated with the outbreak on Pangkor Island.[1034] The DNA sequences for *Sarcocystis* from these patients varied 1% from each other, and a BLAST result found they shared 99% homology with *S. nesbitti* from the muscle of *M. fascicularis*.

BIBLIOGRAPHY

Information pertinent to the subject matter of this chapter may be found in the following references:

Intestinal sarcocystosis: **40, 64, 124, 152, 210, 211, 213, 218, 314, 597, 610, 667, 678, 679, 807, 821, 866, 897, 919, 1011, 1012, 1035, 1038, 1051, 1053, 1057, 1059, 1181, 1278, 1382, 1394, 1398, 1399, 1401, 1478, 1480, 1638, 1639, 1645, 1685, 1707, 1714, 1720, 1764, 1782, 1828, 1865, 1884, 1885, 1888, 1954, 2095, 2222, 2314, 2332.**

Muscular sarcocystiosis: **2, 14, 20, 36, 60, 84, 559, 560, 595, 596, 736, 877, 1024, 1034, 1059, 1100, 1179, 1264, 1361, 1374–1377, 1410, 1602, 1607, 1624, 1725, 1726, 1738, 1740, 1743, 1762, 1773, 1781, 1838, 1859, 1891, 1916, 2104, 2161, 2264, 2324, 2348, 2355, 2389, 2402.**

Miscellaneous: **40, 1897, 1908, 1909, 1935, 1954, 2012, 2066, 2077, 2078, 2096, 2134, 2140, 2309, 2334.**

CHAPTER 5

Sarcocystosis in Nonhuman Primates

5.1 INTRODUCTION

Although there are numerous reports of muscular *Sarcocystis* infections in New World monkeys[943,961,1428] and Old World monkeys,[477,795,939,943,1104,1112,1175,1737] only 3 species have been named (Table 5.1). Morphological studies of sarcocysts indicate that there are more. In a review of histologic sections from 744 monkeys, including 374 captive wild monkeys, 4 structurally distinct sarcocysts were found in 21% of the wild captive monkeys, but none was found in laboratory-raised monkeys.[943] Identification of the species of the sarcocysts in primates is difficult because the definitive hosts are not known. Mehlhorn et al.[1175] reported 3 species of *Sarcocystis*, with thin-walled cysts found in rhesus monkeys (*Macaca mulatta*), baboons (*Papio cynocephalus*), and tamarins (*Sanuinus oedipus*), were ultrastructurally distinct (Table 5.1).

5.2 *SARCOCYSTIS KORTEI* CASTELLANI AND CHALMERS 1909

Intermediate hosts include *Macaca mulatta*, *Erythrocebus patas*, *Cercopithecus mitis*, and probably others.

The sarcocysts in skeletal muscles are up to 800 μm. The sarcocyst wall, which appears hairy, is 5- to 6-μm thick. The bradyzoites are $9–14 \times 3–4$ μm in size.[1104]

5.3 *SARCOCYSTIS NESBITTI* MANDOUR 1969

This species was found in of 2 of 44 *Macaca mulatta* from India.[1104] Sarcocysts were found in skeletal muscles but not in the heart. They were up to 1100 μm long. The serrated cyst wall was 1.2- to 1.6-μm thick. A similar parasite was found in 12 (24%) of 51 *M. mulata* imported into New York, United States, but the origin of the monkeys was not stated.[987] Sarcocysts were found in all muscles examined, but in all instances no more than 17 sarcocysts were found. Sarcocysts were thin-walled by LM. By TEM, vp of the sarcocyst wall were up to 60 nm long. Bradyzoites were 4.3–7.8 μm long.

In China, sarcocysts were found in 1 of 69 *M. fascicularis* that were born on a farm.[1859] Sarcocysts were 2.3–14 mm long, and up to 0.1 mm wide. By LM the sarcocyst wall was thin and smooth. By TEM the sarcocyst wall had small vesicular invaginations, 19–380 nm in diameter and 65–120 nm in depth, type 1.[1859] Bradyzoites were $7.5–10 \times 1.25$ μm by LM, and $3.3–6.5 \times 1.0–2.79$ by TEM. Yang et al.[1859] reviewed the morphology of sarcocysts found in monkeys presented by other authors, and concluded that the species they found was *S. nesbitti*; they did not mention references.[987] Because there are no photographs of the parasite named by Mandour,[1104] it is impossible to know if the same parasite was present in the above reports.

Table 5.1   Prevalence of *Sarcocystis* in Nonhuman Primates

Country	Host[a]	Origin	No. Tested	No. Positive (%)	Remarks	Reference[b]
Africa	*Macaca mulatta, Erythrocebus patas, Cercopithecus mitis*	NS	3	3	*S. kortei*	1104
China	*Macaca fascicularis*	Monkey farm	69	1 (1.4%)	*S. nesbitti*	1743, 1859
Germany	*Macaca mulatta*	Wild caught	20	2 (10.0%)	Cysts up to 1 mm, thin walled, TEM type 1, 6 μm long bradyzoites	1175
Germany	*Papio cynocephalus*	Wild caught	2	1 (50.0%)	Cysts up to 800 μm, TEM type 1a, folded vp, 6.4–9.0 μm long bradyzoites	1175
Germany	*Saguinus oedipus*	Wild caught	9	2 (22.2%)	Cysts up to 350 μm, TEM type 1a, papillary vp, 4–5 μm long bradyzoites	1175
India	*Macaca mulatta*	NS	44	2 (4.5%)	*S. nesbitti*	1104
Japan	*Saimiri sciureus*	Imported from animal dealer	37	7 (18.9%)	Histology in femoral muscle	1695
Japan	*Saimiri sciureus*	NS	10	8 (10.0%)	SEM, thin-wall sarcocysts	961
Malaysia	*Macaca fascicularis, Macaca irus*	NS, used in animal experimentation	50	7 (14.0%)	TEM, wavy and invaginated cyst wall	939, 1423
Mexico	*Macaca mulatta*	NS	1	1	Eosinophilic myocarditis, no description of cysts	795
South Africa	*Papio ursinus*	Wild caught	94	44 (46.8%)	Thick and palisade-like wall	1153
South Africa	*Cercopithecus pygerythrus*	NS	134	11 (8.2%)	*S. markusi*	1112
USA	*Macaca mulatta*	NS, used in animal experimentation	4	3 (75.0%)	Different stages of development	477
USA	*Macaca mulatta*	Imported from animal dealer	370	4 (1.1%)	*S. kortei*	765
USA	*Erythrocebus patas*	NS	1	1	Thick and striated wall	1428
USA	14 different species from Old and New World monkeys	Wild caught and laboratory born	774	0% in laboratory born, 21% in wild captive	4 structurally distinct sarcocysts	943
USA	*Macaca mulatta*	Imported from animal dealer	51	12 (24.0%)	TEM, description of the sarcocysts	987

NS = not stated.

[a] Host species: Rhesus macaque (*Macaca mulatta*), crab-eating macaque (*Macaca fascicularis*), cynomolgus monkey (*Macaca irus*), baboon (*Papio cynocephalus*), chacma baboon (*Papio ursinus*), Patas monkey (*Erythrocebus patas*), common squirrel monkey (*Saimiri sciureus*), blue monkey (*Cercopithecus mitis*), velvet monkey (*Cercopithecus pygerythrus*), tamarin (*Saguinus oedipus*).

[b] There are several other reports on the presence of *Sarcocystis* in monkeys.[943]

DNA was extracted[1743] from 2 unfixed *S. nesbitti* sarcocysts from the monkey from China.[1859] Characterization of the 18S rRNA gene indicated a close relationship to *S. singaporensis*, which cycles between pythons and rodents, suggesting that snakes are likely a definitive host for *S. nesbitti*. The genetic sequences from *S. nesbitti* matched the sequences derived from a sarcocyst from 3 symptomatic cases of *Sarcocystis* in a human from Malaysia.[14,877] It is probable that snakes may act as definitive hosts for *S. nesbitti*, because recently, based on amplification of the 18S rRNA genes, sequences were found matching this species in sporocysts excreted by reticulated python (*Braghammerus reticulatus* syn. *Python reticulatus*) and monocled cobra (*Naja kaouthia*) from Malaysia.[1033] No bioassay was carried out. Evidence—or lack of it—to link *S. nesbitti* to human infections is discussed in Chapter 4.

5.4 *SARCOCYSTIS MARKUSI* (MARKUS, KAISER, AND DALY 1981) ODENING 1997

5.4.1 Intermediate Host

South African Vervet monkey (*Cercopithecus pygerythrus*).

5.4.2 Definitive Host

Unknown.

5.4.3 Structure and Prevalence

Sarcocysts were found in muscles of 11 of 134 *C. pygerythrus* in South Africa.[1112] The sarcocysts' size was not stated. Bradyzoites were $10–12 \times 2–3$ μm. Sarcocysts had unique vp, type 9i. Although the size of the vp was not stated by the authors, measurements from the illustrations indicate that vp were approximately 10×2.5 μm. The vp contain many smooth microtubules. The PVM was lined by an electron-dense layer that was invaginated.

5.5 CLINICAL SARCOCYSTOSIS IN NATURALLY INFECTED MONKEYS

There are 4 clinical reports of sarcocystosis in monkeys, all in *Macaca mulatta*, all from the United States, and all with different clinical presentations. The first case was a young male rhesus weighing 3.6 kg that was asymptomatic before being introduced into the colony.[1737] Three months later, the monkey developed myositis, with stiffness of all limbs and edema of the scrotum. A necropsy was performed because of the deteriorating condition. Severe myositis was associated with numerous sarcocysts. There were lesions similar to eosinophilic myositis. The sarcocysts were up to 650 μm long. The sarcocyst walls were 4–5 μm thick with perpendicular vp, similar to those described for *S. kortei*. Bradyzoites were 9–12 μm long. These authors also reviewed literature and concluded that clinical sarcocystosis is relatively rare in monkeys.

The second case was a 22-month-old rhesus that became anorexic and dull.[1020] The monkey was euthanized in moribund condition. Gross lesions included stippling and streaking on the myocardium and with facial ecchymosis and edema. The most striking findings were numerous intravascular schizonts in many organs. Some schizonts occluded capillaries in the brain (Figure 5.1). The presence of schizonts was confirmed by TEM. Immature sarcocysts with metrocytes were seen in skeletal muscles. The sarcocyst wall appeared to be thick (Figure 5.1). PCR amplification placed the organism in the genus *Sarcocystis* but final identity was not confirmed.

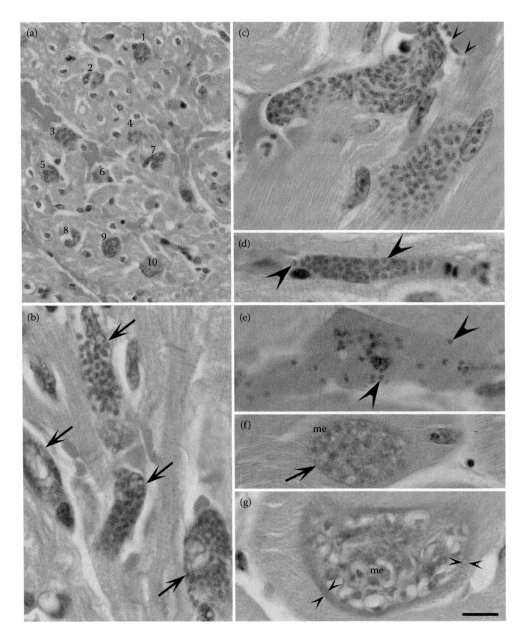

Figure 5.1   *Sarcocystis* sp. schizonts in heart (a–e) and sarcocysts in skeletal muscle (f, g) of a rhesus
monkey (*Macaca mulata*), H and E stain. (a) Ten schizonts that appear like sarcocysts at this
magnification. (b) Four intravascular schizonts (arrows). (c) Two schizonts (arrowheads point to a
longitudinally cut merozoite). (d) Elongated schizont (arrowheads) in a capillary (all merozoites in
cross-section). (e) Blood vessel with free intravascular merozoites (arrowheads). (f) Early immature sarcocyst with a very thin sarcocyst wall (arrow) enclosing metrocytes (me) that stain faintly
with H and E stain. (g) Immature sarcocyst enclosing me with a thick sarcocyst wall (between
arrowheads). Bar applies to all parts; bar in (a) = 50 µm, bars in (b–g) = 10 µm. (Courtesy of the
authors in Lane, J.H. et al., 1998. *Vet. Pathol.* 35, 499–505.)

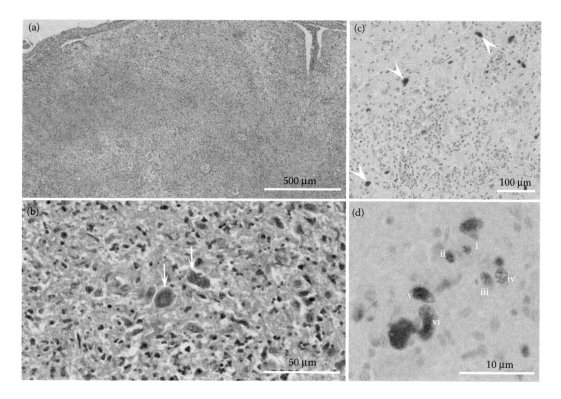

Figure 5.2 *Sarcocystis* myelitis in spinal cord of a *Macaca mulatta*. (a) Severe, extensive inflammation in grey and white matter and obliterating the spinal canal. H and E stain. (b) Higher magnification to indicate necrosis and infiltration with mixed leukocytes. Arrows point to schizonts. There are many free merozoites that are not visible at this magnification. H and E stain. (c) Vasculitis and glial cell proliferation. Arrowheads point to schizonts. Immunohistochemical staining with anti-*S. cruzi* rabbit antibodies. (d) Higher magnification of Figure 5.2c. Note the developing meronts (i–vi, in presumed order of development). The meront nucleus enlarges and becomes lobed (Adapted from Klumpp, S.A. et al., 1994. *Am. J. Trop. Med. Hyg.* 51, 332–338.)

The third case was a 2-year-old monkey euthanized as part of a clinical trial. Diffuse, pale streaking of several muscles was noted.[723] Enormous numbers of sarcocysts were seen in skeletal muscles, some associated with myositis (Figure 5.2). The sarcocyst wall was thick, with vp 2.7–6.3 μm long and 0.5–1.1 μm wide. Bradyzoites were 10–13 μm long. PCR amplification of the 18S rRNA gene showed a 91% similarity to *S. hominis*, but sequencing of the ITS-1 fragment showed little homology to any previously published sequences; therefore, the authors suggested a new/unique *Sarcocystis* species.

The fourth case was a rhesus born in captivity and used for research in simian immunodeficiency virus (SIDV). It was inoculated with SIDV at 18 months of age. Seven months PI with SIDV, the animal became lethargic and had a head tilt. After euthanasia, a necropsy was performed. Histologically, the most prominent lesion was in the lumbar spinal cord, with severe inflammation and obliteration of the central canal (Figure 5.3), marked by necrosis, mixed leukocyte cell infiltrations, perivasculitis, and proliferation of astrocytes (Figure 5.3). Both immature and mature schizonts were in the lesions. Ultrastructurally, schizonts divided by endopolygeny, and merozoites, lacked rhoptries.[968] The parasite reacted to polyclonal *S. cruzi* antibodies but not to *S. neurona*, *T. gondii*, and *N. caninum* antibodies.

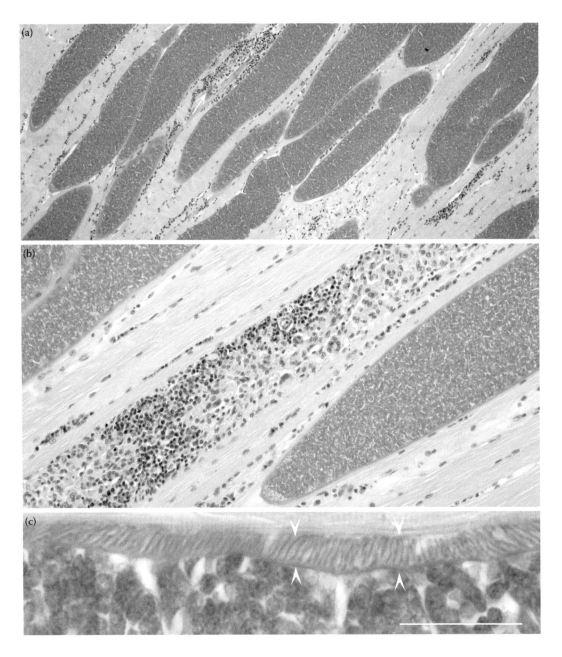

Figure 5.3 *Sarcocystis*-associated myositis in a rhesus monkey. (a) Numerous sarcocysts in skeletal muscle. Arrow points to an inflammatory focus. (b) Inflammatory focus with mononuclear cells and giant cells. (c) Sarcocyst wall with elongated villar protrusions on the sarcocyst wall (arrowheads). H and E stain. Bar applies to all parts; bar in (a) = 500 μm, in (b) = 200 μm, and (c) = 20 μm. (Courtesy of Gozalo, A.S. et al., 2007. *Vet. Pathol.* 44, 695–699.)

BIBLIOGRAPHY

Information pertinent to the subject matter of this chapter may be found in references: **14, 477, 723, 765, 795, 877, 939, 943, 961, 968, 987, 1020, 1033, 1104, 1112, 1153, 1175, 1423, 1428, 1695, 1737, 1743, 1859**.

CHAPTER 6

Sarcocystosis in Pigs (*Sus scrofa*)

6.1 INTRODUCTION

There are 2 valid species in pigs: *S. miescheriana* and *S. suihominis* (Figure 6.1). *S. miescheriana* is type species of the genus.

6.2 *SARCOCYSTIS MIESCHERIANA* (KÜHN 1865) LABBÉ 1899

6.2.1 Definitive Hosts

Dog (*Canis familiaris*), raccoon (*Procyon lotor*), wolf (*Canis lupus*), red fox (*Vulpes vulpes*), and jackal (*Canis aureus*).[1188]

6.2.2 Structure and Life Cycle

Sarcocysts are up to 1500 µm long and 200 µm wide and are found in skeletal and cardiac muscles. The sarcocyst wall is 3–6 µm thick and appears radially striated. The vp on the sarcocyst wall are up to 5 µm long and 1.3 µm wide, type 10b (Figure 6.1a and c).[547,1176] Bradyzoites are up to 20 µm long.

Information on endogenous development is summarized in Table 6.1. There are 2 known generations of schizogony, which develop much more rapidly than those of cattle and sheep. First- and second-generation schizonts mature within 13 DPI; immature sarcocysts are found as early as 27 DPI.[75,814,832]

6.2.3 Pathogenicity

S. miescheriana can cause weight loss and purpura of the skin, especially of the ears and buttocks, dyspnea, muscle tremors, abortion, and death, depending on the number of sporocysts ingested.[74,75,547,549,550,1880] Ingestion of less than 1 million sporocysts generally results in subclinical infection. However, 1 pig fed 200,000 sporocysts died at 60 DPI, probably related to heart failure.[301] Stress of pregnancy may modify the severity of infection. Of 5 pregnant sows experimentally infected with 50,000 sporocysts each, all became ill: 2 aborted 12–14 DPI, 1 died, and 2 became moribund and were euthanatized.[549] A similar number of sporocysts fed to finishing pigs caused no clinical signs, but resulted in weight gains that were 11%–27% less than uninfected controls.[118] Administration of 25,000 or 15,000 sporocysts neither reduced weight gains nor produced clinical signs.

The effect of subclinical sarcocystosis on meat quality has been studied.[296–298] Pigs inoculated with 50,000 *S. miescheriana* sporocysts weighed 5–12 kg less than uninfected pigs, when examined 3 months PI; however, the quality of meat as determined by water-absorbing and water-binding capacity, pH, rigor values, color brightness, and back-fat ratios, was not reduced, compared with

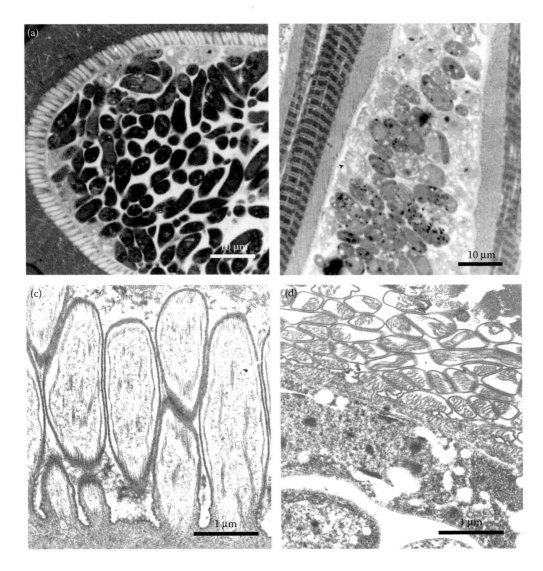

Figure 6.1   Sarcocysts of *S. miescheriana* (a and c) and *S. suihominis* (b and d) in skeletal muscles of pigs. The vp of *S. suihominis* are thinner and longer than those of *S. miescheriana*. a and c, Toluidine blue stain. c and d, TEM.

that of uninfected pigs.[296,298] Host genetics could affect susceptibility to *Sarcocystis* infection. The number of bradyzoites were 20 times higher in European Pietrain (n = 14) than in the Chinese Meishan (n = 11) pigs that were fed 50,000 *S. miescheriana* sporocysts and monitored for severity of disease.[1457] The Pietrain pigs developed more severe disease than the Meishan pigs and 2 Pietrain pigs died, versus no deaths in the Chinese pigs. In a subsequent study, the authors provided evidence for genes regulating resistance to *S. miescheriana*.[1459]

6.2.4 Protective Immunity

Pigs immunized once with 1000–50,000 *S. miescheriana* sporocysts orally become refractive to lethal challenge with large numbers of *S. miescheriana* sporocysts.[1488,1559,1560,1813,1880] This protection was found when the challenge infection was initiated at 80 DPI, but not at 120 DPI.[1880] Pigs fed

Table 6.1   Comparison of Development at Stages of *Sarcocystis* Species in Pigs

	S. miescheriana	*S. suihominis*
First-Generation Schizonts		
Location	Liver endothelium	Liver endothelium
Duration (DPI)	6–7	5–6
Size of meronts (µm)	Unknown	Unknown
No. of merozoites	Unknown	Unknown
Size of merozoites (µm)	Unknown	Unknown
Second-Generation Schizonts		
Location	Many organs	Many organs
Duration (DPI)	7–13	7–20
Size of meronts (µm)	7–17 × 11–79	Unknown
No. of merozoites	20–94	50–90
Size of merozoites (µm)	Unknown	6.5 × 3.2
Parasitemia		
Duration (DPI)	Unknown	Unknown
Blood-cell multiplication	Unknown	Unknown
Sarcocysts		
Wall (µm)	Thick, striated (2.7–6)	Thick, striated (4–9)
First seen (DPI)	27	27
Maximum length (µm)	Up to 1500	Up to 1500
Infective (d)	58	56
Bradyzoites (µm)	17.1 × 3.8	15 × 4.5
Metrocytes	10.8 × 4.6	9.8 × 4.9
Main references	**51, 75, 707, 814**	**51, 806, 809, 1174**

multiple low doses of sporocysts (trickle infections) also developed protection.[1559] This protective effect resulting from feeding of *S. miescheriana* sporocysts was not protective against a challenge infection with *S. suihominis* sporocysts.[550]

6.3 *SARCOCYSTIS SUIHOMINIS* (TADROS AND LAARMAN 1976) HEYDORN 1977

6.3.1 Distribution

Europe, India, Japan.

6.3.2 Definitive Hosts

Humans (*Homo sapiens*) and nonhuman primates (*Macaca mulatta, Macaca irus, Pan troglodytes*, and *Papio cynocephalus*).

6.3.3 Structure and Life Cycle

Sarcocysts are up to 1500 µm long.[547] The sarcocyst wall is 4–9 µm thick[547] and appears hirsute, with vp up to 13 µm long, type 31 (Figure 6.1b and d).[1174] The bradyzoites are approximately 15 µm long.

Life-cycle data is summarized in Table 6.1. The 2 generations of schizogony develop at about the same time as—and are structurally similar to—those of *S. miescheriana*.[806,808,809] Immature sarcocysts were first seen 27 DPI, and bradyzoites were seen 56 DPI.[1174]

6.3.4 Pathogenicity

S. suihominis is pathogenic for pigs. Pigs fed 50,000 or more sporocysts became ill, and half of those fed 1 million sporocysts died.[806] Clinical signs in pigs were similar to those in pigs infected with *S. miescheriana*.[806]

6.3.5 Genetic Diversity

There are few reports on the genetic diversity of *S. miescheriana*, which appears to be very low. Thus, minimal differences have been detected for the 18S rRNA gene in isolates from domestic pigs and wild boars in Iran, Lithuania, China, Switzerland, Germany, and the United States.[161,168,956,1854]

6.4 *SARCOCYSTIS PORCIFELIS* DUBEY 1976

The validity of this species is uncertain. This species was named[349] for the parasite described from the former USSR (Union of Soviet Socialist Republics).[716] *Sarcocystis*-infected esophagi from pigs were fed to 4 cats.[716] The cats excreted sporulated oocysts 5–10 days after ingesting the infected swine tissues. The sporocysts were 13.5 × 7.6 μm and each contained 4 sporozoites measuring 9.5 × 3.8 μm. The sporocysts from cat feces were fed to 8 littermate pigs. All pigs became ill and 1 died 89 DPI. Sarcocysts were found in skeletal muscles and the heart. Other pigs were killed 3.5 months after ingestion of sporocysts, and sarcocysts were found in them.[716] The structure of the sarcocysts was not described.

In Romania, sarcocysts were found in the muscles of a wild pig during examination for *Trichinella*.[1694] Sarcocysts were 800 μm long. Infected meat was fed to 2 cats. Both cats excreted sporocysts from day 9–13 after ingesting pork. The sporocysts were 14.5 × 8.9 μm.

A few sarcocysts of a third type, different from *S. miescheriana* and *S. suihominis*, were reported in wild boars from Lithuania.[33] But no data on morphological description was provided. Authors comment that, most likely, this finding could be an artifact (Kutkiené, L., pers. comm.).

In Japan, 1 case of *S. porcifelis* was reported in slaughtered pigs.[1323] The sarcocysts had thin and smooth walls, as can be seen in Figure 3 provided by authors; no additional details are given.

In India, a macroscopic sarcocyst was found in the diaphragm of a naturally infected domestic pig.[791] The sarcocyst wall was broken during separation from the host tissue. The sarcocyst was 4.1 mm long, and the wall was smooth and thin, with uniform radial striations that were 0.85 μm long. The bradyzoites were 7 × 3 μm in size. The authors called it *S. porcifelis*. Until now, this is the best morphological description of the sarcocyst.

Note: *Sarcocystis* sporocysts have never been reported in feces of cats that have been fed pork. In a national survey, in the United States, for *Toxoplasma gondii* infection, 2049 samples of pork were fed to cats and feces of cats were examined for oocysts[458]; *Sarcocystis* sporocysts were not found (unpublished). In other investigations, sporocysts were not found in feces of cats fed *Sarcocystis*-infected pork from domestic pigs[353] or from wild pigs.[73]

6.5 PREVALENCE OF *SARCOCYSTIS* AND ECONOMIC IMPACT

The prevalence of sarcocysts in domestic pigs (Table 6.2) is lower than in wild pigs (Table 6.3). The prevalence in finishing pigs or market-age pigs is lower than in adult culled pigs.

Table 6.2   Surveys of *Sarcocystis* spp. in Domestic Pigs

Country	Year	No.	Method	No. Positive	% Positive	Species	Reference
Australia	1974	169	Histology	7	1.4 (<1 year) 16.7 (>1 year)	NS	1244
Austria	1978–79	712	Trypsin digestion	130	18.3 5.5 (fattening) 31.6 (sows)	*S. suihominis* (4.1%) "*S. suicanis*" (13.6%) Mixed (0.7%)	832
Chile	NS	100	Pepsin digestion	79 (esophagus)	85.0 73.2 (<1 year) 100.0 (>1 year)	NS	721
China, Guangxi	2004	42	NS	30	71.4%	NS	1720
China	2011	4	Histology, PCR	4 isolates	—	*S. miescheriana*	1854
Czech Republic	1979–80	2000	Trypsin digestion, trichinoscopy	0	0.0	NS	764
Czech Republic	NS	335	Trypsin digestion	8	4.0	NS	1086
Czech Republic	NS	1409	Inspection, pepsin digestion	Not clear	0.9 Not clear	NS	1101
Egypt	NS	105	Bioassay in dogs	NS	NS	*S. miescheriana*	1873
Germany	1977	1000	Trichinoscopy	419	41.9	NS	810
Germany	1975–77	1175	Trypsin digestion	219	18.6 9.7 (fattening) 35.5 (breeders)	*S. suihominis* (60.7%) "*S. suicanis*" (48.3%) Mixed (n = 11 cases)	115
Germany	NS	NS	NS	NS	3.0–30.0	*S. suihominis*	156
Germany	1997	2041 sows	ELISA	592	29.0	*Sarcocystis* spp.	291
Ghana	1997	60	Trichinoscopy	17	28.3	NS	1387
India, Haldwani, Rudrapur and Aligarh	NS	890	Pepsin digestion	605	67.9	*S. suihominis* (43.14%) *S. miescheriana* (47.11%)	1526
India, Haryana	1981–82	157	Pepsin digestion, trichinoscopy, bioassay in dog and cat	108	68.8	*S. miescheriana* after successful bioassay in dog	761
India	NS	37	Histology	2 (cysts in brain)	5.4	NS	759
India, Bihar	1968–80	170	Pepsin digestion	91	53.5	NS	1508
India, Bombay	1966	125	Histology	23	18.4	*S. miescheriana*	315

(Continued)

Table 6.2 (Continued) Surveys of Sarcocystis spp. in Domestic Pigs

Country	Year	No.	Method	No. Positive	% Positive	Species	Reference
India, Assam	1991–92	372	Pepsin digestion	283	76.1 25.0 (<1 year) 79.6 (>1 year)	S. suihominis (56.06%) S. miescheriana (34.85%)	319
India, Punjab	NS	229	Pepsin digestion	168	73.3 33.3 (<1 year) 83.9 (>1 year)	S. suihominis (53.6%) S. miescheriana (31.8%) Mixed infections (14.5%)	50
India, Uttar Pradesh	NS	NS	Pepsin digestion	NS	NS	S. miescheriana-like	18
India, Madhya Pradesh	1987–88	200	Trypsin digestion	168	84.0	S. suihominis (49.5%) S. miescheriana (34.5%) Mixed (4.5%)	1638
India, Uttar Pradesh	NS	296	Trypsin digestion, microscopic examination	228	77.0	S. suihominis (48.9%) S. miescheriana (40.2%), S. porcifelis	790, 791
Italy, Reggio Emiglia	NS	272 fattening	Pepsin digestion	15	5.5	NS	143
Italy, Sardegna	NS	328	Pepsin digestion	271	82.6	NS	37
Japan, Saitama	1996–97	600 breeding	Trichinoscopy	5	0.83	S. suihominis	1520
Japan, Hokkaido	1992	144	Trichinoscopy	17	0.0 (fattening) 16.5 (sows)	"S. suicanis"	1334
Japan, Okinawa	1991–92	300	Pepsin digestion, pathology	107	35.7 13.0 (fattening) 47.0 (breeding)	S. porcifelis (1 case) S. miescheriana (106 cases)	1323
Japan, Saitama	Unknown	300	Trypsin digestion, pathology	17	0.0 (fattening) 8.5 (older)	S. miescheriana	1510
Philippines	1994	225	Histology	1	0.4	Unidentified, immature cysts	219
Philippines	NS	33 (2–5 years)	Histology	9	27.0	S. miescheriana	224
Poland	1967–77	NS	Trypsin digestion	NS	Incidence 0.328 in 1976 0.363 in 1977	NS	317
Thailand, Bangkok	NS	839	Trypsin digestion	99	11.8	NS	1234
Serbia	NS	788 (6 months)	Trypsin digestion	193	24.5	NS	1388

(Continued)

Table 6.2 (Continued) Surveys of *Sarcocystis* spp. in Domestic Pigs

Country	Year	No.	Method	No. Positive	% Positive	Species	Reference
Slovak Republic	Unknown	1409	Pepsin digestion, trichinoscopy	13	0.9	NS	1101
Spain, Northwest	NS	100	Pepsin digestion, trichinoscopy	43	43.0	NS	1384
Spain, Santiago de Compostela	NS	100	Pepsin digestion, ELISA, HAI	5	5.0	NS	1385
Spain, Zaragoza	1978–80	484	Pepsin digestion, trichinoscopy	178–484	36.7–100	NS	1530
Switzerland	NS	1 two-year-old boar	Histology, PCR	1 isolate	–	*S. miescheriana*	168
Thailand	NS	295	Pepsin digestion	35	11.8	NS	1234
Uruguay	NS	269 (90–140 kg)	Pepsin digestion, pathology	154	57.2	"*S. suicanis*"	635
USA, Georgia	NS	168 sows	Pepsin digestion, bioassay in dog, cat, raccoon and opossum	28	16.6	"*S. suicanis*"	1424
USA, Iowa	1990–92	893 sows	Pepsin digestion, pathology	163	18.2	*S. miescheriana*	409
USA, Ohio	1977	236 sows	Trypsin digestion, bioassay in dogs and cats	8	3.4	No species with dog and cats as definitive host	353
USA, Maryland	NS	39	Pepsin digestion	17	43.6	NS	878
USA, Michigan	1973	103	Pepsin digestion, pathology	7	6.8 0.0 (<1 year) 12.7 (>1 year)	NS	1586

Note: *S. suicanis* is now called *S. miescheriana*.

Table 6.3 Surveys of *Sarcocystis* spp. in Wild Swine

Country	Year	No.	Method	No. Positive	% Positive	Species	Reference
Bulgaria	1972–75	NS	Trichinoscopy	NS	66.7	NS	1187
Germany	NS	200	Trichinoscopy	140	70.0	NS	335
Germany	NS	103	Trichinoscopy	46	45.0	NS	546
Iran	2007	–	Pathology, PCR	1 isolate	–	*S. miescheriana*	956
Lithuania	2005–11	51	Trichinoscopy	45	88.2	*S. miescheriana*	1414
Lithuania	1981–86	715	? (In Lithuanian)	353	49.4	"*S. suicanis*" *S. suihominis*	33
Lithuania	1998–2001	55	Histology	49	89.1	NS	1102
New Zealand	NS	50	Pepsin digestion, pathology	5	10.0	NS	238
Poland	NS	30	Trichinoscopy	20	66.0	NS	546
Poland	NS	166	Trichinoscopy		24.7	–	1763
Romania	1995	NS	Trichinoscopy, bioassay in cats	NS	NS	*S. porcifelis*	1694
Slovak Republic	2005–07	20	Trypsin digestion, Histology	17	85.0 (0.5–2 years)	NS	869
Slovak Republic	2006–09	30	Trypsin digestion	25	83.3	NS	869
USA, Southeast	NS	192	Pepsin digestion	62	32.0 40.0 (adults) 10.0 (immature)	*S. suicanis*	73
USA, Texas	1973–74	5 (6 mo–4 years)	Histology	3	60.0	NS	250
USA, 29 states	2012–14	1006	Histology, pepsin digestion, PCR	251	25.0%	*S. miescheriana*	161

Note: S. suicanis is now called *S. miescheriana*.

Several food-safety institutions (e.g., European Food Safety Authority) encourage monitoring and characterization of zoonotic *Sarcocystis* species in animals and foodstuffs.[1727] Although carcass condemnation for *Sarcocystis* has been reported, there is no documentation of any species of *Sarcocystis* in swine, with a macroscopic sarcocyst, so it is unclear if sarcocysts are actually present or whether there has been misidentification. Greve[735] reported association between *Sarcocystis* infection and lymphadenopathy in the pelvic cavity during inspections in Danish abattoirs; it has been suggested that calcified sarcocysts are found only in domestic pigs and not in wild boars.[335] The basis for documentation is uncertain.

Only 1 report was found of clinical sarcocystosis in a naturally infected pig. Clinical fatal sarcocystosis attributed to *S. miescheriana* was reported in a pig from Switzerland.[168] The affected pig was a 2-year-old boar that had fever, anorexia, and died after a short period of illness. Histopathological examination revealed severe myocarditis associated with *S. miescheriana*-like schizonts; sarcocysts were not seen. Molecular characterization of DNA extracted from paraffin embedded myocardium indicated *S. miescheriana* infection.[168]

BIBLIOGRAPHY

Information pertinent to the subject matter of this chapter may be found in References: **75, 74, 79, 97, 119, 120, 296–298, 317, 349, 350, 353, 546, 547, 549, 550, 567, 583, 707, 716, 734–736, 806, 808–810, 812, 814, 819, 960, 1057, 1174, 1177, 1188, 1244, 1286, 1287, 1289, 1383, 1478, 1485, 1488, 1559, 1560, 1688, 1727, 1813, 1826, 1827, 1880, 1881, 1892, 1911, 1914, 1918, 1982, 2011, 2069, 2109, 2149, 2259, 2274, 2276–2278, 2290, 2320, 2323, 2363.**

Prevalence: **18, 33, 37, 40, 49, 50, 73, 115, 143, 156, 168, 219, 224, 238, 250, 291, 315, 317, 319, 335, 353, 409, 546, 635, 721, 735, 759, 761, 764, 790, 791, 810, 832, 869, 878, 956, 1086, 1101, 1102, 1187, 1234, 1244, 1323, 1334, 1384, 1385, 1387, 1388, 1414, 1424, 1508, 1510, 1520, 1526, 1530, 1586, 1638, 1694, 1720, 1763, 1873, 1916, 1930, 1938, 1953, 1967, 1982, 2037, 2038, 2054, 2075, 2091, 2124, 2231, 2242, 2297, 2311, 2325.**

Experimental: **74, 75, 120, 297, 298, 299–303, 322, 806, 954, 989, 1052, 1053, 1057, 1168, 1413, 1425, 1426, 1457, 1458, 1459, 1637, 1640, 1641, 1900, 1906, 1925, 1928, 2001, 2021, 2046, 2052, 2067, 2084, 2087, 2129, 2189, 2232, 2275, 2374.**

Morphology and pathology: **51, 52, 209, 707, 814, 809, 989, 1174, 2152, 2216.**

Molecular: **161, 719, 956, 1854.**

Miscellaneous: **209, 567, 1892, 1906, 1911.**

CHAPTER 7

Sarcocystosis in Cattle (*Bos taurus*)

There are 4 species of *Sarcocystis* in cattle: *S. cruzi, S. hirsuta,* and *S. hominis* (Figure 7.1), with canids, felids, and primates, respectively, as definitive hosts; and a fourth species, *S. rommeli,* with unknown life cycle.[471]

7.1 *SARCOCYSTIS CRUZI* (HASSELMANN 1926) WENYON 1926

7.1.1 Intermediate Hosts

Cattle (*Bos taurus*), North American bison (*Bison bison*),[361,162] European bison (*Bison bonasus*),[608] and banteng (*Bos javaticus*).[1307]

7.1.2 Definitive Hosts

Dog (*Canis familiaris*), coyote (*Canis latrans*), red fox (*Vulpes vulpes*), crab-eating fox (*Cerdocyon thous*),[1475] raccoon dog (*Procyon lotor, Nyctereutes procyonoides*),[1512] and wolf (*Canis lupus*).

7.1.3 Structure and Life Cycle

Sarcocysts are microscopic, <1000 μm long, and are formed in virtually all striated muscles, Purkinje fibers of the heart, and in the CNS.[360] The sarcocyst wall is thin (<1 μm) and its surface is covered with long, narrow, ribbon-like vp (Figures 7.1a and 7.2a).[127,360,1171,1348] The vp are up to 3.5 μm long and up to 0.3 μm wide at the base.[638] They taper towards the tip, which is blunt. There are no supporting tubules or microtubules in the villar core; therefore, the vp are often folded over the surface (Figure 7.2a), type 7a. The sarcocyst wall surface has numerous small invaginations, which are hexagonal when viewed from the surface. The primary cyst wall shows small invaginations (contributing to a dentate appearance), and zipper-like structures.[638]

The life cycle that was used as an example in Chapter 1, is summarized in Table 7.1.

7.1.4 Pathogenicity

S. cruzi is the most pathogenic species found in cattle.[366,572,577,918,1269] It can cause fever, anorexia, anemia, weight loss, hair loss, weakness, muscle twitching, prostration, abortion, reduced milk yield, hypersalivation, neurologic signs, and death, depending on the isolate and the number of sporocysts ingested.[63,167,232,681] Ingestion of 1000 or fewer sporocysts under experimental conditions caused little or no clinical signs. Doses of 5 million or more are uniformly fatal. Cattle fed 200,000 sporocysts became clinically ill, some died of acute sarcocystosis, and some survivors did not grow to their full potential.[591] In experimental infections, cattle had no clinical signs until the fourth week, irrespective

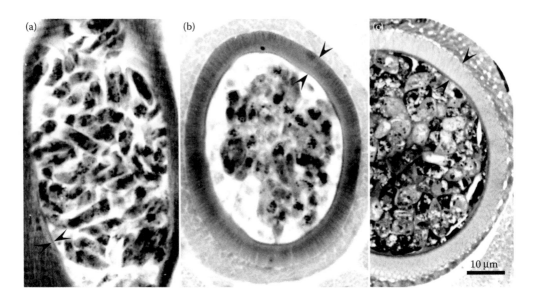

Figure 7.1 Sarcocysts of *S. cruzi* (a), *S. hirsuta* (b), and *S. hominis* (c) in skeletal muscles of cattle. Note outer and inner layers of sarcocyst walls (arrowheads). (a and c), Toluidene blue stain and (b), PAS reaction.

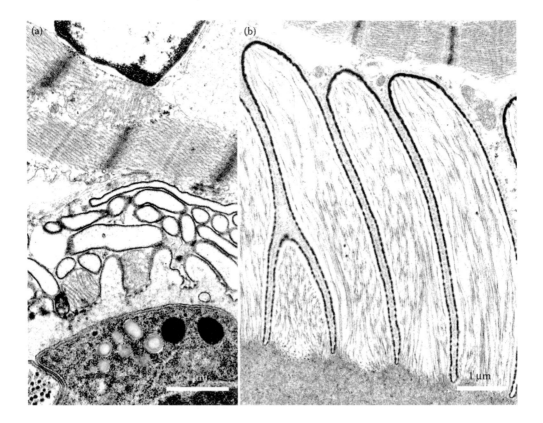

Figure 7.2 TEM of sarcocyst walls of *S. cruzi* (a) and *S. hominis* (b) in experimentally infected calves. (a) Filamentous villar protrusions, without microtubules, folded over the primary sarcocyst wall. (b) Finger-like vp with microtubules.

Table 7.1 Comparison of Developmental Stages of *S. cruzi*, *S. hirsuta*, and *S. hominis* in Cattle

	S. cruzi	*S. hirsuta*	*S. hominis*
First-Generation Schizonts			Unknown
Location	Arterioles of several organs	Mesenteric and intestinal arterioles	—
Duration (DPI)	7–26	7–23	—
Peak development (DPI)	15	15	—
Size of meronts (μm)	41.0 × 17.5	37.2 × 22.3	—
No. of merozoites	>100	>100	—
Size of merozoites (μm)	6.3 × 1.5	5.1 × 1.2	—
Second-Generation Schizonts			Unknown
Location	Capillaries of several organs	Capillaries of striated muscles, heart	—
Duration (DPI)	19–46	15–23	—
Peak development (DPI)	24–28	16	—
Size of meronts (μm)	19.6 × 11.0	13.9 × 6.5	—
No. of merozoites	4–37	3–35	—
Size of merozoites (μm)	7.9 × 1.5	4 × 1.5	—
Parasitemia			Unknown
Duration (DPI)	24–46	11	—
Intraleukocytic			—
Multiplication	Yes	No	—
Sarcocysts			—
First seen (DPI)	45	30	—
Maturation time (DPI)	86	75	—
Size	183–417 × 20–98 μm	>8 mm × 1 mm	0.7–2.6 mm × 50–150 μm
Wall (μm)	Thin (<1.0), type 6/7	Thick (3–7), type 28	Thick (>6), type 10/15/16
Villar protrusions	Ribbon-like, wide at the base. Without microtubules, folded over the surface, >3.5 μm × 0.3 μm	Narrow stalk, expanded laterally and tapered distally, 8 μm long, bent at 45–90°	Finger-like, 0.7–3.0 × 6.8–8.1 μm
Metrocytes	Globular, 4.5 × 9.6 μm	Globular, 3 × 5 μm	Globular or oval, 4–5.5 × 8–9 μm
Bradyzoites	Crescentic	Banana shape	Banana shape, 2.5 × 7–9.0 μm
Location	Mostly in heart, high frequency in skeletal muscle, also found in CNS	Not in heart or CNS	Not in heart
Main references	**225,360**	**126,127,360,395**	**391,395,884,1840**

of dose. Beginning as early as 24 DPI they developed persistent fever, anorexia, weight loss, weakness, muscle twitching, prostration, and some died. Pregnant cows aborted,[844,1152] and milking cows had reduced yield and poor quality milk. Cattle that did not recover lapsed into chronic sarcocystosis, in which there was continued inappetance, weight loss, hypersalivation, hyperexcitability, and loss of hair, especially on the neck, rump, and tail switch (Figure 1.84). Some months after infection, a few animals developed signs of CNS involvement, including nystagmus, opisthotonus, and lateral recumbency with a running gait. All cattle that developed CNS disorders eventually died.

Gross and microscopic lesions were described in Chapter 1.

7.1.5 Natural Outbreaks

In 1961, a febrile illness was noticed in a dairy herd.[256] Within 8 weeks, a total of 25 animals were affected and 17 died. Acutely ill cows had intermittent anorexia, drop in milk yield, diarrhea, transient fever, nasal discharge, hypersalivation, and hemorrhagic vaginitis. Of 17 pregnant cows, 10 aborted in the last trimester of pregnancy and only 1 of the 10 survived. Erosions were found on the tongue and buccal mucosa of the salivating cows. Chronic cases were marked by emaciation, pale or icteric mucous membranes, mandibular edema, exophthalmos, cessation of lactation, and sloughing of the tail switch. Cachectic animals became recumbent, with muscle tremors over the body resembling those of hypocalcemia. Gross and histologic lesions resembled those described for acute experimental sarcocystosis. Numerous schizonts were present in the endothelium of many organs in 11 of 16 animals. Because neither the stage nor clinical signs of sarcocystosis had been described, this outbreak was called Dalmeny disease, after the small Canadian town in which it occurred. In retrospect, however, it was an outbreak of acute sarcocystosis.[256]

Subsequently, cases of natural clinical sarcocystosis in cattle have been reported in Canada,[1161] England,[227,604,1016] Ireland,[232] Norway,[1019] Australia,[167] and the United States.[626,627,681,1238] Unlike Dalmeny disease, these were infections in yearlings or younger calves. A well-documented outbreak was reported in Kentucky, United States.[681] Approximately 30 of 41 heifers in a 1-hectare lot had diarrhea, were losing weight, and consuming less feed, and had rough coats. The condition worsened to include severe laminitis, hypersalivation, extreme nervousness, and marked hair loss from the ears, lower limbs, and the tail switch, giving a "rat tail" appearance (Figure 1.84). Within a month, more than half the animals were severely debilitated and eight died. In the next 5 weeks, two-thirds died or were euthanatized, only a third were salvaged for the market.

Macroscopic erosions were found in several tissues. The skin of the carpus and tarsus, just above the coronary band, was denuded and abraded; interdigital erosions were present, and the hoof wall was cracked, necrotic, and about to separate from the coronary band; the cornea had erosions and opacities; erosions and shallow ulcers were found in the muzzle, lips, tongue, hard palate, esophageal mucosa, and omasum; the lungs were congested. Histologic examination revealed intense mononuclear cell infiltrates in the muscles and muscular organs, with accompanying myodegeneration and numerous immature sarcocysts (Figure 1.96). The *Sarcocystis* IHA antibody titers in affected heifers were fourfold greater than in unaffected cows on the same farm.[681] Hay contaminated with feces from farm dogs was implicated as the source of infection.[626,681]

The association between sarcocystosis and ulcerations in the mouth and foot, seen in naturally infected cattle, has not been explained. Neither buccal nor pedal lesions were seen in over 100 experimentally inoculated cattle. Naturally infected cattle may be more susceptible to secondary infections.

Sporadic cases of bovine abortion and neonatal mortality have been reported in New Zealand,[1246,1741,1791] Australia,[908,1152,1242,1246] Canada,[256] the United States,[30,365,844,988,1371,1558] and Cuba,[251] in which fetal lesions and/or protozoa were found. Fetal encephalitis, myocarditis, and hepatitis were often associated with the outbreaks. Intravascular schizonts resembling those of *Sarcocystis* were found in some of the fetuses in all of the reported outbreaks. Apart from focal placentitis and hypersalivation in a few cows, no clinical abnormalities were identified.

An unusual necrotizing encephalitis was found in 2 cattle from South Africa.[1774] A 2-year-old heifer developed convulsions and died suddenly. Eleven of 15 young calves developed severe clinical signs over 5 days, and 1 calf died of neural signs. Both of these animals were necropsied. Multiple areas of necrosis and vasculitis were found in both animals. Schizonts, considered to be first generations, were identified in blood vessels in affected areas.

7.1.6 Protective Immunity

Cattle are susceptible to clinical sarcocystosis from birth and throughout their life, despite repeated natural infections. However, it has been possible to establish a protective immunity under experimental conditions.[590] Calves infected with 50,000 or 100,000 *S. cruzi* sporocysts were protected from illness and death that would have resulted from challenge infections with large numbers of *S. cruzi* sporocysts given 70–252 days later. Calves treated prophylactically with amprolium from 21 to 35 days after infection to reduce second-generation schizonts were also protected.[590] Infection with the nonpathogenic *S. hirsuta* did not protect calves against *S. cruzi* infection.

7.1.7 Molecular Studies and Genetic Diversity

Several studies were focused on molecular diagnostic and characterization of *S. cruzi* infecting cattle. Initial genomic probes[957] were later applied[1272] for the diagnosis of acute sarcocystiosis in experimentally infected cattle. An RAPD assay was developed for specific diagnosis of *S. cruzi* infections.[1095]

Methodologies focused on the amplification of 18S rRNA, 28S rRNA, ITS-1, rRNA and *cox*1 genes allow differentiation of species.[608,1218,1477,1493,1856] Epidemiological surveys based on molecular identification of *Sarcocystis* species are accumulating; most are compiled in Table 7.2.

The most comprehensive study on genetic variability of *S. cruzi* found a very low/restricted genetic diversity in isolates from the United States and Uruguay, and also minimal differences with sequences available in international databases, belonging to isolates from remote locations like Australia, China, or Germany.[1493] These findings suggested that the dispersion of parasites of veterinary importance is closely related to human movements; in this sense, by domesticating dogs and cattle, humans may have profoundly influenced the ecology and evolution of *S. cruzi*.

7.2 *SARCOCYSTIS HIRSUTA* MOULÉ 1888

7.2.1 Definitive Host

Cat (*Felis catus*).

7.2.2 Structure and Life Cycle

Sarcocysts as large as 8 mm long and 1 mm wide (Figures 7.1b and 7.3) have been found in naturally infected cattle,[126] and others as large as 800 μm long and 80 μm wide have been found in experimentally infected cattle.[362] Sarcocyst walls as thick as 7 μm have been seen, some radially striated, others hirsute.[362,673] The vp are up to 7.0 μm long and 1.5 μm wide, and have been seen to contain numerous microtubules.[127,128] They can be closely packed (0.03 μm apart) or widely spaced (1 μm) (Figure 7.4), type 28. By SEM, wart-like structures were on the vp.[126]

Information on the life cycle and structure is summarized in Table 7.1. There are 2 known generations of schizogony. First-generation schizonts have been found only in small arteries of the intestine and mesentery, and second-generation schizonts have been found only in capillaries of the heart and skeletal muscles.

7.2.3 Pathogenicity

S. hirsuta is mildly pathogenic.[371] Calves fed 100,000 or more sporocysts became febrile, had diarrhea, and were mildly anemic; none died. At necropsy, hemorrhagic lesions were not seen.[371]

Table 7.2   Prevalence of *Sarcocystis* Sarcocysts in Muscle of Cattle

Country	Year	Source	Tissue	No. Tested	Total Positive	Method Digestion/ Squeezing	Method PCR	Method LM, SEM, TEM	Species Identified	Remarks	Refer- ence
Argentina	2004	Abattoir	H, D, E	90	90 (100.0%) IFAT	100.0% dissection	—	93.3% histology	*S. cruzi*	—	**1213**
Argentina, Buenos Aires	2006	Farm	Serum	172 calves at birth and their mothers	IFAT: 3 calves (1.7%) before ingestion of colostrum; 172 cows (100.0%)	—	—	—	*S. cruzi*	Vertical trans- mission: 1.7%	**1215**
Argentina	2005	2 farms	Blood	112	IFAT: close to 100.0% in both farms	—	—	—	*S. cruzi* antigen used	Increment of titers in calves with time	**1216**
Argentina	NS	Abattoir	H (10 g), loin (10 g)	380	378 (99.5% in heart)	Muscle squash	135 (35.5%)	LM, TEM	99.5% of thin walled (*S. cruzi*), 32.1% of thick-walled (*S. hominis*, *S. hirsuta*, *S. sp.*)	IFAT antibodies in 99.7% (379/380)	**1217**
Australia	1989– 90	Abattoir	E	714	371	52.0%	—	—	—	—	**1542**
Australia	1989	Abattoir	E, D, masseter	202	125 (62%)	Pepsin digestion	—	LM, histology	1 thick-walled, S. cruzi in all infected cattle	D: 53% E: 57% Masseter: 52%	**1541**
Belgium, Gent	1987	Abattoir	H, E, D	100	97 (97.0%)	Pepsin digestion: 97% E, H: 89, D: 75	—	LM, histology	Thick-walled: 56%, Thin-walled (*S. cruzi*) in all	—	**1787**

(Continued)

Table 7.2 (*Continued*) Prevalence of *Sarcocystis* Sarcocysts in Muscle of Cattle

Country	Year	Source	Tissue	No. Tested	Total Positive	Method			Species Identified	Remarks	Reference
						Digestion/ Squeezing	PCR	LM, SEM, TEM			
Belgium	2006	Retail stores	Mince loin	67	63 (94.0%)	—	63 positives	—	*S. cruzi*: 49%; thick wall: 91%, of those, 97.4% was *S. hominis*	—	1776
Belgium	1994–2007	Abattoir	Carcasses	97 carcasses affected with eosinophilic myositis	—	—	PCR and sequencing	LM	Extralesional: 80% of *S. hominis* and 20% unknown *S.* sp. Intralesional: 81.8% *S. hominis*, 4.5% *S. cruzi*, and 9% each *S. hirsuta* and unknown *S.* sp.	—	1778
Brazil	NS	NS	H (50 g)	50	50 (100.0%)	Homogenization in saline	—	LM	*S. cruzi* (but no morphologic study was done)	—	275
Cambodia	1994–95	Abattoir	Not clear (in Vietnamese)	Not clear	36.4%	Not clear	Not clear	Not clear	Not clear	Not clear	857
Czech Republic, Bohemia	NS	NS	E	200	174 (87%)	Trypsin (87.0%)	—	Gross examination (29.0%), trichinoscopy (62.5%)	*S. cruzi, S. hominis, S. hirsuta*	40% single infections, 42% two species, and 18% three species	764
Czech Republic, Bohemia	NS	Abattoir	E	208	167 (80.3%)	Trichinoscopy	—	LM, H and E	*S. cruzi* (40.0%), *S. hominis* (5.7%), *S. hirsuta* (19.2%), mixed (4 cases)	—	332

(*Continued*)

Table 7.2 (Continued) Prevalence of Sarcocystis Sarcocysts in Muscle of Cattle

Country	Year	Source	Tissue	No. Tested	Total Positive	Method: Digestion/ Squeezing	Method: PCR	Method: LM, SEM, TEM	Species Identified	Remarks	Reference
Egypt	1989–90	Abattoir	H, D, T	10	4 (40.0%)	—	—	Histology	NS	—	564
Egypt, Assiut	NS	Abattoir	H, E, D, ocular muscle	100	94 (94.0%)	Muscle squash	—	LM	S. cruzi, by morphology and the success in bioassay in dog; cats did not shed sporocysts. Macroscopic cysts (4.0%) are suspected to be S. hirsuta	By ELISA using S. fusiformis antigen: 98.0%.	1551
Egypt, Sharkia	NS	Abattoir	E	81	24 (29.6%)	Gross, muscle squash	RAPD-PCR	LM, TEM	Only microcysts: thin (S. cruzi, 29.6%) and thick (S. hominis, 4.9%) walled	—	54
Estonia	NS	NS	NS	2892	1537 (53.1%)	NS	NS	NS	NS	NS	1193
Germany, South	1977–78	Abattoir	E, D, T	1020	1007 (99.7%)	Trypsin (99.7%)	—	Trichi-noscopy (57.9%)	S. cruzi (65.6%), S. hominis (63.6%), S. hirsuta (34.5%)	—	116
Germany	NS	Retail stores	100 g beef	257	174 (67.7%)	—	67.7% by PCR, 69.6% multiplex	40% (52/130) by LM	28.5% thin walled (S. cruzi) and 24.6% thick walled (S. hominis, S. hirsuta, S. sp.)	Mixed infection: 28%	1221
Hungary	2014	Abattoir	E, H	151	100 (66.0%)	—	PCR	—	S. cruzi (64%), S. hominis (19%), S. sp., S. rommeli (17%)	—	1143

(Continued)

Table 7.2 (Continued) Prevalence of *Sarcocystis* Sarcocysts in Muscle of Cattle

Country	Year	Source	Tissue	No. Tested	Total Positive	Digestion/ Squeezing	Method PCR	Method LM, SEM, TEM	Species Identified	Remarks	Refer- ence
India	NS	Abattoir	E, T, D, H, ocular muscle, M	238	191 (80.25%)	Trypsin digestion	—	LM	*S. cruzi, S. hominis, S. hirsuta*	—	884
India	NS	Animals submitted previously to experi- mental infection with *Clostridium chauvoei*	Brain	30	12 (40.0%)	—	—	Histology	NS	NS	1616
India	1994– 95	Abattoir	H	200	98 (49.0%)	—	—	Gross, histology	*S. cruzi, S. hominis*	NS	323
India, Bangalore	NS	NS	E, H, D, T	100	84 (84.0%)	Pepsin digestion	—	LM, Histology	*S. cruzi* (78.5%), *S. hominis* (21.4%), *S. hirsuta* (34.5%)	E: 84.5%, H: 79.7%	277
Iran, Kerman	2005– 06	Abattoir	H, D, T, M	480	480 (100%)	Pepsin digestion of 10 g	—	—	NS	No macro- scopic cysts	1283
Iran, South West	2009	Abattoir	H, D, T, M	344	480 (100%)	Pepsin digestion of 50 g: 100%, Muscle squash: 94.7%	—	—	NS	The most frequently infected tissue was heart: 93.3%	773
Iran, Isfahan	2008	Abattoir	E, D	100	92 (92.0%)	Muscle squash	—	Gross inspection, LM, TEM	*S. cruzi* in 89%, *S. hominis* or *S. hirsuta* (identified as thick-walled cysts) in 21%	—	1282

(Continued)

Table 7.2 (Continued) Prevalence of *Sarcocystis* Sarcocysts in Muscle of Cattle

Country	Year	Source	Tissue	No. Tested	Total Positive	Method			Species Identified	Remarks	Reference
						Digestion/ Squeezing	PCR	LM, SEM, TEM			
Iran, Babol	NS	Retail stores	5 g muscle	1 isolate	—	—	PCR, sequencing	LM, histology	*S. cruzi*	—	928
Iran	NS	Abattoir	Intercostalis, D	1 cow	—	—	—	Gross inspection, LM, TEM	3–5 mm length cysts of *S. hirsuta*	First report in Iran	1601
Iran, North Khorasan	2013	Abattoir	H, D, meat, liver	32	32 (100.0%)	Pepsin digestion	—	—	NS	Bradyzoites were found in 100% of 32 livers	1524
Iran, Kurdistan	NS	Abattoir	E	130	56 (43.1%)	Muscle squash	—	LM	—	—	1449
Italy, Milan	NS	Abattoir	H, masseter	154	91 (59.1%)	—	—	LM, histology	*S. cruzi*, *S. hominis*, *S. hirsuta*	—	1553
Italy	2008–09	Abattoir	22 different muscles	50	48 (96%)	—	—	LM, histology	Thin-walled, *S. cruzi*	—	150
Italy	2008	Abattoir	H, E, D	384	300 (78.1%) by histology	—	PCR	LE, TEM	By PCR: *S. cruzi* (74.2%), *S. hirsuta* (1.8%), *S. hominis* (42.7%), *S. hominis*-like (18.5) present 0.7%–1.4% divergence with *S. hominis*	E: 11.7%, E+H+D: 62.0%, E+H: 10.0%, E+D: 9.0%, D: 3.7%, H: 2.3%, H+D: 1.3%	331
Japan	Not clear (in Japanese)	Not clear (in Japanese)	H, D	105	44 (41.9%)	—	—	Histology Heart: 97.7%, Diaphragm: 18.2%	*S. cruzi*	Bioassay in dog and cat; only sporocysts were shed by dogs	1223

(Continued)

Table 7.2 (Continued) Prevalence of Sarcocystis Sarcocysts in Muscle of Cattle

Country	Year	Source	Tissue	No. Tested	Total Positive	Digestion/ Squeezing	PCR	LM, SEM, TEM	Species Identified	Remarks	Reference
Japan, Niigata	1989	Abattoir	H, T, masseter	25	25 (100.0%)	Trypsin digestion	—	LM, SEM	*S. cruzi*	H: 100.0%, T: 4.0%, masseter: 8.0%	**1852**
Japan, Aichi	1989	Abattoir	H, masseter	30	17 (56.7%)	Trypsin digestion, squeezing	—	LM, histology	*S. cruzi* (94.0%); *S. hirsuta* (23.6%)	H: 56.7%; Masseter: 53.3%	**872**
Japan, Hokkaido	1993	Abattoir	H	83 autoch-thonous, 185 imported	13 (15.7%) in autoch-thonous, and 92 (49.7%) in imported	Muscle squash	—	LM	*S. cruzi*	Bioassay in dogs	**1335**
Japan	1996–98	Abattoir, retail stores	Loin	482	72 (14.9%)	—	—	LM, histology	*S. cruzi*, only one case presented thick wall	Origin of cattle: Japan (6.3%), America (36.8%), Australia (29.5%)	**1337**
Japan, Saitama	NS	Abattoir	D	5	4 (80.0%)	—	—	LM, TM, SEM	*S. hominis* (other species not specified)	Bioassay in two cynomolgus monkeys: shed sporocysts 10 DPI	**1521**
Japan, Saitama	1999–00	Abattoir	D (100 g)	840 (3–10 yr old)	2 (0.23%)	—	—	LM, SEM, TEM	Thick-walled (7–10 μm) but not recognized species. Possibly, *S. hirsuta*	—	**1522**

(*Continued*)

Table 7.2 (Continued) Prevalence of *Sarcocystis* Sarcocysts in Muscle of Cattle

Country	Year	Source	Tissue	No. Tested	Total Positive	Digestion/ Squeezing	PCR	LM, SEM, TEM	Species Identified	Remarks	Refer- ence
								Method			
Lithuania	Not clear (in Lithu- anian)	Abattoir	H, D, chest muscle	NS	Not clear (100.0%)	NS	—	NS	*S. cruzi* (100% hearts, 34.6% diaphragm, 12.1% chest muscle), *S. hominis* (0% hearts, 65.4% diaphragm, 87.9% chest muscle), *S. hirsuta* (1 case)	—	738
Malaysia	2011– 12	Abattoir	E, H, T, M, D	102	37 (36.2%)	Squeezing: 37 (36.2%)	7 cases sequen- ced	Histology	*S. cruzi*	Skeletal muscle and diaphragm (27%), tongue and esophagus (24.3%), and heart (8%)	1029
Mexico, Mexico DF	NS	Retail meat store	H	100	94 (94.0%)	—	—	LM	NS	—	1185
Mongolia	1998– 99	Abattoir	H, D, T	30	27 (90.0%)	Squeezing	—	—	NS	H (100%), D (61.1%), T (90.0%)	640
New Zealand	1983– 84	Abattoir	E, D	500	498 (99.6%)	E: 97.4%, D: 51.8%	—	LM, TEM, SEM	*S. cruzi* in 98.0%, thick-walled cysts in 79.8%	Bioassay in cat and human volunteer; only cat shed sporocysts	127

(Continued)

Table 7.2 (Continued) Prevalence of Sarcocystis Sarcocysts in Muscle of Cattle

Country	Year	Source	Tissue	No. Tested	Total Positive	Digestion/ Squeezing	Method PCR	Method LM, SEM, TEM	Species Identified	Remarks	Refer- ence
New Zealand	1985	Abattoir	E, D, T, H	100	64 (64.0%)	—	—	Gross	—	Macro sarcocysts in 64. No microcysts investigated	1200
Nigeria, Zaria	2011	Abattoir	E, D	200	85 (42.5%)	Pepsin digestion	—	LM	S. cruzi in 99.0%, S. hominis in 4.0%	E (88.2%), D (65.9%)	1300
Peru	NS	Abattoir	H, E	85	H: 76 (89.5%), E: 2 (2.4%)	Trichi- noscopy	—	LM	NS	—	170
Philippines, Manila	1994	Abattoir	E, H, M, D	608	99 (16.3%)	Muscle squash	—	LM	NS	—	219
Philippines, Manila	NS	Abattoir	E, H, muscle from limbs and cervix	370 (3–11 yr)	40 (10.8%)	Muscle squash	—	LM, TEM (unsuc- cessful)	S. cruzi, S. hominis	—	221
South Korea	1988	Abattoir	H	1442	419 (29.1%)	Trysin digestion	—	LM	Apparently S. cruzi	—	896
South Korea	1991	Abattoir	H	330	144 (43.6%)	Trysin digestion	—	LM	S. cruzi	Bioassay in dog and cat (300 g meat)	1372
Sudan	NS	Abattoir	E, H, M, D	180	176 (97.8%)	Pepsin digestion	—	–	NS	E: 56.7% H: 95.6% M: 64.4% D: 63.9%	868
Thailand	NS	Abattoir	E	328	326 (99.4%)	Pepsin digestion	—	Gross, histology	S. cruzi, S. hominis, S. hirsuta	Macroscopic: 2 (0.61%) Microscopic: 57 (17.38%)	1233

(*Continued*)

Table 7.2 (Continued)   Prevalence of Sarcocystis Sarcocysts in Muscle of Cattle

Country	Year	Source	Tissue	No. Tested	Total Positive	Digestion/ Squeezing	Method PCR	LM, SEM, TEM	Species Identified	Remarks	Reference
Turkey	1985–86	Abattoir	D	1546	1416 (91.5%)	Not clear (in Turkish)	—	Not clear (in Turkish)	S. cruzi, S. hominis, S. hirsuta	No macroscopic cysts	1343
Uruguay	NS	Abattoir	H, E, T	15	15 (100.0%)	—	—	Histology	S. cruzi 15 (100%) of 15, S. hirsuta 2 (13.3%) of 15	Bioassay in dog and cat	634
USA, Texas	NS	Retail meat store	D	4	4 (100%)	Pepsin digestion	—	—	NS	—	132
USA, Uruguay	2004–05	Retail meat store	Sirloin	110	60 (54.5%)	—	54.5%: 41 sequenced	LM: 44 (40.0%)	Only S. cruzi	—	1427
Venezuela	2008	Abattoir	H, myocardium of fetuses	35	34 cows (97%), 0 fetuses (0.0%)	Muscle squash, histology	—	LM	NS	Myocarditis and pericarditis was detected in fetuses; presence of schizonts was not reported	1523
Vietnam	2003	Abattoir	D, E, T, muscle of neck	101	64 (63.3%)	Muscle squash: 25.0%	RFLP–PCR: 63%	LM	S. cruzi, S. hominis, S. hirsuta S. sp (1 S. sibirica-like sarcocyst)	—	905

D = diaphragm; E = esophagus; H = heart; M = skeletal muscle; T = tongue; NS = not stated.

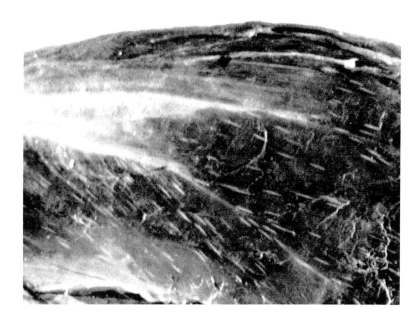

Figure 7.3 Macroscopic sarcocysts (arrow), probably *S. hirsuta*, in beef. (From Collins, G.H., 1980. *Sarcocystis and the Meat Industry*. Monograph, Massey University, New Zealand, 1–15. With permission.)

7.2.4 Other Aspects

There are numerous surveys in which *S. hirsuta* has been detected, in most of cases, as mixed infections (Table 7.2). Although this species does not represent a zoonotic risk, its frequent macroscopic character has resulted in economic loss by condemnation of carcasses during meat inspections.[397] *S. hirsuta* has been detected in meats and meat products destined for human comsumption.[771,1221,1382,1776]

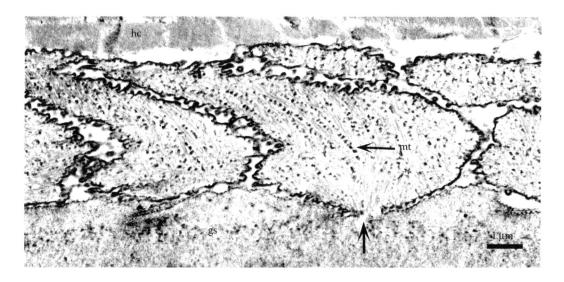

Figure 7.4 TEM of type 28 sarcocyst wall of *S. hirsuta*. Note conical VP, constricted at the base (arrow) and fine microtubules (mt).

7.3 *SARCOCYSTIS HOMINIS* (RAILLIET AND LUCET 1891) DUBEY 1976

7.3.1 Distribution

Argentina, Belgium, Brazil, Canada, England, Germany, Italy, Japan, United States, and Vietnam.

7.3.2 Definitive Hosts

Humans (*Homo sapiens*), rhesus monkey (*Macaca mulatta*), cynomolgus monkey (*Macaca fascicularis*), baboon (*Papio cynocephalus*), and possibly chimpanzee (*Pan troglodytes*).

7.3.3 Structure and Life Cycle

The sarcocyst wall is up to 6 μm thick and appears radially striated because of numerous vp, which are up to 7 μm long and 0.7 μm wide, type 10b[1135] (Figures 7.1c, 7.2b, 7.5 and 7.6). The villar

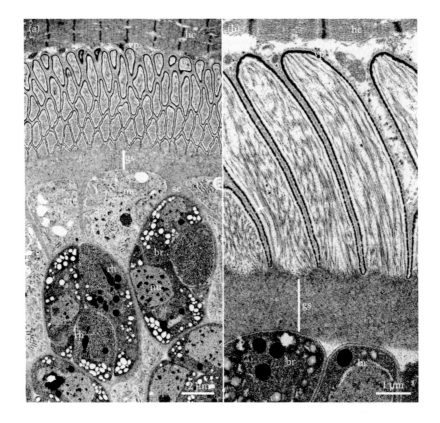

Figure 7.5   TEM of *S. hominis* sarcocyst in tongue of 2 calves after oral inoculation with sporocysts from the feces of a human volunteer who ate raw beef. Note elongated vp with thin microtubules mt. The parasitophorous vacuolar membrane is lined by electron dense layer that is thinned out at places (arrowheads). The gs is thicker at the point of origination of se. (a) 111 days p.i. and (b) 222 days p.i. The vp are of the same length in both sarcocysts but thinner at day 111 p.i. (Adapted from Dubey, J.P., Fayer, R., Speer, C.A., 1988. *J. Parasitol.* 74, 875–879.)

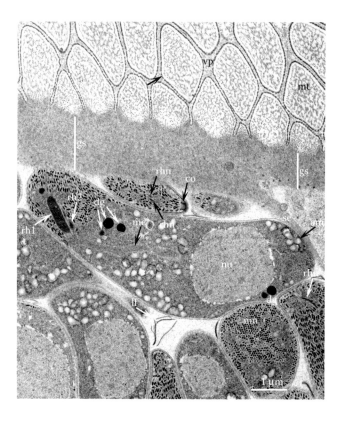

Figure 7.6   TEM of *S. hominis* sarcocyst in tongue of a calf. The vp are cut at an angle. The pvm in vp is lined by electron dense layer that is thinned out at places (arrow). The mt are filamentous. The gs is thicker at the point of origination of se. Also note bradyzoites with a co, 2 rh, few dg, numerous am, a nu, an elongated mc, and leaflet–like structures (Lf) between bradyzoites.

core contains numerous microtubules that run from the base to the apex. Bradyzoites are 7–9 μm long and are arranged in packets (Figure 7.6).[1135]

S. hominis is only mildly pathogenic.[391] Calves inoculated with 1 million sporocysts develop mild anemia, but survive.[391]

A novel *S. hominis*-like parasite was detected in 18.5% of the cattle from northwestern Italy.[331] Hook-like structures were observed at the base and at the apex of the vp; further studies will be needed to confirm its identity.

7.4 *SARCOCYSTIS ROMMELI* DUBEY, MORÉ, VAN WILPE, CALERO-BERNAL, VERMA, SCHARES

This species was recently named[2392]; it was initially reported as *S. sinensis,* which is now regarded as *nomen nudum*.[211,471,701,1218,1221,1858]

7.4.1 Distribution

China, Vietnam, Germany, Hungary, Austria, Argentina.

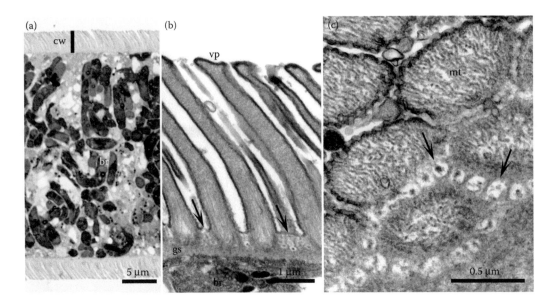

Figure 7.7 *Sarcocystis rommeli* from cattle. (a) Part of thick-walled sarcocyst in loin sample from Argentina. (b) TEM of the sarcocyst wall. Finger-like villar protrusions with microtubules bent at an angle. Note the thin ground substance (0.5 μm). Sarcocysts collected from a minced beef sample from Germany. (c) TEM image of the base of the villar protrusions. Note the presence of numerous microtubules (mt) and vesicles (arrowheads). Sarcocysts collected from a cattle loin sample from Argentina. Bar in (a) = 40 μm, (b) = 1.5 μm, (c) = 0.5 μm. (Courtesy of G. Moré.)

7.4.2 Definitive Host

Unknown, but not humans, dog, racoon dog, nor red fox.

7.4.3 Structure and Life Cycle

Sarcocysts microscopic, sarcocyst wall 4.5–5.2 μm thick. Vp 4–5 μm long, wavy, bent at an angle with mt extending from the tip to the middle of the gs, with vesicles at the base. The gs thin (<1 μm thick). Bradyzoites 10–12 μm long (Figure 7.7a–c).

7.4.4 Molecular studies

Based on 18S rRNA and cox1 gene sequences,[701,1143,1218,1221,2392] *S. rommeli* is molecularly different than *S. cruzi*, *S. hirsuta*, and *S. hominis*.

7.5 OTHER UNCONFIRMED SPECIES

Novak et al.[1284] described a species of *Sarcocystis* from cattle in Alma-Ata, former USSR. Sarcocysts were macroscopic, up to 20 mm long. By LM, cyst wall was up to 5 μm thick. By TEM, the PVM had mushroom-like protrusions. Odening[1317] named this parasite as *S. novaki*. In our opinion, this is not a valid species.

7.6 PREVALENCE OF SARCOCYSTS IN CATTLE

Based on examination of tissues obtained at abattoirs, most cattle in the United States and throughout the world are infected with *Sarcocystis*. Recent surveys indicate that *S. cruzi* is the most prevalent species (Table 7.2), and this species is easily recognized in histologic sections, whereas *S. hirsuta* is difficult to distinguish from *S. hominis* microscopically. However, the sarcocyst walls of *S. hirsuta* and *S. hominis* can be differentiated ultrastructurally. The vp of the sarcocyst wall of *S. hirsuta* are approximately 8 μm long, constricted at their base, expanded laterally in the midregion, and tapered distally. The vp core contains numerous filaments and rows of electron-dense granules oriented parallel to the longitudinal axis of the protrusion. The sarcocyst wall appears only approximately 4 μm thick, because the distal region of each protrusion is folded over neighboring protrusions. The villar protrusions of *S. hominis* measure 5.7×0.6 and 7×1.4 μm at 111 and 222 DPI, respectively, are cylinder shaped and contain scattered granules and filaments that extend from the base to the tip of the projection.[391]

7.7 EOSINOPHILIC MYOSITIS

Eosinophilic myositis was discussed in Chapter 1 (Figures 1.5 and 1.6). After the acceptance that *Sarcocystis* spp. was the etiological agent of this process,[309,652,730,1472] several studies were focused on identifying which species of the genus were involved in the process. In this sense, while analyzing cattle carcasses affected by EM, Jensen et al.[907] detected, for the first time, *S. hominis* infecting cattle in the United States. Lesions were also associated with *S. cruzi* and *S. hirsuta*. Several studies report that *S. cruzi* is the main cause of EM,[526,654,725,1511] but *S. hominis* is the principal agent.[12,1778,1840] It is hypothesized that degeneration or rupture of the sarcocysts is the trigger for the development of EM.[652,1777]

7.8 SARCOCYSTOSIS-LIKE ENCEPHALITIS IN CATTLE

Isolated cases of fatal encephalitis were reported from Canada,[389] Germany,[1718] England,[751] and the United States (Dubey, unpublished). In both published instances, affected cattle developed neural signs before death. The case from Canada was in an 18-month-old steer. The steer became ataxic, recumbent, and blind. Grossly, a 1-cm-diameter, gray discolored area was seen in the cerebellum. Microscopically, the grossly visible lesion consisted of focal malacia, vasculitis, and infiltration of mononuclear cells, mainly macrophages (Figure 7.8). Numerous immature and mature schizonts and extracellular merozoites were seen; these stages resembled *S. cruzi* in structure. However, such lesions have never been seen in experimentally infected cattle. No clinical infection in the brain stem was reported by Singh and Parihar[1616] in 12 bullocks from India.

7.9 FOOD SAFETY

S. hominis is a well-known health risk for humans[210,807,1382]; the European Food Safety Authority is concerned about the risk associated with the presence of *Sarcocystis* sarcocysts in beef and stimulates efforts on its control.[1727] Generally, carcass condemnation by meat inspection services is based on gross examination and is associated with nonzoonotic species *S. hirsuta*.[397] This effort is not enough, and complementary assays based on microscopical examination[150,1427] are needed; related to this, *S. hominis*-infected meat and meat products had been detected in high percentages in Belgium, Brazil, Italy, Germany, and Japan.[331,675,1221,1337,1382,1776]

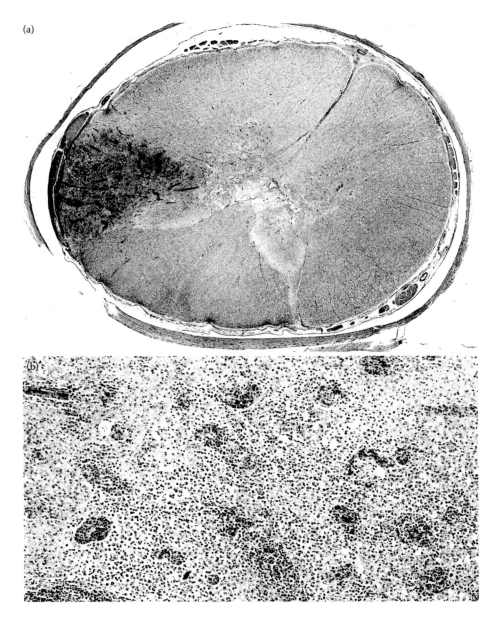

Figure 7.8 Encephalomyelitis due to a *Sarcocystis*-like organism in a naturally infected calf. H and E stain. (a) An area of necrosis in the spinal cord. (b) High-power magnification of (a), showing severe vasculitis and infiltration of macrophages.

BIBLIOGRAPHY

Complementary information pertinent to the subject matter of this chapter may be found in the following references:

Eosinophilic myositis: **12, 139, 151, 309, 526, 652, 654, 724, 725, 730, 907, 1472, 1511, 1777, 1778, 1840, 1943, 2014**.

Experimental transmission: **304, 507, 508, 666, 673, 676, 882, 883, 925, 948, 1059, 1084, 1268, 1276, 1280, 1475, 1512, 1548, 1549, 1550, 1667, 1785, 1814, 1966, 1978, 1983, 2004, 2006–2008,**

2047–2050, 2057, 2087, 2118, 2135, 2163, 2164, 2171, 2172, 2202, 2233, 2234, 2269, 2273, 2307, 2318, 2329, 2350, 2356, 2370.

Food safety/zoonosis: **150, 210, 211, 397, 675, 771, 807, 1221, 1337, 1382, 1427, 1727, 1776, 1943, 1947, 1982, 1984, 2022, 2042, 2043, 2304.**

Miscellaneous: **31, 54, 55, 56, 58, 125, 154, 177, 255, 594, 874, 1228, 1271, 1362, 1516, 1545, 1547, 1920, 1935, 1938, 1950, 1965, 2017, 2076, 2090, 2105, 2136, 2137, 2138, 2148, 2162, 2166, 2186, 2188, 2217, 2220, 2258, 2259, 2260, 2315, 2339, 2351, 2364, 2366.**

Molecular biology: **608, 719, 771, 747, 905, 957, 1056, 1095, 1218, 1272, 1283, 1477, 1493, 1846, 1856, 1858.**

Pathogenesis and clinical disease: **30, 55, 56, 70, 71, 167, 227, 232–236, 251, 318, 323, 505, 506, 507, 508, 681, 751, 782, 844, 1152, 1242, 1246, 1444, 1529, 1582, 1774, 2359.**

Prevalence/epidemiology: **37, 54, 116, 126, 127, 132, 139, 150, 170, 219, 221, 275, 277, 318, 323, 331, 332, 340, 471, 564, 634, 640, 738, 747, 764, 773, 857, 868, 872, 884, 896, 905, 928, 1029, 1143, 1185, 1193, 1200, 1213, 1215–1217, 1221, 1223, 1233, 1241, 1282, 1283, 1300, 1335, 1337, 1343, 1372, 1417, 1427, 1446, 1449, 1521–1524, 1529, 1541, 1542, 1546, 1551, 1553, 1601, 1616, 1776, 1784, 1786, 1787, 1852 1916, 1930, 1931, 1933, 1944, 1964, 1965, 1967, 1968, 1973, 1978, 2003, 2012, 2014, 2017, 2036, 2040, 2056, 2061, 2064, 2070, 2075, 2081, 2101, 2119, 2126, 2151, 2190, 2215, 2226, 2231, 2242, 2243, 2249, 2253, 2265, 2297, 2302, 2306, 2308, 2325, 2338, 2354, 2357, 2358, 2360, 2373.**

Morphology and ultrastructure: **126, 127, 225, 391, 395, 884, 1135, 1382, 1441, 1785, 1840, 2065, 2392, 2396.**

CHAPTER 8

Sarcocystosis in Sheep (*Ovis aries*)

There are 4 species of *Sarcocystis* in sheep (Figure 8.1). *S. tenella* and *S. arieticanis* are trans-mitted via canids; *S. gigantea* and *S. medusiformis* are transmitted via cats. *Sarcocystis* species are prevalent in sheep worldwide (Table 8.1).

8.1 *SARCOCYSTIS TENELLA* (RAILLIET 1886) MOULÉ 1886

8.1.1 Definitive Hosts

Dog (*Canis familiaris*), Coyote (*Canis latrans*), and Red fox (*Vulpes vulpes*).

8.1.2 Structure and Life Cycle

Sarcocysts are microscopic, up to 700 µm long, and are found in striated muscles, including heart and tongue, and in the CNS.[367] The sarcocyst wall is 1–3 µm thick and has vp up to 3.5 µm long and up to 0.5 µm wide (Figure 8.1a).[367,1796] There are no microtubules in the villi, but plaques are present (Figure 8.2a).

Information on the life cycle is summarized in Table 8.2. There are 2 well-documented genera-tions of schizogony in the blood vessels; a possible third generation was reported in 2 lambs at 36 DPI in macrophages in visceral lymph nodes and Küpffer cells.[1288] Schizonts were small (7.4 × 5.1 µm) and contained 6–9 merozoites. Differences in size, parasitemia, and a number of phases of schizog-ony reported by various authors[367,804,818,1036,1037,1245,1288] might reflect the techniques used, the number of stages measured, or the possibility that *S. arieticanis* was present as a contaminant.

Sarcocysts first seen at 35 DPI had a 1-µm-wide sarcocyst wall that appeared nonstriated until over 52 DPI, when the wall was up to 2 µm thick and appeared cross-striated, type 14. At 75 DPI, sarcocysts were up to 700 µm long and contained both metrocytes and bradyzoites.[367,1170]

8.1.3 Pathogenicity

S. tenella can cause anorexia, weight loss, fever, anemia, loss of wool, abortion, pre-mature birth, neural signs, myositis, and death, depending on the number of sporocysts ingested.[3,390,551,672,867,1243,1245,1250,1254,1258,1261,1393] Doses of 1000–100,000 sporocysts have been tested for their effect on lambs. Data using doses lower than 50,000 sporocysts are reviewed.

8.1.3.1 Clinical Disease

In Australia, 288 (179 inoculated, 109 uninoculated) 2- to 4-week-old *Sarcocystis*-free lambs were examined at 1 and 4 months PI.[1250,1261] At low doses of 1000–5000 sporocysts, as few as

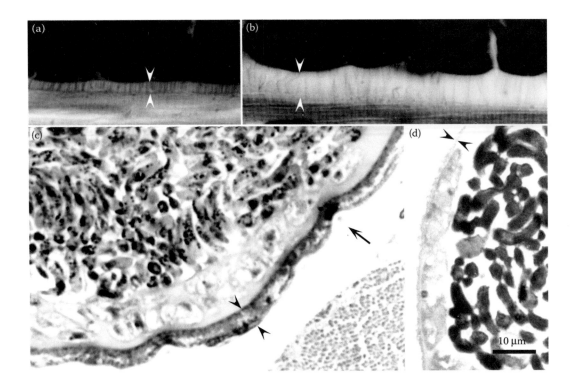

Figure 8.1 A comparison of sarcocyst walls of the 4 species of *Sarcocystis* in sheep. Arrowheads point to the tips of villi or outer margin of the sarcocyst wall. (a) *S. tenella* with thick vp. (b) *S. arieticanis* with long, hirsute protrusions. (c) *S. gigantea* with thin sarcocyst wall (arrow). A connective tissue capsule (arrowhead) surrounds the sarcocyst wall. (d) *S. medusiformis* with thin sarcocyst wall with small protrusions. (a and b) H and E stain. (d) Toluidene blue stain. Bar applies to all parts.

1000 sporocysts reduced hematocrits at 1 month PI, but both hematocrits and weights at 4 months were the same as in uninoculated lambs (Table 8.3). Ingestion of 10,000 sporocysts significantly depressed wool-growth in lambs at 1 month of age, but not at 5 months of age.[1258]

In Germany, ingestion of 25,000 sporocysts affected the health and growth of lambs.[551] The mean carcass weight and mean daily weight of 40 infected lambs in 4 groups of 10 lambs each, based on identical weights, was compared with those of 40 noninfected control lambs; all lambs were reared and managed on a commercial farm. Three infected lambs died and 1 lamb was condemned at slaughter because of carcass appearance. The carcass weights of the 4 groups of the inoculated lambs were 27%, 15.9%, 18.5%, and 6.5% less than the weights of the uninfected controls.[551]

8.1.3.2 *Effect on Parturition and Reproductivity*

S. tenella can cause abortion in sheep.[1037] Of 12 ewes bred 83–85 days earlier, 4 were each fed with 50,000, 100,000, or 500,000 sporocysts. Of those fed 50,000 sporocysts, 1 was not pregnant, 1 aborted 33 DPI, and 2 lambed normally; all survived the acute phase of sarcocystosis. Of those fed 100,000 sporocysts, 1 aborted 40 DPI, 1 lambed normally, and 2 ewes carrying dead fetuses became moribund and were euthanized. Of those fed 500,000 sporocysts, 3 died 27 and 39 DPI, and 1 became moribund and was euthanized 39 DPI; 2 of these ewes aborted 28 and 30 DPI and 2 others had dead fetuses in utero. *Sarcocystis* was not found in tissue sections of any fetus or lamb.[1037]

In another study to determine the effects of sporocyst dosage on reproduction, 20 *Sarcocystis*-free ewes were allotted to 5 groups, A–E (Table 8.4).[1254] Ewes in Group A (6 ewes) were grazed on

Table 8.1 Reports of *Sarcocystis* Prevalence in Sheep

Country	Year	Source	No. Tested	Tissue	No. Positive	Species	Method Histology	Digestion	PCR	Bioassay	Reference
Australia, South	1974–75	Abattoir	864	E, H, D, M	805 (93.2%) microscopic, 54 (6.7%) macroscopic	*S. gigantea, S. medusiformis*	Gross, trichinoscopy	—	—	—	1290
Australia, North	NS	Abattoir	51 lambs	H	47 (92.1%)	*S. tenella, S. arieticanis*	Histology	Pepsin digestion	—	—	1405
Australia, Perth	1989–92	Abattoir	146	H, D (200 g)	138 (95.2%)	*S. tenella* 9.8%, *S. arieticanis* 48.8%, mixed 41.5%	Yes	Pepsin digestion	—	—	1544
Austria	1991–92	Abattoir	500	E	313 (62.6%)	*S. tenella* 25.6%, *S. arieticanis* 24.0%, *S. gigantea* 2.4%	Gross, trichinoscopy	Trypsin digestion	—	—	834
Brazil	NS	Abattoir	602	H, D, Brain	2 (of 22 serologically positive for *Toxoplasma*)	*S. tenella*	—	Pepsin digestion	PCR-RFLP	—	276
Czech Republic	NS	Abattoir	342 lambs	Blood, muscle NS	35.7% microcysts, 94.4% antibodies by IFAT	NS	NS	NS	—	—	1699
Czech Republic	1979–80	NS	100	E	92 (92.0%)	*S. tenella* in all positive animals, *S. gigantea* in 37.0%	Gross, trichinoscopy	Pepsin digestion	—	—	764
Czech Republic	1985–88	Abattoir	1014 different ages	E, D	611 (60.3%)	*S. gigantea* in 3 sheep	—	Trypsin digestion	—	—	1698
Ethiopia, Northwest	NS	Abattoir	208	E, D, H, M	195 (93.7%)	NS	Histology	—	—	—	1835

(Continued)

Table 8.1 (Continued) Reports of *Sarcocystis* Prevalence in Sheep

Country	Year	Source	No. Tested	Tissue	No. Positive	Species	Histology	Digestion	PCR	Bioassay	Reference
								Method			
Germany, Bavaria	1978	Abattoir	500	E, D	427 (85.4%)	*S. tenella* in 63.%, and *S. arieticanis* 84.8% of 166 sheep	Gross examination, trichinoscopy	Trypsin digestion	—	—	117
India, Uttar Pradesh	1990	Abattoir	32	E, D	23 (71.8%)	*S. tenella* (56.5%), *S. arieticanis* (26.0%)	Gross examination, muscle squash	—	—	—	1528
India, Orissa	NS	Abattoir	102	E, T, H, D, M	83 (81, 3%); H: 90.0%, T: 85.8%, E: 90.2%, D: 81.4%, M: 78.4%	*S. tenella*	Muscle squash	Pepsin digestion	—	—	1207
India, Hissar	NS	Veterinary laboratory and abattoir	32	D, H	26 (81.2%)	NS	—	Pepsin digestion	—	—	760
India, Andhra Pradesh	1961–82	Veterinary laboratory	3335	NS	1295 (37.3%)	NS	Histology	—	—	—	1448
India, Andhra Pradesh	NS	Abattoir	197	T, E, H, D, M	174 (88.3%)	*S. tenella*	—	Pepsin digestion	—	—	1784
Iraq	1992–96	Abattoir	605	E, H, D, M	587 (97.0%), macroscopic cyst in 25 (4.1%)	NS	Gross examination, muscle squash	Pepsin digestion	—	—	1030
Iraq, Duhok	NS	Abattoir	175	E	25 (14.3%)	*S. gigantea*	Gross examination	—	—	—	24
Iran, North Khorasan	2012–13	Abattoir	40	H, M	40 (100%)	*S. tenella*, *S. gigantea*	Gross examination	Pepsin digestion	PCR-RFLP, sequencing	—	58
Iran	NS	Abattoir	110	E	78 (70.9%)	NS	—	Muscle squash	—	—	1449
Iran, Ardabil	2004–05	Abattoir	2110	E, D, M	716 (33.9%)	NS	Gross examination	—	—	—	294

(Continued)

Table 8.1 (Continued)   Reports of *Sarcocystis* Prevalence in Sheep

Country	Year	Source	No. Tested	Tissue	No. Positive	Species	Method			Bioassay	Refer-ence
							Histology	Digestion	PCR		
Iran, Fars	1992–94	Abattoir	1362	E, D, T, H, M	786 (57.7%)	*S. tenella, S. arieticanis, S. medusiformis, S. gigantea*	Gross examination, histology	Pepsin digestion	—	Dog, cat, hamster and Guinea pig	1338
Iran, North Khorasan	2013	Abattoir	32	H, D, M	32 (100.0%)	NS	—	Pepsin digestion	—	—	1524
Iran, Khouzestan and Lourestan	NS	Abattoir	NS	E, D, M and macroscopic cysts	NS	*S. gigantea*	Gross examination	Pepsin digestion	PCR-RFLP	—	775
Iran, Urmia	2011-12	Abattoir	638	E, D, M	235 (36.8%)	*S. medusiformis, S. gigantea*	Gross examination	Pepsin digestion	RFLP-PCR	—	2393
Japan, Saitama	1994–96	Abattoir	100	H, D-50 g	45 (45.0%)	12% *S. arieticanis* 45% *S. tenella*	Histology, TEM	—	—	Yes, dogs shed sporocysts	1517
Japan, Saitama	NS	Abattoir	16	D, masseter, M	2 (12.5%)	*S. mihoensis* in both cases	LM, TEM, SEM	—	—	Dog and cat; only dogs shed sporocysts	1519
Jordan	1986–88	Abattoir	620	E, D	Macroscopic: 70 (11.3%), microscopic E: 44.1%, D: 50.1%.	*S. tenella, S. gigantea, S. medusiformis*	Gross examination, trichinoscopy	—	—	—	13
Mongolia	1998–99	Abattoir	777	D, H, T	753 (96.9%); D: 93.2%, H: 94.2%, T: 100.0%	NS	Trichinoscopy	—	—	—	640
Nigeria, North	NS	Abattoir	400	D, E	36 (9.0%)	No macroscopic, *S. tenella*	Gross examination, dissection	Pepsin digestion	—	—	984

(Continued)

Table 8.1 (Continued)   Reports of Sarcocystis Prevalence in Sheep

Country	Year	Source	No. Tested	Tissue	No. Positive	Species	Histology	Digestion	PCR	Bioassay	Reference
Peru, Lima	NS	Abattoir	134	H, E	122 (91.0%); E: 39.6%, H: 91.0%	NS	Trichinoscopy	—	—	—	170
Romania			48	E, H, D	44 (91.7%); E: 81.6%, H: 79.9%, D: 68.8%	S. tenella	Gross examination, trichinoscopy	—	—	Dog, cat. Only dog shed sporocysts	1753
Saudi Arabia	2002–03	Abattoir	3050	T, E, D, H, M	2552 (83.7%)	"S. moulei"	Histology, TEM	—	—	Dog, cat. Only cat shed sporocysts	26
Saudi Arabia	2011	Abattoir	3250	E, T, D, H, M	2790 (85.8%); E: 79.0%, T: 68.7%, D: 81.3%, H: 46.8%, M: 67.7%.	S. arieticanis, S. gigantea	Gross examination, histology	—	—	Dog, cat. Dog shed sporocysts	23
Senegal, Dakar	NS	Abattoir	100	E	82 (82.0%)	S. tenella	Histology	Homogenization	—	—	1786
Slovakia	NS	Abattoir	134-Sample A 30-sample B	E, H, D, T	134 (100%) in sample A 30 (100%) in sample B	S. tenella-96.3%, S. arieticanis 3.7%, S. gigantea 5.2% No macrocysts in sample B	—	Trypsin digestion	—	—	1557
Spain, Zaragoza	1978–80	Abattoir	770	E	486 (63.1%) by trichinoscopy, 756 (98.2%) by digestion, 416 (54.0%) by gross examination	NS	Gross examination, trichinoscopy	Pepsin digestion	—	—	1530

(Continued)

Table 8.1 (*Continued*)   Reports of *Sarcocystis* Prevalence in Sheep

Country	Year	Source	No. Tested	Tissue	No. Positive	Species	Method Histology	Method Digestion	Method PCR	Method Bioassay	Reference
Spain, Center and South	2009–10	Abattoir	5720	Carcasses	712 (12.4%)	*S. gigantea* and *S. medusiformis*	Gross examination	—	—	—	1123
Spain, North West	NS	Abattoir	100	E	36 (78.3%) by digestion, 54 (54.4%) macroscopic cysts by gross examination	NS	Gross examination, trichinoscopy	Pepsin digestion	—	—	1384
Sudan	NS	Abattoir	128	E, H, M, D	123 (96.1%); E: 46.1%, H: 84.4%, M: 89.8%, D: 66.4%	NS	—	Pepsin digestion	—	—	868
Turkey, Kirikkale	2005–06	Abattoir	112	E, D, M	66 (58.9%)	*S. tenella* (47.3%), *S. arieticanis* (1.23%)	Gross examination, histology	—	—	—	1345
USA, Maryland	NS	Abattoir	86	D, E	84 (97.7%)	NS	Dissection	Pepsin digestion	—	—	878
USA, Detroit, Michigan	1973	Abattoir	789 sheep, 306 lambs	E (20 g)	10.8% of 306 lambs 75.3% of 789 adult	NS	Gross examination, histology	Pepsin digestion	—	—	1586
USA, North-West and Texas	1983	Abattoir	512 ewes	T, E, D	430 (84.4%); T: 82.1%, E: 44.4%, D: 51.7%	84.0% *S. tenella*, 3.5% *S. arieticanis*, *S. gigantea* (% NS)	Gross examination, histology, TEM	—	—	Dog, cat, both shed sporocysts	392

E = esophagus, H = heart, D = diaphragm, T = tongue, M = skeletal muscle.

NS = not specified.

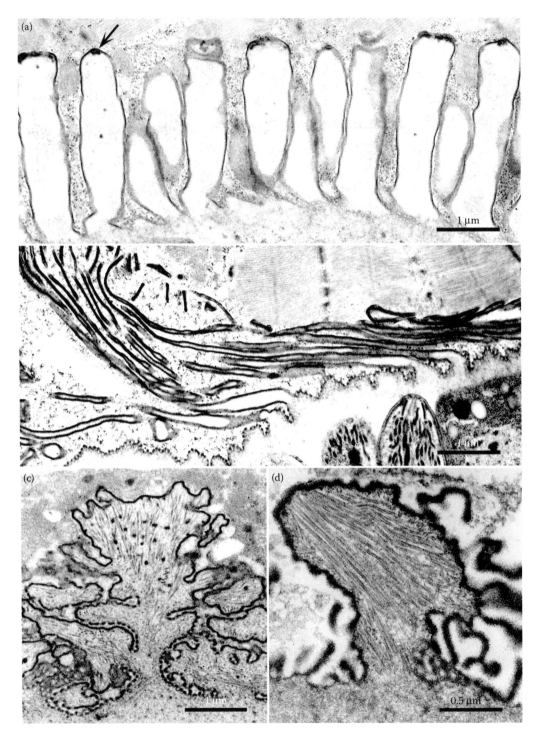

Figure 8.2   TEM showing a comparison of villar protrusions on the sarcocyst wall of the 4 species of *Sarcocystis* in sheep. (a) *S. tenella* with finger-like vp without microtubules. Unusual plaque structures are visible (arrow) at the tip of the vp. (b) *S. arieticanis* with long filamentous vp tapering at the distal end. (c) *S. gigantea* with cauliflower-like vp (arrow). (d) *S. medusiformis* with short, broad villi and numerous convoluted filaments arising from the main villus.

Table 8.2   Comparison of Developmental Stages of *S. arieticanis*, *S. tenella* in Sheep

	S. arieticanis[a]	*S. tenella*[a]	*S. tenella*[b]
First-Generation Schizonts			
Location	Mesenteric and mesenteric lymph node arteries	Arterioles and arteries throughout the body	Arterioles
Duration (DPI)	14–19	9–21	6–19
Peak development (DPI)	Unknown	16	Unknown
Size of meronts (μm)	45–80 × 35–50	29–45 × 24–32	13.4–35.5 × 12.2–15.1
No. of merozoites	100	Up to 168	18–28
Size of merozoites (μm)	6.8–7.5 × 2.3–3.0	7 × 1.5	Unknown
Second-Generation Schizonts			
Location	Capillaries	Capillaries	Capillaries
Duration (DPI)	26–31	16–40	21–34
Peak development (DPI)	Unknown	25	Unknown
Size of meronts (μm)	Unknown	10.5–42 × 7–17.5	15.2 × 10.6 (12.0–27.4 × 7.5–15.3)
No. of merozoites	Unknown	Up to 54	18–38
Size of merozoites (μm)	6–7.5 × 2.3–3	6 × 15	Unknown
Parasitemia			
Duration (DPI)	Unknown	14–16, 25–32	Unknown
Intraleukocytic multiplication	Yes	Yes	Unknown
Sarcocysts	Microscopic	Microscopic	Microscopic
First seen (DPI)	31	35	41
Maturation time (DPI)	70	75	Unknown
Location	Heart	CNS, heart	Heart
Sarcocyst Wall			
Light microscopy	Thin	Thick	–
TEM	Hair-like protrusions of variable shape,[c] type 7	Palisade-like and cylindrical with smooth surface, type 14	–
Bradyzoites	4.6–6.4 × 1.9–2.3 μm	6.6–7.8 × 1.1–2.2 μm	–
References	**787, 817, 1565**	**367, 818, 1565**	**1288**

[a] In conventionally reared lambs.
[b] In SPF lambs.
[c] Snake-like 0.5–1.2 × 0.07–0.21 μm and trapezoidal 0.87–1.30 × 0.14–0.24 μm in size.[787]

a 0.23-hectare field 1 day after it had been sprayed with 1 million *S. tenella* sporocysts. In Group B, 3 ewes were fed 60,000 sporocysts each; in Group C, 3 ewes were fed 10,000 sporocysts each; and in Group D, 4 ewes were fed with 2500 sporocysts each. In Group E, 4 ewes were not given sporocysts. All 3 ewes fed with 60,000 sporocysts became severely ill and were euthanized 48, 50, and 84 DPI. No others became ill or died. All lambs were euthanized soon after birth, and their tissues were examined histologically. 2 of the 3 lambs born to ewes in Group B were very weak and were estimated to be 1–3 weeks premature. 1 lamb from an ewe in Group A was stunted and also appeared premature; focal areas of leukoencephalomalacia were found in the cerebrum and midbrain; the accompanying placenta had small foci of necrosis. Lesions were not seen in other lambs. *Sarcocystis* was not seen in tissues of any lambs.

Table 8.3 Comparison of Hematocrits and Weights of Lambs Infected with 1000–10,000 S. tenella Sporocysts and of Uninoculated Controls

Expt. No.	No. of Lambs	Dose	Hematocrits (%) 1 Month PI	Hematocrits (%) 4 Months PI	Weights (kg) 0 Day	Weights (kg) 1 Month PI	Weights (kg) 4 Months PI
1[a]	43	0	36	37	9	22	32
	38	5000	28	35	11	22	28
	37	10,000	25	34	10	20	27
2[b]	32	0	36	38	17	25	30
	32	1000	32	35	18	27	31
3[b]	34	0	37	37	13	20	28
	35	2500	31	35	12	19	27
	37	5000	29	35	12	19	26

Source: Data from Munday, B.L., 1979. *Vet. Parasitol.* 5, 129–135; Munday, B.L., 1986. *Vet. Parasitol.* 21, 21–24.
[a] Number of sarcocysts per cubic millimeter were <1, 117, and 102, respectively, in 3 groups.
[b] Sarcocysts were not found in control lambs, but were found in inoculated lambs.

Table 8.4   Effect of S. tenella Sporocysts on Lambing Weights

Group	No. of Sporocysts	No. of Ewes	No. of Sarcocysts/ mm³ [a]	Lamb Weights (kg) and Number of Lambs Singles	Lamb Weights (kg) and Number of Lambs Twins
A	Grazed infected paddock	6	333	2.25 (2)	3.21 (8)
B	60,000	3	1250	2.38 (3)	None known
C	10,000	3	530	3.95 (2)	2.5 (2)
D	2500	4	315	4.0 (3)	2.38 (2)
E	None	4	6	3.6 (3)	3.25 (2)

Source: Data from Munday, B.L., 1981. *Vet. Parasitol.* 9, 17–26.
[a] Biopsied 1 day before and 90 DPI. No sarcocysts were seen in preinoculation biopsies.

8.1.4 Immunity and Protection

Oral inoculation of low numbers of *S. tenella* sporocysts induces protective immunity in sheep.[612] In 20 *Sarcocystis*-free 1-year-old sheep divided into 5 groups of 4 sheep each, sheep in each group were fed 0, 5, 50, 500, or 5000 sporocysts and were challenged with 50,000 homologous sporocysts apiece 13 weeks later.[612] The sheep were euthanized 10 weeks after the challenge infection. In the sheep fed 50 or 500 sporocysts, there were fewer sarcocysts, lower body temperatures, and higher hematocrits than in the sheep fed 0 or 5 sporocysts. The sheep fed 5000 sporocysts became ill.[612]

Protection was reported in sheep inoculated with 150,000 *S. tenella* sporocysts attenuated with ultraviolet irradiation (UV-30 to UV-60 lamp for 30 min) weekly for 3 weeks and then challenged with 15,000 untreated sporocysts.[3] Clinical signs were milder in vaccinated *versus* unvaccinated infected sheep later challenged with 15,000 sporocysts.

Prophylactically administered anticoccidial drugs reduce clinical sarcocystosis.[1040,1041] Amprolium® or Salinomycin® was administered to sheep from 1 day before until 29 days after 100,000 or 1 million *S. tenella* sporocysts were fed. Amprolium premix® (50 or 100 mg/kg body weight) and Salinomycin (1 or 2 mg/kg) were mixed in the daily grain ration. Both drugs reduced clinical sarcocystosis, but did not prevent the completion of the life cycle for some parasites. Then, 63 days after sporocysts were fed, lambs given Salinomycin were challenged with 1 million *S. tenella* sporocysts, and were found to have developed a protective immunity.[1041]

8.2 *SARCOCYSTIS ARIETICANIS* HEYDORN 1985

8.2.1 Definitive Host

Dog (*Canis familiaris*).

8.2.2 Structure and Life Cycle

Sarcocysts are found in striated muscles, but not in the CNS. They are up to 900 μm long, thin-walled (<1 μm), and have hirsute projections 5–9 μm long (Figure 8.1b). The vp lack microtubules and, thus, are folded over the sarcocyst wall (Figure 8.2b), type 7b.[23,392,787,819,940,1557]

Salient life cycle features are summarized in Table 8.2.

8.2.3 Pathogenicity

S. arieticanis is less pathogenic than *S. tenella*.[817] A group of 5 sheep fed 50,000, 1 million, 10 million, or 30 million sporocysts developed 2 peaks of fever. The first peak of fever occurred 14–16 DPI, and the second peak was 26–31 DPI.[817] Sheep fed 50,000 or 1 million sporocysts survived the acute phase of sarcocystosis. Sheep fed 2, 10, or 30 million sporocysts died between 16 and 31 DPI.

8.3 *SARCOCYSTIS GIGANTEA* (RAILLIET 1886) ASHFORD 1977

8.3.1 Definitive Host

Domestic Cat (*Felis catus*).

8.3.2 Structure and Life Cycle

S. gigantea sarcocysts are found primarily in the muscles of the esophagus, larynx (Figures 1.5b and 8.3), and tongue, and, to a lesser extent, in the diaphragm and the rest of the

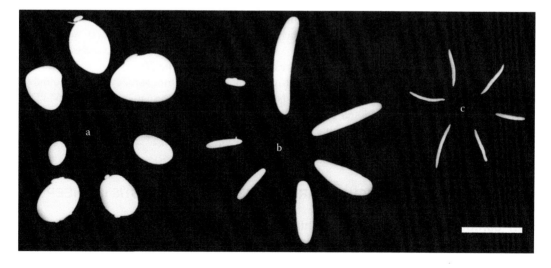

Figure 8.3 Macroscopic sarcocysts in naturally infected sheep: (a) fat *S. gigantea* sarcocysts from the esophagus, (b) slender *S. gigantea* from the diaphragm, and (c) slender *S. medusiformis* from the abdomen. (Courtesy of S. More.)

carcass.[388,712,1258,1259,1572] Sarcocysts have not been found in the heart or the CNS. Macroscopic sarcocysts are found mainly in old sheep. Such sarcocysts are up to 1 cm long, dull white, round, oval, or pear-shaped, sometimes resembling rice grains (Figures 1.5b and 8.3). The sarcocyst wall is thin (<2 μm), smooth, and often surrounded by a PAS-positive connective tissue secondary cyst wall (Figure 8.1c). Ultrastructurally, the sarcocyst wall has cauliflower-like vp, type 21[1260] (Figure 8.2c). The septa are thin (<1 μm wide). In older sarcocysts, live bradyzoites are located peripherally, and the centers of the sarcocysts are often empty.

Little is known of the endogenous development of *S. gigantea* in sheep. Only 1 generation of schizogony is known.[1298] First-generation schizonts were found 7 and 14 DPI, but not 21, 28, and 35 DPI. They were 21–44 × 5–12 μm and were located in capillaries and arterioles in the lung, kidney, and brain.[1298] A sarcocyst was first seen 40 DPI; it measured 71.5 × 8 μm and contained only metrocytes. Bradyzoites were first seen 119 DPI.[1260] Between 10 and 14 months, PI sarcocysts grew up to 1 mm.[388,1260] At 47 months, PI sarcocysts were up to 7.5 × 5 mm. Sarcocysts first became infectious between 230 and 265 DPI. Sarcocysts grew for 4 years or more, and bradyzoites were not highly infectious to cats.[231,388,613,614,674,1259] Cats excreted sporocysts after a prepatent period of 9–11 days.[745]

8.3.3 Pathogenicity

S. gigantea is mildly pathogenic for sheep. Other than fever, lambs inoculated with 1 million sporocysts did not develop clinical signs.[388,1260] The only economic losses reported were due to carcass condemnation after sanitary inspections.[241]

8.4 *SARCOCYSTIS MEDUSIFORMIS* COLLINS, ATKINSON, CHARLESTON 1979

8.4.1 Definitive Host

Cat (*Felis catus*).

8.4.2 Structure and Life Cycle

S. medusiformis sarcocysts are up to 8 mm long and 0.2 mm wide and are found primarily in the diaphragm, abdominal muscles, and the carcass (Figure 8.3). The sarcocyst wall is thin (<2 μm) (Figure 8.1d), and there is no secondary wall.[1211] Ultrastructurally, the vp are trapezoidal. Serpentine filaments arise from the villar surfaces[240] (Figure 8.2d), type 20.

The endogenous development up to 188 DPI is not known. Immature sarcocysts were found in sheep at between 188 and 1132 DPI.[1299] Sarcocysts at 188 and 260 DPI were microscopic; at 443 DPI they were 2–3.5 mm long; at 765 and 1132 DPI they were 4–5 mm long. Cats fed these sarcocysts did not shed sporocysts; therefore, maturation time is unknown.

Cats excreted sporocysts with a prepatent period of 15–30 days.[240,1156]

8.5 REPORTS OF UNUSUAL/RARE SPECIES OF *SARCOCYSTIS* IN SHEEP

8.5.1 *Sarcocystis mihoensis* Saito, Shibata, Kubo, Itagaki 1997

This species was reported only once in 2 of 16 sheep from a farm in Miho, Ibaraki Prefecture, Japan.[1519] Sarcocysts were detected in skeletal muscles. They were up to 2.1 mm long and up to 300 μm wide. The sarcocyst wall was 10–12 μm thick and contained palisade-like vp. Ultrastructurally, the vp were unique type 39 (Figure 1.56a), wide at base, and gradually tapered towards the free end. The microtubules in vp were electron-dense with a mushroom-shaped end. 2 mongrel 6-month-old dogs fed infected muscles excreted sporocysts that were 15–16 × 8–9 µm in size; 2 cats fed the same material did not excrete oocysts.

8.5.2 *Sarcocystis gracilis* Rátz 1909

S. gracilis-like sarcocysts were found in 1 of 50 sheep in Italy.[677] Sarcocysts were 800 µm long and had a 5-µm-thick cyst wall. By TEM, the sarcocyst wall had 3.0–4.5 µm long villar protrusions with type 10 cyst wall characteristic of *S. gracilis* from roe deer (see Chapter 18).

8.5.3 *Sarcocystis microps* Wang, Wei, Wang, Li, Zhang, Dong, Xiao 1988

Oval-shaped sarcocysts were found in myocardium of 21 sheep from Qinghai, China.[1807] Sarcocysts were 150–300 × 50–80 µm. By TEM, the cyst wall was 1.13–1.4 µm, with 0.6–0.69 µm long T-shape vp. Bradyzoites were 8.5 × 3 µm. Also, bioassay in dog and cat was done, only dogs shed sporocyts measuring 14 × 9.1 µm.

8.6 CLINICAL SARCOCYSTOSIS IN NATURALLY INFECTED SHEEP

Scott[1572,1573] spent his career studying sarcocystosis in sheep in Wyoming, United States, and although his efforts were hampered because the life cycle was unknown, he established strong circumstantial evidence that heavy infections retard the growth of sheep. Reports of clinical sarcocystosis in sheep are summarized in Table 8.5.

There are 3 reports of neonatal sarcocystosis in naturally infected sheep. In Brazil, *Sarcocystis*-associated mortality was reported in sheep.[1390] In a flock of Corriedale sheep, 8 ewes aborted or had stillborn lambs. One stillborn lamb was necropsied. Encephalitis and myocarditis were associated with intravascular schizonts in these organs and in kidneys.

A similar case of *Sarcocystis*-associated mortality was observed in a stillborn lamb born in Denmark.[17] The predominant lesions were encephalitis and placentitis associated with schizonts. The encephalitis was characterized by multiple inflammatory foci throughout the brain, predominantly in the cerebrum. Foci of necrosis were surrounded by a zone of microglia and mononuclear cells. Immature and mature schizonts were seen in vascular endothelium, sometimes occluding the lumen (Figure 8.4). Placental lesions consisted of multiple foci of necrosis, inflammation, and mild calcification; numerous schizonts and merozoites were seen within the lesions.

In Australia, neonatal sarcocystosis was found in a 3-week-old lamb with generalized gliosis, and immature and mature schizonts were seen in the brain.[386]

There are several reports of sarcocystosis in weaned and adult sheep. In the United States, debilitation and lymphadenopathy have been found in association with a *Sarcocystis*-like organism.[1324] A 3-year-old ewe had been debilitated for 2 months and for 2 weeks had enlarged prescapular lymph nodes. Microscopically, the lymph node enlargement was found to be due to proliferation of connective tissue. Multiple foci of degeneration and calcified myofibers were seen in the heart, along with sarcocysts and schizonts. Schizonts were also seen in endothelial cells of lymphatic and blood vessels in lymph nodes. Mild arteritis and schizonts were seen in the adventitia. The schizonts were PAS-negative and contained merozoites 5–6 × 1.5–2 µm. Although it was stated that merozoites had rhoptries, none were visible in 3 illustrations. Rhoptries were absent from *Sarcocystis* merozoites (see Chapter 1). In Brazil, schizont-like structures were found in the brain of an adult ataxic sheep.[274]

Table 8.5 Clinical Sarcocystosis in Naturally-Infected Sheep

Country	No., Age	History/Clinical Signs	Diagnosis	Reference
Australia	2, 5–6-month-old lambs	Neurological	Encephalitis, schizonts in brain	396, 784
	3-week-old lamb	Neurological	Encephalitis, schizonts in brain	386
Brazil	1 adult	Sudden death	Encephalitis, schizonts in brain	274
	1 stillborn	8 sheep aborted	Encephalitis, myocarditis, schizonts in several organs	1390
Bulgaria	1 sheep	Neurological	Encephalitis, sarcocysts in brain	89
Canada	4 adult ewes	Neurological	Encephalitis, sarcocysts in brain	793
Denmark	1 stillborn	Abortion	Encephalitis, pneumonia, schizonts	17
France	3-month-old lamb	Neurological	Meningitis	746
Germany	2 sheep submitted to inhalational anesthetic	Sudden death	Myocarditis	148
Hungary	6 adult ewes	Neurological	Encephalitis, sarcocysts in brain	772
Iran	2, 18- and 21-month-old sheep	Heart failure, atrioventricular block	Myocarditis, sarcocysts in heart	1466
Ireland	2 lambs	Neurological	No encephalitis, schizonts in spinal cord	1154
Spain	6 adult sheep	Edema, ataxia	Sarcocysts in muscle	1608
Turkey	1, 10-month-old lamb	Neurological	Encephalitis, sarcocysts in brain	1861
	10 sheep	Neurological	Encephalitis, schizonts in brain	1346
United Kingdom	1 adult	Neurological	Encephalitis, schizonts in brain	1882
	2 young lambs	Neurological	Encephalitis, schizonts and sarcocysts in brain	1222
	9 lambs	Neurological	Encephalitis, schizonts in brain	899, 901, 1689
	3 lambs	Neurological	Encephalitis, schizonts in brain	1297
	3 lambs	Neurological	Encephalitis, schizonts in brain	1534
	Several lambs within 3 different flocks	Neurological	Encephalitis, schizonts in brain	160
	1, 18-month ram	Heart failure	Myocarditis, sarcocysts in heart	1575
	1 lamb	Pneumonia, ascite	Schizonts in lung	1563
	20, 2–5 years ewes	Neurological	Sarcocysts in muscles	900
USA	3-year-old ewe	Neurological	Encephalitis, schizonts in brain	1324
	3 lambs	Neurological	Encephalitis, schizonts in brain, sarcocysts in muscles	609

Severe encephalomyelitis has been reported in sheep from Australia, Bulgaria, Canada, England, Hungary, Turkey, and the United States, and it is not clear if these were caused by *S. tenella* or another unknown species of *Sarcocystis*, because schizonts were located extravascularly in neural cells.[396,784] In Australia, severe encephalomyelitis was found in two 5- or 6-month-old lambs associated with an unidentified *T. gondii*-like organism.[784] Both lambs had noncoordination of the limbs, and flaccid paralysis. Both lambs were necropsied; their brains and spinal cords were fixed in formalin and studied histologically. Schizonts were seen in many cells, predominantly in white matter of the spinal cord. Merozoites had typical apicomplexan structures, including rhoptries. Reevaluation of the electronmicrographs indicated that dense granules were likely misidentified as rhoptries. A

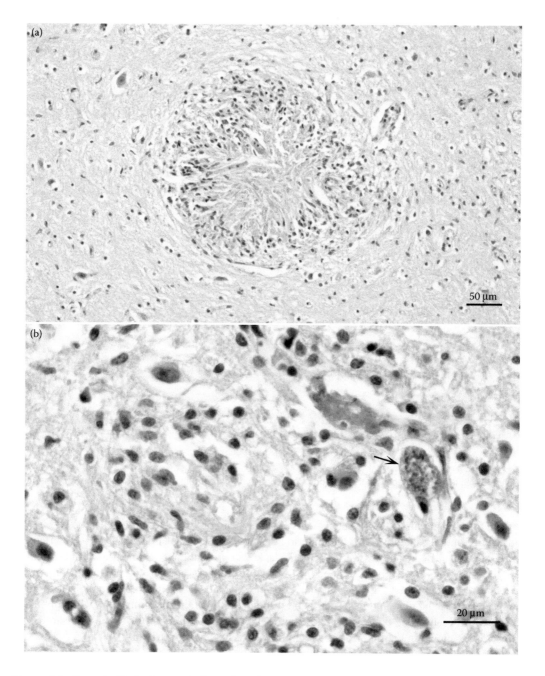

Figure 8.4 Encephalitis in the cerebellum of a stillborn ovine fetus with sarcocystosis. H and E stain. (a) Central necrosis surrounded by inflammatory cells. (b) A schizont in an inflammatory focus. (Adapted from Agerholm, J.S., Dubey, J.P., 2014. *Reprod. Dom. Anim.* 49, e60–e63.)

mature *S. tenella* sarcocyst was found in the brain of 1 of the lambs. Muscles were not examined. One of these cases was restudied ultrastructurally[396]; schizonts were located directly in host-cell cytoplasm, merozoites were formed by endopolygeny, and merozoites lacked rhoptries. An almost identical *Sarcocystis* infection was reported in 2 young lambs in England.[1222] Schizonts and mature sarcocysts were identified in the brains of both lambs.

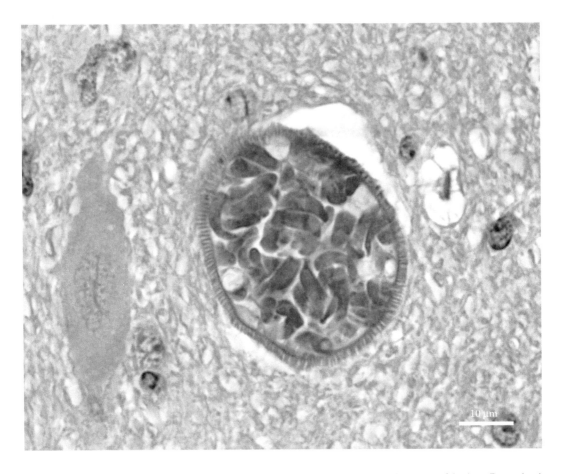

Figure 8.5 *S. tenella* sarcocyst in the cerebrum of a lamb with encephalitis. (Courtesy of Andrew Pregerine.)

On a farm in Indiana, United States, with 300 sheep, an outbreak of clinical sarcocystosis was reported in 5-month-old lambs.[609] Over a period of 4 weeks, 32 of 48 lambs in 1 pen were affected, and 22 died. The affected lambs were anorectic, lacked coordination, had stiff limbs, and exhibited trembling and general weakness. Three lambs were examined at necropsy; 2 had been euthanized and the third had died. Meningoencephalomyelitis and myositis were the predominant lesions in all 3 lambs. Necrosis and glial nodules were found throughout the brain and spinal cord. One immature schizont was found in the pons of 1 lamb. Numerous immature and mature *S. tenella*, sarcocysts similar to that shown in Figure 8.5, were seen in the muscles, particularly in the heart.

Two episodes of sarcocystosis occurred in a group of 1510-month-old Bluefaced Leicester rams in England[1297]; 12 of the 15 rams had high antibody titers to *Sarcocystis* antigen in an ELISA. In the first episode, 7 of 8 homegrown rams became ill, while seven purchased rams remained healthy. Paresis was the main clinical sign. Two mildly affected rams recovered. Three severely affected rams were euthanized and examined at necropsy. The predominant lesions in all 3 rams were edema, gliosis, and nonsuppurative myelitis. Extravascular schizonts were seen in the spinal cords of 2 of the 3 rams. Sarcocysts of *S. tenella* and *S. arieticanis* were detected in all 3 rams. In the second episode, in another flock of sheep in England, 15 ten-month-old lambs developed acute illness.[1534] Three lambs examined at necropsy had encephalitis, but schizonts were not seen. Numerous *S. tenella* sarcocysts were present in muscles. Similar outbreaks were observed in Black-faced sheep, less than 1 year old, from 3 unrelated flocks in England.[160] In each flock, sheep were recumbent, and from each flock 2 to 4 sheep were necropsied. All had myeloencephalitis, and schizonts were detected in the spinal cord or brain from at least 1 sheep. These schizonts reacted with a *Sarcocystis*-labeled antibody, and were negative for *Toxoplasma* and *Neospora*. A heavy infection with sarcocysts was seen in the muscles of all sheep examined.[160]

Neurological signs were observed in 10 of 350 sheep from a flock in Turkey.[1346] All 10 were examined at necropsy and had encephalitis. Schizonts were present in encephalitic lesions, and sarcocysts were seen in the muscles. Prophylactic treatment with Amprolium in another report 1 of 170 sheep flock in Izmir, Turkey became ataxic. Intravascular schizonts and mature sarcocysts were seen in sections of the brain.[1861]

In another unusual case, sarcocystosis associated with pneumonitis was reported in a lamb from a flock of 150 5- to 7 month-old sheep in England.[1563] The lamb had respiratory distress. The pathology report indicated pneumonitis and myocarditis with demonstrable *Sarcocystis* merozoies. Additionally, sarcocysts were seen in the heart. PCR-based on the DNA from the paraffin-embedded block of lung indicated *S. tenella*.

Severe *Sarcocystis*-associated myositis was reported in sheep in England. Heavy infection with sarcocysts was thought to be the cause of congestive heart failure in an 18-month-old Blue-faced Leicestershire ram.[1575] Clinical myositis was found in 20 of 600 ewes, associated with heavy infection with sarcocysts.[900] Four ewes were examined at necropsy. Muscle degeneration and myositis were found in all 4 ewes. Inadvertent ingestion of monensin sodium was considered to have aggravated myodegeneration.

In Germany, *Sarcocystis*-associated myocarditis was suspected as a contributory factor in mortality in sheep that failed to recover from inhalation anesthesia.[148]

In Hungary, muscle weakness, ataxia, loss of fleece, and 6.5%–10.8% mortality was reported in a flock of 210 Suffolk sheep.[772] Six affected sheep were examined at necropsy. Schizonts were seen in the heart and the brain. Sarcocysts of *S. arieticanis* were also identified in the hearts.

In Bulgaria, encephalitis was reported in a sheep, in association with *Sarcocystis* in the brain.[89]

Encephalitis associated with *S. capracanis* merozoites was reported from England,[622] but this report was questioned because of the mistaken identity of bradyzoites with merozoites in CSF.[469] In another case, *Sarcocystis* zoite-like structures, measuring 10.5–11.5 × 4–5 μm were seen in cerebrospinal fluid smears of 2 sheep with chronic coenurosis in Italy[1883]; it appears that these were probably bradyzoites released from intramuscular cysts while obtaining the CSF, because they were too large to be merozoites.

A case of laryngeal hemiplegia in an 18-month-old ram with respiratory distress and poor body condition associated with *Sarcocystis* spp. was reported in Spain.[1505] Histopathological examination of the laryngeal nerve revealed demyelination, degenerate axons, and atrophy. Multiple *Sarcocystis* spp. microcysts were found in the larynx musculature.

Whole carcasses or parts of adult sheep are condemned because of the presence of macroscopic sarcocysts of *S. gigantea* and *S. medusiformis*.[241] Carcasses are also condemned for eosinophilic myositis in lambs and in adult sheep.[563,907]

8.7 MOLECULAR STUDIES

Initially, *S. gigantea* was genetically characterized based on isoenzyme markers[616]; patterns were identical in all isolates from North American and Australasian origin. Genetic diversity seems to be very low for *S. gigantea*; RFLP–PCR detected null variation in the ITS-1 region among cysts within the same animal, and from different animals and countries (namely, Germany and Australia).[903]

Specific oligonuclotide probes for 18S rRNA of *S. tenella* were useful for diagnosis of acute sarcocystiosis.[840] A PCR based on 18S rRNA fragments was able to differentiate *S. tenella, S.*

arieticanis, and *S. gigantea.*[1736] Other methodologies like RAPD–PCR,[910] nested-PCR,[788] or RFLP–PCR[58,276,775,1858] can diagnose *Sarcocystis* spp. infecting ovine livestock.

Based on the 18S rRNA sequencing, established phylogenetic relationships were established among ovine species (except *S. medusiformis*), and other major *Sarcocystis* spp.[843] Data from the mitochondrial *cox1* gene also contributed to this phylogeny.[701]

Little is known about genetic diversity of *Sarcocystis* spp. in sheep. The pioneering study of Jeffries et al.[903] exposed null divergence. Further studies targeting new markers for 18S rRNA or *cox1*[701] hopefully will provide data that demonstrates the real genetic variability in *Sarcocystis* spp. in sheep.

BIBLIOGRAPHY

Clinical disease: **17, 89, 148, 160, 274, 386, 396, 551, 563, 609, 622, 746, 772, 784, 793, 899, 900, 901, 907, 1154, 1222, 1297, 1324, 1346, 1390, 1466, 1505, 1534, 1563, 1575, 1608, 1689, 1861, 1882**.

Experimental/immunology: **3, 231, 334, 553, 674, 745, 846, 867, 946, 1156–1160, 1206, 1285, 1295, 1461, 1468, 1606, 1632, 1656, 1700, 1731, 1734, 1745–1748, 1789, 1902, 1904, 1906, 1926–1928, 1945, 2015, 2016, 2023, 2039, 2053, 2183, 2185, 2188, 2192, 2223, 2229, 2230, 2263, 2266, 2295, 2300, 2330, 2352**.

Molecular biology: **58, 276, 616, 701, 775, 788, 840, 843, 903, 910, 1186, 1736, 1858**.

Morphology/ultrastructure: **96, 149, 240, 273, 367, 432, 677, 787, 817, 818, 940, 1288, 1390, 1442, 1518, 1552, 1649, 1652, 1796, 1807, 1899, 1936, 1970, 2165, 2184, 2221, 2227**.

Prevalence/biology: **13, 16, 23, 24, 26, 58, 65, 117, 170, 276, 294, 392, 640, 677, 760, 764, 775, 834, 868, 878, 984, 1030, 1123, 1207, 1290, 1338, 1345, 1384, 1405, 1448, 1449, 1517, 1519, 1524, 1528, 1530, 1544, 1557, 1586, 1587, 1698, 1699, 1753, 1784, 1786, 1835, 1895, 1910, 1912, 1916, 1919, 1930, 1934, 1936, 1944, 1963, 1968, 1970, 1976, 1977, 2000, 2003, 2033, 2034, 2055, 2098, 2106, 2107, 2108, 2143, 2179, 2210, 2011, 2213, 2231, 2243, 2250, 2312, 2360, 2361, 2376**.

Miscellaneous: **47, 57, 65, 214, 216, 240, 242, 243, 1899, 1902, 1904, 1912, 1945, 1970, 1976, 2015**.

CHAPTER 9

Sarcocystosis in Goats (*Capra hircus*)

There are 3 species of *Sarcocystis* in domestic goats: *Sarcocystis capracanis*, *Sarcocystis hircicanis* (Figure 9.1), and *Sarcocystis moulei*.

9.1 *SARCOCYSTIS CAPRACANIS* FISCHER 1979

9.1.1 Definitive Hosts

Dog (*Canis familiaris*), coyote (*Canis latrans*), red fox (*Vulpes vulpes*), and crab-eating fox (*Cerdocyon thous*).

9.1.2 Structure and Life Cycle

Sarcocysts are up to 1000 μm long and 100 μm wide. The sarcocyst wall is up to 3 μm thick with radial striations and finger-like vp, type 14 (Figure 9.1a and c). Sarcocysts are found in virtually all skeletal muscles, in the CNS, and in the heart.[382] Information on the life cycle is summarized in Table 9.1 and Figure 9.2. There are 2 generations of schizogony in blood vessels.

9.1.3 Pathogenicity

S. capracanis is the most pathogenic species of *Sarcocystis* in goats.[239,359,815] It can cause fever, weakness, anorexia, weight loss, tremors, irritability, abortion, and death, depending on the number of sporocysts ingested. As few as 5000 sporocysts cause clinical disease, and 100,000 sporocysts are generally fatal.[359] Goats that recover from acute sarcocystosis remain unthrifty, have a dull, dry hair coat, and are predisposed to other infections.[372] Dosing with 1 million sporocysts cause severe acute sarcocytosis and goats die between 18 and 21 days.

9.1.4 Protective Immunity

Goats immunized once with 100 or 1000 sporocysts develop subclinical infections and become refractive to lethal challenge with large numbers of *S. capracanis* sporocysts.[358,372,373,381] This protective immunity persisted at least 274 DPI. Immunization with 1000 sporocysts provided better protection than with 100, and ingestion of 10 sporocysts induced no protection.[381]

9.1.5 Clinical Disease in Naturally Infected Goats

In Victoria, Australia, 8 of 38 Saanen goats aborted or gave birth to stillborn kids over a period of 3 months.[1094] 1 stillborn goat was necropsied. The predominant lesion was encephalitis characterized

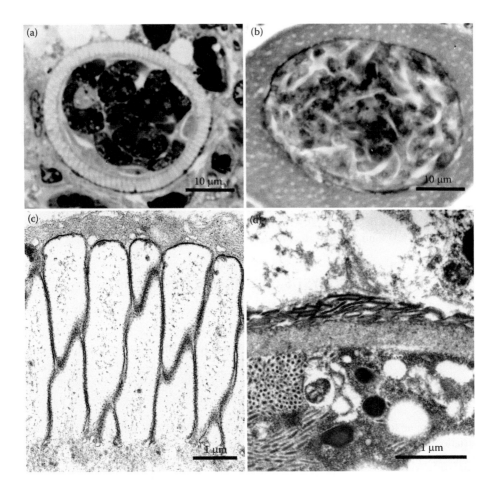

Figure 9.1   Sarcocyst walls of *S. capracanis* (a and c) and *S. hircicanis* (b and d) in skeletal muscles of
goats. Arrows point to thick villi in *S. capracanis* and hirsute protrusions in *S. hircicanis*. (a)
Toluidine blue stain, (b) H and E stain.

by necrosis, perivascular cuffing, and nonsuppurative encephalitis. Schizonts were seen in the blood
vessels in the brain. Occasional schizonts were also identified in the lungs and kidneys.[1094]

9.2 *SARCOCYSTIS HIRCICANIS* HEYDORN AND UNTERHOLZNER 1983

9.2.1 Definitive Host

Dog (*Canis familiaris*).

9.2.2 Structure and Life Cycle

Sarcocysts are up to 2.5 mm long. The sarcocyst wall is thin (<1 μm) and has hirsute protrusions (Figure 9.1b and d).[815] The vp are of the type 7a, long, filamentous, without tubules, and thus
folded over the cyst wall; they were up to 5 μm long.[1514] Little is known of the pathogenicity of this
species. Of 2 goats fed 10 million sporocysts, 1 died 43 DPI and the other was euthanatized while
moribund 35 DPI.

Table 9.1 Comparison of Developmental Stages of *Sarcocystis* spp. in Goats

	S. capracanis	*S. hircicanis*	*S. moulei*
First-generation Schizonts			Unknown
Location	Gut-associated arteries	Arterioles, intestinal lymph nodes, liver	—
Duration (DPI)	15–17	17–18	—
Peak development (DPI)	10–12	—	—
Size of meronts (μm)	26.1 × 17.4 (sections)	—	—
No. of merozoites	Up to 80	Up to 100 or more	—
Size of merozoites (μm)	5.5–7.1 × 2.5 (sections)	6–7.5 × 2.5–3.0 (smear)	—
Second-generation Schizonts			—
Location	Capillary endothelium of all organs	Capillary endothelium of several organs	—
Duration (DPI)	14–24	28–35	—
Peak development (DPI)	19	—	—
Size of meronts (μm)	18.8 × 10.1	—	—
No. of merozoites	4–36	—	—
Size of merozoites (μm)	5.5 × 1.5 (sections)	6–7.5 × 2.5–3.0 (smear)	—
Parasitemia			—
Duration (DPI)	17–24	—	—
Intraleukocytic	Yes	—	—
Sarcocysts			
Size (μm)	Up to 1000	Up to 2500	Up to 1000
Earliest seen (DPI)	30	43	—
Maturation time (DPI)	64	84	—
Wall (μm)	2–3 thick, striated	<1, thin, hairy projections	3–4, cauliflower-like
Location	Striated muscle, CNS	Striated muscles	Striated muscles
Main references	**382, 815, 2013**	**816**	**7, 820**

— = unknown.

9.3 *SARCOCYSTIS MOULEI* NEVEU-LEMAIRE 1912

9.3.1 Definitive Host

Cat (*Felis catus*).

9.3.2 Structure and Life Cycle

Sarcocysts are dull white, ovoid, up to 17 mm long, and up to 7 mm wide, and typically have been found in the esophagus.[7,69] The sarcocyst wall is branched, cauliflower-like (type 21), and contains 13–15 μm long bradyzoites.[7,820]

Two sizes of macroscopic sarcocysts were found in domestic goats from Iraq.[69] Fat sarcocysts had thick rounded ends and were found only in the esophageal muscle. Bradyzoites in smears were 14.1 × 4.5 (11–17 × 3–6) μm in size. The thin sarcocysts were small, slender, and were also found in diaphragm and skeletal muscles. The contained bradyzoites were 12.2 × 2.9 (9–13 × 2–4) μm in size. However, it is uncertain if both types of sarcocysts represent the same

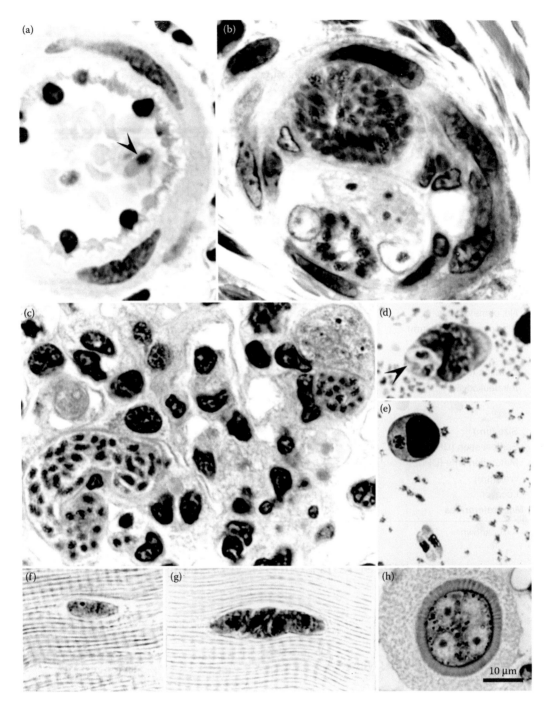

Figure 9.2 Developmental stages of *S. capracanis* in tissues of experimentally infected goats. (a) One sporo-
zite (arrowhead) in the lumen of an artery at 5 DPI. (b) Three developing first-generation schizonts
in an artery. The vascular lumen is occluded by hypertrophied endothelial cells and parasites at 10
DPI. (c) Second-generation schizonts and merozoites in renal glomerulus at 20 DPI. (d) Dividing
merozoite in a monocyte-like cell in peripheral blood at 20 DPI. (e) An intracellular merozoite and
2 extracellular merozoites in peripheral blood at 20 DPI. (f) Binucleate sarcocyst at 35 DPI. (g)
Sarcocyst with four nuclei at 35 DPI. (h) Sarcocyst with well-developed wall at 68 DPI. Bar applies
to all parts. (From Dubey, J.P. et al., 1984. *Int. Goat Sheep Res.* 2, 252–265. With permission.)

Table 9.2 Prevalence of *Sarcocystis* spp. in Domestic Goats

Country	Province	Year	Source	No	No. Positive	Species	Method	Bioassay	Reference
Brazil	Bahia	NS	Abattoir	250	E: 198 (79.2%)	*S. capracanis*	Gross examination, pepsin digestion	Dog and cat	**68**
Chile	Región Metropolitana	NS	Abattoir	266	244 (91.7%)	—	IFAT	—	**720**
Egypt	Cairo	2009–10	Abattoir	680	540 (79.4%)	*S. capracanis*	Gross examination, muscle squash, histology, TEM	Dog and cat	**1226**
India	Bihar	1985	Abattoir	288	146 (50.69%)	*S. hircicanis, S. capracanis*	Dissection	—	**1665, 1668**
India	Madhya Pradesh	1986–87	NS	76	51 (67.1%)	NS	Muscle squash	—	**1615**
India	Uttar Pradesh	1988–90	Abattoir	675	495 (73.33%)	*S. hircicanis, S. capracanis*	Muscle squash, pepsin digestion	—	**1589**
India	Maharashtra	1987–88	Not for human consumption	35	35 (100%)	*S. capracanis*	Muscle squash	—	**1593, 1595**
India	Maharashtra	1989	Abattoir	372	209 (56.2%)	*S. hircicanis, S. capracanis*	NS	—	**1801**
India	Uttar Pradesh	NS	NS	170	111 (65.3%)	*S. capracanis*	Muscle squash, pepsin digestion	—	**920**
India	Uttar Pradesh	1988	Abattoir	310	118 (58.4%); D: 60.0%, E: 58.9%, H: 55.1%	NS	Gross examination, pepsin digestion	—	**1525**
India	Jabalpur	NS	Abattoir	790	528 (66.9%)	*S. capracanis*	Gross examination, homogenization	—	**19**
India	Tripura	NS	Abattoir	80	23 (28.7%)	NS	Muscle squash	—	**1507**
India	Madhya Pradesh	NS	Abattoir	750	NS	*S. capracanis* and *S. hircicanis*	Isolation by washing with saline	Dog	**1592**
India	Akola	1994–95	Abattoir	100	25(25.0%)	All macroscopic	Gross examination, histology	—	**323**

(Continued)

Table 9.2 (Continued)   Prevalence of *Sarcocystis* spp. in Domestic Goats

Country	Province	Year	Source	No	No. Positive	Species	Method	Bioassay	Reference
India	Andhra Pradesh	NS	NS	111	80 (72.1%)	*S. capracanis*	Pepsin digestion	—	1784
India	Bengaluru	NS	Abattoir	100	72 (72.0)% E: 69.4%, H: 36.1%, D: 62.5%, T: 12.5%	72.0% *S. capracanis*; 21.0% *S. hircicanis*	Muscle squash, pepsin digestion	—	278
India	Orissa	NS	Abattoir	120	92 (76.6%); H: 76.6%, T: 85.8%, E: 80.4%, D: 76.6%, M: 68.3%	—	Gross examination, pepsin digestion	—	1207
India	Rajasthan	1989	Abattoir	703	480 (68.28%)	*S. capracanis*-like	Muscle squash	—	1618
India	Nagpur, Maharashtra	1996–97	NS	800	277 (34.62%)	NS	Muscle squash	—	916
India	Punjab	NS	Abattoir	45	9 (20.0%)	*S. capracanis*	Muscle squash, histology	—	48
Iran	Shiraz	NS	Abattoir	169	168 (99.4%)	Macroscopic in 28 (16.6%)	Pepsin digestion	—	1599
Iran	Tehran, Ghazvin	NS	Abattoir	NS	NS	All macroscopic: *S. moulei*	Muscle squash, pepsin digestion, PCR, TEM	—	1229
Iraq	Baghdad	1992–96	NS	826	278 (33.6%)	NS	Pepsin digestion: 97.4%; IFAT: 91.0%; muscle squash: 89.7%	Dog	1030
Iraq	North	1991–92	Abattoir	826	Macroscopic: 278 (33.6%); microscopic: 805 (97.4%)	*S. caprifelis*, *S. capracanis*	Gross examination, muscle squash, pepsin digestion, histology	Dog	69

(Continued)

Table 9.2 (*Continued*) Prevalence of *Sarcocystis* spp. in Domestic Goats

Country	Province	Year	Source	No	No. Positive	Species	Method	Bioassay	Reference
Japan	Niigata, Kanagawa	1986–89	Abattoir	15	2 (13.3%)	*S. capracanis*	Typsin digestion, histology, TEM	—	873
Japan	Saitama	1995	Abattoir	21	10 (47.6%)	*S. capracanis*, *S. hircicanis*	Dissection, histology, SEM	Dog	1514
Jordan	Central	1986–88	Abattoir	1261	Macrocysts in 46 (11.7%) of 393; microcysts in 711 (56.4%) of 1261	*S. capracanis*, *S. hircicanis*, *S. moulei*	Gross examination, trichinoscopy, IHA	—	13
Malaysia	Selangor	2014	Abattoir	105	55 (52.3%) by LM; 95 (90.0%) by PCR	*S. capracanis*	Muscle squash, histology, PCR	—	1400
New Zealand	Hawke's Bay	1977	Feral goats	60	17 (28.3%)	NS	Pepsin digestion, histology	—	237
Peru	Lima	NS	Abattoir	63	33 (52.4%); E: 76.2%; H: 52.4%	NS	Trichinoscopy	—	170
Philippines	Several	NS	NS	26	11 (42.3%)	*S. capracanis*	Gross examination, histology, TEM	—	226
Saudi Arabia	NS	NS	NS	NS	77.0%	NS	Gross examination, electron microscopy. Uncomplete	Cat and presumably dog	26
Senegal	Dakar	NS	Abattoir	75	66 (88.0%)	*S. capracanis*	Pepsin digestion, TEM	—	1786
Sudan	Several	NS	Abattoir	116	101 (87.1%); E: 32.8%; H: 35.3%; M: 53.4%; D: 50.0%	NS	Pepsin digestion, histology	—	868

D = diaphragm; E = esophagus; H = heart; M = skeletal muscle; T = tongue.
IFAT = immunofluorescent antibody test; IHA = indirect hemagglutination; LM = light microscopy.

Sarcocystis species. The 18S rRNA gene of fat and thin macrocysts collected from the esophagus and other striated muscle of goats in Iran was sequenced, and it was found that both sizes of cysts were the same species.[7] Results matched with previously sequences for *S. moulei* reported in international databases.

Reports on macroscopic sarcocysts in goats[1231,1274,1275] were reviewed, and part of the life cycle of *S. moulei* was described.[820] In experimentally infected goats, sarcocysts were not macroscopic and were not infectious to cats at 6 months PI. At 19 months PI sarcocysts were up to 2×1 mm in size and infective to cats. At 86 months PI sarcocysts reached a size of $7–13 \times 5–8$ mm, and remained infective for cats. Cats fed sarcocysts shed sporocysts with a prepatent period of 10 days. Sporocysts were $11.6–13.1 \times 8.7–9.4$ μm.[820]

Review of all evidence[491,554,820] indicates that there is only 1 species of *Sarcocystis*, *S. moulei*, with macroscopic sarcocysts and transmitted by cats.

9.4 NATURAL INFECTIONS IN GOATS

To investigate the presence of *Sarcocystis* spp. in goat carcasses, gross examination is useful in many cases, but artificial digestion methods showed higher sensitivity yielding very high rates of prevalence of *Sarcocystis* spp. Table 9.2 summarizes surveys carried out on goats.

BIBLIOGRAPHY

Epidemiology/prevalence: **13, 19, 26, 48, 68, 69, 170, 214, 226, 237, 278, 288, 323, 491, 554, 720, 868, 873, 916, 920, 1030, 1206, 1207, 1226, 1229, 1400, 1507, 1514, 1525, 1589, 1592, 1593, 1595, 1599, 1615, 1618, 1665, 1668, 1766, 1784, 1786, 1801, 1839, 1955, 1985, 2013, 2060, 2071, 2191, 2218, 2255, 2280, 2302, 2312, 2345.**

Experimental infections: **39, 199, 239, 320, 321, 717, 718, 798, 820, 921, 986, 1349–1353, 1613, 1614, 1617, 1619, 1802, 1959, 1960, 2117, 2205, 2214, 2219, 2266, 2352.**

Life cycle and ultrastructure: **7, 39, 199, 334, 382, 815, 816, 820, 1226, 1514, 1617, 1618, 1896, 2305.**

Miscellaneous: **24, 733, 846, 1094, 1229, 1336, 1608, 1614, 1735.**

CHAPTER **10**

Sarcocystosis in Water Buffalo (*Bubalus bubalis*)

10.1 INTRODUCTION

There are 4 valid species of *Sarcocystis* in water buffalo (Table 10.1, Figure 10.1).[471] Endogenous development of schizonts and sarcocysts of any of the 4 species in buffalo is unknown or uncertain. None of the 4 species are considered pathogenic for buffalo, and there is no clinical report of acute sarcocystosis in buffalo.

Sarcocystis levinei (Dissanaike and Kan 1978) Huong, Dubey and Uggla 1997

Sarcocystis fusiformis (Railliet 1897) Bernard and Bauche 1912: The sarcocyst of this species was redescribed[474]

Sarcocystis buffalonis Huong, Dubey, Nikkila and Uggla 1997

Sarcocystis dubeyi Huong and Uggla 1999

10.2 PREVALENCE

Natural infections by *Sarcocystis* spp. in water buffalo are frequent; data are summarized in Table 10.2.

Table 10.1 Species of *Sarcocystis* in Water Buffalo (*Bubalus bubalis*)

Character	Sarcocystis fusiformis	Sarcocystis levinei	Sarcocystis buffalonis	Sarcocystis dubeyi
Sarcocyst	Macroscopic	Microscopic	Macroscopic	Microscopic
Length	Up to 25 mm	Microscopic, <1 mm	Up to 8 mm	Up to 600 µm
Cyst wall	Thin, 2–5.2 µm	Thin	Thick, 3–7.7 µm	Thick
Cyst wall type—TEM	Type 21b, vp 6 µm long, with common stalk-like branches of a dead tree Microfilaments in vp	Type 7a, vp 0.5 µm at base, 0.08 µm at tip	Type 28, vp 12 µm long, narrow at base, expanded laterally, tapered distally. Microfilaments in vp	Type 10c, pallisade, bent at 45°, 4.5–8 µm long and 50–300 nm wide
Metrocytes	9.0–17.5 × 5.0–8.5 µm	Large	NS	NS
Bradyzoites	14.5–19.0 × 4.0–4.5 µm	9–12 × 2.5–4.0 µm	10.3–12.8 × 2.6 × 3.2 µm	8–11 × 2–2.5 µm
Definitive host	Cat	Dog	Cat	Unknown (not dog or cat)
Prepatent period—days	8–14	16–18	10	Unknown
Sporocyst (µm)	NS by Dissanaike and Kan (1978)[1654]	9.5–10.5 × 14.0–16.5 µm	7.5–9.5 × 12.5–13.5 µm	Unknown
Main references	327,474	327,862	863	865

NS = not stated.

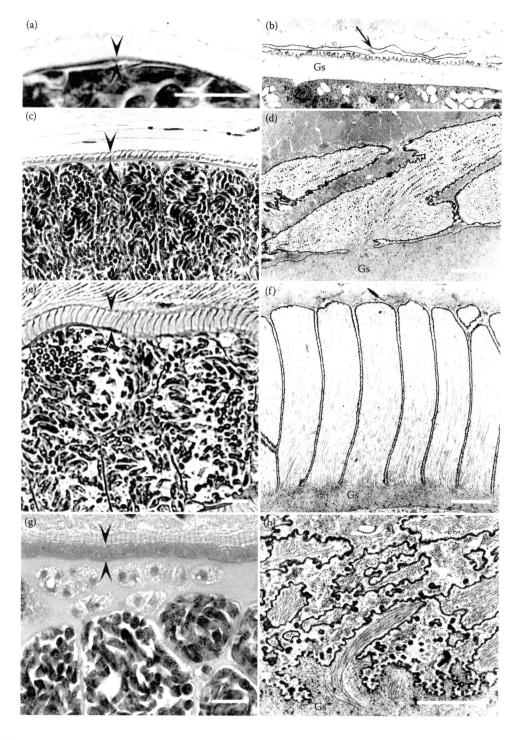

Figure 10.1 Sarcocyst walls of *S. levinei* (a, b), *S. buffalonis* (c, d), *S. dubeyi* (e, f), and *S. fusiformis* (g, h) from the water buffalo from Vietnam. (a, c, e, g), light micrographs, bar = 10 μm. (b, d, f, h) transmission electron micrographs, bar = 2 μm. Arrowheads point to the thickness of the sarcocyst walls. The vp (arrow) in *S. levinei* are slender, and folded over the sarcocyst wall, in *S. buffalonis*, have a narrow stalk, expanded laterally, in *S. dubeyi* they are elongated, and in *S. fusiformis* they are highly branched. (Courtesy of Arvid Uggla and Lam T. Huong.)

Table 10.2 Prevalence of *Sarcocystis* Sarcocysts in Water Buffaloes (*B. bubalis*)

Country	Province	Year	No. Tested	No. Positive (%) and Species Involved	Method (Tissue)	Reference
Argentina	Northeast	NS	500	254 (50.8%)	IFAT	971
China	Yunnan	1988	50	S. fusiformis: (80%) S. levinei: (75%) S. sp: (58%)	Muscle squash	1886
Egypt	Cairo	NS	130	S. fusiformis: 65 (50.0%)	Muscle examination (O)	488
	El-Gharbia	2009	35	11 (31.4%), S. dubeyi	Histology, TEM	827
	Beni-Suef	NS	379	299 (78.9%), S. fusiformis and S. levinei	Gross examination, histology, pepsin digestion	487
	Alexandria	NS	300	203 (67.6%)	ELISA	43
	El-Gharbia	2011–12	528	S. fusiformis: 280 (85.4%) S. buffalonis: 65 (12.3%) Mixed infection: 358 (67.8%)	Histology, PCR, and sequencing	492
	Assiut	NS	90	23 (25.5%) by digestion and 85 (94.4%) by ELISA. S. fusiformis, S. cruzi, S. levinei and S. hominis	Histology, SEM, digestion, ELISA	1190
India	Mumbai	NS	50792	S. fusiformis: 7776 (15.31%)	Gross examination	312
	West Bengal (Kolkata)	1987	530	276 (52.08%) S. fusiformis: 200 (37.7%) S. levinei: 76 (14.3%)	Gross examination, pepsin digestion (H, T, D, O)	106
Iraq	Baghdad	1992–96	580	Macroscopic: 91 (15.6%) Microscopic: 481 (82.9%) by pepsin digestion and 464 (80.0%) by IFAT	Muscle examination, IFAT, pepsin digestion, bioassay in dogs (O, T, H)	1030
Iran	Khuzestan	NS	100	S. levinei and S. dubeyi: 83 (83.0%) 3 carcasses with macroscopic cysts (S. buffalonis)	Histology	1339
	Khuzestan	2009	NS	Macroscopic and microscopic sarcocysts of S. fusiformis: NS (NS)	Muscle examination, PCR-RFLP	1340

(Continued)

Table 10.2 (*Continued*) Prevalence of *Sarcocystis* Sarcocysts in Water Buffaloes (*B. bubalis*)

Country	Province	Year	No. Tested	No. Positive (%) and Species Involved	Method (Tissue)	Reference
Italy	Northern	1994–95	249	82 (32.9%), *S. fusiformis*, *S. levinei*	Muscle squash, H	163
Malaysia	Selangor	2011–12	18	12 (66.7%), *S. levinei* and *S. cruzi*	Muscle squash, H	1029
Philippines	Luzon, Visayas, Mindanao	NS	142	92 (64.8%), *S. fusiformis*, *S. levinei*	Histology, TEM	220, 222, 223
Romania	Bucarest	NS	NS	*S. fusiformis*	Bioassay in dog and cats	1429
Sri Lanka	Colombo	1999	123	93 (75.6%)	ELISA	932
Thailand	Ratchaburi	NS	211	81 (38.4%), *S. cruzi*	Histology, digestion	1233
Turkey	Elazig	1985–86	183	Microscopic cyst (*S. dubeyi*): 174 (95.1%) Macroscopic cysts (*S. fusiformis*): 48 (26.2%)	Muscle examination, histology	1343
	Ankara	1993–94	125	*S. fusiformis*: 49 (39.2%) *S. levinei*: 92 (73.6%) *S.* sp.: 39 (31.2%) Mixed: 58 (46.4%)	Trypsin digestion, bioassay (O and others)	481
Vietnam	Ho Chi Minh City	1996–97	502	396 (79%), of these *S. levinei* in 74%, *S. fusiformis* in 41%, *S. buffalonis* in 33%, and *S. dubeyi* in 12%; 8% were infected with all 4 species	Gross examination, histopathology	864
	Son La	2003	30	*S. fusiformis*, *S. cruzi*, *S. hominis* and *S. hirsuta*	Histology, PCR, TEM	905

D = diaphragm; H = heart; O = oesophagus; T = tongue; NS = not specified.

BIBLIOGRAPHY

Information pertinent to the subject matter of this chapter may be found in references grouped by topics as:

Biochemistry: **754–758, 955, 1527, 1948, 1949, 1974, 2240, 2271**.

Molecular biology: **57, 492, 701, 864, 1056, 1340, 1846, 1856–1858**.

Epidemiology: **43, 106, 222, 223, 294, 312, 327, 481, 487, 932, 971, 1029, 1030, 1190, 1221, 1233, 1339, 1343, 1452, 1476, 1893, 1903, 1916, 1937, 1952, 1969, 1973, 1975, 2024, 2025, 2059, 2070, 2071, 2114, 2116, 2173, 2243, 2248, 2278, 2342, 2373**.

Experimental infections: **211, 213, 488, 881, 1059, 1429, 1809, 1811, 1847, 1849, 1893, 1951, 1971, 1972, 1974, 1979, 2026–2028, 2030, 2031, 2062, 2063, 2326, 2327**.

Morphology and speciation: **5, 163, 220, 326, 327, 394, 471, 474, 481, 651, 862, 863, 865, 937, 1058, 1369, 1438, 1808, 1810, 1846, 1848, 1867, 1871, 1878, 1886, 1887, 1889, 1890, 1903, 1973, 1975, 2029, 2059, 2065, 2130, 2272, 2343, 2344**.

Miscellaneous: **213, 326, 488**.

CHAPTER 11

Sarcocystosis in Horses, Mules, and Donkeys (*Equus* spp.)

11.1 INTRODUCTION

There is considerable confusion concerning the validity of different species of *Sarcocystis* in horses and other equids.[833,1306,1481] Early records of *Sarcocystis* infections in horses have been reviewed.[1481] The name *Sarcocystis bertrami* was first proposed as a new species in the horse[329] in a book. Sarcocysts were 9–10 mm long and were found in the esophagus and leg muscles. No other information was given. In retrospect, it is impossible to determine the validity of the species because there are no archived specimens.

Dogs fed naturally infected esophagus from horses in Germany excreted sporocysts.[1481] The species was named *Sarcocystis equicanis* following the nomenclature proposed.[803] The sarcocysts were thin walled (Figure 11.1b).

The life cycle of a species of *Sarcocystis* in horses in the United States was determined and it was named *Sarcocystis fayeri* based on morphological features.[351] Sarcocysts were thick walled. The name *S. bertrami* was used[1136] for the parasite originally described as *S. equicanis*[1481] and its life cycle was further described.

Based on limited observations, the parasite in the muscles of a donkey was named *Sarcocystis asinus*.[650] The validity of *S. asinus* has been questioned,[1048,1306] and we regard it as invalid. Experimental evidence indicated the similarity of the *Sarcocystis* species in horses and donkeys.[1131] Tissues from 20 naturally infected horses were fed to a dog, and those of 10 donkeys were fed to another dog. Both dogs excreted sporocysts, but cats and raccoons fed the same infected tissues did not excrete sporocysts. Experimental infections were conducted with 4 ponies raised in captivity using the sporocysts derived from the infected donkeys (inoculum A) and sporocysts derived from the inocula from horses (inoculum B). A foal fed 100,000 sporocysts of inoculum A developed 2 peaks of fever on days 10 and 11, and 19–21 DPI. Sarcocysts were detected histologically in muscles of the foal when killed 138 DPI; sarcocysts contained only metrocytes. 3 foals were fed inoculum B. Two foals fed 100,000 sporocysts of inoculum B developed 2 peaks of fever between 10 and 21 DPI; these foals were killed 21, and 197 DPI. Mature sarcocysts were found in the foal killed 197 DPI, but no parasites were detected in the foal killed 21 DPI. The fourth foal fed only 10,000 sporocysts did not develop fever; mature sarcocysts were found in its muscles when killed 212 DPI. A dog fed muscles of the experimentally infected foal excreted sporocysts. Sporocysts were structurally similar in 3 dogs that excreted sporocysts; the sporocysts were 12.2–13.8 × 9.2–9.9 μm in size and the prepatent period was 9–10 days.[1131]

Review of the literature indicates there are 2 distinct species of *Sarcocystis*, the thick-walled species (*S. fayeri*) with an unique type 11a cyst wall, and the thin-walled species (*S. equicanis*/*S. bertrami*) with a type 11c cyst wall[552,1853,1306] (Figure 11.1). The identity of the original *S. bertrami* will never be known because there are no archived specimens. Therefore, to avoid further confusion, only 2 species are identified and discussed, *S. bertrami* and *S. fayeri*.

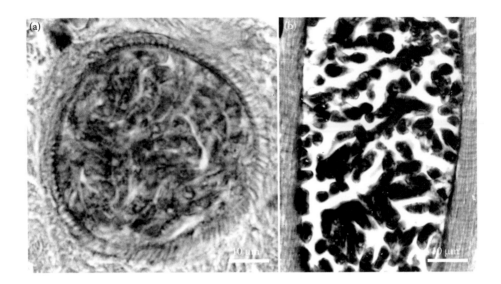

Figure 11.1 Sarcocysts of *S. fayeri* and *S. bertrami* in skeletal muscles. (a) *S. fayeri* from a horse in the United States. (b) *S. bertrami* from a horse in Germany. Note the thick, striated wall in *S. fayeri* and thin wall in *S. bertrami*. H and E stain.

11.2 *SARCOCYSTIS BERTRAMI* DOFLEIN 1901
(SYN. *S. EQUICANIS* ROMMEL AND GEISEL 1975)

Sarcocysts are up to 15 mm long.[53,552,833] By LM the thickness of the wall is variable but generally thin. By TEM, the cyst wall has villar protrusions that are up to 11 μm long and less than 0.5 μm wide. The vp are folded on the cyst wall giving it a thin-walled appearance, type 11c.[708,1306] Bradyzoites are 8–10 μm long and 2.5–3.5 μm wide.

The dog, but not the cat, raccoon or human, is the definitive host.[833] The prepatent period is 8–10 days. Sporocysts were 15–16.3 × 8.8–11.3 μm[552] and slightly smaller.[1131]

In experiments conducted in Germany, only mild clinical signs were noted in ponies. In 1 experiment, 5 ponies were each fed 200,000 sporocysts.[1562] One pony was febrile on day 12 and all ponies had fever 20–22 DPI; these 2 peaks of fever probably indicate the 2 schizogonic phases of the parasite. Three of the 5 ponies had neurological signs, apathy, and anorexia between 51 and 70 DPI. Muscle enzymes (creatinine kinase) values were elevated between 63 and 147 DPI. Sarcocysts were 2 mm long at 378 DPI and up to 9 mm long at 1040 DPI.[1136]

In another experiment, 2 ponies, each fed 100,000 sporocysts, remained asymptomatic[552]; these ponies were killed 167 and 189 DPI. Both thick-walled (*S. fayeri*) and thin-walled (*S. bertrami/S. equicanis*) sarcocysts were seen.

11.3 *SARCOCYSTIS FAYERI* DUBEY, STREITEL,
STROMBERG, AND TOUSSANT 1977

Sarcocysts were found primarily in skeletal muscles and rarely in the heart. Sarcocysts in histologic sections of naturally infected horses were up to 990 μm long and 136 μm wide with radially striated cyst walls 1–3 μm thick.[351,1752] In experimentally infected ponies sarcocysts were up to 1 mm long at 79 DPI.[1539] The cyst wall has vp 2.2–3.1 μm long, with microtubules that extend from tip of vp to the zoite plasmalemma (Figure 3.10b), type 11a (Figures 11.1a and 11.2). Bradyzoites are

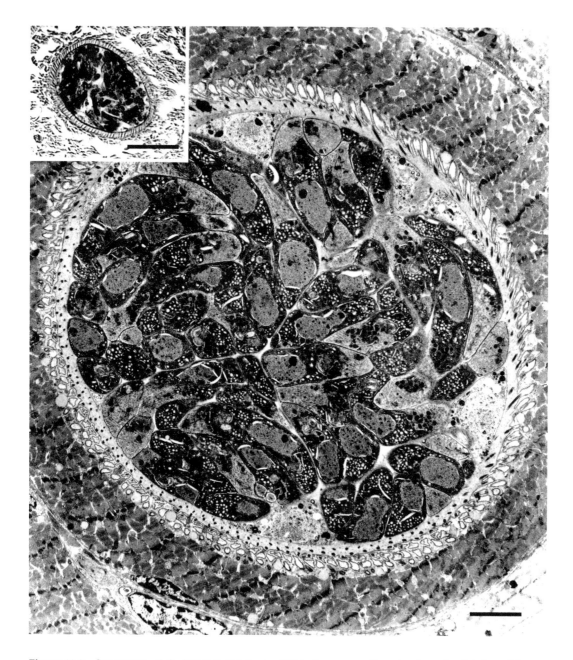

Figure 11.2 Sarcocysts of *S. fayeri* in the horse. TEM of cross section showing microtubules of the vp extending into the interior of the sarcocyst. Bar = 5 μm. Inset shows a cyst seen by light microscopy. H and E stain. Bar = 10 μm. (From Tinling, S.P. et al. 1980. *J. Parasitol.* 66, 458–465. With permission.)

12–16 μm long in sections and 15–20 μm in smears. The dog (but not the cat) excreted sporocysts, 11–13 × 7–8.5 μm, with a prepatent period of 12–15 days.

Endogenous development was studied in 1 horse and 12 ponies. In the first experiment, 8 ponies and 1 horse were fed 10,000 to 10 million sporocysts and were necropsied 10–156 DPI.[585] Two generations of schizonts were found in arteries or capillaries of the heart, brain, and kidney between 10 and 25 DPI. Immature sarcocysts were seen from 55 DPI, and some sarcocysts at 77 DPI had

Figure 11.3 A depressed pony with hair loss, 6 months after inoculation with *S. fayeri*. (From Fayer, R. et al. 1983. *Vet. Rec.* 113, 216–217. With permission.)

bradyzoites and were infectious for dogs. At 156 DPI, the sarcocysts were up to 338 μm long in histologic sections. In the second experiment, 3 ponies were fed 100,000 to 10 million sporocysts and necropsied 79 DPI;[1539] only immature sarcocysts were found in muscles.

The *Sarcocystis* species of equids are only mildly pathogenic. In 1 study, ponies fed 1 million sporocysts of *S. fayeri* developed mild anemia and fever. A horse fed 10 million sporocysts developed a stiff gait, but was otherwise clinically normal.[1651] There are indications that pathogenicity may vary among different isolates of *Sarcocystis* or among horses. For example, a pony fed 2 million sporocysts of a Texas isolate of *Sarcocystis* became lethargic, stiff-legged, tired quickly, and lost hair on its body, especially on its head and neck by 150 DPI (Figure 11.3). This pony had severe myositis associated with sarcocysts and developed an autoimmune anemia. In another study, 3 ponies fed 100,000 to 10 million sporocysts of the Texas isolate remained asymptomatic.[1539] Two cases of remarkable granulomatous and eosinophilic myositis associated with *S. fayeri* have been diagnosed in USA. At gross examination, disseminated granulomas were very patent. In 1 horse, refractory muscle pain and recurrent esophageal dysphagia were observed.[2394]

Recently, several outbreaks of food poisoning were reported after consumption of raw horse meat in Japan.[781,934] This has been related to a 15-kDa protein present in *S. fayeri* sarcocysts.[934] Freezing of horse meat seems to control the toxicity of *S. fayeri*[781] associated with raw meat consumption.

11.4 PREVALENCE OF NATURAL INFECTIONS AND CLINICAL DISEASE

Sarcocysts have been found in horses, mules, and donkeys (Table 11.1). All but 2[625,750] were based on samples collected from slaughterhouses and thus do not represent prevalence in well-kept

Table 11.1 Prevalence of Sarcocysts in Equids

Location	Year	Tissues	Methods	Results	Species	Reference
Austria	1979–81	E	Histology, bioassay in dog, cat and human	Sarcocysts in 89 (32.4%) of 275 horses and 1 of 2 donkeys. Only dog shed sporocysts	*S. bertrami*	833
Belgium	NS	M, biopsy	Histology	Sarcocysts in 12 (13.2%) of 91 horses with chronic muscle problems	NS	625
Egypt, Giza	NS	H	Histology, TEM	Sarcocysts in 18 (90.0%) of 20 donkeys	NS, similar to those described in horses	824
Egypt, Giza	NS	E, D, H	Bioassay in dog	After prepatent period of 11 days, 4 dogs, fed a pool of donkey tissues (500 g), shed sporocysts	*S. bertrami*	1873
Germany, Munich	NS	E	Trypsin digestion, histology, bioassay in dog and cat	Bradyzoites in 21 (23.3%) of 90 horses. 3 dogs, but not 6 cats, shed sporocysts	*S. equicanis*	1481
Germany, Munich	1978–79	E, H, D	Trypsin digestion, histology, bioassay in dog	Bradyzoites in 31(15.5%) of 200 horses. No bradyzoites in heart digests Dogs shed sporocysts	*S. equicanis, S. fayeri*	552
Great Britain	1981	E	Trypsin digestion, muscle squash, TEM	Sarcocysts in 245 (62.2%) of 394 horses, and 1 of 2 donkeys	Sarcocysts thin walled, *S. bertrami*	485
Great Britain and Ireland	NS	Postural, propulsive and respiratory muscles	Histology	Sarcocysts in 3 (4.0%) of 74 horses	NS	750
India, Madras	NS	E, H, D, M	Muscle squash, histology	Sarcocysts in 1 pony; no. tested not stated	NS	15
India, Punjab	NS	E, D	Muscle squash, bioassay in dog	Sarcocysts in 1 mare; no. tested not stated. Dog shed sporocysts 7 DPI	NS	922

(Continued)

Table 11.1 *(Continued)* Prevalence of Sarcocysts in Equids

Location	Year	Tissues	Methods	Results	Species	Reference
India, Punjab	NS	E, D, T	Muscle squash, histology	Sarcocysts in 1 mare; no. tested not stated	NS	924
Japan	1981–93	E, D, T, H, M	Trypsin digestion, histology, SEM, TEM	Thin-walled sarcocysts in 6 (6.4%) of 93 horses	*S. equicanis*	1853
Mongolia	1998–99	D, H, T	Trichinoscopy	Sarcocysts in 3 of 4 horses	NS	640
Mongolia, Ulaan Baatar	1998	D, H, T	Trichinoscopy	Sarcocysts in 40 (93.0%) of 43 horses	NS	641
Morocco	NS	E	Trichinoscopy, histology	Sarcocysts in 55 (46.2%) of 119 horses, 9 (50.0%) of 18 mules, 9 (21.9%) of 41 donkeys	NS	965
Turkey, Ankara	NS	E, D	Gross examination, muscle squash, trypsin digestion	Sarcocystis in 22 (68.8%) of 32 horses, and 60 (84.5%) of 71 donkeys	*S. equicanis*, *S. fayeri*, and mixed infections in both hosts	1344
USA: Ohio, Indiana and Wisconsin	1976	E, D, H	Trypsin digestion, histology, bioassay	Sarcocysts in H and E sections of 4 esophagi, and 2 hearts of 97 horses. Bradyzoites found in digestion of 17 (17.5%) of 97 horses 10 dogs, but not 10 cats, shed sporocysts	*S. fayeri*	351

E = esophagus; D = diaphragm; H = heart; T = tongue; M = skeletal muscle.
NS = not stated.

companion horses. In 36 thoroughbred horses, both types of myocytes (dark, light) were infected.[750] Infections are more prevalent in older horses.[485,1481] The finding of mature sarcocysts in the heart of a 3-day-old foal from Canada[269] and in a 6-week-old foal in Britain[485] indicates transplacental infection.

Clinical myositis has been reported to be associated with the presence of sarcocysts in histologic sections of horses in Germany,[631,1556] in a mule in Israel,[1702] in a malnourished horse in Canada,[178] in an ataxic horse in California,[1752] and in a horse with pronounced muscle wasting in Colorado,[1760] both in the United States. In a controlled study of 91 horses with clinical myositis in Belgium, sarcocysts were found in 12 muscle biopsies, but there was no association between the presence of sarcocysts and clinical myositis.[625] Eosinophilic myositis in a horse from New Zealand is illustrated in Figure 11.4.

There are only few reports of *Sarcocystis* in wild Equidae. These include *Sarcocystis neurona* infection in a Grant's zebra (*Equus burchelli bohmi*) from the United States[1117] (see Chapter 3), and a report by Odening et al.[1306] who detected TEM wall type 11 sarcocysts in 5 species of wild equids (Przewalski's feral horse, *Equus caballus przewalskii*; Chapman's plain zebra, *Equus burchellii chapmani*; Damara plain zebra, *Equus burchellii antiquorum*; kulan, *Equus onager kulan*; and kiang, *Equus kiang holdereri*) in zoological gardens in Germany. A dog fed

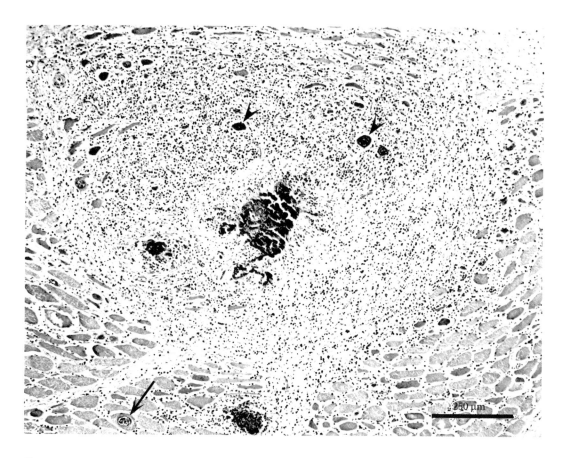

Figure 11.4 Eosinophilic myositis in skeletal muscle of a naturally infected horse. The central mineralized area is surrounded by numerous eosinophils and fibroblasts. A few giant cells (arrowheads) are present. Cross section of a sarcocyst (arrow) is in the lower left area. H and E stain. (Courtesy of late W. J. Hartley.)

Sarcocystis species from mountain zebra (*Equus zebra hartmannae*) from South Africa excreted sporocysts.[2145]

BIBLIOGRAPHY

15, 53, 178, 269, 329, 351, 485, 552, 585, 588, 625, 631, 640, 641, 650, 665, 708, 750, 781, 824, 833, 922–924, 934, 965, 1117, 1131, 1136, 1306, 1344, 1481, 1518, 1539, 1556, 1562, 1702, 1752, 1760, 1853, 1873, 1916, 2145, 2292, 2394.

CHAPTER 12

Sarcocystosis in Camels (*Camelus dromedarius* and *Camelus bactrianus*)

12.1 INTRODUCTION

In Egypt, Mason[1125] first reported sarcocysts in striated muscles, including the heart, of a camel. He saw 2 types of sarcocysts, 1 with a striated wall 1–2 μm thick, and the other with a smooth, thin wall (<1 μm). Mason thought that they were both different stages of the same parasite and named them *Sarcocystis cameli*. Therefore, in retrospect, it is not possible to decide which type of sarcocyst is *S. cameli*. Because the thick-walled striated sarcocyst has been found repeatedly, Dubey et al.[393] called it *S. cameli* (Figure 12.1) and did not name the thin-walled species, which was subsequently named *Sarcocystis ippeni* by Odening.[1314] Another unnamed species of *Sarcocystis* with hirsute protrusions on the sarcocyst wall was seen occasionally in *Camelus bactrianus* born in the zoo, but additional details were not provided.[1314]

Macroscopic sarcocysts (up to 15 mm long) were reported in camels from Russia[990] and Egypt[2399] but structural details are missing. Surveys of *Sarcocystis* in camels have been summarized in Table 12.1.

The presence of thick and thin-walled sarcocysts was confirmed in camels from Saudi Arabia,[568] Somalia,[770] and Sudan.[876] Ishag et al.[875] found sporocysts of 2 sizes (13.2–13.6 × 6.5–9.5 and 16.0 × 9.9–11.5 μm) in feces of dogs fed camel meat; the larger sporocyst was called *Sarcocystis camelocanis*.

Thick-walled sarcocysts from camels were transmitted to dogs and named *S. camelicanis* without any explanation for this new name.[11] It is evident from data in Table 12.2, that the size of sporocysts is variable and species should not be based on sporocyst size alone. To add to this confusion, another new species from the camel was named *Sarcocystis miescheri*, based on finding oocysts in feces of dogs fed naturally infected camel meat.[1106] Illustrations provided by the authors resemble *Cystoisospora ohioensis* oocysts measuring 20.8–26.7 × 18.5–20.7 μm with a thick wall containing 2 sporoblasts and bearing no resemblance to other species of *Sarcocystis*. The bradyzoites measuring 21.5–32.8 × 7.7–17.7 μm appear to be artifacts misidentified as bradyzoites.

Recently, Dubey et al.[2391] reviewed all reports of sarcocystosis in camels and concluded that there are 2 structurally distinct species of *Sarcocystis* (both with microscopic sarcocysts), *S. cameli*, and *S. ippeni*. The status of species forming macroscopic sarcocysts is unclear.

The first molecular characterization of *S. cameli* was reported by Motamedi et al.[1230] who developed a restriction map for the species. This information will be useful for further speciation studies.

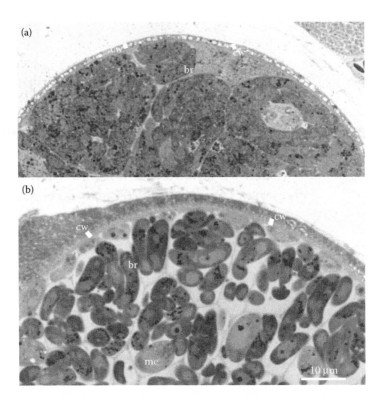

Figure 12.1 LM of *Sarcocystis* from camels; they are difficult to speciate based on light microscopy. Toluidine blue stain. (a) *S. cameli* sarcocyst. Vp are very thin and barely visible, and whitish areas are probably degenerated host tissue (hc) between vp. (b) *S. ippeni* based on tringular vp. (a, b) Note pale metrocytes (me) and banana shaped bradyzoites (br).

12.2 *SARCOCYSTIS CAMELI* (MASON 1910) AMENDED DUBEY, HILALI, VAN WILPE, CALERO-BERNAL, VERMA, AND ABBAS 2015 (SYN. *S. CAMELICANIS, S. CAMELOCANIS, S. MIESCHERI*)

12.2.1 Distribution

Egypt, Iran, Sudan, Afghanistan, Morocco, and the former USSR.

12.2.2 Definitive Host

Most likely dog.

12.2.3 Structure and Life Cycle

The sarcocysts found in striated muscles including the tongue, esophagus, and heart are microscopic. By LM, the sarcocyst wall appears thin (Figure 12.1a). The vp are conical, sloping to straight, and 3 μm long and 0.5 μm wide at the base, with knob-like structures on villar pvm, type 9j (Figure 12.2a). The bradyzoites are 14–15 × 3–4 μm long in histologic sections.

Table 12.1 Prevalence of *Sarcocystis* Sarcocysts in Camels

Country	Year	No	Method	No. Positive	% Positive	Thick Walled	Thin Walled	Bioassay in Dog and Cat	Reference
Afghanistan	1984	192	Muscle squash, histology	118	61.4	Yes	–	NS	966
Egypt	NS	112	Muscle squash, histology	41	36.6	Yes	–	Dog and cat	822
	NS	13	Histology	3	23.1	NS	NS	NS	530
	2008	180	Muscle squash, histology, trypsin digestion	116	64.0	Yes	–	Dog and cat	11
	2009–10	156	Muscle squash	66	42.3	Yes	–	Dog	1106
	NS	180	Gross examination, trypsin digestion	120	66.6	–	Yes	Dog and cat	2399
Ethiopia	1998–99	121	Histology	55	45.5	–	Yes	NS	1836
India	NS	1	Histology	1	–	NS	NS	NS	1447
Iran	NS	400	Muscle squash	209	52.3	NS	NS	NS	1600
	2002–05	250	Histology	209	83.6	–	Yes	No	1772
	2009	130	Pepsin digestion	67	51.5	NS	NS	NS	774
Iraq	1992–96	36	Pepsin digestion	33	91.6	NS	NS	No	1030
Jordan	NS	110	Muscle squash	24	21.8	Yes	Yes	NS	1031
Mongolia	1998–99	5	Muscle squash	5	100.0	NS	NS	NS	640
Saudi Arabia	1992–93	103	Trypsin digestion	91	88.3	Yes	Yes, more common	Dog and cat	568
	NS	40	Histology	31	77.5	–	Yes	–	2401
	2002–03	624	Trypsin digestion	399	64.0	–	Yes	NS	25
Somalia	1987	200	Trypsin digestion	165	82.5	Yes	Yes	NS	770
			Histology	120	60.0				
Sudan	NS	100	Pepsin digestion	81	81.0	NS	NS	NS	868
	NS	NS	Bioassay	NS	NS	Yes	Yes	Dog	876
Former USSR (Russia)	NS	NS	Unknown	6	NS	Yes	Yes	Dog	990

NS = not stated.

Table 12.2   Shedding of Sporocysts in Feces of Dogs Fed Camel Meat

Country	Prepatent Period (Days)	Size of Sporocysts (µm)	Reference
Egypt	Not stated	12 × 9	822
Egypt	10–11	12.0–14.0 × 8.9–11.3	825
Egypt	11	13.7–15.6 × 7.8–10.7 ($n = 50$)	11
Egypt	13–15	10.1–13.9 × 8.59–9.94 (type A) 8.7–14.3 × 11.5–10.0 (type B)	1106
Egypt	11	13.1–15.2 × 7.4–10.1	2399
Russia	Not stated	16.4 × 8.3	990
Saudi Arabia	9–10	10.7–14.3 × 8.3–10.7 ($n = 20$)	568, 826
Sudan	9–13	13.2–13.6 × 6.5–9.5 (type A) 16.0 × 9.9–11.5 (type B)	876

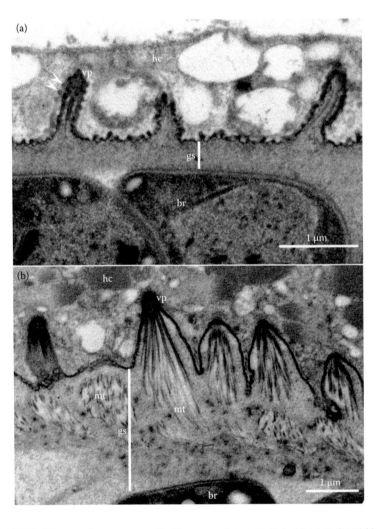

Figure 12.2   (a) TEM of *S. cameli* sarcocyst walls. The vp are interspersed with vacuolated (degenerated) hc. Note corrugations (arrowheads) at the villar tips and prominent mt. (b) TEM of *S. ippeni* sarcocyst walls. Note the vp are cut at different angles with prominent mt in conical vp. The gs is mostly electron lucent and not well demarcated. The mt in vp are more electron dense towards the villar tips and blunt tip of vp.

12.3 *SARCOCYSTIS IPPENI* (ODENING 1997) AMENDED DUBEY, HILALI, VAN WILPE, CALERO-BERNAL, VERMA, AND ABBAS 2015

12.3.1 Structure and Life Cycle

Sarcocysts are microscopic (Figure 12.1b). By TEM, sarcocyst wall has unique type 32 vp with an electron dense knob (Figure 12.2b). The vp are approximately 1.0 μm long, 1.2 μm wide at the base, and 0.25 μm at the tip; microtubules in vp originate at midpoint of gs and continue up to tip, criss-crossed, smooth and without granules or dense areas. The total thickness of the sarcocyst wall (from the base of gs to vp tip) is 2.3–3.0 μm. Bradyzoites are 12.0–13.5 × 2.0–3.0 μm in size.[2391]

BIBLIOGRAPHY

6, 11, 25, 53, 393, 530, 568, 640, 770, 774, 822, 823, 825, 826, 868, 875, 876, 966, 990, 1030, 1031, 1106, 1125, 1230, 1314, 1447, 1600, 1772, 1836, 2092, 2241, 2391, 2399, 2401.

CHAPTER 13

Sarcocystosis in South American Camelids (Alpaca, Guanaco, Llama, and Vicugna/Vicuña)

13.1 INTRODUCTION

A sarcocyst observed in a llama (*Lama glama*) was named *Sarcocystis aucheniae* but no other detail was provided.[149] In Argentina, sarcocysts found in a guanaco, *Lama guanicoe*, were named *Sarcocystis tilopodi* based on having found it in a different species of *Lama*.[1435] In Chile, macroscopic and microscopic sarcocysts were found in *L. guanicoe* which transmitted infection to dogs.[722] Four dogs fed sarcocysts excreted sporocysts but 4 cats fed the same infected meat did not excrete sporocysts. Additionally, 1 dog fed isolated macrocysts also excreted sporocysts. Based on these results, the author suggested that the parasite in *L. glama* should be named *S. lamacanis* and the parasite in the species in *L. guanicoe* should be called *S. guanicoe-canis*.[1435] In Germany, both microscopic and macroscopic sarcocysts were found in *L. glama*.[1561] 1 dog and 1 cat were fed isolated macroscopic sarcocysts; only the dog excreted oocysts. Moreover, another cat fed infected meat also did not excrete sporocysts. These experiments confirmed the findings that sarcocysts from llamas are transmissible to dogs but not cats.[722] The idea that 2 species of *Sarcocystis* (*S. aucheniae* and "*Sarcocystis lamacenis*," possibly *S. lamacanis*) infect alpacas was supported by others.[1043,1127]

Critical examination of the taxonomy of the *Sarcocystis* species led to the conclusion that there is only 1 valid name, *S. aucheniae*, and that there is no valid reason to separate species infecting llamas and alpacas (*Vicugna pacos*).[1317] Because the sarcocyst that Brumpt first recognized was macroscopic it seems reasonable to designate macroscopic species as *S. aucheniae*. A new name can be proposed for the microscopic species when its structure is described. Some authors have called the microscopic sarcocyst *S. lamacanis*,[649] others have called it *S. lamacensis*.[1490] There is no morphological description of the microcyst sarcocyst of the llama. The closest description is an illustration of the microcyst in a histologic section of the heart of a llama in figure 1 from Reference **393** reproduced here (Figure 13.1). Therefore, the names *S. guanicoe-canis*, *S. lamacanis*, *S. lamacensis* should be *nomina dubia*.

In the first morphological description of the macroscopic sarcocyst in llamas, the parasitized myocyte is located in a vacuole.[1561] By light microscopy, the sarcocyst wall is thin. By TEM, the sarcocyst wall is type 21b with highly branched cauliflower-like vp. The dog, and not the cat, is the definitive host for *S. aucheniae*.[164] Macroscopic sarcocysts, appearing like rice grains, 3–5 mm long, were isolated from the meat of 3 naturally infected llamas in the Andean flatlands region of Argentina. DNA was obtained from each of 6 clean sarcocysts that were 17.3–33.1 mg and contained more than 10 million bradyzoites that were 17.6 μm long. Full-length 18S rRNA gene sequences from individual sarcocysts had 99.1%–100% similarity, indicating a cloned parasite. In comparison, the Australian isolate of *S. aucheniae* from *V. pacos* had 96% similarity. Whether these differences were related to methods used to compare the Australian or Argentinian isolates or were related to different *Sarcocystis* species could not be determined. Based on short sequences published for the "micro-cyst producing species

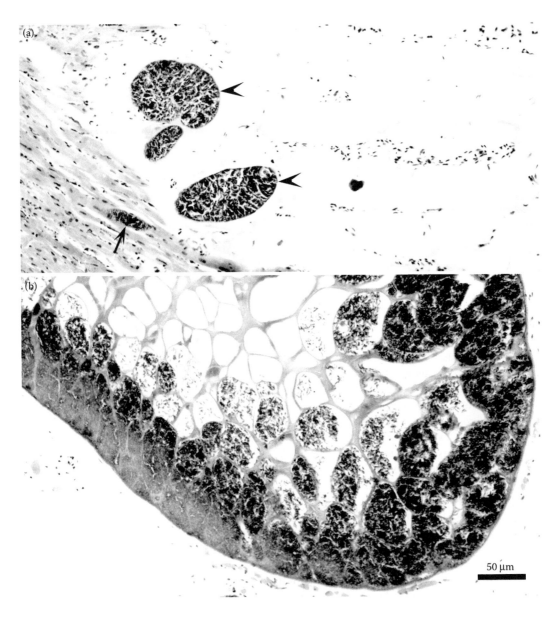

Figure 13.1   Sarcocysts in tissues of llamas (*Lama glama*). (a) Microscopic sarcocysts of an unnamed species in myocytes (arrow) and in Purkinje fibers (arrowheads) in the myocardium. (b) A macroscopic *S. aucheniae* sarcocyst in the esophagus. Note the prominent septa, zoites at the periphery, and empty areas in the center without zoites. H and E stain.

S lamacanis" 18S rRNA gene,[1162,1163] it was published and concluded that there were enough phylogenetic differences such that *S. aucheniae* should be considered a separate species.

13.2 CLINICAL AND SUBCLINICAL SARCOCYSTOSIS IN LLAMA AND ALPACA

Llamas are important to the economy of several South American countries. Little is known of clinical sarcocystosis in llamas. Severe myositis was diagnosed in a 6-year old female llama imported to the United States from Peru.[1010] Necrosis, hemorrhage, and eosinophilic myositis were associated with macroscopic sarcocysts; schizonts were not found. The llama aborted an 8-month-old fetus but parasites were not found in the placenta or in the fetal tissues.

Severe myositis was reported in an alpaca born in Australia.[649] This alpaca had numerous nodules and subcutaneous abscesses on head and neck. Examination of needle biopsy tissue revealed infiltration with leukocytes, predominantly eosinophils. Because of its deteriorating condition the alpaca was euthanized. At necropsy multiple caseating, pale white streaks were found in the head and neck area. Histologically, the lesions resembled eosinophilic myositis similar to that discussed in Chapter 1. Sarcocysts were found in the lesions. The authors indicated that the sarcocysts were ultrastructurally distinct from those reported in llamas but details were insufficient.

Sarcocysts are common in asymptomatic llamas and other similar camelidae (Table 13.1)[129,710,1043] and result in the down grading of carcasses at meat inspection. In an extensive survey of llamas in local abattoirs from 2006 to 2011 data, 1196 slaughtered llamas in Bolivia were analyzed with respect to *Sarcocystis* infection,[1490] and the following conclusions were drawn: (1) Yearly prevalence of sarcocysts varied from 23.4% to 50.3% and infection rates were higher in females and in older animals. (2) Prevalence in younger llamas (2.4 years) was 19.9% (96 of 482) versus 90.4% (160 of 177) in animals older than 4.5 years. (3) Down grading of *Sarcocystis*-carcasses resulted in between 13% and 20% economic loss to the farmer.

13.3 OTHER ASPECTS ON SARCOCYSTOSIS IN SOUTH AMERICAN CAMELIDS

13.3.1 Experimental Infections

Host specificity is discussed for the *Sarcocystis* spp. mentioned in the literature; thus, experimental infections have been carried out in different hosts: 1 investigation erroneously searched for sarcocysts in dogs, mice, and rats after feeding with guanaco meat but did not search for sporocysts in feces.[1435] Dogs, cats, rats, and mice were fed guanaco meat containing sarcocysts, but only dogs excreted sporocysts.[722] This study also tested the negative effect of temperature (cooking and freezing) on the infectivity of such sarcocysts for dogs. In Peru, macrocysts (considered *S. aucheniae*) of 2 different sizes (>5 and 1–3 mm) were fed to dogs.[254] Dogs that ingested the smaller sarcocysts excreted more sporocysts, possibly due to higher viability of bradyzoites in younger cysts.

Sixty-three young alpacas (5 months old) were infected with sporocysts of "*S. lamacanis*" obtained from dogs fed myocardium of alpacas containing *Sarcocystis* microcysts.[200] After feeding 1000, 2500, and 5000 sporocysts, 92.3% of the alpacas fed 5000 sporocysts died between 30 and 90 days later. However, there were no data on lesions or the stage of *Sarcocystis* (schizonts/sarcocysts) and no strong correlation between infection and loss of weight or wool production.

Five dogs were infected with 400 macrocysts obtained from naturally infected alpacas.[1866] Large numbers of oocysts of *S. aucheniae* were retrieved from the mucosa intestinalis of all dogs. Also, asexual stages of *S. aucheniae* have been cultured in different cell lines.[1769]

Experimental infections of alpacas with "*S. lamacanis*" in Peru studied the possible influence of sarcocysts located in the myocardium on electrocardiographical parameters.[129] After histological examination, sarcocysts were detected only in myocites and not in Purkinje cells.

13.3.2 Food Safety

The importance of gastrointestinal symptoms resulting from the ingestion of infected meat, due to the toxin present in the sarcocysts was highlighted.[1043] The effects of *S. aucheniae* sarcocystine toxin and its immunogenic activity were tested in rabbits.[215] Several experiments have been carried out on the control of toxic risks derived from ingesting meats containing *Sarcocystis* sarcocysts.

Table 13.1 Prevalence of *Sarcocystis* Sarcocysts in South American Camelids

Country	Year	Host	No	Method	No. Positive	% Positive	Microscopic cysts	Macroscopic cysts	Bioassay in dog and cat	Species	References
Argentina	1966–68	Guanaco[a]	119	Gross examination, squeezing, histology	48	40.3%	NS	Yes	Dogs, mice, rats	*S. tilopodi*	1435
Argentina, Patagonia	2000	Guanaco[a]	12	Gross examination, histology	8	66.7	No	Yes (T, H, M)	–	NS	90
Argentina, Jujuy	2005–07	Llama[b]	308	IFAT	295	96.0% (92.5%: *S. cruzi* antigen; 77.0%: *S. aucheniae* antigen)	–	–	–	NS	1214
Bolivia	2006–11	Llama	1196	Gross examination	408	34.1%	Macroscopic	No	–	–	1490
Peru	NS	Alpaca[c]	200	Gross examination, histology	200	100.0%	Yes (E: 99.5%; M: 95.5%; neck: 87.5%)	Yes (E: 16.5%)	–	NS	748
Peru	NS	Alpaca[c]	15	ELISA, PCR	15	100.0%	Yes	Yes	–	By sequencing of 18S rRNA gene: *S. aucheniae* and *S. lamacanis*	1162
Peru, Junin	NS	Alpaca[c]	941	ELISA	844	89.7%	–	–	–	–	169

Hosts: [a] *Lama guanicoe*, [b] *Lama glama*, [c] *Vicugna pacos*.
Tissues: E = esophagus, H = heart, M = skeletal muscle, T = tongue. NS = not stated.

Four different physical treatments for *S. aucheniae* infected meat included boiling, baking, frying, and freezing.[710] Treated meats were not toxic or infectious for rabbits and dogs, respectively. Different methods of curing and smoking failed to detoxify alpaca meat.[1792]

BIBLIOGRAPHY

Information pertinent to the subject matter of this chapter may be found in references:
Indentations Epidemiology/prevalence: **25, 90, 169, 748, 1043, 1127, 1214, 1490, 1962**.
Food safety: **215, 710, 1792**.
Molecular biology: **164, 1162, 1163**.
Morphology and speciation: **149, 393, 1317, 1435, 1561**.
Pathology and experimental infections: **129, 200, 254, 649, 722, 1010, 1477, 1769, 1866**.

CHAPTER 14

Sarcocystosis in Dogs (*Canis familiaris*)

14.1 MUSCULAR SARCOCYSTOSIS

Until 2005, there were few reports of sarcocysts in dogs and these were considered incidental (Table 14.1). Recently, severe myositis was diagnosed in 4 dogs and 2 species, *Sarcocystis caninum*, and *Sarcocystis svanai* were named.[476]

14.1.1 *Sarcocystis caninum* Dubey, Sykes, Shelton, Sharp, Verma, Calero-Bernal, Viviano, Sundar, Khan, and Grigg 2015

14.1.1.1 Structure and Life Cycle

Sarcocysts are microscopic, up to 1.2 mm long; the sarcocyst wall is 1–2 μm thick and is type 9c (Figure 14.1a). By TEM, vp are approximately 1 μm long, lacking microtubules. Bradyzoites are 6.0–7.5 × 2–3 μm in size. The life cycle is unknown.

14.1.2 *Sarcocystis svanai* Dubey, Sykes, Shelton, Sharp, Verma, Calero-Bernal, Viviano, Sundar, Khan, and Grigg 2015

14.1.2.1 Structure and Life Cycle

The sarcocyst wall is <0.5 μm thick and type 1a. Bradyzoites contain up to 4 rhoptries (Figure 14.1b).

14.2 CLINICAL DISEASE

Four (case numbers 5–8, Table 14.1) unrelated dogs (from Canada and the United States) had fever, apathy, anorexia, muscle weakness, ataxia, and elevated liver and muscle enzymes. One dog had dysphagia. Initially, the dogs were suspected to have neosporosis. Examination of muscle biopsies revealed severe myositis associated with numerous sarcocysts (Figure 14.2). Treatment with clindamycin was seemingly effective.

Table 14.1 Details of Dogs with Muscular Sarcocystosis

Case No.	Year	Location	Breed	Age	Sex	Main Signs	Location of Sarcocysts	TEM	References
1	1964–65	Madhya Pradesh, India	Not known	Not known	Not known	Not known	Esophagus	No	1509
2	Not stated	Georgia, USA	Doberman pinscher	2 yr	Female	Adenocarcinoma	Heart	No	828
3	Not stated	Alabama, USA	Not known	4 yr	Male	Ataxia	Leg, biceps	Yes, irregularly spaced vp	107
4	1991	Kenya	German shepherd	Adult	Female	Adenocarcinoma	Heart	No	159
5	2003	British Columbia, Canada	Labrador cross	5 yr	Male	a	Shoulder, biceps	Yes, S. caninum	196, 476
6	2010	Cañon City, Colorado, USA	Golden retriever	11 yr	Male	b	Digital flexor	Yes, S. caninum + S. svanai	476, 1701
7	2010	Rural Montana, USA	Rottweiler	4 yr	Female	c	Biceps, triceps	Yes, S. caninum	476, 1701
8	2012	British Columbia, Canada	Rhodesian ridgeback	1.5 yr	Female	d	Quadriceps	Yes, S. caninum	476

a Lethargy, anorexia, fever, ataxia, inability to walk, dysphagia.
b Lethargy, ptyalism, panting, anorexia, ataxia, diarrhea then stiff gait and generalized muscle pain 28 days after the onset of illness.
c Lethargy, ptyalism, panting, anorexia, diarrhea then stiff gait and generalized muscle pain 28 days after the onset of illness.
d Anorexia, diarrhea, head tremors, fever, muscle weakness, ataxia.

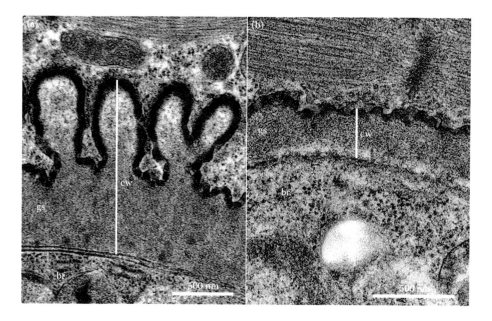

Figure 14.1 Comparison of the sarcocyst walls of two species of *Sarcocystis*. (a) *S. caninum* and (b) *S. svanai*. The cw of *S. caninum* is more than twice as thick as that of *S. svanai*. (Adapted from Dubey, J.P. et al., 2015. *J. Eukaryot. Microbiol.* 62, 307–317.)

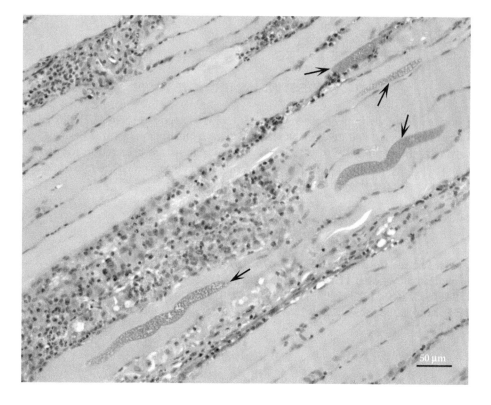

Figure 14.2 Severe myositis in a naturally infected dog. Note 4 elongated immature sarcocysts (arrows). H and E stain. (Adapted from Chapman, J., Mense, M., Dubey, J.P., 2005. *J. Parasitol.* 91, 187–190.)

Table 14.2 *Sarcocystis* **Species Sporocysts in Feces of Dogs**

Species	Intermediate Host	Size of Sporocysts (μm)
S. arieticanis	Sheep	15–16.5 × 9.8–10.5
S. tenella	Sheep	14–15 × 9–10.5
S. mihoensis	Sheep	15–16 × 8–9
S. cruzi	Cattle	14.5–17 × 9–11
S. capracanis	Goat	12–16 × 9–11
S. hircicanis	Goat	15.0–17.3 × 10.5–11.3
S. bertrami	Horse	11–14.4 × 8–10.1
S. equicanis	Horse	15–16.3 × 8.8–11.3
S. fayeri	Horse	11–13 × 7.0–8.5
S. miescheriana	Pig	12.7 × 10.1 (12.2–13.2 × 9.8–10.4)
S. poephagicanis	Yak	15.4–27.8 × 14.2–25.2 (oocysts)
S. alceslatrans	Moose	14–17 × 8.5–11.5
S. wapiti	Wapiti-elk	15.7–17.4 × 9.6–11.9
S. sybillensis	Wapiti-elk	15–17 × 10.5–12.0
S. cervicanis	Wapiti-elk	15.5–16.5 × 10.5–11.3
S. grueneri	Reindeer and wapiti-elk	12.4–15.7 × 9.2–11.2
S. capreolicanis	Roe deer	12.0–18.0 × 9.0–11.4
S. hemionilatrantis	Mule deer	14–16 × 9.5–11
S. odocoileocanis	White-tailed deer	13.2–15.7 × 8.8–12.1
S. levinei	Water buffalo	9.5–10.5 × 14.0–16.5
S. cameli	Camel	12.0–14.0 × 8.9–11.3
S. gracilis (?)	Roe deer	12–18 × 9–11.4
S. sp.	Grant's gazelle	10.8–15.6 × 8.4–12.0
S. sp.	Chicken	11–12.2 × 8.5–10
S. sp.	Pheasant	13–15 × 9–11
S. sp.	Fallow deer	15.4 × 8.8
S. sp. (S. aucheniae)	Llama	14.6 × 10.6

14.3 INTESTINAL SARCOCYSTOSIS

The dog is a major definitive host for many *Sarcocystis* species (Table 14.2). Sexual stages of those species have been tested in experimental infections by feeding parasitized tissues to coccidia-free dogs (and other canids).[11,383,392,585] *Sarcocystis* sporocysts are excreted in the feces in high number and generally for a sustained period; several reports have shown the role of domestic and stray dogs in excreting sporocysts.[80,451,880,1060,1265,1281,1506,1543,1570,1594,1612,1686,1783]

BIBLIOGRAPHY

Muscular sarcocystiosis: **107, 159, 196, 476, 828, 1509, 1701**.

Intestinal sarcocystiosis: **11, 16, 80, 383, 392, 451, 585, 880, 1060, 1265, 1281, 1506, 1543, 1570, 1594, 1612, 1686, 1783, 1905, 1913, 1914, 1920, 1922, 1924, 1962, 1990, 2002, 2102, 2139, 2157, 2158, 2200, 2228, 2255, 2256, 2261, 2310**.

CHAPTER 15

Sarcocystosis in Cats (*Felis catus*)

Cats are both intermediate and definitive hosts for *Sarcocystis*.

15.1 MUSCULAR SARCOCYSTOSIS

There are several reports of sarcocysts in cats, all except 1 were from the United States (Table 15.1). The presence of sarcocysts in cats is considered incidental. Only 1 morphologic type, *Sarcocystis felis*, has been found.

15.1.1 *Sarcocystis felis* Dubey, Hamir, Kirkpatrick, Todd, and Rupprecht 1992

15.1.1.1 *Structure and Life Cycle*

Sarcocysts were up to 22 mm long and up to 150 μm wide.[407] Viewed by LM, the sarcocyst wall has minute protrusions with a hobnail appearance (Figure 15.1). Ultrastructurally, vp are up to

Table 15.1   Reports of Muscular Sarcocystosis in Cats

Country	No. of Cats	Remarks	Location of Sarcocysts	TEM	References
USA, Illinois	4	4 of 10 asymptomatic cats killed after termination of unrelated study	Sarcocysts in skeletal muscle of 3, and heart of 1	Yes	**561**
USA	3	5-, 5-, and 10-yr-old, hemangioendothelioma, cardiomyopathy, adenocarcinoma	Skeletal muscle, heart, abdominal muscle	Yes	**963**
USA, Texas	1	1.5-yr-old, lymphosarcoma	Sarcocysts in muscles	Yes	**486**
USA, Georgia	2	5-yr-old, lymphosarcoma, 5-yr-old leukopenia	Sarcocysts in skeletal muscle of 1 and heart of the other	Yes	**828**
USA	2	Stray cats	Sarcocysts in leg muscles of 2 cats	Yes	**607**
Chile, Valdivia	1	Domestic cats	Sarcocysts in 1 of 24 diaphragms	No	**1757**
USA, Florida	5	5 of 50 cats from animal shelter	Sarcocysts in sections of muscles of 5, up to 20 mm long. Molecular characterization	Yes	**684**
USA	3	4 yr with neurological signs, 2 yr with undefined illness, 3 yr with sudden death	Skeletal muscle, heart. Molecular characterization	Yes	**521, 519**

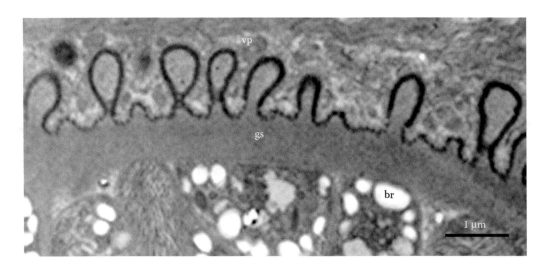

Figure 15.1   TEM of a mature *S. felis* sarcocyst in skeletal muscle of a naturally infected domestic cat. Note pleomorphic vp, thick, smooth gs and bradyzoites with amylopectin granules (br).

1.5 µm long, spaced unevenly on the sarcocyst wall. The vp are lined with a thick electron-dense layer, microtubules are absent from the vp and the ground substance layer, type 9c. Bradyzoites are slender 7–10 µm long and have numerous micronemes.[407]

The life cycle is unknown. Cats fed infected muscle did not excrete sporocysts.

15.2 INTESTINAL SARCOCYSTOSIS

Cats are definitive hosts for numerous *Sarcocystis* species (Table 15.2). There are a few surveys on the prevalence of sporocyts excreted by domestic or stray cats,[80,1021,1060,1800] nevertheless the rates

Table 15.2   Cats as Definitive Hosts for *Sarcocystis* Species

Species	Intermediate Host	Size of Sporocysts (µm)
S. gigantea	Sheep	10.5–14.0 × 8.0–9.7
S. medusiformis	Sheep	10.3–13.0 × 7.3–8.8
S. moulei	Goat	11.6–13.1 × 8.7–9.4
S. hirsuta	Cattle	11–14 × 7–9
S. fusiformis	Water buffalo	11.5–14 × 9–10
S. sp.	Grant's gazelle	12.6–18.0 × 8.4–12.0
S. muris	House mouse	8.7–11.7 × 7.5–9.0
S. cuniculi	Rabbit	11.6–14.5 × 8.7–10
S. cymruensis	Rat	10.5 × 7.9
S. sp.	Chicken	11.2–12.2 × 8.5–9.3
S. leporum	Cottontail rabbit	13–16.7 × 9.3–11.1
S. odoi	White-tailed deer	11–15 × 9–11
S. porcifelis	Pig	13.2–13.5 × 7.2–8.0
S. poephagi	Yak	20.8–25.7 × 15.9–22.4

seem to be highly related to the high prevalence of sarcocysts found in domestic livestock (as shown in other chapters in this book).

BIBLIOGRAPHY

80, 407, 486, 519, 521, 561, 607, 684, 828, 963, 1021, 1060, 1757, 1800, 1905, 1923, 1956, 1958, 1962, 2139, 2156, 2158–2160, 2201, 2228, 2255, 2256, 2259, 2367.

CHAPTER 16

Sarcocystosis in Chickens (*Gallus gallus*)

16.1 INTRODUCTION

There is considerable confusion concerning the species of *Sarcocystis* in chickens. The first reports of *Sarcocystis* infection in chickens were from Hungary, and the parasite was named *Sarcocystis horvathi*.[1450,1451] Sarcocysts were up to 1 mm long, the sarcocyst wall 1.8–2.7 μm thick, but the bradyzoites were not described.

In Bulgaria, *Sarcocystis* was found in chickens and a new name, *Sarcocystis gallinarum*, was proposed.[977] Sarcocysts were up to 10 mm long, bradyzoites were banana shaped, 7–13 μm long and 2–4.5 μm wide. The name was synonymized with *S. horvathi*.[1048]

In Russia, sarcocysts from naturally infected chickens with *S. horvathi* and from ducks infected with *Sarcocystis rileyi* were transmitted to dogs and cats.[715] There were 4 groups of dogs and cats: (1 and 3) 3 dogs 1.5–2 years of age; (2 and 4) 3 kittens 1.5–2 months of age. Each had 2 infected and 1 control animal. Groups 1 and 2 were fed 250–300 g of infected chicken meat and groups 3 and 4 were fed infected duck meat. Both dogs and cats fed infected meat excreted sporocysts with a prepatent period of 10 or 11 days; the patent period was 23 days for dogs, and 19 days for cats. The sporocysts were 13.3–15.4 × 10.6 μm. The author did not indicate whether dogs or cats were positive with *S. horvathi* or *S rileyi* or if both were positive with both species and whether dogs and cats had never ingested meat before the experiment.

Sarcocysts were seen in histological sections of 35 (45.5%) of 78 chickens from Papua New Guinea and in 2 chickens from Tasmania, Australia.[1248] Sarcocysts were up to 2 mm long and had a 1.5 μm thick striated wall. The ultrastructure of sarcocysts in muscles from a fowl from Tasmania and another fowl from Papua New Guinea were also described.[1173] The vp on the sarcocyst wall were 1.5 μm long and contained criss-crossing microtubules. Bradyzoites were not described. A dog fed infected muscles from a fowl excreted sporocysts but the sporocysts were not infectious for chickens. Thus, dogs could not be proven to be the definitive host,[1248] or the species found in the fowl was not infectious for domesticated chickens.

In Germany, sarcocysts were found in 45 of 241 free-range chickens but were not found in any of 207 battery raised chickens.[1822] Two types of sarcocysts were recognized based on the shape of bradyzoites. The banana-shaped bradyzoite sarcocysts were considered *S. horvathi* described in 1909[1451]; dog, cat, polecat, marten, and goshawk fed infected chicken muscle did not excrete sporocysts. They found a second type of sarcocyst containing lancet-shaped bradyzoites that were 12–15 × 2.5–3.0 μm in size. Both dogs and cats fed these sarcocysts shed sporocysts.[1822] The sporocysts were 11.6 × 9.2 μm in size. They fed 90 chickens with 1000–10,000 sporocysts (either dog or cat) and killed chickens at intervals for histological study. Schizonts were not found in chickens killed 1–15 DPI. Sarcocysts were detected starting 16 DPI. This seems very early compared with other species. At 40 DPI sarcocysts were immature, but they were mature at 71 DPI. Both dogs and cats fed chickens 88 DPI excreted sporocysts.[1822]

Reports of undefined species have been published more recently. In the Czech Republic, sarcocysts were found in 3 (10.0%) of 30 hens; and a dog fed infected tissues did not excrete sporocysts.[1380,1381] In China, sarcocysts were found in 6 (2.1%) of 284 free-range chickens from 3 shops in Kunming.[1109] Sarcocysts were found in most muscles of the carcass but not in the heart. Sarcocysts contained lancet-shaped bradyzoites (14.3 × 2.9 μm), and the cyst wall was identical to that from sarcocysts from Australia and Papua New Guinea.[1173] Four of 6 dogs that were 2 months old and 5 of 8 cats excreted sporocysts after consuming naturally infected muscles of chickens; sporocysts were 13.0 × 8.6 μm in size. Sporocysts (500,000; source not stated) were fed to 5 chickens that were 2 weeks old, and the chickens were killed at 20, 33, 63, 116, and 128 DPI; 3 chickens were controls but results were not given. Sarcocysts were not observed at 20 DPI. Sarcocysts at 33 DPI were immature. Mature sarcocysts were found in chickens at 63, 116, and 128 DPI.[1109]

The species with lancet-shaped bradyzoites[1822] was named *Sarcocystis wenzeli*.[1314]

In China, *S. wenzeli* was found in 17 (8.9%) of 191 free-range chickens from Yunnan and the morphological description of the species was added.[212]

In Azerbaijan, sarcocysts were found in 24 (27.9%) of 86 hens.[1182] Macroscopic sarcocysts found in pectoral and esophageal muscles were considered to be *S. horvathi*.

It appears that there are at least 2 named species, *S. horvathi* and *S. wenzeli* in chickens, with cats and dogs as definitive hosts. However, it is uncertain if both cats and dogs are hosts for the same parasite. In contrast to the foregoing reports, *Sarcocystis* sporocysts were not found in the feces of cats fed breast muscles from over 2000 chickens from grocery stores in the United States[458] and hearts from more than 1000 free-range chickens from different countries[465] although the muscles were not examined microscopically for sarcocysts or bradyzoites.

16.2 *SARCOCYSTIS WENZELI* (WENZEL, ERBER, BOCH, SCHELLNER 1982) ODENING 1997

16.2.1 Distribution

Australia, People's Republic of China, Papua New Guinea, Czech Republic, Germany.

16.2.2 Definitive Hosts

Dog, cat.

16.2.3 Structure and Life Cycle

Sarcocysts up to 3.1 mm long, thread-like, sarcocyst wall up to 3.5 μm thick, vp up to 1.9 μm long with prominent criss-crossing microtubules extending the entire length of the vp and embedded in the ground substance, villar tips with thick edl, type 9 k (Figure 16.1). Bradyzoites are lanceolate and measure 12.2–17.7 × 1.8–2.9 μm.[212] Sarcocysts in experimentally infected chickens mature by 63 DPI.[1109]

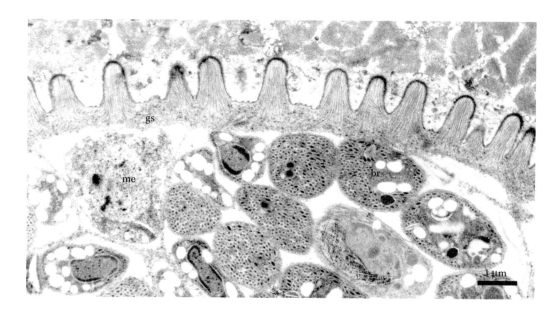

Figure 16.1 TEM of sarcocyst of *S. wenzeli.* Note stubby vp with criss-crossing microtubules. (From Mehlhorn, H., Hartley, W.J., Heydorn, A.O., 1976. *Protistologica* 12, 451–467. With permission.)

16.3 *SARCOCYSTIS HORVATHI* RÁTZ 1908

16.3.1 Distribution

Hungary, Bulgaria, Russia, Germany, Papua New Guinea, Australia, USA.

16.3.2 Structure and Life Cycle

Sarcocysts up to 980 μm long with indistinct septa. The sarcocyst wall is striated, up to 2.5–3.0 μm thick. Bradyzoites are banana shaped, 9.0–12.5 × 2.5–3.0 μm in size. No other details were found.

16.4 CLINICAL DISEASE IN NATURALLY INFECTED CHICKENS

Severe myositis was seen in chickens from Papua New Guinea and Australia[1248]; affected chickens had muscular weakness and adopted a duck-sitting posture. Numerous immature and mature sarcocysts were found in lesions of nonsuppurative myositis.

Severe encephalitis (Figure 16.2) was seen in free-range adult chickens in Mississippi, USA where many chickens had died with neurological signs.[1266] Three chickens were examined at necropsy, and all 3 had severe encephalitis associated *Sarcocystis* schizonts in various stages of development. The schizonts appeared to be in neural cells. Retrospectively, schizonts did not react to the *S. neurona* antibody (Dubey, J.P., unpublished data).

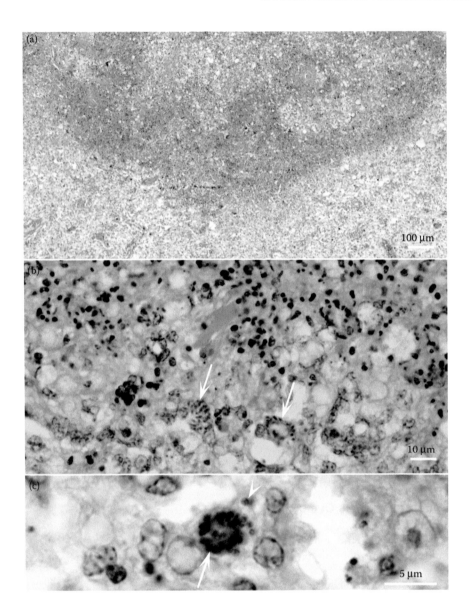

Figure 16.2   *Sarcocystis*-associated encephalitis in the cerebrum of a chicken. H and E stain. (a) Focal area of necrosis and perivascular cuffing. Numerous schizonts are present but not visible at this magnification. (b) Higher magnification of the edge of the lesion, showing schizonts (arrows). (c) Higher magnification to show a schizont (arrow) and individual merozoites (arrowhead). (Adapted from Mutalib, A. et al., 1995. *Avian Dis*. 39, 436–440.)

Myositis associated with undiagnosed *Sarcocystis/Leucocytozoon*-like parasite has been reported from Malaysia.[2203]

BIBLIOGRAPHY

212, 458, 465, 715, 977, 1048, 1109, 1173, 1182, 1248, 1266, 1314, 1380, 1381, 1450, 1451, 1822, 1916, 2372.

CHAPTER 17

Sarcocystosis in Other Avian Species

17.1 INTRODUCTION

There are numerous species of *Sarcocystis* in birds. In this chapter, they are arbitrarily discussed in 6 groups (A–F): groups A–E, species with birds (other than raptorial) as intermediate hosts; and group F, species with raptors as definitive hosts. Group A, constituting species with an unusually wide host range and marked pathogenicity; group B constituting *Sarcocystis* species with a known definitive host; group C, *Sarcocystis* species with an unknown life cycle; group D, species with uncertain taxonomic status; group E, sarcocysts in raptorial birds; and group F, species with raptors as definitive hosts.

17.2 GROUP A

There are 2 species, *Sarcocystis falcatula* and *Sarcocystis calchasi* (Table 17.1). These species are biologically and morphologically distinct.

17.2.1 *Sarcocystis falcatula* Stiles 1893

There is considerable uncertainty concerning this parasite. It was first named for the sarcocyst in muscles of the rose-breasted grosbeck (*Pheucticus ludovicianus*). Box et al.[137] redescribed it. Sarcocysts are thick-walled, type 11b (Figure 17.1). The opossum (*Didelphis virginiana*) is the definitive host[133] and several species of birds act as intermediate hosts. Subsequently, 2 other species of South American opossums (*Didelphis albiventris* and *Didelphis marsupialis*) were found to be its natural definitive hosts.[418,441] The type specimens of *S. falcatula* deposited by Stiles have been re-examined by us, and there is no way to determine whether the parasite Box described is the same as described by Stiles. It is most likely that *S. falcatula* constitutes a group of several species. Recently, a new species *Sarcocystis lindsayi* was recognized with *D. albiventris* and *D. aurita* as definitive hosts.[433,1670] The natural intermediate host/hosts for *S. lindsayi* is unknown; budgerigars (*Melopsittacus undulatus*) were found experimentally to be an intermediate host. Budgerigars fed sporocysts from feces of opossum that died of acute sarcocystosis, similar to infection with *S. falcatula*. Another *S. falcatula*-like parasite was recognized in the carmine bee-eater (*Merops nubicus*) in the United States.[436] *In vitro* culture, life cycle studies with cloned parasites, and molecular assays will be needed to resolve the speciation issue.

Numerous outbreaks of acute sarcocystosis have been reported in passerine and psittacine birds in captivity in the Americas (Table 17.1). These outbreaks were presumed to be as *S. falcatula* infections. Some affected birds die acutely without prior signs. Depression and dyspnea are the most

Table 17.1   Comparison of 2 Most Pathogenic Avian Species of *Sarcocystis*, *S. falcatula*, and *S. calchasi*

Character	*S. falcatula* (STILES 1893) BOX, MEIER, AND SMITH 1984	*S. calchasi* OLIAS, GRUBER, HAFEZ, HEYDORN, MEHLHORN AND LIERZ 2010
	Intermediate Hosts	
Natural	Numerous passeriformes, psittaciformes, columbiformes, strigiformes, and falconiformes including great horned owl (*Bubo virginianus*), bald eagles (*Haliaeetus leucocephalus*), golden eagle (*Aquila chrysaetos*), brown-headed cowbird (*Molothrus ater*). Luzon bleeding-heart pigeon (*Gallicolumba luzonica*), Pinon's imperial pigeon (*Ducula pinon*), nutmeg imperial pigeon (*Ducula bicolor bicolor*), Patagonian conure (*Cyanoliseus patagonus*), grackle (*Quiscalus quiscula*) and many other new and old world avian species[262,436,484,711,1305,1354,1630,1631,1691,1795]	Domestic pigeon (*Columbia livia* f. *dom.*), psittaciformes in captivity as Alexander parrots (*Polytelis alexandrae*), long-billed corella (*Cacatua tenuirostris*) and rose-breasted cockatoos (*Eolophus roseicapilla*)
Type host	Rose-breasted grosbeck (*P. ludovicianus*)	Domestic pigeon
Experimental	Canary (*S. canaria*), zebra finch (*Poephila guttata*), budgerigar (*M. undulatus*), domestic pigeon (*Columbia livia* f. *dom.*), Guinea fowl (*Numida meleagris*), house sparrows (*Passer domesticus*)	Domestic pigeon, cockatiel (*Nymphicus hollandicus*), great tit (*Parus major*)
Distribution	Americas	Europe, North America
Original	North America	Europe
	Schizonts	
Persistence	Extended period, schizonts and sarcocysts in the same animal	Extended period, schizonts and sarcocysts in the same animal
Duration	Extended	7–12 days, experimental
Main locale	Lung, liver	Liver
	Sarcocysts	
Size	Up to 3.2 mm long	Up to 2.0 mm long
Wall type	Striated, TEM type 11b	Smooth, wavy, TEM type 1a
Bradyzoites	6.8×2.1 (4.8–8.4×1.2–3.6) µm (muscle digest)	Lancet-shaped, 7.5×1.5 µm (fixed smears, Giemsa)
Molecular data	18S, ITS-1	18S, 28S, ITS-1
	Definitive Hosts	
Natural	Opossums (*D. virginiana, D. albiventris, D. marsupialis*)	Northern goshawk (*Accipiter g. gentilis*), European sparrowhawk (*Accipiter nisus*)
Experimental		Northern goshawk
Sporocysts size	11.2×7.4 (9.6–12.0×6.0–8.0) µm	11.9×7.9 µm
Reports of outbreaks and clinical cases	101, 122, 228, 229, 436, 484, 711, 831, 879, 1032, 1189, 1354, 1631, 1691, 1795, 1841, 1842	1325, 1328, 1329, 1473, 1841, 1843
General references	131, 133, 134, 135, 262, 412, 414, 418, 421, 441, 792, 1070, 1088, 1116, 1173, 1273, 1305, 1409, 1626, 1627, 1629, 1630, 1631, 1660, 1932, 2209	1099, 1326, 1327, 1330, 1331, 1332, 1696, 2395

Figure 17.1   TEM of *S. falcatula* sarcocyst in skeletal muscle of a budgerigar (*M. undulatus*). Note the elongated vp on the sarcocyst wall. Few microtubules are present in the vp, and they extend into the gs (arrowheads). The bradyzoites (br) are juxtaposed with the gs. (Adapted from rom Box, E.D., Meier, J.L., Smith, J.H., 1984. *J. Protozool*. 31, 521–524; Courtesy of late Edith Box.)

prominent signs. At post mortem, edema and congestion of the lungs and enlargement of the spleen and liver are the most prominent findings. Microscopically, lesions are associated with intravascular schizonts, often occluding capillaries (Figure 17.2).

The host range of *S. falcatula* is wide; in other taxonomic groups, there is a report (other than those reported in Table 17.4 in this chapter) of severe encephalitis in a free-ranging great horned owl (*Bubo virginianus*) from Minnesota, USA.[1841] Pathogenic *Sarcocystis* species add a new concern in conservation.

17.2.2 *Sarcocystis calchasi* Olias, Gruber, Hafez, Heydorn, Mehlhorn, and Lierz 2010

Unlike *S. falcatula*, *S. calchasi* sarcocysts are thin-walled, type 1a (Figure 17.3). They cause neurological signs in pigeons and psittacine birds (Table 17.1). The affected birds have signs of depression, torticollis, opisthotonus, trembling, paralysis, and polyuria.[1325] Necrotic and granulomatous meningoencephalitis is the predominant lesion associated with schizogonic development of the parasite in neural cells.

17.3 GROUP B

There are 6 species in this group (Table 17.2).

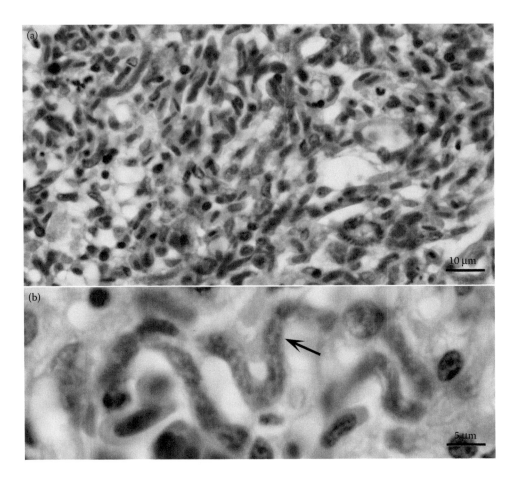

Figure 17.2 *S. falcatula.* (a) Section of a lung of a budgerigar fed sporocysts from *D. albiventris* from Argentina. Note severe parasitization of pulmonary blood vessels. H and E stain. (b) Higher magnification to show serpentine schizonts (arrow).

17.4 GROUP C

Sarcocyst morphology has been described for 23 species (Table 17.3).

Unnamed sarcocysts have been found in the straw-necked ibis (*Threskiornis spinicollis* [*Carphibis spinicollis*]),[437] coot (*Fulica atra*),[1822] wild turkey (*Meleagris gallopavo*),[1728] Northern gannet (*Morus bassanus*),[1647] American woodcock (*Scolopax minor*),[829] goldeneye (*Bucephala clangula*)[1004] and in several species of wading birds (Ciconiiformes).[1646] Seven avian families including Anatidae, Icteridae, and Corvidae in Canada were surveyed.[341] Sarcocysts were found in 11 host species but parasites were not named. The same year, in Australia, *Sarcocystis* was detected in 44 species representing 25 families (e.g., Pelecanidae, Anatidae, Falconidae, Rallidae, Corvidae, etc.).[1251] Thin- and thick-walled cysts were recorded, but final identity was not provided.

17.5 GROUP D

These species are considered invalid/species *inquirendae* as there are no differential morphological descriptions of the sarcocyst. Namely, *Sarcocystis ammodrami* (Splendore, 1907) Babudieri,

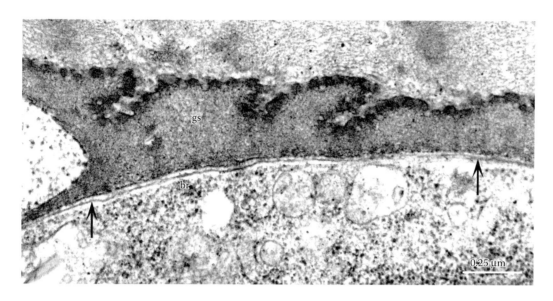

Figure 17.3 TEM of *S. calchasi* sarcocyst in skeletal muscle of a pigeon. Note the thin sarcocyst wall, without villi, and microtubules. Arrows point to the bradyzoite (br) pellicle. (Courtesy of A.G. Armién, Diagnostic Ultrastructural Pathology Service, College of Veterinary Medicine, University of Minnesota.)

1932 from the Fringillid bird (*Pheucticus ludovicianus*)[1049]; *Sarcocystis aramidis* Spendore, 1907 from the Rallid bird (*Aramides saracura*)[1049]; *Sarcocystis jacarinae* Barreto, 1940 from the Fringillid bird (*Volatinia jacarina*)[1049]; *Sarcocystis oliverioi* Pessõa, 1935 from the Psittacid bird (*Forpus passerinus*)[1391]; *Sarcocystis turdi* Brumpt, 1913 from the European blackbird (*Turdus merula*), *S. nontenella* Eble, 1961 from the common buzzard (*Buteo buteo*)[1049] and *Sarcocystis debonei* (Vogelsang, 1929) Duszynski and Box 1978 from the cowbird (*Molothrus ater*).[482]

There are 3 additional species that need validation, *Sarcocystis garzettae* Odening 1997 from the little egret (*Egretta garzetta*), *Sarcocystis kaiserae* Odening 1997 from the laughing dove (*Streptopelia senegalensis*), and *Sarcocystis spaldingae* Odening 1997 from Ciconiiformes (*Ardea Herodias*, *Egretta caerulea*, etc.).[1314,1317]

17.6 GROUP E

This group includes sarcocysts in the muscle of raptors. Sarcocysts are apparently common in raptors.[179] *Sarcocystis*-like bradyzoites were found in digests of 52 (45.6%) of 114 samples of muscles from 11 species of raptorial birds.[1069] Sarcocysts were identified in the histological sections of muscles of 28 (37.8%) of 74; 12 of 21 red-tailed hawks (*Buteo jamaicensis*), 2 of 5 red-shouldered hawks (*Buteo lineatus*), 4 of 4 Cooper's hawks (*Accipiter cooperi*), 2 of 3 sharp-shinned hawks (*Accipiter striatus*), 1 of 2 American kestrels (*Falco sparverius*), 2 of 4 turkey vultures (*Cathartes aura*), 1 of 2 black vultures (*Coragyps atratus*), 1 of 10 barred owls (*Strix varia*), and 3 of 8 screech owls (*Megascops* spp.).[1069]

Two types of sarcocysts with thin and thick walls have been reported in bald eagles (*Haliaeetus leucocephalus*). Thin-walled microscopic sarcocysts were found in a female bald eagle in Alabama.[261] Both thin- (variety A) and thick-walled (variety B) sarcocysts were seen in histologic sections of muscles of a female bald eagle from Missouri.[1333] Variety A sarcocysts were microscopic, had a thin

Table 17.2 Sarcocystis Species with Birds as Intermediate Host and Known Definitive Hosts

Species	Intermediate Host	Sarcocysts			Schizonts	Definitive Host	Molecular Data	Location	References
		Size (μm)	TEM Wall Type	Bradyzoites (μm)					
S. albifronsi Kutkiené, Prakas, Sruoga, and Butkauskas 2012	White-fronted goose (*Anser albifrons*)	Microscopic, ribbon-like, up to 4 mm long, 750 μm wide	Thick, up to 2.3 μm long vp, type 9a	Lanceolate, 10.0–13.5 × 1.5–2.5	–	Arctic fox (*Alopex lagopus*), sporocysts 10.0–12.8 × 6.8–8.6 μm	18S, 28S, ITS-1	Lithuania	158, 1002, 1003, 1008
S. alectorivulpes Pak, Sklyarova, and Pak 1989	Chukar partridge (*A. chukar*)	Macroscopic. Thick, striated wall	–	Cigar-shaped, 2.1–4.2 × 9.8–14.0	No data	Foxes (*Vulpes vulpes, Alopex corsac*)	No data	Kazakhstan	1284, 1357
S. lindsayi Dubey, Rosenthal, and Speer 2001	Experimental budgerigar (*Melopsittacus undulatus*)	Up to 600 μm, up to 2 μm	Thick, 2.0 × 0.3 μm vp like, type 11d	6 × 1.5 μm	In cell culture, merozoites 4.8–6.4 × 1.6–2.4	*D. albiventris, D. aurita*; sporocysts 12 × 7 μm	ITS-1	Brazil	425, 433, 1669, 1670
S. peckai (Pecka 1988) Odening 1997	Pheasant (*Phasianus colchicus*)	–	–	Lancet-shaped, 14–16 × 2–3	–	Dog	No data	Former Czechoslovakia	1314, 1381, 1822
S. rileyi (Stiles 1893) Dubey, Cawthorn, Speer, and Wobeser 2003	Shoveler duck (*Anas clypeata*), mallard duck (*Anas platyrhynchos*), common eider (*Somateria mollissima*), Eurasian wigeon (*Anas penelope*), common teal (*Anas crecca*)[a]	Macroscopic, up to 12 mm, smooth	7.5 μm long vp, type 23	Slender, 10.7–13.0 × 2.5–5.1	–	Skunk (*Mephitis mephitis*), sporocysts 10–14 × 5.5–9.5 μm. Molecular identification in feces of red fox (*Vulpes vulpes*) and raccoon dog (*Nyctereutes procyonoides*) from Europe	18S, 28S, ITS-1, cox1	North America, Lithuania, Finland, Norway, other countries[b]	452, 466, 703, 1007, 1421, 2398
S. sp. Wenzel, Erber, Boch, and Schellner 1982	Pheasant (*Phasianus colchicus*)	Up to 2 mm. Thick, striated wall, with 3–4 μm long finger-like vp	–	Lanceolate, 12.5–17 × 2–3	No data	Dog, sporocysts 13–15 × 9–11	No data	Czech Republic, Germany	191, 1380, 1822

a Cornwell[257] and Chabreck[194] reported *S. rileyi*-like in Anatidae as redhead (*Aythya americana*), canvasback (*Aythya valisineria*), and common goldeneye (*Bucephala clangula*) and reviewed other previous reports in the England, Mexico, Panama, and the United States.

b Bulgaria, Germany, Poland, Russia,[715,931,933] and in other European countries.[1421]

Table 17.3 Sarcocystis Species with Birds as Intermediate Host and with Unknown Definitive Hosts

| Species | Intermediate Host | Sarcocysts | | | Molecular Data | Country-Location | References |
		Size (μm)	Wall Type	Bradyzoites (μm)			
S. anasi Kutkiené, Prakas, Sruoga, and Butkauskas 2012	Mallard duck (*Anas platyrhynchos*)	Microscopic, up to 5 mm long	Thick, palisade-like vp, type 9a	Blunt ends, 13–16.1 × 1.8–2.5	18S, 28S, ITS-1	Lithuania	**1002, 1004, 1008**
S. colii Fantham 1913	Red-faced African mouse bird (*Colius erythromelon*)	Macroscopic, 2.5 mm	No data	5–7 in smears	No data	Origin unknown, Liverpool museum, UK	565
S. columbae Olias, Olias, Liez, Mehlhorn, and Gruber 2010	Wood pigeon (*Columba palumbus*)	Microscopic, ribbon-like, up to 7 mm long	Thin, type 1a	Banana-shaped, 6.3–7.3	18S, 28S, ITS-1	Germany, Lithuania	**1326, 1416**
S. cornixi Kutkiené, Prakas, Sruoga, and Butkauskas 2009	Hooded crow (*Corvus cornix*)	Microscopic, ribbon-shaped, up to 6 mm	2.1 μm thick, stump-like vp, type 1 g	Banana-shaped, 6.1–7.9 × 1.4–1.8	18S, 28S, ITS-1	Lithuania	1005
S. corvusi Prakas, Kutkiené, Butkauskas, Sruoga, and Žalakevičius 2013	Jackdaw (*Corvus monedula*)	Microscopic, thick thread-like, up to 6 mm long	Thin, type 1d	Lancet-shaped, 5.9–7.3	18S, 28S, ITS-1	Lithuania	1419
S. chloropusae El-Morsey, El-Seify, Desouky, Abdel-Aziz, Sakai, and Yanai 2015	Common moorhen (*Gallinula chloropus*)	Microscopic, up to 650 μm	2.0–4.5 μm thick wall, type 10	6–12 × 1–2	18S, 28S, ITS-1	Egypt	**489, 490**
S. kaiserai (Kaiser 1983) Odening 1997	Laughing dove (*Streptopelia senegalensis*)	No data	Type 9j	No data	No data	South Africa	926, 927, 1314
S. kirmsei Garnham, Duggan, and Sinden 1979	Hill mynah (*Gracula religiosa*), Siamese fire-backed pheasant (*Lophura diardi*)	Cyst in brain, round to oval, up to 1100 μm. 1–4 μm thick wall with spines	—	7–10 × 1.9–2.6	No data	Thailand, Panama, Germany	664, 895

(Continued)

Table 17.3 (Continued) Sarcocystis Species with Birds as Intermediate Host and with Unknown Definitive Hosts

Species	Intermediate Host	Sarcocysts				Molecular Data	Country-Location	References
		Size (µm)	Wall Type	Bradyzoites (µm)				
S. lari Prakas, Kutkienė, Butkauskas, Sruoga, and Žalakevičius 2014	Black-baked gull (*Larus marinus*)	Microscopic, thread-like, up to 6 mm long	Thin, type 1d	Lancet-shaped, 6.3–7.9 × 1.2–1.5		18S, 28S, ITS-1	Lithuania	**1420**
S. phoeniconaii (Murata 1986) Gobel, Erber, and Grimm 1996	Lesser flamingo (*Phoeniconaias minor*)	Macroscopic, rice grain-like, up to 7 mm	Thick, vp up to 7 µm long, type 9l	Lancet-shaped, 15–20 × 1.8–2.5		No data	Germany, Tanzania	**47, 709, 1319**
S. ramphastosi Dubey, Lane, and Van Wilpe 2004	Keel-billed toucan (*Ramphastos sulfuratus*)	Macroscopic, up to 3 mm long	Thin, smooth, vp up to 6.5 µm, type 9e	4.0–4.5 × 1.3–1.6		18S	South Africa, Costa Rica	**457, 463**
S. setophagae Crawley 1914	Redstart (*Setophaga ruticilla*)	Macroscopic 2.5 mm	No data	4–5		No data	USA	**260**
S. sulfuratusi Dubey, Lane, and Van Wilpe 2004	Keel-billed toucan (*Ramphastos sulfuratus*)	Microscopic, up to 900 µm	Thin, smooth, vp up to 4.3 µm long, type 9f	5.0–7.0 × 1.5–2.2		No data	South Africa (Bird was imported from Central America)	**457**
S. turdusi Kutkiene, Prakas, Butkauskas, and Sruoga 2012	Common blackbird (*Turdus merula*)	Microscopic, ribbon-like, up to 7 mm long	Thick, vp branched, similar to type 18b	Orange segment-shaped, 5.5–7.2 × 1.2–1.5		18S, 28S, ITS-1	Lithuania	**1009**
S. wobeseri Kutkiené, Prakas, Sruoga, and Butkauskas 2010	Barnacle goose (*Branta leucopsis*), mallard duck (*Anas platyrhynchos*), herring gull (*Larus argenticus*)	Microscopic, ribbon-like, up to 6 mm long	Thin, type 1d	Banana-shaped, 6.4–7.9		18S, 28S, ITS-1	Lithuania	**1834, 1002, 1006, 1415**
S. sp. Wenzel, Erber, Boch, and Schellner 1982	Coot (*Fulica atra*)	Microscopic, up to 360 µm long. Smooth, not clear	—	Banana-shaped, 7.5–8 × 1.5–2		No data, dog, cat not hosts	Germany	**1822**

(Continued)

Table 17.3 (Continued) Sarcocystis Species with Birds as Intermediate Host and with Unknown Definitive Hosts

Species	Intermediate Host	Sarcocysts			Molecular Data	Country-Location	References
		Size (μm)	Wall Type	Bradyzoites (μm)			
S. sp. Teglas, Little, Latimer, and Dubey 1998	Turkey (*Meleagris gallopavo*)	Sarcocyst in cross section 60 μm diameter. 5 μm thick sarcocyst wall with leaf-like vp with a stalk	—	NS	No data	USA	1728
S. sp. Dubey, Johnson, Bermudez, Suedmeyer, and Fritz 2001	Straw-necked ibis (*Carphibis spinicollis*)	Microscopic, up to 350 μm long	2 μm thick, striated, vp 4.5 μm long, type 11d	4.5–5.0 × 1.5	No data	USA	437
S. sp. Spalding, Yowell, Lindsay, Greiner, and Dame 2002	Northern gannet (*Morus bassanus*)	NS	Palisade-like, type 11d	NS	ITS-1 (for schizonts in brain, not done in sarcocysts)	USA	1647
S. sp. Dubey and Morales 2006	Buffon's macaw (*Ara ambigua*)	Up to 950 μm long	Vp up to 4 μm long, folded over, type 11c	4.0–5.9 × 0.8–2.8	No data	Costa Rica	1989
S. sp. Hiller, Sidor, De Guise, and Barclay 2007	American woodcock (*Scolopax minor*)	Microscopic, up to 125 μm	Thin, 2 μm long finger-like vp, type 9b	7 × 2	No data	USA	829
S. sp. Kutkienė, Sruoga, and Butkauskas 2008	Goldeneye (*Bucephala clangula*)	Microscopic, up to 1.4 mm	Irregular surface, wavy, type 1a	Banana-shaped, 8.5–7.0	No data	Lithuania	1002, 1004

NS = not stated

sarcocyst wall with spines, and contained bradyzoites that measured 5×1 μm. Variety B sarcocysts had a 2 μm thick striated cyst wall, and were immature.

Only *Sarcocystis otus* has been proposed as a species name for *Sarcocystis* in raptors. Three different types of sarcocysts were detected in the Eurasian buzzard (*B. buteo*) and long-eared owl (*Asio otus*) from Germany; the third species (proposed as *S. otus*) had hirsute vp and 5.1–2.9 × 1.9–2.8 μm long bradyzoites.[979]

17.7 GROUP F

In addition to raptors acting as definitive hosts for species of *Sarcocystis* in rodents (see Chapter 21), the sizes of oocysts and sporocysts detected in 9 species within the orders Accipitriformes and Strigiformes from the Berlin–Brandenburg area in Germany were reviewed.[1302] *Sarcocystis* sporocysts were found in 1 of 2 turkey vultures from Alabama, USA, but sporocysts were not infectious for cotton rats and mice.[1066] Sporocysts were found in an Indian pariah kite (*Milvus migrans*) that were not infective to rats and mice.[1591] Molecular methods distinguished 8 different genetic sequences of *Sarcocystis* spp. present in fecal matter from 4 species of hawks in Georgia, USA.[1851] Intermediate hosts need to be defined. In addition, dissemination of *Sarcocystis* sporocysts is also possible in aegagropiles of diurnal and nocturnal birds.[1531]

Three species in this group need additional comments:

Sarcocystis accipitris[192] was proposed as a new species between goshawk (*Accipiter gentilis*, Accipitriformes) and Passeriformes, experimentally canaries (*Serinus canaria*) acting as intermediate hosts. The sizes of the sporocysts in definitive hosts are 15–17 × 13–15 μm. Under experimental conditions, canaries developed sarcocysts up to 900 μm long, thin-walled (<1 μm) and with small

Table 17.4   Neurological Sarcocystosis in Birds in Groups B–F

Host	Country	Clinical Signs	Main Lesion	Immuno-staining	References
Northern gannet (*Morus bassanus*)[a]	USA, Florida	Neurologic	Meningoen-cephalitis	Not reacted against *S. neurona*	1647
Bald eagle (*Haliaeetus leucocephalus*)[b]	USA, Missouri	Neurologic	Meningoen-cephalitis	*S. neurona*	1333
Northern goshawk (*Accipiter gentilis atricapillus*)[b]	USA, Minnesota	Paralysis	Encephalitis	*S. cruzi*	21
Golden eagle (*Aquila chrysaetos*)[a]	USA, Virginia	Neurological, blindness	Necrotizing encephalitis	*S. cruzi*	403
Straw-necked ibis (*Threskiornis spinicollis*)[b]	USA, Missouri	Neurological	Encephalitis	*S. falcatula, S. neurona, S. cruzi*	437
Turkey (*Meleagris gallopavo*)[a]	USA, West Virginia	Ataxia, lethargy	Myocarditis, encephalitis	*S. cruzi*	1728
Capercaillie (*Tetrao urogallus*)[a]	Sweden	53 dead	Meningoen-cephalitis	*S. cruzi*	763
	Finland	Found dead	Myocarditis, encephalitis, other organs affected	*S. cruzi, S. neurona*	413

a   Caught in wild.
b   In captivity.

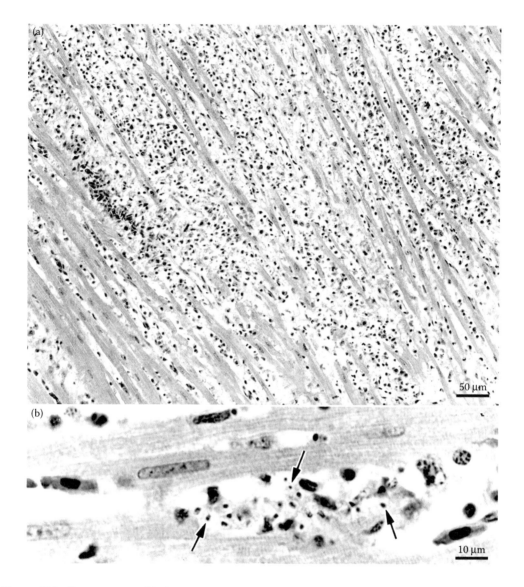

Figure 17.4 Severe myocarditis in a naturally infected capercaillie from Finland. (a) Massive mononuclear cell infiltration. (b) Degenerating merozoites (arrows) among inflammatory cells. (Adapted from Dubey, J.P., Rudbäck, E., Topper, M.J., 1998. *J. Parasitol.* 84, 104–108.)

(5–7 × 1.5–2.0 μm) crescent-shaped bradyzoites. TEM information has not been reported. This is similar to *Sarcocystis* sp. reported by Svobodová[1696] in a great tit, *Parus major*, with a goshawk as definitive host, and is also similar to *S. calchasi*.

Sarcocystis alectoributeonis completes its life cycle between the common buzzard (*B. buteo*, Accipitriformes) as the definitive host and the chukar partridge (*Alectoris chukar*, Galliformes) as the intermediate host.[1357] It was originally detected in Kazakhstan. Sarcocysts are macroscopic, thin-walled (0.5–1.0 μm) and smooth. TEM type was not reported. Bradyzoites are 1.2–2.1 × 7.0–8.4 μm. Sporocysts shed by definitive hosts are 8.4–10.5 × 11.2–14.7 μm.

The life cycle of *Sarcocystis ovalis* is completed between the magpie (*Pica pica*, Passeriformes) as the definitive host,[699] this is not a raptor but an opportunistic scavenger, and the moose (*Alces alces*) as the intermediate host. Information related to this species is summarized in Chapter 18.

17.8 CLINICAL OBSERVATIONS OF SARCOCYSTOSIS IN BIRDS IN GROUPS B–F

Mature schizonts and free merozoites of *Sarcocystis* as well as *Chlamydia* sp. were found in pulmonary lesions in an immature wild red-tailed hawk (*B. jamaicensis*) in Louisiana, USA that died following respiratory distress.[1199]

Neurological sarcocystosis has been reported in the eagle, ibis, wild turkey, gannet, and goshawk from the United States (Table 17.4). Numerous *Sarcocystis* schizonts and merozoites were found in the lesions. The 4 raptors listed in Table 17.4 had been caught in the wild but were in captivity for different reasons. Therefore, it is uncertain if they acquired sarcocystosis in the wild or in captivity. The wild turkey was killed because it was ill. Grossly, there were pale areas on the myocardium suggestive of necrosis. Schizonts were seen in areas of myocardial necrosis and in areas of encephalitis.[1728]

Systemic sarcocystosis has been reported in a capercaillie (*Tetrao urogallus*) in Finland,[413] and severe encephalitis has been seen in these galliform birds in Sweden.[763] In the bird from Finland, severe lesions were found in the heart, spleen, lung, brain, and other organs. Mortality in capercaillie has been recognized in Scandinavia since 1947, but the disease was misdiagnosed as toxoplasmosis.[763] Among the 87 capercaillie examined at post mortem, nonsuppurative encephalitis was diagnosed in 53 birds in association with *Sarcocystis*; lesions and protozoa were confined to the brain. Until now, only schizonts, no sarcocysts, of *Sarcocystis* have been found in capercaillie. Most schizonts were extravascular and up to 55 μm long (Figures 17.4).

BIBLIOGRAPHY

Group A: **101, 122, 131, 133, 134, 135, 137, 228, 229, 261, 262, 412, 414, 418, 421, 433, 436, 441, 484, 510, 711, 792, 831, 879, 1032, 1070, 1088, 1099, 1116, 1189, 1273, 1305, 1325–1332, 1354, 1409, 1473, 1626, 1627, 1629–1631, 1660, 1670, 1691, 1696, 1795, 1841–1843, 1932, 2209.**

Group B: **158, 171, 179, 191, 194, 413, 425, 433, 452, 466, 703, 715, 763, 931, 933, 1003, 1007, 1008, 1357, 1380, 1421, 1669, 1670, 2182, 2328.**

Group C: **47, 155, 257, 259, 260, 341, 342, 437, 457, 463, 489, 490, 565, 570, 599, 664, 709, 829, 845, 895, 926, 927, 931, 1002, 1004–1009, 1126, 1173, 1182, 1212, 1251, 1314, 1319, 1326, 1380, 1381, 1415, 1416, 1419, 1420, 1646, 1647, 1728, 1822, 1834, 1989, 2382.**

Group D: **393, 482, 1049, 1314, 1317, 1391, 1917, 1929, 1957, 1981, 1999, 2073, 2187, 2204, 2237, 2238, 2244, 2251, 2279, 2299, 2316, 2333, 2341, 2365.**

Group E: **179, 261, 979, 1069, 1199, 1333.**

Group F: **21, 192, 403, 699, 1066, 1302, 1333, 1357, 1531, 1591, 1696, 1851, 2141, 2146, 2147, 2150, 2347, 2353.**

Miscellaneous: **122, 192, 794, 1917, 1981.**

CHAPTER 18

Sarcocystosis in Wild Ruminants and Other Large Animals

18.1 *SARCOCYSTIS* INFECTIONS IN CERVIDS

Sarcocysts are ubiquitous in cervids, and numerous species of *Sarcocystis* in cervids have been named. Whether *Sarcocystis* species are host-specific for cervid species is not known. Cross transmission in animals raised in captivity and/or DNA sequence data will be necessary to prove or disprove host specificity.[285,354,380,586,704,858] Available information on wild Cervidae, other large animals, and marine mammals are summarized in Tables 18.1 through 18.3, but even morphological information is incomplete for most species. Examples of species detected but considered *inquirendae* are given in Table 18.4.

18.1.1 *Sarcocystis hemionilatrantis* Hudkins and Kistner 1977

Among all cervids, the life cycle is known for only *S. hemionilatrantis* which has been studied in fawns of mule deer (*Odocoileus hemionus*) raised in captivity. The definitive hosts include the dog (*Canis familiaris*)[379,858] and the coyote (*Canis latrans*).[858] In histologic sections, the sarcocysts are microscopic, measuring up to 525 μm long. The sarcocyst wall is 1–2 μm thick (Figure 1.10f). The vp are 1.8–2.4 μm long and 1.9–2.8 μm wide and are highly convoluted and branched at the base, type 17 (Figure 1.41a). The bradyzoites seen in tissue sections are 13.3–16 × 2.6–3.2 μm. The sporocysts are 14–16 × 9.5–11.0 μm.

S. hemionilatrantis has at least 3 generations of schizogony.[379] First- and second-generation schizonts were identified in arteries and capillaries of the lung, heart, spleen, and several other organs. First-generation schizonts at 14 DPI were 14–39 × 14–25 μm and contained around 100 nuclei. The second-generation schizonts, seen 24–39 DPI, were 14–32 × 10–20 μm and contained 20–35 nuclei. Merozoites were seen in peripheral blood 24 DPI; 1 merozoite was binucleated, suggesting multiplication in the bloodstream.

A terminal generation of schizonts, characteristic of *S. hemionilatrantis*, was found in macrophages in the muscles (Figure 1.20). Schizonts were 10–28 × 7–14 μm and contained up to 40 nuclei.[379,970]

The sarcocysts at 63 DPI measured up to 350 μm long and were immature; the sarcocyst wall was thin (<1 μm) and not striated. Mature sarcocysts, containing bradyzoites, were seen 90 DPI, measured up to 525 μm long, and were striated.[379]

Experimentally, *S. hemionilatrantis* is pathogenic for mule deer.[970] Of 12 animals fed 50,000, 250,000, or 1 million sporocysts (4 deer for each dosage), all became anorectic and uncoordinated. 9 died between 29 and 65 DPI, and 3 were euthanatized 41, 63, and 90 DPI.[970]

The lesions and clinical signs were similar to those in cattle inoculated with 150,000 or more *S. cruzi* sporocysts, but the clinical disease was more protracted in mule deer than in cattle.

Table 18.1 Sarcocystis Species in Wild Cervids

Species	Sarcocyst					Definitive Host, Sporocysts (µm)	Molecular Markers Studied	Reference
	Shape, Size	Wall			Bradyzoites			
		SEM	TEM					
Moose (Alces alces)								
S. alces Dahlgren and Gjerde 2008	Spindle-shaped, up to 5.5 mm long, smooth with no visible vp	Square-platform-like vp	No data		No data	Red fox (Vulpes vulpes), arctic fox (V. lagopus)	18S, cox1	282, 286, 702
S. alceslatrans Dubey 1980	7 mm, thread-like, smooth, thin wall	Hair-like, 10–12 µm vp	Type 7b		12.6–15.8 × 3.0–3.5 µm	Coyote (C. latrans), dog, 14.4–15.8 × 10.8–11.5	18S, cox1	244, 245, 355, 702, 999, 1576
S. hjorti (Dahlgren and Gjerde 2008) Dahlgren and Gjerde 2010	Up to 2 mm long, thin-walled, hair-like vp 10 µm long	—	No data		No data	Red fox (Vulpes vulpes), arctic fox (V. lagopus)	18S, cox1	282, 285, 286, 702
S. ovalis Dahlgren and Gjerde 2008	Ovoid, up to 1.5 mm, vp 13 µm long, secondary encapsulation	Tongue-like vp in rows	No data		No data	Magpie (Pica pica), 12.5 × 7.9	18S, cox1	282, 699, 702
S. scandinavica Dahlgren and Gjerde 2008	Spindle-shaped, 1 mm long, 10 µm long finger-like vp	No data	No data		No data	Unknown	18S, cox1	282, 702
S. silva (Type D, Dahlgren and Gjerde 2008) Gjerde 2012	Sac-like, rounded end, smooth, no visible vp	No data	No data		No data	Unknown	18S, cox1	282, 700, 702
S. taeniata Gjerde 2014	Sac-like, 1.1 mm, thin smooth wall	Ribbon-like vp in parallel rows, 2 µm, sparse	No data		No data	Unknown	18S, cox1	702
European Roe Deer (Capreolus capreolus)								
S. capreolicanis Erber, Boch, and Barth 1978	Thread-like, up to 1.7 mm, thin, hair-like vp	Hair-like, up to 9.5 µm long vp, aligned in parallel rows	Thread or tube-like vp, type 6		No data	Dog, red fox, (not felids, raccoon, ferret)	18S	532, 548, 700, 1533, 1577

(Continued)

Table 18.1 (Continued) Sarcocystis Species in Wild Cervids

Species	Sarcocyst				Bradyzoites	Definitive Host, Sporocysts (µm)	Molecular Markers Studied	Reference
	Shape, Size	Wall						
		SEM	TEM					
S. gracilis Rátz 1908	Up to 2.5 mm, hirsute vp	Densely packed, knob-like or cube-like vp, up to 0.5 µm long	—		15.6–12.5 × 3.9–3.1	Dog, red fox, silver fox (*V. vulpes*) and blue fox (*V. lagopus*), 15 × 10	18S	**284, 548, 700, 1577, 1683**
S. oviformis Dahlgren and Gjerde 2009	Ovoid, up to 600 µm long	Tongue-like vp, 10 µm long, pointed distal tip	No data		No data	Unknown	18S	**284, 700**
S. silva Gjerde 2012	Cigar-shaped, thick striated, microscopic, 0.6 mm long, upright vp 8 µm long	No data	Tightly packed finger-like, up to 7 long, granules and microfilaments, type 10		15.6–11.7 × 3.9–3.1	Unknown	18S	**532, 548, 700, 1577**

Siberian Roe Deer (*Capreolus pygargus*)

Species	Shape, Size	SEM	TEM		Bradyzoites	Definitive Host, Sporocysts (µm)	Molecular Markers Studied	Reference
S. sibirica Machul'skii 1947	Macroscopic, up to 7.5 mm, thin wall	No data	Vp with stalk, 0.4 wide tip, type 27		8.6–11.5 × 2.1–2.6	Unknown	No data	**1089, 1692**

Sika Deer (*Cervus nippon centralis*)

Species	Shape, Size	SEM	TEM		Bradyzoites	Definitive Host, Sporocysts (µm)	Molecular Markers Studied	Reference
S. sp. Saito, Itagaki, Shibata, and Itagaki 1995	Up to 600 mm thick hairy 6–10 µm thick wall	No data	Hair-like vp with microtubules, type 8b		12.9 × 5.0	Raccoon dog (*Nycterectes procyonides*), domestic dog, 16 × 11.5	No data	**1515**

Fallow Deer (*Dama dama*)

Species	Shape, Size	SEM	TEM		Bradyzoites	Definitive Host, Sporocysts (µm)	Molecular Markers Studied	Reference
S. jorrini Hernández- Rodriguez, Acosta, and Navarrete 1992	Macroscopic, up to 1.4 mm	No data	Secondary cyst wall, vp 2–10 long, type 15		8.6–13 × 3–5	Unknown	No data	**799**

(Continued)

Table 18.1 (Continued) Sarcocystis Species in Wild Cervids

Species	Sarcocyst				Definitive Host, Sporocysts (µm)	Molecular Markers Studied	Reference
	Shape, Size	Wall					
		SEM	TEM	Bradyzoites			
S. sp. Poli, Mancianti, Marconcini, Nigro, and Colagreco 1988	Up to 360 µm	No data	Secondary absent, 20–30 nm thick, sheet-like vp, 2–3 µm long, type 8a	13.2–21.0 × 3.3–5.5	Dog, not cat, 15.4 × 8.8	No data	1404
S. cf. hofmanni Wesemeier and Sedlaczek 1995	Probably, S. jorrini [799]						1823
S. cf. grueneri Wesemeier and Sedlaczek 1995	Probably, S. sp [1404]						1823
White-Tailed Deer (Odocoileus virginianus)							
S. odoi Dubey and Lozier 1983	1050 × 260 µm long, 6.5–12 µm thick wall	No data	Vp 5–10 thick, type 10a	9.5–12.5 × 2.5–3.5	Cat, 11–15 × 9–11	No data	378
S. odocoileocanis Crum, Fayer, and Prestwood 1981	Up to 620 µm long, 2–3 thick, striated	No data	Vp up to 4 long, convoluted invaginations at the base, type 17	9.5–14 × 2.8–4.6	Dog, cat, red fox, coyote, gray fox (Urocyon cinereoargenteus), wolf	No data	46, 266, 378
S. sp. Dubey and Lozier 1983	7–11 µm thick wall	No data	Vp 3.2–4.4, type 15	11–14.5 × 2.4–3.4	Unknown	No data	46, 378
S. sp. type II Atkinson, Wright, Telford, Mclaughlin, Forrester, Roelke, and Mccown 1993	77.6 µm wide, thick wall, striated	No data	Two types IIA and IIB, both with narrow vp, similar to type 36	No data	Unknown	No data	46
Mule Deer (Odocoileus hemionus)							
S. americana Dubey and Speer 1986	Microscopic, up to 1500, 2–10 thick, hairy	No data	No data	No data	Unknown	No data	383, 385

(Continued)

Table 18.1 (Continued) Sarcocystis Species in Wild Cervids

Species	Sarcocyst Shape, Size	Wall SEM	Wall TEM	Bradyzoites	Definitive Host, Sporocysts (µm)	Molecular Markers Studied	Reference
S. hemioni Dubey and Speer 1986	Microscopic, up to 800 µm, wall 5.0–8.5 thick	No data	Vp 3.6–7.1 × 06–1.3, type 10d	10–13 × 2–4	Unknown	No data	**383, 385**
S. hemionilatrantis Hudkins and Kistner 1977	Up to 535 µm, 1–2 µm thick wall	No data	Tee-shaped vp, type 17	13.3–16.0 × 2.6–3.2	Dog, coyote, 14–16 × 9.5–11.0	No data	**383, 858**
S. youngi Dubey and Speer 1986	Microscopic, up to 800, wall 4–8 thick	No data	Vp 8.1–9.0 × 3.5–5.6, type 16	10.7–15.8 × 3.2–3.6	Unknown	No data	**383, 385**
Reindeer (*Rangifer tarandus tarandus*)							
S. grueneri Yakimoff and Sokoloff 1934	Rounded ends, sack-like, 580 × 140	Thin strip-like vp	Ribbon-like up to 4.6 × 0.5 × 0.05, type 8a	13.9–16.5 × 2.4–4.8	Fox (silver and blue), raccoon, dog	No data	**281, 283, 685, 690, 698**
S. hardangeri Gjerde 1984	Ovoid, macroscopic, up to 2 mm	Prominent, slanting tongue-like vp	Tongue-like 25 × 5 × 1, type 26	11.2–12.9 × 2.8–3.5	Unknown	18S rRNA	**281, 283, 686, 698**
S. rangi Gjerde 1984	Thread-like up to 16 mm	Long hair-like vp	Hair-like up to 12.6 × 0.3–0.6, type 7a	12.4–16.5 × 2.4–4.6	Blue fox (*Alopex lagopus*)	18S	**281, 283, 687, 697, 693, 698**
S. rangiferi Gjerde 1984 (also found in red deer)	Cigar-like shaped 400 × 2106	Upright finger-like 7 µm long vp	Villiform 13.2 × 6.7, type 15	10.4–13.0 × 2.9–4.2	Unknown	18S, ITS-1, *cox1*	**281, 283, 685, 692, 698, 1682**
S. tarandi Gjerde 1984	Spindle shaped 1000 × 80	Vp upright, polygonal to oval	Villiform 9.2 × 2.2, type 10e	9.7–13.0 × 2.5–4.0	Unknown	18S	**281, 283, 685, 696, 698**
S. tarandivulpes Gjerde 1984	Spindle shaped up to 3 mm	Short knob-like vp interconnected by microfolds	Knob-like 0.6–1.2 type 17	—	Fox (blue and silver), raccoon (*Nyctereutes procyonides*), dog, 13.6 × 9.8	18S	**281, 283, 688, 695, 698**

(Continued)

Table 18.1 (Continued)   Sarcocystis Species in Wild Cervids

Species	Sarcocyst					Definitive Host, Sporocysts (µm)	Molecular Markers Studied	Reference
	Shape, Size	Wall			Bradyzoites			
		SEM	TEM					

*Red Deer (**Cervus elaphus hippelaphus and hispanicus**)*

Species	Shape, Size	SEM	TEM	Bradyzoites	Definitive Host, Sporocysts (µm)	Molecular Markers Studied	Reference
S. cervicanis Hernández-Rodríguez, Navarrete, and Martínez-Gómez 1981	150 to 200 µm long, wall smooth and thin (0.5 µm)	No data	Hair-like vp up to 1.4 µm long and 32 nm wide and bifurcate at the tip, type 8a	15–17 × 4–6	Dog, 15.1–17.1 × 10.3–11.9	No data	797
S. elongata (Gjerde 1984) Gjerde (2014)	Up to 2.3 mm, slender and spindle-shaped, striated, thick walled, 7–8 × 1.5–2.0 µm	Upright finger-like hexagonal vp, 7–8 µm long, hexagonal	No data from this host	No data	Unknown	18S, ITS-1, cox1	**285, 704**
S. hardangeri Gjerde 1984 (also found in reindeer)	Macroscopic, ovoid up to 1.9 mm, fibrous covering	Loosely packed flattened protrusions 10–18 × 2	No data from this host	No data	Unknown	18S, ITS-1	285, 704
S. hjorti Dahlgren and Gjerde 2010 (also found in moose)	Slender and spindle-shaped, up to 3.1 mm long	Slender, thin hair-like protrusions 10–12 µm, narrow at base, tapered	No data	No data	Unknown	18S, ITS-1, cox1	285, 704
S. ovalis Dahlgren and Gjerde 2008	Macroscopic, ovoid up to 1.9 mm, fibrous covering	Loosely packed flattened tongue shaped protrusions 10–18 × 2	No data from this host	No data	Unknown	18S, ITS-1, cox1	**282, 285, 704**
S. tarandi Gjerde 1984 (also found in reindeer)	Spindle shaped, up to 2.3 mm, tapering pointed ends	Polygonal 6 µm long vp, aligned in rows	Villiform vp 9.2 × 2.2, type 10e	No data	Unknown	18S, ITS-1, cox1	**280, 285, 1477**

(Continued)

Table 18.1 (Continued) Sarcocystis Species in Wild Cervids

| Species | Sarcocyst | | | Bradyzoites | Definitive Host, Sporocysts (μm) | Molecular Markers Studied | Reference |
| | Shape, Size | Wall | | | | | |
		SEM	TEM				
S. truncata (Gjerde 1984) Gjerde (2014)	Cigar-shaped to ellipsoidal, up to 1.7 mm	Upright 7 × 3 vp, hexagonal base, minute pits on tips	No data from this host	No data	Unknown	18S, ITS-1, cox1	**285, 704**
Wapiti or North American Elk (Cervus elaphus canadensis)							
S. sybillensis Dubey, Jolley, and Thorne 1983	Up to 637 μm long, wall up to 8 μm thick and hairy	No data	Up to 6.1 μm long vp, 0.3 μm wide at the base and 0.1 μm at the apex, have a central core, and the villar tips bifurcate, type 12	9.5–12 × 2.5–4	Dog, 15–17 × 10.5–12	No data	**380**
S. wapiti Speer and Dubey 1982	Up to 837 μm long. The sarcocyst wall is thin (<1 μm)	No data	Stubby, finger-like vp, folded over the PVM, slender invaginations, type 2	14–18 × 1.5–4	Coyote, dog, 15.7–17.4 × 9.6–11.9	No data	**1651**
S. sp. Stolte, Bockhardt, and Odening 1996	Thick wall, striated	Ear-like vp	Unknown	No data	Unknown	No data	**1680**

Table 18.2 Sarcocystis Species in Other Wild Ruminants and Large Animals

Species	Sarcocyst Shape, Size	Wall SEM	Wall TEM	Bradyzoites	Definitive Host, Sporocysts (µm)	Molecular Markers Studied	Reference
African Buffalo (Syncerus caffer)							
S. cafferi Dubey, Lane, Wilpe, Suleman, Reininghaus, Verma, Rosenthal and Mtshali 2014	Macrocysts, up to 12 mm long, cashew nut-like	Mesh-like structure with irregular shaped vp	3.6 µm thick walled, highly branched vp, up to 3 µm long, type 23b	12.1 × 2.7 µm	Unknown	18S, cox1	473, 1434
Springbok (Antidorcas marsupialis)							
S. gazellae Balfour 1913 (originally described from Grant's gazelle, Gazella rufifrons)	Macroscopic, up to 2.8 mm, smooth	No data	Vp finger-like, up to 11 µm wide, type 14	12–16 × 5–6	Unknown	No data	62, 1320
S. woodhousei (Dogiel 1915) Wenyon 1926	850 µm long, microscopic	No data	Tooth-like vp 0.9–7.0 × 1.3–4.5 µm, type 29a	10–12 × 3.0–3.4	Unknown	No data	1320
S. sp. Odening, Rudolph, Quandt, Bengis, Bockhardt, and Viertel 1998 Also described in Aepyceros melampus	Microscopic, 1.6 mm long, smooth	No data	Hair-like vp, 0.1–0.6 µm wide, type 7a	10.5–17.0 × 3.2–4.5	Unknown	No data	1320
Impala (Aepyceros melampus)							
S. melampi Odening, Rudolph, Quandt, Bengis, Bockhardt, and Viertel 1998	Microscopic, 1.6 mm	No data	Wart-like vp, flat, 4.6–7.0 µm wide, 1.0–2.5 µm high, type 40	11.0–13.5 × 3.2–3.4	Unknown	No data	1320
Defassa Waterbuck (Kobus ellipsiprymnus defassa)							
S. nelsoni Mandour and Keymer 1970	5 mm in esophagus, microscopic in heart, 4–5 µm, thin wall	No data	No data	No data	Unknown	No data	1105

(Continued)

Table 18.2 (Continued) Sarcocystis Species in Other Wild Ruminants and Large Animals

Species	Shape, Size	Sarcocyst Wall SEM	Sarcocyst Wall TEM	Bradzoites	Definitive Host, Sporocysts (µm)	Molecular Markers Studied	Reference
Saiga (*Saiga tatarica*)							
S. saiga Pak, Sklyarova, and Dymkova 1991	Thread-like, up to 3 mm long, striated, 4.2–6.5 µm long vp	No data	No data	11.2–14.0 × 4.2–5.6	Unknown	No data	**1359**
Greater Kudu (*Tragelaphus strepsiceros*)							
S. cf. hominis	Microscopic, 2.5 mm	No data	Tombstone- or cone-like vp, type 15	12–15 × 2.6–3.0	Unknown	No data	**1320**
Mongolian Gazelle (*Procapra gutturosa*)							
S. danzani Odening, Stolte, Lux, and Bockhardt 1996	Microscopic, 1.5 mm long	Molar-like vp, with a hollow from the top view	Molar tooth-like vp, 1.2–2.2 wide, type 29	11–16 × 3–4	Unknown	No data	**1313**
S. mongolica Machul'skii 1947	Macroscopic, secondary wall, rice grain like, up to 5.1 mm long	Visible small invaginations	Unusual small invaginations, type 1c	16–21 × 3–6	Unknown	No data	**1089, 1313**
S. sp. Odening, Stolte, Lux, and Bockhardt 1996	Microscopic, up to 3.8 mm long, thin wall	Hair-like vp	Thin wall (<0.5 µm), 8–15 µm long hair-like vp, type 7a	12.5–14.5 × 3.0–3.5	Unknown	No data	**1313, 1680**
Addax (*Addax nasomaculatus*)							
S. cf. medusiformis Stolte, Odening, and Bockhardt 1996	Up to 2.8 mm long	No data	Scale-like or trapezoidal vp, 0.8–2.6 µm long, type 20	14.0–17.8 × 2.8–4.0	Unknown	No data	**1679**
Blackbuck (*Antilope cervicapra*)							
S. cf. capracanis Stolte, Odening, and Bockhardt 1996	Up to 1.0 mm long	No data	Palisade-like texture with finger-like vp, 3.4–4.4 µm long, type 14	13.5 × 4.5	Unknown	No data	**1679**

(Continued)

Table 18.2 (Continued) Sarcocystis Species in Other Wild Ruminants and Large Animals

Species	Sarcocyst					Definitive Host, Sporocysts (μm)	Molecular Markers Studied	Reference
	Shape, Size	Wall			Bradyzoites			
		SEM	TEM					
Common Eland (Taurotragus oryx)								
S. cf. hominis Stolte, Odening, and Bockhardt 1996	Up to 800 μm long	No data	Tombstone- or cone-like vp, 6.8–8.2 μm long, type 15		No data	Unknown	No data	1679
Nilgai (Boselaphus tragocamelus)								
S. cf. cruzi (also found in eland and blackbuck)	Up to 3.5 μm long	No data	Long hair-like villar protrusions, 13 μm, type 7a		15.0–17.8 × 3.0–4.2	Unknown	No data	1679
Mountain Goat (Oreamnos americanus)								
S. sp. Foreyt 1989	Microscopic, up to 248 μm long	No data	Mushroom-shape vp, type 24		No data	Not canids, felids, bears, several Mustelidae, pigtail macaque, hawk, and owl	No data	619
Siberian Ibex (Capra sibirica)								
S. orientalis Machul'skii and Miskaryan 1958	Macroscopic, up to 7.5 mm long	No data	No data		14.1–9.2 × 6.2–2.3	Unknown	No data	1089
Chamois (Rupicapra rupicapra)								
S. cornagliai Odening, Stolte, and Bockhardt 1996	Microscopic, up to 700 μm long	No data	Thin wall (<0.77 μm), mushroom-shaped, type 24		11.5–13.5 × 3–4	Unknown	No data	1311
S. sp. Odening, Stolte, and Bockhardt 1996	Microscopic, up to 1.9 mm log	No data	0.2–1.3 μm thick wall, finger-shaped vp, type 14		12.0–12.5 × 3.1–4.3	Unknown	No data	1311

(Continued)

Table 18.2 (Continued) Sarcocystis Species in Other Wild Ruminants and Large Animals

| Species | Sarcocyst | | | | Definitive Host, Sporocysts (μm) | Molecular Markers Studied | Reference |
| | Shape, Size | Wall | | Bradyzoites | | | |
		SEM	TEM				
Alpine Ibex or Steinbock (Capra ibex)							
S. arieticanis-like (type III) Cornaglia, Giaccherino, and Peracino 1998	No data	No data	Discontinuous elongate digitations, similar to type 7a	No data	No data	No data	253
S. capracanis-like (type II) Cornaglia, Giaccherino, and Peracino 1998	No data	No data	Villus-like projections in palisade position, similar to type 14	No data	No data	No data	253
S. cornagliai-like (type I) Cornaglia, Giaccherino, and Peracino 1998	No data	No data	Sessile and polyp-shaped (mushroom) vp, similar to type 24	No data	No data	No data	253, 103
Hispanic Ibex (Capra pyrenaica hispanica)							
S. capricornis-like (type IV) Cornaglia, Giaccherino, and Peracino 1998	No data	No data	Finger-like projections with indented edges, fire flame-like, similar to type 41	No data	No data	No data	253
Himalayan Blue Sheep (Pseudois nayaur)							
S. pseudois (Syn. S. cf. capracanis) Novak, Fedossenko, and Orazalinova 1987) Odening, Stolte, and Brockhardt 1996	Up to 2.2 mm long, thick wall, striated	No data	1.0–2.4 μm thick wall, 2.0–2.7 μm long irregular finger-like vp, type 14	12.5–13.5 × 3.3–5.0	Unknown	18S	608, 1284, 1309, 1317
Himalayan Goral (Naemorhedus goral)							
S. sp. Agrawal, Chauhan, and Ahluwalia 1982	Macroscopic, up to 5 mm long, 2 mm wide. 3.7–5.6 μm thick wall	No data	No data	7.5–13.1 × 1.8–2.8	Unknown	No data	1898

(Continued)

Table 18.2 (Continued) Sarcocystis Species in Other Wild Ruminants and Large Animals

Species	Sarcocyst			Bradyzoites	Definitive Host, Sporocysts (µm)	Molecular Markers Studied	Reference
	Shape, Size	Wall					
		SEM	TEM				
Japanese Serow (Capricornis crispus)							
S. capricornis Odening, Stolte, and Bockhardt 1996	Up to 4.8 mm long, striated	No data	3.3–6.4 µm thick wall, T-shaped or short, thick nails with angular head vp, 2.5–3.5 µm long, type 27	14–17 × 3–4	Unknown	No data	**1309**
Mouse Deer (Moschiola meminna)							
S. sp. Kannangara 1970	Up to 345 µm long, thick, striated	No data	No data	7.2–4.3 × 2.4–1.7	No data	No data	**941**
Bighorn Sheep (Ovis canadensis)							
S. ferovis Dubey 1983	Microscopic, up to 780 µm, thin <1 µm, smooth	No data	Flattened and mushroom-like vp, type 3a	10–14 × 3–3.5	Coyote 13–15 × 9–11 (not transmissible to sheep, goat, cattle)	No data	**368**
European Mouflon (Ovis ammon musimon)							
S. tenella Raillet 1886	(See Chapter 8)						**1277, 1308**
S. arieticanis Heydorn 1985	(See Chapter 8)						**1277, 1308**
Hippopotamus (Hippopotamus amphibius)							
S. africana Odening, Quandt, Bengis, Stolte, and Bockhardt 1997	Microscopic, round to oval, up to 720 µm wide	No data	Rectangular or semicircular vp, 2.5–5.0 × 3.5–5.0, type 15b	7.0–11.0 × 2.6–3.7	Not known	No data	**1316**
S. hippopotami Odening, Quandt, Bengis, Stolte, and Bockhardt 1997	Macroscopic, up to 10 mm, fibrous capsule	Fibrous capsule visible	Thumb-like vp, 1.2–2.4 × 0.7–1.8, type 33	15.6 × 2.2	Not known	No data	**1316**

(Continued)

Table 18.2 (Continued) Sarcocystis Species in Other Wild Ruminants and Large Animals

Species	Sarcocyst				Definitive Host, Sporocysts (µm)	Molecular Markers Studied	Reference
	Shape, Size	Wall		Bradyzoites			
		SEM	TEM				
Warthog (Phacochoerus aethiopicus)							
S. dubeyella Stolte, Odening, Quandt, Bengis, and Bockhardt 1998	Macrocysts, up to 12 mm, fibrous capsule	Indented margins	Rectangular vp, 2–10 µm long, 2.8–11.0 µm wide, type 30	12–19 × 2.5–3.5	Unknown	No data	1684
S. phacochoeri Stolte, Odening, Quandt, Bengis, and Bockhardt 1998	Macrocysts, up to 4 mm, thin cyst wall	Hollow present on one side of the vp	Molar-like vp, 1.6–3.3 × 1.7–3.3, type 29b	12–16 × 2.0–3.5	Unknown	No data	1684
Yak (Poephagus grunniens)							
S. poephagi Wei, Chang, Duong, Wang, and Xia 1985	Macroscopic, filiform, 4 mm, thick wall 7.4 µm, secondary cyst wall	Tree-bark-like surface	Thick	No data	Cat, oocysts were 13.7 × 9.4	No data	1815, 1817
S. poephagicanis Wei, Chang, Duong, Wang and Xia 1985	Macroscopic, oval, 0.28 mm long, thin wall, secondary cyst wall	Honeycomb-like surface, smooth	Thin, spines	No data	Dog, oocysts were 14.6 × 10.6	No data	1815, 1817
Giraffe (Giraffa camelopardalis)							
S. camelopardalis Bengis, Odening, Stolte, Quandt, and Bockhadt 1998	Microscopic, <1 mm long, 120 µm wide	No data	Strap-like vp, 2–3 µm long, type 36	9–14 × 2–4	Unknown	No data	91
S. giraffae Bengis, Odening, Stolte, Quandt, and Bockhadt 1998	Macroscopic, slim, up to 11 mm long, collagen capsule	No data	Finger-shaped vp, 6–13 µm long, type 34	14–18 × 2.5–3.5	Unknown	No data	91
S. klaseriensis Bengis, Odening, Stolte, Quandt, and Bockhadt 1998	Microscopic, filiform, <1 mm long, 105 µm wide	No data	Finger-like vp, 4–7 µm long, type 35	9–14 × 2.6–4.0	Unknown	No data	91

Table 18.3 Sarcocystis Species in Marine Mammals

Intermediate Host	Species	Shape, Size	TEM	Bradyzoites	Definitive Host, Sporocysts (μm)	Molecular Assays	References
Sea lion (Zalophus californianus)	S. hueti (Moulé, 1888) Labbé 1899	Up to 4 mm long	No data	4–5	No data	No data	1013, 1232, 1317
	S. sp. Mense, Dubey, and Homer 1992	Up to 100 μm wide	Thin, 0.5–1.2 μm, sloping-to-straight 1.5 μm vp, folded down the surface, type 11e	No data	No data	No data	1184
Harbor seal (Phoca vitulina richardsi)	S. richardi Hadwen 1922	Macroscopic, up to 2 cm long	No data	10 × 2.5	No data	No data	766
Antarctic leopard seal (Hydrurga leptonyx)	S. hydrurgae Odening and Zipper 1986	Up to 13 mm long	Thin, type 1e	6.1 × 2.2	No data	No data	2195
Bearded seal (Erignathus barbatus)	S. sp. Bishop 1979	Location in tongue. No data	No data	No data	No data	No data	105
Ringed seal (Pusa hispida)	S. sp. Migaki and Albert 1980	Up to 550 μm long	Thin wall (0.8–1.0 μm)	Comma-shaped, 12–10 × 3–2	No data	No data	1192
Northern fur seal (Callorhinus ursinus)	S. sp. Brown, Smith, and Keyes 1974	Numerous cysts in several muscles. No data	No data	No data	No data	No data	147
Whale (Balaenoptera borealis)	S. balaenopterialis Akao (1970)	Thread-like, macroscopic, 10–20 cm long, 0.5 mm wide	Thin, 2–3 μm, type 1a	7.7 × 2.3	No data	No data	22
Beluga whale (Delphinapterus leucas)	S. sp. De Guise, Lagacé, Girard, and Béland 1993	Microscopic, up to 280 μm wide, >1 mm long	Thin, type 1a	No data	No data	No data	310
Sperm whale (Physeter catodon)	S. sp. Munday and Mason 1978	2.5 mm long × 2.5 mm wide	Thin and smooth, type 1a	No data	No data	No data	1249
Atlantic white-sided dolphin (Lagenorhyncus acutus)	S. sp. De Guise, Lagacé, Girard, and Béland 1993	Microscopic, up to 115 μm wide	Thin	No data	No data	No data	310, 562
Delphinus sp.	S. sp. Munday and Mason 1978	750 × 80 μm	Thin	No data	No data	No data	1249
Striped dolphin (Stenella coeruleoalba)	S. sp. Dailey and Stroud (1978)	No data	No data	No data	No data	No data	287
Sea otter (Enhydra lutris)	S. sp. Dubey, Lindsay, Rosenthal, and Thomas 2003	Microscopic, 110 μm long	Thin, 0.5–0.7 μm, type 1a	5.0–5.7 × 1.6–1.9	No data	rpoB	453

Data on natural epizootiology of *Sarcocystis* infections in mule deer indicated that *S. hemionilatrantis* affects the growth of fawns and, coupled with winter conditions, probably predisposes infected deer to predation.[384]

18.2 *SARCOCYSTIS* INFECTIONS IN OTHER WILD RUMINANTS AND OTHER LARGE ANIMALS

Numerous *Sarcocystis* species have been named from animals in zoological gardens, wild ruminants, herbivores, and several omnivores. Some are summarized in Tables 18.1 and 18.2. It is likely that wild goats and feral sheep share species that are also present in domestic animals. This is the case in mouflon (*S. tenella*)[1308] and wood bison (*S. cruzi*).[162] Some *Sarcocystis* species of wild animals are host specific, such as *S. ferovis* in bighorn sheep (Table 18.2) which is morphologically different from any species in domestic animals and is not transmissible to ox, sheep, or goats.[368] Another structurally distinct, unnamed, and enigmatic species was described from the North American mountain goat (*Oreamnos americanus*).[619] It was found in 24 (43%) of 56 goats from Washington state, USA. None of the 4 coyotes, 8 dogs, 4 domestic cats, 3 bears, 2 raccoons, 2 martens, 2 fishers, 3 skunks, 5 minks, 5 ferrets, 1 pigtail macaque, 2 red tailed hawks, or 1 great horned owl fed infected meat excreted sporocysts.[619] A structurally similar sarcocyst, *Sarcocystis*

Table 18.4   Species *inquirenda*/Uncertain Status

Species	Intermediate host	Morphology	References
S. ruandae Chiwy and Colback 1926	Unnamed antelope	Microscopic cysts, striated wall, 9–12 × 1–5 µm banana-shaped bradyzoites	216
S. sp. Dubey 1980	Pronghorn (*Antilocapra americana*)	Microscopic cysts, 127–297 × 63–95 µm, thin walled and smooth	355
S. bubalis Diegel 1915	Coke's hartebeest (*Alceslaphus cokei*)	Up to 2 mm long, striated wall	1105
S. woodhousei Diegel 1915	Grant's gazelle (*Nanger granti*)	Up to 1.5 mm long, with rods or rhombic prisms, striated wall	1105
S. gusevi (Blyth 1841) Krylov and Saponzhnikov 1965	Arkhar (*Ovis ammon polii*)	Only description of bradyzoites was provided, they were lunate- or bean-shape, 10.8–14.4 × 4.8–8.4 µm	981
S. sp. Krylov and Saponzhnikov 1965	Bactrian deer (*Cervus elaphus bactrianus*)	No description of sarcocysts. Bean shaped, and orange slice-shaped 10.8–15.6 × 4.8–8.4 µm bradyzoites	981
S. sp. Rioseco, Cubillos, Gonzalez and Diaz 1976	Pudu (*Pudu pudu*)	Microscopic	2252
S. sp. Samuel and Gray 1974	Muskox (*Ovibos moschatus*)	Microscopic	2281
S. sp. Gangadharan, Valsala, Nair and Rajan 1992	Sambar deer (*Cervus unicolor*)	Microscopic	660
S. sp. Shrivastav, Sharma, Chaudhry, and Malik 1999	Barasingha or hard-ground swamp deer (*Cervus duvauceli branderi*)	Microscopic	1605

cornagliai was described from the European chamois (*Rupicapra rupicapra*)[1311]; it was concluded that sarcocysts in the North American mountain goat were morphologically identical to *S. cornagliai*. Another example of the complexity of the epidemiology of *Sarcocystis* genus in wildlife was the finding of sarcocysts resembling *S. gracilis* of roe deer (Cervidae) in muscles of dwarf zebu (Bovidae) born and housed in a German zoo.[1312]

18.3 *SARCOCYSTIS* INFECTIONS IN MARINE MAMMALS

Information concerning *Sarcocystis* infections in marine mammals is limited. Several species have been proposed and a high prevalence was found.[562] Further studies should focus on detailed descriptions of the ultrastructure. Information regarding *Sarcocystis* spp. in marine mammals is summarized in Table 18.3.

18.4 *SARCOCYSTIS* SPP. WITH UNCERTAIN STATUS

Several species detected in wild animals are considered *inquirendae* based on poor descriptions or because there is only one report.[1317] Several examples are given in Table 18.4.

BIBLIOGRAPHY

Taxonomy and prevalence of *Sarcocystis* species in wild ruminants and other large animals can be found in the following papers:

Alpine ibex and Hispanic ibex: **103, 253, 1311**.

Antelopes: **62, 1105, 1205, 1320, 1679**.

Fallow deer: **337, 338, 542, 799, 1404, 1455, 1823**.

Mule deer: **379, 383, 384, 385, 858, 859**.

Sika, sambar, Barasingha deer: **121, 660, 1454, 1515, 1605**;

Red deer (wapiti): **61, 121, 238, 285, 337, 338, 704, 796, 797, 1000, 1102, 1664, 1676, 1763, 1824, 1961, 1997, 2100, 2132, 2212, 2291, 2322**.

Roe deer: **97, 110, 284, 336, 339, 531, 532, 534, 538, 548, 700, 969, 998, 1085, 1102, 1386, 1533, 1577, 1664, 1682, 1683, 1692, 1763, 1997, 2054, 2088, 2212, 2268, 2322**.

Reindeer (caribou): **280, 281, 283, 685–698, 953, 1682, 2083, 2349**.

White-tailed deer: **46, 266, 267, 480, 533, 1995, 1996, 2079**.

Elk: **380, 620, 1014, 2125**.

European mouflon: **1277, 1308, 1664, 2266**.

Moose: **244, 245, 282, 284–286, 699, 700, 702, 704, 999, 1477, 2080**.

Other wild ruminants: **103, 110, 253, 368, 529, 619, 660, 894, 929, 1277, 1308, 1309, 1311, 1312, 1342, 1389, 1434, 1446, 1815, 1816, 1915, 1980, 1995, 2074, 2086, 2128, 2146, 2147, 2180, 2197–2199, 2206, 2207, 2224, 2225, 2235, 2266, 2338, 2340**.

Marine mammals: **22, 105, 108, 147, 206, 265, 287, 310, 453, 562, 766, 930, 1184, 1192, 1249, 1317, 1588, 2193–2196, 2208**.

Miscelanea: **62, 91, 432, 1028, 1102, 1314, 1315, 1317, 1321, 1418, 1664, 1680, 1898, 1915**.

CHAPTER 19

Sarcocystosis in Wild Terrestrial Carnivores

19.1 INTRODUCTION

As seen in previous chapters wild carnivores serve as definitive hosts for numerous *Sarcocystis* species.[286,374,383] Prevalence ranges from 4% to 92%.[94,246,364,504,953,1122,1837] Carnivores can also act as intermediate hosts harboring sarcocysts (Table 19.1).

19.1.1 Prevalence and Clinical Disease

The prevalence of sarcocysts in the muscles of wild carnivores might be higher than previous estimates. Muscle infections have been reported in more than 21 terrestrial species. Sarcocysts were reported in 6 of 21 wild dogs (*Lycaon pictus*) from Kenya,[159] and the prevalence in raccoons (*Procyon lotor*) ranges between 47.5% and 66% in different U.S. states.[776,964,1633] Sarcocysts have been found in 4 of 5 feral skunks (*Mephitis mephitis*) from Oregon,[446] in 80.4% of wolverines (*Gulo gulo*) from Canada,[467] in 83% of fishers (*Martes pennanti*) from Pennsylvania,[1025] and in 2 coyotes from Oklahoma.[268]

The first record of *Sarcocystis* in bears reported a prevalence of 11% in black bears (*Ursus americanus*) from southeastern and northeastern United States.[265] Several cases of infection have been reported in North Carolina, Oregon, Florida, and Pennsylvania.[206,415,462,621] Sarcocysts were also reported in brown bears (*Ursus arctos*) from Alaska.[461]

Among wild Felidae the prevalence of *Sarcocystis* is very high. A prevalence of 50% was found in Florida bobcats (*Felis rufus floridanus*),[29] and infections were found in 11 of 14 Florida panthers and 4 of 4 cougars (*Felis concolor*).[732] Infection was reported in a young leopard (*Panthera pardus*)[1644] and in 2 lions (*Panthera leo*), both from India.[100] Sarcocysts were detected in a lion from Kenya and in 4 lions from Namibia,[410,962] as well as 7 of 10 cheetahs (*Acinonyx jubatus*) in a breeding colony in Oregon, USA.[142]

Clinical sarcocystosis has been described in carnivores. Numerous reports attributed to *S. neurona* and *S. canis* infections have been summarized, respectively, in Chapter 3. A case of generalized eosinophilic myositis was reported in a raccoon from Pennsylvania, and although numerous sarcocysts were present, *Sarcocystis* could not be confirmed as the cause.[776] In a case of meningoencephalitis in a young male Japanese raccoon dog (*Nyctereutes procyonoides viverrinus*), *Sarcocystis* schizonts were detected in the brain, causing severe encephalitis specially in the cerebellum.[983] The protozoa reacted weakly with *S. cruzi* antiserum, and the analysis of the partial 18S rRNA gene sequence revealed its close relationship with a *Sarcocystis* sp. isolated from a white-fronted goose (*Anser albifrons*) presumably *S. albifronsi*, with a high degree of similarity (99.3%–99.7%) to *S. neurona* and *S. canis*.

Table 19.1 Sarcocystis Species in Wild Terrestrial Carnivores

Species	Sarcocyst				Definitive Host, Sporocysts (µm)	Molecular	References
	Shape, Size	SEM	TEM, Wall	Bradyzoites			
Raccoon (*Procyon lotor*)[a]							
S. kirkpatricki Synder, Sanderson, Toivio-Kinnucan, and Blagburn 1990	Microscopic, up to 801 µm long, 2–3 µm thick, striated	No data	2–3 µm thick, striated, type 11b	5.2–3.9 × 2.0–1.2	Not dog or cat	No data	1633
S. sp. 1 Stolte, Odening, Walter, and Bockhardt 1996	Microscopic, 99.5 µm wide, only 1 cyst seen, palisade-like wall	No data	4.6–6.9 µm thick wall, vp 5.8–7.4 µm long, type 17	No data	Unknown	No data	1681
S. sp. 2 Stolte, Odening, Walter, and Bockhardt 1996	2 sarcocysts detected, 1.8 and 1.9 mm	No data	Thin, type 7a	14.0–17.1 × 3.4–3.9	Unknown	No data	1681
S. sp. 3 (sf. *S. sebeki*) Stolte, Odening, Walter, and Bockhardt 1996	Up to 18 mm, thin wall, smooth	No data	Thin, smooth, type 1a	6.6–8.5 × 1.3–1.8	Unknown	No data	1681
Arctic Fox (*Vulpes lagopus*)							
S. arctica Gjerde and Schulze 2014	Up to 12 mm, serrated, knob-like protrusions on wall	No data	No data	No data	Unknown	18S, ITS-1, cox1	705
S. lutrae Gjerde and Josefsen 2015	Up to 970 µm long, thin wall (0.5 µm)	No data	No data	No data	Unknown	18S, 28S, ITS-1, cytb, coxI	706
Red Fox (*Vulpes vulpes*)							
S. sp. Pak 1991	Up to 3.9 mm long, 2.0–2.8 µm, thick wall, smooth	No data	No data	2.8–1.4 × 9.8–8.4	Unknown	No data	1359
Corsac Fox (*Alopex corsac*)							
S. corsaci Pak 1979	Up to 8.2 mm, cyst wall with two layers. Wall 2.8–2.1 µm thick	No data	-	2.8–1.4 × 8.4–7.0	No data	No data	1360

(Continued)

Table 19.1 (Continued) Sarcocystis Species in Wild Terrestrial Carnivores

Species	Shape, Size	Sarcocyst SEM	Sarcocyst TEM, Wall	Bradyzoites	Definitive Host, Sporocysts (µm)	Molecular	Reference
Jackal (*Canis mesomelas*)							
S. sp. 1 Wesemeier, Odening, Walter, and Bockhardt 1995	Up to 393 µm, thick wall	No data	Up to 6.7 µm palisade-like vp, type 10f	No data	Unknown	No data	1825
S. sp. 2 Wesemeier, Odening, Walter, and Bockhardt 1995	Up to 695 µm, thin wall	No data	Flat, mushroom-like vp, type 3b	No data	Unknown	No data	1825
Coyote (*Canis latrans*)							
S. sp. Cummings, Kocan, Barker, and Dubey 2000	Microscopic, 99 µm, thin cyst wall	No data	Type 1b	5.0 long	Unknown	No data	268
Wolverine (*Gulo gulo*)							
S. kalvikus Dubey, Reichard, Torretti, Garvon, Sundar, and Grigg 2010	Macroscopic, 900 µm, smooth, thin wall	No data	Type 1a	5 × 1	Unknown	18S, ITS-1	467
S. kitikmeotensis Dubey, Reichard, Torretti, Garvon, Sundar, and Grigg 2010	Microscopic, 1100, thick wall	No data	vp 1 µm long, type 9 h	9–13 × 2–3	Unknown	18S, ITS-1	467
Wild Dog (*Lycaon pictus*)							
S. sp. Bwangamoi, Ngatia, and Richardson 1993	Microscopic, up to 830 µm, 3 µm thick wall	No data	No data	No data	Unknown	No data	159
European Badger (*Meles meles*)							
S. hofmanni Odening, Stolte, Walter, and Bockhardt 1994	Microscopic, up to 1.2 mm long	No data	vp up to 6.6 µm long, finger-like, type 10b	8.3–9.9 × 2.6–3.6	Unknown	No data	1303, 1304
S. melis Odening, Stolte, Walter, and Bockhardt 1994	2–5 mm long, thin wall	No data	Thin, 1f	8.2–6.9 × 1.7–0.8	Unknown	No data	1303, 1304

(Continued)

Table 19.1 (Continued) Sarcocystis Species in Wild Terrestrial Carnivores

| Species | Sarcocyst | | | Bradyzoites | Definitive Host, Sporocysts (μm) | Molecular | Reference |
	Shape, Size	SEM	TEM, Wall				
S. cf. *sebeki* Odening, Stolte, Walter, and Bockhardt 1994	Up to 9 mm long	No data	Thin, wavy, type 1a	6.1–7.2 × 1.6–2.2	Unknown	No data	1303, 1304
S. cf. *gracilis* Odening, Stolte, Walter, Bockhardt, and Jakob 1994	Up to 1.2 mm long	No data	Thin, 1.7 μm thick wall, type 1g	13.2–11.0 × 3.5–2.7	Unknown	No data	1304
Japanese Badger (*Meles meles anakuma*)							
S. sp. Kubo, Okano, Ito, Tsubota, Sakai, and Yanai 2009	Up to 380 μm long	No data	2.3–1.8 μm thick, type 12	No data	Unknown	No data	982
Otter (*Lutra lutra*)							
S. lutrae Gjerde and Josefsen 2015	Up to 970 μm long, thin wall (0.5 μm)	No data	No data	No data	Unknown	18S, 28S, ITS-1, cytb, coxI	706
S. sp. Wahlström, Nikkila, and Uggla 1999	Up to 2.3 mm long, thin wall	No data	Thin 2 μm, type 1b	No data	Unknown	No data	1803
Skunk (*Mephitis mephitis*)[a]							
S. erdmanae (Erdman 1978) Odening 1997	Microscopic, up to 300 μm long	No data	No data	No data	Dog, 12 × 10	No data	555, 1314
S. mephitisi Dubey, Hamir, and Topper 2002	Microscopic, up to 400 μm long, >6 μm thick wall	No data	vp up to 5 long, constricted at the base, leaf-like, type 9 m.	11 × 3.2	Unknown	No data	446
Mink (*Mustela vison*)							
S. sp. (most likely *S. neurona*) Ramos-Vara, Dubey, Watson, Winn-Elliot, Patterson, and Yamini 1997	Microscopic, up to 400 μm long, thick wall	No data	vp up to 2 long, constricted at the base, leaf-like, type 11b	4.5–5.0	Unknown	No data	1445

(Continued)

Table 19.1 (Continued) Sarcocystis Species in Wild Terrestrial Carnivores

Species	Sarcocyst				Definitive Host, Sporocysts (μm)	Molecular	Reference
	Shape, Size	SEM	TEM, Wall	Bradyzoites			
Steppe Polecat (Mustela eversmanii)							
S. eversmanni Pak, Sklyarova, and Dymkova 1991	No data	No data	No data	2.8–3.0 × 7.0–8.4	Unknown	No data	**1314, 1359**
Common European Weasel (Mustela nivalis)							
S. sp. Tadros and Laarman 1979	Macroscopic, up to several mm long, smooth	No data	No data	9 × 2.5	Possibly tawny owl (*Strix aluco*)	No data	**1711**
Fisher (Martes pennanti)							
S. sp. Larkin, Gabriel, Gerhold, Yabsley, Wester, Humphreys, Beckstead, and Dubey 2011	Microscopic, 1.6 μm long vp	No data	No data	No data	Unknown	18S	**1025**
Black Bear (Ursus americanus)[b]							
S. ursusi Dubey, Humphreys, and Fritz 2008	Microscopic, thin wall, <1 μm, serrations	No data	vp up to 3.8 μm long, type 9g.	4.8–6.0 × 1.4–1.8	Unknown	No data	**415, 462**
S. sp. Dubey, Humphreys, and Fritz 2008	Microscopic, 700 μm long, 2 μm palisade-like wall, thick wall	No data	No data	No data	Unknown	No data	**462**
S. sp. Cheadle, Cunningham, and Greiner 2002 (From Florida black bear, *Ursus americanus floridanus*)	Microscopic, 181 μm long, thin wall	No data	No data	No data	Unknown	No data	**206**

(Continued)

Table 19.1 (Continued)   Sarcocystis Species in Wild Terrestrial Carnivores

Species	Sarcocyst Shape, Size	SEM	TEM, Wall	Bradyzoites	Definitive Host, Sporocysts (µm)	Molecular	Reference
Brown Bear (*Ursus arctos*)[b]							
S. arctosi Dubey, Rosenthal, Sundar, Velmurugan, and Beckman 2007	Microscopic, 75 µm long, thin wall, 1 µm	No data	Type 1a	5.6–6.8 × 0.7–1.8	Unknown	No data	461
Wild Felids, Namely Bobcat (*Felis rufus*)[a,c], Cheetah (*A. jubatus*)[c], Lion (*P. leo*)[c], Florida Panther and Cougar (*Felis concolor*)[c]							
S. felis Dubey, Hamir, Kirkpatrick, Todd, and Rupprecht 1992	2.1 mm, cyst wall minute striations	No data	vp hob-nail, no microtubules	7.0–10.0 × 1.5	Unknown	No data	407
Leopard (*Panthera pardus*)							
S. sp. Somvanshi, Koul, and Biswas 1987	Thin wall	No data	No data	No data	No data	No data	1644

[a] Intermediate host for *S. neurona*.[204,446,1788]
[b] Intermediate host for *S. canis*.[305,663,1874]
[c] Wild Felidae as hosts for *S. felis*: bobcat,[29,407,732,1788] cheetah,[142] lion,[410,962] Florida panther and cougar.[732]

19.2 *SARCOCYSTIS* SPECIES IN WILD CARNIVORES

In a survey of 10 wild carnivore species in Japan, 4 were found to be positive, namely, Japanese raccoon dog (*Nyctereutes procyonoides viverrinus*), Japanese red fox (*Vulpes vulpes japonica*), Japanese martens (*Martes melampus melampus*), and Japanese badger (*Meles meles anakuma*).[982] Two types of sarcocysts were detected; 1 with a thin wall and minute undulations (presumably TEM type 1) was common in canids and martens, and type 2, with a thick wall and finger-like villous projections containing microtubules, was detected in Japanese badgers (Table 19.1).

The species present in wild Felidae comprising lions, cheetahs, and bobcats, among others was named *Sarcocystis felis*[407] (Table 19.1). This species has been reported in the domestic cat[521,684] (see Chapter 15).

Data on *Sarcocystis* species infecting other frequently observed terrestrial carnivores as foxes, badgers, or coyotes are summarized in Table 19.1.

BIBLIOGRAPHY

29, 94, 100, 103, 142, 153, 159, 204, 206, 246, 265, 268, 286, 305, 364, 374, 383, 407, 410, 415, 446, 461, 462, 467, 504, 521, 546, 555, 621, 663, 684, 705, 706, 732, 776, 933, 953, 962, 964, 982, 983, 1025, 1122, 1187, 1303, 1304, 1314, 1359, 1360, 1445, 1633, 1644, 1681, 1711, 1788, 1803, 1825, 1837, 1874, 1894, 1905, 1907, 1939, 1940, 1958, 1996, 2002, 2035, 2045, 2102, 2168, 2181.

CHAPTER 20

Sarcocystosis in Marsupials

20.1 INTRODUCTION

Opossums act as definitive and intermediate hosts for *Sarcocystis* species.

20.2 AMERICAN OPOSSUMS AS DEFINITIVE HOSTS FOR *SARCOCYSTIS* SPECIES

The North American opossum (*Didelphis virginiana*) and the 3 species of South American opossums (*Didelphis albiventris, Didelphis marsupialis,* and *Didelphis aurita*) are hosts for at least 4 species of *Sarcocystis: S. neurona* (discussed in Chapter 3), *S. falcatula,* and *S. lindsayi* (Chapter 17), and *S. speeri* (discussed here, Table 20.1).

20.2.1 Prevalence of *Sarcocystis* Sporocysts in Opossums

Sporocysts are common in opossums and the estimated prevalence varies with methods employed. Because sporocysts are trapped in the lamina propria, they are excreted in feces irregularly and sometimes none are detectable in feces although millions are present in the intestines. Based on the examination of intestinal digests, sporocysts were found in as many as 54.4% of

Table 20.1 American Opossums as Definitive Hosts for *Sarcocystis* Species

Opossum Species (*Didelphis* spp.)	*Sarcocystis* Species (Sporocysts)[a]	Place of Report	Main Reference
D. albiventris (White-eared opossum)	*S. falcatula*	Argentina	**418**
	S. lindsayi	Brazil	**433**
	S. neurona	Brazil	**442**
	S. speeri	Argentina	**426**
D. aurita[b] (Big-eared opossum)	*S. lindsayi*	Brazil	**1670**
D. marsupialis (Common opossum)	*S. falcatula*	Brazil	**441**
	S. neurona	Brazil	**442**
	S. speeri	Brazil	**427**
D. virginiana (Virginia opossum)	*S. falcatula*	USA	**137**
	S. neurona	USA	**601**
	S. speeri	USA	**416**

[a] KO mice are intermediate host for *S. speeri* under experimental conditions. Its natural intermediate host is unknown.
[b] Monteiro et al.[1209] reported DNA of *S. neurona*- and *S. falcatula*-like sporocysts in *D. aurita* and *D. albiventris* from Brazil.

opossums (see Table 3.3). Identification of *Sarcocystis* species in opossum feces is difficult and expensive. Several methods are available for species identification, including morphology of sporocysts, molecular, and bioassays (see Chapter 2). Among these methods, sporocyst morphology is unreliable and DNA of not all *Sarcocystis* species of opossums has been characterized for definitive identification. Currently, bioassay is the most definitive method for sporocyst identification. Bioassay in KO (knockout) mice and budgerigars can distinguish *S. falcatula* from *S. neurona* and *S. speeri*; *S. falcatula* is not infective to KO mice and *S. neurona* and *S. speeri* are not infective to budgerigars.[421] As indicated in Chapter 3, *S. neurona* does not produce sarcocysts in KO mice whereas *S. speeri* does (Figure 20.1). Based on the bioassays in KO mice and budgerigars, of the 44 opossums examined the prevalences were *S. speeri* 18.8%, *S. falcatula* 43.1%, and *S. neurona* 31.8%.[421] It has been suggested this diversity of *Sarcocystis* species in the intestine of opossums may enable sexual recombination contributing to their already allelic variability.[1209]

20.2.2 *Sarcocystis speeri* Dubey and Lindsay 1999

Opossums (*D. virginiana, D. albiventris,* and *D. marsupialis*) are definitive hosts and a parasite has been found in the United States, Argentina, and Brazil.[414,416,423,426,427] The parasite has been transmitted from the South American opossum to the North American opossum via infections induced in the KO mouse experimental intermediate host.[428] Sporocysts from experimentally infected opossums are 12–15 × 8–10 μm in size. The natural intermediate host for *S. speeri* remains unknown.

Sarcocysts of *S. speeri* in skeletal muscles of KO mice are up to 5 mm long and filiform (Figures 1.18 and 20.1). By LM, the sarcocyst wall is thin (<1 μm thick); by TEM, the cyst wall is up to 1.8 μm thick and has characteristic steeple-shaped vp surmounted by a spire, type 38 (Figure 1.52).[416,1432]

Schizonts are found in many organs, including liver, brain, and uterus of KO mice. Schizonts have also been grown in cell cultures seeded with merozoites from the liver of a KO mouse infected with an isolate from *D. albiventris* from Argentina. Merozoites from cell culture are not infective to KO mice, thus, *S. speeri* has obligatory 2-host life cycle.[423]

The endogenous development of *S. speeri* in tissues of KO mice is illustrated in Figure 20.1 to facilitate further studies using this parasite as laboratory model for *Sarcocystis*.

20.3 AMERICAN OPOSSUMS AS INTERMEDIATE HOSTS FOR *SARCOCYSTIS* SPECIES

Opossums serve as intermediate hosts for 6 *Sarcocystis* species (Table 20.2). There are 2 prevalence surveys involving *Didelphis virginiana* in which 24 of 240 opossums were found infected with *S. greineri*[59]; and, in at least 3 of 137 cases there was coinfection with *S. neurona*,[516] but overall prevalence was not clear.

20.4 SARCOCYSTS OF *SARCOCYSTIS* SPECIES IN OTHER MARSUPIALS

In a vast survey of Australian mammals, sarcocysts were found in the muscle of individuals from 8 different families (Dasyuridae, Peramelidae, Phalangeridae, Burramyidae, Petauridae, Macropodidae, Phascolarctidae, and Vombatidae) including infections in Bennett's wallaby (*Macropus rufogriseus*), swamp wallaby (*Wallabia bicolor*), pademelon (*Thylogale billardierii*), eastern grey kangaroo (*Macropus giganteus*), red kangaroo (*Macropus rufus*), and Tasmanian devil (*Sarcophilus harrisii*).[1249] Two types of sarcocysts were observed: type A, thick walls and/or large

Table 20.2   Sarcocystis Species in Marsupials

Intermediate Host	Sarcocystis Species	Sarcocysts			Definitive Host	Molecular Data	References
		Maximum Length (μm or Stated)	Wall Thickness (μm) and TEM Type	Bradyzoites (μm)			
Common opossum (*Didelphis marsupialis*)[a]	*S. didelphidis* Scorza, Torrealba, and Dagert 1957	Macroscopic, 935 μm long, 5.2 μm long vp	–	6.5 × 1.5	Unknown	No data	1571
	S. garnhami Mandour 1965 Four-eyed opossum (*Philander opossum*) is also a host for this species	Macroscopic, 3.3 mm long, spiny cyst wall	8–6 μm long vp, type 9 m	5.3–6.8 × 1.3–1.9	Unknown	No data	1103, 1567, 1596
Virginia opossum (*Didelphis virginiana*)[a]	*S. greineri* Cheadle 2001	Macroscopic, up to 6 mm long	4.0–2.8 μm long vp, stumpy, type 9b	6.2–9.0 × 1.8–3.0	Unknown, not dog, cat nor opossum	No data	201
	S. inghami Elsheikha, Fitzgerald, Mansfield, and Saeed 2003	Microscopic, up to 700 μm long	7 μm thick wall, type 9n	9.7–12.0 × 2.5–3.0	Unknown	ITS-1	509, 516
	S. sp. Scholtyseck, Entzeroth, and Chobotar 1982	Microscopic, 140 μm long	7–3.4 μm long, type 9 m	7.0–10.0 × 2.5–3.0	Unknown	No data	1567
Murine opossum (*Marmosa murina*)	*S. marmosae* Shaw and Lainson 1969	Macroscopic, up to 2 mm. Spiny, rose-thorn 11.5–13.0 μm long vp	–	6.2–9.0 × 1.8–3.0	Unknown	No data	1596
Wallabies (*Macrops rufogriseus*, *Petrogale assimilis* and pademelon (*Thylogale billardierii*)	*S. mucosa* (Blanchard 1885) Labbe 1889	Macroscopic, ovoid, up to 2 mm diameter, secondary cyst wall	Thin, type 13	8.5 × 2	Unknown	No data	1293, 893
Bettong or rat kangaroo (*Bettongia* sp.)	*S. bettongiae* Bourne 1932	No data	No data	No data	Unknown	No data	128

[a] Common and Virginia opossum are also hosts for *S. neurona*, *S. falcatula*, *S. speeri*, and *S. lindsayi*.

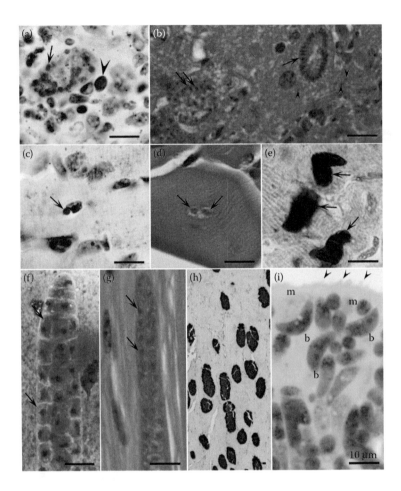

Figure 20.1   Development of *S. speeri* in experimentally infected KO mice. Schizonts (a, b) in the liver, and sarcocysts (c–i) in skeletal muscles of KO mice fed sporocysts. (b, d, and g) stained with H and E, I stained with Toluidine blue. (a) Immunohistochemical (IHC) staining with anti-*S. speeri* merozoite antibodies. (c, e, and h) IHC staining with anti-*S. speeri* bradyzoite antibodies. (f) Smear stained with Giemsa, the rest are tissue sections. (a) Immature schizont (arrow), and a uninucleate merozoite/schizont (arrowhead) at 11 DPI. (b) A schizont with peripherally arranged developing merozoites (arrow), a mature schizont (double arrows), and numerous individual merozoites (arrowheads) scattered in liver parenchyma at 11 DPI. (c) Myocyte containing 1 metrocyte inside a vacuole at 14 DPI. (d) A myocyte with 2 metrocytes (arrows), each in a separate vacuole. (e) Three immature sarcocysts. Day 17 PI. (f, g) Immature sarcocysts with metrocytes (arrows) at 25 DPI. The metrocytes in the smear are much bigger than metrocytes in tissue section. (h) Numerous sarcocysts at 52 DPI. (i) Mature sarcocyst. Note the villar protrusions (arrowheads) on the sarcocyst wall, few metrocytes (me), and longitudinally cut bradyzoites (br) at 222 DPI.

zoites and/or trabeculation, and type B, thin wall with small zoites. The prevalence of macroscopic sarcocysts of *Sarcocystis mucosa* was found to be 13% and 16% for Bennett's wallabies and pademelon, respectively.[893]

BIBLIOGRAPHY

Opossums: **59, 134, 201, 202, 416, 421, 423, 426–428, 433, 509, 516, 1103, 1209, 1432, 1492, 1567, 1571, 1596, 1670, 1722, 1942**.

Other marsupials: **128, 893, 941, 1249, 1293**.

CHAPTER 21

Sarcocystosis in Rodents, Lagomorphs, and Other Small Mammals

This chapter summarizes *Sarcocystis* species detected in rodents, namely mice (Table 21.1), rats (Table 21.2), voles (Table 21.3), and in lagomorphs/leporids (Table 21.4), and other small mammals (Table 21.5). In most small animals the clinical significance of infection with *Sarcocystis* spp. is unknown. Three species with rodents as intermediate hosts, *S. muris*, *S. cymruensis* (Syn. *S. rodentifelis*), and *S. singaporensis*, are described in detail because of their interest for researchers.

21.1 *SARCOCYSTIS MURIS* (RAILLIET 1886) LABBÉ 1899

Sarcocystis muris is one of the first species named in the genus.[1191] The house mouse (*Mus musculus*) is the intermediate host, and the domestic cat and the ferret (*Putorius putorius furo*) are definitive hosts[1484,1503] (Table 21.1). In Egypt, sporocysts from a naturally infected cat were infectious for immunocompetent and immunocompromised mice.[27] The definitive host was required to complete the life cycle even in the case of immunodeficient mice, although theoretically *S. muris* possesses both diheteroxenous and dihomoxenous life cycles enabling transmission by cannibalism among mice.[974]

S. muris (Figure 1.1) does not cause clinical disease and is not zoonotic. The main interest in this species is its usefulness as a biological model for research on cyst forming coccidia.[1486] Many studies have been done on *in vitro* cultivation and immunology of *S. muris* infection (see additional references summarized in the Bibliography of this chapter).

Asexual stages of *S. muris* were found only in *Mus musculus*. Schizonts were seen in hepatocytes 11–17 DPI, but details of schizogony are unknown.[1503] Sarcocysts in muscles were first seen at 28 DPI, and first attained infectivity for cats at 76 DPI. As sarcocysts grew in size, the structure of the sarcocyst wall (type 1a) did not change.[1597] Mice remained asymptomatic, except for an abnormal gait, despite observations that 80% of their muscles were infected (Figures 1.1, 21.1, and 21.2). As indicated in Table 21.1, *S. muris* sarcocysts are thread-like, and several centimeters long. Bradyzoites in stained smears are 14–16 × 4–6 μm in size. Sporocysts are 8.7–11.7 × 7.5–9.0 μm.[1503]

Most international research on *S. muris* from 1973 to 2010 was performed using the Costa Rican isolate of *S. muris*, originally obtained by feeding a naturally infected house mouse in Costa Rica to a kitten.[1597] Both outbred and inbred *Mus musculus* are susceptible to infection, but results are inconsistent. After testing oral doses of 1–100,000 sporocysts, either as a single dose or as repeated doses, the most consistent results were obtained using 50 sporocysts.[1486] Infections were produced more consistently in immunosuppressed or immunodeficient mice (including nude, SCID, KO). In 2010, *S. muris* was isolated from the feces of a naturally infected *Felis catus* in Egypt.[27] Attempts to find schizonts in KO or Swiss mice inoculated orally with the Egyptian isolate of

Table 21.1 Sarcocystis Species in Mice

| Intermediate Host | Sarcocystis Species | Sarcocysts | | Definitive Host | References |
		Maximum Length (μm, or Stated)	Wall Thickness (μm) and Type		
House mouse (Mus musculus)	S. atheridis Šlapeta, Modrý, and Koudela 1999	Macroscopic, 30 × 0.9 mm	Thin wall, type 1b	Nitsche's bush viper (Atheris nitschei), 10.0–11.0 × 7.0–8.5	1620, 1623
	S. crotali Entzeroth, Chobotar, and Scholtyseck 1985	Up to 4 mm long	Smooth, type 1a	Mojave rattlesnake (Crotalus scutulatus)	540
	S. dispersa Černá, Kolářova, and Šulc 1978	Microscopic	Smooth, <1 μm, type 18a	Barn owl (Tyto alba), masked owl (Tyto novaehollandiae)	187, 1585
	S. muris (Railliet 1886) Labbe 1899	Several centimeters	Smooth, <1 μm, type 1a	Cat (Felis catus), ferret (Putorius putorius furo)	1484, 1503, 1597
	S. muriviperae Matuschka, Heydorn, Mehlhorn, Abd-al-Aal, Diesing, and Biehler 1987	Up to 8 mm, thick wall	3.5 μm long vp, type 18b	Palestinian viper (Vipera palaestinae) and Caspian whipsnake (Coluber jugularis)	1142, 1363
	S. scotti Levine and Tadros 1980	No data	No data	Tawny owl (Strix aluco)	1048, 1247, 1715, 1716
	S. sebeki Tadros and Laarman 1976	Several centimeters, smooth	<1 μm, type 1b	Tawny owl (Strix aluco)	1705, 1708, 1712
Deer mouse (Peromyscus maniculatus)	S. espinosai (Espinosa, Sterner, and Blixt 1988) Odening 1997	Microscopic, thin wall	<1 μm, type 1h	Northern saw-whet owl (Aegolius acadicus)	558
	S. idahoensis Bledsoe 1980	Up to 9.4 mm long	Thin wall (<1 μm)	Gopher snake (Pituophis melanoleucus)	111, 112, 113, 375
	S. peromysci Dubey 1983	Up to 1.8 mm	Thick wall (2–5.5 μm), hairy vp	Unknown	375

Table 21.2 Sarcocystis Species in Rats

Intermediate Host	Sarcocystis Species	Sarcocysts			References
		Maximum Length (μm, or Stated)	Wall Thickness (μm) and Type	Definitive Host	
Rice rat (*Oryzomys capito*)	*S. azevedo* Shaw and Lainson 1969	Up to 2.7 mm long	1–1.8 μm thick wall, striated	Unknown	1596
	S. oryzomyos Shaw and Lainson 1969	Up to 24.7 mm long	3–3.2 μm thick wall, striated	Unknown	1596
Cotton rat (*Sigmodon hispidus*)	*S. sigmodontis* Dubey and Sheffield 1988	More than 1 cm long	Thin wall, vp up to 1 μm, type 4	Unknown	1996
Moon rat (*Echinosorex gymnurus*)	*S. booliati* Dissanaike and Poopalachelvam 1975	Up to 5 mm, thin wall	<0.5 μm long vp, type 1b	Unknown	325, 935
Multimammate rat (*Mastomys natalensis*)	*S. disrumpens* Häfner and Matuschka 1984	Macroscopic, up to 2.5 cm	Thin, smooth, type 1b	Vipers (*Bitis* spp.) (see Table 22.2)	768, 769, 1138
Experimentally: rodents of genera *Mesocricetus*, *Phodopus*, *Gerbillus*, *Meriones*, and *Mus*	*S. hoarensis* Matuschka and Häfner 1984	Macroscopic, up to 2.5 mm, smooth	Secondary cyst wall, type 42	Vipers (*Bitis* spp.) (see Table 22.2)	1133, 1141
Indonesian rats (*Bunomys chrysocomus*, *Bunomys fratrorum*, *Paruromys dominator*)	*S. sulawesiensis* O'Donoghue, Watts, and Dixon 1987	Up to 120 μm long, smooth	<1 μm thick, type 5	Unknown	1292
Spiny rat (*Proechimys guyannensis*)	*S. proechimyos* Shaw and Lainson 1969	Up to 3.9 mm	3.5–4 μm thick wall, striated	Unknown	1596
Norway rat (*Rattus norvegicus*)	*S. cymruensis* Ashford 1978 (Syn. *S. rodentifelis*) Grikienienė, Arnastauskienė, and Kutkienė 1993 Also in Oriental house rat (*Rattus flavipectus*). Experimental hosts: laboratory mouse, bank vole	Macroscopic, up to 50 mm, thin wall, smooth	Thin <1 μm thick, smooth, type 1c	Cat (*Felis catus*)	41, 741, 854
	S. murinotechis Munday and Mason 1980	Up to 300 μm	Thick, 6 μm long vp, type 9b	Tiger snake (*Notechis ater*) (see Table 23.2)	1252
	S. singaporensis Zaman and Colley (1975) 1976	1 mm long, striated, and thick wall	3–5 μm thick, type 19	Python (*Python reticulatus*)	1868, 1870

(Continued)

Table 21.2 (Continued) Sarcocystis Species in Rats

| Intermediate Host | Sarcocystis Species | Sarcocysts | | Definitive Host | References |
		Maximum Length (μm, or Stated)	Wall Thickness (μm) and Type		
	S. villivillosi Beaver and Maleckar 1981	1.1 mm long	1.5 μm thick, type 22	Python (Python reticulatus)	85
	S. zamani Beaver and Maleckar 1981	2 mm long	1–3 μm thick, type 18c	Python (Python reticulatus)	85
	S. zuoi Hu, Ma, and Li 2005	Up to 2.3 mm, striated wall	8.5–9.6 μm long vp, type 17	King ratsnake (Elaphe carinata)	853, 855
	S. dispersa-like Munday 1983	Microscopic	Smooth, <1 μm, type 18	Masked owl (Tyto novaehollandiae)	1256
Malaysian house rats (Rattus rattus diardii)	S. sp.	Microscopic	5.7 μm long finger-like and hexagonal in section vp, type 19	Unknown	936
Wood rat (Neotoma micropus)	S. neotomafelis Galaviz-Silva, Mercado-Hernández, Ramírez-Bon, Arredondo-Cantú, and Lazcano-Villareal 1991	Macroscopic, up to 2 mm long	2–4 μm thick, type 1c	Domestic cat (Felis catus)	658
Bush rat (Rattus fuscipes)	S. sp. type B Rzepczyk and Scholtyseck 1976	Macroscopic, up to 1 mm long, thin wall	Type 1b	Unknown	1504
Bandicoot (Bandicota indica)	S. sp. Kannangara 1970	Up to 440 μm long, smooth wall	No data	Unknown	941

Table 21.3 Sarcocystis Species in Voles

Intermediate Host	Sarcocystis Species	Sarcocysts		Definitive Host	References
		Maximum Length (μm, or Stated)	Wall Thickness (μm) and type		
Common European vole (*Microtus arvalis*)	*S. cernae* Levine 1977	Microscopic, up to 900 μm, smooth	<1 μm thick	Kestrel (*Falco tinnunculus*)	182, 188, 185, 187, 1047
Common European vole (*Microtus arvalis*), short-tailed vole (*Microtus agrestis*)	*S. putorii* (Railliet and Lucet 1891) Tadros and Laarman 1978	Macroscopic, several mm, smooth	Thin, bristly, type 9b	Ferret (*Mustela putorius var. furo*), common European weasel (*Mustela nivalis*), stoat (*Mustela nivalis, Mustela erminea*), mink (*Mustela lutreola*)	1703, 1704, 1708, 1710, 1716
European voles (*Microtus arvalis, Microtus oconomus, Microtus guentheri, Clethrionomys glareolus*)	*S. clethrionomyelaphis* Matuschka 1986	Up to 4.5 mm long, striated	4–5 μm long vp, type 9	Aesculapian snake (*Zamenis longissimus*) and others under experimental conditions.	830, 1135, 1178
	S. dirumpens Häfner and Matuschka 1984	Up to 25 mm, smooth	3–5 μm thick wall	Vipers (*Bitis arietans, Bitis gabonica, Bitis caudalis, Bitis nasicornis*)	768
	S. muriviperae Matuschka, Heydorn, Mehlhorn, Abd-al-Aal, Diesing, and Biehler 1987	Up to 8 mm long	3.5 μm long vp, type 18b	Palestinian viper (*Vipera palestinae*)	1142
Meadow vole (*Microtus pennsylvanicus*) and long-tailed vole (*Microtus longicaudus*)	*S. jaypeedubeyi* (Dubey 1983) Modry, Votypka, and Svobodova 2004	Macroscopic, 1.2 mm, thick wall	1.8–3.3 μm thick wall, type 9o	Unknown	370, 1204
	S. montanaensis Dubey 1983	Up to 648 mm, thin wall	<1 μm thick, smooth, type 1a	Unknown	370
Prairie vole (*Microtus ochrogaster*)	*S. montanaensis*-like Lindsay, Upton, Blagburn, Toivio-Kinnucan, McAllister, and Trauth 1991	Up to 800 μm	<1 μm thick, smooth, type 1a	Southern copperhead (*Agkistrodon contortrix*)	1065
Vole (*Microtus savii*)	*S. pitymysi* Splendore 1918	No data	No data	No data	1317
Large oriental vole (*Eothenomys miletus*)	*S. eothenomysi* Hu, Liu, Yang, Esch, Guo, and Zou 2014	Up to 3 mm long, smooth	1–2 μm thick, striated, type 18a	Unknown	856

(Continued)

Table 21.3 (Continued) Sarcocystis Species in Voles

| Intermediate Host | Sarcocystis Species | Sarcocysts | | Definitive Host | References |
		Maximum Length (μm, or Stated)	Wall Thickness (μm) and type		
Japanese vole (*Clethrionomys rufocanus bedfordiae*)	*S. clethrionomysi* Ohbayashi and Kitamura 1959	Up to 1.5 mm long	3–7 μm thick wall	Unknown	1322
Voles (*Arvicola sapidus*, *Clethrionomys rufocanus*, *Clethrionomys rutilus*)	*S. glareoli* (Erhardová 1955) Odening 1997	Up to 400 μm	Type 1a	Common buzzard (*Buteo buteo*) and rough-legged buzzard (*Buteo lagopus*)	1314, 1317
Northern mole vole (*Ellobius talpinus*)	*S. talpini* Pak, Sklyarova, and Dymkova 1991	Up to 3.6 mm long	1.0–1.4 μm thick wall	Unknown	1359
European water vole (*Arvicola terrestris*)	*S. terrestri* Pak, Sklyarova, and Dymkova 1991	Up to 3.0 mm long	Thin wall (0.3–0.4 μm), hirsute	Unknown	1359

Table 21.4 Sarcocystis Species in Lagomorphs

Intermediate Host	Sarcocystis Species	Sarcocysts		Definitive Host	References
		Maximum Length (μm, or Stated)	Wall Thickness (μm) and Type		
European rabbit (Oryctolagus cuniculus)	S. cuniculorum (Brumpt 1913) Odening, Wesemeier, and Bockhardt 1996	Up to 4.5 mm	8–11 μm, striated, type 10b	Cat (Felis catus)	525, 1305, 1310, 1706
Cottontail rabbit (Sylvilagus floridanus, S. nuttalli, S. pallistris)	S. leporum Crawley 1914	Macroscopic, 2 mm	5–6 μm, striated, type 10b	Cat (Felis catus), raccoon (Procyon lotor)	260, 525, 581
European hare (Lepus europaeus)	S. cf. cuniculi Brumpt 1913	Up to 2.2 mm	10.3-8.6 μm, packed long slim-looking vp, type 10b	Unknown	1305
	S. sp. 1 Odening, Wesemeier, Pinkowski, Walter, Sedlaczek, and Bockhardt 1994	800 μm	2.7-0.9 μm, hair-like vp, type 7a	Unknown	1305
	S. sp. 2 Odening, Wesemeier, Pinkowski, Walter, Sedlaczek, and Bockhardt 1994	4.3 mm	4.0-3.2 μm, cauliflower-like vp, type 26	Unknown	1305
	S. cf. bertrami Odening, Wesemeier, Pinkowski, Walter, Sedlaczek, and Bockhardt 1994	Up to 4.0 mm	4.4-1.9 μm, elongated leaflet-like vp, type 11b	Unknown	1305
	S. sp. 1 Odening, Wesemeier, and Bockhardt 1996	Oval to cigar-shaped, up to 835 μm	No vp, type 1a	Unknown	1310
	S. sp. 2 Odening, Wesemeier, and Bockhardt 1996	Elliptical, up to 385 μm	Stub-like vp, type 29	Unknown	1310
Pika (Ochotona spp.)	S. ochotonae Odening, Aue, Ochs, and Stolte 1998	Up to 0.6 mm long Sarcocysts in brain	Thin wall, type 1a	Unknown	1318
Alpine pika (Ochotona alpina)	S. galuzoi Levit, Orlov, and Dymkova 1984	Macroscopic, up to 2 mm long. Smooth	3–5 μm thick wall	Unknown	1050, 1317
Daurian pika (Ochotona daurica)	S. dogeli Machul'skii 1947	Fat macrocysts	—	Unknown	1317

Table 21.5 Sarcocystis Species in other Small Mammals

Intermediate Host	Sarcocystis Species	Sarcocysts		Definitive Host	References
		Maximum Length (µm, or Stated)	Wall Thickness (µm) and Type		
Striped hamster (*Cricetulus griseus*)	*S. cricetuli* Patton and Hindle 1926	Macroscopic, 1.5 mm	No data	Unknown	1378
Large gerbil (*Rhombomys opimus*)	*S. fedoseenkoi* (Fedoseenko and Romanova 1983) Odening 1997	No data	70 nm thick, type 9	Unknown	598, 1314, 1317
Gerbil (*Gerbillus gerbillus, Gerbillus perpallidus, Psammomys obesus, Pachyuromys duprasi*)	*S. gerbilliechis* Jäkel 1995	Macroscopic, 11.7 × 0.19 mm	Smooth, type 1b	Arabian saw-scaled viper (*Echis coloratus*), 12.3–13.3 × 10.7–11.4	885
Guinea pig (*Cavia porcellus*)	*S. caviae* Almeida 1928	No data	No data	Unknown	1048
Porcupine (*Erethizon dorsatum*)	*S. sehi* Dubey, Hamir, Brown, and Rupprecht 1992	Up to 2.3 mm long	Thin wall (<1.2 µm), type 1e	Unknown	406
9-banded armadillo (*Dasypus novemcinctus*)[a]	*S. dasypi* Howells, Carvalho, Mello, and Rangel 1975	Up to 1.53 mm, moderately thick wall	3.7–4.3 µm, spinose, type 11	Unknown	851
	S. diminuta Howells, Carvalho, Mello, and Rangel 1975	Up to 224 µm, thick wall	2.6–3.5 µm, hair-like projections, type 11	Unknown	851
	S. sp. Lindsay, McKown, Upton, McAllister, Toivio-Kinnucan, Veatch, and blagburn 1996	Only one cyst. No data	Vp narrow at the base and expand as they grow outwardly. Unclassified	Unknown	1067
Marmot (*Marmota baibacina*)	*S. baibacinacanis* syn. S. *baibacina* Umbetaliev 1979	Up to 17 mm	No data	Dog (*Canis familiaris*), fox (*Vulpes vulpes*), wolf (*Canis lupus*)	1767
Collared anteater (*Tamandua tetradactyla*)	*S. tamanduae* Artigas and Oria 1932	No data	No data	Unknown	38, 1048
Richardson's ground squirrel (*Spermophilus richardsonii*)	*S. bozemanensis* Dubey 1983	300 µm, thin-walled	Smooth, <1 µm, type 1a	Unknown	369

(Continued)

Table 21.5 (Continued) Sarcocystis Species in other Small Mammals

Intermediate Host	Sarcocystis Species	Sarcocysts Maximum Length (μm, or Stated)	Sarcocysts Wall Thickness (μm) and Type	Definitive Host	References
	S. campestris Cawthorn, Wobeser, and Gajadhar 1983 Also in 13-lined ground squirrel (Spermophilus tridecemlineatus)	Macroscopic, 4 mm, thick-walled, 3.5–6.4 μm	3.5 μm long vp, type 9a	American badger (Taxidea taxus)	172, 369, 418
Red squirrel (Tamiasciurus hudsonicus)	S. sp. Entzeroth, Chobotar, and Scholtyseck 1983	Microscopic, total length not known	<1 μm thick wall, type 1b	Unknown	535
Yellow suslik (Spermophilus fulvus)	S. citellibuteonis Pak, Pak, and Sklyarova 1989	Up to 10 mm, thin wall	1.0–0.7 μm long vp	Common buzzard (Buteo buteo)	1355,1358
	S. citellivulpes Pak, Perminora, and Yeshtokina 1979	Up to 8 mm, thick wall, striated	1.4–3.5 μm long vp	Red fox (Vulpes vulpes) and Corsac fox (V. corsac)	1356
Varying lemming (Dicrostonyx richardsoni)	S. rauschorum Cawthorn, Gajadhar, and Brooks 1984	Up to 122 μm long, thin wall, smooth	<1 μm thick, type 1h	Snowy owl (Bubo scandiacus syn. Nyctea scandiaca)	173, 174, 175, 636, 1671
Treeshrew (Tupaia belangeri chinensis)	S. tupaia Xiang, Rosenthal, He, Wang, Wang, Song, Shen, Li, and Yang 2010	Thread-like, up to 10 mm long	Smooth, type 1a	Unknown	1845
Greater white-toothed shrew (Crocidura russula)	S. russuli Pak, Sklyarova, and Dymkova 1991	Up to 6.0 mm long	Thin wall (0.5–1.0 μm)	Unknown	1359
Chipmunk (Eutamias asiaticus)	S. eutamias Tanabe and Okinami 1940	Up to 530 μm long	Not striated	Unknown	1721
Eastern chipmunk (Tamias striatus)	S. sp. Entzeroth, Scholtyseck, and Chobotar 1983	Microscopic, up to 75.6 μm long	Thin wall, type 1b	Unknown	536

a Nine-banded armadillo is also intermediate host for Sarcocystis neurona.[205,1723]

Figure 21.1   *Sarcocystis muris*-associated myositis in a mouse. Note the numerous intact (a) and degenerating (b) sarcocysts. H and E stain.

S. muris failed, although the mice had hepatitis (unpublished). The Costa Rican and the Egyptian isolates of *S. muris* are no longer available.

21.2 *SARCOCYSTIS CYMRUENSIS* ASHFORD 1978 (SYN. *S. RODENTIFELIS* GRIKIENIENÉ, ARNASTAUSKIENÉ, AND KUTKIENÉ 1993)

It has been proposed that the parasite from the bank vole and the Norway rat is capable of circulating among some rodents without the participation of the definitive host (domestic cat), based on a series of observations.[34,739,740,741,992] This species exhibited low specificity, able to infect multiple families such as Cricetidae and Muridae. Of principal interest was the possibility of transmission by cannibalism and transplacental transmission.[993,996] This second idea was almost discarded after comprehensive research[1001]; no sarcocysts were detected in 83 offspring born from experimentally infected rats. Norway rats are both definitive and intermediate hosts for *S. rodentifelis* under laboratory conditions[742]; a similar situation has been detected for *S. muris* and *S. cymruensis*. Thus, based on indistinguishable morphological and biological characteristics we have synonymized *S. rodentifelis* with *S. cymruensis* (Table 21.2).

21.3 *SARCOCYSTIS SINGAPORENSIS* ZAMAN AND COLLEY (1975) 1976

The rats (*Rattus norvegicus*) are the intermediate hosts and the pythons (*Python reticulatus*) are the definitive hosts (Table 21.2).[85,140,1868,1870] The ultrastructural development of *S. singaporensis* sarcocysts in rats has been described.[1367,1368] 2 subpopulations (type 1 and 2) of merozoites were detected in lungs from acutely infected rats.[888] Developmental stages in the definitive host (*Python reticulatus*) have been described.[1365]

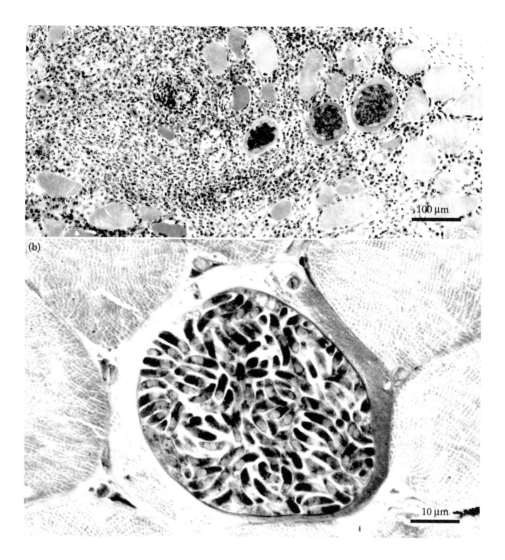

Figure 21.2 *S. muris* infection in skeletal muscle of a mouse. H and E stain. (a) Severe inflammation around degenerating sarcocysts. (b) An intact sarcocyst within a myocyte without host reaction.

Unlike other rodent *Sarcocystis* species, schizogony of *S. singaporensis* occurs in blood vessels.[85,140,1368] The dose and the strain of the parasite might affect the organ parasitized. In rats fed 200,000 sporocysts in the United States,[85] schizonts were seen only in the liver at 7 DPI but at 10 days schizonts were also present in the muscles, including cardiomyocytes, but never in the liver. Two generations of schizonts, first at 6 DPI and the second at 16 DPI, were found in rats infected in Germany.[140] In rats euthanized at 10 days after feeding them 40 million sporocysts, schizonts were seen in brain, kidneys, lungs, and the heart. Heavy infection in these rats facilitated the detailed description of schizogonic development of the parasite, including an occasional parasitization of erythrocytes and blood vessel occlusion due to hypertrophied parasitized endothelial cells.[1368]

The use of *S. singaporensis* has been proposed for the control of wild populations of rodents. It is known that natural infections in wild rats do not induce protective immunity, but a significant immune response can be induced experimentally in laboratory rats using about 1000 sporocysts.[886] A specific immunoglobulin subclass response suggests the existence of an immune evasion strategy for *S. singaporensis*,[890] and a natural selection for intermediate degrees of virulence of parasite strains has been confirmed.[892] The use of *S. singaporensis* as a biocontrol agent was preceded by studies on host specificity and pathogenicity.[886] Under controlled conditions,[889] bait-pellets containing high numbers of sporocysts resulted in mortalities ranging from 58% to 92% in 3 species of rodents (*Rattus norvegicus*, *Rattus tiomanicus*, and *Bandicota indica*). Using species-specific monoclonal antibodies against merozoites of *S. singaporensis* in lungs, the parasitic etiology of mortality was confirmed.[889,891]

Highly variable 18S rDNA sequences were found in *S. singaporensis* from pythons,[1622] which also seem to have a phylogenetic close relationship with *Sarcocystis atheridis* from the African viper and *Sarcocystis* sp. from the American rattlesnake. *Sarcocystis* sporocysts from the green python (*Morelia viridis*) have 95% of identity with *S. singaporensis* 18S rRNA gene sequences.[1219] Further molecular investigations will be needed in order to confirm the identity of such genotypes and the possibilities of additional hosts for *S. singaporensis*.

21.4 *SARCOCYSTIS* SPECIES WITH SARCOCYSTS IN BRAIN (PREVIOUSLY CALLED *FRENKELIA*)

21.4.1 Introduction

In 1934, Findlay and Middleton[605] reported large parasitic cysts in the brains of voles in Wales, United Kingdom, and they called it *Toxoplasma microti*. In 1953, a similar organism was found in a vole in Montana, USA (Figure 21.3) and was called "M" organism.[630] A similar parasite was described in a bank vole and the name *Toxoplasma glareolus* was proposed.[556] A new genus, *Frenkelia* was also proposed for these organisms.[102] When the life cycle of a similar parasite was completed it was named *Frenkelia clethrionomyobuteonis* (Figure 1.3).[1483] With the development of molecular tools during the last 3 decades, several investigators indicated that the genus *Frenkelia* is a synonym of *Sarcocystis*.[76,1236,1317,1799]

21.4.2 *Sarcocystis microti* (Findlay and Middleton 1934) Modrý, Votypka, and Svobodová 2004

21.4.2.1 Distribution

Europe and North America.

21.4.2.2 Intermediate Hosts

Microtus agrestis,[605] *Microtus arvalis*,[976] *Apodemus sylvaticus*,[976] *Apodemus flavicollis*,[976] *Apodemus agrarius*,[976] *Mesocricetus auratus*,[976] *Rattus norvegicus*,[1483] *Mus musculus*,[976] *Mastomys natalensis*,[1482] *Cricetus cricetus*,[1482] *Chinchilla laniger*,[424,1180,1482] *Oryctolagus cuniculus*,[1482] and probably *Lemmus lemmus*,[528] *Microtus modestus*,[630] *Ondatra zibethica*,[944] *Rattus* sp.,[786] and *Erethizon dorsatum*.[951]

21.4.2.3 Definitive Hosts

The common buzzard (*Buteo buteo*) is the definitive host but remains unknown for *S. microti* in North America; hawks and owls are suspected.[794,1063] European workers[1716] found that *F. microti* can cycle through the American red-tailed hawk, *Buteo borealis*.

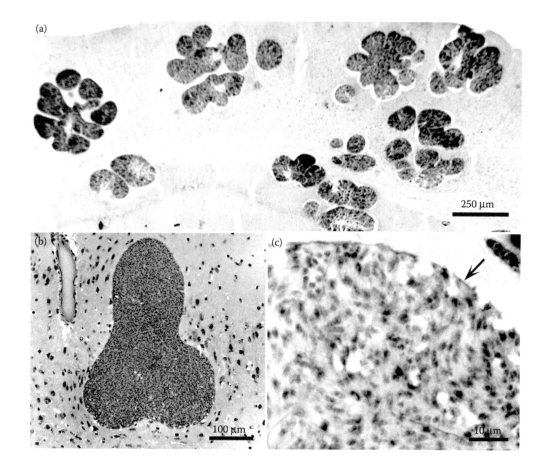

Figure 21.3 *Sarcocystis microti*-like cysts from rodents in the United States. (a) Lobulated cysts in the brain of *Microtus modestus*. (Courtesy of late J. K. Frenkel.) (b and c) A cyst in the brain of a chinchilla; arrow points to the thin cyst wall. (Adapted from Dubey, J.P. et al., 2000. *J. Parasitol.* 86, 1149–1150.)

21.4.2.4 Structure

The sarcocysts are thin-walled, lobulated, measure up to 1 mm in diameter, and are located in the brain (Figure 21.3). The sarcocyst wall is type 1a. Cysts were first observed in the brain 23 DPI.[669]

The sporocysts are 12–15 × 9–12 (mean 12 × 10) μm.[976] The schizonts are in the liver parenchymal cells 6 and 7 DPI.

21.4.3 *Sarcocystis glareoli* Erhardová 1955

21.4.3.1 Distribution

Europe and Japan.[639]

21.4.3.2 Intermediate Hosts

The European bank vole (*Clethrionomys glareolus*) is its intermediate host.[528,556,1482]

21.4.3.3 Definitive Host

The common buzzard (*Buteo buteo*).[975]

The sporocysts are 11.3–13.8 × 7.8–10.0 (mean 12.5 × 8.8) μm.[1482,1483] The schizonts are in the liver parenchymal cells between 5 and 8 DPI; merozoites are 7.6 × 2.2 μm.[975] The cysts are found in the brain, beginning 18 DPI. They are microscopic,[952,1717] and up to 400 μm at 120 DPI.[975] Congenital transmission is suspected, but not proven.[975,1709]

21.5 PREVALENCES OF SARCOCYSTS IN RODENTS, LAGOMORPHS, AND OTHER SMALL MAMMALS

Small animals, and specially rodents, are the intermediate hosts for numerous species of *Sarcocystis* involving snakes, raptors, and carnivorous mammals as definitive hosts. Thus, the prevalence of infections in those animals is high. Infection rates of 88% for *S. dasypi* and 21% for *S. diminuta* have been found in armadillos (*Dasypus novemcinctus*) from 4 southern states in the United States.[1067] Similar infection rates (*S. dasypi*, 60.3% and *S. diminuta*, 9.5%) were found in Florida.[313] Mixed infections were also recorded in these surveys. In lagomorphs, macroscopic and microscopic sarcocysts were found in 54.1% of Eastern cottontails from Pennsylvania (*Sylvilagus floridanus*).[258] Risk factors such as the presence of cats in the environment are important; the absence of such definitive hosts seems to be the cause of the null presence of sarcocysts in 555 farmed rabbits from Lithuania.[991] The prevalence of thick-walled sarcocysts in rabbits (*Oryctolagus cuniculus*) from Australia was 21.1%.[1249]

In Richardson's ground squirrels (*Spermophilus richardsonii*) from Montana 11.1% had sarcocysts.[369] In the same state, 40.8% and later 25.5% of deer mice (*Peromyscus maniculatus*) were infected.[112,375] In woodrats (*Neotoma microtus*) from Mexico, 28.7% were infected.[658] Infection with *Sarcocystis* has also been reported in 25% of woodrats from Texas.[197,198] In large oriental voles (*Eothenomys miletus*) from China, 25% were infected with *Sarcocystis*.[856]

Wide surveys involving several host species had been carried out. The presence of sarcocysts in the brain and muscle was detected in 13 species of micromammals from different habitats in Italy.[252] Species of *Sarcocystis* were not stated. *Sarcocystis* infection was absent in 59 common moles (*Talpa europaea*) from Lithuania.[1872] Previously, in the same country, infection was reported in an Insectivora, the common shrew (*Sorex araneus*), and in 3 additional species of rodents.[743] Factors such as habitat or host weight influenced the prevalences of *Sarcocystis* in populations of small mammals and buzzards (*Buteo buteo*) in the Czech Republic.[1697]

BIBLIOGRAPHY

References on mice: 111–113, 180, 181, 183, 184, 186, 187, 375, 540, 1048, 1142, 1247, 1363, 1484, 1585, 1597, 1705, 1708, 1712, 1715, 1716, 2010, 2041, 2058, 2089, 2093, 2131, 2240, 2257, 2286, 2313.

References on S. muris: 1, 27, 42, 86, 316, 537, 539, 541, 543, 544, 545, 557, 632, 633, 653, 682, 683, 780, 789, 839, 914, 967, 973, 974, 1001, 1027, 1093, 1183, 1191, 1403, 1411, 1439, 1440, 1462, 1469–1471, 1484, 1486, 1503, 1625, 1628, 1642, 1643, 1687, 1729, 1730, 1733, 1749, 1794, 1804, 1805, 1839, 1914, 1921, 1970, 1998, 2058, 2082, 2103, 2131, 2142, 2174, 2239, 2245, 2254, 2293, 2296, 2298, 2346, 2362, 2368, 2369.

References on rats: 41, 85, 197, 198, 325, 658, 741, 768, 769, 853–855, 935, 936, 1133, 1138, 1141, 1252, 1256, 1292, 1504, 1596, 1868, 1870, 2270.

References on *S. singaporensis*: **140, 886–892, 1365, 1367, 1368, 1622, 2018**.

References on *S. cymruensis* (Syn. *S. rodentifelis*): **34, 739–742, 893, 992–997, 1001**.

References on voles: **35, 182, 185, 187, 188, 370, 768, 830, 856, 1047, 1065, 1135, 1142, 1178, 1204, 1317, 1322, 1703, 1704, 1708, 1710, 1716, 2032, 2294**.

References on *Sarcocystis* species located in brain: **99, 102, 527, 528, 556, 605, 630, 639, 669, 786, 794, 944, 949, 951, 952, 975, 976, 1063, 1166, 1167, 1180, 1204, 1270, 1483, 1565, 1584, 1709, 1717, 2019, 2112, 2287, 2288, 2301, 2331, 2353**.

Other references in rodents: **34, 35, 180, 181, 183, 188, 190, 193, 252, 558, 743, 744, 767, 959, 1247, 1255, 1256, 1363, 1364, 1379, 1697, 1872, 2094**.

References on lagomorphs: **189, 258, 260, 264, 525, 581, 991, 1050, 1249, 1253, 1305, 1310, 1318, 1706, 1708, 1713, 1751, 1901, 1970, 2054**.

Other small mammals: **38, 172–175, 325, 369, 406, 418, 535, 536, 598, 636, 851, 1048, 1067, 1355, 1356, 1358, 1378, 1671, 1721, 1767, 1845, 2113, 2127, 2236**.

CHAPTER 22

Sarcocystosis in Poikilothermic Animals

22.1 REPTILES

Reptiles can act as intermediate or definitive hosts for numerous *Sarcocystis* species. Information is summarized in Tables 22.1 and 22.2, respectively.

22.1.1 Reptiles as Intermediate Hosts for *Sarcocystis*

There are 23 *Sarcocystis* species in which reptiles serve as intermediate hosts (Table 22.1). Few definitive hosts are known, but 7 snakes play this role, especially for species parasitizing lizards and geckoes (*S. stenodactylicolubris, S. gongyli, S. chalcidicolubris, S. acanthocolubri, S. ameivamastigodryasi, S. lacerate, S. podarcicolubris*).

S. gallotiae has an unusual life cycle.[1139] It can complete its life cycle in 1 host, the Gran Canaria giant lizard *Gallotia galloti*. Its sarcocysts are found in the tails of the lizards. Cannabalism is common in these reptiles, and sporocysts are produced in the intestines of lizards after eating the tails of other lizards.[1139] This particular dihomoxenous life cycle has been tested for some other *Sarcocystis* species of lizards.[67]

None of the species of *Sarcocystis* are known to cause clinical illness in reptiles. Lainson and Shaw[1015] reported granulomatous myositis in several naturally infected tortoises; lesions seemed to be associated with cyst rupture.

Epidemiological surveys on *Sarcocystis* in reptiles are scarce. In Australia, sarcocysts were reported in the muscles of 2 goannas (1 *Varanus gouldii* and 1 *V. varius*) and in 1 metallic skink (*Niveoscincus metallicus* syn. *Leiolopisma metallica*).[1251] No morphological details were given. Only 3 reports were found of sarcocysts in snakes from different locations (Table 22.1). *S. pythonis* was found in the Australian carpet snake (*Morelia spilota* syn. *Python spilotes*) in Australia.[1744] Sarcocysts in muscle were microscopic, 1.1 mm long, and thin-walled. *S. atractaspidis* was named from 2 Ogaden burrowing asps (*Atractaspis leucomelas*) from Kenya.[1370] Sarcocysts were macroscopic and were not reported in striated muscle but in mesentery and lungs. No illustrations accompanied the article, so this species should be regarded as *species inquirendae*. A massive number of microscopic sarcocysts were observed in striated muscle, not in the heart, of a wild caught female South American rattlesnake (*Crotalus durissus terrificus*) of unknown origin but confiscated in Tennessee.[1474] The sarcocyst wall was type 1h; the 18S rRNA gene sequence was closely related to that of *S. mucosa* from Australian marsupials. The authors hypothesize that an opossum is a final host.

22.1.1.1 *Species Inquirendae*

The following species need further revision, because there are no differential morphological descriptions of the sarcocysts and no DNA data: (1) *S. atractaspidis* Parenzan, 1947 from the

Table 22.1 Reptiles as Intermediate Hosts for Sarcocystis

Species	Intermediate Host	Size of Sarcocyst (μm)	Size of Bradyzoites (μm)	Sarcocyst Wall	Definitive Host, Sporocysts (μm)	Molecular Data	Reference
S. acanthocolubri Morsy, Bashtar, Abdel-Ghaffar, Mehlhorn, Quraishy, Al-Ghamdi, Koura, and Maher 2012	Lizards (Acanthodactylus boskianus, A. scutellatus, A. pardalis)	Macroscopic, 0.95 × 10.1 mm	4–7	2–5.8 μm thick, finger-like and stalkless vp, type 9d	Colubrid snake (Spalerosophis diadema), 10.6 × 9.1	No data	1227
S. ameivamastigodryasi Lainson and Paperna 2000	Giant ameiva lizard (Ameiva ameiva)	1000–2000 × 80	7.2 × 2.5	Similar to type 1b	Colubrid snake (Mastigodryas bifossatus), 10–7.5 × 11.2–8.7	No data	1018
S. atractaspidis Parenzan 1947	Ogaden burrowing asp (Atractaspis leucomelas)	7 × 1 mm	Piriform, 15–25	No data	Unknown	No data	1370
S. bunopusi Abdel-Baki, Abdel-Haleem, and Al-Quraishy 2012	Rock gecko (Bunopus tuberculatus)	42–45 × 22–25	5–7 × 1.5–2.0	Smooth, thin, type 1a	Unknown	No data	4
S. chalcidicolubris Matuschka 1987	Ocellated skink (Chalcides ocellatus)	Microscopic, 124–275 × 27–53	10–12 × 2	Thin, irregular, type 1e	Spotted whip snake (Hemorrhois ravergieri), 9.4–11.7 × 8.2–9.9	No data	1137
S. chamaeleonis Frank 1966	Chameleon (Chamaeleo fischeri)	500–1000 × 15,000	10–13 × 2	Thin (<1 μm), short vp	Unknown	No data	623
S. dugesii Matuschka and Mehlhorn 1984	Madeira wall lizard (Lacerta dugesii)	40–125 × 220–605	12–16 × ?	Thick, 4 μm long vp, type 17	Autophagy, 10.2–10.7 × 8.7–9.2	No data	1132, 1144
S. gallotiae Matuschka and Mehlhorn 1984	Western Canaries lizard (Gallotia galloti)	43–140 × 110–370	16 × 12	Thick, 7 μm long vp, type 37a	Unknown	18S	1132, 1621
S. gongyli Trinci 1911	Ocellated skink (Chalcides ocellatus)	60–100 × 200–900	4.1–5.3 × 0.8–2.0	Striated, 2–6 μm thick, leaf-like 1.9–3.0 μm long vp, type 9d	Diadem snake (Spalerosophis diadema), 9.1 × 10.6	No data	8, 81

(Continued)

Table 22.1 (Continued) Reptiles as Intermediate Hosts for Sarcocystis

Species	Intermediate Host	Size of Sarcocyst (μm)	Size of Bradyzoites (μm)	Sarcocyst Wall	Definitive Host, Sporocysts (μm)	Molecular Data	Reference
S. kinosterni (Lainson and Shaw 1971) 1972	Scorpion mud turtle (Kinosternon scorpioides)	170–230 × 8000	18.4 × 1.7	Smooth, thin-walled (<1 μm)	Unknown	No data	1015, 1017
S. lacertae Babudieri 1932	Eurasian wall lizard (Podarcis muralis)	Macroscopic, 1000 × 1800- 2000	Banana-shaped, 6.5–7.3 × 1.5–2.0	Smooth, 2.5–3.2 μm, type 37a	Smooth snake (Coronella austriaca), 9.0–10.0 × 7.0–8.0	18S	53, 1621, 1798
S. lacertautae (Syn. S. utae) Ball (1944) Lainson and Shaw 1971	Side-blotched lizard (Uta stansburiana)	120 × 950	5.5–7 × 1.5–2	Smooth, thin (<1 μm)	Unknown	No data	1015
S. mitrani Abdel-Ghaffar and AL-Johany 2002	Eastern skink or sand-fish lizard (Scincus mitranus)	Macroscopic, 0.05–0.3 × 0.5–1.8	5.2 × 2.0	Thick, vp up to 3.1 μm long vp, serrated pvm, microtubules in center of vp, type 9d	Unknown	No data	9
S. platydactyli Bertram 1892	Common wall gecko (Tarentola mauritanica)	Macroscopic, 400 × 2000	5–6 × 1.5–2	Striated, thick wall, 7–10 μm	Unknown	No data	1138
S. platydactyliscelopori (Ball 1944) Odening 1998	San Joaquin fence lizard (Sceloporus occidentalis biseriatus)	600 × 180	5.2–6 × 1.5–2	Thick wall (2–8 μm)	Unknown	No data	63, 1317, 1015
S. podarcicolubris Matuschka 1981	Eurasian wall lizards (Podarcis spp., Lacerta spp., Algyroides nigropunctatus)	90–290 × 430–1300	7.7–10 × 2	Striated, type 9m	Whip snakes (Coluber viridiflavus, C. gemonensis, C. jugularis, C. najadum, Macropodon cucullatus), 6.9 × 9.6	No data	1130, 1132, 1134
S. pythonis Tiegs 1931	Python (Morelia spilota)	1.1 mm	Crescent-shaped, 4–7 × ?	Thin	Unknown	No data	1744
S. scelopori Ball 1944 (Lainson and Shaw 1971)	Western fence lizard (Sceloporus occidentalis)	180 × 600 thickness	5.2–6.0 × 1.5–2.0	Striated, 2–8 μm thick wall	Unknown	No data	63, 1015

(Continued)

Table 22.1 (Continued) Reptiles as Intermediate Hosts for Sarcocystis

Species	Intermediate Host	Size of Sarcocyst (µm)	Size of Bradyzoites (µm)	Sarcocyst Wall	Definitive Host, Sporocysts (µm)	Molecular Data	Reference
S. schneideri Bashtar, Al Aal, Maarouf, Morsy, and Quraishy 2014	Barber skink (*Eumeces schneideri*)	Microscopic, 250–900 × 50–100	Banana-shaped, 3–5 × 1.5–2.5	Thin wall (0.28 µm), type 1b	Unknown	No data	82
S. simonyi Bannert 1992	Hierro giant lizard (*Gallotia simonyi*)	Microscopic, 400–720 × 130–220	5.5–6.3 × 2–3	Thin-walled, vp up to 2 µm long, unusual spine-like shape 37a	Autophagy, 7.8–9.4 × 6.2–7.8	No data	66, 67
S. stehlinii Matuschka and Bannert 1989	Gran Canarian giant lizard (*Gallotia stehlini*)	90–400 × 60–160	6–7 × 1.5–2.3	Thick, 11.8–14.7 µm long vp, type 7c	Dihomoxenous, 8.2–9.4 × 5.9–7.0	No data	67, 1145
S. stenodactylicolubris Modrý, Koudela, and Šlapeta 2000	Fan-fingered gecko (*Ptyodactylus guttatus*) (Also: Jordan short-fingered gecko, *Stenodactylus grandiceps*)	175–200 × 35–50	5–6 × 1.5	1–2 µm thick wall, spine-like and lobulated vp, type 4	Dahl's whip snake (*Coluber najadum*), 9–10 × 7–8 (Also: *Coluber rogersi*)	No data	1203
S. turcicii Abdel-Ghaffar, Bashtar, Al-Quraishy, Al Nasr, and Mehlhorn 2009	Mediterranean house gecko (*Hemidactylus turcicus*)	250–800 × 70–90	5–7	Thick, serrated, up to 4 µm long vp, without stalk and folded, type 9d	Unknown	No data	10
S. sp. Roberts, Wellehan, Weisman, Rush, Childress, and Sibley 2014	South American rattlesnake (*Crotalus durissus terrificus*)	? × 50–150	Unknown	Thin, type 1h	Unknown	18S	1474

Table 22.2 Species of Sarcocystis Utilizing Reptiles as Definitive Hosts

Sarcocystis Species	Intermediate Host	Schizonts	Sarcocysts Length, Cyst Wall	Bradyzoites	Maturation Time	Clinical Disease in IH	Definitive Host, Sporocysts (μm)	Molecular Data	References
S. atheridis Šlapeta, Modrý, and Koudela 1999	House mouse (Mus musculus) and Barbary striped mouse (Lemniscomys barbarus)	Liver	Macroscopic, 30 × 0.9 mm, thin wall, type 1b	6–8 × 1–1.5	61–63 dpi	Yes	Nitsche's bush viper (Atheris nitschei), 10.0–11.0 × 7.0–8.5	18S, 28S	**1620, 1623**
S. clethrionomyelaphis Matuschka 1986	Voles (Microtus arvalis, M. oeconomus, M. guentheri, Clethrionomys glareolus)	Hepatocytes, 7–9 dpi	Up to 4.5 mm, 3 μm long vp, smooth, type 9	10–12 × 2 μm	68 dpi	Yes	Aesculapian snake (Zamenis longissimus, and others in the genus), 8.2–9.4 × 11.1–12.3	No data	**1135, 1178**
S. crotali Entzeroth, Chobotar, and Scholtyseck 1985	House mouse (Mus musculus)	Unknown	Up to 4 mm, thin, smooth type 1a	Unknown	Up to 67 dpi	No data	Mojave's rattle-snake (Crotalus scutulatus scutulatus), 7.6–10.4 × 10.4–11.7	No data	**540**
S. dirumpens (Hoare 1933) Häfner and Matuschka 1984	Natal multimammate rat (Mastomys natalensis) and others (Gerbillus perpallidus, Meriones unguiculatus, Mesocricetus auratus, Phodopus sungorus, Mus musculus)	No data	Macroscopic, up to 2.5 cm long, smooth, type 1b	8–9 × 2	Over 60 dpi	No	Snakes of genus Bitis (B. arietans, B. caudalis, B. gabonica, B. nasicornis), 10.6–11.6 × 7.9–8.3	No data	**768**
S. gerbilliechis Jäkel 1995	Gerbils (Gerbillus gerbillus, G. perpallidus, Psammomys obesus, Pachyuromys duprasi)	Endothelium, several organs	Macroscopic, 11.7 × 0.19 mm, smooth, type 1b	8.1 × 2.2	60 dpi	Yes	Arabian saw-scaled viper (Echis coloratus), 12.3–13.3 × 10.7–11.4	No data	**885**

(Continued)

Table 22.2 (Continued)   Species of Sarcocystis Utilizing Reptiles as Definitive Hosts

Sarcocystis Species	Intermediate Host	Schizonts	Sarcocysts Length, Cyst Wall	Bradyzoites	Maturation Time	Clinical Disease in IH	Definitive Host, Sporocysts (µm)	Molecular Data	References
S. hoarensis Matuschka, Mehlhorn, and Abd-al-Aal 1987	Natal multimammate rat (*Mastomys natalensis*) and others (*Gerbillus perpallidus, Meriones unguiculatus, Mesocricetus auratus, Phodopus sungorus, Mus musculus*)	No data	Macroscopic, up to 2.5 mm long, secondary cyst wall, type 42	11–12 × 2	No data	No data	Snakes of genus *Bitis* (*B. arietans, B. caudalis, B. gabonica, B. nasicornis*), 11.7–13.1 × 7.8–9.6	No data	**1129, 1133, 1141**
S. idahoensis Bledsoe 1980	Deer mouse (*Peromyscus maniculatus*)	Hepatocytes	Up to 9.4 mm, thin wall (<l µm)	Crescent-shaped, 2.0 × 6–7	50 dpi	Yes	Gopher snake (*Pituophis melanoleucus*), 11–12 × 13–14	No data	**111, 112, 113, 375**
S. murinotechis Munday and Mason 1980	Norway rat (*Rattus norvegicus*) and other rats of genus *Rattus, Pseudomys* and *Mastacomys*	Endothelium, several organs	Up to 300 µm long, thick-walled, 6 µm long vp, type 9b	No data	90 dpi	Yes	Tiger snake (*Notechis ater*), 10.5–11.5 × 7–8	No data	**1252**
S. muriviperae Matuschka, Heydorn, Mehlhorn, Abd-al-Aal, Diesing, and Biehler 1987	House mouse (*Mus musculus*)	Hepatocytes	Up to 8 mm long, 3.5 µm long vp, type 18b	No data	36 dpi	Yes	Palestinian viper (*Vipera palaestinae*) and Caspian whip snake (*Coluber jugularis*), 8.8–10.5 × 11.7–12.9	No data	**1142, 1363, 1364**

(Continued)

Table 22.2 (Continued)   Species of Sarcocystis Utilizing Reptiles as Definitive Hosts

Sarcocystis Species	Intermediate Host	Schizonts	Sarcocysts Length, Cyst Wall	Bradyzoites	Maturation Time	Clinical Disease in IH	Definitive Host, Sporocysts (μm)	Molecular Data	References
S. nesbitti Mandour 1969	See Chapter 5						Possibly reticulated python (*Braghammerus reticulatus* syn. *Python reticulatus*) and monocled cobra (*Naja kaouthia*)	18S	**1033** For other references, see Chapters 4 and 5
S. singaporensis Zaman and Colley (1975) 1976	Norway rat (*Rattus norvegicus*), bandicoot rats (*Neosokia indica*), roof rats (*R. rattus frugivorous*), Nile grass rats (*Arvicanthis niloticus*)	Endothelium, several organs	Up to 4.4 mm long, striated and thick wall, 3–5 μm thick, type 19	4.5–6.0 × 1.0–1.5	60 dpi	Yes	Python (*Python reticulatus*), 8.4–10.2 × 7.2–8.0	18S	**886, 1033, 1365, 1868, 1870**
S. villivillosi Beaver and Maleckar 1981	Norway rat (*Rattus norvegicus*)	Endothelium, several organs	Up to 1.1 mm long, 1.5 μm thick wall, striated, type 22	Fusiform, 5 × 1	60 dpi	No	Pythons (*Python reticulatus, P. timorensis, P. sebae, Aspidites melanocephalus*), size of sporocysts could not be stated	No data	**85, 767**
S. zamani Beaver and Maleckar 1981	Norway rat (*Rattus norvegicus*)	Endothelium, several organs	Up to 2 mm long, 1–3 μm thick wall, smooth, type 18c	Banana-shaped, 10–12 × 1.5	60 dpi	No	Python (*Python reticulatus*), size of sporocysts could not be stated	No data	**85**
S. zuoi Hu, Ma, and Li 2005	Norway rat (*Rattus norvegicus*)	No data	Up to 2.3 mm, striated wall, 8.5–9.6 μm long vp, type 17	7–10 × 2.0–2.5	No data	No data	King ratsnake (*Elaphe carinata*), 10.8 × 8.0	18S	**853, 855, 1033**

Ogaden burrowing asp (*Atractaspis leucomelas*).[1370] In this case, the author provided information: up to 7 mm long, piriform, and 15–25 μm long bradyzoites. This species was considered a *nomina dubia*[1138] cited that (2) *S. pythonis* Tiegs 1931,[1744] (3) *S. gallotiae* Matuschka and Mehlhorn 1984,[1132] (4) *S. scelopori* Ball 1944, (Lainson and Shaw 1971),[63,1015] and (5) *S. lacertautae* (Syn. *S. utae*) Ball, 1944, Lainson and Shaw 1971.[63,1015]

22.1.2 Reptiles as Definitive Hosts for *Sarcocystis*

Snakes serve as the definitive host for 13 species of *Sarcocystis* which have mammals as intermediate hosts (Table 22.2). The importance of *S. singaporensis*, *S. nesbitti*, and other species is discussed in Chapter 21.

There are 4 additional reports of *Sarcocystis* sporocysts in the feces of snakes, whose final identity was not stated. Sporocysts (no information about size and shape was given) were found in 1 Australian copperhead snake (*Austrelaps superba*) and 4 tiger snakes (*Notechis ater*).[1251] In another case, sporocysts measuring 12.0–13.6 × 10.0–11.2 μm were detected in a Western green rat snake, *Senticolis triaspis intermedia*, from New Mexico.[1149] Oocysts measuring 9–16 × 5.5–12 μm, suspected to be *S. idahoensis*, were found in a bullsnake, *Pituophis melanoleucus sayi*, in London via Tennessee.[295] Sporocysts measuring 11.9 × 10.3 μm were detected in 2 New Mexico ridgenose rattlesnakes, *Crotalus willardi obscurus*, from Sonora, Mexico.[1150]

22.1.3 Molecular Tools for *Sarcocystis* Species in Reptiles

Initial molecular assays on *Sarcocystis* from reptiles were carried out by Doležel et al.,[330] who discussed the coevolution of *Sarcocystis* spp. in their final hosts. It was later hypothesized that despite similar 18S rRNA gene sequences for *S. gallotiae* and *S. lacertae*, their dihomoxenous and heteroxenous life cycles differ because *S. gallotiae* evolved in a snake-free environment in the Canary Islands.[1621] Evolutionary relationships were shown among *Sarcocystis* spp. in 7 endemic African tree vipers (*Atheris* spp., and *Bitis* spp.) that are all monophyletic.[1623] Bioassays in different rodent species were most successful in laboratory mice. In a captive Indian cobra (*Naja naja*) from China, ITS-1 sequencing detected a *Sarcocystis* species closely related to species with rodents as intermediate hosts (*S. muris*, *S. singaporensis*, and *S. zuoi*).[942] Based on 18S rRNA gene sequences from sporocysts excreted by different Malaysian snakes, *S. nesbitti* was detected in a monocled cobra, *Naja kaouthia*, and a reticulated python, *Braghammerus reticulatus* syn. *Python reticulatus*; *S. singaporensis* was identified in a reticulated python and a Malayan brown pit viper, *Ovophis convictus*; *S. zuoi* was identified in a Malayan keeled rat snake, *Ptyas carinata*; and an unidentified *Sarcocystis* sp. was detected in a reticulated python.[1033] Based on 18S rRNA gene sequences from sporocysts from green pythons (*Morelia viridis*), 2 different *Sarcocystis* spp. were detected with 95%–97% sequence identity with *S. singaporensis*.[1219]

22.2 FISHES

S. salvelini Fantham and Porter, 1943, was reported from the trout, *Salvelinus fontinalis*.[566] Sarcocysts were observed only once among several hundred trout examined in eastern Canada.[566] Sarcocysts up to 0.5 mm long were seen as whitish threads in the abdominal muscles. The bradyzoites were 5.2–8.8 × 1.5–2.5 μm. No other information is available.

Also in Canada, sarcocysts of an unnamed species were found in an eel (*Zoarces angularis*).[566] They were up to 1 mm long and zoites were up to 15 μm long and up to 3.5 μm wide. No other information is available. In 7 fish species examined in Australia, *Sarcocystis* was not found.[1251]

BIBLIOGRAPHY

Reptiles as intermediate hosts: **4, 8, 9, 10, 53, 63, 66, 67, 81, 82, 623, 1015, 1017, 1018, 1130, 1132, 1134, 1137, 1138, 1144, 1145, 1203, 1227, 1370, 1474, 1621, 1744, 1798**.

Reptiles as definitive hosts: **85, 111, 112, 113, 295, 375, 540, 767, 768, 853, 855, 885, 886, 1129, 1133, 1135, 1138, 1141, 1142, 1149, 1150, 1178, 1251, 1252, 1363, 1364, 1365, 1620, 1868, 1870, 2270**.

Other references: **330, 566, 942, 1033, 1219, 1621, 1623, 2175**.

CHAPTER 23

Genetics, Phylogeny, and Molecular Epidemiology of *Sarcocystis*

23.1 INTRODUCTION

Sarcocystis species, especially *S. neurona*, are morphologically and biologically related to the coccidians *T. gondii* and *N. caninum* that are economically important pathogens. For every paper indexed in PubMed (http://www.ncbi.nlm.nih.gov/pubmed) from the last 10 years referencing any species of *Sarcocystis*, more than 10 appeared that referenced *T. gondii*. As explained in Chapter 3, the vast amount of research on *T. gondii* has facilitated investigations on genetics of *S. neurona*. The progress with the genetics of *S. neurona* is summarized in Chapter 3.

23.2 GENE REGULATION OF METABOLIC PATHWAYS

The first successful genetic transfection of *S. neurona* was accomplished using electroporation of a plasmid-encoded transgene encoding a selectable marker (pyrimethamine resistance) and reporter molecules (beta-galactosidase and yellow fluorescent protein).[656] This system was not, however, subsequently embraced by a large research community. However, this achievement was built upon by an approach exploiting the dependency of such parasites on host-derived purines to construct a system to systematically study gene function through positive and negative selection. By rendering parasites deficient in an enzyme (hypoxanthine-xanthine-guanine phosphoribosyltransferase) that mediates purine salvage, a line of *S. neurona* absolutely dependent on a second enzyme was made that performed this essential metabolic function.[293] Inhibiting this second pathway renders parasites entirely dependent on functional complementation with a plasmid encoding the first, providing a means to select for parasites containing this "rescue" plasmid. By altering the plasmid to convey other genes, this system can be flexibly deployed to create and isolate parasites harboring specific genetic traits. The phenotypic effects of any such traits are thereby amenable for study, opening new vistas in our biological understanding.

The ability to manipulate and study the effects of particular genes could not have come at a more opportune moment, given newfound knowledge of the composition of the *S. neurona* genome. For example, 6 rhoptry kinases were identified in the *S. neurona* genome when surveying members of this important family of virulence factors.[1719] This is an important observation because the merozoites of *Sarcocystis*, including *S. neurona*, lack rhoptries. Until now, it has not been easy to study, in any species of *Sarcocystis,* the biological functions mediated by any such molecule; the magnitude of this missed opportunity is clear in view of experiments using *T. gondii* that have implicated such molecules as key regulators of host metabolism and immunology and essential to establishing intracellular parasitism. Differential expression of certain rhoptry kinases explain a great deal of the variation in virulence resulting from infection in mice and may underlie evident differences in the clinical spectrum of *S. neurona* infections in animals.

Developing insights in model systems are needed, given how difficult and expensive it is to perform clinical research in horses or sea otters (see Chapter 3). Pigeon protozoal myeloencephalitis (PPE), caused by infections with *S. calchasi* (Chapter 17), may provide a valuable basis for exploring the pathobiology of neurological sarcocystosis. In a recent PPE study, immune evasion and delayed-type hypersensitivity were identified in hosts developing extensive cerebral lesions.[1331] During the early phases of schizogony, the parasites appeared to suppress IL-12, IL-18 and interferon in their hosts. Later, as disease progressed, cytokines induced by interferon gamma and tumor necrosis factor alpha increased accompanied by increased expression of MHC-II genes and the infiltration of various T cells, B cells, and macrophages.[1331]

Species of *Sarcocystis* modulate host immunity. For example, studies[1110,1240] have resolved structures for homologous microneme proteins secreted, during host cell invasion, by *T. gondii* and *S. muris*, respectively. In *T. gondii,* the molecule plays an important role in binding glycans expressed on the surface of host cells that enable the parasite to "pull itself" into the host cell by engaging the host cell's actin/myosin motor. Because this molecule may compete with host lectins (termed galectins because they preferentially bind terminal galactose residues) that regulate cell adhesion, cell signaling, and apoptosis, Marchant et al.[1110] proposed that "proteolytic maturation of TgMIC4 provides a mechanism to liberate a soluble galectin-like lectin, which could subsequently contribute independently to parasite dissemination or down-regulation of the host immune response." Indeed, galectins and their ligands have been termed "master regulators of immune cell homeostasis."[1436] Related data from *Neospora caninum* suggest that galactose recognition by these microneme proteins may be a common strategy by which tissue-cyst forming coccidia regulate host immunity and, specifically, forestall destruction and elimination of parasitized cells by cytotoxic lymphocytes.[1864] Similarly, enzymes that degrade extracellular nucleoside triphosphates (NTPases) regulate purigenic signaling, thereby modulating immune responses,[1532] and homologous versions of 1 such gene (TgNTP1 and SnNTP1) have been identified and shown to be functional in *T. gondii* and *S. neurona*.[1876]

The function of other structural elements of species of *Sarcocystis* is suggested by experiments in the model system (*T. gondii*) and may now be subjected to direct examination in a species of *Sarcocystis*. For example, a dense granule protein (DGP) was identified that may be important in the formation and/or maintenance of the parasitophorous vacuole in which *T. gondii* lives.[1124] Sequence homology suggests that this protein (TgGRA23) shares common origins (and, perhaps, a common function) with a DGP (SmDG32) previously characterized in *S. muris*. The availability of a comprehensive catalogue of such DGPs (furnished by emerging genome projects) and the newfound ability to study the independent phenotypic effects of specified genes illustrate the opportunity that awaits those curious to understand the cell biology, molecular biology, and pathobiology of sarcocystosis. In another example, stable fluorescent fusion proteins were used to localize a protein (named TgRNG1 in *T. gondii*, SnRNG1 in *S. neurona*) to the apical polar ring, a structure which organizes microtubules at the terminus of the inner membrane complex.[1758] Beyond this apical ring, characteristic of and unique to coccidian parasites, the parasite is bounded by only 1 membrane; apical secretion of molecules vital to host cell invasion occur only through the "extreme apex."

Defining surface proteins, as has been done in *S. neurona*, in other pathogenic *Sarcocystis* species might help in developing diagnostic tests and immunoprophylaxis (see Chapter 3).[263,498,668,850] Differences in the expression of shared genes contribute to the distinctive attributes of various species of *Sarcocystis*. For example, antibodies raised to the enolase present in various stages of *T. gondii* also recognize the enolase of *S. neurona* but highlight distinct patterns of stage-specific expression in these parasites.[123] This underscores the need not only to identify which genes are (or are not) shared among various species of parasites, but to also study their differential expression and regulation as they define myriad and unique biological outcomes.

Molecular approaches seeking to understand what makes each parasitic species unique will also undoubtedly encounter genes that are unique to the particular taxon. For example, libraries of expressed sequence tags from *S. neurona* include numerous sequences with no known homolog in *T. gondii* or elsewhere. One such protein is expressed prominently on the surface of merozoites (SnSPR1) and mediates attachment to host cells. It has no known homolog in any other parasite yet examined; in particular, it should not be confused with surface antigens collectively ascribed to the SAG family.[1877] Logically, sequencing the genomes of more species of *Sarcocystis* will be necessary in order to understand how many genes, and what kinds of genes, are truly unique to each.

Comparing the fully sequenced genomes of *S. neurona* and *T. gondii* identifies broad conservation of metabolism, GRA proteins, ROP kinases, and SRS family adhesions, with interesting differences such as the absence in *S. neurona* of certain ROP kinases especially implicated as virulence factors in *T. gondii* (i.e., ROP 5, 16, and 18). When considering other genomes, including that for *N. caninum,* another tissue cyst-forming coccidia closely related to *T. gondii,* and *Eimeria tenella,* a coccidia that completes its life cycle in a single host and does not form tissue cysts, 18% of the predicted proteins were found to be unique to *S. neurona,* another 18% were unique to tissue cyst forming coccidia, and 10% were shared with these and with more distantly related apicomplexans (belonging to *Plasmodium, Theileria,* and/or *Cryptosporidium*).[109] As genomes of additional species accumulate, we will gain a better understanding of the drivers and consequences of parasite diversification and uncover new means of understanding and managing these infections.

23.3 GENETIC MARKERS THAT HAVE BEEN EMPLOYED FOR PHYLOGENETIC AND DIAGNOSTIC PURPOSES

23.3.1 18S rDNA

For most species of *Sarcocystis,* at most 1 gene has been sequenced (the gene encoding the small subunit ribosomal RNA [18S rDNA]) and for 1 purpose (to ascribe a genetic signature so that the species might be compared to others). Ribosomal RNA has many attractive attributes, and a few drawbacks, for use in characterizing and comparing biological lineages. Advantages include the fact that it is ubiquitous, portions of the molecule are so highly conserved among related groups of organisms as to allow easy application of conserved PCR primers, and portions of the molecule are more variable, allowing distinction among related biological species and lineages. Disadvantages include the possibility that closely related species, although phenotypically distinct, may differ little or not at all in this locus, and the possibility that distinctions among versions of the gene harbored by a given lineage may confound attempts to compare 1 parasite to another. Finally, the vastly differing rates of evolution among portions of the molecule raise questions as to how to model the evolutionary process when seeking to infer phylogenetic relationships from extant variation in such sequences.

In spite of such complications, 18S remains obvious choice when first characterizing an isolate of *Sarcocystis* owing to its relative ease of amplification and, moreover, owing to the richness of the available reference set of sequences. Table 23.1 summarizes the literature employing this molecule to render diagnoses and to infer phylogenetic relationships. In spite of benefits that might have been offered by alternative loci, the history of the field dictates that new species descriptions should, whenever possible, include 18S rDNA. So too, new 18S sequences stand the greatest chance of being employed in future comparative studies. Nonetheless, science advances when hypotheses withstand independent validation, and phylogenetic hypotheses are no exception. To date, that concept has been evaluated in *Sarcocystis* primarily by subjecting the same 18S data to independent tree-building approaches. Sampling other data, however, would provide even more compelling tests of such relationships, because results from any given set of data can engender an undeserved sense of certainty.

Table 23.1   Studies of 18S rDNA Variation in Species of *Sarcocystis*

Hosts	Parasite	Year	Reference
Mouse (*Mus musculus*)	*S. muris*	1991	653
Mouse (*Mus musculus*)	*S. muris*	1991	1732
Mouse (*Mus musculus*)	*S. muris, S. gigantea*	1992	655
Various	Various	1993	144
Cattle (*Bos taurus*), sheep (*Ovis aries*)	*S. cruzi, S. tenella, S. fusiformis, S. gigantea*	1993	840
Cattle (*Bos taurus*)	*S. cruzi*	1994	1095
Cattle (*Bos taurus*)	*Neospora caninum*	1994	494
Various	*Neospora caninum* and *T. gondii*	1994	841
Sheep (*Ovis aries*)	*S. tenella, S. arieticanis, S. gigantea,* and *T. gondii*	1994	1736
Horse (*Equus caballus*)	*S. neurona*	1994	600
Opossum (*Didelphis virginiana*)	*S. neurona*	1995	601
Brown-headed cowbird (*Molothrus ater*) and Moluccan cockatoo (*Cacatua moluccensis*)	*S. neurona, S. falcatula*	1995	290
Sheep (*Ovis aries*)	*S. arieticanis, S. gigantea, S. tenella*	1995	497
Various	Various	1995	495
Horse (*Equus caballus*)	*S. neurona*	1996	1113
Various	Various, compared to *Eimeria*	1997	77
Goat (*Capra hircus*)	*S. capricanis* and *S. moulei*	1997	904
Cattle (*Bos taurus*), dwarf zebu (*B. taurus*), and bison (*Bison bison*)	*S. hirsuta, S. cruzi, S. hominis*-like	1998	608
Rodents	Synonymize *Frankelia* with *Sarcocystis*	1998	1799
Various	Sarcocystidae include *Sarcocystis* spp. and *Isospora* spp., *Neospora* spp., and *Toxoplasma* sp.	1998	166
Macaque (*Macaca mulatta*)	*S.* sp.	1998	1020
Various ruminants	*Sarcocystis* infecting ruminants are divided between those having cats and dogs as definitive hosts	1999	843
Mouse (*Mus musculus*), barn owl (*Tyto alba*), and Nitsche's bush viper (*Atheris nitschei*)	*Sarcocystis* sharing a type of final host are especially closely related	1999	330
Various	*Hammondia hammondi* is similar to *T. gondii, Frenkelia* are *Sarcocystis*	1999	906
Nitsche's bush viper (*Atheris nitschei*)	*S. atheridis*	1999	1620
Sheep (*Ovis aries*)	*S. tenella, S. arieticanis*	1999	788
Various	Effects of 18S sequence alignment	2001	1237
Various	Relationship of *Lankesterella* and *Caryospora* to other Sarcocystidae	2001	78
Various	Dihomoxenous life cycle evolved more than once	2001	1621
Cattle (*Bos taurus*) and water buffalo (*Bubalus bubalis*)	*S. cruzi*	2001	1856
Water buffalo (*Bubalus bubalis*)	*S. hominis*-like	2001	1857
Cat (*Felis catus*)	*S. felis*	2003	684
Snakes and rodents	*S. singaporensis*	2002	1622
Water buffalo (*Bubalus bubalis*)	*S. cruzi*	2002	1056
Various	Various: PCR followed by RFLP	2002	1858

(Continued)

Table 23.1 (*Continued*) Studies of 18S rDNA Variation in Species of *Sarcocystis*

Hosts	Parasite	Year	Reference
Vipers	Various	2003	**1623**
Opossum (*Didelphis virginiana*)	*S. neurona, S. inghami*	2004	**513**
Dog (*Canis familiaris*)	*S. neurona*-like	2005	**1780**
Horses and opossums (*Didelphis virginiana*)	*S. neurona*	2005	**518**
White-fronted goose (*Anser albifrons*)	*S.* sp.	2006	**1003**
Reindeer (*Rangifer tarandi*)	Various	2007	**280**
Cattle (*Bos taurus*)	*S. hominis*	2007	**1776**
Reindeer (*Rangifer tarandi*)	Various	2007, 2008	**281, 283**
Moose (*Alces alces*)	*S. alces, S. ovalis, S. scandinavica, S. alceslatrans, S. silva,* and *S.* sp.	2008	**282**
Cattle (*Bos taurus*)	*S. cruzi*	2008	**1427**
Cattle (*Bos taurus*) and water buffalo (*Bubalus bubalis*)	*S. cruzi, S. hominis, S. hirsuta, S. fusiformis*	2009	**905**
Southern sea otters (*Enhydra lutris*)	*S. neurona*	2009	**1196**
Human (*Homo hominis*), cat (*Felis catus*), and dog (*Canis familiaris*) definitive hosts	*S. hominis, S. fusiformis,* and *S. cruzi*	2009	**1844**
Raccoons (*Procyon lotor*)	*S.* sp.	2009	**909**
Racing pigeon (*Columbia livia*)	*S.* sp.	2009	**1325**
Buteo spp. buzzards and *Accipiter* spp. hawks	*S.* sp.	2009	**1851**
Hooded crow (*Corvus cornix*)	*S. cornixi*	2009	**1005**
Roe deer (*Capreolus capreolus*)	*S. gracilis, S. oviformis, S. silva*	2008, 2014	**284, 969**
Raccoon dogs (*Nyctereutes procyonoides*)	*S.* sp.	2010	**983**
Sparrow hawks (*Accipiter nisus*)	*S. calchasi*	2010	**1327**
Red deer (*Cervus elaphus*)	*S. hjorti, S. hardangeri, S. ovalis, S. rangiferi,* and *S. tarandi*	2010	**285**
Wolverine (*Gulo gulo*)	*S. kalvikus, S. kitikmeotensis*	2010	**467**
Red (*Vulpes vulpes*) and arctic fox (*V. lagopus*)	*S. alces, S. hjorti*	2010	**286**
Striped skunk (*Mephitis mephitis*)	*S. neurona*	2010	**153**
Water buffalo (*Bubalus bubalis*)	*S. cruzi*	2010	**1846**
Mallard duck (*Anas platyrhynchos*)	*S. rileyi*	2010	**466**
Corvid birds (Corvidae)	*S. ovalis*	2010	**699**
Racing pigeons (*Columbia livia*) and goshawks (*Accipiter* spp.)	*S. calchasi*	2010	**1329**
Racing pigeons (*Columbia livia*) and goshawks (*Accipiter* spp.)	*S. calchasi, S. columbae,* S. sp.	2010	**1326**
Ferret (*Mustela putorius furo*)	*S. neurona*	2010	**145**
Black bear (*Ursus americanus*)	*S. canis*-like	2011	**305**
Herring gull (*Larus argentatus*)	*S. wobeseri*	2011	**1415**
Mallard duck (*Anas platyrhynchos*)	*S. rileyi*	2011	**1007**
Wood pigeon (*Columba palumbus*)	*S. columbae*	2011	**1416**
Water buffalo (*Bubalus bubalis*)	*S. rommeli*	2011	**211**
Cattle (*Bos taurus*)	*S. hominis*-like	2011	**331**
Water buffalo (*Bubalus bubalis*)	*S. fusiformis*	2011	**1340**

(*Continued*)

Table 23.1 (*Continued*)   Studies of 18S rDNA Variation in Species of *Sarcocystis*

Hosts	Parasite	Year	Reference
Fischer *Martes pennanti*	*S.* sp.	2011	**1025**
Camel *Camelus dromedarius*	*S.* sp.	2011	**1230**
Swine (*Sus scrofa*)	*S. miescheriana*	2011	**168**
Swine (*Sus scrofa*)	*S. miescheriana*	2011	**956**
Goat (*Capra hircus*)	*S. moulei*	2010	**1229**
Rhesus monkey (*Macaca mulatta*)	*S. nesbitti*	2012	**1743**
Roe deer (*Capreolus capreolus*)	*S. capreolicanis, S. silva*	2012	**700**
Wall lizard (*Podarcis* sp.)	*S. gallotiae*	2012	**783**
Woodrat (*Neotoma micropus*)	*S. neotomafelis*	2012	**198**
Jackdaw (*Corvus monedula*)	*S. corvusi*	2013	**1419**
Various psittacine species	*S. calchasi*	2013	**1473**
Cat (*Felis catus*), house mouse (*Mus musculus*)	*S. muris*	2013	**27**
Swine (*Sus scrofa*)	*S. miescheriana*	2013	**1854**
Llama (*Llama glama*) and alpaca (*Vicugna pacos*)	*S. aucheniae*	2013	**164**
Water buffalo (*Bubalis bubalis*)	*S. cruzi*	2013	**1029**
Human (*Homo sapiens*)	*S. nesbitti*	2013	**14**
Great black-backed gull (*Larus marinus*)	*S. lari*	2014	**1420**
Dog (*Canis familiaris*)	*S. caninum, S. svanai*	2014	**476**
Oriental vole (*Eothenomys miletus*)	*S. eothenomysi*	2014	**856**
Sheep (*Ovis aries*)	*S. gigantea*	2014	**775**
Beef (*Bos taurus*)	*S.* sp.	2014	**1221**
Feral swine (*Sus scrofa*)	*S. miesheriana*	2014	**161**
African buffalo (*Syncerus caffer*)	*S. cafferi*	2014	**473**
Stellar sea lion (*Eumetopias jubatus*)	*S. canis*	2014	**1818**
Human (*Homo sapiens*)	*S. nesbitti*	2014	**1034**
Malaysian snakes (*Naja* sp., *Braghammerus* sp., *Ovophis* sp., *Ptyas* sp.).	*S. nesbitti, S. singaporensis, S. zuoi,* and *S.* sp.	2014	**1033**
Ducks (*Anas* spp.)	*S. rileyi*	2014	**1421**
Sheep (*Ovis aries*)	*S. tenella*	2014	**2402**
Eurasian otter (*Lutra lutra*)	*S. lutrae*	2015	**706**
Common moorhen (*Gallinula chloropus*)	*S. chloropusae*	2015	**490**
Wood bison (*Bison bison athabascae*)	*S. cruzi*	2015	**162**
Bobcat (*Felis rufus*)	*S. neurona*	2015	**1788**
Cattle (*Bos taurus*)	*S. cruzi, S. hominis, S.* sp.	2015	**1143**
Goat (*Capra hircus*)	*S. capracanis*	2015	**1400**
Cattle (*Bos taurus*)	*S. rommeli*	2015	**2392**
Opossum (*Didelphis* spp.)	*S. speeri*	2015	**1432**

Overreliance on the 18S marker, in particular, masks true differences between closely related lineages and confounds attempts to resolve their order of evolutionary branching. Controversies regarding the validity of distinctions between phenotypically and biologically distinctive parasites in this genus (see, e.g., *S. neurona* vs. *S. falcatula*) illustrates how 18S can fail to provide researchers sufficient resolution when attempting to differentially diagnose lineages that deserve independent recognition (based, e.g., on distinct host ranges).[271,290,1722] Moreover, ribosomal DNA is subjected to vastly different functional constraints in its active sites, its structure-defining and base-paired stems, and certain variable loops. Such complexities justify meaningful debate as to how best to align homologous positions and model the process of evolutionary change.[495,1237] Structural models of the ribosomal RNA can help ensure the accurate comparison of evolutionarily homologous nucleotides and have been shown to be capable of improving phylogenetic analyses of *Sarcocystis* rDNA; until that process can be validated and automated, algorithmic approaches to sequence alignment that do not take into account structural information are likely to continue to play an important role in phylogenetic inference with only limited reduction in the power to resolve well-supported relationships.

What should be used to augment the information provided by sequencing 18S? The publication of genome data provides a nearly limitless menu of possibilities. However, only several loci have so far been used as a basis by which to compare species, populations, or individual isolates (reviewed in Reference **1678**). Table 23.2 summarizes research characterizing variation in loci other than the small subunit rDNA gene. These include genes encoded in the apicoplast genome (its ribosomal RNA and the beta subunit of the RNA polymerase II gene), in the mitochondrial genome (cytochrome b), and the nuclear genome (the large subunit ribosomal RNA gene/28S, various microsatellite markers, and cloned fragments of anonymous DNA derived initially from RAPD DNA profiling). With some justification, the extant variation among those markers is presumed not to influence the survival of the parasites (because functionally consequential mutations likely so disfavor survival and reproduction as to lead to their hasty demise). In contrast, variation in surface antigens likely reflects not only historical subdivision but also ongoing selection for particular function, increasing its interest for those studying the host–parasite interaction but decreasing its reliability as a surrogate for genome evolution, writ-large. None of these is as easy to initially characterize as 18S rRNA, but each affords access to more rapidly evolving portions of the genome.

23.3.2 Other Nuclear Genes

Few other genes have been characterized from multiple species of *Sarcocystis*. Bands originally obtained from RAPD PCRs were cloned and sequenced in efforts to help discriminate *S. neurona* from *S. falcatula*.[1722] Limited work was done to characterize surface antigens and microneme proteins from *S. muris*.[633] The internal transcribed spacers of rDNA (ITS1 and ITS2) have been employed in several instances to discriminate among species of *Sarcocystis* that are identical in 18S rDNA (see Table 23.2). The larger subunit rDNA (28S) encompasses phylogenetically informative variation that augments information derived from the small subunit.[1235] With the publication of the *S. neurona* genome, this limited repertoire has been inundated with thousands of annotated genes, any subset of which may prove useful in understanding the identities of and relationships among species of *Sarcocystis*.

23.3.3 Plastid Genes

A defining feature of the phylum Apicomplexa is its plastid, an organelle (related to the chloroplasts of plants) derived from alpha-proteobacteria. This organelle contributes to the distinctive metabolism of these parasites and possesses a genome derived from the formerly free-living ancestor presumed to have been acquired via an ancient symbiosis. That genome encodes its own transcriptional machinery, as well as certain enzymes, in species of *Sarcocystis*, and these have been explored for their contribution to understanding relationships among diverse species. When seeking to describe phylogenetic relationships among a wide sample of Apicomplexa, a study concluded that the plastid small subunit RNA afforded more resolution than did the corresponding sequences derived from the nucleus.[1301] Similarly, the beta subunit of the plastid-encoded RNA polymerase gene (rpoB) has been used a basis to explore the diversification history of parasite species.[453,902]

Table 23.2   Analyses of Variation in Other Molecular Markers

Host	Parasite	Locus	Year	Reference
Cattle (*Bos taurus*)	*Neospora caninum* and *T. gondii*	ITS-1, 5.8S rDNA	1996	**842**
Various	Various	RNA Polymerase B (RPOb)	1996	**902**
Sheep (*Ovis aries*)	*S. gigantea*	ITS-1	1996	**903**
Cattle (*Bos taurus*)	*Neospora caninum*	ITS-1	1997	**1675**
Various	*S. neurona, S. falcatula*	ITS-1	1999	**1116**
Various	*S. neurona, S. falcatula*	Cloned RAPD bands	1999	**1722**
Opossums (*Didelphis marsupialis* and *Didelphis albiventris*)	*S. falcatula*	Cloned RAPD band	2001	**441**
Various	*Sarcocystis* and *Frenkelia* should be synonymized	28S	1999	**1236**
Cattle (*Bos taurus*), wildebeest (*Connochaetes* spp.) and goats (*Capra hircus*)	Relationship of *Besnoitia* spp. to Toxoplasmatinae	18S, ITS-1	2000	**496**
Pacific harbor seal (*Phoca vitulina richardsi*)	*S. neurona*	18S, ITS-1	2001	**1194**
Opossum (*Didelphis albiventris*)	*S. lindsayi*	28S, ITS-1	2001	**433**
Opossum (*Didelphis virginiana*)	*S. neurona*	Cloned RAPD band	2001	**442**
Opossum (*Didelphis virginiana, D. albiventris,* and *D. marsupialis*)	*S. neurona*	Cloned RAPD band	2001	**1492**
Various	Various	Apicoplast rDNA	2002	**1301**
Southern sea otter (*Enhydra lutris*)	*S.* sp.	Plastid-encoded RNA polymerase	2003	**453**
African tree viper (*Atheris* spp., *Bitis* spp., *Proatheris* sp.)	*S. atheridis* and other unknown genotypes	18S, 28S	2003	**1623**
Cell cultures	Various	18S, hsp70	2006	**520**
Various	*S. neurona, S. falcatula*	Microsatellites	2006	**45, 44**
Dog (*Canis familiaris*)	Various	18S, ITS-1	2006	**460**
African gray parrot (*Psittacus erithacus*)	*S.* sp.	18S, ITS-1	2006	**459**
Pig (*Sus scrofa*) and cattle (*Bos taurus*)	*S. hominis, S. suihominis*	ITS1, ITS-2	2006	**719**
Cat (*Felis catus*)	*S. felis*	18S, ITS-1	2006	**521, 519**
Rhesus macaque (*Macaca mulatta*)	*S.* sp.	18S, ITS-1	2007	**723**
Brown-headed cowbird (*Molothrus ater*)	*S. neurona, S. falcatula*	RAPD bands	2008	**1108**
Various	*S. neurona*	Microsatellites	2008	**1693**
Cattle (*Bos taurus*)	*S. cruzi*	18S, ITS-1	2008	**1493**
Goldeneye (*Bucephala clangula*) and mallard duck (*Anas platyrhyncos*)	Various	28S	2008	**1004**
Hooded crow (*Corvus cornix*)	*S. cornixi*	18S, 28S	2009	**1005**
Australian glider (*Petaurus australis*)	*S.* sp.	18S, 28S	2009	**1879**
Racing pigeon (*Columbia livia*)	*S. calchasi*	28S, ITS-1	2009	**1325**
Southern sea otter (*Enhydra lutris*)	*S. neurona*	18S, RPOb, CO1, SnSAG 1–5, microsatellite Sn9	2010	**1821**

(Continued)

Table 23.2 (*Continued*)   Analyses of Variation in Other Molecular Markers

Host	Parasite	Locus	Year	Reference
Southern sea otter (*Enhydra lutris*)	*S. neurona*	ITS-1, 25/396 RAPD marker, SnSAG 2, 3, 4, microsatellites	2010	**1464**
Southern sea otter (*Enhydra lutris*)	*S. neurona*	ITS-1	2009, 2010	**1197, 1196**
Wolverine (*Gulo gulo*)	*S. kalvikus, S. kitikmeotensis*	18S, ITS-1	2010	**467**
Wood pigeon (*Columbia palumbus*) and sparrowhawk (*Accipiter nisus*)	*S. calchasi, S. columbae*	18S, 28S, ITS-1	2010	**1327**
Racing pigeon (*Columbia livia*) and goshawk (*Accipiter* spp.)	*S. calchasi, S. columbae, S. sp.*	18S, 28S, ITS-1	2010	**1326**
Mallard (*Anas platyrhyncos*)	*S. rileyi*	18S, 28S, ITS-1	2010	**466**
Racing pigeon (*Columbia livia*) and goshawk (*Accipiter* spp.)	*S. calchasi*	18S, 28S	2010	**1329**
Mallard (*Anas platyrhyncos*)	*S. wobeseri*	18S, 28S, ITS-1	2010	**1006**
Domestic pigeon (*Columbia livia*)	*S. calchasi*	28S, ITS-1	2011	**1843**
Herring gull (*Larus argentatus*)	*S. wobeseri*	18S, 28S, ITS-1	2011	**1415**
Accipiter hawk (*Accipiter g. gentilis, Accipiter n. nisus*)	*S. calchasi*	ITS-1	2011	**1330**
Mallard duck (*Anas platyrhynchos*)	*S. rileyi*	18S, 28S, ITS-1	2011	**1007**
Common blackbird (*Turdus merula*)	*S. turdusi*	18S, 28S, ITS-1	2012	**1009**
White-fronted goose (*Anser albifrons*) and mallard duck (*Anas platyrhynchos*)	*S. albifronsi, S. anasi*	18S, 28S, ITS-1	2012	**1008**
Jackdaw (*Corvus monedula*)	*S. corvusi*	18S, 28S, ITS-1	2013	**1419**
Various psittacine species	*S. calchasi*	18S, ITS-1	2013	**1473**
Duck (*Anas* spp.)	*S. rileyi*	18S, 28S, ITS-1	2014	**1421**
Cockatiel (*Nymphicus hollandicus*) and North American rock pigeon (*Columba livia f. dom.*)	*S. calchasi*	ITS-1	2014	**1332**
Great black-backed gull (*Larus marinus*)	*S. lari*	18S, 28S, ITS-1	2014	**1420**
Moose (*Alces alces*)	*S. alces, S. alceslatrans, S. ovalis, S. taeniata*	18S, cox1	2014	**702**
Arctic fox (*Vulpes lagopus*)	*S. arctica*	18S, 28S, ITS-1, and cox1	2014	**705**
Red deer (*Cervus elaphus*)	*S. elongata, S. truncata*	cox1	2014	**704**
Common eider (*Somateria mollissima*)	*S. rileyi*	18S, 28S, ITS-1, cox1	2014	**703**
Dog (*Canis familiaris*)	*S. caninum, S. svanai*	18S, RPOb	2014	**476**
African buffalo (*Syncerus caffer*)	*S. cafferi*	18S, cox1	2014	**473**
Eurasian otter (*Lutra lutra*)	*S. lutrae*	18S, 28S, ITS-1, cytb, cox1	2015	**706**
Common moorhen (*Gallinula chloropus*)	*S. chloropusae*	18S, 28S, ITS-1	2015	**490**
Bobcat (*Felis rufus*)	*S. neurona*	18S, ITS-1	2015	**1788**
Fox (*Vulpes vulpes*), racoon dog (*Nyctereutes procyonoides*)	*S. rileyi*	ITS-1	2015	**2398**
Opossum (*Didelphis* spp.)	*S. speeri*	18S, 28S, ITS-1,	2015	**1432**

23.3.4 Mitochondrial Genes

In recent years, the cytochrome c oxidase subunit 1 (cox1) has been promoted as a means to differentiate among various species of *Sarcocystis*.[473,702,1477,1941] This approach mirrors broader efforts in the biological community to "barcode" species with greater resolution than that which is possible using ribosomal RNA alone. Initiated by entomologists and subsequently embraced by a broad spectrum of biologists, this essential gene (whose product mediates a step in the respiratory chain of oxidative phosphorylation in all aerobic life forms) is comprised of codons whose variation tends to be restricted to synonymous sites (that leave the resulting amino acid sequences unchanged, preserving enzymatic function). Variation in this locus has been used to examine intra- and interpopulation subdivision in metazoa, as well as descent relationships among closely related organisms; mutational saturation at the limited subset of mutable sites, however, obscures efforts to recover evolutionarily deeper relationships at such loci.

23.3.5 Surface Antigens

The surface of *Sarcocystis* parasites includes gpi-anchored glycoproteins that engender immunological responses. It has been established that a given surface antigen in the SAG family can harbor any 1 of several, markedly distinct alleles, and that, when considering the SAG family as a whole, these alleles combine to form distinctive assemblages of surface antigens.[1209] "Swapping" of alleles at these loci has evidently diversified the surfaces of parasites derived from *S. falcatula*. Although these differ one to another, the pool of alleles from which they sample appears distinct from the pool of alleles used by sympatric parasites attributed, by contrast, to *S. neurona*. These, in turn, consistently differed (but to a lesser extent) from alleles characteristic of *S. neurona* in North America. Taken together, these data illustrate a dynamic process of allele swapping and undermine variation in any given SAG locus as the basis by which to derive inferences about parasite phylogeny.

23.3.6 Microsatellites

Short nucleotide repeats can provide a rich source of allelic variation with which to characterize the structure of biological populations and have been used to some extent in examining the connections between populations of *S. neurona*.[44,45,1693 1821] These have identified only limited differentiation among candidate populations in the United States and have affirmed that infections in terrestrial and marine mammals share a common source of infection.

23.4 EPIDEMIOLOGICAL INSIGHTS DERIVED FROM MOLECULAR MARKERS

Molecular markers have contributed greatly to our understanding of the diversity and transmission of species of *Sarcocystis* and have become established as essential to future progress in this area. Three important examples illustrate the point. A molecular signature (18S rDNA) was essential to unlocking the transmission cycle responsible for EPM. Although a protozoal agent was identified in this neurological disease of horses, it was not clear how horses contracted the infection. Because no natural predator of horses exists where this disease is prevalent, there was little *a priori* basis to suspect any particular carnivore as responsible. Neither cats nor dogs (the most readily tested and abundant of carnivores) could be demonstrated as competent definitive hosts. The serendipitous discovery of oocysts bearing 18S rDNA matching the agent of EPM[601] solved the mystery, and engendered a flurry of productive research into the biology and epidemiology of this organism. In addition to linking prey species with the predators (or scavengers) that support the sexual development of species of *Sarcocystis*, molecular data have been employed to establish the range of intermediate hosts susceptible to infection. The case of *S. neurona* is notable for an uncommon and unexpected range of hosts (including raccoons, cats, armadillos, horses, birds, and sea otters). Without this sequence (and additional markers that provided greater specificity), there is little chance that we would understand the vast range of intermediate hosts that would be shown susceptible to infection with this agent, including a variety of domesticated and wildlife species. Certainly a strong connection between terrestrial systems and the marine aquatic environment could not have been predicted, but *S. neurona* was subsequently confirmed as a major infectious cause of mortality in Southern sea otters *Enhydra lutris*.[453] Similarly, genetic signatures have been crucial to differentiating among sporocysts excreted by a given host, for example *S. neurona* and *S. falcatula* in the feces of opossums.[1722]

A similar application of molecular epidemiology established an unexpected cycle of transmission of *S. ovalis*.[699] Here, the definitive host of a parasite of a moose (*Alces alces*) was sought. Noting that corvid birds often feed on moose carcasses, droppings were examined and found to contain sporocysts. These matched the signature sequence of 18S rDNA, and the mucosal scrapings of a magpie (*Pica pica*) were thereby demonstrated to harbor parasites bearing this same sequence.

A precise "match" between sequences, whether identified through serendipity or hypothesis testing, need not occur in order to derive epidemiologically valuable clues. A good example of this concerns successful prediction of the definitive host responsible for human infections with *S. nesbitti* (see Chapters 4 and 5). Infections in a macaque[1859] were characterized at the 18S rDNA. On the basis of the resulting phylogenetic tree, the entity was shown to share an especially close relationship to a species of *Sarcocystis* that complete their life cycle in predatory snakes.[1743] When human cases of sarcocystosis[14,877,559,560] were subsequently attributed to a parasite identical in 18S sequence, a presumptive diagnosis was provided with a presumptive transmission mode implied. Focused sampling efforts subsequently confirmed the occurrence of related sporocysts in predatory snakes,[1033,1034] and ongoing efforts are searching for such parasites in the monitor lizards that abound in the vicinity of an important ongoing outbreak. Without a phylogenetic clue as to the likely source of such sporocysts, the rapid progress in understanding the source(s) of human infection would not have been possible.

23.5 EVOLUTIONARY INFERENCES FROM MOLECULAR MARKERS

Molecular phylogeny provides the context for understanding parasite history and evolution in ways too numerous to mention. Important landmarks in our understanding of *Sarcocystis* and the evolution of tissue cyst parasites, however, deserve mention here:

a. *Sarcocystis* comprises an evolutionary ancient and diverse group. They parasitize all tetrapod groups, and likely have been diversifying over that entire period. Unlike the *Eimeria,* which show signs of early diversification in fish hosts,[1208] the ultimate origins of tissue cyst forming coccidia remains ill-defined. However, their extant diversity (over 150 named species) seems certain to grow as more hosts are examined and the tools of modern genetics are applied to their diagnosis.

b. Not all genera are created equal. As opposed to *Sarcocystis*, which signifies a broad and ecologically diverse assemblage of parasites, other generic names demarcate far less diverse biological lineages. We speak of strains of *Toxoplasma* but not of other coccidia; the paucity of known species of *Hammondia* and *Neospora* also stand in stark contrast to the diverse and deeply branching *Sarcocystis*. Together, the Toxoplasmatinae (including *T. gondii, N. caninum,* and species of *Hammondia* and *Besnoitia*) constitute a sister group to *Sarcocystis,* which itself comprises greater numbers of species and probably has existed for far longer. Molecular phylogenies also served as the basis for subsuming *Frenkelia* within *Sarcocystis* on grounds of common evolutionary heritage. These genera were initially separated by location of tissue cysts in brain (*Frenkelia*) and in muscle (*Sarcocystis*).

c. Definitive hosts have bequeathed parasites to the prey upon which they feed. It was noted that, among parasites encysting in ruminant intermediate hosts, those cycling through cats descended from a lineage distinct from the one encompassing those cycling through dogs.[843] A reasonable conclusion was that an ancestral parasite in a feline ancestor gave rise to those now specific to each intermediate host, and the same was true for the ancestor in an ancestral canine. That pattern generally holds up when considering the deep divergences in the group's evolutionary history: the parasites of a given definitive host share an especially close evolutionary history. The successful search for definitive snake hosts of *S. nesbitti*, summarized above, also illustrates this notion. Although it is also true that related intermediate hosts can share related species of parasite, instances of host shifts are notable (e.g., *S. neurona* and *S. falcatula* infect mammals and birds but share a common definitive host). From an evolutionary perspective, feline transmitted *Sarcocystis* species will eventually disappear from livestock because the sarcocysts stage of feline transmitted species takes several months or years to mature, and by that time the livestock are slaughtered for meat.

d. Phylogenies have also illustrated the flexibility of life-history attributes. The evolutionary histories of parasites have, of necessity, evolved the means to use the same species as its definitive and intermediate host. It was shown that these species are closely related to parasites that complete their life cycles in the expected 2-host cycle, indicating that this life-cycle has originated at least twice,

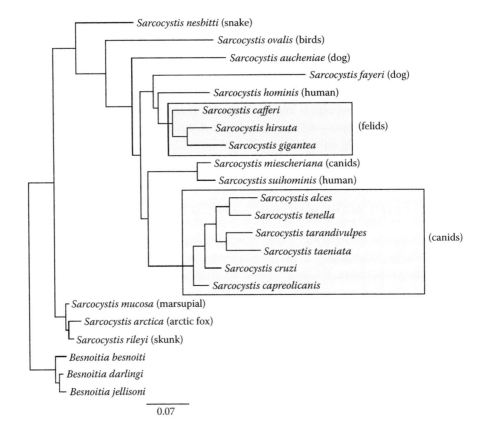

Figure 23.1   Phylogenetic relationships among species of *Sarcocystis* with known or suspected definitive hosts, reconstructed from 18S rDNA sequences aligned using MAFFT[947] under the criterion of maximum likelihood as implemented by PhyML[749] using Geneious 8.2.5. Parasites of felids appear to derive from a single common ancestor, which might have been acquired from a human or canid host. Parasites of more diverse definitive hosts (mustelids, marsupials, reptiles, and birds) represent distinct lineages.

independently.[1623] This result is interesting in light of the major life-history departure that evidently sets apart *T. gondii* from (so far as we know) all other tissue cyst forming coccidians: its ability to circumvent (at least for a time) definitive hosts altogether.[1690]

The transmission of each species of *Sarcocystis* requires a definitive host to feed upon contaminated prey and to contaminate the environment inhabited by competent intermediate hosts. It is therefore interesting to consider the phylogenetic evidence suggesting how new parasite species have become established in new hosts. Although all infected prey species of a given predator or scavenger might conceivably be suited to becoming a reservoir of infection and although all animals imbibing water or consuming vegetation contaminated with the feces of each infected carnivore might acquire new infections, actual cycles of transmission tend to be much more circumscribed. Until the definitive host of many more parasites has been established, our view of this process will remain clouded. But even the fragmentary data now available (Figure 23.1) suggest that certain species transmitted by felid-definitive hosts had a common ancestor (presumably also transmitted by cats) and that these in turn may have derived from an ancestor in human or canine hosts. Greater evolutionary diversity characterizes species of *Sarcocystis* transmitted by other hosts (including marsupials, mustelids, reptiles, and birds), indicating that an important constraint in the elaboration of diversity in this parasite group has been imposed by adaptations to particular definitive host lineages.

BIBLIOGRAPHY

General references: **109, 123, 263, 293, 498, 656, 668, 850, 1110, 1124, 1240, 1331, 1436, 1532, 1719, 1758, 1864, 1876, 1877.**

18S rDNA: **14, 27, 77, 78, 144, 145, 153, 161, 162, 164, 166, 168, 198, 211, 271, 280–286, 290, 305, 330, 331, 466, 467, 473, 476, 490, 494, 495, 497, 513, 518, 600, 601, 608, 653, 655, 684, 699, 700, 706, 775, 783, 788, 840, 841, 843, 856, 904–906, 909, 956, 969, 983, 1003, 1005, 1007, 1020, 1025, 1029, 1033, 1034, 1056, 1095, 1113, 1196, 1221, 1229, 1230, 1237, 1325–1327, 1329, 1340, 1415, 1416, 1419–1421, 1427, 1432, 1473, 1620–1623, 1678, 1722, 1732, 1736, 1743, 1776, 1780, 1788, 1799, 1818, 1844, 1846, 1851, 1854, 1856–1858.**

Other nuclear genes: **44, 45, 433, 441, 442, 453, 459, 460, 466, 467, 473, 476, 490, 496, 519, 520, 521, 633, 702–706, 719, 723, 842, 902, 903, 1004–1009, 1108, 1116, 1194, 1196, 1197, 1235, 1236, 1301, 1325–1327, 1329, 1330, 1332, 1415, 1419–1421, 1432, 1464, 1473, 1492, 1493, 1623, 1675, 1693, 1722, 1788, 1821, 1843, 1879.**

Plastid genes: **453, 902, 1301.**

Mitochondrial genes: **473, 702, 1477, 1941.**

Surface antigens: **1209.**

Microsatellites: **44, 45, 1693, 1821.**

Epidemiological insights: **14, 453, 559, 560, 601, 699, 877, 1033, 1034, 1722, 1743, 1859.**

Evolutionary inferences: **843, 1208, 1623, 1690.**

CHAPTER 24

Current Status of *Sarcocystis* Species

Initial revisions on species-systematic of genus *Sarcocystis* by Levine and Tadros[1048] reported 93 species in 1980; later, Levine[1049] listed 122 in 1986; and more recently, Odening[1317] listed 189 in 1998. Here, we summarize the current status in 7 categories A–G (Tables 24.1 and 24.2): **A**, species with full life cycle known, including endogenous stages; **B**, with definitive and intermediate hosts known; **C**, with definitive host unknown; **D**, partial description of sarcocysts; **E**, *species inquirendae*; **F**, species invalid; and **G**, species synonymized here.

Table 24.3 summarizes 85 unnamed *Sarcocystis* species reported mostly once in different hosts. All species are listed alphabetically in Table 24.4.

Table 24.1   List of Valid *Sarcocystis* Species

Category	*Sarcocystis* Species	Chapter
A	*S. arieticanis*	9
	S. bertrami	11
	S. calchasi	17
	S. capracanis	9
	S. cernae	21
	S. clethrionomyelaphis	22
	S. cruzi	7
	S. dispersa	21
	S. falcatula	17
	S. fayeri	11
	S. glareoli	21
	S. hemionilatrantis	18
	S. hircicanis	9
	S. hirsuta	7
	S. idahoensis	21, 22
	S. microti	21
	S. miescheriana	6
	S. murinotechis	21, 22
	S. muris	21
	S. muriviperae	21, 22
	S. neurona	3A
	S. rauschorum	21
	S. singaporensis	21
	S. tenella	8

(Continued)

Table 24.1 (*Continued*)   List of Valid *Sarcocystis* Species

Category	*Sarcocystis* Species	Chapter
	S. villivillosi	21, 22
	S. zamani	21
B	*S. acanthocolubri*	22
	S. accipitris	17
	S. albifronsi	17
	S. alces	18
	S. alceslatrans	18
	S. alectoributeonis	17
	S. alectorivulpes	17
	S. ameivamastigodryasi	22
	S. atheridis	21, 22
	S. aucheniae	13
	S. baibacinacanis	21
	S. buffalonis	10
	S. cameli	12
	S. campestris	21
	S. capreolicanis	18
	S. cervicanis	18
	S. chalcidicolubris	22
	S. citellibuteonis	21
	S. crotali	21, 22
	S. cuniculorum	21
	S. cymruensis	21
	S. disrumpens	21
	S. dugesii	22
	S. espinosai	21
	S. ferovis	18
	S. fusiformis	10
	S. gerbilliechis	21
	S. gigantea	8
	S. gongyli	22
	S. gracilis	18
	S. grueneri	18
	S. hjorti	18
	S. hoarensis	21
	S. hominis	4, 7
	S. ippeni	12
	S. lacertae	22
	S. leporum	21
	S. levinei	10
	S. lindsayi	17
	S. medusiformis	8
	S. moulei	9
	S. neotomafelis	21
	S. odocoileocanis	18
	S. odoi	18

(*Continued*)

Table 24.1 (*Continued*) List of Valid *Sarcocystis* Species

Category	*Sarcocystis* Species	Chapter
	S. ovalis	18
	S. podarcicolubris	22
	S. poephagi	18
	S. poephagicanis	18
	S. putorii	21
	S. rangi	18
	S. rileyi	17
	S. rommeli	7
	S. sebeki	21
	S. simonyi	22
	S. stehlinii	22
	S. stenodactylicolubris	22
	S. suihominis	4, 6
	S. sybillensis	18
	S. tarandivulpes	18
	S. wapiti	18
	S. wenzeli	16
	S. zuoi	21
C	*S. africana*	18
	S. americana	18
	S. anasi	17
	S. arctica	19
	S. arctosi	19
	S. azevedoi	22
	S. balaenopteralis	18
	S. booliati	21
	S. bozemanensis	21
	S. bunopusi	21
	S. cafferi	18
	S. camelopardalis	18
	S. caninum	14
	S. canis	3B
	S. capricornis	18
	S. chamaeleonis	22
	S. chloropusae	17
	S. clethrionomysi	21
	S. columbae	17
	S. cornagliai	18
	S. cornixi	17
	S. corsaci	19
	S. corvusi	17
	S. danzani	18
	S. dasypi	21
	S. didelphidis	20
	S. diminuta	21
	S. dubeyella	18

(*Continued*)

Table 24.1 (*Continued*)   List of Valid *Sarcocystis* Species

Category	*Sarcocystis* Species	Chapter
	S. dubeyi	10
	S. elongata	18
	S. eothenomysi	21
	S. felis	15
	S. gallotiae	22
	S. garnhami	20
	S. gazellae	18
	S. giraffae	18
	S. greineri	20
	S. hardangeri	18
	S. hemioni	18
	S. hippopotami	18
	S. hofmanni	19
	S. horvathi	16
	S. hydrurgae	18
	S. inghami	20
	S. jaypeedubeyi	21
	S. jorrini	18
	S. kalkivus	19
	S. kinosterni	22
	S. kirkpatricki	19
	S. kirmsei	17
	S. kitikmeotensis	19
	S. klaseriensis	18
	S. kortei	5
	S. lari	17
	S. lutrae	19
	S. markusi	5
	S. marmosae	20
	S. melampi	18
	S. melis	19
	S. mephitisi	19
	S. mitrami	22
	S. mongolica	18
	S. montanaensis	21
	S. mucosa	20
	S. nesbitti	22
	S. ochotonae	21
	S. oryzomyos	21
	S. oviformis	18
	S. peromysci	21
	S. phacochoeri	18
	S. phoeniconaii	17

(*Continued*)

Table 24.1 (*Continued*)   List of Valid *Sarcocystis* Species

Category	*Sarcocystis* Species	Chapter
	S. proechimyos	21
	S. ramphastosi	17
	S. rangiferi	18
	S. scandinavica	18
	S. scelopori	22
	S. schneideri	22
	S. sehi	21
	S. sibirica	18
	S. sigmodontis	21
	S. silva	18
	S. speeri	20
	S. sulawesiensis	21
	S. sulfuratusi	17
	S. svanai	14
	S. taeniata	18
	S. tarandi	18
	S. truncata	18
	S. tupaia	21
	S. turcicii	22
	S. turdusi	17
	S. ursusi	19
	S. wobeseri	17
	S. woodhousei	18
	S. youngi	18
D	S. dogeli	21
	S. erdmanae	19
	S. eutamias	21
	S. eversmanni	19
	S. fedoseenkoi	21
	S. galuzoi	21
	S. hueti	18
	S. orientalis	18
	S. platydactyliscelopori	22
	S. pseudois	18
	S. rhombomys	21
	S. richardii	21
	S. undulati	21

Table 24.2   List of Invalid *Sarcocystis* Species

Category	*Sarcocystis* Species	Chapter
E	S. capreoli	18
	S. colii	17
	S. ctenosauris	NR
	S. garzettae	17
	S. kaiserae	17
	S. lacertaeutae	22
	S. lampropeltis	22
	S. microps	8
	S. mihoensis	8
	S. nelsoni	18
	S. novaki	18
	S. peckai	17
	S. porcifelis	6
	S. russuli	21
	S. saiga	18
	S. salvelini	22
	S. setophagae	17
	S. spaldingae	17
	S. talpini	21
	S. terrestri	NR
	S. vulpis	19
F	S. ammodrami	17
	S. aramidis	17
	S. asinus	11
	S. atlanticae	22
	S. atractaspidis	22
	S. bettongiae	21
	S. bubalis	NR
	S. caviae	21
	S. cervi	18
	S. cheeli	NR
	S. citellivulpes	21
	S. cricetuli	21
	S. gusevi	18
	S. jacarinae	17
	S. miescheri	12
	S. nontenella	17
	S. oliverioi	NR
	S. otus	17
	S. pitymysi	21
	S. platydactyli	22
	S. pythonis	22
	S. roudabushi	NR
	S. scotti	NR

(Continued)

Table 24.2 (*Continued*)   List of Invalid *Sarcocystis* Species

Category	*Sarcocystis* Species	Chapter
	S. tamanduae	21
	S. tropicalis	NR
	S. turdi	17
G	S. buteonis	21
	S. cuniculi	21
	S. equicanis	11
	S. rodentifelis	21

NR = not reported in this book.

Table 24.3   List of the 85 Unnamed *Sarcocystis* Species

Sarcocystis Species	Intermediate Host	Main Reference	Chapter
Nonhuman Primates			
S. sp. Mehlhorn, Heydorn, and Janitschke 1977	Rhesus monkey (*Macaca mulatta*)	1175	5
S. sp. Mehlhorn, Heydorn, and Janitschke 1977	Baboon (*Papio cynocephalus*)	1175	5
S. sp. Mehlhorn, Heydorn, and Janitschke 1977	Tamarin (*Saguinus oedipus*)	1175	5
Wild Ruminants			
S. sp. Saito, Itagaki, Shibata, and Itagaki 1995	Sika deer (*Cervus nippon centralis*)	1515	18
S. sp. Gangadharan, Valsala, Nair, and Rajan 1992	Sambar deer (*Cervus unicolor*)	660	18
S. sp. Shrivastav, Sharma, Chaudhry, and Malik 1999	Barasingha or hard-ground swamp deer (*Cervus duvauceli branderi*)	1605	18
S. sp. Poli, Mancianti, Marconcini, Nigro, and Colagreco 1988	Fallow deer (*Dama dama*)	1404	18
S. cf. *hofmanni* Wesemeier and Sedlaczek 1995	Fallow deer (*Dama dama*)	1823	18
S. cf. *grueneri* Wesemeier and Sedlaczek 1995	Fallow deer (*Dama dama*)	1823	18
S. sp. Dubey and Lozier 1983	White-tailed deer (*Odocoileus virginianus*)	378	18
S. sp. type II Atkinson, Wright, Telford, Mclaughlin, Forrester, Roelke, and Mccown 1993	White-tailed deer (*Odocoileus virginianus*)	46	18
S. sp. Stolte, Bockhardt, and Odening 1996	Wapiti or North American elk (*Cervus elaphus canadensis*)	1680	18
S. sp. Odening, Rudolph, Quandt, Bengis, Bockhardt, and Viertel 1998	Springbok (*Antidorcas marsupialis*), Impala (*Aepyceros melampus*)	1320	18
S. sp. cf. *hominis* Odening, Rudolph, Quandt, Bengis, Bockhardt, and Viertel 1998	Greater kudu (*Tragelaphus strepsiceros*)	1320	18

(*Continued*)

Table 24.3 (*Continued*)   List of the 85 Unnamed *Sarcocystis* Species

Sarcocystis Species	Intermediate Host	Main Reference	Chapter
S. sp. cf. *medusiformis* Stolte, Odening, and Bockhardt 1996	Addax (*Addax nasomaculatus*)	1679	18
S. sp. Odening, Stolte, Lux, and Bockhardt 1996	Mongolian gazelle (*Procapra gutturosa*)	1313	18
S. sp. cf. *capracanis* Stolte, Odening, and Bockhardt 1996	Blackbuck (*Antilope cervicapra*)	1679	18
S. sp. cf. *hominis* Stolte, Odening, and Bockhardt 1996	Common eland (*Taurotragus oryx*)	1679	18
S. sp. cf. *cruzi* Stolte, Odening, and Bockhardt 1996	Nilgai (*Boselaphus tragocamelus*), Blackbuck (*Antilope cervicapra*)	1679	18
S. sp. Foreyt 1989	Mountain goat (*Oreamnos americanus*)	619	18
S. sp. Dubey 1980	Pronghorn (*Antilocapra americana*)	355	18
S. sp. Odening, Stolte, and Bockhardt 1996	Chamois (*Rupicapra rupicapra*)	1311	18
S. sp. Kannangara 1970	Mouse deer (*Moschiola meminna*)	941	18
S. cornagliai-like (type I) Cornaglia, Giaccherino, and Peracino 1998	Alpine ibex or steinbock (*Capra ibex*)	253	18
S. capracanis-like (type II) Cornaglia, Giaccherino, and Peracino 1998	Alpine ibex or steinbock (*Capra ibex*)	253	18
S. arieticanis-like (type III) Cornaglia, Giaccherino, and Peracino, 1998	Alpine ibex or steinbock (*Capra ibex*)	253	18
S. capricornis-like (type IV) Cornaglia, Giaccherino, and Peracino 1998	Hispanic ibex (*Capra pyrenaica hispanica*)	253	18
S. sp. Agrawal, Chauhan, and Ahluwalia 1982	Himalayan goral (*Naemorhedus goral*)	1898	18
S. sp. Rioseco, Cubillos, Gonzalez, and Diaz 1976	Pudu (*Pudu pudu*)	2252	18
S. sp. Samuel and Gray 1974	Musk ox (*Ovibos moschatus*)	2281	18
Marine Mammals			
S. sp. Mense, Dubey, and Homer 1992	Sea lion (*Zalophus californianus*)	1184	18
S. sp. Bishop 1979	Bearded seal (*Erignathus barbatus*)	105	18
S. sp. Migaki and Albert 1980	Ringed seal (*Pusa hispida*)	1192	18
S. sp. Brown, Smith, and Keyes 1974	Northern fur seal (*Callorhinus ursinus*)	147	18
S. sp. De Guise, Lagacé, Girard, and Béland 1993	Beluga whale (*Delphinapterus leucas*)	310	18
S. sp. Munday and Mason 1978	Sperm whale (*Physeter catodon*)	1249	18
S. sp. De Guise, Lagacé, Girard, and Béland 1993	Atlantic white-sided dolphin (*Lagenorhyncus acutus*)	310, 562	18
S. sp. Munday and Mason 1978	*Delphinus* sp.	1249	18

(*Continued*)

Table 24.3 (*Continued*)   List of the 85 Unnamed *Sarcocystis* Species

Sarcocystis Species	Intermediate Host	Main Reference	Chapter
S. sp. Dailey and Stroud 1978	Striped dolphin (*Stenella coeruleoalba*)	287	18
S. sp. Dubey, Lindsay, Rosenthal, and Thomas 2003	Sea otter (*Enhydra lutris*)	453	18
Wild Carnivores			
S. sp. 1 Stolte, Odening, Walter, and Bockhardt 1996	Raccoon (*Procyon lotor*)	1681	19
S. sp. 2 Stolte, Odening, Walter, and Bockhardt 1996	Raccoon (*Procyon lotor*)	1681	19
S. sp. 3 (sf. *S. sebeki*) Stolte, Odening, Walter, and Bockhardt 1996	Raccoon (*Procyon lotor*)	1681	19
S. sp. Pak 1991	Red fox (*Vulpes vulpes*)	1359	19
S. sp. 1 Wesemeier, Odening, Walter, and Bockhardt 1995	Jackal (*Canis melomelas*)	1825	19
S. sp. 2 Wesemeier, Odening, Walter, and Bockhardt 1995	Jackal (*Canis melomelas*)	1825	19
S. sp. Cummings, Kocan, Barker, and Dubey 2000	Coyote (*Canis latrans*)	268	19
S. sp. Bwangamoi, Ngatia, and Richardson 1993	Wild dog (*Lycacon pictus*)	159	19
S. sp. cf. *sebeki* Odening, Stolte, Walter, and Bockhardt 1994	European badger (*Meles meles*)	1303, 1304	19
S. sp. cf. *gracilis* Odening, Stolte, Walter, Bockhardt, and Jakob 1994	European badger (*Meles meles*)	1304	19
S. sp. Kubo, Okano, Ito, Tsubota, Sakai, and Yanai 2009	Japanese badger (*Meles meles anakuma*)	982	19
S. sp. Wahlström, Nikkila, and Uggla 1999	Otter (*Lutra lutra*)	1803	19
S. sp. (most likely *S. neurona*) Ramos-Vara, Dubey, Watson, Winn-Elliot, Patterson, and Yamini 1997	Mink (*Mustela vison*)	1445	19
S. sp. Tadros and Laarman 1979	Common European weasel (*Mustela nivalis*)	1711	19
S. sp. Larkin, Gabriel, Gerhold, Yabsley, Wester, Humphreys, Beckstead, and Dubey 2011	Fisher (*Martes pennanti*)	1025	19
S. sp. Dubey, Humphreys, and Fritz 2008	Black bear (*Ursus americanus*)	462	19
S. sp. Cheadle, Cunningham, and Greiner 2002	Florida black bear (*Ursus americanus floridanus*)	206	19
S. sp. Somvanshi, Koul, and Biswas 1987	Leopard (*Panthera pardus*)	1644	19
Small Mammals			
S. dispersa-like Munday 1983	Norway rat (*Rattus norvegicus*)	1256	21
S. sp. cf. *singaporensis* Kan and Dissanaike 1977	Malaysian house rats (*Rattus rattus diardii*)	936	21
S. sp. type B Rzepczyk and Scholtyseck 1976	Bush rat (*Rattus fuscipes*)	1504	21

(*Continued*)

Table 24.3 (*Continued*)   List of the 85 Unnamed *Sarcocystis* Species

Sarcocystis Species	Intermediate Host	Main Reference	Chapter
S. montanaensis-like Lindsay, Upton, Blagburn, Toivio-Kinnucan, Mcallister, and Trauth 1991	Prairie vole (*Microtus achrogaster*)	1065	21
S. cf. cuniculi Brumpt 1913	European hare (*Lepus europaeus*)	1305	21
S. sp. 1 Odening, Wesemeier, Pinkowski, Walter, Sedlaczek, and Bockhardt 1994	European hare (*Lepus europaeus*)	1305	21
S. sp. 2 Odening, Wesemeier, Pinkowski, Walter, Sedlaczek, and Bockhardt 1994	European hare (*Lepus europaeus*)	1305	21
S. cf. bertrami Odening, Wesemeier, Pinkowski, Walter, Sedlaczek, and Bockhardt 1994	European hare (*Lepus europaeus*)	1305	21
S. sp. 1 Odening, Wesemeier, and Bockhardt 1996	European hare (*Lepus europaeus*)	1310	21
S. sp. 2 Odening, Wesemeier, and Bockhardt 1996	European hare (*Lepus europaeus*)	1310	21
S. sp. Lindsay, Mckown, Upton, Mcallister, Toivio-Kinnucan, Veatch, and Blagburn 1996	9-banded armadillo (*Dasypus novemcinctus*)	1067	21
S. sp. Scholtyseck, Entzeroth, and Chobotar 1982	Virginia opossum (*Didelphis virginiana*)	1567	21
S. sp. Entzeroth, Chobotar, and Scholtyseck 1983	Red squirrel (*Tamiasciurus hudsonicus*)	535	21
S. sp. Entzeroth, Scholtyseck, and Chobotar 1983	Eastern chipmunk (*Tamias striatus*)	536	21
S. sp. Kannangara 1970	Bandicoot (*Bandicota malabarica*)	941	21
Birds			
S. sp. type A Crawley, Ernst, and Milton 1982	Bald eagle (*Haliaeetus leucocephalus*)	261	17
S. sp. type B Olson, Wunschmann, and Dubey 2007	Bald eagle (*Haliaeetus leucocephalus*)	1333	17
S. sp. Wenzel, Erber, Boch, and Schellner 1982	Pheasant (*Phasianus colchicus*)	1822	17
S. sp. Hiller, Sidor, De Guise, and Barclay 2007	American woodcock (*Scolopax minor*)	829	17
S. sp. Dubey, Johnson, Bermudez, Suedmeyer, and Fritz 2001	Straw-necked ibis (*Carphibis spinicollis*)	437	17
S. sp. Wenzel, Erber, Boch, and Schellner 1982	Coot (*Fulica atra*)	1822	17
S. sp. Teglas, Little, Latimer, and Dubey 1998	Turkey (*Meleagris gallopavo*)	1728	17
S. sp. Spalding, Yowell, Lindsay, Greiner, and Dame 2002	Northern gannet (*Morus bassanus*)	1647	17
S. sp. Dubey and Morales 2006	Buffon's macaw (*Ara ambigua*)	1989	17
S. sp. Kutkiené, Sruoga, and Butkauskas 2008	Goldeneye (*Bucephala clangula*)	1004	17

(Continued)

Table 24.3 (*Continued*)   List of the 85 Unnamed *Sarcocystis* Species

Sarcocystis Species	Intermediate Host	Main Reference	Chapter
Reptiles			
S. sp. Roberts, Wellehan, Weisman, Rush, Childress, and Sibley 2014	South American rattlesnake (*Crotalus durissus terrificus*)	**1474**	22
Fish			
S. sp. Fantham and Porter 1943	Eel (*Zoarces angularis*)	**566**	22

Table 24.4   List of Named *Sarcocystis* Species (Alphabetized)

S. acanthocolubri	*S. cervicanis*
S. accipitris	*S. chalcidicolubris*
S. africana	*S. chamaeleonis*
S. albifronsi	*S. chloropusae*
S. alces	*S. citellibuteonis*
S. alceslatrans	*S. clethrionomyelaphis*
S. alectoributeonis	*S. clethrionomysi*
S. alectorivulpes	*S. columbae*
S. ameivamastigodryasi	*S. cornagliai*
S. americana	*S. cornixi*
S. anasi	*S. corsaci*
S. arctica	*S. corvusi*
S. arctosi	*S. crotali*
S. arieticanis	*S. cruzi*
S. atheridis	*S. cuniculorum*
S. aucheniae	*S. cymruensis*
S. azevedoi	*S. danzani*
S. baibacinacanis	*S. dasypi*
S. balaenopteralis	*S. didelphidis*
S. bertrami	*S. diminuta*
S. booliati	*S. dispersa*
S. bozemanensis	*S. disrumpens*
S. buffalonis	*S. dogeli*
S. bunopusi	*S. dubeyella*
S. cafferi	*S. dubeyi*
S. calchasi	*S. dugesii*
S. cameli	*S. elongata*
S. camelopardalis	*S. eothenomysi*
S. campestris	*S. erdmanae*
S. caninum	*S. espinosai*
S. canis	*S. eutamias*
S. capracanis	*S. eversmanni*
S. capreolicanis	*S. falcatula*
S. capricornis	*S. fayeri*
S. cernae	*S. fedoseenkoi*

(*Continued*)

Table 24.4 (*Continued*)   List of Named *Sarcocystis* Species (Alphabetized)

S. felis	S. melampi
S. ferovis	S. melis
S. fusiformis	S. mephitisi
S. gallotiae	S. microti
S. galuzoi	S. miescheriana
S. garnhami	S. mitrami
S. gazellae	S. mongolica
S. gerbilliechis	S. montanaensis
S. gigantea	S. moulei
S. giraffae	S. mucosa
S. glareoli	S. murinotechis
S. gongyli	S. muris
S. gracilis	S. muriviperae
S. greineri	S. neotomafelis
S. grueneri	S. nesbitti
S. hardangeri	S. neurona
S. hemioni	S. ochotonae
S. hemionilatrantis	S. odocoileocanis
S. hippopotami	S. odoi
S. hircicanis	S. orientalis
S. hirsuta	S. otus
S. hjorti	S. oryzomyos
S. hoarensis	S. ovalis
S. hofmanni	S. oviformis
S. hominis	S. peromysci
S. horvathi	S. phacochoeri
S. hueti	S. phoeniconaii
S. hydrurgae	S. platydactyliscelopori
S. idahoensis	S. podarcicolubris
S. inghami	S. poephagi
S. ippeni	S. poephagicanis
S. jaypeedubeyi	S. proechimyos
S. jorrini	S. pseudois
S. kalkivus	S. putorii
S. kinosterni	S. ramphastosi
S. kirkpatricki	S. rangi
S. kirmsei	S. rangiferi
S. kitikmeotensis	S. rauschorum
S. klaseriensis	S. rhombomys
S. kortei	S. richardii
S. lacertae	S. rileyi
S. lari	S. rommeli
S. leporum	S. scandinavica
S. levinei	S. scelopori
S. lindsayi	S. schneideri
S. markusi	S. sebeki
S. marmosae	S. sehi
S. medusiformis	S. sibirica

(Continued)

Table 24.4 (*Continued*)   List of Named *Sarcocystis* Species (Alphabetized)

S. silva	S. sigmodontis
S. simonyi	S. truncata
S. singaporensis	S. tupaia
S. speeri	S. turcicii
S. stehlinii	S. turdusi
S. stenodactylicolubris	S. undulati
S. suihominis	S. ursusi
S. sulawesiensis	S. villivillosi
S. sulfuratusi	S. wapiti
S. svanai	S. wenzeli
S. sybillensis	S. wobeseri
S. taeniata	S. woodhousei
S. tarandi	S. youngi
S. tarandivulpes	S. zamani
S. tenella	S. zuoi

References

1. Abbas, M.K., Powell, E.C., 1983. Identification of surface antigens of *Sarcocystis muris* (Coccidia). *J. Protozool.* 30, 356–361.
2. Abdel Mawla, M.M., 1990. Ultrastructure of the cyst wall of *S. lindemanni* with pathological correlations. *J. Egypt. Soc. Parasitol.* 20, 319–325.
3. Abdel-Baki, A.-A., Allam, G., Sakran, T., El-Malah, E.-M., 2009. Lambs infected with UV-attenuated sporocysts of *Sarcocystis ovicanis* produced abnormal sarcocysts and induced protective immunity against a challenge infection. *Korean J. Parasitol.* 47, 131–138.
4. Abdel-Baki, A.S., Abdel-Haleem, H.M., Al-Quraishy, S., 2012. A new *Sarcocystis* species (Apicomplexa: Sarcocystidae) from the rock gecko *Bunopus tuberculatus* in Saudi Arabia. *J. Parasitol.* 98, 951–953.
5. Abdel Ghaffar, F.A., Hilali, M., Scholtyseck, E., 1978. Ultrastructural study of *Sarcocystis fusiformis* (Railliet, 1897) infecting the Indian water buffalo (*Bubalus bubalis*) of Egypt. *Tropenmed. Parasitol.* 29, 289–294.
6. Abdel Ghaffar, F., Entzeroth, R., Chobotar, B., Scholtyseck, E., 1979. Ultrastructural studies of *Sarcocystis* sp. from the camel (*Camelus dromedarius*) in Egypt. *Tropenmed. Parasitol.* 30, 434–438.
7. Abdel Ghaffar, F.A., Heydorn, A.O., Mehlhorn, H., 1989. The fine structure of cysts of *Sarcocystis moulei* from goats. *Parasitol. Res.* 75, 416–418.
8. Abdel-Ghaffar, F., Bashtar, A.R., Ashour, M.B., Sakran, T., 1990. Life cycle of *Sarcocystis gongyli* Trinci 1911 in the skink *Chalcides ocellatus ocellatus* and the snake *Spalerosophis diadema*. A light and electron microscopic study. *Parasitol. Res.* 76, 444–450.
9. Abdel-Ghaffar, F., Al-Johany, A.M., 2002. A light and electron microscope study of *Sarcocystis mitrani* (sp. nov.) infecting the skink *Scincus mitranus* in the central region of Saudi Arabia. *Parasitol. Res.* 88, 102–106.
10. Abdel-Ghaffar, F., Bashtar, A.R., Al-Quraishy, S., Al Nasr, I., Mehlhorn, H., 2009. *Sarcocystis* infecting reptiles in Saudi Arabia. 1–Light and electron microscopic study on sarcocysts of *Sarcocystis turcicii* sp. nov. infecting the gecko *Hemidactylus turcicus* Linnaeus. *Parasitol. Res.* 104, 503–508.
11. Abdel-Ghaffar, F., Mehlhorn, H., Bashtar, A.R., Al-Rasheid, K., Sakran, T., El-Fayoumi, H., 2009. Life cycle of *Sarcocystis camelicanis* infecting the camel (*Camelus dromedarius*) and the dog (*Canis familiaris*), light and electron microscopic study. *Parasitol. Res.* 106, 189–195.
12. Abdirahman, O.M., Finazzi, M., Scanziani, E., Sironi, G., 1991. Reperti de miosite eosinofilica in bovini regolarmente macellati. *Arch. Vet. Ital.* 42, 97–104.
13. Abo-Shehada, M.N., 1996. Age variations in the prevalence of sarcocystosis in sheep and goats from northern and central Jordan. *Prev. Vet. Med.* 27, 135–140.
14. AbuBakar, S., Teoh, B.T., Sam, S.S., Chang, L.Y., Johari, J., Hooi, P.S., Lakhbeer-Singh, H.K. et al. 2013. Outbreak of human infection with *Sarcocystis nesbitti*, Malaysia, 2012. *Emerg. Infect. Dis.* 19, 1989–1991.
15. Achuthan, H.N., Raja, E.E., 1990. Occurrence of *Sarcocystis* sp. in horse (*Equus caballus*). *Indian Vet. J.* 67, 472.
16. Aganga, A.O., Aganga, A.A., Umoh, J.U., Kudi, A.C., 1988. Isolation of *Sarcocystis ovicanis* sporocysts in dogs in Zaria area: Its significance to small ruminant production. *Bull. Anim. Health Prod. Afr.* 36, 138–140.
17. Agerholm, J.S., Dubey, J.P., 2014. Sarcocystosis in a stillborn lamb. *Reprod. Dom. Anim.* 49, e60–e63.
18. Agnihotri, R.K., Juyal, P.D., Bhatia, B.B., 1987. Microsarcocysts in pigs in Uttar Pradesh, India. *J. Vet. Parasitol.* 1, 69–70.
19. Agrawal, M.C., Singh, K.P., Shah, H.L., 1991. Caprine sarcocystosis in Jabalpur area. *J. Vet. Parasitol.* 5, 108–112.
20. Agarwal, P.K., Srivastava, A.N., 1983. *Sarcocystis* in man: A report of two cases. *Histopathology* 7, 783–787.
21. Aguilar, R.F., Shaw, D.P., Dubey, J.P., Redig, P., 1991. *Sarcocystis*-associated encephalitis in an immature northern goshawk (*Accipiter gentilis atricapillus*). *J. Zoo Wildl. Med.* 22, 466–469.
22. Akao, S., 1970. A new species of *Sarcocystis* parasitic in the whale *Balaenoptera borealis*. *J. Protozool.* 17, 290–294.

23. Al Quraishy, S., Morsy, K., Bashtar, A.R., Ghaffar, F.A., Mehlhorn, H., 2014. *Sarcocystis arietica-nis* (Apicomplexa: Sarcocystidae) infecting the heart muscles of the domestic sheep, *Ovis aries* (Artiodactyla: Bovidae), from K. S. A. on the basis of light and electron microscopic data. *Parasitol. Res.* 113, 3823–3831.

24. Al-Bayati, S.M.H., 2012. Studying some biochemical parameters of *Sarcocystis* parasites isolated from local sheep and goat in Duhok area. *Al-Anbar J. Vet. Sci.* 5, 94–97.

25. Al-Goraishy, S.A.R., Bashtar, A.R., Al-Rasheid, K.A.S., Abdel-Ghaffar, F.A., 2004. Prevalence and ultrastructure of *Sarcocystis* species infecting camels (*Camelus dromedarius*) slaughtered in Riyadh City, Saudi Arabia. *Saudi J. Biol. Sci.* 11, 135–142.

26. Al-Hoot, A.S., Al-Qureishy, S.A., Al-Rashid, K., Bashtar, A.R., 2005. Microscopic study on *Sarcocystis moulei* from sheep and goats in Saudi Arabia. *J. Egypt. Soc. Parasitol.* 35, 295–312.

27. Al-Kappany, Y.M., Abu-Elwafa, S.A., Hilali, M., Rosenthal, B.M., Dunams, D.B., Dubey, J.P., 2013. Experimental transmission of *Sarcocystis muris* (Apicomplexa: Sarcocystidae) sporocysts from a naturally infected cat (*Felis catus*) to immunocompetent and immunocompromised mice. *J. Parasitol.* 99, 997–1001.

28. Allison, R., Williams, P., Lansdowne, J., Lappin, M., Jensen, T., Lindsay, D., 2006. Fatal hepatic sarcocystosis in a puppy with eosinophilia and eosinophilic peritoneal effusion. *Vet. Clin. Pathol.* 35, 353–357.

29. Anderson, A.J., Greiner, E.C., Atkinson, C.T., Roelke, M.E., 1992. Sarcocysts in the Florida bobcat (*Felis rufus floridanus*). *J. Wildl. Dis.* 28, 116–120.

30. Anderson, M.L., Blanchard, P.C., Barr, B.C., Dubey, J.P., Hoffman, R.L., Conrad, P.A., 1991. *Neospora*-like protozoan infection as a major cause of abortion in California dairy cattle. *J. Am. Vet. Med. Assoc.* 198, 241–244.

31. Andrews, C.D., Fayer, R., Dubey, J.P., 1990. Continuous *in vitro* cultivation of *Sarcocystis cruzi*. *J. Parasitol.* 76, 254–255.

32. Arias, M., Yeargan, M., Francisco, I., Dangoudoubiyam, S., Becerra, P., Francisco, R., Sánchez-Andrade, R., Paz-Silva, A., Howe, D.K., 2012. Exposure to *Sarcocystis* spp. in horses from Spain determined by Western blot analysis using *Sarcocystis neurona* merozoites as heterologous antigen. *Vet. Parasitol.* 185, 301–304.

33. Arnastauskienė, T., 1989. Sarcosporidia of wild boars in the Lithuanian SSR. *Acta Parasitol. Lituanica* 23, 51–58. (In Russian).

34. Arnastauskienė, T., Grikienienė, J., 1989. Possibility of the development of *Sarcocystis* sp. taken from the bank vole in the organism of laboratory rats without the definitive host's participation. *Lietuvos TSR Mokslu Akademijos DarbaiC Serija* 4, 51–60. (In Russian).

35. Arnastauskienė, T., Grikienienė, J., 1993. Infection of small mammals with sarcosporidians in the south-eastern Baltic region. *Ekologija* 2, 47–56.

36. Arness, M.K., Brown, J.D., Dubey, J.P., Neafie, R.C., Granstrom, D.E., 1999. An outbreak of acute eosinophilic myositis attributed to human *Sarcocystis* parasitism. *Am. J. Trop. Med. Hyg.* 61, 548–553.

37. Arru, E., Cosseddu, A.M., 1976. Diffusione e distribuzione dei sarcosporidi negli animali da macello in sardegna. *Clinica Veterinaria* 99, 322–327.

38. Artigas, P.T., Oria, J., 1932. Sobre urn novo *Sarcocystis* parasito do *Tamandua tetradactyla*. *Anais da Faculdade de Medicina de São Paulo* 6, 37–40.

39. Aryeetey, M., Mehlhorn, H., Heydorn, A.O., 1980. Electron microscopic studies on the development of *Sarcocystis capracanis* in experimentally infected goats. *Zentralbl. Bakteriol. Hyg. I. Abt. Orig. A* 247, 543–556.

40. Aryeetey, M.E., Piekarski, G., 1976. Serologische *Sarcocystis*-Studien an Menschen und Ratten. *Z. Parasitenkd.* 50, 109–124.

41. Ashford, R.W., 1978. *Sarcocystis cymruensis* n. sp., a parasite of rats *Rattus norvegicus* and cats *Felis catus*. *Ann. Trop. Med. Parasitol.* 72, 37–43.

42. Ashford, R.W., 1978. Who named *Sarcocystis muris*? *Ann. Trop. Med. Parasitol.* 72, 95.

43. Ashmawy, K.I., Abu-Akkada, S.S., Ghashir, M.B., 2014. Prevalence and molecular characterization of *Sarcocystis* species in water buffaloes (*Bubalus bubalis*) in Egypt. *Trop. Anim. Health Prod.* 46, 1351–1356.

44. Asmundsson, I.M., Rosenthal, B.M., 2006. Isolation and characterization of microsatellite markers from *Sarcocystis neurona*, a causative agent of equine protozoal myeloencephalitis. *Mol. Ecol. Notes* 6, 8–10.

45. Asmundsson, I.M., Dubey, J.P., Rosenthal, B.M., 2006. A genetically diverse but distinct North American population of *Sarcocystis neurona* includes an overrepresented clone described by 12 microsatellite alleles. *Infect. Genet. Evol.* 6, 352–360.

46. Atkinson, C.T., Wright, S.D., Telford, S.R., McLaughlin, G.S., Forrester, D.J., Roelke, M.E., McCown, J.W., 1993. Morphology, prevalence, and distribution of *Sarcocystis* spp. in white-tailed deer (*Odocoileus virginianus*) from Florida. *J. Wildl. Dis.* 29, 73–84.

47. Atkinson, E.M., Collins, G.H., 1981. Electrophoretic studies on three enzymes from *Sarcocystis* species in sheep. *Syst. Parasitol.* 2, 213–216.

48. Aulakh, R.S., Joshi, D.V., Juyal, P.D., 1997. Prevalence of *Sarcocystis capracanis* in naturally infected goats: A preliminary report. *J. Vet. Parasitol.* 11, 99–100.

49. Avapal, R.S., Singh, B.S., Sharma, J.K., Juyal, P.D., 2002. Seroprevalence of sarcocystosis in swine. *J. Parasit. Dis.* 26, 109–110.

50. Avapal, R.S., Sharma, J.K., Juyal, P.D., 2003. Prevalence of *Sarcocystis* species infection in slaughtered pigs. *J. Vet. Parasitol.* 17, 151–153.

51. Avapal, R.S., Sharma, J.K., Juyal, P.D., 2003. Comparative morphology of *Sarcocystis suihominis* and *S. miescheriana* in domestic pigs. *Indian J. Anim. Sci.* 73, 392–393.

52. Avapal, R.S., Sharma, J.K., Juyal, P.D., 2004. Pathological changes in *Sarcocystis* infection in domestic pigs (*Sus scrofa*). *Vet. J.* 168, 358–361.

53. Babudieri, B., 1932. I Sarcosporidi e le sarcosporidiosi (Studio monografico). *Arch. Protistenkd.* 76, 421–580.

54. Badawy, A.I.I., Abouzaid, N.Z., Ahmed, H.A., 2012. *Sarcocystis hominis* and other *Sarcocystis* species infecting cattle in Sharkia Province, Egypt. *J. Amer. Sci.* 8, 271–275.

55. Baetz, A.L., Barnett, D., Bryner, J.H., Cysewski, S.J., 1980. Plasma progesterone concentration in the bovine before abortion or parturition in pregnant animals exposed to *Sarcocystis cruzi*, *Campylobacter fetus*, or *Aspergillus fumigatus*. *Am. J. Vet. Res.* 41, 1767–1768.

56. Baetz, A.L., Crandell, S.E., Schmerr, M.J.F., Barnett, D., Bryner, J.H., 1981. Plasma α-fetoprotein concentrations in pregnant cows exposed to *Sarcocystis cruzi*, *Campylobacter fetus*, or *Aspergillus fumingatus*. *Am. J. Vet. Res.* 42, 2146–2148.

57. Bagir-Zade, S.S., 1989. DNA content in the cysts of *Sarcocystis fusiformis* and *S. gigantea*. *Izvestiia Akademii Nauk Azerbaidzhanskoi SSR, Biologichkie Nauki* 2, 43–45. (In Russian).

58. Bahari, P., Salehi, M., Seyedabadi, M., Mohammadi, A., 2014. Molecular identification of macroscopic and microscopic cysts of *Sarcocystis* in sheep in north Khorasan Province, Iran. *Int. J. Mol. Cell Med.* 3, 51–56.

59. Baird, K.L., Cheadle, M.A., Greiner, E.C., 2002. Prevalence and site specificity of *Sarcocystis greineri* sarcocysts in Virginia opossum (*Didelphis virginiana*) in Florida. *J. Parasitol.* 88, 624–625.

60. Balakrishna, J.P., Chacko, G., Manipadam, M.T., Ramyal, 2013. Glomerulopathy in a patient with *Sarcocystis* infestation. *Indian J. Pathol. Microbiol.* 56, 285–287.

61. Balbo, T., Rossi, L., Lanfranchi, P., Meneguz, P.G., Peirone, B., 1985. Sarcosporidiosis in red deer (*Cervus elaphus*) in regional park of "La Mandria." Erkrankungen der Zootiere. Verhandlungsbercht Dec. 27. Internationalen Symposiums Uber Die Erkrankungen der zootiere 9–13 June, 1985, St. Vincent, Torino. Akademie Verlag, Berlin, pp. 159–162.

62. Balfour, A., 1913. A sarcocyst of a gazelle (*G. rufifrons*) showing differentiation of spores by vital staining. *Parasitology* 6, 52–56.

63. Ball, G.H., 1944. Sarcosporidia in southern California lizards. *Trans. Am. Micro. Soc.* 63, 144–148.

64. Banerjee, P.S., Bhatia, B.B., Pandit, B.A., 1994. *Sarcocystis suihominis* infection in human beings in India. *J. Vet. Parasitol.* 8, 57–58.

65. Banerjee, P.S., 1998. Studies on pathogenesis of *Sarcocystis tenella* in sheep. *J. Vet. Parasitol.* 12, 65.

66. Bannert, B., 1992. *Sarcocystis simonyi* sp. nov. (Apicomplexa: Sarcocystidae) from the endangered Hierro giant lizard *Gallotia simonyi* (Reptilia: Lacertidae). *Parasitol. Res.* 78, 142–145.

67. Bannert, B., 1994. Investigations on the host specificity of dihomoxenous sarcosporidia in the intermediate and definitive host. *J. Eukaryot. Microbiol.* 41, 183–188.

68. Barci, L.A.G., do Amaral, V., Santos, S.M., Rebouças, M.M., 1983. Sarcocistose caprina: Prevalência em animais provenientes do estado da Bahia-Brasil, com identificação do agente etiológico. *Biológico, São Paulo* 49, 97–102.

69. Barham, M., Stützer, H., Karanis, P., Latif, B.M., Neiss, W.F., 2005. Seasonal variation in *Sarcocystis* species infections in goats in northern Iraq. *Parasitology* 130, 151–156.

70. Barnett, D., Carter, J.K.Y., Hughes, D.E., Baetz, A.L., Fayer, R., 1977. Practicable diagnosis of acute bovine sarcocystosis causally related to bovine abortion. *20th Annual Proceedings of the American Association of Veterinary Laboratory Diagnosticians*, Minneapolis, USA, pp. 131–138.

71. Barr, B.C., Anderson, M.L., Blanchard, P.C., Daft, B.M., Kinde, H., Conrad, P.A., 1990. Bovine fetal encephalitis and myocarditis associated with protozoal infections. *Vet. Pathol.* 27, 354–361.

72. Barr, S.C., Warner, K., 2003. Characterization of a serine protease activity in *Sarcocystis neurona* merozoites. *J. Parasitol.* 89, 385–388.

73. Barrows, P.L., Smith, H.M. Jr., Prestwood, A.K., Brown, J., 1981. Prevalence and distribution of *Sarcocystis* sp. among wild swine of southeastern United States. *J. Am. Vet. Med. Assoc.* 179, 1117–1118.

74. Barrows, P.L., Prestwood, A.K., Green, C.E., 1982. Experimental *Sarcocystis suicanis* infections: Disease in growing pigs. *Am. J. Vet. Res.* 43, 1409–1412.

75. Barrows, P.L., Prestwood, A.K., Adams, D.D., Dykstra, M.J., 1982. Development of *Sarcocystis suicanis* Erber, 1977 in the pig. *J. Parasitol.* 68, 674–680.

76. Barta, J.R., 1997. Investigating phylogenetic relationships within the Apicomplexa using sequence data: The search for homology. *Methods* 13, 81–88.

77. Barta, J.R., Martin, D.S., Liberator, P.A., Dashkevicz, M., Anderson, J.W., Feighner, S.D., Elbrecht, A. et al. 1997. Phylogenetic relationships among eight *Eimeria* species infecting domestic fowl inferred using complete small subunit ribosomal DNA sequences. *J. Parasitol.* 83, 262–271.

78. Barta, J.R., Martin, D.S., Carreno, R.A., Siddall, M.E., Profous-Juchelkat, H., Hozza, M., Powles, M.A., Sundermann, C., 2001. Molecular phylogeny of the other tissue coccidia: *Lankesterella* and *Caryospora*. *J. Parasitol.* 87, 121–127.

79. Barutzki, D., Erber, M., Boch, J., 1981. Möglichkeiten der Desinfektion bei Kokzidiose (*Eimeria, Isospora, Toxoplasma, Sarcocystis*). *Berl. Münch. Tierärztl. Wochenschr.* 94, 451–455.

80. Barutzki, D., Schaper, R., 2003. Endoparasites in dogs and cats in Germany 1999–2002. *Parasitol. Res.* 90, S148–S150.

81. Bashtar, A.R., Abdel-Ghaffar, F.A., Ashour, M.B., Sakran, T., 1991. Endodyogony and cyst formation of *Sarcocystis gongyli* (Trinci 1911) from the skink *Chalcides ocellatus*. *J. Egypt. Soc. Parasitol.* 21, 383–390.

82. Bashtar, A.R., Al Aal, Z.A., Maarouf, W., Morsy, K., Al Quraishy, S., 2014. *Sarcocystis schneideri* n. sp. (Sarcocystidae) infecting the barber skink *Eumeces schneideri schneideri* (Scincidae) Daudin, 1802. A light and ultrastructural study. *Parasitol. Res.* 113, 2153–2159.

83. Baszler, T.V., Shkap, V., Mwangi, W., Davies, C.J., Mathison, B.A., Mazuz, M., Resnikov, D., Fish, L., Leibovitch, B., Staska, L.M., Savitsky, I., 2008. Bovine immune response to inoculation with *Neospora caninum* surface antigen SRS2 lipopeptides mimics immune response to infection with live parasites. *Clin. Vaccine Immunol.* 15, 659–667.

84. Beaver, P.C., Gadgil, R.K., Morera, P., 1979. *Sarcocystis* in man: A review and report of five cases. *Am. J. Trop. Med. Hyg.* 28, 819–844.

85. Beaver, P.C., Maleckar, J.R., 1981. *Sarcocystis singaporensis* Zaman and Colley, (1975) 1976, *Sarcocystis villivillosi* sp. n., and *Sarcocystis zamani* sp. n.: Development, morphology, and persistence in the laboratory rat, *Rattus norvegicus*. *J. Parasitol.* 67, 241–256.

86. Becker, B., Mehlhorn, H., Heydorn, A.O., 1979. Light and electron microscopic studies on gamogony and sporogony of 5 *Sarcocystis* species *in vivo* and in tissue cultures. *Zentralbl. Bakteriol. Hyg. I. Abt. Orig. A* 244, 394–404.

87. Beech, J., 1974. Equine protozoan encephalomyelitis. *Vet. Med. Small Anim. Clin.* 69, 1562–1566.

88. Beech, J., Dodd, D.C., 1974. *Toxoplasma*-like encephalomyelitis in the horse. *Vet. Pathol.* 11, 87–96.

89. Belchev, L., Arnaoudov, D., 1987. A rare case of sarcocystic encephalitis in sheep. *Veterinarnomeditsinski* 24, 33–37. (In Russian).

90. Beldomenico, P.M., Uhart, M., Bono, M.F., Marull, C., Baldi, R., Peralta, J.L., 2003. Internal parasites of free-ranging guanacos from Patagonia. *Vet. Parasitol.* 118, 71–77.

91. Bengis, R.G., Odening, K., Stolte, M., Quandt, S., Bockhardt, I., 1998. Three new *Sarcocystis* species, *Sarcocystis giraffae, S. klaseriensis*, and *S. camelopardalis* (Protozoa: Sarcocystidae) from the giraffe (*Giraffa camelopardalis*) in South Africa. *J. Parasitol.* 84, 562–565.

92. Bentz, B.G., Granstrom, D.E., Stamper, S., 1997. Seroprevalence of antibodies to *Sarcocystis neurona* in horses residing in a county of southeastern Pennsylvania. *J. Am. Vet. Med. Assoc.* 210, 517–518.

93. Bentz, B.G., Ealey, K.A., Morrow, J., Claypool, P.L., Saliki, J.T., 2003. Seroprevalence of antibodies to *Sarcocystis neurona* in equids residing in Oklahoma. *J. Vet. Diagn. Invest.* 15, 597–600.

94. Berentsen, A.R., Becker, M.S., Stockdale-Walden, H., Matandiko, W., McRobb, R., Dunbar, M.R., 2012. Survey of gastrointestinal parasite infection in African lion (*Panthera leo*), African wild dog (*Lycaon pictus*) and spotted hyaena (*Crocuta crocuta*) in the Luangwa Valley, Zambia. *African Zoology* 47, 363–368.

95. Bergler, K.-G., Erber, M., Boch, J., 1980. Untersuchungen zur Überlebensfähigkeit von Sporozysten bzw. Oozysten von *Sarcocystis*, *Toxoplasma*, *Hammondia* und *Eimeria* unter Labor-und Freilandbedingungen. *Berl. Münch. Tierärztl. Wochenschr.* 93, 288–293.

96. Bergmann, V., Kinder, E., 1975. Unterschiede in der Struktur der Zystenwand bei Sarkozysten des Schafes. *Monatsh. Veterinärmed.* 30, 772–774.

97. Bergmann, V., Kinder, E., 1976. Elektronenmikroskopische Untersuchungen zur Wandstruktur von Sarkozysten in der Skelettmuskulatur von Wildschwein und Reh. *Mh. Vet. Med.* 31, 785–788.

98. Berrocal, A., López, A., 2003. Pulmonary sarcocystosis in a puppy with canine distemper in Costa Rica. *J. Vet. Diagn. Invest.* 15, 292–294.

99. Bestetti, G., Fankhauser, R., 1978. Doppelinfektion des Gehirns mit *Frenkelia* und *Toxoplasma* bei einem Chinchilla. Licht-und elektronenmikroskopische Untersuchung. *Schweiz. Arch. Tierheilkd.* 120, 591–601.

100. Bhatavdekar, M.Y., Purohit, B.L., 1963. A record of sarcosporidiosis in lion. *Indian Vet. J.* 40, 44–45.

101. Bicknese, E.J., Murnane, R.D., Rideout, B.A., Bunte, R.M., Miller, L.A., 1993. A pathologic muscular form of *Sarcocystis* in two species of exotic columbiformes. *1992 Proceedings of Joint Meeting Am. Assoc. Zoo Vet. and Am. Assoc. Wildl. Vet.*, St. Louis, Missouri, pp. 186–190.

102. Biocca, E., 1968. Class Toxoplasmatea: Critical review and proposal of the new name *Frenkelia* gen. n. for M-organism. *Parassitologia* 10, 89–98.

103. Biocca, E., Balbo, T., Guarda, E., Costantini, R., 1975. L'importanza della volpe (*Vulpes vulpes*) nella transmissione della sarcosporidiosi dello stambeco (*Capra ibex*) nel Parco Nazionale del Gran Paradiso. *Parassitologia* 17, 17–24.

104. Bisby, T.M., Holman, P.J., Pitoc, G.A., Packer, R.A., Thompson, C.A., Raskin, R.E., 2010. *Sarcocystis* sp. encephalomyelitis in a cat. *Vet. Clin. Pathol.* 39, 105–112.

105. Bishop, L., 1979. Parasite-related lesions in a bearded seal, *Erignathus barbatus*. *J. Wildl. Dis.* 15, 285–293.

106. Biswas, S., Chakrabarti, A., Roy, S., 1990. *Sarcocystis* infection in buffalo meat (carabeef) in West Bengal. *J. Vet. Anim. Sci.* 21, 29–31.

107. Blagburn, B.L., Braund, K.G., Amling, K.A., Toivio-Kinnucan, M., 1989. Muscular *Sarcocystis* in a dog. *Proc. Helminthol. Soc. Wash.* 56, 207–210.

108. Blanchard, R., 1885. Note sur les sarcosporidies et sur un essai de classification de ces sporozoaires. *Soc. Zool. France. Bull.* 10, 244–276.

109. Blazejewski, T., Nursimulu, N., Pszenny, V., Dangoudoubiyam, S., Namasivayam, S., Chiasson, M.A., Chessman, K. et al. 2015. Systems based analysis of the *Sarcocystis neurona* genome identifies pathways that contribute to a heteroxenous life cycle. *mBio.* 6, e02445–14.

110. Blazek, K., Schramlová, J., Ippen, R., Kotrlý, A., 1978. Die sarkosporidiose des rehwildes (*Capreolus capreolus* L.). *Folia Parasitol. (Praha)* 25, 99–102.

111. Bledsoe, B., 1979. Sporogony of *Sarcocystis idahoensis* in the gopher snake, *Pituophis melanoleucus* (Daudin). *J. Parasitol.* 65, 875–879.

112. Bledsoe, B., 1980. *Sarcocystis idahoensis* sp. n. in deer mice *Peromyscus maniculatus* (Wagner) and gopher snakes *Pituophis melanoleucus* (Daudin). *J. Protozool.* 27, 93–102.

113. Bledsoe, B., 1980. Transmission studies with *Sarcocystis idahoensis* of deer mice (*Peromyscus maniculatus*) and gopher snakes (*Pituophis melanoleucus*). *J. Wildl. Dis.* 16, 195–200.

114. Blythe, L.L., Granstrom, D.E., Hansen, D.E., Walker, L.L., Bartlett, J., Stamper, S., 1997. Seroprevalence of antibodies to *Sarcocystis neurona* in horses residing in Oregon. *J. Am. Vet. Med. Assoc.* 210, 525–527.

115. Boch, J., Mannewitz, U., Erber, M., 1978. Sarkosporidien bei Schlachtschweinen in Süddeutschland. *Berl. Münch. Tierärztl. Wochenschr.* 91, 106–111.

116. Boch, J., Laupheimer, K.E., Erber, M., 1978. Drei Sarkosporidienarten bei Schlachtrindern in Süddeutschland. *Berl. Münch. Tierärztl. Wochenschr.* 91, 426–431.

117. Boch, J., Bierschenck, A., Erber, M., Weiland, G., 1979. *Sarcocystis*- und *Toxoplasma*-Infektionen bei Schlachtschafen in Bayern. *Berl. Münch. Tierärztl. Wochenschr.* 92, 137–141.

118. Boch, J., Hennings, R., Erber, M., 1980. Die wirtschaftliche Bedeutung der Sarkosporidiose (*Sarcocystis suicanis*) in der Schweinemast. Auswertung eines Feldversuches. *Berl. Münch. Tierärztl. Wochenschr.* 93, 420–423.

119. Boch, J., Erber, M., 1981. Vorkommen sowie wirtschaftliche und hygienische Bedeutung der Sarkosporidien von Rind, Schaf und Schwein. *Fleischwirtschaft* 61, 1–5.

120. Bogush, A.A., Pyshko, I., 1976. I. Experimental infection of the piglets with *Sarcocystis* [In Russian]. *Zhivotnovodstvu* 2, 46–50.

121. Böhm, M., White, P.C.L., Daniels, M.J., Allcroft, D.J., Munro, R., Hutchings, M.R., 2006. The health of wild red and sika deer in Scotland: An analysis of key endoparasites and recommendations for monitoring disease. *Vet. J.* 171, 287–294.

122. Bolon, B., Greiner, E.C., Calderwood Mays, M.B., 1989. Microscopic features of *Sarcocystis falcatula* in skeletal muscle from a patagonian conure. *Vet. Pathol.* 26, 282–284.

123. Bolten, K.E., Marsh, A.E., Reed, S.M., Dubey, J.P., Toribio, R.E., Saville, W.J.A., 2008. *Sarcocystis neurona*: Molecular characterization of enolase domain I region and a comparison to other protozoa. *Exp. Parasitol.* 120, 108–112.

124. Boonjaraspinyo, S., Boonmars, T., Kaewsamut, B., Ekobol, N., Laummaunwai, P., Aukkanimart, R., Wonkchalee, N., Juasook, A., Sriraj, P., 2013. A cross-sectional study on intestinal parasitic infections in rural communities, Northeast Thailand. *Korean J. Parasitol.* 51, 727–734.

125. Botelho, G.G., Lopes, C.W.G., 1984. Esporocistos de *Sarcocystis cruzi* (Apicomplexa: Sarcocystidae) em linfonodos mesentericos de cães. *Arq. Univ. Fed. Rur. Rio de Janeiro* 7, 87–88.

126. Bottner, A., Charleston, W.A.G., Hopcroft, D., 1987. The structure and the identity of macroscopically visible *Sarcocystis* cysts in cattle. *Vet. Parasitol.* 24, 35–45.

127. Böttner, A., Charleston, W.A.G., Pomroy, W.E., Rommel, M., 1987. The prevalence and identity of *Sarcocystis* in beef cattle in New Zealand. *Vet. Parasitol.* 24, 157–168.

128. Bourne, G., 1932. Sarcosporidia. *J. Roy. Soc. Western Aust.* 19, 1–8.

129. Bowler, B., Grandez, R., 2008. Electrocardiographical parameters in alpacas infected with *Sarcocystis lamacanis*. *Intern. J. Appl. Res. Vet. Med.* 6, 87–92.

130. Bowman, D.D., Iordanescu, M.A., Chou, H.H., Horton, K.H., 2001. *Sarcocystis neurona* and *Sarcocystis falcatula*: Monitoring of schizogony in cell culture using fluorescent nuclear labeling. *Vet. Parasitol.* 95, 353–356.

131. Box, E.D., Duszynski, D.W., 1977. Survey for *Sarcocystis* in the brown-headed cowbird (*Molothrus ater*): A comparison of macroscopic, microscopic and digestion techniques. *J. Wildl. Dis.* 13, 356–359.

132. Box, E.D., McGuinness, T.B., 1978. *Sarcocystis* in beef from retail outlets demonstrated by digestion technique. *J. Parasitol.* 64, 161–162.

133. Box, E.D., Duszynski, D.W., 1978. Experimental transmission of *Sarcocystis* from icterid birds to sparrows and canaries by sporocysts from the opossum. *J. Parasitol.* 64, 682–688.

134. Box, E.D., Marchiondo, A.A., Duszynski, D.W., Davis, C.P., 1980. Ultrastructure of *Sarcocystis* sporocysts from passerine birds and opossums: Comments on classification of the genus *Isospora*. *J. Parasitol.* 66, 68–74.

135. Box, E.D., Smith, J.H., 1982. The intermediate host spectrum in a *Sarcocystis* species of birds. *J. Parasitol.* 68, 668–673.

136. Box, E.D., 1983. Recovery of *Sarcocystis* sporocysts from feces after oral administration. *Proc. Helminthol. Soc. Wash.* 50, 348–350.

137. Box, E.D., Meier, J.L., Smith, J.H., 1984. Description of *Sarcocystis falcatula* Stiles, 1893, a parasite of birds and opossums. *J. Protozool.* 31, 521–524.

138. Boy, M.G., Galligan, D.T., Divers, T.J., 1990. Protozoal encephalomyelitis in horses: 82 cases (1972–1986). *J. Am. Vet. Med. Assoc.* 196, 632–634.

139. Bradley, J.A., Taylor, C.M., 1993. Disposition of feedlot heifer and beef carcasses under a Canadian streamlined inspection system. *Can. Vet. J.* 34, 38–40.

140. Brehm, H., Frank, W., 1980. Der Entwicklungskreislauf von *Sarcocystis singaporensis* Zaman und Colley, 1976 im End- und Zwischenwirt. *Z. Parasitenkd.* 62, 15–30.

141. Brewer, B.D., Mayhew, I.G., 1988. Multifocal neurologic disease in a horse. *Equine Vet. Sci.* 8, 302–304.

142. Briggs, M.B., Leathers, C.W., Foreyt, W.J., 1993. *Sarcocystis felis* in captive cheetahs (*Acinonyx jubatus*). *J. Helminthol. Soc. Wash.* 60, 277–279.

143. Brindani, F., Perini, S., Cabassi, E., Marastoni, G., 1982. Diffusione della sarcosporidiosi nei suini macellati in provincia di Reggio Emilia 2a. Nota: suini all'ingrasso. *Sel. Vet. Ist. Zooprofil. Sper. Lomb. Emilia* 24, 57–60.

144. Brindley, P.J., Gazzinelli, R.T., Denkers, E.Y., Davis, S.W., Dubey, J.P., Belfort, R., Jr., Martins, M.C., Silveira, C., Jamra, L.M.F., Waters, A.P., Sher, A., 1993. Differentiation of *Toxoplasma gondii* from closely related coccidia by riboprint analysis and a surface antigen gene polymerase chain reaction. *Am. J. Trop. Med. Hyg.* 48, 447–456.

145. Britton, A.P., Dubey, J.P., Rosenthal, B.M., 2010. Rhinitis and disseminated disease in a ferret (*Mustela putorius furo*) naturally infected with *Sarcocystis neurona*. *Vet. Parasitol.* 169, 226–231.

146. Brown, C.M., Morrow, J.K., Carleton, C.L., Ramanathan, B., Reddy, R., Vaidya, V., Karthikeyan, S.M., Zulfikar, A.A., Kannadkar, V.S., 2006. Persistence of serum antibodies to *Sarcocystis neurona* in horses moved from North America to India. *J. Vet. Intern. Med.* 20, 994–997.

147. Brown, R.J., Smith, A.W., Keyes, M.C., 1974. *Sarcocystis* in the northern fur seal. *J. Wildl. Dis.* 10, 53.

148. Brumloop, A., Sager, M., 1993. Sarkosporidien als Ursache von plötzlichen Todesfällen in der Narkose beim Schaf. *Berl. Münch. Tierärztl. Wschr.* 106, 80–83.

149. Brumpt, E., 1936. *Precis de parasitologie. Collection de Précis Médieux.* Masson et cie, Paris.

150. Bucca, M., Brianti, E., Giuffrida, A., Ziino, G., Cicciari, S., Panebianco, A., 2011. Prevalence and distribution of *Sarcocystis* spp. cysts in several muscles of cattle slaughtered in Sicily, Southern Italy. *Food Control* 22, 105–108.

151. Bundza, A., Feltmate, T.E., 1989. Eosinophilic myositis/lymphadenitis in slaughter cattle. *Can. Vet. J.* 30, 514–516.

152. Bunyaratvej, S., Bunyawongwiroj, P., Nitiyanant, P., 1982. Human intestinal sarcosporidiosis: Report of six cases. *Am. J. Trop. Med. Hyg.* 31, 36–41.

153. Burcham, G.N., Ramos-Vara, J.A., Vemulapalli, R., 2010. Systemic sarcocystosis in a striped skunk (*Mephitis mephitis*). *Vet. Pathol.* 47, 560–564.

154. Burgess, D.E., Speer, C.A., Reduker, D.W., 1988. Identification of antigens of *Sarcocystis cruzi* sporozoites, merozoites and bradyzoites with monoclonal antibodies. *J. Parasitol.* 74, 828–832.

155. Burtscher, H., 1983. Große Coccidien-Gewebszysten im Gehirn von *Gracula religiosa* (Aves: Sturnidae). *Zbl. Vet. Med. B.* 30, 590–599.

156. Bussieras, J., 1994. An example of holozoonosis: Human coccidiosis due to *Sarcocystis* spp. *Bull. Acad. Natle. Méd.* 178, 613–622; discussion 622-3.

157. Butcher, M., Lakritz, J., Halaney, A., Branson, K., Gupta, G.D., Kreeger, J., Marsh, A.E., 2002. Experimental inoculation of domestic cats (*Felis domesticus*) with *Sarcocystis neurona* or *S. neurona*-like merozoites. *Vet. Parasitol.* 107, 1–14.

158. Butkauskas, D., Sruoga, A., Kutkiené, L., Prakas, P., 2007. Investigation of the phylogenetic relationships of *Sarcocystis* spp. from greylag (*Anser anser*) and white-fronted (*Anser albifrons*) geese to other cyst forming coccidia using 18S and 28S rRNA gene sequences. *Acta Zool. Lituanica* 17, 124–128.

159. Bwangamoi, O., Ngatia, T.A., Richardson, J.D., 1993. *Sarcocystis*-like organisms in musculature of a domestic dog (*Canis familiaris*) and wild dogs (*Lycaon pictus*) in Kenya. *Vet. Parasitol.* 49, 201–205.

160. Caldow, G.L., Gidlow, J.R., Schock, A., 2000. Clinical, pathological and epidemiological findings in three outbreaks of ovine protozoan myeloencephalitis. *Vet. Rec.* 146, 7–10.

161. Calero-Bernal, R., Verma, S.K., Oliveira, S., Yang, Y., Rosenthal, B.M., Dubey, J.P., 2015. In the United States, negligible rates of zoonotic sarcocystosis occur in feral swine that, by contrast, frequently harbor infections with *Sarcocystis meischeriana*, a related parasite contracted from canids. *Parasitology* 142, 549–556.

162. Calero-Bernal, R., Verma, S.K., Seaton, C.T., Sinnett, D., Ball, E., Dunams, D., Rosenthal, B.M., Dubey, J.P., 2015. *Sarcocystis cruzi* infection in wood bison (*Bison bison athabascae*). *Vet. Parasitol.* 210, 102–105.

163. Camisasca, S., Corsico, G., Tessuto, L., Scanziani, E., Genchi, C., Benedetti, G., Alfonsi, R., Crippa, L., 1996. Sarcocistosi in bufali allevati e macellati in Italia. *Ingegneria Alimentare* 12, 9–12.

164. Carletti, T., Martin, M., Romero, S., Morrison, D.A., Marcoppido, G., Florin-Christensen, M., Schnittger, L., 2013. Molecular identification of *Sarcocystis aucheniae* as the macrocyst-forming parasite of llamas. *Vet. Parasitol.* 198, 396–400.

165. Carlson-Bremer, D.P., Gulland, F.M.D., Johnson, C.K., Colegrove, K.M., Van Bonn, W.G., 2012. Diagnosis and treatment of *Sarcocystis neurona*-induced myositis in a free-ranging California sea lion. *J. Am. Vet. Med. Assoc.* 240, 324–328.

166. Carreno, R.A., Schnitzler, B.E., Jeffries, A.C., Tenter, A.M., Johnson, A.M., Barta, J.R., 1998. Phylogenetic analysis of Coccidia based on 18S rDNA sequence comparison indicates that *Isospora* is most closely related to *Toxoplasma* and *Neospora*. *J. Eukaryot. Microbiol.* 45, 184–188.

167. Carrigan, M.J., 1986. An outbreak of sarcocystosis in dairy cattle. *Aust. Vet. J.* 63, 22–24.

168. Caspari, K., Grimm, F., Kühn, N., Caspari, N.C., Basso, W., 2011. First report of naturally acquired clinical sarcocystosis in a pig breeding stock. *Vet. Parasitol.* 177, 175–178.

169. Castro, C.E., Sam, T.R., López, U.T., González, Z.A., Silva, I.M., 2004. Evaluación de la edad como factor de riesgo de seropositividad a *Sarcocystis* sp. en alpacas. *Rev. Inv. Vet. Perú* 15, 83–86.

170. Castro, J., Lequía, G., 1992. Prevalencia de *Sarcocystis* sp. en vacunos, ovinos y caprinos beneficiados en los camales de Lima. *Rev. Per. Biol.* 4, 21–24.

171. Cawthorn, R.J., Rainnie, D., Wobeser, G., 1981. Experimental transmission of *Sarcocystis* sp. (Protozoa: Sarcocystidae) between the shoveler (*Anas clypeata*) duck and the striped skunk (*Mephitis mephitis*). *J. Wildl. Dis.* 17, 389–394.

172. Cawthorn, R.J., Wobeser, G.A., Gajadhar, A.A., 1983. Description of *Sarcocystis campestris* sp. n. (Protozoa: Sarcocystidae): A parasite of the badger *Taxidea taxus* with experimental transmission to the Richardson's ground squirrel, *Spermophilus richardsonii*. *Can. J. Zool.* 61, 370–377.

173. Cawthorn, R.J., Gajadhar, A.A., Brooks, R.J., 1984. Description of *Sarcocystis rauschorum* sp. n. (Protozoa: Sarcocystidae) with experimental cyclic transmission between varying lemmings (*Dicrostonyx richardsoni*) and snowy owls (*Nyctea scandiaca*). *Can. J. Zool.* 62, 217–225.

174. Cawthorn, R.J., Brooks, R.J., 1985. Histological observations on precystic merogony and metrocyte formation of *Sarcocystis rauschorum* (Apicomplexa: Sarcocystidae) in varying lemmings, *Dicrostonyx richardsoni*. *Can. J. Zool.* 63, 2907–2912.

175. Cawthorn, R.J., Brooks, R.J., 1985. Light microscopical observations on sporogony of *Sarcocystis rauschorum* (Protozoa: Sarcocystidae) in snowy owls (*Nyctea scandiaca*). *Can. J. Zool.* 63, 1455–1458.

176. Cawthorn, R.J., Reduker, D.W., Speer, C.A., Dubey, J.P., 1986. *In vitro* excystation of *Sarcocystis capracanis*, *Sarcocystis cruzi* and *Sarcocystis tenella* (Apicomplexa). *J. Parasitol.* 72, 880–884.

177. Cawthorn, R.J., Markham, R.J.F., Hitt, N.D., Despres, D., 1990. *In vitro* cultivation of the vascular phase of *Sarcocystis hirsuta* (Apicomplexa). *Can. J. Zool.* 68, 1068–1070.

178. Cawthorn, R.J., Clark, M., Hudson, R., Friesen, D., 1990. Histological and ultrastructural appearance of severe *Sarcocystis fayeri* infection in a malnourished horse. *J. Vet. Diagn. Invest.* 2, 342–345.

179. Cawthorn, R.J., 1993. Cyst-forming Coccidia of raptors: Significant pathogens or not? In: Reding, P.T., Cooper, J.E., Remple, J.D., Hunter, D.B. (Eds.), *Raptor Biomedicine*. University of Minnesota Press, Minneapolis, pp. 14–20.

180. Černà, Z., 1976. Relationship of oocysts of "*Isospora buteonis*" from the barn-owl (*Tyto alba*) to muscle cysts of sarcosporidians from the house mouse (*Mus musculus*). *Folia Parasitol. (Praha)* 23, 285.

181. Černà, Z., 1976. Sarcocystosis of mice (*Mus musculus*) acquired from oocysts isolated from the intestines of the owl (*Tyto alba*). *J. Protozool.* 23, 6A.

182. Černà, Z., Loučková, M., 1976. *Microtus arvalis* as the intermediate host of a coccidian from the kestrel (*Falco tinnunculus*). *Folia Parasitol. (Praha)* 23, 110.

183. Černà, Z., 1977. Cycle de dévelopment sarcosporidien d'une coccidie, chez la souris, après infestation des animaux par des oocystes- sporocystes isolés de l'intestin de la chouette effraie (*Tyto alba*). *Protistologica* 13, 401–405.

184. Černà, Z., Senaud, J., 1977. Sur un type nouveau de multiplication asexuée d'une sarcosporidie, dans le foie de la Souris. *C. R. Seances Acad. Sci.* 285, 347–349.

185. Černà, Z., Louckova, M., 1977. *Microtus arvalis*, the intermediate host of a coccidian from the kestrel (*Falco tinnunculus*). *Vest. Cs. Spolec. Zool.* 41, 1–4.

186. Černà, Z., Kolářová, I., 1978. Contribution to the serologic diagnosis of sarcocystosis. *Folia Parasitol. (Praha)* 25, 289–292.

187. Černà, Z., Kolářová, I., Sulc, P., 1978. *Sarcocystis cernae* Levine, 1977, excystation, life-cycle and comparison with other heteroxenous coccidians from rodents and birds. *Folia Parasitol. (Praha)* 25, 201–207.

188. Černà, Z., Kolářová, I., Sulc, P., 1978. Contribution to the problem of cyst-producing coccidians. *Folia Parasitol. (Praha)* 25, 9–16.

189. Černà, Z., Loučková, M., Nedvedová, H., Vavra, J., 1981. Spontaneous and experimental infection of domestic rabbits by *Sarcocystis cuniculi* Brumpt, 1913. *Folia Parasitol. (Praha)* 28, 313–318.

190. Černà, Z., 1983. Multiplication of merozoites of *Sarcocystis dispersa* Černà, Kolářová et Šulc, 1978 and *Sarcocystis cernae* Levine, 1977 in the blood stream of the intermediate host. *Folia Parasitol. (Praha)* 30, 5–8.

191. Černà, Z., Pecka, Z., 1984. Muscle sarcocystosis in pheasants and first records of the genus *Sarcocystis* in *Phasianus colchicus* Linné, 1758 in Czechoslovakia. *Folia Parasitol. (Praha)* 31, 85–88.

192. Černà, Z., Kvašnovská, Z., 1986. Life cycle involving bird-bird relation in *Sarcocystis* coccidia with the description of *Sarcocystis accipitris* sp. n. *Folia Parasitol. (Praha)* 33, 305–309.

193. Červa, L., Černà, Z., 1982. Indirect haemagglutination reaction with *Sarcocystis dispersa* antigen. *Folia Parasitol. (Praha)* 29, 219–225.

194. Chabreck, R.H., 1965. Sarcosporidiosis in ducks in Louisiana. *Trans. North Am. Wildl. Conf.* 30, 174–184.

195. Challis, J.R.G., Lye, S.J., 1986. Parturition. In: Clark, J.R. (Ed.), *Oxford Reviews of Reproductive Biology*. Clarendon Press, Oxford, UK, pp. 61–112.

196. Chapman, J., Mense, M., Dubey, J.P., 2005. Clinical muscular sarcocystosis in a dog. *J. Parasitol.* 91, 187–190.

197. Charles, R.A., Ellis, A.E., Dubey, J.P., Barnes, J.C., Yabsley, M.J., 2011. Besnoitiosis in a southern plains woodrat (*Neotoma micropus*) from Uvalde, Texas. *J. Parasitol.* 97, 838–841.

198. Charles, R.A., Kjos, S., Ellis, A.E., Dubey, J.P., Shock, B.C., Yabsley, M.J., 2012. Parasites and vector-borne pathogens of southern plains woodrats (*Neotoma micropus*) from southern Texas. *Parasitol. Res.* 110, 1855–1862.

199. Chaudhry, R.K., Shah, H.L., 1988. Role of man in the life-cycle of *Sarcocystis* sp. of the goat (*Capra hircus*). *Indian Vet. J.* 65, 742.

200. Chávez, V.A., Leyva, V.V., Panez, L.S., Ticona, S.D., García, V.W., Pezo, C.D., 2008. Sarcocistiosis y la eficiencia productiva de la alpaca. *Rev. Inv. Vet. Perú* 19, 160–167.

201. Cheadle, M.A., 2001. *Sarcocystis greineri* n. sp. (Protozoa: Sarcocystidae) in the Virginia opossum (*Didelphis virginiana*). *J. Parasitol.* 87, 1085–1089.

202. Cheadle, M.A., Dame, J.B., Greiner, E.C., 2001. Sporocyst size of isolates of *Sarcocystis* shed by the Virginia opossum (*Didelphis virginiana*). *Vet. Parasitol.* 95, 305–311.

203. Cheadle, M.A., Tanhauser, S.M., Scase, T.J., Dame, J.B., MacKay, R.J., Ginn, P.E., Greiner, E.C., 2001. Viability of *Sarcocystis neurona* sporocysts and dose titration in gamma-interferon knockout mice. *Vet. Parasitol.* 95, 223–231.

204. Cheadle, M.A., Yowell, C.A., Sellon, D.C., Hines, M., Ginn, P.E., Marsh, A.E., Dame, J.B., Greiner, E.C., 2001. The striped skunk (*Mephitis mephitis*) is an intermediate host for *Sarcocystis neurona*. *Int. J. Parasitol.* 31, 843–849.

205. Cheadle, M.A., Tanhauser, S.M., Dame, J.B., Sellon, D.C., Hines, M., Ginn, P.E., MacKay, R.J., Greiner, E.C., 2001. The nine-banded armadillo (*Dasypus novemcinctus*) is an intermediate host for *Sarcocystis neurona*. *Int. J. Parasitol.* 31, 330–335.

206. Cheadle, M.A., Cunningham, M.W., Greiner, E.C., 2002. Prevalence of sarcocysts in Florida black bears (*Ursus americanus floridanus*). *J. Parasitol.* 88, 629–630.

207. Cheadle, M.A., Ginn, P.E., Lindsay, D.S., Greiner, E.C., 2002. Neurologic disease in gamma-interferon gene knockout mice caused by *Sarcocystis neurona* sporocysts collected from opossums fed armadillo muscle. *Vet. Parasitol.* 103, 65–69.

208. Cheadle, M.A., Lindsay, D.S., Greiner, E.C., 2006. Lack of *Sarcocystis neurona* antibody response in Virginia opossums (*Didelphis virginiana*) fed *Sarcocystis neurona*-infected muscle tissue. *J. Parasitol.* 92, 652–654.

209. Chen, L.Y., Zhou, B.J., Yang, Z.Q., Li, C.Y., Attwood, S.W., Wang, W.L., 2007. Effects of frozen storage on the structure of sarcocysts in pig muscle and implications in taxonomic studies. *Exp. Parasitol.* 115, 393–398.

210. Barbosa, L., Johnson, C.K., Lambourn, D.M., Gibson, A.K., Haman, K.H., Huggins, J.L., Sweeny, A.R., Sundar, N., Raverty, S.A., Grigg, M.E., 2015. A novel *Sarcocystis neurona* genotype XIII is associated with severe encephalitis in an unexpectedly broad range of marine mammals from the northeastern Pacific Ocean. *Int. J. Parasitol.*, in press.

211. Chen, X., Zuo, Y., Rosenthal, B.M., He, Y., Cui, L., Yang, Z., 2011. *Sarcocystis sinensis* is an ultra-structurally distinct parasite of water buffalo that can cause foodborne illness but cannot complete its life-cycle in human beings. *Vet. Parasitol.* 178, 35–39.

212. Chen, X., He, Y., Liu, Y., Olias, P., Rosenthal, B.M., Cui, L., Zuo, Y., Yang, Z., 2012. Infections with *Sarcocystis wenzeli* are prevalent in the chickens of Yunnan Province, China, but not in the flocks of domesticated pigeons or ducks. *Exp. Parasitol.* 131, 31–34.

213. Chen, X.W., Zuo, Y.X., Hu, J.J., 2003. Experimental *Sarcocystis hominis* infection in a water buffalo (*Bubalus bubalis*). *J. Parasitol.* 89, 393–394.

214. Chhabra, M.B., Samantaray, S., 2013. *Sarcocystis* and sarcocystosis in India: Status and emerging perspectives. *J. Parasit. Dis.* 37, 1–10.

215. Chileno, M.M., Chávez, V.A., Casas, A.E., Chavera, C.A., Puray, C.N., 2011. Efectos tóxicos del contenido de dos tamaños de quistes de *Sarcocystis aucheniae* en conejos inoculados experimentalmente. *Rev. Inv. Vet. Perú* 22, 360–368.

216. Chiwy, M.M., 1926. Sur la sarcosporidiose. *Ann. Méd. Vét.* 71, 64–67.

217. Clark, E.G., Townsend, H.G.G., McKenzie, N.T., 1981. Equine protozoal myeloencephalitis: A report of two cases from western Canada. *Can. Vet. J.* 22, 140–144.

218. Clavel, A., Doiz, O., Varea, M., Morales, S., Castillo, F.J., Rubio, M.C., Gómez-Lus, R., 2001. Molestias abdominales y heces blandas en consumidor habitual de carne de vacuno poco cocinada. *Enferm. Infecc. Microbiol. Clin.* 19, 29–30.

219. Claveria, F.G., Petersen, B., Macabagdal, M.R., Farolan, R.J., Farrol, M.A., Gonzalvo, F., Cadiz, R., Ajero, R., Roque, R., Lozano, G., 1997. A survey of bovine, bubaline and swine sarcocystosis in the Philippines. *Southeast Asian J. Trop. Med. Pub. Health* 28(Suppl 1), 173–178.

220. Claveria, F.G., Cruz, M.J., 1999. Light microscopic and ultrastructural studies on *Sarcocystis* spp. infection in Philippine water buffaloes (*Bubalus bubalis*). *J. Protozool. Res.* 9, 17–25.

221. Claveria, F.G., Farolan, R.J., Macabagdal, M.R., Criss, A., Salvador, R., Lim, R.S., 1999. Light microscopic studies of *Sarcocystis* spp. infection in cattle slaughtered in three different abattoirs in metro Manila. *J. Protozool. Res.* 9, 26–31.

222. Claveria, F.G., Cruz, M.J., 2000. *Sarcocystis levinei* infection in Philippine water buffaloes (*Bubalus bubalis*). *Parasitol. Int.* 48, 243–247.

223. Claveria, F.G., Cruz, M.J., Lim, R.S., 2000. *Sarcocystis* spp. infection in Philippine water buffaloes (*Bubalus bubalis*). *Southeast Asian J. Trop. Med. Pub. Health* 31(Suppl 1), 44–47.

224. Claveria, F.G., De La Peña, C., Cruz-Flores, M.J., 2001. *Sarcocystis meischeriana* infection in domestic pigs (*Sus scrofa*) in the Philippines. *J. Parasitol.* 87, 938–939.

225. Claveria, F.G., San-Pedro Lim, R., Cruz-Flores, M.J., Nagasawa, H., Suzuki, N., De La Peña, C., 2001. Ultrastructural studies of *Sarcocystis cruzi* (Hasselmann, 1926) Wenyon, 1926 infection in cattle (*Bos taurus*): Philippine cases. *Parasite* 8, 251–254.

226. Claveria, F.G., San Pedro-Lim, M.R., Tan, J.E., Flores-Cruz, M.J., 2004. *Sarcocystis capracanis* infection in Philippine domestic goats (*Capra hircus*): Ultrastructural studies. *Philippine J. Sci.* 133, 33–37.

227. Clegg, F.G., Beverley, J.K.A., Markson, L.M., 1978. Clinical disease in cattle in England resembling Dalmeny disease associated with suspected *Sarcocystis* infection. *J. Comp. Pathol.* 88, 105–114.

228. Clubb, S.L., Frenkel, J.K., 1992. *Sarcocystis falcatula* of opossums: Transmission by cockroaches with fatal pulmonary disease in Psittacine birds. *J. Parasitol.* 78, 116–124.

229. Clubb, S.L., Frenkel, J.K., Gardiner, C.H., Graham, D.L., 1996. An acute fatal illness in old world psittacine birds associated with *Sarcocystis falcatula* of opossums. *Ann. Proc. Assoc. Avian Vet.* 139–149.

230. Cohen, N.D., McKay, R.J., 1997. Interpreting immunoblot testing of cerebrospinal fluid for equine protozoal myeloencephalitis. Compendium of Continuing Education. *Pract. Vet. Equine Forum*, 1176–1181.

231. Cole, D.J.W., 1982. Attempted transmission of ovine *Sarcocystis* from cats to SPF lambs. *New Zealand J. Zool.* 9, 48.

232. Collery, P., Weavers, E., 1981. An outbreak of *Sarcocystis* in calves in Ireland. *Irish Vet. J.* 35, 159–162.

233. Collery, P., 1987. The pathogenesis of acute bovine sarcocystosis I. Clinical signs and anaemia. *Irish Vet. J.* 41, 273–280.

234. Collery, P., 1988. The pathogenesis of acute bovine sarcocystosis II. Bone marrow and peripheral blood white cell response. *Irish Vet. J.* 43, 33–39.

235. Collery, P., 1989. The pathogenesis of acute bovine sarcocystosis III. *In vitro* studies on the interactions between erythrocytes and mononuclear phagocytes. *Irish Vet. J.* 42, 69–74.

236. Collery, P., 1989. The pathogenesis of acute bovine sarcocystosis IV: The effects of corticosteroid therapy on the course of the anaemia. *Irish Vet. J.* 42, 85–92.

237. Collins, G.H., Crawford, S.J.S., 1978. *Sarcocystis* in goats: Prevalence and transmission. *N. Z. Vet. J.* 26, 288.

238. Collins, G.H., Charleston, W.A.G., 1979. Studies on *Sarcocystis* species II. Infection in wild and feral animals–prevalence and transmission. *N. Z. Vet. J.* 27, 134–135.

239. Collins, G.H., Charleston, W.A.G., 1979. Studies on *Sarcocystis* species IV. A species infecting dogs and goats; development in goats. *N. Z. Vet. J.* 27, 260–262.

240. Collins, G.H., Atkinson, E., Charleston, W.A.G., 1979. Studies on *Sarcocystis* species III. The macrocystic species of sheep. *N. Z. Vet. J.* 27, 204–206.

241. Collins, G.H., 1980. *Sarcocystis and the Meat Industry.* Monograph Massey University, New Zealand, pp. 1–15.

242. Collins, G.H., Charleston, W.A.G., 1980. Studies on *Sarcocystis* species VII. The effect of temperature on the viability of macrocyst (*Sarcocystis gigantea*) of sheep. *N. Z. Vet. J.* 28, 189–191.

243. Collins, G.H., Charleston, W.A.G., Wiens, B.G., 1980. Studies on *Sarcocystis* species VI. A comparison of three methods for the detection of *Sarcocystis* species in muscle. *N. Z. Vet. J.* 28, 173.

244. Colwell, D.D., Mahrt, J.L., 1981. Ultrastructure of the cyst wall merozoites of *Sarcocystis* from moose (*Alces alces*) in Alberta, Canada. *Z. Parasitenkd.* 65, 317–329.

245. Colwell, D.D., Mahrt, J.L., 1983. Development of *Sarcocystis alceslatrans* Dubey, 1980, in the small intestine of dogs. *Am. J. Vet. Res.* 44, 1813–1818.

246. Conder, G.A., Loveless, R.M., 1978. Parasites of the coyote (*Canis latrans*) in central Utah. *J. Wildl. Dis.* 14, 247–249.

247. Cook, A.G., Buechner-Maxwell, V., Morrow, J.K., Ward, D.L., Parker, N.A., Dascanio, J.J., Ley, W.B., Cooper, W., 2001. Interpretation of the detection of *Sarcocystis neurona* antibodies in the serum of young horses. *Vet. Parasitol.* 95, 187–195.

248. Cook, A.G., Maxwell, V.B., Donaldson, L.L., Parker, N.A., Ward, D.L., Morrow, J.K., 2002. Detection of antibodies against *Sarcocystis neurona* in cerebrospinal fluid from clinically normal neonatal foals. *J. Am. Vet. Med. Assoc.* 220, 208–211.

249. Cooley, A.J., Barr, B., Rejmanek, D., 2007. *Sarcocystis neurona* encephalitis in a dog. *Vet. Pathol.* 44, 956–961.

250. Coombs, D.W., Springer, M.D., 1974. Parasites of feral pig x European wild boar hybrids in Southern Texas. *J. Wildl. Dis.* 10, 436–441.

251. Cordovés Céspedes, C.O., Mernio, M., Fernández Alvarez, B., 1981. Primer reporte de *Sarcocystis* en fetos bovinos. *Acad. Ciencias Cuba Ser. Biol.* 155, 2–6.

252. Cornaglia, E., Misciattelli, M.E., Guarda, F., Lanfranchi, P., 1980. Aspetti istopatologici ed ultrastrutturali di protozoi della famiglia Sarcocystidae (Poche 1913) a sede cardiaca, muscolare ed encefalica in micromammiferi di differenti habitat d'Italia. *Ann. Fac. Med. Vet. Torino* 27, 355–374.

253. Cornaglia, E., Giaccherino, A.R., Peracino, V., 1998. Ultrastructural morphology of sarcosporidiosis in Alpine ibex (*Capra ibex*). *Vet. Parasitol.* 75, 21–32.

254. Cornejo, B.R., Chávez, V.A., Leyva, V.V., Falcón, P.N., Panez, L.S., Ticona, S.D., 2007. Relación entre el tamaño de los macroquistes de *Sarcocystis aucheniae* y su viabilidad en *Canis familiaris*. *Rev. Inv. Vet. Perú* 18, 76–83.

255. Cornelissen, A.W.C.A., Overdulve, J.P., van der Ploeg, M., 1984. Determination of nuclear DNA of five eucoccidian parasites, *Isospora* (*Toxoplasma*) *gondii*, *Sarcocystis cruzi*, *Eimeria tenella*, *E. acervulina* and *Plasmodium berhei*, with special reference to gamontogenesis and meiosis in *I.* (*T.*) *gondii*. *Parasitology* 88, 531–553.

256. Corner, A.H., Mitchell, D., Meads, E.B., Taylor, P.A., 1963. Dalmeny disease. An infection of cattle presumed to be caused by an unidentified protozoon. *Can. Vet. J.* 4, 252–264.

257. Cornwell, G., 1963. New waterfowl host records for *Sarcocystis rileyi* and a review of sarcosporidiosis in birds. *Avian Dis.* 7, 212–216.

258. Cosgrove, M., Wiggins, J.P., Rothenbacher, H., 1982. *Sarcocystis* sp. in the eastern cottontail (*Sylvilagus floridanus*). *J. Wildl. Dis.* 18, 37–40.

259. Costanzo, G.R., 1990. *Sarcocystis* in American black ducks wintering in New Jersey. *J. Wildl. Dis.* 26, 387–389.

260. Crawley, H., 1914. Two new Sarcosporidia. *Proc. Natl. Acad. Sci.* (*Philadelphia*) 66, 214–218.

261. Crawley, R.R., Ernst, J.V., Milton, J.L., 1982. *Sarcocystis* in the bald eagle (*Haliaeetus leucocephalus*). *J. Wildl. Dis.* 18, 253–255.

262. Cray, C., Zielezienski-Roberts, K., Bonda, M., Stevenson, R., Ness, R., Clubb, S., Marsh, A., 2005. Serologic diagnosis of sarcocystosis in psittacine birds: 16 cases. *J. Avian Med. Surg.* 19, 208–215.

263. Crowdus, C.A., Marsh, A.E., Saville, W.J., Lindsay, D.S., Dubey, J.P., Granstrom, D.E., Howe, D.K., 2008. SnSAG5 is an alternative surface antigen of *Sarcocystis neurona* strains that is mutually exclusive to SnSAG1. *Vet. Parasitol.* 158, 36–43.

264. Crum, J.M., Prestwood, A.K., 1977. Transmission of *Sarcocystis leporum* from a cottontail rabbit to domestic cats. *J. Wildl. Dis.* 13, 174–175.

265. Crum, J.M., Nettles, V.F., Davidson, W.R., 1978. Studies on endoparasites of the black bear (*Ursus americanus*) in the southeastern United States. *J. Wildl. Dis.* 14, 178–186.

266. Crum, J.M., Fayer, R., Prestwood, A.K., 1981. *Sarcocystis* spp. in white-tailed deer I. Definitive and intermediate host spectrum with a description of *Sarcocystis odocoileocanis* n. sp. *J. Wildl. Dis.* 17, 567–579.

267. Crum, J.M., Prestwood, A.K., 1982. Prevalence and distribution of *Sarcocystis* spp. among white-tailed deer of the southeastern United States. *J. Wildl. Dis.* 18, 195–203.

268. Cummings, C.A., Kocan, A.A., Barker, R.W., Dubey, J.P., 2000. Muscular sarcocystosis in coyotes from Oklahoma. *J. Wildl. Dis.* 36, 761–763.

269. Cunningham, C.C., 1973. Sarcocysts in the heart muscle of a foal. *Vet. Rec.* 92, 684.

270. Cusick, P.K., Sells, D.M., Hamilton, D.P., Hardenbrook, H.J., 1974. Toxoplasmosis in two horses. *J. Am. Vet. Med. Assoc.* 164, 77–80.

271. Cutler, T.J., MacKay, R.J., Ginn, P.E., Greiner, E.C., Porter, R., Yowell, C.A., Dame, J.B., 1999. Are *Sarcocystis neurona* and *Sarcocystis falcatula* synonymous? A horse infection challenge. *J. Parasitol.* 85, 301–305.

272. Cutler, T.J., MacKay, R.J., Ginn, P.E., Gillis, K., Tanhauser, S.M., LeRay, E.V., Dame, J.B., Greiner, E.C., 2001. Immunoconversion against *Sarcocystis neurona* in normal and dexamethasone-treated horses challenged with *S. neurona* sporocysts. *Vet. Parasitol.* 95, 197–210.

273. D'Haese, J., Mehlhorn, H., Peters, W., 1977. Comparative electron microscope study of pellicular structures in coccidia (*Sarcocystis, Besnoitia,* and *Eimeria*). *Int. J. Parasitol.* 7, 505–518.

274. da Cruz Brandão, F., Chaplin, E.L., Fernandes, R.E., Nunes, P.C.S., 1989. Encefalite ovina: Registro de um caso em carneiro importado, no municipio de Marau, RS, Brasil. *Arquivos da Faculdade de Veterinaria UFRGS* 17, 29–34.

275. da Silva, N.R.S., Rodrigues, R.J.D., Araújo, F.A.P., Beck, C., Olicheski, A.T., 2002. Detection of bovine *Sarcocystis cruzi* cysts in cardiac muscles: a new technique of concentration for diagnostic. *Acta Sci. Vet.* 30, 127–129.

276. da Silva, R.C., Su, C., Langoni, H., 2009. First identification of *Sarcocystis tenella* (Railliet, 1886) Moulé, 1886 (Protozoa: Apicomplexa) by PCR in naturally infected sheep from Brazil. *Vet. Parasitol.* 165, 332–336.

277. Dafedar, A., D'Souza, P.E., Mamatha, G.S., 2011. Prevalence and morphological studies on *Sarcocystis* species infecting cattle in Bangalore. *J. Vet. Parasitol.* 25, 183–184.

278. Dafedar, A.M., D'Souza, P.E., Ananda, K.J., Puttalakshmamma, G.C., 2008. Prevalence of sarcocystosis in goats slaughtered at an abattoir in Bangalore, Karnataka state. *Veterinary World* 1, 335–337.

279. Daft, B.M., Barr, B.C., Gardner, I.A., Read, D., Bell, W., Peyser, K.G., Ardans, A., Kinde, H., Morrow, J.K., 2002. Sensitivity and specificity of western blot testing of cerebrospinal fluid and serum for diagnosis of equine protozoal myeloencephalitis in horses with and without neurologic abnormalities. *J. Am. Vet. Med. Assoc.* 221, 1007–1013.

280. Dahlgren, S.S., Gjerde, B., 2007. Genetic characterisation of six *Sarcocystis* species from reindeer (*Rangifer tarandus tarandus*) in Norway based on the small subunit rRNA gene. *Vet. Parasitol.* 146, 204–213.

281. Dahlgren, S.S., Gjerde, B., Skirnisson, K., Gudmundsdottir, B., 2007. Morphological and molecular identification of three species of *Sarcocystis* in reindeer (*Rangifer tarandus tarandus*) in Iceland. *Vet. Parasitol.* 149, 191–198.

282. Dahlgren, S.S., Gjerde, B., 2008. *Sarcocystis* in moose (*Alces alces*): molecular identification and phylogeny of six *Sarcocystis* species in moose, and a morphological description of three new species. *Parasitol. Res.* 103, 93–110.

283. Dahlgren, S.S., Gouveia-Oliveira, R., Gjerde, B., 2008. Phylogenetic relationships between *Sarcocystis* species from reindeer and other Sarcocystidae deduced from ssu rRNA gene sequences. *Vet. Parasitol.* 151, 27–35.

284. Dahlgren, S.S., Gjerde, B., 2009. *Sarcocystis* in Norwegian roe deer (*Capreolus capreolus*): Molecular and morphological identification of *Sarcocystis oviformis* n. sp. and *Sarcocystis gracilis* and their phylogenetic relationship with other *Sarcocystis* species. *Parasitol. Res.* 104, 993–1003.

285. Dahlgren, S.S., Gjerde, B., 2010. Molecular characterization of five *Sarcocystis* species in red deer (*Cervus elaphus*), including *Sarcocystis hjorti* n. sp., reveals that these species are not intermediate host specific. *Parasitology* 137, 815–840.

286. Dahlgren, S.S., Gjerde, B., 2010. The red fox (*Vulpes vulpes*) and the arctic fox (*Vulpes lagopus*) are definitive hosts of *Sarcocystis alces* and *Sarcocystis hjorti* from moose (*Alces alces*). *Parasitology* 137, 1547–1557.

287. Dailey, M., Stroud, R., 1978. Parasites and associated pathology observed in cetaceans stranded along the Oregon coast. *J. Wildl. Dis.* 14, 503–511.

288. Dalimi, A., Arshad, M., Ghaffari Far, F., 2009. Asessment of *Sarcocystis* infection in slaughtered goats by different methods. *Pajouhesh and Sazandegi* 21, 38–42.

289. Daly, T.J.M., Markus, M.B., Biggs, H.C., 1983. *Sarcocystis* of domestic and wild equine hosts. *Proc. Electron Microsc. Soc. S. Afr.* 13, 71–72.

290. Dame, J.B., MacKay, R.J., Yowell, C.A., Cutler, T.J., Marsh, A., Greiner, E.C., 1995. *Sarcocystis falcatula* from passerine and psittacine birds: Synonymy with *Sarcocystis neurona*, agent of equine protozoal myeloencephalitis. *J. Parasitol.* 81, 930–935.

291. Damriyasa, I.M., Bauer, C., Edelhofer, R., Failing, K., Lind, P., Petersen, E., Schares, G., Tenter, A.M., Volmer, R., Zahner, H., 2004. Cross-sectional survey in pig breeding farms in Hesse, Germany: Seroprevalence and risk factors of infections with *Toxoplasma gondii*, *Sarcocystis* spp. and *Neospora caninum* in sows. *Vet. Parasitol.* 126, 271–286.

292. Dangoudoubiyam, S., Oliveira, J.B., Víquez, C., Gómez-García, A., González, O., Romero, J.J., Kwok, O.C.H., Dubey, J.P., Howe, D.K., 2011. Detection of antibodies against *Sarcocystis neurona*, *Neospora* spp., and *Toxoplasma gondii* in horses from Costa Rica. *J. Parasitol.* 97, 522–524.

293. Dangoudoubiyam, S., Zhang, Z., Howe, D.K., 2014. Purine salvage in the apicomplexan *Sarcocystis neurona*, and generation of hypoxanthine-xanthine-guanine phosporibosyltransferase-deficient clones for positive-negative selection of transgenic parasites. *Parasitology* 141, 1399–1405.

294. Daryani, A., Alaei, R., Dehghan, M.H., Arab, R., Sharif, M., Ziaei, H., 2006. Survey of *Sarcocystis* infection in slaughtered sheep and buffaloes in Ardabil, Iran. *J. Anim. Vet. Adv.* 5, 60–62.

295. Daszak, P., Cunningham, A.A., 1995. A report of intestinal sarcocystosis in the bullsnake (*Pituophis melanoleucus sayi*) and a re-evaluation of *Sarcocystis* sp. from snakes of the genus *Pituophis*. *J. Wildl. Dis.* 31, 400–403.

296. Daugschies, A., Rommel, M., Schnieder, T., Henning, M., Kallweit, E., 1987. Effects of *Sarcocystis miescheriana* infection on carcass weight and meat quality of halothane-tested fattening pigs. *Vet. Parasitol.* 25, 19–31.

297. Daugschies, A., Schnieder, T., Rommel, M., Bickhardt, K., 1988. The effects of *Sarcocystis miescheriana* infections on blood enzymes and weight gain of stress-sensitive and stress-insensitive pigs. *Vet. Parasitol.* 27, 221–229.

298. Daugschies, A., Rommel, M., Schnieder, T., Henning, M., Kallweit, E., 1988. Effects of *Sarcocystis miescheriana* infection on carcass quality and on the water-binding capacity of the meat of halothane-tested fattening pigs. *Vet. Parasitol.* 27, 231–237.

299. Daugschies, A., Altfeld, E., Rommel, M., 1989. Hemostatic alterations in pigs fed sublethal doses of *Sarcocystis miescheriana*. *Vet. Parasitol.* 34, 1–13.

300. Daugschies, A., Rommel, M., Hoppen, H.O., 1989. Prostanoids during acute sarcocystosis in growing pigs. *Parasitol. Res.* 76, 115–118.

301. Daugschies, A., Jacobs, M., Rommel, M., 1989. Herztod bei einem chronisch mit *Sarcocystis miescheriana* infizierten Schwein-Ein Fallbericht. *Berl. Münch. Tierärztl. Wochenschr.* 102, 184–187.

302. Daugschies, A., Hasche, H.-O., Rommel, M., 1990. Aktivitäten ausgewählter Enzyme im Blutplasma und in der Muskulatur Sarkosporidien-infizierter Schweine. *Mitt. Österr. Ges. Tropenmed. Parasitol.* 12, 131–140.

303. Daugschies, A., Bode, H.-C., Rommel, M., 1992. Balance trial in pigs sublethally infected with *Sarcocystis miescheriana*. In: *In memorian al Profesor Doctor D. F. de P. Martínez Gómez*. Hernández-Rodríguez, S. (Ed.), Universidad de Cordoba, Spain, pp. 51–64.

304. Daugschies, A., Rupp, U., Rommel, M., 1998. Blood clotting disorders during experimental sarcocystosis in calves. *Int. J. Parasitol.* 28, 1187–1194.

305. Davies, J.L., Haldorson, G.J., Bradway, D.S., Britton, A.P., 2011. Fatal hepatic sarcocystosis in a captive black bear (*Ursus americanus*) associated with *Sarcocystis canis*-like infection. *J. Vet. Diagn. Invest.* 23, 379–383.

306. Davis, C.R., Barr, B.C., Pascoe, J.R., Olander, H.J., Dubey, J.P., 1999. Hepatic sarcocystosis in a horse. *J. Parasitol.* 85, 965–968.

307. Davis, S.W., Daft, B.M., Dubey, J.P., 1991. *Sarcocystis neurona* cultured *in vitro* from a horse with equine protozoal myelitis. *Equine Vet. J.* 23, 315–317.

308. Davis, S.W., Speer, C.A., Dubey, J.P., 1991. *In vitro* cultivation of *Sarcocystis neurona* from the spinal cord of a horse with equine protozoal myelitis. *J. Parasitol.* 77, 789–792.

309. De Bosschere, H., Ducatelle, R., 2001. Inverse correlation between myositis eosinophilica and number of *Sarcocystis* cystozoites in heart tissue of cattle. *Vlaams Diergeneesk. Tijdsch.* 70, 118–123.

310. De Guise, S., Lagacé, A., Girard, C., Béland, P., 1993. Intramuscular *Sarcocystis* in two beluga whales and an Atlantic white-sided dolphin from the St. Lawrence estuary, Québec, Canada. *J. Vet. Diagn. Invest.* 5, 296–300.

311. Deepe, G.S., 1994. Role of CD8+ T cells in host resistance to systemic infection with *Histoplasma capsulatum* in mice. *J. Immunol.* 152, 3491–3500.

312. Degloorkar, N.M., Kulkarni, G.B., Deshpande, B.B., Digraskar, S.U., 1993. Incidence of *Sarcocystis fusiformis* in buffaloes (*Bubalus bubalis*). *Indian J. Comp. Microbiol. Immunol. Infect. Dis.* 14, 29–30.

313. DeLucia, P.M., Cheadle, M.A., Greiner, E.C., 2002. Prevalence of *Sarcocystis* sarcocysts in nine-banded armadillos (*Dasypus novemcinctus*) from Florida. *Vet. Parasitol.* 103, 203–205.

314. Deluol, A.M., Mechali, D., Cenac, J., Savel, J., Coulaud, J.P., 1980. Incidence et aspects cliniques des coccidioses intestinales dans une consultation de médecine tropicale. *Bull. Soc. Pathol. Exot.* 73, 259–266.

315. Deodhar, N.S., Narsapur, V.S., Ajinkya, S.M., 1968. A survey on the incidence of *Trichinella* and sarcosporidium in pigs at Bombay. *Bull. Ind. Soc. Mal. Com. Dis.* 5, 267–269.

316. Derothe, J.M., Le Brun, N., Loubes, C., Perriat-Sanguinet, M., Moulia, C., 2001. Susceptibility of natural hybrids between house mouse subspecies to *Sarcocystis muris*. *Int. J. Parasitol.* 31, 15–19.

317. Derylo, A., Kinka, R., 1978. Skutki ekonomiczne wystepowania sarkosporydiozy u świń. *Medycyna Wet.* 34, 729–731.

318. Désilets, A., 1989. Sarcosporidiose chez un veau de trois semaines. *Méd. Vét. Québec* 19, 145–146.

319. Devi, P.P., Sarmah, P.C., Saleque, A., 1998. *Sarcocystis* spp. infection in carcasses of domestic pigs in Guwahati, Assam. *J. Vet. Parasitol.* 12, 56–57.

320. Dey, S., Gupta, S.L., Singh, R.P., 1991. Effect on intensity of coccidiosis in goats due to concurrent infection with *Sarcocystis* infection–an experimental study. *Indian Vet. Med. J.* 15, 272–274.

321. Dey, S., Gupta, S.L., Singh, R.P., 1995. Clinico-haematologic findings in caprine sarcocystosis. *Indian J. Anim. Sci.* 65, 44–46.

322. Dey, S., Khatkar, S.K., Ghosh, J.D., Akbar, M.A., 1995. Clinicohaematologic and serum biochemical changes in piglets experimentally infected with *Sarcocystis miescheriana*. *Indian Vet. J.* 72, 126–130.

323. Dhage, V.N., Bhandarkar, A.G., Kurkure, N.V., Joshi, M.V., 1997. Preliminary haematological observations on cardiac sarcocystosis in cattle and goat. *J. Maharashtra Agric. Univ.* 22, 250–252.

324. Dirikolu, L., Karpiesiuk, W., Lehner, A.F., Hughes, C., Woods, W.E., Harkins, J.D., Boyles, J., Atkinson, A., Granstrom, D.E., Tobin, T., 2006. New therapeutic approaches for equine protozoal myeloencephalitis: Pharmacokinetics of diclazuril sodium salts in horses. *Veterinary Therapeutics* 7, 52–63.

325. Dissanaike, A.S., Poopalachelvam, M., 1975. *Sarcocystis booliati* n. sp. and a parasite of undetermined taxonomic position, *Octoplasma garnhami* n. gen. n. sp., from the moonrat, *Echinosorex gymnurus*. *Southeast Asian J. Trop. Med. Pub. Health* 6, 175–185.

326. Dissanaike, A.S., Kan, S.P., Retnasabapathy, A., Baskaran, G., 1977. Developmental stages of *Sarcocystis fusiformis* (Railliet, 1897) and *Sarcocystis* sp., of the water buffalo, in the small intestines of cats and dogs respectively. *Southeast Asian J. Trop. Med. Pub. Health* 8, 417.

327. Dissanaike, A.S., Kan, S.P., 1978. Studies on *Sarcocystis* in Malaysia. I. *Sarcocystis levinei* n. sp. from the water buffalo *Bubalus bubalis*. *Z. Parasitenkd.* 55, 127–138.

328. Divers, T.J., Bowman, D.D., Lahunta, A., 2000. Equine protozoal myeloencephalitis: Recent advances in diagnosis and treatment. *Vet. Med. Supplement*, 3–17.

329. Doflein, F., 1901. *Sarcocystis bertrami* n. sp. In: *Die Protozoen als Parasiten und Krankheitserreger, nach biologischen Gesichtspunkten dargestellt*. Gustav Fischer., Jena, pp. 219–220.

330. Dolezel, D., Koudela, B., Jirku, M., Hypsa, V., Obornik, M., Votypka, J., Modry, D., Slapeta, J.R., Lukes, J., 1999. Phylogenetic analysis of *Sarcocystis* spp. of mammals and reptiles supports the coevolution of *Sarcocystis* spp. with their final hosts. *Int. J. Parasitol.* 29, 795–798.

331. Domenis, L., Peletto, S., Sacchi, L., Clementi, E., Genchi, M., Felisari, L., Felisari, C. et al. 2011. Detection of a morphogenetically novel *Sarcocystis hominis*-like in the context of a prevalence study in semi-intensively bred cattle in Italy. *Parasitol. Res.* 109, 1677–1687.

332. Donat, K., 1989. Sarcosporidia in cattle slaughtered at the abattoir at Ceske Budejovice. *Veterinarstvi* 39, 114–115. (In Czech).

333. Dorr, T.E., Higgins, R.J., Dangler, C.A., Madigan, J.E., Witham, C.L., 1984. Protozoal myeloencephalitis in horses in California. *J. Am. Vet. Med. Assoc.* 185, 801–802.

334. Drössigk, U., Hiepe, T., Pötzsch, F., Scholz, D., Tietz, H.J., 1998. Stimulation of human immunodeficiency virus expression in permanent monocytic cells by *Sarcocystis gigantea* extract. *Parasitol. Res.* 84, 455–458.

335. Drost, S., Graubmann, H.-D., 1972. Der sarkosporidienbefall beim Wildschwein. *Eingegangen* 20, 870–872.

336. Drost, S., Graubmann, H.D., 1974. Der Sarkosporidienbefall beim Rehwild. *Monatsh. Veterinärmed.* 29, 620–621.

337. Drost, S., Graubmann, H.D., 1975. Der Sarkosporidienbefall de Rot- und Damwildes. *Monatsh. Veterinärmed.* 30, 587–589.

338. Drost, S., 1977. Die Sarkosporidien des Schalenwildes. III. Sarkosporidien beim Rot- und Damwild. *Angew. Parasitol.* 18, 219–225.

339. Drost, S., 1977. Die Sarkosporidien des Schalenwildes. II. Sarkosporidien beim Rehwild. *Angew. Parasitol.* 18, 121–131.

340. Drost, S., Brackmann, H., 1978. Zum Sarkosporidienvorkommen bei Rindern eines Sanitätsschlachtbetriebes. *Manatsh. Veterinardmed.* 33, 175–178.

341. Drouin, T.E., Mahrt, J.L., 1979. The prevalence of *Sarcocystis* Lankester, 1882, in some bird species in western Canada, with notes on its life cycle. *Can. J. Zool.* 57, 1915–1921.

342. Drouin, T.E., Mahrt, J.L., 1980. The morphology of cysts of *Sarcocystis* infecting birds in western Canada. *Can. J. Zool.* 58, 1477–1482.

343. Duarte, P.C., Daft, B.M., Conrad, P.A., Packham, A.E., Gardner, I.A., 2003. Comparison of a serum indirect fluorescent antibody test with two Western blot tests for the diagnosis of equine protozoal myeloencephalitis. *J. Vet. Diagn. Invest.* 15, 8–13.

344. Duarte, P.C., Conrad, P.A., Wilson, W.D., Ferraro, G.L., Packham, A.E., Bowers-Lepore, J., Carpenter, T.E., Gardner, I.A., 2004. Risk of postnatal exposure to *Sarcocystis neurona* and *Neospora hughesi* in horses. *Am. J. Vet. Res.* 65, 1047–1052.

345. Duarte, P.C., Conrad, P.A., Barr, B.C., Wilson, W.D., Ferraro, G.L., Packham, A.E., Carpenter, T.E., Gardner, I.A., 2004. Risk of transplacental transmission of *Sarcocystis neurona* and *Neospora hughesi* in California horses. *J. Parasitol.* 90, 1345–1351.

346. Duarte, P.C., Daft, B.M., Conrad, P.A., Packham, A.E., Saville, W.J., MacKay, R.J., Barr, B.C., Wilson, W.D., Ng, T., Reed, S.M., Gardner, I.A., 2004. Evaluation and comparison of an indirect fluorescent antibody test for detection of antibodies to *Sarcocystis neurona*, using serum and cerebrospinal fluid of naturally and experimentally infected, and vaccinated horses. *J. Parasitol.* 90, 379–386.

347. Duarte, P.C., Ebel, E.D., Traub-Dargatz, J., Wilson, W.D., Conrad, P.A., Gardner, I.A., 2006. Indirect fluorescent antibody testing of cerebrospinal fluid for diagnosis of equine protozoal myeloencephalitis. *Am. J. Vet. Res.* 67, 869–876.

348. Dubey, J.P., Davis, G.W., Koestner, A., Kiryu, K., 1974. Equine encephalomyelitis due to a protozoan parasite resembling *Toxoplasma gondii*. *J. Am. Vet. Med. Assoc.* 165, 249–255.

349. Dubey, J.P., 1976. A review of *Sarcocystis* of domestic animals and of other coccidia of cats and dogs. *J. Am. Vet. Med. Assoc.* 169, 1061–1078.

350. Dubey, J.P., Streitel, R.H., 1976. Shedding of *Sarcocystis* in feces of dogs and cats fed muscles of naturally infected food animals in the midwestern United States. *J. Parasitol.* 62, 828–830.

351. Dubey, J.P., Streitel, R.H., Stromberg, P.C., Toussant, M.J., 1977. *Sarcocystis fayeri* sp. n. from the horse. *J. Parasitol.* 63, 443–447.

352. Dubey, J.P., Fayer, R., Seesee, F.M., 1978. *Sarcocystis* in feces of coyotes from Montana: Prevalence and experimental transmission to sheep and cattle. *J. Am. Vet. Med. Assoc.* 173, 1167–1170.

353. Dubey, J.P., 1979. Frequency of *Sarcocystis* in pigs in Ohio and attempted transmission to cats and dogs. *Am. J. Vet. Res.* 40, 867–868.

354. Dubey, J.P., 1980. Coyote as a final host for *Sarcocystis* species of goats, sheep, cattle, elk, bison, and moose in Montana. *Am. J. Vet. Res.* 41, 1227–1229.

355. Dubey, J.P., 1980. *Sarcocystis* species in moose (*Alces alces*), bison (*Bison bison*), and pronghorn (*Antilocapra americana*) in Montana. *Am. J. Vet. Res.* 41, 2063–2065.

356. Dubey, J.P., Speer, C.A., Douglass, T.G., 1980. Development and ultrastructure of first-generation meronts of *Sarcocystis cruzi* in calves fed sporocysts from coyote feces. *J. Protozool.* 27, 380–387.

357. Dubey, J.P., 1981. Abortion and death in goats inoculated with *Sarcocystis* sporocysts from coyote feces. *J. Am. Vet. Med. Assoc.* 178, 700–703.

358. Dubey, J.P., 1981. Development of immunity to sarcocystosis in dairy goats. *Am. J. Vet. Res.* 42, 800–804.

359. Dubey, J.P., Weisbrode, S.E., Speer, C.A., Sharma, S.P., 1981. Sarcocystosis in goats: Clinical signs and pathologic and hematologic findings. *J. Am. Vet. Med. Assoc.* 178, 683–699.

360. Dubey, J.P., 1982. Development of ox-coyote cycle of *Sarcocystis cruzi*. *J. Protozool.* 29, 591–601.

361. Dubey, J.P., 1982. Sarcocystosis in neonatal bison fed *Sarcocystis cruzi* sporocysts derived from cattle. *J. Am. Vet. Med. Assoc.* 181, 1272–1274.

362. Dubey, J.P., 1982. Development the ox-cat cycle of *Sarcocystis hirsuta*. *Proc. Helminthol. Soc. Wash.* 49, 295–304.

363. Dubey, J.P., 1982. Quantitative parasitemia in calves fed *Sarcocystis cruzi* sporocysts from coyotes. *Am. J. Vet. Res.* 43, 1085–1086.

364. Dubey, J.P., 1982. *Sarcocystis* and other coccidia in foxes and other wild carnivores from Montana. *J. Am. Vet. Med. Assoc.* 181, 1270–1271.

365. Dubey, J.P., Bergeron, J.A., 1982. *Sarcocystis* as a cause of placentitis and abortion in cattle. *Vet. Pathol.* 19, 315–318.

366. Dubey, J.P., Speer, C.A., Epling, G.P., 1982. Sarcocystosis in newborn calves fed *Sarcocystis cruzi* sporocysts from coyotes. *Am. J. Vet. Res.* 43, 2147–2164.

367. Dubey, J.P., Speer, C.A., Callis, G., Blixt, J.A., 1982. Development of sheep-canid cycle of *Sarcocystis tenella*. *Can. J. Zool.* 60, 2464–2477.

368. Dubey, J.P., 1983. *Sarcocystis ferovis* sp. n. from the bighorn sheep (*Ovis canadensis*) and coyote (*Canis latrans*). *Proc. Helminthol. Soc. Wash.* 50, 153–158.

369. Dubey, J.P., 1983. *Sarcocystis bozemanensis* sp. nov. (Protozoa: Sarcocystidae) and *S. campestris* from the Richardson's ground squirrel (*Spermophilus richardsonii*), in Montana, USA. *Can. J. Zool.* 61, 942–946.

370. Dubey, J.P., 1983. *Sarcocystis montanaensis* and *S. microti* sp. n. from the meadow vole (*Microtus pennsylvanicus*). *Proc. Helminthol. Soc. Wash.* 50, 318–324.

371. Dubey, J.P., 1983. Clinical sarcocystosis in calves fed *Sarcocystis hirsuta* sporocysts from cats. *Vet. Pathol.* 20, 90–98.

372. Dubey, J.P., 1983. Immunity to sarcocystosis: Modifications of intestinal coccidiosis, and disappearance of sarcocysts in dairy goats. *Vet. Parasitol.* 13, 23–34.

373. Dubey, J.P., 1983. Impaired protective immunity to sarcocystosis in pregnant dairy goats. *Am. J. Vet. Res.* 44, 132–134.

374. Dubey, J.P., 1983. Experimental infections of *Sarcocystis cruzi*, *Sarcocystis tenella*, *Sarcocystis capracanis* and *Toxoplasma gondii* in red foxes (*Vulpes vulpes*). *J. Wildl. Dis.* 19, 200–203.

375. Dubey, J.P., 1983. *Sarcocystis peromysci* n. sp. and *S. idahoensis* in deer mouse (*Peromyscus maniculatus*) in Montana. *Can. J. Zool.* 61, 1180–1182.

376. Dubey, J.P., 1983. Microgametogony of *Sarcocystis hirsuta* in the intestine of the cat. *Parasitology* 86, 7–9.

377. Dubey, J.P., 1983. Immunity to sarcocystosis: Modifications of intestinal coccidiosis, and disappearance of sarcocysts in dairy goats. *Vet. Parasitol.* 13, 23–34.

378. Dubey, J.P., Lozier, S.C., 1983. *Sarcocystis* infection in the white-tailed deer (*Odocoileus virginianus*) in Montana: Intensity and description of *Sarcocystis odoi* n. sp. *Am. J. Vet. Res.* 44, 1738–1743.

379. Dubey, J.P., Kistner, T.P., Callis, G., 1983. Development of *Sarcocystis* in mule deer transmitted through dogs and coyotes. *Can. J. Zool.* 61, 2904–2912.

380. Dubey, J.P., Jolley, W.R., Thorne, E.T., 1983. *Sarcocystis sybillensis* sp. nov. from the North American elk (*Cervus elaphus*). *Can. J. Zool.* 61, 737–742.

381. Dubey, J.P., 1984. Protective immunity to *Sarcocystis capracanis*-induced abortion in dairy goats. *J. Protozool.* 31, 553–555.

382. Dubey, J.P., Speer, C.A., Epling, G.P., Blixt, J.A., 1984. *Sarcocystis capracanis*: Development in goats, dogs and coyotes. *Int. Goat Sheep Res.* 2, 252–265.

383. Dubey, J.P., Speer, C.A., 1985. Prevalence and ultrastructure of three types of *Sarcocystis* in mule deer, *Odocoileus hemionus* (Rafinesque), in Montana. *J. Wildl. Dis.* 21, 219–228.

384. Dubey, J.P., Kistner, T.P., 1985. Epizootiology of *Sarcocystis* infections in mule deer fawns in Oregon. *J. Am. Vet. Med. Assoc.* 187, 1181–1186.

385. Dubey, J.P., Speer, C.A., 1986. *Sarcocystis* infections in mule deer (*Odocoileus hemionus*) in Montana and the description of three new species. *Am. J. Vet. Res.* 47, 1052–1055.

386. Dubey, J.P., Towle, A., 1986. Toxoplasmosis in sheep: A review and annotated bibliography. In: Commonwealth Institute of Parasitology, St. Albans, United Kingdom, pp. 1–152.

387. Dubey, J.P., Miller, S., 1986. Equine protozoal myeloencephalitis in a pony. *J. Am. Vet. Med. Assoc.* 188, 1311–1312.

388. Dubey, J.P., Leek, R.G., Fayer, R., 1986. Prevalence, transmission, and pathogenicity of *Sarcocystis gigantea* of sheep. *J. Am. Vet. Med. Assoc.* 188, 151–154.

389. Dubey, J.P., Perry, A., Kennedy, M.J., 1987. Encephalitis caused by a *Sarcocystis*-like organism in a steer. *J. Am. Vet. Med. Assoc.* 191, 231–232.

390. Dubey, J.P., 1988. Lesions in sheep inoculated with *Sarcocystis tenella* sporocysts from canine feces. *Vet. Parasitol.* 26, 237–252.

391. Dubey, J.P., Fayer, R., Speer, C.A., 1988. Experimental *Sarcocystis hominis* infection in cattle: Lesions and ultrastructure of sarcocysts. *J. Parasitol.* 74, 875–879.

392. Dubey, J.P., Lindsay, D.S., Speer, C.A., Fayer, R., Livingston, C.W., 1988. *Sarcocystis arieticanis* and other *Sarcocystis* species in sheep in the United States. *J. Parasitol.* 74, 1033–1038.

393. Dubey, J.P., Speer, C.A., Fayer, R., 1989. *Sarcocystosis of Animals and Man.* CRC Press, Boca Raton, Florida.

394. Dubey, J.P., Speer, C.A., Shah, H.L., 1989. Ultrastructure of sarcocysts from water buffalo in India. *Vet. Parasitol.* 34, 149–152.

395. Dubey, J.P., Speer, C.A., Charleston, W.A.G., 1989. Ultrastructural differentiation between sarcocysts of *Sarcocystis hirsuta* and *Sarcocystis hominis*. *Vet. Parasitol.* 34, 153–157.

396. Dubey, J.P., Speer, C.A., Munday, B.L., Lipscomb, T.P., 1989. Ovine sporozoan encephalomyelitis linked to *Sarcocystis* infection. *Vet. Parasitol.* 34, 159–163.

397. Dubey, J.P., Udtujan, R.M., Cannon, L., Lindsay, D.S., 1990. Condemnation of beef because of *Sarcocystis hirsuta* infection. *J. Am. Vet. Med. Assoc.* 196, 1095–1096.

398. Dubey, J.P., Hamir, A.N., Hanlon, C.A., Topper, M.J., Rupprecht, C.E., 1990. Fatal necrotizing encephalitis in a raccoon associated with a *Sarcocystis*-like protozoon. *J. Vet. Diagn. Invest.* 2, 345–347.

399. Dubey, J.P., 1991. Sarcocystosis of the skeletal and cardiac muscle, mouse. In: Jones, T.C., Mohr, U., Hunt, R.D. (Eds.), *Cardiovascular and Musculoskeletal Systems*. Springer-Verlag, Berlin, Germany, pp. 165–169.

400. Dubey, J.P., Speer, C.A., 1991. *Sarcocystis canis* n. sp. (Apicomplexa: Sarcocystidae), the etiologic agent of generalized coccidiosis in dogs. *J. Parasitol.* 77, 522–527.

401. Dubey, J.P., Slife, L.N., Speer, C.A., Lipscomb, T.P., Topper, M.J., 1991. Fatal cutaneous and visceral infection in a Rottweiler dog associated with a *Sarcocystis*-like protozoon. *J. Vet. Diagn. Invest.* 3, 72–75.

402. Dubey, J.P., Speer, C.A., Hamir, A.N., Topper, M.J., Brown, C., Rupprecht, C.E., 1991. Development of a *Sarcocystis*-like apicomplexan protozoan in the brain of a raccoon (*Procyon lotor*). *J. Helminthol. Soc. Wash.* 58, 250–255.

403. Dubey, J.P., Porter, S.L., Hattel, A.L., Kradel, D.C., Topper, M.J., Johnson, L., 1991. Sarcocystosis-associated clinical encephalitis in a golden eagle (*Aquila chrysaetos*). *J. Zoo Wildl. Med.* 22, 233–236.

404. Dubey, J.P., Davis, S.W., Speer, C.A., Bowman, D.D., de Lahunta, A., Granstrom, D.E., Topper, M.J., Hamir, A.N., Cummings, J.F., Suter, M.M., 1991. *Sarcocystis neurona* n. sp. (Protozoa: Apicomplexa), the etiologic agent of equine protozoal myeloencephalitis. *J. Parasitol.* 77, 212–218.

405. Dubey, J.P., Cosenza, S.F., Lipscomb, T.P., Topper, M.J., Speer, C.A., Hoban, L.D., Davis, S.W., Kincaid, A.L., Seely, J.C., Marrs, G.E., 1991. Acute sarcocystosis-like disease in a dog. *J. Am. Vet. Med. Assoc.* 198, 439–443.

406. Dubey, J.P., Hamir, A.N., Brown, C., Rupprecht, C.E., 1992. *Sarcocystis sehi* sp. n. (Protozoa: Sarcocystidae) from the porcupine (*Erethizon dorsatum*). *J. Helminthol. Soc. Wash.* 59, 127–129.

407. Dubey, J.P., Hamir, A.N., Kirkpatrick, C.E., Todd, K.S., Rupprecht, C.E., 1992. *Sarcocystis felis* sp. n. (Protozoa: Sarcocystidae) from the bobcat (*Felis rufus*). *J. Helminthol. Soc. Wash.* 59, 227–229.

408. Dubey, J.P., Hedstrom, O.R., 1993. Meningoencephalitis in mink associated with a *Sarcocystis neurona*-like organism. *J. Vet. Diagn. Invest.* 5, 467–471.

409. Dubey, J.P., Powell, E.C., 1994. Prevalence of *Sarcocystis* in sows from Iowa. *Vet. Parasitol.* 52, 151–155.

410. Dubey, J.P., Bwangamoi, O., 1994. *Sarcocystis felis* (Protozoa: Sarcocystidae) from the African lion (*Panthera leo*). *J. Helminthol. Soc. Wash.* 61, 113–114.

411. Dubey, J.P., Higgins, R.J., Barr, B.C., Spangler, W.L., Kollin, B., Jorgensen, L.S., 1994. *Sarcocystis*-associated meningoencephalomyelitis in a cat. *J. Vet. Diagn. Invest.* 6, 118–120.

412. Dubey, J.P., Lindsay, D.S., 1998. Isolation in immunodeficient mice of *Sarcocystis neurona* from opossum (*Didelphis virginiana*) faeces, and its differentiation from *Sarcocystis falcatula*. *Int. J. Parasitol.* 28, 1823–1828.

413. Dubey, J.P., Rudbäck, E., Topper, M.J., 1998. Sarcocystosis in capercaillie (*Tetrao urogallus*) in Finland: Description of the parasite and lesions. *J. Parasitol.* 84, 104–108.

414. Dubey, J.P., Speer, C.A., Lindsay, D.S., 1998. Isolation of a third species of *Sarcocystis* in immunodeficient mice fed feces from opossums (*Didelphis virginiana*) and its differentiation from *Sarcocystis falcatula* and *Sarcocystis neurona*. *J. Parasitol.* 84, 1158–1164.

415. Dubey, J.P., Topper, M.J., Nutter, F.B., 1998. Muscular *Sarcocystis* infection in a bear (*Ursus americanus*). *J. Parasitol.* 84, 452–454.

416. Dubey, J.P., Lindsay, D.S., 1999. *Sarcocystis speeri* n. sp. (Protozoa: Sarcocystidae) from the opossum (*Didelphis virginiana*). *J. Parasitol.* 85, 903–909.

417. Dubey, J.P., Kerber, C.E., Granstrom, D.E., 1999. Serologic prevalence of *Sarcocystis neurona*, *Toxoplasma gondii*, and *Neospora caninum* in horses in Brazil. *J. Am. Vet. Med. Assoc.* 215, 970–972.

418. Dubey, J.P., Venturini, L., Venturini, C., Basso, W., Unzaga, J., 1999. Isolation of *Sarcocystis falcatula* from the South American opossum (*Didelphis albiventris*) from Argentina. *Vet. Parasitol.* 86, 239–244.

419. Dubey, J.P., Venturini, M.C., Venturini, L., McKinney, J., Pecoraro, M., 1999. Prevalence of antibodies to *Sarcocystis neurona*, *Toxoplasma gondii*, and *Neospora caninum* in horses from Argentina. *Vet. Parasitol.* 86, 59–62.

420. Dubey, J.P., Mattson, D.E., Speer, C.A., Baker, R.J., Mulrooney, D.M., Tornquist, S.J., Hamir, A.N., Gerros, T.C., 1999. Characterization of *Sarcocystis neurona* isolate (SN6) from a naturally infected horse from Oregon. *J. Eukaryot. Microbiol.* 46, 500–506.

421. Dubey, J.P., 2000. Prevalence of *Sarcocystis* species sporocysts in wild caught opossums (*Didelphis virginiana*). *J. Parasitol.* 86, 705–710.

422. Dubey, J.P., Hamir, A.N., 2000. Immunohistochemical confirmation of *Sarcocystis neurona* infections in raccoons, mink, cat, skunk and pony. *J. Parasitol.* 86, 1150–1152.

423. Dubey, J.P., Speer, C.A., Lindsay, D.S., 2000. *In vitro* cultivation of schizonts of *Sarcocystis speeri* Dubey and Lindsay, 1999. *J. Parasitol.* 86, 671–678.

424. Dubey, J.P., Clark, T.R., Yantist, D., 2000. *Frenkelia microti* infection in a chinchilla (*Chinchilla laniger*) in the United States. *J. Parasitol.* 86, 1149–1150.

425. Dubey, J.P., Lindsay, D.S., Rezende, P.C.B., Costa, A.J., 2000. Characterization of an unidentified *Sarcocystis falcatula*-like parasite from the South American opossum, *Didelphis albiventris* from Brazil. *J. Eukaryot. Microbiol.* 47, 538–544.

426. Dubey, J.P., Venturini, L., Venturini, M.C., Speer, C.A., 2000. Isolation of *Sarcocystis speeri* Dubey and Lindsay, 1999 from the South American opossum (*Didelphis albiventris*) from Argentina. *J. Parasitol.* 86, 160–163.

427. Dubey, J.P., Kerber, C.E., Lindsay, D.S., Kasai, N., Pena, H.F.J., 2000. The South American opossum, *Didelphis marsupialis*, from Brazil as another definitive host for *Sarcocystis speeri* Dubey and Lindsay, 1999. *Parasitology* 121, 589–594.

428. Dubey, J.P., Speer, C.A., Bowman, D.D., Horton, K.M., Venturini, C., Venturini, L., 2000. Experimental transmission of *Sarcocystis speeri* Dubey and Lindsay, 1999 from the South American opossum (*Didelphis albiventris*) to the North American opossum (*Didelphis virginiana*). *J. Parasitol.* 86, 624–627.

429. Dubey, J.P., Saville, W.J.A., Lindsay, D.S., Stich, R.W., Stanek, J.F., Speer, C.A., Rosenthal, B.M., Njoku, C.J., Kwok, O.C.H., Shen, S.K., Reed, S.M., 2000. Completion of the life cycle of *Sarcocystis neurona*. *J. Parasitol.* 86, 1276–1280.

430. Dubey, J.P., 2001. Parasitemia and early tissue localization of *Sarcocystis neurona* in interferon gamma gene knockout mice fed sporocysts. *J. Parasitol.* 87, 1476–1479.

431. Dubey, J.P., 2001. Migration and development of *Sarcocystis neurona* in tissues of interferon gamma knockout mice fed sporocysts from a naturally infected opossums. *Vet. Parasitol.* 95, 341–351.

432. Dubey, J.P., Odening, K., 2001. Toxoplasmosis and related infections. In: Samuel, W.M., Pybus, M.J., Kocan, A.A. (Eds.), *Parasitic Diseases of Wild Mammals.* Iowa State University Press, Ames, pp. 478–519.

433. Dubey, J.P., Rosenthal, B.M., Speer, C.A., 2001. *Sarcocystis lindsayi* n. sp. (Protozoa: Sarcocystidae) from the South American opossum, *Didelphis albiventris* from Brazil. *J. Eukaryot. Microbiol.* 48, 595–603.

434. Dubey, J.P., Lindsay, D.S., Kwok, O.C.H., Shen, S.K., 2001. The gamma interferon knockout mouse model for *Sarcocystis neurona*: Comparison of infectivity of sporocysts and merozoites and routes of inoculation. *J. Parasitol.* 87, 1171–1173.

435. Dubey, J.P., Lindsay, D.S., Fritz, D., Speer, C.A., 2001. Structure of *Sarcocystis neurona* sarcocysts. *J. Parasitol.* 87, 1323–1327.

436. Dubey, J.P., Garner, M.M., Stetter, M.D., Marsh, A.E., Barr, B.C., 2001. Acute *Sarcocystis falcatula*-like infection in a carmine bee-eater (*Merops nubicus*) and immunohistochemical cross reactivity between *Sarcocystis falcatula* and *Sarcocystis neurona*. *J. Parasitol.* 87, 824–832.

437. Dubey, J.P., Johnson, G.C., Bermudez, A., Suedmeyer, K.W., Fritz, D.L., 2001. Neural sarcocystosis in a straw-necked ibis (*Carphibis spinicollis*) associated with a *Sarcocystis neurona*-like organism and description of muscular sarcocysts of an unidentified *Sarcocystis* species. *J. Parasitol.* 87, 1317–1322.

438. Dubey, J.P., Lindsay, D.S., Saville, W.J.A., Reed, S.M., Granstrom, D.E., Speer, C.A., 2001. A review of *Sarcocystis neurona* and equine protozoal myeloencephalitis (EPM). *Vet. Parasitol.* 95, 89–131.

439. Dubey, J.P., Fritz, D., Lindsay, D.S., Shen, S.K., Kwok, O.C.H., Thompson, K.C., 2001. Diclazuril preventive therapy of gamma interferon knockout mice fed *Sarcocystis neurona* sporocysts. *Vet. Parasitol.* 94, 257–263.

440. Dubey, J.P., Rosypal, A.C., Rosenthal, B.M., Thomas, N.J., Lindsay, D.S., Stanek, J.F., Reed, S.M., Saville, W.J.A., 2001. *Sarcocystis neurona* infections in sea otter (*Enhydra lutris*): Evidence for natural infections with sarcocysts and transmission of infection to opossums (*Didelphis virginiana*). *J. Parasitol.* 87, 1387–1393.

441. Dubey, J.P., Lindsay, D.S., Rosenthal, B.M., Kerber, C.E., Kasai, N., Pena, H.F.J., Kwok, O.C.H., Shen, S.K., Gennari, S.M., 2001. Isolates of *Sarcocystis falcatula*-like organisms from South American opossums *Didelphis marsupialis* and *Didelphis albiventris* from São Paulo, Brazil. *J. Parasitol.* 87, 1449–1453.

442. Dubey, J.P., Lindsay, D.S., Kerber, C.E., Kasai, N., Pena, H.F.J., Gennari, S.M., Kwok, O.C.H., Shen, S.K., Rosenthal, B.M., 2001. First isolation of *Sarcocystis neurona* from the South American opossum, *Didelphis albiventris*, from Brazil. *Vet. Parasitol.* 95, 295–304.

443. Dubey, J.P., Black, S.S., Rickard, L.G., Rosenthal, B.M., Lindsay, D.S., Shen, S.K., Kwok, O.C.H., Hurst, G., Rashmir-Raven, A., 2001. Prevalence of *Sarcocystis neurona* sporocysts in opossums (*Didelphis virginiana*) from rural Mississippi. *Vet. Parasitol.* 95, 283–293.

444. Dubey, J.P., Mattson, D.E., Speer, C.A., Hamir, A.N., Lindsay, D.S., Rosenthal, B.M., Kwok, O.C.H., Baker, R.J., Mulrooney, D.M., Tornquist, S.J., Gerros, T.C., 2001. Characteristics of a recent isolate of *Sarcocystis neurona* (SN7) from a horse and loss of pathogenicity of isolates SN6 and SN7 by passages in cell culture. *Vet. Parasitol.* 95, 155–166.

445. Dubey, J.P., Saville, W.J.A., Stanek, J.F., Lindsay, D.S., Rosenthal, B.M., Oglesbee, M.J., Rosypal, A.C. et al. 2001. *Sarcocystis neurona* infections in raccoons (*Procyon lotor*): Evidence for natural infection with sarcocysts, transmission of infection to opossums (*Didelphis virginiana*), and experimental induction of neurologic disease in raccoons. *Vet. Parasitol.* 100, 117–129.

446. Dubey, J.P., Hamir, A.N., Topper, M.J., 2002. *Sarcocystis mephitisi* n. sp. (Protozoa: Sarcocystidae), *Sarcocystis neurona*-like and *Toxoplasma*-like infections in striped skunk (*Mephitis mephitis*). *J. Parasitol.* 88, 113–118.

447. Dubey, J.P., Lindsay, D.S., Saville, W.J.A., 2002. Serologic responses of cats against experimental *Sarcocystis neurona* infections. *Vet. Parasitol.* 107, 265–269.

448. Dubey, J.P., Lindsay, D.S., Hill, D., Romand, S., Thulliez, P., Kwok, O.C.H., Silva, J.C.R., Oliveira-Camargo, M.C., Gennari, S.M., 2002. Prevalence of antibodies to *Neospora caninum* and *Sarcocystis neurona* in sera of domestic cats from Brazil. *J. Parasitol.* 88, 1251–1252.

449. Dubey, J.P., Saville, W.J., Sreekumar, C., Shen, S.K., Lindsay, D.S., Pena, H.F., Vianna, M.C., Gennari, S.M., Reed, S.M., 2002. Effects of high temperature and disinfectants on the viability of *Sarcocystis neurona* sporocysts. *J. Parasitol.* 88, 1252–1254.

450. Dubey, J.P., Benson, J., Larson, M.A., 2003. Clinical *Sarcocystis neurona* encephalomyelitis in a domestic cat following routine surgery. *Vet. Parasitol.* 112, 261–267.

451. Dubey, J.P., Ross, A.D., Fritz, D., 2003. Clinical *Toxoplasma gondii, Hammondia heydorni,* and *Sarcocystis* spp. infections in dogs. *Parassitologia* 45, 141–146.

452. Dubey, J.P., Cawthorn, R.J., Speer, C.A., Wobeser, G.A., 2003. Redescription of the sarcocysts of *Sarcocystis rileyi* (Apicomplexa: Sarcocystidae). *J. Eukaryot. Microbiol.* 50, 476–482.

453. Dubey, J.P., Lindsay, D.S., Rosenthal, B.M., Thomas, N.J., 2003. Sarcocysts of an unidentified species of *Sarcocystis* in sea otter (*Enhydra lutris*). *J. Parasitol.* 89, 397–399.

454. Dubey, J.P., Mitchell, S.M., Morrow, J.K., Rhyan, J.C., Stewart, L.M., Granstrom, D.E., Romand, S., Thulliez, P., Saville, W.J., Lindsay, D.S., 2003. Prevalence of antibodies to *Neospora caninum, Sarcocystis neurona,* and *Toxoplasma gondii* in wild horses from central Wyoming. *J. Parasitol.* 89, 716–720.

455. Dubey, J.P., Zarnke, R., Thomas, N.J., Wong, S.K., Van Bonn, W., Briggs, M., Davis, J.W. et al. 2003. *Toxoplasma gondii, Neospora caninum, Sarcocystis neurona,* and *Sarcocystis canis*-like infections in marine mammals. *Vet. Parasitol.* 116, 275–296.

456. Dubey, J.P., 2004. Equine protozoal myeloencephalitis and *Sarcocystis neurona.* In: Coetzer, J.A.W., Thomson, G.R., Tustin, R.C., Kriek, N.P.J. (Eds.), *Infectious Diseases of Livestock with Special Reference to Southern Africa.* Oxford University Press, Ni City, South Africa, pp. 394–403.

457. Dubey, J.P., Lane, E., van Wilpe, E., 2004. *Sarcocystis ramphastosi* sp. nov. and *Sarcocystis sulfuratusi* sp. nov. (Apicomplexa, Sarcocystidae) from the keel-billed toucan (*Ramphastos sulfuratus*). *Acta Parasitologica* 49, 93–101.

458. Dubey, J.P., Hill, D.E., Jones, J.L., Hightower, A.W., Kirkland, E., Roberts, J.M., Marcet, P.L. et al. 2005. Prevalence of viable *Toxoplasma gondii* in beef, chicken, and pork from retail meat stores in the United States: Risk assessment to consumers. *J. Parasitol.* 91, 1082–1093.

459. Dubey, J.P., Rosenthal, B.M., Morales, J.A., Alfaro, A., 2006. Morphologic and genetic characterization of *Sarcocystis* sp. from the African grey parrot, *Psittacus erithacus,* from Costa Rica. *Acta Parasitologica* 51, 161–168.

460. Dubey, J.P., Chapman, J.L., Rosenthal, B.M., Mense, M., Schueler, R.L., 2006. Clinical *Sarcocystis neurona, Sarcocystis canis, Toxoplasma gondii,* and *Neospora caninum* infections in dogs. *Vet. Parasitol.* 137, 36–49.

461. Dubey, J.P., Rosenthal, B.M., Sundar, N., Velmurugan, G.V., Beckmen, K.B., 2007. *Sarcocystis arctosi* sp. nov. (Apicomplexa, Sarcocystidae) from the brown bear (*Ursus arctos*), and its genetic similarity to schizonts of *Sarcocystis canis*-like parasite associated with fatal hepatitis in polar bears (*Ursus maritimus*). *Acta Parasitologica* 52, 299–304.

462. Dubey, J.P., Humphreys, G.J., Fritz, D., 2008. A new species of *Sarcocystis* (Apicomplexa; Sarcocystidae) from the black bear (*Ursus americanus*). *J. Parasitol.* 94, 496–499.

463. Dubey, J.P., Morales, J.A., Leandro, D., Rosenthal, B.M., 2008. Molecular phylogeny implicates new world opossums (Didelphidae) as the definitive host of *Sarcocystis ramphastosi,* a parasite of the keel-billed toucan (*Ramphasotos sulfuratus*). *Acta Protozool.* 47, 55–61.

464. Dubey, J.P., 2010. *Toxoplasmosis of Animals and Humans.* CRC Press, Boca Raton, FL.

465. Dubey, J.P., 2010. *Toxoplasma gondii* infections in chickens (*Gallus domesticus*): Prevalence, clinical disease, diagnosis, and public health significance. *Zoonoses Public Health* 57, 60–73.

466. Dubey, J.P., Rosenthal, B.M., Felix, T.A., 2010. Morphologic and molecular characterization of the sarcocysts of *Sarcocystis rileyi* (Apicomplexa: Sarcocystidae) from the mallard duck (*Anas platyrhynchos*). *J. Parasitol.* 96, 765–770.

467. Dubey, J.P., Reichard, M.V., Torretti, L., Garvon, J.M., Sundar, N., Grigg, M.E., 2010. Two new species of *Sarcocystis* (Apicomplexa: Sarcocystidae) infecting the wolverine (*Gulo gulo*) from Nunavut, Canada. *J. Parasitol.* 96, 972–976.

468. Dubey, J.P., Thomas, N.J., 2011. *Sarcocystis neurona* retinochoroiditis in a sea otter (*Enhydra lutris kenyoni*). *Vet. Parasitol.* 183, 156–159.

469. Dubey, J.P., Rosenthal, B.M., 2013. *Sarcocystis capracanis*-associated encephalitis in sheep. *Vet. Parasitol.* 197, 407–408.

470. Dubey, J.P., Sundar, N., Kwok, O.C.H., Saville, W.J.A., 2013. *Sarcocystis neurona* infection in gamma interferon gene knockout (KO) mice: Comparative infectivity of sporocysts in two strains of KO mice, effect of trypsin digestion on merozoite viability, and infectivity of bradyzoites to KO mice and cell culture. *Vet. Parasitol.* 196, 212–215.

471. Dubey, J.P., Fayer, R., Rosenthal, B.M., Calero-Bernal, R., Uggla, A., 2014. Identity of *Sarcocystis* species of the water buffalo (*Bubalus bubalis*) and cattle (*Bos taurus*) and the suppression of *Sarcocystis sinensis* as a *nomen nudum*. *Vet. Parasitol.* 205, 1–6.

472. Dubey, J.P., Black, S.S., Verma, S.K., Calero-Bernal, R., Morris, E., Hanson, M.A., Cooley, A.J., 2014. *Sarcocystis neurona* schizonts-associated encephalitis, chorioretinitis, and myositis in a two-month-old dog simulating toxoplasmosis, and presence of mature sarcocysts in muscles. *Vet. Parasitol.* 202, 194–200.

473. Dubey, J.P., Lane, E.P., van Wilpe, E., Suleman, E., Reininghaus, B., Verma, S.K., Rosenthal, B.M., Mtshali, M.S., 2014. *Sarcocystis cafferi*, n. sp. (Protozoa: Apicomplexa) from the African buffalo (*Syncerus caffer*). *J. Parasitol.* 100, 817–827.

474. Dubey, J.P., Hilali, M., van Wilpe, E., Verma, S.K., Calero-Bernal, R., Abdel-Wahab, A., 2015. Redescription of *Sarcocystis fusiformis* sarcocysts from the water buffalo (*Bulbalus bubalis*). *Parasitology.* 142, 385–394.

475. Dubey, J.P., Howe, D.K., Furr, M., Grigg, M.E., Saville, W.J., Marsh, A.E., Reed, S.M., 2015. An update on *Sarcocystis neurona* infections in animals and equine protozoal myeloencephalitis (EPM). *Vet. Parasitol.* 209, 1–42.

476. Dubey, J.P., Sykes, J.E., Shelton, G.D., Sharp, N., Verma, S.K., Calero-Bernal, R., Viviano, J., Sundar, N., Khan, A., Grigg, M.E., 2015. *Sarcocystis caninum* and *Sarcocystis svanai* n. spp. (Apicomplexa: Sarcocystidae) associated with severe myositis and hepatitis in the domestic dog (*Canis familiaris*). *J. Eukaryot. Microbiol.* 62, 307–317.

477. Dubin, I.N., Wilcox, A., 1947. *Sarcocystis* in *Macaca mulatta*. *J. Parasitol.* 33, 151–153.

478. Dubremetz, J.F., Porchet-Henneré, E., Parenty, M.D., 1975. Croissance de *Sarcocystis tenella* en culture cellulaire. *C. R. Seances Acad. Sci.* 280, 1793–1795.

479. Dubremetz, J.F., Dissous, C., 1980. Characteristic proteins of micronemes and dense granules from *Sarcocystis tenella* zoites (protozoa, coccidia). *Mol. Biochem. Parasitol.* 1, 279–289.

480. Duncan, R.B., Fox, J.H., Lindsay, D.S., Dubey, J.P., Zuccaro, M.E., 2000. Acute sarcocystosis in a captive white-tailed deer (*Odocoileus virginianus*). *J. Wildl. Dis.* 36, 357–361.

481. Dündar, B., Özer, E., 1996. Mandalarda bulunan *Sarcocystis* türleri ve gelismeleri. *Etlik Vet. Mikrob. Derg.* 8, 58–69.

482. Duszynski, D.W., Box, E.D., 1978. The opossum (*Didelphis virginiana*) as a host for *Sarcocystis debonei* from cowbirds (*Molothrus ater*) and grackles (*Cassidix mexicanus, Quiscalus quiscula*). *J. Parasitol.* 64, 326–329.

483. Dzierszinski, F., Mortuaire, M., Cesbron-Delauw, M.F., Tomavo, S., 2000. Targeted disruption of the glycosylphosphatidylinositol-anchored surface antigen *SAG3* gene in *Toxoplasma gondii* decreases host cell adhesion and drastically reduces virulence in mice. *Mol. Microbiol.* 37, 574–582.

484. Ecco, R., Luppi, M.M., Malta, M.C.C., Araújo, M.R., Guedes, R.M.C., Shivaprasad, H.L., 2008. An outbreak of sarcocystosis in psittacines and a pigeon in a zoological collection in Brazil. *Avian Diseases* 52, 706–710.

485. Edwards, G.T., 1984. Prevalence of equine *Sarcocystis* in British horses and a comparison of two detection methods. *Vet. Rec.* 115, 265–267.

486. Edwards, J.F., Ficken, M.D., Luttgen, P.J., Frey, M.S., 1988. Disseminated sarcocystosis in a cat with lymphosarcoma. *J. Am. Vet. Med. Assoc.* 193, 831–832.

487. El-Dakhly, K.M., El-Nesr, K.A., El-Nahass, E.S., Hirata, A., Sakai, H., Yanai, T., 2011. Prevalence and distribution patterns of *Sarcocystis* spp. in buffaloes in Beni-Suef, Egypt. *Trop. Anim. Health Prod.* 43, 1549–1554.

488. El-Kelesh, E.A.M., Abdel-Maogood, S.Z., Abdel-Wahab, A.M., Radwan, I.G.H., Ibrahim, O., 2011. The effect of freezing and heating on the infectivity of *Sarcocystis fusiformis* to cats and evaluation of ELISA for its diagnosis in water buffaloes (*Bubalus bubalis*). *J. Amer. Sci.* 7, 55–57.

489. El-Morsey, A., El-seify, M., Desouky, A.R.Y., Abdel-Aziz, M.M., Sakai, H., Yanai, T., 2014. Morphologic identification of a new *Sarcocystis* sp. in the common moorhen (*Gallinula chloropus*) (Aves: Gruiformes: Rallidae) from Brolos Lake, Egypt. *Parasitol. Res.* 113, 391–397.

490. El-Morsey, A., El-seify, M., Desouky, A.R.Y., Abdel-Aziz, M.M., Sakai, H., Yanai, T., 2015. *Sarcocystis chloropusae* n. sp. infecting the common moorhen (*Gallinula chloropus*): Molecular characterization based on 18S rRNA, 28S rRNA genes and ITS-1 region analytical investigations. *Parasitology*. 31, 1–3.

491. El-Refaii, A.H., Abdel-Baki, G., Selim, M.K., 1980. Sarcosporidia in goats of Egypt. *J. Egypt. Soc. Parasitol.* 10, 471–472.

492. El-Seify, M., El-Morsey, A., Hilali, M., Zayed, A., El-Dakhly, K., Haridy, M., Sakai, H., Yanai, T., 2014. Molecular characterization of *Sarcocystis fusiformis* and *Sarcocystis buffalonis* infecting water buffaloes (*Bubalus bubalis*) from Egypt. *Am. J. Anim. Vet. Sci.* 9, 95–104.

493. Elitsur, E., Marsh, A.E., Reed, S.M., Dubey, J.P., Oglesbee, M.J., Murphy, J.E., Saville, W.J.A., 2007. Early migration of *Sarcocystis neurona* in ponies fed sporocysts. *J. Parasitol.* 93, 1222–1225.

494. Ellis, J., Luton, K., Baverstock, P.R., Brindley, P.J., Nimmo, K.A., Johnson, A.M., 1994. The phylogeny of *Neospora caninum*. *Mol. Biochem. Parasitol.* 64, 303–311.

495. Ellis, J., Morrison, D., 1995. Effects of sequence alignment on the phylogeny of *Sarcocystis* deduced rom 18S rDNA sequences. *Parasitol. Res.* 81, 696–699.

496. Ellis, J.T., Holmdahl, O.J.M., Ryce, C., Njenga, J.M., Harper, P.A.W., Morrison, D.A., 2000. Molecular phylogeny of *Besnoitia* and the genetic relationships among *Besnoitia* of cattle, wildebeest and goats. *Protistologica* 151, 329–336.

497. Ellis, T.J., Luton, K., Baverstock, P.R., Whitworth, G., Tenter, A.M., Johnson, A.M., 1995. Phylogenetic relationships between *Toxoplasma* and *Sarcocystis* deduced from a comparison of 18S rDNA sequences. *Parasitology* 110, 521–528.

498. Ellison, S., Witonsky, S., 2009. Evidence that antibodies against recombinant SnSAG1 of *Sarcocystis neurona* merozoites are involved in infection and immunity in equine protozoal myeloencephalitis. *Can. J. Vet. Res.* 73, 176–183.

499. Ellison, S.P., Omara-Opyene, A.L., Yowell, C.A., Marsh, A.E., Dame, J.B., 2002. Molecular characterization of a major 29 kDa surface antigen of *Sarcocystis neurona*. *Int. J. Parasitol.* 32, 217–225.

500. Ellison, S.P., Kennedy, T., Brown, K.K., 2003. Development of an ELISA to detect antibodies to rSAG1 in the horse. *Int. J. Appl. Res. Vet. Med.* 1, 318–327.

501. Ellison, S.P., Kennedy, T., Brown, K.K., 2003. Early signs of equine protozoal myeloencephalitis. *Int. J. Appl. Res. Vet. Med.* 1, 272–278.

502. Ellison, S.P., Greiner, E., Brown, K.K., Kennedy, T., 2004. Experimental infection of horses with culture-derived *Sarcocystis neurona* merozoites as a model for equine protozoal myeloencephalitis. *Int. J. Appl. Res. Vet. Med.* 2, 79–89.

503. Ellison, S.P., Lindsay, D.S., 2012. Decoquinate combined with levamisole reduce the clinical signs and serum SAG 1, 5, 6 antibodies in horses with suspected equine protozoal myeloencephalitis. *Intern. J. Appl. Res. Vet. Med.* 10, 1–7.

504. Elmore, S.A., Lalonde, L.F., Samelius, G., Alisauskas, R.T., Gajadhar, A.A., Jenkins, E.J., 2013. Endoparasites in the feces of arctic foxes in a terrestrial ecosystem in Canada. *Int. J. Parasitol.: Parasites and Wildlife* 2, 90–96.

505. Elsasser, T.H., Hammond, A.C., Rumsey, T.S., Fayer, R., 1986. Perturbed metabolism and hormonal profiles in calves infected with *Sarcocystis cruzi*. *Dom. Anim. Endocrinol.* 3, 277–287.

506. Elsasser, T.H., Rumsey, T.S., Hammond, A.C., Fayer, R., 1988. Influence of parasitism on plasma concentrations of growth hormone, somatomedin-c and somatomedin-binding proteins in calves. *J. Endocrinol.* 116, 191–200.

507. Elsasser, T.H., Fayer, R., Rumsey, T.S., Hammond, A.C., 1990. Plasma and tissue concentrations and molecular forms of somatostatin in calves infected with *Sarcocystis cruzi*. *Dom. Anim. Endocrinol.* 7, 537–550.

508. Elsasser, T.H., Sartin, J.L., McMahon, C., Romo, G., Fayer, R., Kahl, S., Blagburn, B., 1998. Changes in somatotropic axis response and body composition during growth hormone adminstraton in progressive cachectic parasitism. *Dom. Anim. Endocrinol.* 15, 239–255.

509. Elsheikha, H.M., Fitzgerald, S.D., Mansfield, L.S., Saeed, A.M., 2003. *Sarcocystis inghami* n. sp. (Sporozoa: Sarcocystidae) from the skeletal muscles of the Virginia opossum *Didelphis virginiana* in Michigan. *Syst. Parasitol.* 56, 77–84.

510. Elsheikha, H.M., Saeed, M.A., Fitzgerald, S.D., Murphy, A.J., Mansfield, L.S., 2003. Effects of temperature and host cell type on the *in vitro* growth and development of *Sarcocystis falcatula*. *Parasitol. Res.* 91, 22–26.

511. Elsheikha, H.M., Murphy, A.J., Fitzgerald, S.D., Mansfield, L.S., Massey, J.P., Saeed, M.A., 2003. Purification of *Sarcocystis neurona* sporocysts from opossum (*Didelphis virginiana*) using potassium bromide discontinuous density gradient centrifugation. *Parasitol. Res.* 90, 104–109.

512. Elsheikha, H.M., Mansfield, L.S., 2004. *Sarcocystis neurona* major surface antigen gene 1 (*SAG1*) shows evidence of having evolved under positive selection pressure. *Parasitol. Res.* 94, 452–459.

513. Elsheikha, H.M., Murphy, A.J., Mansfield, L.S., 2004. Prevalence of *Sarcocystis* species sporocysts in Northern Virginia opossums (*Didelphis virginiana*). *Parasitol. Res.* 93, 427–431.

514. Elsheikha, H.M., Murphy, A.J., Mansfield, L.S., 2004. Prevalence of and risk factors associated with the presence of *Sarcocystis neurona* sporocysts in opossum (*Didelphis virginiana*) from Michigan: A retrospective study. *Vet. Parasitol.* 125, 277–286.

515. Elsheikha, H.M., Murthy, A.J., Mansfield, L.S., 2004. Viability of *Sarcocystis neurona* sporocysts after long-term storage. *Vet. Parasitol.* 123, 257–264.

516. Elsheikha, H.M., Fitzgerald, S.D., Rosenthal, B.M., Mansfield, L.S., 2004. Concurrent presence of *Sarcocystis neurona* sporocysts, *Besnoitia darlingi* tissue cysts, and *Sarcocystis inghami* sarcocysts in naturally infected opossums (*Didelphis virginiana*). *J. Vet. Diagn. Invest.* 16, 352–356.

517. Elsheikha, H.M., Murphy, A.J., Mansfield, L.S., 2005. Phylogenetic congruence of *Sarcocystis neurona* Dubey et al., 1991 (Apicomplexa: Sarcocystidae) in the United States based on sequence analysis and restriction fragment length polymorphism (RFLP). *Syst. Parasitol.* 61, 191–202.

518. Elsheikha, H.M., Lacher, D.W., Mansfield, L.S., 2005. Phylogenetic relationships of *Sarcocystis neurona* of horses and opossums to other cyst-forming coccidia deduced from SSU rRNA gene sequence. *Parasitol. Res.* 97, 345–357.

519. Elsheikha, H.M., Soltan, D.M., el-Garhy, M.F., 2006. Inference of molecular phylogeny of *Sarcocystis felis* (Sarcocystidae) from cats based on nuclear-encoded ribosomal gene sequences. *J. Egypt. Soc. Parasitol.* 36, 441–453.

520. Elsheikha, H.M., Schott, H.C., Mansfield, L.S., 2006. Genetic variation among isolates of *Sarcocystis neurona*, the agent of protozoal myeloencephalitis, as revealed by amplified fragment length polymorphism markers. *Infect. Immun.* 74, 3448–3454.

521. Elsheikha, H.M., Kennedy, F.A., Murphy, A.J., Soliman, M., Mansfield, L.S., 2006. Sarcocystosis of *Sarcocystis felis* in cats. *J. Egypt. Soc. Parasitol.* 36, 1073–1088.

522. Elsheikha, H.M., Rosenthal, B.M., Murphy, A.J., Dunams, D.B., Neelis, D.A., Mansfield, L.S., 2006. Generally applicable methods to purify intracellular coccidia from cell cultures and to quantify purification efficacy using quantitative PCR. *Vet. Parasitol.* 135, 223–234.

523. Elsheikha, H.M., Mansfield, L.S., 2007. Molecular typing of *Sarcocystis neurona*: Current status and future trends. *Vet. Parasitol.* 149, 43–55.

524. Elsheikha, H.M., 2009. Has *Sarcocystis neurona* Dubey et al., 1991 (Sporozoa: Apicomplexa: Sarcocystidae) cospeciated with its intermediate hosts? *Vet. Parasitol.* 163, 307–314.

525. Elwasila, M., Entzeroth, R., Chobotar, B., Scholtyseck, E., 1984. Comparison of the structure of *Sarcocystis cuniculi* of the European rabbit (*Oryctolagus cuniculus*) and *Sarcocystis leporum* of the cottontail rabbit (*Sylvilagus floridanus*) by light and electron microscopy. *Acta Vet. Hung.* 32, 71–78.

526. Ely, R.W., Fox, J.C., 1989. Elevated IgG antibody to *Sarcocystis cruzi* associated with eosinophilic myositis in cattle. *J. Vet. Diagn. Invest.* 1, 53–56.

527. Enemar, A., 1963. Studies on a parasite resembling *Toxoplasma* in the brain of the bank-vole, *Clethrionomys glareolus*. *Arkiv för Zoologi* 15, 381–382.

528. Enemar, A., 1965. M-organisms in the brain of the Norway lemming, *Lemurs lemurs*. *Arkiv för Zoologi* 18, 9–16.

529. Entzeroth, R., 1981. Untersuchungen an Sarkosporidien (Mieschersche Schläuche) des einheimischen Rehwildes (*Capreolus capreolus* L.). *Z. Jagdwiss.* 27, 247–257.

530. Entzeroth, R., Abdel Ghaffar, F., Chobotar, B., Scholtyseck, E., 1981. Fine structural study of *Sarcocystis* sp. from Egyptian camels (*Camelus dromedarius*). *Acta Vet. Acad. Sci. Hung.* 29, 335–339.

531. Entzeroth, R., 1982. Ultrastructure of gamonts and gametes and fertilization of *Sarcocystis* sp. from the roe deer (*Capreolus capreolus*) in dogs. *Z. Parasitenkd.* 67, 147–153.

532. Entzeroth, R., 1982. A comparative light and electron microscope study of the cysts of *Sarcocystis* species of roe deer (*Capreolus capreolus*). *Z. Parasitenkd.* 66, 281–292.

533. Entzeroth, R., Chobotar, B., Scholtyseck, E., 1982. Ultrastructure of *Sarcocystis* sp. from the muscle of a white-tailed deer (*Odocoileus virginianus*). *Z. Parasitenkd.* 68, 33–38.

534. Entzeroth, R., 1983. Electron microscope study of merogony preceding cyst formation of *Sarcocystis* sp. in roe deer (*Capreolus capreolus*). *Z. Parasitenkd.* 69, 447–456.

535. Entzeroth, R., Chobotar, B., Scholtyseck, E., 1983. Ultrastructure of a *Sarcocystis* species from the red squirrel (*Tamiasciurus hudsonicus*) in Michigan. *Protistologica* 19, 91–94.

536. Entzeroth, R., Scholtyseck, E., Chobotar, B., 1983. Ultrastructure of *Sarcocystis* sp. from the eastern chipmunk (*Tamias striatus*). *Z. Parasitenkd.* 69, 823–826.

537. Entzeroth, R., 1984. Electron microscope study of host-parasite interactions of *Sarcocystis muris* (Protozoa, Coccidia) in tissue culture and *in vivo*. *Z. Parasitenkd.* 70, 131–134.

538. Entzeroth, R., 1985. Light-, scanning-, and transmission electron microscope study of the cyst wall of *Sarcocystis gracilis* Ràtz, 1909 (Sporozoa, Coccidia) from the roe deer (*Capreolus capreolus* L.). *Arch. Protistenk.* 129, 183–186.

539. Entzeroth, R., 1985. Invasion and early development of *Sarcocystis muris* (Apicomplexa, Sarcocystidae) in tissue cultures. *J. Protozool.* 32, 446–453.

540. Entzeroth, R., Chobotar, B., Scholtyseck, E., 1985. *Sarcocystis crotali* sp. n. with Mojave rattlesnake (*Crotalus scutulatus scutulatus*)—mouse (*Mus musculus*) cycle. *Arch. Protistenkd.* 129, 19–23.

541. Entzeroth, R., Chobotar, B., Scholtyseck, E., 1985. Electron microscope study of gamogony of *Sarcocystis muris* (Protozoa, Apicomplexa) in the small intestine of cats (*Felis catus*). *Protistologica* 21, 399–408.

542. Entzeroth, R., Chobotar, B., Scholtyseck, E., Neméséri, L., 1985. Light and electron microscope study of *Sarcocystis* sp. from the fallow deer (*Cervus dama*). *Z. Parasitenkd.* 71, 33–39.

543. Entzeroth, R., Dubremetz, J.F., Hodick, D., Ferreira, E., 1986. Immunoelectron microscopic demonstration of the exocytosis of dense granule contents into the secondary parasitophorous vacuole of *Sarcocystis muris* (Protozoa, Apicomplexa). *Eur. J. Cell Biol.* 41, 182–188.

544. Entzeroth, R., Chobotar, B., 1989. A freeze-fracture study of the host cell-parasite interface during and after invasion of cultured cells by cystozoites of *Sarcocystis muris*. *Eur. J. Protistol.* 25, 89–99.

545. Entzeroth, R., Konig, A., Dubremetz, J.F., 1991. Monoclonal antibodies identify micronemes and a new population of cytoplasmic granules cross-reacting with micronemes of cystozoites of *Sarcocystis muris*. *Parasitol. Res.* 77, 59–64.

546. Erber, M., Boch, J., 1976. Untersuchungen über Sarkosporidien des Schwarzwildes. Sporozystenausscheidung durch Hund, Fuchs und Wolf. *Berl. Münch. Tierärztl. Wochenschr.* 89, 449–450.

547. Erber, M., 1977. Möglichkeiten des Nachweises und der Differenzierung von zwei *Sarcocystis*-Arten des Schweines. *Berl. Münch. Tierärztl. Wochenschr.* 90, 480–482.

548. Erber, M., Boch, J., Barth, D., 1978. Drei Sarkosporidienarten des Rehwildes. *Berl. Münch. Tierärztl. Wochenschr.* 91, 482–486.

549. Erber, M., Meyer, J., Boch, J., 1978. Aborte beim Schwein durch Sarkosporidien (*Sarcocystis suicanis*). *Berl. Münch. Tierärztl. Wochenschr.* 91, 393–395.

550. Erber, M., Geisel, O., 1979. Untersuchungen zur Klinik und Pathologie der *Sarcocystis-suicanis*-Infektion beim Schwein. *Berl. Münch. Tierärztl. Wochenschr.* 92, 197–202.

551. Erber, M., Burgkart, M., 1981. Wirtschaftliche Verluste durch Sarkosporidiose (*Sarcocystis ovicanis* und *S.* spec.) bei der Mast von Schaflämmern. *Praktische Tierarzt* 62, 422–427.

552. Erber, M., Geisel, O., 1981. Vorkommen und Entwicklung von 2 Sarkosporidienarten des Pferdes. *Z. Parasitenkd.* 65, 283–291.

553. Erber, M., 1982. Life cycle of *Sarcocystis tenella* in sheep and dog. *Z. Parasitenkd.* 68, 171–180.

554. Erber, M., Göksu, K., 1984. Sarcosporidia in goats in Turkey and the differentiation of species. *Deutsche Forschungsgemeinschaft*, 21–29.

555. Erdman, L.F., 1978. *Sarcocystis* in striped skunks. *Iowa St. Univ. Vet.* 40, 112.

556. Erhardova, B., 1955. Nález cizopasníkú podobnych toxoplasme v mozku norníka rudého–*Clethrionomys glareolus* Schr. *Ceskoslovenska Biologie* 4, 251–252.

557. Eschenbacher, K.H., Sommer, I., Meyer, H.E., Mehlhorn, H., Rüger, W., 1992. Cloning and expression in *Escherichia coli* of cDNAs encoding a 31-kilodalton surface antigen of *Sarcocystis muris*. *Mol. Biochem. Parasitol.* 53, 159–167.

558. Espinosa, R.H., Sterner, M.C., Blixt, J.A., Cawthorn, R.J., 1988. Description of a species of *Sarcocystis* (Apicomplexa: Sarcocystidae), a parasite of the northern saw-whet owl, *Aegolius acadicus*, and experimental transmission to deer mice *Peromyscus maniculatus*. *Can. J. Zool.* 66, 2118–2121.

559. Esposito, D.H., Freedman, D.O., Neumayr, A., Parola, P., 2012. Ongoing outbreak of an acute muscular *Sarcocystis*-like illness among travellers returning from Tioman Island, Malaysia, 2011–2012. *Euro Surveill.* 17(45), pii = 20310.

560. Esposito, D.H., Stich, A., Epelboin, L., Malvy, D., Han, P.V., Bottieau, E., da Silva, A. et al. 2014. Acute muscular sarcocystosis: An international investigation among ill travelers returning from Tioman Island, Malaysia, 2011–2012. *Clin. Infect. Dis.* 59, 1401–1410.

561. Everitt, J.I., Bagsall, E.J., Hooser, S.B., Todd, K.S., 1987. *Sarcocystis* sp. in the striated muscle of domestic cats, *Felis catus. Proc. Helminthol. Soc. Wash.* 54, 279–281.

562. Ewing, R., Zaias, J., Stamper, M.A., Bossart, G.D., Dubey, J.P., 2002. Prevalence of *Sarcocystis* sp. in stranded Atlantic white-sided dolphins (*Lagenorhynchus acutus*). *J. Wildl. Dis.* 38, 291–296.

563. Ezzi, A., Gholami, M.R., Ahourai, P., 1992. Eosinophilic myositis associated with *Sarcocystis* in sheep. *Arch. Inst. RAZI* 42/43, 65–68.

564. Fahmy, M.F.M., 1992. Studies on muscle-infection with cysticercosis and sarcosporidiosis in cattle. *Egypt. J. Comp. Pathol. Clin. Pathol.* 5, 1–9.

565. Fantham, H.B., 1913. *Sarcocystis colii* n. sp. a sarcosporidian occurring in the red-faced African mouse bird, *Colius erythromelon. Proceedings of the Cambridge Philosophical Society* 17, 221–224.

566. Fantham, H.B., Porter, A., 1943. *Plasmodium struthionis*, sp. n. from Sudanese ostriches and *Sarcocystis salvelini*, sp. n., from Canadian speckled trout (*Salvelinus fontinalis*), together with a record of a *Sarcocystis* in the eel pout (*Zoarces angularis*). *Proc. Zool. Soc. London B* 113, 25–30.

567. Farooqui, A.A., Adams, D.D., Hanson, W.L., Prestwood, A.K., 1987. Studies on the enzymes of *Sarcocystis suicanis*: Purification and characterization of an acid phosphatase. *J. Parasitol.* 73, 681–688.

568. Fatani, A., Hilali, M., Al-Atiya, S., Al-Shami, S., 1996. Prevalence of *Sarcocystis* in camels (*Camelus dromedarius*) from Al-Ahsa, Saudi Arabia. *Vet. Parasitol.* 62, 241–245.

569. Fayer, R., 1970. *Sarcocystis*: Development in cultured avian and mammalian cells. *Science* 168, 1104–1105.

570. Fayer, R., Kocan, R.M., 1971. Prevalence of *Sarcocystis* in grackles in Maryland. *J. Protozool.* 18, 547–548.

571. Fayer, R., 1972. Gametogony of *Sarcocystis* sp. in cell culture. *Science* 175, 65–67.

572. Fayer, R., Johnson, A.J., 1973. Development of *Sarcocystis fusiformis* in calves infected with sporocysts from dogs. *J. Parasitol.* 59, 1135–1137.

573. Fayer, R., Leek, R.G., 1973. Excystation of *Sarcocystis fusiformis* sporocysts from dogs. *Proc. Helminthol. Soc. Wash.* 40, 294–296.

574. Fayer, R., 1975. Effects of refrigeration, cooking, and freezing on *Sarcocystis* in beef from retail food stores. *Proc. Helminthol. Soc. Wash.* 42, 138–140.

575. Fayer, R., Johnson, A.J., 1975. *Sarcocystis fusiformis* infection in the coyote (*Canis latrans*). *J. Infect. Dis.* 131, 189–192.

576. Fayer, R., Johnson, A.J., 1975. Effect of amprolium on acute sarcocystosis in experimentally infected calves. *J. Parasitol.* 61, 932–936.

577. Fayer, R., Johnson, A.J., Lunde, M., 1976. Abortion and other signs of disease in cows experimentally infected with *Sarcocystis fusiformis* from dogs. *J. Infect. Dis.* 134, 624–628.

578. Fayer, R., Johnson, A.J., Hildebrandt, P.K., 1976. Oral infection of mammals with *Sarcocystis fusiformis* bradyzoites from cattle and sporocysts from dogs and coyotes. *J. Parasitol.* 62, 10–14.

579. Fayer, R., 1977. Production of *Sarcocystis cruzi* sporocysts by dogs fed experimentally infected and naturally infected beef. *J. Parasitol.* 63, 1072–1075.

580. Fayer, R., Lunde, M.N., 1977. Changes in serum and plasma proteins and in IgG and IgM antibodies in calves experimentally infected with *Sarcocystis* from dogs. *J. Parasitol.* 63, 438–442.

581. Fayer, R., Kradel, D., 1977. *Sarcocystis leporum* in cottontail rabbits and its transmission to carnivores. *J. Wildl. Dis.* 13, 170–173.

582. Fayer, R., 1979. Multiplication of *Sarcocystis bovicanis* in the bovine bloodstream. *J. Parasitol.* 65, 980–982.

583. Fayer, R., Heydorn, A.O., Johnson, A.J., Leek, R.G., 1979. Transmission of *Sarcocystis suihominis* from humans to swine to nonhuman primates (*Pan troglodytes, Macaca mulatta, Macaca irus*). *Z. Parasitenkd.* 59, 15–20.

584. Fayer, R., Prasse, K.W., 1981. Hematology of experimental acute *Sarcocystis bovicanis* infection in calves. I. Cellular and serologic changes. *Vet. Pathol.* 18, 351–357.

585. Fayer, R., Dubey, J.P., 1982. Development of *Sarcocystis fayeri* in the equine. *J. Parasitol.* 68, 856–860.

586. Fayer, R., Dubey, J.P., Leek, R.G., 1982. Infectivity of *Sarcocystis* spp. from bison, elk, moose, and cattle for cattle via sporocysts from coyotes. *J. Parasitol.* 68, 681–685.

587. Fayer, R., Leek, R.G., Lynch, G.P., 1982. Attempted transmission of *Sarcocystis bovicanis* from cows to calves via colostrum. *J. Parasitol.* 68, 1127–1129.

588. Fayer, R., Hounsel, C., Giles, R.C., 1983. Chronic illness in a *Sarcocystis* infected pony. *Vet. Rec.* 113, 216–217.

589. Fayer, R., Lynch, G.P., Leek, R.G., Gasbarre, L.C., 1983. Effects of sarcocystosis on milk production of dairy cows. *J. Dairy Sci.* 66, 904–908.

590. Fayer, R., Dubey, J.P., 1984. Protective immunity against clinical sarcocystosis in cattle. *Vet. Parasitol.* 15, 187–201.

591. Fayer, R., Dubey, J.P., 1986. Bovine sarcocystosis. *Comp. Cont. Edu. Pract. Vet.* 8, F130–F142.

592. Fayer, R., 1988. Influence of parasitism on growth of cattle possibly mediated through tumor necrosis factor. In: *Proceedings of 12th Beltsville Symp. Biomechanisms Regulating Growth and Development*, Kluwer Academic Publishers, Beltsville, USA, p. 437.

593. Fayer, R., Dubey, J.P., 1988. *Sarcocystis* induced abortion and fetal death. In: Scarpelli, D.G., Migaki, G. (Eds.), *Transplacental Effects on Fetal Health*. Alan R. Liss, Inc., New York, NY, pp. 153–164.

594. Fayer, R., Elsasser, T.H., 1991. Bovine sarcocystosis: How parasites negatively affect growth. *Parasitol. Today* 7, 250–255.

595. Fayer, R., 2004. *Sarcocystis* spp. in human infections. *Clin. Microbiol. Rev.* 17, 894–902.

596. Fayer, R., Esposito, D.H., Dubey, J.P., 2015. Human infections with *Sarcocystis* species. *Clin. Microbiol. Rev.* 28, 295–311.

597. Fazly, Z.A., Nurulaini, R., Shafarin, M.S., Fariza, N.J., Zawida, Z., Muhamad, H.Y., Adnan, M. et al. 2013. Zoonotic parasites from exotic meat in Malaysia. *Trop. Biomed.* 30, 535–542.

598. Fedoseenko, V.M., Romanova, V.A., 1981. The occurrence of *Sarcocystis* sp. cysts in the muscle tissue of the intermediate host, *Rhomeomys opimus*. *Parazity* 1, 172–180. (In Russian).

599. Fedynich, A.M., Pence, D.B., 1992. *Sarcocystis* in mallards on the southern high plains of Texas. *Avian Dis.* 36, 1067–1069.

600. Fenger, C.K., Granstrom, D.E., Langemeier, J.L., Gajadhar, A., Cothran, G., Tramontin, R.R., Stamper, S., Dubey, J.P., 1994. Phylogenetic relationship of *Sarcocystis neurona* to other members of the family Sarcocystidae based on the sequence of the small ribosomal subunit gene. *J. Parasitol.* 79, 966–975.

601. Fenger, C.K., Granstrom, D.E., Langemeier, J.L., Stamper, S., Donahue, J.M., Patterson, J.S., Gajadhar, A.A., Marteniuk, J.V., Xiaomin, Z., Dubey, J.P., 1995. Identification of opossums (*Didelphis virginiana*) as the putative definitive host of *Sarcocystis neurona*. *J. Parasitol.* 81, 916–919.

602. Fenger, C.K., Granstrom, D.E., Langemeier, J.L., Stamper, S., 1997. Epizootic of equine protozoal myeloencephalitis on a farm. *J. Am. Vet. Med. Assoc.* 210, 923–927.

603. Fenger, C.K., Granstrom, D.E., Gajadhar, A.A., Williams, N.M., McCrillis, S.A., Stamper, S., Langemeier, J.L., Dubey, J.P., 1997. Experimental induction of equine protozoal myeloencephalitis in horses using *Sarcocystis* sp. sporocysts from the opossum (*Didelphis virginiana*). *Vet. Parasitol.* 68, 199–213.

604. Ferguson, H.W., Ellis, W.A., 1979. Toxoplasmosis in a calf. *Vet. Rec.* 104, 392–393.

605. Findlay, G.M., Middleton, A.D., 1934. Epidemic disease among voles (*Microtus*) with special reference to *Toxoplasma*. *J. Anim. Ecol.* 3, 150–160.

606. Finno, C.J., Packham, A.E., Wilson, W.D., Gardner, I.A., Conrad, P.A., Pusterla, N., 2007. Effects of blood contamination of cerebrospinal fluid on results of indirect fluorescent antibody tests for detection of antibodies against *Sarcocystis neurona* and *Neospora hughesi*. *J. Vet. Diagn. Invest.* 19, 286–289.

607. Fiori, M.G., Lowndes, H.E., 1988. Histochemical study of *Sarcocystis* sp. intramuscular cysts in gastrocnemius and soleus of the cat. *Parasitol. Res.* 75, 123–131.

608. Fischer, S., Odening, K., 1998. Characterization of bovine *Sarcocystis* species by analysis of their 18S ribosomal DNA sequences. *J. Parasitol.* 84, 50–54.

609. Fitzgerald, S.D., Janovitz, E.B., Kazacos, K.R., Dubey, J.P., Murphy, D.A., 1993. Sarcocystosis with involvement of the central nervous system in lambs. *J. Vet. Diagn. Invest.* 5, 291–296.

610. Flentje, B., Jungmann, R., Hiepe, Th., 1975. Vorkommen von *Isospora-hominis*-Sporozysten beim Menschen. *Dt. Gesundh. -Wesen* 30, 523–525.

611. Ford, G.E., 1974. Prey–predator transmission in the epizootiology of ovine sarcosporidiosis. *Aust. Vet. J.* 50, 38–39.

612. Ford, G.E., 1985. Immunity of sheep to homologous challenge with dog-borne *Sarcocystis* species following varying levels of prior exposure. *Int. J. Parasitol.* 15, 629–634.

613. Ford, G.E., 1986. Role of the dog, fox, cat and human as carnivore vectors in the transmission of the sarcosporidia that affect sheep meat production. *Aust. J. Agric. Res.* 37, 79–88.

614. Ford, G.E., 1986. Completion of the cycle for transmission of sarcosporidiosis between cats and sheep reared specific pathogen free. *Aust. Vet. J.* 63, 42–44.

615. Ford, G.E., 1987. Hosts of two canid genera, the red fox and the dog, as alternate vectors in the transmission of *Sarcocystis tenella* from sheep. *Vet. Parasitol.* 26, 13–20.

616. Ford, G.E., Fayer, R., Adams, M., O'Donoghue, P.J., Dubey, J.P., Baverstock, P.R., 1987. Genetic characterization by isoenzyme markers of North American and Australian isolates of species of *Sarcocystis* (Protozoa: Apicomplexa) from mice, sheep, goats and cattle. *Syst. Parasitol.* 9, 163–167.

617. Forest, T.W., Abou-Madi, N., Summers, B.A., Tornquist, S.J., Cooper, B.J., 2001. *Sarcocystis neurona*-like encephalitis in a Canada lynx (*Felis lynx canadensis*). *J. Zoo Wildl. Med.* 31, 383–387.

618. Foreyt, W.J., 1986. Evaluation of decoquinate, lasalocid, and monensin against experimentally induced sarcocystosis in calves. *Am. J. Vet. Res.* 47, 1674–1676.

619. Foreyt, W.J., 1989. *Sarcocystis* sp. in mountain goats (*Oreamnos americanus*) in Washington: Prevalence and search for the definitive host. *J. Wildl. Dis.* 25, 619–622.

620. Foreyt, W.J., Baldwin, T.J., Lagerquist, J.E., 1995. Experimental infections of *Sarcocystis* spp. in rocky mountain elk (*Cervus elaphus*) calves. *J. Wildl. Dis.* 31, 462–466.

621. Foreyt, W.J., High, W.A., Green, R.L., 1999. Search for the trematode *Prouterina wescotti* in black bears in Oregon. *J. Wildl. Dis.* 35, 622–623.

622. Formisano, P., Aldridge, B., Alony, Y., Beekhuis, L., Davies, E., Del Pozo, J., Dunn, K. et al. 2013. Identification of *Sarcocystis capracanis* in cerebrospinal fluid from sheep with neurological disease. *Vet. Parasitol.* 193, 252–255.

623. Frank, W., 1966. Eine *Sarcocystis*-infektion mit Pathologischen Veränderungen bei *Chamaeleo fischeri* durch *Sarcocystis chamaeleonis* n. spec. (Protozoa, Sporozoa). *Z. Parasitenkd.* 27, 317–335.

624. Franklin, R.P., MacKay, R.J., Gillis, K.D., Tanhauser, S.M., Ginn, P.E., Kennedy, T.J., 2003. Effect of a single dose of ponazuril on neural infection and clinical disease in *Sarcocystis neurona*-challenged interferon-gamma knockout mice. *Vet. Parasitol.* 114, 123–130.

625. Fransen, J.L.A., Degryse, A.-D.A.Y., van Mol, K.A.C., Ooms, L.A.A., 1987. *Sarcocystis* und chronische Myopathien bei Pferden. *Berl. Münch. Tierärztl. Wochenschr.* 100, 229–232.

626. Frelier, P., Mayhew, I.G., Fayer, R., Lunde, M.N., 1977. Sarcocystosis: A clinical outbreak in dairy calves. *Science* 195, 1341–1342.

627. Frelier, P.F., Mayhew, I.G., Pollock, R., 1979. Bovine sarcocystosis: Pathologic features of naturally occurring infection with *Sarcocystis cruzi*. *Am. J. Vet. Res.* 40, 651–657.

628. Frelier, P.F., 1980. Experimentally induced bovine sarcocystosis: Correlation of *in vitro* lymphocyte function with structural changes in lymphoid tissue. *Am. J. Vet. Res.* 41, 1201–1207.

629. Frelier, P.F., Lewis, R.M., 1984. Hematologic and coagulation abnormalities in acute bovine sarcocystosis. *Am. J. Vet. Res.* 45, 40–48.

630. Frenkel, J.K., 1953. Infections with organisms resembling *Toxoplasma*, together with the description of a new organism: *Besnoitia jellisoni*. *Atti del VI Congreso Internazionale di Microbiologia* 5, 426–434.

631. Freudenberg, Fr., 1956. Zur klinik der Myositis sarcosporidica des Pferdes. *Tierärztl. Umschau* 11, 91–93.

632. Freyer, B., Eschenbacher, K.H., Mehlhorn, H., Rueger, W., 1998. Isolation and characterization of cDNA clones encoding a 32-kDa dense- granule antigen of *Sarcocystis muris* (Apicomplexa). *Parasitol. Res.* 84, 583–589.

633. Freyer, B., Hansner, T., Mehlhorn, H., Rüger, W., 1999. Characterization of a genomic region encoding the 32-kDa dense granule antigen of *Sarcocystis muris* (Apicomplexa). *Parasitol. Res.* 85, 923–927.

634. Freyre, A., Pedroff, E., Mattos, J., 1988. *Sarcocystis bovicanis* y *S. bovifelis* en ganado vacuno; identificación, patogénesis y colonización preferencial. *Rev. Latinoam. Microbiol.* 30, 253–258.

635. Freyre, A., Chifflet, L., Mendez, J., 1992. Sarcosporidian infection in pigs in Uruguay. *Vet. Parasitol.* 41, 167–171.

636. Friesen, D.L., Cawthorn, R.J., Speer, C.A., Brooks, R.J., 1989. Ultrastructural development of the sarcocyst of *Sarcocystis rauschorum* (Apicomplexa: Sarcocystidae) in the varying lemming *Dicrostonyx richardsoni*. *J. Parasitol.* 75, 422–427.

637. Fritz, D.L., Dubey, J.P., 2002. Pathology of *Sarcocystis neurona* in interferon-gamma gene knockout mice. *Vet. Pathol.* 39, 137–140.

638. Fujino, T., Koga, S., Ishii, Y., 1982. Stereoscopical observations on the sarcosporidian cyst wall of *Sarcocystis bovicanis. Z. Parasitenkd.* 68, 109–111.

639. Fujita, O., Oku, Y., Ohbayashi, M., 1988. *Frenkelia* sp. from the red-backed vole, *Clethrionomys rufocanus bedfordiae*, in Hokkaido, Japan. *Jpn. J. Vet. Res.* 36, 69–71.

640. Fukuyo, M., Battsetseg, G., Byambaa, B., 2002. Prevalence of *Sarcocystis* infection in meat-producing animals in Mongolia. *Southeast Asian J. Trop. Med. Pub. Health* 33, 490–495.

641. Fukuyo, M., Battsetseg, G., Byambaa, B., 2002. Prevalence of *Sarcocystis* infection in horses in Mongolia. *Southeast Asian J. Trop. Med. Pub. Health* 33, 718–719.

642. Furr, M., Pontzer, C., 2001. Transforming growth factor beta concentrations and interferon gamma responses in cerebrospinal fluid of horses with equine protozoal myeloencephalitis. *Equine Vet. J.* 33, 721–725.

643. Furr, M., Kennedy, T., 2001. Cerebrospinal fluid and serum concentrations of ponazuril in horses. *Vet. Therapeut.* 2, 232–237.

644. Furr, M., Pontzer, C., Gasper, P., 2001. Lymphocyte phenotype subsets in the cerebrospinal fluid of normal horses and horses with equine protozoal myeloencephalitis. *Vet. Therapeut.* 2, 317–324.

645. Furr, M., Kennedy, T., MacKay, R., Reed, S., Andrews, F., Bernard, B., Bain, F., Byars, D., 2001. Efficacy of ponazuril 15% oral paste as a treatment for equine protozoal myeloencephalitis. *Vet. Therapeut.* 2, 215–222.

646. Furr, M., MacKay, R., Granstrom, D., Schott, H., Andrews, F., 2002. Clinical diagnosis of equine protozoal myeloencephalitis (EPM). *J. Vet. Intern. Med.* 16, 618–621.

647. Furr, M., McKenzie, H., Saville, W.J.A., Dubey, J.P., Reed, S.M., Davis, W., 2006. Prophylactic administration of ponazuril reduces clinical signs and delays seroconversion in horses challenged with *Sarcocystis neurona. J. Parasitol.* 92, 637–643.

648. Furr, M., Howe, D., Reed, S., Yeargan, M., 2011. Antibody coefficients for the diagnosis of equine protozoal myeloencephalitis. *J. Vet. Intern. Med.* 25, 138–142.

649. Gabor, M., Gabor, L.J., Srivastava, M., Booth, M., Reece, R., 2010. Chronic myositis in an Australian alpaca (*Llama pacos*) associated with *Sarcocystis* spp. *J. Vet. Diagn. Invest.* 22, 966–969.

650. Gadaev, A., 1978. On sarcocysts of ass (*Equus asinus*). *Akad. Nauk. Uzbecks. SSR.* 1, 47–48. (In Russian).

651. Gaibova, G.D., Radchenko, A.I., 1990. Electron microscopic study of macro- and microcysts of *Sarcocystis* (Coccidia, Sporozoa, Apicomplexa) from the buffalo. *Tsitologiya* 32, 801–805. (In Russian).

652. Gajadhar, A.A., Yates, W.D.G., Allen, J.R., 1987. Association of eosinophilic myositis with an unusual species of *Sarcocystis* in a beef cow. *Can. J. Vet. Res.* 51, 373–378.

653. Gajadhar, A.A., Marquardt, W.C., Hall, R., Gunderson, J., Ariztia-Carmona, E.V., Sogin, M.L., 1991. Ribosomal RNA sequences of *Sarcocystis muris*, *Theileria annulata* and *Crypthecodinium cohnii* reveal evolutionary relationships among apicomplexans, dinoflagellates, and ciliates. *Mol. Biochem. Parasitol.* 45, 147–154.

654. Gajadhar, A.A., Marquardt, W.C., 1992. Ultrastructural and transmission evidence of *Sarcocystis cruzi* associated with eosinophilic myositis in cattle. *Can. J. Vet. Res.* 56, 41–46.

655. Gajadhar, A.A., Blair, C.D., 1992. Development of a model ribosomal RNA hybridization assay for the detection of *Sarcocystis* and other coccidia. *Can. J. Vet. Res.* 56, 208–213.

656. Gaji, R.Y., Zhang, D., Breathnach, C.C., Vaishnava, S., Striepen, B., Howe, D.K., 2006. Molecular genetic transfection of the coccidian parasite *Sarcocystis neurona. Mol. Biochem. Parasitol.* 150, 1–9.

657. Gaji, R.Y., Howe, D.K., 2009. The heptanucleotide motif GAGACGC is a key component of a *cis*-acting promoter element that is critical for SnSAG1 expression in *Sarcocystis neurona. Mol. Biochem. Parasitol.* 166, 85–88.

658. Galavíz-Silva, L., Mercado-Hernández, R., Ramírez-Bon, E., Arredondo-Cantú, J.M., Lazcano-Villarreal, D., 1991. *Sarcocystis neotomafelis* sp. n. (Protozoa: Apicomplexa) from the woodrat *Neotoma micropus* in Mexico. *Rev. Lat-amer. Microbiol.* 33, 313–322.

659. Galfre, G., Howe, S.C., Milstein, C., Butcher, G.W., Howard, J.C., 1977. Antibodies to major histocompatibility antigens produced by hybrid cell lines. *Nature* 266, 550–552.

660. Gangadharan, B., Valsala, K.V., Nair, M.G., Rajan, A., 1992. Sarcocystosis in a Sambar deer (*Cervus unicolor*). *Indian J. Anim. Sci.* 62, 127–128.

661. Gargala, G., Baishanbo, A., Favennec, L., Francois, A., Ballet, J.J., Rossignol, J.F., 2005. Inhibitory activities of epidermal growth factor receptor tyrosine kinase-targeted dihydroxyisoflavone and tri-hydroxydeoxybenzoin derivatives on *Sarcocystis neurona, Neospora caninum*, and *Cryptosporidium parvum* development. *Antimicrob. Agents Chemother.* 49, 4628–4634.

662. Gargala, G., Le Goff, L., Ballet, J.J., Favennec, L., Stachulski, A.V., Rossignol, J.F., 2009. *In vitro* efficacy of nitro- and halogeno-thiazolide/thiadiazolide derivatives against *Sarcocystis neurona. Vet. Parasitol.* 162, 230–235.

663. Garner, M.M., Barr, B.C., Packham, A.E., Marsh, A.E., Burek-Huntington, K.A., Wilson, R.K., Dubey, J.P., 1997. Fatal hepatic sarcocystosis in two polar bears (*Ursus maritimus*). *J. Parasitol.* 83, 523–526.

664. Garnham, P.C.C., Duggan, A.J., Sinden, R.E., 1979. A new species of *Sarcocystis* in the brain of two exotic birds. *Ann. Parasitol.* 54, 393–400.

665. Gasbarre, L.C., 1982. Demonstration of anti-horse red blood cell antibodies in a *Sarcocystis* infected pony. *Vet. Rec.* 111, 15–16.

666. Gasbarre, L.C., Suter, P., Fayer, R., 1984. Humoral and cellular immune responses in cattle and sheep inoculated with *Sarcocystis. Am. J. Vet. Res.* 45, 1592–1596.

667. Gauert, B., Jungmann, R., Hiepe, T., 1983. Beziehungen zwischen Intestinalstörungen und para-sitären Darminfektionen des Menschen unter besonderer berücksichtigung von *Sarcocystis*-befall. *Dt. Gesundh. Wesen* 38, 62–66.

668. Gautam, A., Dubey, J.P., Saville, W.J., Howe, D.K., 2011. The SnSAG merozoite surface antigens of *Sarcocystis neurona* are expressed differentially during the bradyzoite and sporozoite life cycle stages. *Vet. Parasitol.* 183, 37–42.

669. Geisel, O., Kaiser, E., Vogel, O., Krampitz, H.E., Rommel, M., 1979. Pathomorphologic findings in short-tailed voles (*Microtus agrestis*) experimentally-infected with *Frenkelia microti. J. Wildl. Dis.* 15, 267–270.

670. Gerhold, R., Newman, S.J., Grunenwald, G.M., Crews, A., Hodshon, A., Su, C., 2014. Acute onset of encephalomyelitis with atypical lesions associaed with dual infection of *Sarcocystis neurona* and *Toxoplasma gondii. Vet. Parasitol.* 205, 697–701.

671. Gerhold, R.W., Howerth, E.W., Lindsay, D.S., 2005. *Sarcocystis neurona*-associated meningoencephali-tis and description of intramuscular sarcocysts in a fisher (*Martes pennanti*). *J. Wildl. Dis.* 41, 224–230.

672. Gestrich, R., Heydorn, A.O., 1974. Untersuchungen zur Überlebensdauer von Sarkosporidienzysten im Fleisch von Schlachttieren. *Berl. Münch. Tierärztl. Wschr.* 87, 475–476.

673. Gestrich, R., Mehlhorn, H., Heydorn, A.O., 1975. Licht- und elektronenmikroskopische Untersuchungen an Cysten von *Sarcocystis fusiformis* in der Muskulatur von Kälbern nach experimenteller Infektion mit Oocysten und Sporocysten der groben Form von *Isopora bigemina* der Katze. *Zbl. Bakt. Hyg., I. Abt. Orig. A* 233, 261–276.

674. Gestrich, R., Heydorn, A.O., Baysu, N., 1975. Beiträge zum Lebenszyklus der Sarkosporidien VI. Untersuchungen zur Artdifferenzierung bei *Sarcocystis fusiformis* und *Sarcocystis tenella. Berl. Münch. Tierärztl. Wochenschr.* 88, 191–197.

675. Ghisleni, G., Robba, S., Germani, O., Scanziani, E., 2006. Identification and prevalence of *Sarcocystis* spp. cysts in bovine canned meat. *Food Control* 17, 691–694.

676. Ghosal, S.B., Joshi, S.C., Shah, H.L., 1987. Cross-transmission studies of *Sarcocystis fusiformis* and *Sarcocystis levinei* of buffalo to cattle calves. *Indian J. Anim. Sci.* 57, 524–525.

677. Giannetto, S., Poglayen, G., Brianti, E., Gaglio, G., Scala, A., 2005. *Sarcocystis gracilis*-like sarcocysts in a sheep. *Vet. Rec.* 156, 322–323.

678. Giboda, M., Rakar, J., 1978. First record of "*Isopora hominis*" in Czechoslovakia. *Folia Parasitol. (Praha)* 25, 16.

679. Giboda, M., Ditrich, O., Scholz, T., Viengsay, T., Bouaphanh, S., 1991. Current status of food-borne parasitic zoonoses in Laos. *Southeast Asian J. Trop. Med. Pub. Health* 22(Suppl), 56–61.

680. Gibson, A.K., Raverty, S., Lambourn, D.M., Huggins, J., Magargal, S.L., Grigg, M.E., 2011. Polyparasitism is associated with increased disease severity in *Toxoplasma gondii*-infected marine sentinel species. *PLoS Negl. Trop. Dis.* 5, issue 5-e1142.

681. Giles, R.C., Tramontin, R., Kadel, W.L., Whitaker, K., Miksch, D., Bryant, D.W., Fayer, R., 1980. Sarcocystosis in cattle in Kentucky. *J. Am. Vet. Med. Assoc.* 176, 543–548.

682. Gill, H.S., Charleston, W.A.G., Moriarty, K.M., 1988. Immunosuppression in *Sarcocystis muris*-infected mice: Evidence for suppression of antibody and cell-mediated responses to a heterologous antigen. Immunol. *Cell Biol.* 66, 209–214.

683. Gill, H.S., Moriarty, K.M., Charleston, W.A.G., 1989. *In vitro* and *in vivo* interaction between sporocysts of *Sarcocystis muris* and mouse peritoneal macrophages. *Vet. Parasitol.* 32, 341–347.

684. Gillis, K.D., MacKay, R.J., Yowell, C.A., Levy, J.K., Greiner, E.C., Dame, J.B., Cheadle, M.A., Hernández, J., Massey, E.T., 2003. Naturally occurring *Sarcocystis* infection in domestic cats (*Felis catus*). *Int. J. Parasitol.* 33, 877–883.

685. Gjerde, B., 1984. A light microscopic comparison of the cysts of four species of *Sarcocystis* infecting the domestic reindeer (*Rangifer tarandus*) in northern Norway. *Acta Vet. Scand.* 25, 195–204.

686. Gjerde, B., 1984. *Sarcocystis* infection in wild reindeer (*Rangifer tarandus*) from Hardangervidda in southern Norway: With a description of the cysts of *Sarcocystis hardangeri* n. sp. *Acta Vet. Scand.* 25, 205–212.

687. Gjerde, B., 1984. *Sarcocystis hardangeri* and *Sarcocystis rangi* n. sp. from the domestic reindeer (*Rangifer tarandus*) in northern Norway. *Acta Vet. Scand.* 25, 411–418.

688. Gjerde, B., 1984. The fox as definitive host for *Sarcocystis* sp. Gjerde, 1984 from skeletal muscle of reindeer (*Rangifer tarandus*) with a proposal for *Sarcocystis tarandivulpes* n. sp. as replacement name. *Acta Vet. Scand.* 25, 403–410.

689. Gjerde, B., 1984. The raccoon dog (*Nyctereutes procyonoides*) as definitive host for *Sarcocystis* spp. of reindeer (*Rangifer tarandus*). *Acta Vet. Scand.* 25, 419–424.

690. Gjerde, B., Bratberg, B., 1984. The domestic reindeer (*Rangifer tarandus*) from northern Norway as intermediate host for three species of *Sarcocystis*. *Acta Vet. Scand.* 25, 187–194.

691. Gjerde, B., 1985. Ultrastructure of the cysts of *Sarcocystis hardangeri* from skeletal muscle of reindeer (*Rangifer tarandus tarandus*). *Can. J. Zool.* 63, 2676–2683.

692. Gjerde, B., 1985. Ultrastructure of the cysts of *Sarcocystis rangiferi* from skeletal muscle of reindeer (*Rangifer tarandus tarandus*). *Can. J. Zool.* 63, 2669–2675.

693. Gjerde, B., 1985. Ultrastructure of the cysts of *Sarcocystis rangi* from skeletal muscle of reindeer (*Rangifer tarandus tarandus*). *Rangifer* 5, 43–52.

694. Gjerde, B., 1985. Ultrastructure of the cysts of *Sarcocystis grueneri* from cardiac muscle of reindeer (*Rangifer tarandus tarandus*). *Z. Parasitenkd.* 71, 189–198.

695. Gjerde, B., 1985. Ultrastructure of the cysts of *Sarcocystis tarandivulpes* from skeletal muscle of reindeer (*Rangifer tarandus tarandus*). *Acta Vet. Scand.* 26, 91–104.

696. Gjerde, B., 1985. Ultrastructure of the cysts of *Sarcocystis tarandi* from skeletal muscle of reindeer (*Rangifer tarandus tarandus*). *Can. J. Zool.* 63, 2913–2918.

697. Gjerde, B., 1985. The fox as definitive host for *Sarcocystis rangi* from reindeer (*Rangifer tarandus tarandus*). *Acta Vet. Scand.* 26, 140–142.

698. Gjerde, B., 1986. Scanning electron microscopy of the sarcocysts of six species of *Sarcocystis* from reindeer (*Rangifer tarandus tarandus*). *Acta Pathol. Microbiol. Immunol. Scand. B* 94, 309–317.

699. Gjerde, B., Dahlgren, S.S., 2010. Corvid birds (Corvidae) act as definitive hosts for *Sarcocystis ovalis* in moose (*Alces alces*). *Parasitol. Res.* 107, 1445–1453.

700. Gjerde, B., 2012. Morphological and molecular characterization and phylogenetic placement of *Sarcocystis capreolicanis* and *Sarcocystis silva* n. sp. from roe deer (*Capreolus capreolus*) in Norway. *Parasitol. Res.* 110, 1225–1237.

701. Gjerde, B., 2013. Phylogenetic relationships among *Sarcocystis* species in cervids, cattle and sheep inferred from the mitochondrial cytochrome c oxidase subunit I gene. *Int. J. Parasitol.* 43, 579–591.

702. Gjerde, B., 2014. Morphological and molecular characteristics of four *Sarcocystis* spp. in Canadian moose (*Alces alces*), including *Sarcocystis taeniata* n. sp. *Parasitol. Res.* 113, 1591–1604.

703. Gjerde, B., 2014. Molecular characterisation of *Sarcocystis rileyi* from a common eider (*Somateria mollissima*) in Norway. *Parasitol. Res.* 113, 3501–3509.

704. Gjerde, B., 2014. *Sarcocystis* species in red deer revisited: With a re-description of two known species as *Sarcocystis elongata* n. sp. and *Sarcocystis truncata* n. sp. based on mitochondrial *cox*1 sequences. *Parasitology* 141, 441–452.

705. Gjerde, B., Schulze, J., 2014. Muscular sarcocystosis in two arctic foxes (*Vulpes lagopus*) due to *Sarcocystis arctica* n. sp.: Sarcocyst morphology, molecular characteristics and phylogeny. *Parasitol. Res.* 113, 811–821.

706. Gjerde, B., Josefsen, T.D., 2014. Molecular characterisation of *Sarcocystis lutrae* n. sp. and *Toxoplasma gondii* from the musculature of two Eurasian otters (*Lutra lutra*) in Norway. *Parasitol. Res.* 114, 873–886.

707. Göbel, E., Katz, M., Erber, M., 1978. Licht- und elektronenmikroskopische Untersuchungen zur Entwicklung von Muskelzysten von *Sarcocystis suicanis* in Hausschweinen nach experimenteller infektion. *Zentralbl. Bakteriol. Hyg. I. Abt. Orig. A* 241, 368–383.

708. Göbel, E., Rommel, M., 1980. Licht- und elektronenmikroskopische Untersuchungen an Zysten von *Sarcocystis equicanis* in der Ösophagusmuskulatur von Pferden. *Berl. Münch. Tierärztl. Wschr.* 93, 41–47.

709. Göbel, E., Erber, M., Grimm, F., 1996. *Sarcocystis phoeniconaii* n. sp. Murata, 1986 (Apicomplexa: Sarcocystidae) des Lesser Flamingo (*Phoeniconaias minor*: Ciconiiformes). *Berl. Münch. Tierärztl. Wschr.* 109, 239–244.

710. Godoy, Z.R., Vilca, L.M., González, Z.A., Leyva, V.V., Sam, T.R., 2007. Saneamiento y detoxificación de carne de llama (*Lama glama*) infectada con *Sarcocystis aucheniae* mediante cocción, horneado, fritura y congelado. *Rev. Inv. Vet. Perú* 18, 51–56.

711. Godoy, S.N., de Paula, C.D., Cubas, Z.S., Matushima, E.R., Catão-Dias, J.L., 2009. Occurrence of *Sarcocystis falcatula* in captive Psittacine birds in Brazil. *J. Avian Med. Surg.* 23, 18–23.

712. Göksu, K., 1975. Koyunlarda sarcosporidiosis'in yayilişi üzerine araştirmalar. *J. Fac. Vet. Med. Istanbul* 1, 110–127.

713. Goldstein, T., Gill, V.A., Tuomi, P., Monson, D., Burdin, A., Conrad, P.A., Dunn, J.L. et al. 2011. Assessment of clinical pathology and pathogen exposure in sea ottters (*Enhydra lutris*) bordering the threatened population in Alaska. *J. Wildl. Dis.* 47, 579–592.

714. Golubkov, I.A., 1979. Infection of dogs and cats with *Sarcocystis* from hens and ducks. *Veterinariia* 77, 55–56. (In Russian).

715. Golubkov, I.A., 1979. Infection of dogs and kittens with *Sarcocystis* from chickens and ducks. *Veterinaria Moscow* 77, 55–56. (In Russian).

716. Golubkov, V.I., Rybaltovskii, D.V., Kislyakova, Z.I., 1974. The source of infection for swine *Sarcocystis*. *Veterinarya* 11, 85–87. (In Russian).

717. Gomes, A.P.M., Vogel, J., Pacheco, R.G., Botelho, G.G., 1992. Aspectos clínicos da sarcocistose caprina. *Arq. Univ. Fed. Rur. Rio de Janeiro* 15, 1–6.

718. Gomes, A.P.M., Vogel, J., Botelho, G.G., Pacheco, R.G., Rodrigues, O.D., 1994. Electrocardiographs evaluation on goats experimentally infected by *Sarcocystis capracanis* Fischer, 1979. *R. Bras. Med. Vet.* 16, 240–247.

719. González, L.M., Villalobos, N., Montero, E., Morales, J., Álamo Sanz, R., Muro, A., Harrison, L.J.S., Parkhouse, R.M.E., Gárate, T., 2006. Differential molecular identification of *Taeniid* spp. and *Sarcocystis* spp. cysts isolated from infected pigs and cattle. *Vet. Parasitol.* 142, 95–101.

720. Gorman, T., Lorca, M., Pereira, S., Thiermann, E., Núñez, F., 1986. Sarcosporidiosis y toxoplasmosis caprina en la region metropolitana (Comunas de San Jose de Maipo y Til-Til). *Arch. Med. Vet.* 18, 87–94.

721. Gorman, T.R., Alcaino, H., Robles, M., 1981. Sarcosporidiosis in especies de abasto de la zona central de Chile. *Arch. Med. Vet.* 13, 39–43.

722. Gorman, T.R., Alcaíno, H.A., Muñoz, H., Cunazza, C., 1984. *Sarcocystis* sp. in guanaco (*Lama guanicoe*) and effect of temperature on its viability. *Vet. Parasitol.* 15, 95–101.

723. Gozalo, A.S., Montali, R.J., St. Claire, M., Barr, B., Rejmanek, D., Ward, J.M., 2007. Chronic polymyositis associated with disseminated sarcocystosis in a captive-born rhesus macaque. *Vet. Pathol.* 44, 695–699.

724. Granstrom, D.E., Ridley, R.K., Baoan, Y., Gershwin, L.J., 1990. Immunodominant proteins of *Sarcocystis cruzi* bradyzoites isolated from cattle affected or nonaffected with eosinophilic myositis. *Am. J. Vet. Res.* 51, 1151–1155.

725. Granstrom, D.E., Ridley, R.K., Yao, B., Gershwin, L.J., Briggs, D.J., 1990. Immunofluorescent localization of *Sarcocystis cruzi* antigens, IgG and IgE, in lesions of eosinophilic myositis in cattle. *J. Vet. Diagn. Invest.* 2, 147–149.

726. Granstrom, D.E., Giles, R.C., Tuttle, P.A., Williams, N.M., Poonacha, K.B., Petrites-Murphy, M.B., Tramontin, R.R. et al. 1991. Immunohistochemical diagnosis of protozoan parasites in lesions of equine protozoal myeloencephalitis. *J. Vet. Diagn. Invest.* 3, 75–77.

727. Granstrom, D.E., Álvarez, O., Dubey, J.P., Comer, P.F., Williams, N.M., 1992. Equine protozoal myelitis in Panamanian horses and isolation of *Sarcocystis neurona*. *J. Parasitol.* 78, 909–912.

728. Granstrom, D.E., Dubey, J.P., Davis, S.W., Fayer, R., Fox, J.C., Poonacha, K.B., Giles, R.C., Comer, P.F., 1993. Equine protozoal myeloencephalitis: Antigen analysis of cultured *Sarcocystis neurona* merozoites. *J. Vet. Diagn. Invest.* 5, 88–90.

729. Granstrom, D.E., MacPherson, J.M., Gajadhar, A.A., Dubey, J.P., Tramontin, R., Stamper, S., 1994. Differentiation of *Sarcocystis neurona* from eight related coccidia by random amplified polymorphic DNA assay. *Mol. Cell. Probes* 8, 353–356.

730. Granstrom, D.E., Ridley, R.K., Baoan, Y., Gershwin, L.J., Nesbitt, P.M., Wempe, L.A., 1994. Type 1 hypersensitivity as a component of eosinophilic myositis (muscular sarcocystosis) in cattle. *Am. J. Vet. Res.* 50, 571–574.

731. Gray, L.C., Magdesian, K.G., Sturges, B.K., Madigan, J.E., 2001. Suspected protozoal myeloencephalitis in a two-month-old colt. *Vet. Rec.* 149, 269–273.

732. Greiner, E.C., Roelke, M.E., Atkinson, C.T., Dubey, J.P., Wright, S.D., 1989. *Sarcocystis* sp. in muscles of free-ranging Florida panthers and cougars (*Felis concolor*). *J. Wildl. Dis.* 25, 623–628.

733. Greiner, M., Ono, K., Horn, K., 1992. Comparative investigation of sporozoites of 4 *Sarcocystis* species by isoenzyme electrophoresis. *J. Protozool. Res.* 2, 123–127.

734. Greve, E., 1973. Forekomsten af *Sarcocystis miescheriana* hos svin. En undersøgelse af infektionsfrekvens og geografisk spredning I et sjællandsk område. *Nord. Vet. Med.* 25, 545–553.

735. Greve, E., 1980. Zur diagnostik der Sarkosporidiose des Schweines bei der routinemäβigen fleischuntersuchung. *Mh. Vet. Med.* 35, 150–151.

736. Greve, E., 1985. Sarcosporidiosis–an overlooked zoonosis. Man as intermediate and final host. *Danish Med. Bull.* 32, 228–230.

737. Grigg, M.E., Sundar, N., 2009. Sexual recombination punctuated by outbreaks and clonal expansions predicts *Toxoplasma gondii* population genetics. *Int. J. Parasitol.* 39, 925–933.

738. Grikienienė, J., Senutaite, Y.A., 1994. Identification of bovine *Sarcocystis* species by the morphological characters of muscle cysts. *Biologija* 1, 31–32. (In Russian).

739. Grikienienė, J., Arnastauskienė, T., 1992. On the possibility of the development of bank vole parasite *Sarcocystis* sp. in the organism of other intermediate hosts, avoiding the definitive host. *Ekologija* 1, 48–58. (In Russian).

740. Grikienienė, J., 1993. New experimental data on development cycle of sarcosporidians (Sarcosporidia) of rodents. *Ekologija* 1, 33–46. (In Russian).

741. Grikienienė, J., Arnastauskienė, T., Kutkienė, L., 1993. On some disregarded ways of sarcosporidians circulation and remarks about systematics of the genus *Sarcocystis* Lankester, 1882 with the description of the new species from rodents. *Ekologija* 1, 16–24. (In Russian).

742. Grikienienė, J., Kutkienė, L., 1998. New experimental data on the laboratory rat as a definitive host of *Sarcocystis rodentifelis*. *Acta Zool. Lituanica. Parasitologia* 8, 121–124. (In Russian).

743. Grikienienė, J., Mažeikytė, R., 2000. Investigation of sarcosporidians (*Sarcocystis*) of small mammals in Kamasta landscape reserve and its surroundings. *Acta Zool. Lituanica* 10, 55–68. (In Russian).

744. Grikienienė, J., 2005. Investgations into endoparasites of small mammals in the environs of Lake Drūkšiai. *Acta Zool. Lituanica* 15, 109–114. (In Russian).

745. Grikienienė, J., 1989. Experimental investigation of the development of *Sarcocystis* from the oesophagus of sheep in the final host. *Acta Parasitol. Lituanica* 23, 59–66. (In Russian).

746. Groulade, P., Vallée, A., 1959. Encéphalite chez un Agneau avec présence de Toxoplasmes. *Bull. Acad. Vét. France* 32, 135–136.

747. Güçlü, F., Aldem, R.O.S., Güler, L., 2004. Differential identifcation of cattle *Sarcocystis* spp. by random amplified polymorphic DNA-polymerase chain reaction (RAPD-PCR). *Revue Méd. Vét.* 155, 8–9.

748. Guerrero, D.C.A., Hernández, D.J., Alva, M.J., 1967. *Sarcocystis* en alpacas. *Revista de la Facultad de Medicina Veterinaria* 21, 69–76.

749. Guindon, S., Dufayard, J.F., Lefort, V., Anisimova, M., Hordijk, W., Gascuel, O., 2010. New algorithms and methods to estimate maximum-likelihood phylogenies: Assessing the performance of PhyML 3.0. *Syst. Biol.* 59, 307–321.

750. Gunn, H.M., Fraher, J.P., 1992. Incidence of sarcocysts in skeletal-muscles of horses. *Vet. Parasitol.* 42, 33–40.

751. Gunning, R.F., Jones, J.R., Jeffrey, M., Higgins, R.J., Williamson, A.G., 2000. *Sarcocystis* encephalomyelitis in cattle. *Vet. Rec.* 146, 328.

752. Gupta, G.D., Lakritz, J., Kim, J.H., Kim, D.Y., Kim, J.K., Marsh, A.E., 2002. Seroprevalence of *Neospora*, *Toxoplasma gondii*, and *Sarcocystis neurona* antibodies in horses from Jeju island, South Korea. *Vet. Parasitol.* 106, 193–201.

753. Gupta, G.D., Lakritz, J., Saville, W.J., Livingston, R.S., Dubey, J.P., Middleton, J.R., Marsh, A.E., 2004. Antigenic evaluation of a recombinant Baculovirus-expressed *Sarcocystis neurona* SAG1 antigen. *J. Parasitol.* 90, 1027–1033.

754. Gupta, R.S., Kushwah, H.S., Kushwah, A., 1992. Some glucose metabolic enzymes in various fractions of sarcocysts of *Sarcocystis fusiformis* of buffalo (*Bubalus bubalis*). *Vet. Parasitol.* 44, 45–50.

755. Gupta, R.S., Kushwah, H.S., Kushwah, A., 1993. *Sarcocystis fusiformis*: Some protein metabolic enzymes in various fractions of sarcocysts of buffalo (*Bubalus bubalis*). *Vet. Parasitol.* 45, 185–189.

756. Gupta, R.S., Kushwah, H.S., Kushwah, A., 1993. Enzymatic studies in various fractions of sarcocysts of *Sarcocystis fusiformis* of buffalo. *Indian J. Anim. Sci.* 63(1), 20–21.

757. Gupta, R.S., Kushwah, H.S., Kushwah, A., 1994. Some enzymes of gluconeogenesis in various fractions of sarcocysts of *Sarcocystis fusiformis* of buffalo (*Bubalus bubalis*) (technical note). *Vet. Parasitol.* 52(1–2), 145–149.

758. Gupta, R.S., Kushwah, H.S., Kushwah, A., 1995. *Sarcocystis fusiformis*: Some Krebs cycle enzymes in various fractions of sarcocysts of buffalo (*Bubalus bubalis*). *Vet. Parasitol.* 56, 1–5.

759. Gupta, S.C., Iyer, P.K.R., 1984. Spontaneous sarcocystosis in the brain of pigs. *Indian Vet. J.* 61, 738–739.

760. Gupta, S.L., Gautam, O.P., Bhardwaj, R.M., 1979. A note on the prevalence of *Sarcocystis* infection in sheep from Hissar area as studied by peptic-digestion technique. *Indian J. Anim. Sci.* 49, 971.

761. Gupta, S.L., Gautam, O.P., 1982. *Sarcocystis* infection in goats of Hissar and its transmission to dogs. *Indian J. Parasitol.* 6, 73–74.

762. Gupta, S.L., Gautam, O.P., 1984. *Sarcocystis* infection in pigs of Hissar, Haryana, India and its transmission to dogs. *Vet. Parasitol.* 13, 1–3.

763. Gustafsson, K., Book, M., Dubey, J.P., Uggla, A., 1997. Meningoencephalitis in capercaillie (*Tetrao urogallus* L.) caused by a *Sarcocystis*-like organism. *J. Zoo Wildl. Med.* 28, 280–284.

764. Gut, J., 1982. Effectiveness of methods used for the detection of sarcosporidiosis in farm animals. *Folia Parasitol. (Praha)* 29, 289–295.

765. Habermann, R.T., Williams, F.P., 1957. Diseases seen at necropsy of 708 *Macaca mulatta* (Rhesus monkey) and *Macaca philippinensis* (Cynomologus monkey). *Am. J. Vet. Res.* 18, 419–426.

766. Hadwen, S., 1922. Cyst-forming protozoa in reindeer and caribou, and a sarcosporidian parasite of the seal (*Phoca richardi*). *J. Am. Vet. Med. Assoc.* 61, 374–382.

767. Häfner, U., Frank, W., 1984. Host specificity and host range of the genus *Sarcocystis* in three snake-rodent life cycles. *Zentralbl. Bakteriol. Mikrobiol. Hyg. A* 256, 296–299.

768. Häfner, U., Matuschka, F.R., 1984. Life cycle studies on *Sarcocystis dirumpens* sp. n. with regard to host specificity. *Z. Parasitenkd.* 70, 715–720.

769. Häfner, U., Frank, W., 1986. Morphological studies on the muscle cysts of *Sarcocystis dirumpens* (Hoare 1933) Häfner and Matuschka 1984 in several host species revealing endopolygeny in metrocytes. *Z. Parasitenkd.* 72, 453–461.

770. Hagi, A.B., Hassan, A.M., Di Sacco, B., 1989. *Sarcocystis* in Somali camel. *Parassitologia* 31, 133–136.

771. Hajimohammadi, B., Eslami, G., Oryan, A., Zohourtabar, A., Pourmirzaei, T.H., Moghaddam, A.M., 2014. Molecular identification of *Sarcocystis hominis* in native cattle of central Iran: A case report. *Trop. Biomed.* 31, 183–186.

772. Hajtós, I., Glávits, R., Pálfi, V., Kovács, T., 2000. Sarcocystosis with neurological signs in breeding-sheep stock. *Magyar Állator Vosok Lapja* 122, 72–78. (In Hungarian).

773. Hamidinejat, H., Jalali, M.H.R., Nabavi, L., 2010. Survey on *Sarcocystis* infection in slaughtered cattle in south-west of Iran, emphasized on evaluation of muscle squash in comparison with digestion method. *J. Anim. Vet. Adv.* 9, 1724–1726.

774. Hamidinejat, H., Hekmatimoghaddam, S., Jafari, H., Sazmand, A., Molayan, P.H., Derakhshan, L., Mirabdollahi, S., 2013. Prevalence and distribution patterns of *Sarcocystis* in camels (*Camelus dromedarius*) in Yazd province, Iran. *J. Parasit. Dis.* 37, 163–165.

775. Hamidinejat, H., Moetamedi, H., Alborzi, A., Hatami, A., 2014. Molecular detection of *Sarcocystis* species in slaughtered sheep by PCR-RFLP from south-western of Iran. *J. Parasit. Dis.* 38, 233–237.

776. Hamir, A.N., Rupprecht, C.E., Ziemer, E.L., 1989. Generalized eosinophilic myositis with eosinophilia of blood and cerebrospinal fluid in a raccoon (*Procyon lotor*). *J. Vet. Diagn. Invest.* 1, 192–194.

777. Hamir, A.N., Moser, G., Rupprecht, C.E., 1992. A five year (1985–1989) retrospective study of equine neurological diseases with special reference to rabies. *J. Comp. Pathol.* 106, 411–421.

778. Hamir, A.N., Dubey, J.P., 2001. Myocarditis and encephalitis associated with *Sarcocystis neurona* infection in raccoons (*Procyon lotor*). *Vet. Parasitol.* 95, 335–340.

779. Hancock, K., Zajac, A.M., Elvinger, F., Lindsay, D.S., 2004. Prevalence of agglutinating antibodies to *Sarcocystis neurona* in raccoons (*Procyon lotor*) from an urban area of Virginia. *J. Parasitol.* 90, 881–882.

780. Hansner, T., Freyer, B., Mehlhorn, H., Rüger, W., 1999. Isolation and characterization of a cDNA clone encoding a thiol proteinase of *Sarcocystis muris* cyst merozoites (Apicomplexa). *Parasitol. Res.* 85, 749–757.

781. Harada, S., Furukawa, M., Tokuoka, E., Matsumoto, K., Yahiro, S., Miyasaka, J., Saito, M. et al. 2013. Control of toxicity of *Sarcocystis fayeri* in horsemeat by freezing treatment and prevention of food poisoning caused by raw consumption of horsemeat. *Shokuhin Eiseigaku Zasshi* 54, 198–203. (In Japanese).

782. Hasselmann, G., 1926. Alterações patholigicas do myocardio na sarcosporideose. *Bol. Inst. Bras. Scien.* 2, 319–326.

783. Harris, D.J., Maia, J.P., Perera, A., 2012. Molecular survey of Apicomplexa in *Podarcis* wall lizards detects *Hepatozoon*, *Sarcocystis*, and *Eimeria* species. *J. Parasitol.* 98, 592–597.

784. Hartley, W.J., Blakemore, W.F., 1974. An unidentified sporozoan encephalomyelitis in sheep. *Vet. Pathol.* 11, 1–12.

785. Haskard, D., Cavender, D., Beatty, P., Springer, T., Ziff, M., 1986. T lymphocyte adhesion to endothelial cells: Mechanisms demonstrated by anti-LFA-1 monoclonal antibodies. *J. Immunol.* 137, 2901–2906.

786. Hayden, D.W., King, N.W., Murthy, A.S.K., 1976. Spontaneous *Frenkelia* infection in a laboratory-reared rat. *Vet. Pathol.* 13, 337–342.

787. Haziroglu, R., Guvenc, T., Tunca, R., 2003. Electron microscopical studies on cysts of *Sarcocystis arieticanis* within cardiac muscle of naturally infected sheep. *Parasitol. Res.* 89, 23–25.

788. Heckeroth, A.R., Tenter, A.M., 1999. Development and validation of species-specific nested PCRs for diagnosis of acute sarcocystosis in sheep. *Int. J. Parasitol.* 29, 1331–1349.

789. Hellwig, M., 1981. *Licht-und Elektronenmikroskopische Untersuchungen zum Lebenszyklus von Sarcocystis muris (Sporozoa, Coccidia) in der Katze*. Inaugural Dissertation, Ludwig-Maximilians-Universität München, Germany, pp. 1–60.

790. Hemaprasanth, 1995. *Sarcocystis* infection in domestic pigs in Uttar Pradesh. *J. Vet. Parasitol.* 2, 119–123.

791. Hemaprasanth, Bhatia, B.B., 1996. On the species of *Sarcocystis* of the pig from Uttar Pradesh, India with a description of a sarcocyst of hitherto unrecognised species. *J. Vet. Parasitol.* 10, 57–61.

792. Hemenway, M.P., Avery, M.L., Ginn, P.E., Schaack, S., Dame, J.B., Greiner, E.C., 2001. Influence of size of sporocyst inoculum upon the size and number of sarcocysts of *Sarcocystis falcatula* which develop in the brown-headed cowbird. *Vet. Parasitol.* 95, 321–326.

793. Henderson, J.M., Dies, K.H., Haines, D.M., Higgs, G.W., Ayroud, M., 1997. Neurologic symptoms associated with sarcocystosis in adult sheep. *Can. Vet. J.* 38, 168–170.

794. Henry, D.P., 1932. *Isopora buteonis* sp. nov. from the hawk and owl, and notes on *Isopora lacazii* (Labbé) in birds. *Univ. Calif. Publ. Zool.* 37, 291–300.

795. Hernández-Jauregui, P., Silva-Lemoine, E., Girón-Rojas, H., 1983. Miocarditis por sarcosporidios en un macaco Rhesus, Estudio di microscopia electrónica y de luz. *Arch. Invest. Méd. (Méx.)* 14, 139–144.

796. Hernández-Rodríguez, S., Navarrete, I., Martínez Gómez, F., 1981. *Sarcocystis cervicanis*, nueva especie parásita del ciervo (*Cervus elaphus*). *Rev. Ibérica Parasitol.* 41, 43–51.

797. Hernández-Rodríguez, S., Martínez-Gómez, F., Navarrete, I., Acosta-García, I., 1981. Estudio al microscopio optico y electronico del quiste de *Sarcocystis cervicanis*. *Rev. Ibérica Parasitol.* 41, 351–361.

798. Hernández-Rodríguez, S., Martínez-Gómez, F., López-Rodríguez, R., Navarrete, I., 1986. Morfología y biología de *Sarcocystis capracanis* Fischer, 1979, primera cita en España. *Rev. Ibérica Parasitol.* 46, 7–12.

799. Hernández-Rodríguez, S., Acosta, I., Navarrete, I., 1992. *Sarcocystis jorrini* sp. nov. from the fallow deer *Cervus dama*. *Parasitol. Res.* 78, 557–562.

800. Heskett, K.A., MacKay, R.J., 2008. Antibody index and specific antibody quotient in horses after intragastric administration of *Sarcocystis neurona* sporocysts. *Am. J. Vet. Res.* 69, 403–409.

801. Heydorn, A.O., Rommel, M., 1972. Beiträge zum Lebenszyklus der Sarkosporidien. II. Hund und Katze als Überträger der Sarkosporidien des Rindes. *Berl. Münch. Tierärztl. Wochenschr.* 85, 121–123.

802. Heydorn, A.O., Mehlhorn, H., Gestrich, R., 1975. Licht- und elektronenmikroskopische Untersuchungen an Cysten von *Sarcocystis fusiformis* in der Muskulatur von Kälbern nach experimenteller Infektion mit Oocysten und Sporocysten von *Isopora hominis* Railliet et Lucet, 1891. 2. Die Feinstruktur der Metrocysten und Merozoiten. *Zbl. Bakt. Hyg., I. Abt. Orig. A* 232, 373–391.

803. Heydorn, A.O., Gestrich, R., Mehlhorn, H., Rommel, M., 1975. Proposal for a new nomenclature of the Sarcosporidia. *Z. Parasitenkd.* 48, 73–82.

804. Heydorn, A.O., Gestrich, R., 1976. Beiträge zum Lebenszyklus der Sarkosporidien. VII. Entwicklungsstadien von *Sarcocystis ovicanis* im Schaf. *Berl. Münch. Tierärztl. Wochenschr.* 89, 1–5.

805. Heydorn, A.O., Gestrich, R., Janitschke, K., 1976. Beiträge zum Lebenszyklus der Sarkosporidien. VIII. Sporozysten von *Sarcocystis bovihominis* in den Fäzes von Rhesusaffen (*Macaca rhesus*) und Pavianen (*Papio cynocephalus*). *Berl. Münch. Tierärztl. Wochenschr.* 89, 116–120.

806. Heydorn, A.O., 1977. Beiträge zum Lebenszyklus der Sarkosporidien. IX. Entwicklungszyklus von *Sarcocystis suihominis* n. spec. *Berl. Münch. Tierärztl. Wochenschr.* 90, 218–224.

807. Heydorn, A.O., 1977. Sarkosporidieninfiziertes Fleisch als mögliche Krankheitsursache für den Menschen. *Arch. Lebensmittelhygiene* 28, 27–31.

808. Heydorn, A.O., Mehlhorn, H., 1978. Light and electron microscopic studies on *Sarcocystis suihominis*. 2. The schizogony preceding cyst formation. *Zbl. Bakt. Hyg., I. Abt. Orig. A* 240, 123–134.

809. Heydorn, A.O., Ipczynski, V., 1978. Zur Schizogonie von *Sarcocystis suihominis* im Schwein. *Berl. Münch. Tierärztl. Wochenschr.* 91(8), 154–155.

810. Heydorn, A.O., Döhmen, H., Funk, G., Pähr, H., Zientz, H., 1978. Zur Verbreitung der Sarkosporidieninfektion beim Hausschwein. *Arch. Lebensmittelhygiene* 29, 184–185.

811. Heydorn, A.O., 1980. Zur Widerstandsfähigkeit von *Sarcocystis bovicanis*-Sporozysten. *Berl. Münch. Tierärztl. Wochenschr.* 93, 267–270.

812. Heydorn, A.O., Matuschka, F.R., 1981. Zur Endwirtspezifität der vom Hund übertragnen Sarkosporidienarten. *Z. Parasitenkd.* 66, 231–234.

813. Heydorn, A.O., Haralambidis, S., Matuschka, F.R., 1981. Zur Chemoprophylaxe und Therapie der akuten Sarkosporidiose. Berl. Münch. *Tierärztl. Wochenschr.* 94, 229–234.

814. Heydorn, A.O., Matuschka, F.R., Ipczynski, V., 1981. Zur Schizogonie von *Sarcocystis suicanis* im Schwein. *Berl. Münch. Tierärztl. Wochenschr.* 94, 49–51.

815. Heydorn, A.O., Haralambidis, S., 1982. Zur Entwicklung von *Sarcocystis capracanis* Fischer, 1979. *Berl. Münch. Tierärztl. Wochenschr.* 95, 265–271.

816. Heydorn, A.O., Unterholzner, J., 1983. Zur Entwicklung von *Sarcocystis hircicanis* n. sp. *Berl. Münch. Tierärztl. Wochenschr.* 96, 275–282.

817. Heydorn, A.O., 1985. Zur Entwicklung von *Sarcocystis arieticanis* n. sp. *Berl. Münch. Tierärztl. Wochenschr.* 98, 231–241.

818. Heydorn, A.O., Karaer, Z., 1986. Zur Schizogonie von *Sarcocystis ovicanis*. *Berl. Münch. Tierärztl. Wochenschr.* 99, 185–189.

819. Heydorn, A.O., Mehlhorn, H., 1987. Fine structure of *Sarcocystis arieticanis* Heydorn, 1985 in its intermediate and final hosts (sheep and dog). *Zentralbl. Bakteriol. Hyg. A* 264, 353–362.

820. Heydorn, A.O., Kirmsse, P., 1996. Isolation und experimentelle Übertragung von *Sarcocystis moulei* Neveu-Lemaire, 1912. *Berl. Münch. Tierärztl. Wochenschr.* 109, 440–445.

821. Hiepe, F., Hiepe, T., Hlinak, P., Jungmann, R., Horsch, R., Weidauer, B., 1979. Experimentelle Infektion des Menschen und von Tieraffen (*Cercopithecus callitrichus*) mit Sarkosporidien-Zysten von Rind und Schwein. *Arch. Exp. Vet. Med.* 33, 819–830.

822. Hilali, M., Mohamed, A., 1980. The dog (*Canis familiaris*) as the final host of *Sarcocystis cameli* (Mason, 1910). *Tropenmed. Parasitol.* 31, 213–214.

823. Hilali, M., Imam, E.S., Hassan, A., 1982. The endogenous stages of *Sarcocystis cameli* (Mason, 1910). *Vet. Parasitol.* 11, 127–129.

824. Hilali, M., Nasser, A.M., 1987. Ultrastructure of *Sarcocystis* spp. from donkeys (*Equus asinus*) in Egypt. *Vet. Parasitol.* 23, 179–183.

825. Hilali, M., Nassar, A.M., El-Ghaysh, A., 1992. Camel (*Camelus dromedarius*) and sheep (*Ovis aries*) meat as a source of dog infection with some coccidian parasites. *Vet. Parasitol.* 43, 37–43.

826. Hilali, M., Fatani, A., Al-Atiya, S., 1995. Isolation of tissue cysts of *Toxoplasma, Isospora, Hammondia* and *Sarcocystis* from camel (*Camelus dromedarius*) meat in Saudi Arabia. *Vet. Parasitol.* 58, 353–356.

827. Hilali, M., El-seify, M., Zayed, A., El-morsey, A., Dubey, J.P., 2011. *Sarcocystis dubeyi* (Huong and Uggla, 1999) infection in water buffaloes (*Bubalus bulalis*) from Egypt. *J. Parasitol.* 97, 527–528.

828. Hill, J.E., Chapman, W.L., Prestwood, A.K., 1988. Intramuscular *Sarcocystis* sp. in two cats and a dog. *J. Parasitol.* 74, 724–727.

829. Hiller, B.J., Sidor, I.F., De Guise, S., Barclay, J.S., 2007. Prevalence and ultrastructure of *Sarcocystis* in American woodcock. *J. Parasitol.* 93, 1529–1530.

830. Hu, J.J., Liu, T.T., Liu, Q., Esch, G.W., Chen, J.Q., 2015. *Sarcocystis clethrionomyelaphis* Matuschla, 1986 (Apicomplexa: Sarcocystidae) infecting the large oriental vole *Eothenomys miletus* (Thomas) (Cricetidae: Microtinae) and its phylogenetic relationships with other *Sarcocystis* Lankester, 1982 species. *Syst. Parasitol.* 91, 273–279.

831. Hillyer, E.V., Anderson, M.P., Greiner, E.C., Atkinson, C.T., Frenkel, J.K., 1991. An outbreak of *Sarcocystis* in a collection of psittacines. *J. Zoo Wildl. Med.* 22, 434–445.

832. Hinaidy, H.K., Supperer, R., 1979. Sarkosporidienbefall des Schweines in Österreich. *Wien. Tierärztl. Mschr.* 66, 281–285.

833. Hinaidy, H.K., Loupal, G., 1982. *Sarcocystis bertrami* Doflein, 1901, ein Sarkosporid des Pferdes, *Equus caballus*. *Zbl. Vet. Med. B* 29, 681–701.

834. Hinaidy, H.K., Egger, A., 1994. Die sarkosporidien des scafes in Österreich. *J. Vet. Med. B* 41, 417–427.

835. Hoane, J.S., Carruthers, V.B., Striepen, B., Morrison, D.P., Entzeroth, R., Howe, D.K., 2003. Analysis of the *Sarcocystis neurona* microneme protein SnMIC10: Protein characteristics and expression during intracellular development. *Int. J. Parasitol.* 33, 671–679.

836. Hoane, J.S., Morrow, J.K., Saville, W.J., Dubey, J.P., Granstrom, D.E., Howe, D.K., 2005. Enzyme-linked immunosorbent assays for the detection of equine antibodies specific to *Sarcocystis neurona* surface antigens. *Clin. Diagn. Lab. Immunol.* 12, 1050–1056.

837. Furr, M., Kennedy, T., 2000. Cerebrospinal fluid and blood concentrations of toltrazuril 5% suspension in the horse after oral dosing. *Vet. Therapeut.* 1, 125–132.

838. Hoane, J.S., Gennari, S.M., Dubey, J.P., Ribeiro, M.G., Borges, A.S., Yai, L.E.O., Aguiar, D.M., Cavalcante, G.T., Bonesi, G.L., Howe, D.K., 2006. Prevalence of *Sarcocystis neurona* and *Neospora* spp. infection in horses from Brazil based on presence of serum antibodies to parasite surface antigen. *Vet. Parasitol.* 136, 155–159.

839. Hofer, J., Boch, J., Erber, M., 1982. Zelluläre und humorale Reaktionen bei Mäusen nach experimenteller *Sarcocystis muris*- und *S. dispersa*-Infektion. *Berl. Münch. Tierärztl. Wochenschr.* 95, 169–175.

840. Holmdahl, O.J.M., Mattsson, J.G., Uggla, A., Johansson, K.E., 1993. Oligonucleotide probes complementary to variable regions of 18S rRNA from *Sarcocystis* species. *Mol. Cell. Probes* 7, 481–486.

841. Holmdahl, O.J.M., Mattsson, J.G., Uggla, A., Johansson, K.E., 1994. The phylogeny of *Neospora caninum* and *Toxoplasma gondii* based on ribosomal RNA sequences. *FEMS Microbiology Letters* 119, 187–192.

842. Holmdahl, O.J.M., Mattsson, J.G., 1996. Rapid and sensitive identification of *Neospora caninum* by *in vitro* amplification of the internal transcribed spacer 1. *Parasitology* 112, 177–182.

843. Holmdahl, O.J.M., Morrison, D.A., Ellis, J.T., Huong, L.T.T., 1999. Evolution of ruminant *Sarcocystis* (Sporozoa) parasites based on small-subunit rDNA sequences. *Mol. Phylogenet. Evol.* 11, 27–37.

844. Hong, C.B., Giles, R.C., Newman, L.E., Fayer, R., 1982. Sarcocystosis in an aborted bovine fetus. *J. Am. Vet. Med. Assoc.* 181, 585–588.

845. Hoppe, D.M., 1976. Prevalence of macroscopically detectable *Sarcocystis* in North Dakota ducks. *J. Wildl. Dis.* 12, 27–29.

846. Horn, K., Ono, K., Heydorn, A.O., 1991. *In vitro* excystation and cryopreservation of ovine and caprine *Sarcocystis* species. *J. Protozool. Res.* 1, 3–21.

847. Houk, A.E., Goodwin, D.G., Zajac, A.M., Barr, S.C., Dubey, J.P., Lindsay, D.S., 2010. Prevalence of antibodies to *Trypanosoma cruzi*, *Toxoplasma gondii*, *Encephalitozoon cuniculi*, *Sarcocystis neurona*, *Besnoitia darlingi*, and *Neospora caninum* in North American opossums, *Didelphis virginiana*, from Southern Louisiana. *J. Parasitol.* 96, 1119–1122.

848. Howe, D.K., 2001. Initiation of a *Sarcocystis neurona* expressed sequence tag (EST) sequencing project: A preliminary report. *Vet. Parasitol.* 95, 233–239.

849. Howe, D.K., Gaji, R.Y., Mroz-Barrett, M., Gubbels, M.J., Striepen, B., Stamper, S., 2005. *Sarcocystis neurona* merozoites express a family of immunogenic surface antigen that are orthologues of the *Toxoplasma gondii* surface antigens (SAGs) and SAG-related sequences. *Infect. Immun.* 73, 1023–1033.

850. Howe, D.K., Gaji, R.Y., Marsh, A.E., Patil, B.A., Saville, W.J., Lindsay, D.S., Dubey, J.P., Granstrom, D.E., 2008. Strains of *Sarcocystis neurona* exhibit differences in their surface antigens, including the absence of the major surface antigen SnSAG1. *Int. J. Parasitol.* 38, 623–631.

851. Howells, R.E., Carvalho, A.D.V., Mello, M.N., Rangel, N.M., 1975. Morphological and histochemical observations on *Sarcocystis* from the nine-banded armadillo, *Dasypus novemcinctus*. *Ann. Trop. Med. Parasitol.* 69, 463–474.

852. Hsu, V., Grant, D.C., Dubey, J.P., Zajac, A.M., Lindsay, D.S., 2010. Prevalence of antibodies to *Sarcocystis neurona* in cats from Virginia and Pennsylvania. *J. Parasitol.* 96, 800–801.

853. Hu, J.J., Ma, T.C., Li, X.R., 2005. A new species of sarcocysts (Sporozoea, Eucoccidiida) from *Rattus norvegicus*. *Acta Zootaxonomica Sinica* 30, 287–290. (In Chinese).

854. Hu, J.J., Liao, J.Y., Meng, Y., Guo, Y.M., Chen, X.W., Zuo, Y.X., 2011. Identification of *Sarcocystis cymruensis* in wild *Rattus flavipectus* and *Rattus norvegicus* from Peoples Republic of China and its transmission to rats and cats. *J. Parasitol.* 97, 421–424.

855. Hu, J.J., Meng, Y., Guo, Y.M., Liao, J.Y., Song, J.L., 2012. Completion of the life cycle of *Sarcocystis zuoi*, a parasite from the Norway rat, *Rattus norvegicus*. *J. Parasitol.* 98, 550–553.

856. Hu, J.J., Liu, Q., Yang, Y.F., Esch, G.W., Guo, Y.M., Zou, F.C., 2014. *Sarcocystis eothenomysi* n. sp. (Apicomplexa: Sarcocystidae) from the large oriental vole *Eothenomys miletus* (Thomas) (Cricetidae: Microtinae) from Anning, China. *Syst. Parasitol.* 89, 73–81.

857. Huan, L.V., Viseth, V., Sophiop, R., 1996. Infection by *Cysticercus cellulosae* in pigs and *C. bovis*, *Sarcocystis* on cattle from Kampuchea. *Khoa Hoc Ky Thuat Thu Y* 3, 61–67. (In Vietnamese).

858. Hudkins, G., Kistner, T.P., 1977. *Sarcocystis hemionilatrantis* (sp. n.) life cycle in mule deer and coyotes. *J. Wildl. Dis.* 13, 80–84.

859. Hudkins-Vivion, G., Kistner, T.P., Fayer, R., 1976. Possible species differences between *Sarcocystis* from mule deer and cattle. *J. Wildl. Dis.* 12, 86–87.

860. Hunyadi, L., Papich, M.G., Pusterla, N., 2015. Pharmacokinetics of a low dose and FDA-labeled dose of diclazuril administered orally as a pelleted topdressing in adult horses. *J. Vet. Pharmacol. Ther.* 38, 243–248.

861. Huong, L.T.T., Uggla, A., Dubey, J.P., Thao, L.P., Doanh, P.T.M., Tien, T., 1995. Study on *Sarcocystis* in water buffalo at some abattoirs in Ho Chi Minh city. In: *Proceeding of The Annual Scientific Conference of the Faculty of Veterinary Medicine and Animal Science*. University of Agriculture and Forestry, Ho Chi Minh city, Vietnam, pp. 93–97.

862. Huong, L.T.T., Dubey, J.P., Uggla, A., 1997. Redescription of *Sarcocystis levinei* Dissanaike and Kan, 1978 (Protozoa: Sarcocystidae) of the water buffalo (*Bubalus bubalis*). *J. Parasitol.* 83, 1148–1151.

863. Huong, L.T.T., Dubey, J.P., Nikkilä, T., Uggla, A., 1997. *Sarcocystis buffalonis* n. sp. (Protozoa: Sarcocystidae) from the water buffalo (*Bubalus bubalis*) in Vietnam. *J. Parasitol.* 83, 471–474.

864. Huong, L.T.T., 1999. Prevalence of *Sarcocystis* spp. in water buffaloes in Vietnam. *Vet. Parasitol.* 86, 33–39.

865. Huong, L.T.T., Uggla, A., 1999. *Sarcocystis dubeyi* n. sp. (Protozoa: Sarcocystidae) in the water buffalo (*Bubalus bubalis*). *J. Parasitol.* 85, 102–104.

866. Husna Maizura, A.M., Khebir, V., Chong, C.K., Azman Shah, A.M., Azri, A., Lokman Hakim, S., 2012. Surveillance for sarcocystosis in Tioman Island, Malaysia. *Malaysian J. Pub. Health Med.* 12, 39–44.

867. Hussain, M.M., Gupta, S.L., Singh, R.P., Verma, P.C., 1986. Pathological changes in lambs infected with *Sarcocystis ovicanis* during experiments. *Indian J. Anim. Sci.* 56, 169–173.

868. Hussein, H.S., Warrag, M., 1985. Prevalence of *Sarcocystis* in food animals in the Sudan. *Trop. Anim. Health Prod.* 17, 100–101.

869. Hvizdošová, H., Goldová, M., 2009. Monitoring of occurrence of sarcocystosis in hoofed game in Eastern Slovakia. *Folia Vet.* 53, 5–7.

870. Hyun, C., Gupta, G.D., Marsh, A.E., 2003. Sequence comparison of *Sarcocystis neurona* surface antigen from multiple isolates. *Vet. Parasitol.* 112, 11–20.

871. Imes, G.D., Migaki, G., 1967. Eosinophilic myositis in cattle-pathology and incidence. *Proceedings of 71st Annual Meeting of US Livestock Sanit. Assoc.* pp. 111–112.

872. Inoue, I., Yamada, C., Yamada, M., Kondo, T., 1990. *Sarcocystis hirsuta* (Protozoa:Apicomplexa) in cattle in Japan. *Jpn. J. Parasitol.* 39, 403–405.

873. Inoue, I., Yamada, M., Fujita, M., Shimizu, H., 1990. *Sarcocystis capracanis* (Protozoa, Apicomplexa) in goats in Japan. *Jpn. J. Parasitol.* 39, 406–409.

874. Inoue, N., Omata, Y., Yonemasu, K., Claveria, F.G., Igarashi, I., Saito, A., Suzuki, N., 1996. Collagen cross-reactive antigen of *Sarcocystis cruzi*. *Vet. Parasitol.* 63, 17–23.

875. Ishag, M.Y., El Amin, E.A., Osman, A.Y., 2001. Camel experimentally infected with *Sarcocystis*. *Sudan J. Vet. Res.* 17, 27–33.

876. Ishag, M.Y., Majid, A.M., Magzoub, A.M., 2006. Isolation of a new *Sarcocystis* species from Sudanese camels (*Camelus dromedarius*). *Int. J. Trop. Med.* 1, 167–169.

877. Italiano, C.M., Wong, K.T., AbuBakar, S., Lau, Y.L., Ramli, N., Syed Omar, S.F., Bador, M.K., Tan, C.T., 2014. *Sarcocystis nesbitti* causes acute, relapsing febrile myositis with a high attack rate: Description of a large outbreak of muscular sarcocystosis in Pangkor Island, Malaysia, 2012. *PLoS Negl. Trop. Dis.* 8, e2876.

878. Jacobs, L., Remington, J.S., Melton, M.L., 1960. A survey of meat samples from swine, cattle, and sheep for the presence of encysted *Toxoplasma*. *J. Parasitol.* 46, 23–28.

879. Jacobson, E.R., Gardiner, C.H., Nicholson, A., Page, C.D., 1984. *Sarcocystis* encephalitis in a cockatiel. *J. Am. Vet. Med. Assoc.* 185, 904–906.

880. Jafari Shoorijeh, S., Sadjjadi, S.M., Asheri, A., Eraghi, K., 2008. *Giardia* spp. and *Sarcocystis* spp. status in pet dogs of Shiraz, Southern part of Iran. *Trop. Biomed.* 25, 154–159.

881. Jain, A.K., Gupta, S.L., Singh, R.P., Mahajan, S.K., 1986. Experimental *Sarcocystis levinei* infection in buffalo calves. *Vet. Parasitol.* 21, 51–53.

882. Jain, P.C., Shah, H.L., 1985. Cross-transmission studies of *Sarcocystis cruzi* of the cattle to the buffalo-calves. *Indian J. Anim. Sci.* 55, 27–28.

883. Jain, P.C., Shah, H.L., 1986. Determination of sporocysts discharged during the patent period by dogs fed with sarcocysts of *Sarcocystis cruzi* of the cattle. *Indian J. Anim. Sci.* 56, 1005–1008.

884. Jain, P.C., Shah, H.L., 1988. A comparative morphological study of sarcocysts of three species of *Sarcocystis* infecting cattle in Madhya Pradesh. *Indian J. Anim. Sci.* 58, 8–13.

885. Jäkel, T., 1995. Cyclic transmission of *Sarcocystis gerbilliechis* n. sp. by the Arabian saw-scaled viper, *Echis coloratus*, to rodents of the subfamily Gerbillinae. *J. Parasitol.* 81, 626–631.

886. Jäkel, T., Burgstaller, H., Frank, W., 1996. *Sarcocystis singaporensis*: Studies on host specificity, pathogenicity, and potential use as a biocontrol agent of wild rats. *J. Parasitol.* 82, 280–287.

887. Jäkel, T., Henke, M., Weingarten, B., Kliemt, D., Seidinger, S., 1997. *In vitro* cultivation of the vascular phase of *Sarcocystis singaporensis*. *J. Eukaryot. Microbiol.* 44, 293–299.

888. Jäkel, T., Archer-Baumann, C., Boehmler, A.M., Sorger, I., Henke, M., Kliemt, D., Mackenstedt, U., 1999. Identification of a subpopulation of merozoites of *Sarcocystis singaporensis* that invades and partially develops inside muscle cells in vitro. *Parasitology* 118, 235–244.

889. Jäkel, T., Khoprasert, Y., Endepols, S., Archer-Baumann, C., Suasa-ard, K., Promkerd, P., Kliemt, D., Boonsong, P., Hongnark, S., 1999. Biological control of rodents using *Sarcocystis singaporensis*. *Int. J. Parasitol.* 29, 1321–1330.

890. Jäkel, T., Khoprasert, Y., Kliemt, D., Mackenstedt, U., 2001. Immunoglobulin subclass responses of wild brown rats to *Sarcocystis singaporensis*. *Int. J. Parasitol.* 31, 273–283.

891. Jäkel, T., Wallstein, E., Müncheberg, F., Archer-Baumann, C., Weingarten, B., Kliemt, D., Mackenstedt, U., 2001. Binding of a monoclonal antibody to sporozoites of *Sarcocystis singaporensis* enhances escape from the parasitophorous vacuole, which is necessary for intracellular development. *Infect. Immun.* 69, 6475–6482.

892. Jäkel, T., Scharpfenecker, M., Jitrawang, P., Rückle, J., Kliemt, D., Mackenstedt, U., Hongnark, S., Khoprasert, Y., 2001. Reduction of transmission stages concomitant with increased host immune responses to hypervirulent *Sarcocystis singaporensis*, and natural selection for intermediate virulence. *Int. J. Parasitol.* 31, 1639–1647.

893. Jakes, K.A., 1998. *Sarcocystis mucosa* in Bennett's wallabies and pademelons from Tasmania. *J. Wildl. Dis.* 34, 594–599.

894. Janitschke, K., Protz, D., Werner, H., 1976. Beitrag zum Entwicklungszyklus von Sarkosporidien der Grantgazelle (*Gazella granti*). *Z. Parasitenkd.* 48, 215–219.

895. Jakob, W., Stolte, M., Odening, K., Bockhardt, I., 1998. *Sarcocystis kirmsei* in the brain of a hill mynah (*Gracula religiosa*). *J. Comp. Pathol.* 118, 75–80.

896. Jang, H., Kang, Y.-B., Wee, S.-H., Choi, S.-H., 1990. Survey of *Sarcocystis* infection in cattle in Korea. *Res. Rept. RDA (V)* 32, 32–37.

897. Janitschke, K., 1975. Neue Erkenntnisse über die Kokzidien-Infektionen des Menschen. II. *Isospora*-Infektion. *Bundesgesundheitsblatt* 18, 419–422.

898. Jantzen, B., Entzeroth, R., 1987. Exocystosis of dense granules of cyst merozoites (cystozoites) of *Sarcocystis cuniculi* (Protozoa, Apicomplexa) in cell cultures. *Parasitol. Res.* 73, 472–474.

899. Jeffrey, M., O'Toole, D., Smith, T., Bridges, A.W., 1988. Immunocytochemistry of ovine sporozoan encephalitis and encephalomyelitis. *J. Comp. Pathol.* 98, 213–224.

900. Jeffrey, M., Low, J.C., Uggla, A., 1989. A myopathy of sheep associated with *Sarcocystis* infection and monensin administration. *Vet. Rec.* 124, 422–426.

901. Jeffrey, M., 1993. Sarcocystosis of sheep. *In Practice* 15, 2, 4–6, 8.

902. Jeffries, A.C., Johnson, A.M., 1996. The growing importance of the plastid-like DNAs of the Apicomplexa. *Int. J. Parasitol.* 26, 1139–1150.

903. Jeffries, A.C., Amaro, N., Tenter, A.M., Johnson, A.M., 1996. Genetic diversity in *Sarcocystis gigantea* assessed by RFLP analysis of the ITS1 region. *Appl. Parasitol.* 37, 275–283.

904. Jeffries, A.C., Schnitzler, B., Heydorn, A.O., Johnson, A.M., Tenter, A.M., 1997. Identification of synapomorphic characters in the genus *Sarcocystis* based on 18S rDNA sequence comparison. *J. Eukaryot. Microbiol.* 44, 388–392.

905. Jehle, C., Dinkel, A., Sander, A., Morent, M., Romig, T., Luc, P.V., De, T.V., Thai, V.V., Mackenstedt, U., 2009. Diagnosis of *Sarcocystis* spp. in cattle (*Bos taurus*) and water buffalo (*Bubalus bubalis*) in northern Vietnam. *Vet. Parasitol.* 166, 314–320.

906. Jenkins, M.C., Ellis, J.T., Liddell, S., Ryce, C., Munday, B.L., Morrison, D.A., Dubey, J.P., 1999. The relationship of *Hammondia hammondi* and *Sarcocystis mucosa* to other heteroxenous cyst-forming coccidia as inferred by phylogenetic analysis of the 18S SSU ribosomal DNA sequence. *Parasitology* 119, 135–142.

907. Jensen, R., Alexander, A.F., Dahlgren, R.R., Jolley, W.R., Marquardt, W.C., Flack, D.E., Bennett, B.W. et al. 1986. Eosinophilic myositis and muscular sarcocystosis in the carcasses of slaughtered cattle and lambs. *Am. J. Vet. Res.* 47, 587–593.

908. Jerrett, I.V., McOrist, S., Waddington, J., Browning, J.W., Malecki, J.C., McCausland, I.P., 1984. Diagnostic studies of the fetus, placenta and maternal blood from 265 bovine abortions. *Cornell Vet.* 74, 8–20.

909. Jinnai, M., Kawabuchi-Kurata, T., Tsuji, M., Nakajima, M., Fujisawa, K., Nagata, S., Koide, H., Matoba, Y., Asakawa, M., Takahashi, K., Ishihara, C., 2009. Molecular evidence for the presence of new *Babesia* species in feral raccoons (*Procyon lotor*) in Hokkaido, Japan. *Vet. Parasitol.* 162, 241–247.

910. Joachim, A., Tenter, A.M., Jeffries, A.C., Johnson, A.M., 1996. A RAPD-PCR derived marker can differentiate between pathogenic and non-pathogenic *Sarcocystis* species of sheep. *Mol. Cell. Probes* 10, 165–172.

911. Johnson, A.J., Hildebrandt, P.K., Fayer, R., 1975. Experimentally induced *Sarcocystis* infection in calves: Pathology. *Am. J. Vet. Res.* 36, 995–999.

912. Johnson, A.L., Burton, A.J., Sweeney, R.W., 2010. Utility of 2 immunological tests for antemortem diagnosis of equine protozoal myeloencephalitis (*Sarcocystis neurona* infection) in naturally occurring cases. *J. Vet. Intern. Med.* 24, 1184–1189.

913. Jolley, W.R., Jensen, R., Hancock, H.A., Swift, B.L., 1983. Encephalitic sarcocystosis in a newborn calf. *Am. J. Vet. Res.* 44, 1908–1911.

914. Jonas, W.E., Roberts, M.G., Fisher, M.G., Stankiewicz, M., 1995. *Sarcocystis muris*: Changes in the number of mice with sarcocysts and median number of cysts observed after a single oral dose of sporocysts. *Acta Parasitologica* 40, 43–46.

915. Jordan, C.N., Kaur, T., Koenen, K., DeStefano, S., Zajac, A.M., Lindsay, D.S., 2005. Prevalence of agglutinating antibodies to *Toxoplasma gondii* and *Sarcocystis neurona* in beavers (*Castor canadensis*) from Massachusetts. *J. Parasitol.* 91, 1228–1229.

916. Jumde, P.D., Bhojne, G.R., Maske, D.K., Kolte, S.W., 2000. Prevalence of sarcocystosis in goats at Nagpur. *Indian Vet. J.* 77, 662–663.

917. Jung, C., Lee, C.Y.F., Grigg, M.E., 2004. The SRS superfamily of *Toxoplasma* surface proteins. *Int. J. Parasitol.* 34, 285–296.

918. Jungmann, R., Bergmann, V., Hiepe, T., Nedjari, T., 1977. Untersuchungen zur septikämisch verlaufenden experimentellen *Sarcocystis-bovicanis*-Infektion des Rindes. *Monatsh. Veterinärmed.* 32, 885–889.

919. Juyal, P.D., Bhatia, B.B., 1989. Sarcocystosis: an emerging zoonosis. *Indian Vet. Med. J.* 13, 66–69.

920. Juyal, P.D., Bhatia, B.B., Saleque, A., 1989. Prevalence and intensity of *Sarcocystis* infection in goats and buffaloes in Tarai area of Uttar Pradesh. *Indian Vet. Med. J.* 13, 269–271.

921. Juyal, P.D., Ruprah, N.S., Chhabra, M.B., 1989. Experimentally induced *Sarcocystis capracanis* infection in pregnant goats. *Indian Vet. Med. J.* 13, 200–202.

922. Juyal, P.D., Kalra, I.S., Bali, H.S., 1991. Occurrence of *Sarcocystis equicanis* in a horse *(Equus caballus)* in India. *J. Vet. Parasitol.* 5, 53–54.

923. Juyal, P.D., Kalra, I.S., Gupta, P.P., 1993. A preliminary report on the development of *Sarcocystis equicanis* cysts from infected equine musculature in a dog. *Indian Vet. Med. J.* 17, 70–71.

924. Juyal, P.D., Gupta, M.P., Gupta, P.P., Singh, H., Kalra, I.S., 1994. Chronic *Sarcocystis* infection in a naturally infected mare. *Indian Vet. Med. J.* 18, 54–55.

925. Kahl, S., Elsasser, T.H., Sartin, J.L., Fayer, R., 2002. Effect of progressive cachectic parasitism and growth hormone treatment on hepatic 5'-deiodinase activity in calves. *Dom. Anim. Endocrinol.* 22, 211–221.

926. Kaiser, I.A., Markus, M.B., 1983. Species of *Sarcocystis* in wild South African birds. *Proc. Electron Microsc. Soc. S. Afr.* 13, 103.

927. Kaiser, I.A., Markus, M.B., 1983. *Sarcocystis* infection in wild southern African birds. *S. Afr. J. Sci.* 79, 470–471.

928. Kalantari, N., Bayani, M., Ghaffari, S., 2013. *Sarcocystis cruzi*: First molecular identification from cattle in Iran. *Int. J. Mol. Cell Med.* 2, 125–130.

929. Kaliner, G., Grootenhuis, J.G., Protz, D., 1974. A survey for sarcosporidial cysts in East African game animals. *J. Wildl. Dis.* 10, 237–238.

930. Kaliner, G., 1975. Observations on the histomorphology of sarcosporidian cysts of some East African game animals (Artiodactyla). *Z. Parasitenkd.* 46, 13–23.

931. Kalisińska, E., Betlejewska, K.M., Schmidt, M., Goździcka-Jozefiak, A., Tomczyk, G., 2003. Protozoal macrocysts in the skeletal muscles of a mallard duck in Poland: The first recorded case. *Acta Parasitologica* 48, 1–5.

932. Kalubowila, D.G.W., Udagama-Randeniya, P.V., Perera, N.A.N.D., Rajapakse, R.P.V.J., 2004. Seroprevalence of *Sarcocystis* spp. in cattle and buffaloes from the wet and dry zones of Sri Lanka: A preliminary study. *J. Vet. Med. B* 51, 89–93.

933. Kalyakin, V.N., Zasukhin, D.N., 1975. Distribution of *Sarcocystis* (Protozoa: Sporozoa) in vertebrates. *Folia Parasitol. (Praha)* 22, 289–307.

934. Kamata, Y., Saito, M., Irikura, D., Yahata, Y., Ohnishi, T., Bessho, T., Inui, T., Watanabe, M., Sugita-Konishi, Y., 2014. A toxin isolated from *Sarcocystis fayeri* in raw horsemeat may be responsible for food poisoning. *J. Food Prot.* 77, 814–819.

935. Kan, S.P., Dissanaike, A.S., 1976. Ultrastructure of *Sarcocystis booliati* Dissanaike and Poopalachelvam, 1975 from the moonrat, *Echinosorex gymnurus* in Malaysia. *Int. J. Parasitol.* 6, 321–326.

936. Kan, S.P., Dissanaike, A.S., 1977. Ultrastructure of *Sarcocystis* sp. from the Malaysian house rat, *Rattus rattus diardii*. *Z. Parasitenkd.* 52, 219–227.

937. Kan, S.P., Dissanaike, A.S., 1978. Studies on *Sarcocystis* in Malaysia II. Comparative ultrastructure of the cyst wall and zoites of *Sarcocystis levinei* and *Sarcocystis fusiformis* from the water buffalo, *Bubalus bubalis*. *Z. Parasitenkd.* 57, 107–116.

938. Kan, S.P., 1979. Ultrastructure of the cyst wall of *Sarcocystis* spp. from some rodents in Malaysia. *Int. J. Parasitol.* 9, 475–480.

939. Kan, S.P., Prathap, K., Dissanaike, A.S., 1979. Light and electron microstructure of a *Sarcocystis* sp. from the Malaysian long-tailed monkey, *Macaca fascicularis*. *Am. J. Trop. Med. Hyg.* 28, 634–642.

940. Kanakoudis, G., Vlemmas, I., Papaioannou, N., Daffas, G., Lekkas, S., 1993. Ultrastructure of cyst wall of *Sarcocystis arieticanis* of sheep after experimental infection. *Bull. Hellen. Vet. Med. Soc.* 44, 107–111. (In Greek).

941. Kannangara, D.W.W., 1970. Two new host records for *Sarcocystis* lankester, 1882. *Ceylon vet. J.* 18, 123–125.

942. Karim, M.R., Yu, F., Li, J., Li, J., Zhang, L., Wang, R., Rume, F.I., Jian, F., Zhang, S., Ning, C., 2014. First molecular characterization of enteric protozoa and the human pathogenic microsporidian, *Enterocytozoon bieneusi*, in captive snakes in China. *Parasitol. Res.* 113, 3041–3048.

943. Karr, S.L., Wong, M.M., 1975. A survey of *Sarcocystis* in nonhuman primates. *Lab. Anim. Sci.* 25, 641–645.

944. Karstad, L., 1963. *Toxoplasma microti* (the M-organism) in the muskrat (*Ondatra zibethica*). *Can. Vet. J.* 4, 249–251.

945. Katayama, Y., Wada, R., Kanemaru, T., Sasagawa, T., Uchiyama, T., Matsumura, T., Anzai, T., 2003. First case report of *Sarcocystis neurona*-induced equine protozoal myeloencephalitis in Japan. *J. Vet. Med. Sci.* 65, 757–759.

946. Katić-Radivojević, S., Gadanski-Omerović, G., 1996. Immuno-chemical and physico-chemical characterization of antigens of *Sarcocystis ovifelis*. *Acta Vet. (Beograd)* 46, 299–306.

947. Katoh, K., Standley, D.M., 2013. MAFFT multiple sequence alignment software version 7: Improvements in performance and usability. *Mol. Biol. Evol.* 30, 772–780.

948. Kenison, D.C., Elsasser, T.H., Fayer, R., 1991. Tumor necrosis factor as a potential mediator of acute metabolic and hormonal responses to endotoxemia in calves. *Am. J. Vet. Res.* 52, 1320–1326.

949. Kennedy, M.J., Frelier, P.F., 1986. *Frenkelia* sp. from the brain of a porcupine (*Erethizon dorsatum*) from Alberta, Canada. *J. Wildl. Dis.* 22, 112–114.

950. Kennedy, T., Campbell, J., Selzer, V., 2001. Safety of ponazuril 15% oral past in horses. *Vet. Therapeut.* 2, 223–231.

951. Kepka, O., Scholtyseck, E., 1970. Weitere Untersuchungen der Feinstruktur von *Frenkelia* spec. (=M-organismus, Sporozoa). *Protistologica* 6, 249–266.

952. Kepka, O., Skofitsch, G., 1979. Zur Epidemiologie von *Frenkelia* (Apicomplexa, Protozoa) der mitteleuropäischen Waldrötelmaus (*Clethrionomys glareolus*). *Mitt. Naturwiss. Ver. Steiermark* 109, 283–307.

953. Khan, R.A., Evans, L., 2006. Prevalence of *Sarcocystis* spp. in two subspecies of caribou (*Rangifer tarandus*) in Newfoundland and Labrador, and foxes (*Vulpes vulpes*), wolves (*Canis lupus*), and husky dogs (*Canis familiaris*) as potential definitive hosts. *J. Parasitol.* 92, 662–663.

954. Khatkar, S.K., Singh, R.P., Gupta, S.L., Verma, P.C., 1993. Histopathology of experimental *Sarcocystis miescheriana* infections in pigs. *Indian J. Anim. Sci.* 63, 932–935.

955. Khulbe, D.C., Kushwah, A., Kushwah, H.S., 1989. Biochemistry of the various fractions of sarcocysts of *Sarcocystis fusiformis* of buffalo (*Bubalus bubalis*). *Vet. Parasitol.* 31, 1–5.

956. Kia, E.B., Mirhendi, H., Rezaeian, M., Zahabiun, F., Sharbatkhori, M., 2011. First molecular identification of *Sarcocystis miescheriana* (Protozoa, Apicomplexa) from wild boar (*Sus scrofa*) in Iran. *Exp. Parasitol.* 127, 724–726.

957. Kibenge, F.S.B., Cawthorn, R.J., Despres, D., McKenna, P.K., Markham, R.J.F., 1991. Development of genomic probes to *Sarcocystis cruzi* (Apicomplexa). *Vet. Parasitol.* 40, 9–20.

958. Kim, S.K., Boothroyd, J.C., 2005. Stage-specific expression of surface antigens by *Toxoplasma gondii* as a mechanism to facilitate parasite persistence. *J. Immunol.* 174, 8038–8048.

959. Kim, T.H., Han, J.H., Chang, S.N., Kim, D.S., Abdelkader, T.S., Seok, S.H., Park, J.H. et al. 2011. Detection of sarcocystic infection in a wild rodent (*Apodemus agrarius chejuensis*) captured on Jeju island. *Lab. Anim. Res.* 27, 357–359.

960. Kimmig, P., Piekarski, G., Heydorn, A.O., 1979. Zur Sarkosporidiose (*Sarcocystis suihominis*) des Menschen (II). *Immun. Infekt.* 7, 170–177.

961. Kimura, T., Ito, J., Suzuki, M., Inokuchi, S., 1987. *Sarcocystis* found in the skeletal muscle of common squirrel monkeys. *Primates* 28, 247–255.

962. Kinsel, M.J., Briggs, M.B., Venzke, K., Forge, O., Murnane, R.D., 1998. Gastric spiral bacteria and intramuscular sarcocysts in African lions from Namibia. *J. Wildl. Dis.* 34, 317–324.

963. Kirkpatrick, C.E., Dubey, J.P., Goldschmidt, M.H., Saik, J.E., Schmitz, J.A., 1986. *Sarcocystis* sp. in muscles of domestic cats. *Vet. Pathol.* 23, 88–90.

964. Kirkpatrick, C.E., Hamir, A.N., Dubey, J.P., Rupprecht, C.E., 1987. *Sarcocystis* in muscles of raccoons (*Procyon lotor* L.). *J. Protozool.* 34, 445–447.

965. Kirmse, P., 1986. Sarcosporidioses in equines of Morocco. *Br. Vet. J.* 142, 70–72.

966. Kirmse, P., Mohanbabu, B., 1986. *Sarcocystis* sp. in the one-humped camel (*Camelus dromedarius*) from Afghanistan. *Br. Vet. J.* 142, 73–74.

967. Klein, H., Löschner, B., Zyto, N., Pörtner, M., Montag, T., 1998. Expression, purification, and biochemical characterization of a recombinant lectin of *Sarcocystis muris* (Apicomplexa) cyst merozoites. *Glycoconjugate J.* 15, 147–153.

968. Klumpp, S.A., Anderson, D.C., McClure, H.M., Dubey, J.P., 1994. Encephalomyelitis due to a *Sarcocystis neurona*-like protozoan in a rhesus monkey (*Macaca mulatta*) infected with simian immunodeficiency virus. *Am. J. Trop. Med. Hyg.* 51, 332–338.

969. Kolenda, R., Ugorski, M., Bednarski, M., 2014. Molecular characterization of *Sarcocystis* species from Polish roe deer based on *ssu rRNA* and *cox1* sequence analysis. *Parasitol. Res.* 113, 3029–3039.

970. Koller, L.D., Kistner, T.P., Hudkins, G.G., 1977. Histopathologic study of experimental *Sarcocystis hemionilatrantis* infection in fawns. *Am. J. Vet. Res.* 38, 1205–1209.

971. Konrad, J.L., Campero, L.M., Caspe, G.S., Brihuega, B., Moore, D.P., Crudeli, G.A., Venturini, M.C., Campero, C.M., 2013. Detection of antibodies against *Brucella abortus*, *Leptospira* spp., and Apicomplexa protozoa in water buffaloes in the Northeast of Argentina. *Trop. Anim. Health Prod.* 45, 1751–1756.

972. Koudela, B., Steinhauser, L., 1984. Evaluation of vitality of sarcocysts in beef by the DAPI fluorescence test. *Acta Vet. Brno* 53, 193–197.

973. Koudela, B., Modrý, D., Svobodová, M., Votýpka, J., Vávra, J., Hudcovic, T., 1999. The severe combined immunodeficient mouse as a definitive host for *Sarcocystis muris*. *Parasitol. Res.* 85, 737–742.

974. Koudela, B., Modrý, D., 2000. *Sarcocystis muris* possesses both diheteroxenous and dihomoxenous characters of life cycle. *J. Parasitol.* 86, 877–879.

975. Krampitz, H.E., Rommel, M., Geisel, O., Kaiser, E., 1976. Beiträge zum Lebenszyklus der Frenkelien II. Die ungeschlechtliche Entwicklung von *Frenkelia clethrionomyobuteonis* in der Rötelmaus. *Z. Parasitenkd.* 51, 7–14.

976. Krampitz, H.E., Rommel, M., 1977. Experimentelle Untersuchungen über das Wirtsspektrum der Frenkelien der Erdmaus. *Berl. Münch. Tierärztl. Wochenschr.* 90, 17–19.

977. Krause, C., Goranoff, S., 1933. Ueber Sarkosporidiosis bei Huhn und Wildente. *Zeitschrift für Infektiionskrankheit en der Haustieren* 43, 261–278.

978. Kreuder, C., Miller, M.A., Jessup, D.A., Lowenstine, L.J., Harris, M.D., Ames, J.A., Carpenter, T.E., Cibrad, P.A., Mazet, J.A.K., 2003. Patterns of mortality in southern sea otters (*Enhydra lutris nereis*) from 1998–2001. *J. Wildl. Dis.* 39, 495–509.

979. Krone, O., Rudolph, M., Jakob, W., 2000. Protozoa in the breast muscle of raptors in Germany. *Acta Protozool.* 39, 35–42.

980. Kruttlin, E.A., Rossano, M.G., Murphy, A.J., Vrable, R.A., Kaneene, J.B., Schott, H.C., Mansfield, L.S., 2001. The effects of pyrantel tartrate on *Sarcocystis neurona* merozoite viability. *Vet. Therapeut.* 2, 268–276.

981. Krylov, M.V., Sapozhnikov, G.N., 1965. *Sarcocystis gusevi* n. sp. from the Marco Polo sheep (*Ovis ammon polii*) and *Sarcocystis* sp. from the Bactrian deer (*Cervus elaphus bactrianus*). *Izvestiya Otdeleniya Biologicheskikh Nauk Akademii Nauk Tadzhikskoi SSR* 2, 74–77. (In Russian).

982. Kubo, M., Okano, T., Ito, K., Tsubota, T., Sakai, H., Yanai, T., 2009. Muscular sarcocystosis in wild carnivores in Honshu, Japan. *Parasitol. Res.* 106, 213–219.

983. Kubo, M., Kawachi, T., Murakami, M., Kubo, M., Tokuhiro, S., Agatsuma, T., Ito, K. et al. 2010. Meningoencephalitis associated with *Sarcocystis* spp. in a free-living Japanese raccoon dog (*Nyctereutes procyonoides viverrinus*). *J. Comp. Pathol.* 143, 185–189.

984. Kudi, A.C., Aganga, A.O., Ogbogu, V.C., Umoh, J.U., 1991. Prevalence of *Sarcocystis* species in sheep and goats in northern Nigeria. *Rev. Élev. Méd. vét. Pays trop.* 44, 59–60.

985. Kuhn, J., 1865. Untersuchungen uber die Trichinenkrankheit der Schweine. *Mittheilungen des Landwirthschaftlichen Institutes der Universität du Halle*, pp. 1–84.

986. Kumar, A., Srivastava, P.S., Sinha, S.R.P., 1988. Chemotherapy of experimental caprine sarcocystosis in goats and yound dogs with salinomycin (Coxistac-Pfizer). *J. Vet. Parasitol.* 2, 129–132.

987. Kuncl, R.W., Richter, W., 1988. Prevalence and ultrastructure of *Sarcocystis* in Rhesus monkeys. *Jpn. J. Vet. Sci.* 50, 519–527.

988. Kunde, J.M., Jones, L.P., Craig, T.M., 1980. Protozoal encephalitis in a bovine fetus. *Southwestern Veterinarian* 33, 231–232.

989. Kunita, T., Inoue, I., Fujita, M., Nogami, S., Oota, K., Nakamura, A., Kojima, Y., 1990. Changes of blood chemistry in pigs experimentally infected with *Sarcocystis miescheriana*. *Bull. Coll. Agr. Vet. Med.* 47, 116–123.

990. Kuraev, G.T., 1981. Morphology of sarcocysts from naturally infected camels. Khimioprofilaktika 1. Patogenez I Epizootologiya Parazitov Sel'Skokhozyaistvennykh Zhivotnykh. Alma-Ata, USSR: Vostochnoe Otdelenie Vaskhnil, pp. 91–92. (In Russian).

991. Kutkiené, L., Grikieniené, J., 1994. Some peculiarities of distribution of sarcosporidians in rabbits. *Ekologija* 3, 37–40.

992. Kutkiené, L., Grikieniené, J., 1993. Transplacental transmission as one of the ways of circulation of some sarcosporidian species in rodents. *Ekologija* 1, 25–32.

993. Kutkienė, L., Grikienienė, J., 1995. New experimental data on transplacental transmission of *Sarcocystis* in laboratory rats. *Acta Zoologica Lituanica. Parasitologia* 25, 42–44.

994. Kutkienė, L., Grikienienė, J., 1995. On transplacental transmission of *Sarcocystis rodentifelis* (Eucoccidiida, Eimeriidae) in laboratory rats. *Ekologija* 3, 75–80.

995. Kutkienė, L., 1998. Circulation of *Sarcocystis rodentifelis* in laboratory rats (*Rattus norvegicus*): facts and hypotheses. *Ekologija* 4, 38–42.

996. Kutkienė, L., Grikienienė, J., 1998. Transmission of *Sarcocystis rodentifelis* from female rats (*Rattus norvegicus*) to their offspring. *Acta Zoologica Lituanica. Parasitologia* 8, 76–83.

997. Kutkienė, L., Grikienienė, J., 1998. Transmission of *Sarcocystis rodentifelis* from female rate (*Rattus norvegicus*) to offspring in post-natal period. *Ekologija* 1, 44–46.

998. Kutkienė, L., 2001. The species composition of European roe deer (*Capreolus capreolus*) *Sarcocystis* in Lithuania. *Acta Zool. Lituanica* 11, 97–101.

999. Kutkienė, L., 2002. On the investigations of *Sarcocystis* (Protista: Coccidia) fauna in moose (*Alces alces*) in Lithuania. *Acta Zool. Lituanica* 12, 82–85.

1000. Kutkienė, L., 2003. Investigations of red deer (*Cervus elaphus*) *Sarcocystis* species composition in Lithuania. *Acta Zool. Lituanica* 13, 390–395.

1001. Kutkienė, L., Grikienienė, J., 2003. The importance of coprophagy and transplacental transmission in spread of *Sarcocystis rodentifelis* in rats. *Acta Zool. Lituanica* 13, 322–326.

1002. Kutkienė, L., Sruoga, A., 2004. *Sarcocystis* spp. in birds of the order Anseriformes. *Parasitol. Res.* 92, 171–172.

1003. Kutkienė, L., Sruoga, A., Butkauskas, D., 2006. *Sarcocystis* sp. from white-fronted goose (*Anser albifrons*): Cyst morphology and life cycle studies. *Parasitol. Res.* 99, 562–565.

1004. Kutkienė, L., Sruoga, A., Butkauskas, D., 2008. *Sarcocystis* sp. from the goldeneye (*Bucephala clangula*) and the mallard (*Anas platyrhynchos*): Cyst morphology and ribosomal DNA analysis. *Parasitol. Res.* 102, 691–696.

1005. Kutkienė, L., Prakas, P., Sruoga, A., Butkauskas, D., 2009. *Sarcocystis* in the birds family Corvidae with description of *Sarcocystis cornixi* sp. nov. from the hooded crow (*Corvus cornix*). *Parasitol. Res.* 104, 329–336.

1006. Kutkienė, L., Prakas, P., Sruoga, A., Butkauskas, D., 2010. The mallard duck (*Anas platyrhynchos*) as intermediate host for *Sarcocystis wobeseri* sp. nov. from the barnacle goose (*Branta leucopsis*). *Parasitol. Res.* 107, 879–888.

1007. Kutkienė, L., Prakas, P., Sruoga, A., Butkauskas, D., 2011. Identification of *Sarcocystis rileyi* from the mallard duck (*Anas platyrhynchos*) in Europe: Cyst morphology and results of DNA analysis. *Parasitol. Res.* 108, 709–714.

1008. Kutkienė, L., Prakas, P., Sruoga, A., Butkauskas, D., 2012. Description of *Sarcocystis anasi* sp. nov. and *Sarcocystis albifronsi* sp. nov. in birds of the order Anseriformes. *Parasitol. Res.* 110, 1043–1046.

1009. Kutkienė, L., Prakas, P., Butkauskas, D., Sruoga, A., 2012. Description of *Sarcocystis turdusi* sp. nov. from the common blackbird (*Turdus merula*). *Parasitology* 139, 1438–1443.

1010. La Perle, K.M.D., Silveria, F., Anderson, D.E., Blomme, E.A.G., 1999. Dalmeny disease in an alpaca (*Lama pacos*): Sarcocystosis, eosinophilic myositis and abortion. *J. Comp. Pathol.* 121, 287–293.

1011. Laarman, J.J., 1962. *Isospora hominis* (Railliet and Lucet 1891) in the Netherlands. *Acta Leiden* 31, 111–116.

1012. Laarman, J.J., Tadros, W., 1982. Sarcosporidiosis of farm animals. In: Walton, J.R., White, E.G., Hall, S.A. (Eds.), *Agriculture: Some Diseases of Emerging Importance to Community Trade*. Luxemburg Commission of the European Communites, pp. 81–88.

1013. Labbé, A., 1899. Sporozoa. *Das Thierreich* I. Gen. *Sarcocystis* Lank Verlag vm R. Friedlaünder und Sohn. Berlin, Germany, pp. 116–119.

1014. Lagerquist, J.E., Foreyt, W.J., 1993. Prevalence of *Sarcocystis* in elk (*Cervus elaphus*) in Oregon. *Northwest. Sci.* 67, 196–198.

1015. Lainson, R., Shaw, J.J., 1971. *Sarcocystis gracilis* n. sp. from the Brazilian tortoise *Kinosternon scorpioides*. *J. Protozool.* 18, 365–372.

1016. Lainson, R., 1972. A note on sporozoa of undetermined taxonomic position in an armadillo and a heifer calf. *J. Protozool.* 19, 582–586.

1017. Lainson, R., Shaw, J.J., 1972. *Sarcocystis* in tortoises: A replacement name, *Sarcocystis kinosterni*, for the homonym *Sarcocystis gracilis*, Lainson and Shaw, 1971. *J. Protozool.* 19, 212.

1018. Lainson, R., Paperna, I., 2000. The life-cycle and ultrastructure of *Sarcocystis ameivamastigodryasi* n. sp., in the lizard *Ameiva ameiva* (Teiidae) and the snake *Mastigodryas bifossatus* (Colubridae). *Parasite* 7, 263–274.

1019. Landsverk, T., 1979. An outbreak of *Sarcocystis* in a cattle herd. *Acta Vet. Scand.* 20, 238–244.

1020. Lane, J.H., Mansfield, K.G., Jackson, L.R., Diters, R.W., Lin, K.C., MacKey, J.J., Sasseville, V.G., 1998. Acute fulminant sarcocystosis in a captive-born Rhesus macaque. *Vet. Pathol.* 35, 499–505.

1021. Langham, N.P.E., Charleston, W.A.G., 1990. An investigation of the potential for spread of *Sarcocystis* spp. and other parasites by feral cats. *N. Z. J. Agr. Res.* 33, 429–435.

1022. Lankester, E.R., 1882. On *Drepanidium ranarum*, the cell-parasite of the frog's blood and spleen (Gaule's Würmschen). *Quart. J. Microscop. Sci.* 85, 53–65.

1023. Lapointe, J.M., Duignan, P.J., Marsh, A.E., Gulland, F.M., Barr, B.C., Naydan, D.K., King, D.P., Farman, C.A., Huntingdon, K.A.B., Lowenstine, L.J., 1998. Meningoencephalitis due to a *Sarcocystis neurona*-like protozoan in Pacific harbor seals (*Phoca vitulina richardsi*). *J. Parasitol.* 84, 1184–1189.

1024. Larbcharoensub, N., Cheewaruangroj, W., Nitiyanant, P., 2011. Laryngeal sarcocystosis accompanying laryngeal squamous cell carcinoma: Case report and literature review. *Southeast Asian J. Trop. Med. Pub. Health* 42, 1072–1076.

1025. Larkin, J.L., Gabriel, M., Gerhold, R.W., Yabsley, M.J., Wester, J.C., Humphreys, J.G., Beckstead, R., Dubey, J.P., 2011. Prevalence to *Toxoplasma gondii* and *Sarcocystis* spp. in a reintroduced fisher (*Martes pennanti*) population in Pennsylvania. *J. Parasitol.* 97, 425–429.

1026. Larsen, R.A., Kyle, J.E., Whitmire, W.M., Speer, C.A., 1984. Effect of nylon wool purification on infectivity and antigenicity of *Eimeria falciformis* sporozoites and merozoites. *J. Parasitol.* 70, 597–601.

1027. Last, M.J., Powell, E.C., 1978. Separation of *Sarcocystis muris* and *Isospora felis* in mice used by Powell and McCarley (1975) in studies on the life cycle of S. *muris*. *J. Parasitol.* 64, 162–163.

1028. Latif, B., Vellayan, S., Omar, E., Abdullah, S., Desa, N.M., 2010. Sarcocystosis among wild captive and zoo animals in Malaysia. *Korean J. Parasitol.* 48, 213–217.

1029. Latif, B., Vellayan, S., Heo, C.C., Kannan Kutty, M., Omar, E., Abdullah, S., Tappe, D., 2013. High prevalence of muscular sarcocystosis in cattle and water buffaloes from Selangor, Malaysia. *Trop. Biomed.* 30, 699–705.

1030. Latif, B.M.A., Al-Delemi, J.K., Mohammed, B.S., Al-Bayati, S.M., Al-Amiry, A.M., 1999. Prevalence of *Sarcocystis* spp. in meat-producing animals in Iraq. *Vet. Parasitol.* 84, 85–90.

1031. Latif, B.M.A., Khamas, W.A., 2007. Light and ultrastructural morphology of sarcocystiosis in one-humped camel (*Camelus dromedarius*) in northern Jordan. *J. Camel Prac. Res.* 14, 45–48.

1032. Latimer, K.S., Perry, R.W., Mo, I.P., Nietfeld, J.C., Steffens, W.L., Harrison, G.J., Ritchie, B.W., 1990. Myocardial sarcocystosis in a grand eclectus parrot (*Eclectus roratus*) and a Moluccan cockatoo (*Cacatua moluccensis*). *Avian Dis.* 34, 501–505.

1033. Lau, Y.L., Chang, P.Y., Subramaniam, V., Ng, Y.H., Mahmud, R., Ahmad, A.F., Fong, M.Y., 2013. Genetic assemblage of *Sarcocystis* spp. in Malaysian snakes. *Parasit. Vectors* 6, 257.

1034. Lau, Y.L., Chang, P.Y., Tan, C.T., Fong, M.Y., Mahmud, R., Wong, K.T., 2014. *Sarcocystis nesbitti* infection in human skeletal muscle: Possible transmission from snakes. *Am. J. Trop. Med. Hyg.* 90, 361–364.

1035. Lee, S.C., Ngui, R., Tan, T.K., Aidil, R.M., Lim, Y.A.L., 2014. Neglected tropical diseases among two indigenous subtribes in peninsular Malaysia: Highlighting differences and co-infection of helminthiasis and sarcocystosis. *PLoS ONE* 9, e107980.

1036. Leek, R.G., Fayer, R., Johnson, A.J., 1977. Sheep experimentally infected with *Sarcocystis* from dogs. I. Disease in young lambs. *J. Parasitol.* 63, 642–650.

1037. Leek, R.G., Fayer, R., 1978. Sheep experimentally infected with *Sarcocystis* from dogs. II. Abortion and disease in ewes. *Cornell Vet.* 68, 108–123.

1038. Leek, R.G., Fayer, R., 1978. Infectivity of *Sarcocystis* in beef and beef products from a retail food store. *Proc. Helminthol. Soc. Wash.* 45, 135–136.

1039. Leek, R.G., Fayer, R., 1979. Survival of sporocysts of *Sarcocystis* in various media. *Proc. Helminthol. Soc. Wash.* 46, 151–154.

1040. Leek, R.G., Fayer, R., 1980. Amprolium for prophylaxis of ovine *Sarcocystis*. *J. Parasitol.* 66, 100–106.

1041. Leek, R.G., Fayer, R., 1983. Experimental *Sarcocystis ovicanis* infection in lambs: Salinomycin chemoprophylaxis and protective immunity. *J. Parasitol.* 69, 271–276.

1042. Leek, R.G., 1986. Infection of sheep with frozen sporocysts of *Sarcocystis ovicanis*. *Proc. Helminthol. Soc. Wash.* 53, 297–298.

1043. Leguía, G., 1991. The epidemiology and economic impact of llama parasites. *Parasitol. Today* 7, 54–56.

1044. Leier, H., Boch, J., Erber, M., 1982. Möglichkeiten der Immunisierung von Mäusen gegen Sarkosporidien (*Sarcocystis muris*) mit abgeschwächten Sporozysten. *Berl. Münch. Tierärztl. Wschr.* 95, 231–235.

1045. Lekutis, C., Ferguson, D.J.P., Grigg, M.E., Camps, M., Boothroyd, J.C., 2001. Surface antigens of *Toxoplasma gondii*: Variations on a theme. *Int. J. Parasitol.* 31, 1285–1292.

1046. Lerche, M., Brochwitz, H., 1957. Sarkosporidienbefall des Rindes und Perimyositis eosinophilica. *Dtsch. Tierärztl. Wschr.* 64, 251–252.

1047. Levine, N.D., 1977. *Sarcocystis cernae* n. sp., replacement name for *Sarcocystis* sp. Cernà and Loucková, 1976. *Folia Parasitol. (Praha)* 24, 316.

1048. Levine, N.D., Tadros, W., 1980. Named species and hosts of *Sarcocystis* (Protozoa: Apicomplexa: Sarcocystidae). *Syst. Parasitol.* 2, 41–59.

1049. Levine, N.D., 1986. The taxonomy of *Sarcocystis* (Protozoa, Apicomplexa) species. *J. Parasitol.* 72, 372–382.

1050. Levit, A.V., Orlov, G.I., Dymkova, N.D., 1984. Sarcosporidians from the Alpine pika (*Ochotona alpina*). In: Panin, V.J. (Ed.), *Sarcosporidians of Animals in Kazakhstan*. Nauka, Alma-Ata, pp. 86–88. (In Russian).

1051. Li, J.H., Lin, Z., Tan, Y.X., Du, J.F., 2004. *Sarcocystis suihominis* infection discovered in Guangxi. *Chin. J. Parasitol. Parasit. Dis.* 22, 82. (In Chinese).

1052. Li, J.H., Lin, Z., Qin, Y.X., Du, J., 2004. *Sarcocystis suihominis* infection in human and pig population in Guangxi. *Chin. J. Parasitol. Parasit. Dis.* 22, 82. (In Chinese).

1053. Li, J.H., Lin, Z., Du, J.F., Qin, Y.X., 2007. Experimental infection of *Sarcocystis suihominis* in pig and human volunteer in Guangxi. *Chin. J. Parasitol. Parasit. Dis.* 25, 466–468. (In Chinese).

1054. Li, L., Brunk, B.P., Kissinger, J.C., Pape, D., Tang, K., Cole, R.H., Martin, J. et al. 2003. Gene discovery in the Apicomplexa as revealed by EST sequencing and assembly of a comparative gene database. *Genome Res.* 13, 443–454.

1055. Li, L., Crabtree, J., Fischer, S., Pinney, D., Stoeckert, C.J., Jr., Sibley, L.D., Roos, D.S., 2004. ApiEST-DB: Analyzing clustered EST data of the apicomplexan parasites. *Nucleic Acids Res.* 32(Suppl. 1), D326–D328.

1056. Li, Q.Q., Yang, Z.Q., Zuo, Y.X., Attwood, S.W., Chen, X.W., Zhang, Y.P., 2002. A PCR-based RFLP analysis of *Sarcocystis cruzi* (Protozoa: Sarcocystidae) in Yunnan Province, PR China, reveals the water buffalo (*Bubalus bubalis*) as a natural intermediate host. *J. Parasitol.* 88, 1259–1261.

1057. Li, Y., Lian, Z., 1986. Study on man-pig cyclic infection of *Sarcocystis suihominis* found in Yunnan Province, China. *Acta Zoologica Sinica* 32, 329–335. (In Chinese).

1058. Li, Y., 1989. *Sarcocystis fusiformis* in *Bubalus bubalis* of South China. *Acta Zoologica Sinica* 35, 256–258. (In Chinese).

1059. Lian, Z., Ma, J., Wang, Z., Fu, L., Zhou, Z., Li, W., Wang, X., 1990. Studies on man-cattle-man infection cycle of *Sarcocystis hominis* in Yunnan. *Chin. J. Parasitol. Parasit. Dis.* 8, 50–53. (In Chinese).

1060. Lian, Z.C., Ma, J.H., 1989. A survey of sarcocystosis in cat and dog in Yunnan. *Chin. J. Vet. Sci. Tech.* 11, 18–19. (In Chinese).

1061. Lubyanetskii, S.A., 1960. Toxins of sarcosporidia. *Trudy vsesoyuz. Inst. Vet. Sanit.* 17, 135–138. (In Russian).

1062. Liang, F.T., Granstrom, D.E., Zhao, X.M., Timoney, J.F., 1998. Evidence that surface proteins Sn14 and Sn16 of *Sarcocystis neurona* merozoites are involved in infection and immunity. *Infect. Immun.* 66, 1834–1838.

1063. Lindsay, D.S., Ambrus, S.I., Blagburn, B.L., 1987. *Frenkelia* sp.-like infection in the small intestine of a red-tailed hawk. *J. Wildl. Dis.* 23, 677–679.

1064. Lindsay, D.S., Blagburn, B.L., Mason, W.H., Frandsen, J.C., 1988. Prevalence of *Sarcocystis odocoileocanis* from white-tailed deer in Alabama and its attempted transmission to goats. *J. Wildl. Dis.* 24, 154–156.

1065. Lindsay, D.S., Upton, S.J., Blagburn, B.L., Toivio-Kinnucan, M., McAllister, C.T., Trauth, S.E., 1991. Sporocysts isolated from the southern copperhead (*Agkistrodon contortrix contortrix*) produce *Sarcocystis montanaensis*-like sarcocysts in prairie voles (*Microtus ochrogaster*). *J. Wildl. Dis.* 27, 148–152.

1066. Lindsay, D.S., Upton, S.J., Sundermann, C.A., McKown, R.D., Blagburn, B.L., 1994. *Caryospora tremula* and *Sarcocystis* sp. from turkey vultures, *Cathartes aura*: Descriptions of oocysts and sporocysts and attempted transmission to rodents. *J. Helminthol. Soc. Wash.* 61, 12–16.

1067. Lindsay, D.S., McKown, R., Upton, S.J., McAllister, C.T., Toivio-Kinnucan, M.A., Veatch, J.K., Blagburn, B.L., 1996. Prevalence and identity of *Sarcocystis* infections in armadillos (*Dasypus novemcinctus*). *J. Parasitol.* 82, 518–520.

1068. Lindsay, D.S., Dubey, J.P., 1999. Determination of the activity of pyrimethamine, trimethoprim, and sulfonamides and combinations of pyrimethamine and sulfonamides against *Sarcocystis neurona* in cell cultures. *Vet. Parasitol.* 82, 205–221.

1069. Lindsay, D.S., Blagburn, B.L., 1999. Prevalence of encysted apicomplexans in muscles of raptors. *Vet. Parasitol.* 80, 341–344.

1070. Lindsay, D.S., Dubey, J.P., Horton, K.M., Bowman, D.D., 1999. Development of *Sarcocystis falcatula* in cell cultures demonstrates that it is different from *Sarcocystis neurona*. *Parasitology* 118, 227–233.

1071. Lindsay, D.S., Dubey, J.P., 2000. Determination of the activity of diclazuril against *Sarcocystis neurona* and *Sarcocystis falcatula* in cell cultures. *J. Parasitol.* 86, 164–166.

1072. Lindsay, D.S., Thomas, N.J., Dubey, J.P., 2000. Biological characterisation of *Sarcocystis neurona* from a Southern sea otter (*Enhydra lutris nereis*). *Int. J. Parasitol.* 30, 617–624.

1073. Lindsay, D.S., Dubey, J.P., Kennedy, T.J., 2000. Determination of the activity of ponazuril against *Sarcocystis neurona* in cell cultures. *Vet. Parasitol.* 92, 165–169.

1074. Lindsay, D.S., Dykstra, C.C., Williams, A., Spencer, J.A., Lenz, S.D., Palma, K., Dubey, J.P., Blagburn, B.L., 2000. Inoculation of *Sarcocystis neurona* merozoites into the central nervous system of horses. *Vet. Parasitol.* 92, 157–163.

1075. Lindsay, D.S., Dubey, J.P., 2001. Determination of the activity of pyrantel tartrate against *Sarcocystis neurona* in gamma-interferon gene knockout mice. *Vet. Parasitol.* 97, 141–144.

1076. Lindsay, D.S., Dubey, J.P., 2001. Direct agglutination test for the detection of antibodies to *Sarcocystis neurona* in experimentally infected animals. *Vet. Parasitol.* 95, 179–186.

1077. Lindsay, D.S., Thomas, N.J., Rosypal, A.C., Dubey, J.P., 2001. Dual *Sarcocystis neurona* and *Toxoplasma gondii* infection in a Northern sea otter from Washington state, USA. *Vet. Parasitol.* 97, 319–327.

1078. Lindsay, D.S., Rosypal, A.C., Spencer, J.A., Cheadle, M.A., Zajac, A.M., Rupprecht, C., Dubey, J.P., Blagburn, B.L., 2001. Prevalence of agglutinating antibodies to *Sarcocystis neurona* in raccoons, *Procyon lotor*, from the United States. *Vet. Parasitol.* 100, 131–134.

1079. Lindsay, D.S., Mitchell, S.M., Vianna, M.C., Dubey, J.P., 2004. *Sarcocystis neurona* (Protozoa: Apicomplexa): Description of oocysts, sporocysts, sporozoites, excystation, and early development. *J. Parasitol.* 90, 461–465.

1080. Lindsay, D.S., Mitchell, S.M., Yang, J., Dubey, J.P., Gogal, R.M., Witonsky, S.G., 2006. Penetration of equine leukocytes by merozoites of *Sarcocystis neurona*. *Vet. Parasitol.* 138, 371–376.

1081. Lindsay, D.S., Nazir, M.M., Maqbool, A., Ellison, S.P., Strobl, J.S., 2013. Efficacy of decoquinate against *Sarcocystis neurona* in cell cultures. *Vet. Parasitol.* 196, 21–23.

1082. Lombardo de Barros, C.S., de Barros, S.S., dos Santos, M.N., 1986. Equine protozoal myeloencephalitis in southern Brazil. *Vet. Rec.* 119, 283–284.

1083. Long, M.T., Mines, M.T., Knowles, D.P., Tanhauser, S.M., Dame, J.B., Cutler, T.J., MacKay, R.J., Sellon, D.C., 2002. *Sarcocystis neurona*: Parasitemia in a severe combined immunodeficient (SCID) horse fed sporocysts. *Exp. Parasitol.* 100, 150–154.

1084. Lopes, C.W., de Sá, W.F., Botelho, G.G., 2005. Lesions in cross-breed pregnant cows, experimentally infected with *Sarcocystis cruzi* (Hasselmann, 1923) Wenyon, 1926 (Apicomplexa: Sarcocytidae). *Rev. Bras. Parasitol. Vet.* 14, 79–83. (In Portuguese).

1085. López, C., Panadero, R., Bravo, A., Paz, A., Sánchez-Andrade, R., Díez-Baños, P., Morrondo, P., 2003. *Sarcocystis* spp. infections in roe deer (*Capreolus capreolus*) from the north-west of Spain. *Z. Jagdwiss.* 49, 211–218.

1086. Lukesova, D., Nevole, M., 1984. Detection of sarcocystosis in heifers and pigs using the indirect immunofluorescence method and trypsinization of muscle tissue. *Vet. Med. (Praha)* 29, 307–312. (In Czech).

1087. Lunde, M.N., Fayer, R., 1977. Serologic tests for antibody to *Sarcocystis* in cattle. *J. Parasitol.* 63, 222–225.

1088. Luznar, S.L., Avery, M.L., Dame, J.B., MacKay, R.J., Greiner, E.C., 2001. Development of *Sarcocystis falcatula* in its intermediate host, the brown-headed cowbird (*Molothrus ater*). *Vet. Parasitol.* 95, 327–334.

1089. Machul'skii, S.N., Miskaryan, N.D., 1958. Sarcosporidiosis in wild Artiodactyla. *Trudy Buryat-Mongol Zoovestinstituta* 13, 297–299. (In Russsian).

1090. MacKay, R.J., Davis, S.W., Dubey, J.P., 1992. Equine protozoal myeloencephalitis. *Comp. Cont. Edu. Pract. Vet.* 14, 1359–1367.

1091. MacKay, R.J., 1997. Equine protozoal myeloencephalitis. *Vet. Clin. N. Am. Equine Pract.* 13, 79–96.

1092. MacKay, R.J., Tanhauser, S.T., Gillis, K.D., Mayhew, I.G., Kennedy, T.J., 2008. Effect of intermittent oral administration of ponazuril on experimental *Sarcocystis neurona* infection of horses. *Am. J. Vet. Res.* 69, 396–402.

1093. Mackenstedt, U., Wagner, D., Heydorn, A.O., Mehlhorn, H., 1990. DNA measurements and ploidy determination of different stages in the life cycle of *Sarcocystis muris*. *Parasitol. Res.* 76, 662–668.

1094. Mackie, J.T., Rahaley, R.S., Nugent, R., 1992. Suspected *Sarcocystis* encephalitis in a stillborn kid. *Aust. Vet. J.* 69, 114–115.

1095. MacPherson, J.M., Gajadhar, A.A., 1994. Specific amplification of *Sarcocystis cruzi* DNA using a randomly primed polymerase chain reaction assay. *Vet. Parasitol.* 55, 267–277.

1096. Madigan, J.E., Higgins, R.J., 1987. Equine protozoal myeloencephalitis. *Vet. Clin. N. Am. Equine Pract.* 3, 397–403.

1097. Mahaffey, E.A., George, J.W., Duncan, J.R., Prasse, K.W., Fayer, R., 1986. Hematologic values in calves infected with *Sarcocystis cruzi*. *Vet. Parasitol.* 19, 275–280.

1098. Mahrt, J.L., Fayer, R., 1975. Hematologic and serologic changes in calves experimentally infected with *Sarcocystis fusiformis*. *J. Parasitol.* 61, 967–969.

1099. Maier, K., Olias, P., Enderlein, D., Klopfleisch, R., Mayr, S.L., Gruber, A.D., Lierz, M., 2015. Parasite distribution and early-stage encephalitis in *Sarcocystis calchasi* infections in domestic pigeons (*Columba livia* f. *domestica*). *Avian Pathol.* 44, 5–12.

1100. Makhija, M., 2012. Histological identification of muscular *Sarcocystis*: A report of two cases. *Indian J. Pathol. Microbiol.* 55, 552–554.

1101. Malá, P., Baranova, M., 1995. Detection of sarcocystosis in slaughterhouse animals during a veterinary inspection. *Vet. Med. (Praha)* 40, 97–100. (In Slovak).

1102. Malakauskas, M., Grikienié, G., 2002. *Sarcocystis* infection in wild ungulates in Lithuania. *Acta Zool. Lituanica* 12, 372–380.

1103. Mandour, A.M., 1965. *Sarcocystis garnhami* n. sp. in the skeletal muscle of an opossum, *Didelphis marsupialis*. *J. Protozool.* 12, 606–609.

1104. Mandour, A.M., 1969. *Sarcocystis nesbitti* n. sp. from the rhesus monkey. *J. Protozool.* 16, 353–354.

1105. Mandour, A.M., Keymer, I.F., 1970. *Sarcocystis* infection in African antelopes. *Ann. Trop. Med. Parasitol.* 64, 513–523.

1106. Mandour, A.M., Rabie, S.A., Mohammed, N.I., Hussein, N.M., 2011. On the presence of *Sarcocystis miescheri* sp. nov. in camels of Qena Governorate. *Egypt. Acad. J. Biol. Sci.* 3, 1–7.

1107. Mansfield, L.S., Schott, H.C., Murphy, A.J., Rossano, M.G., Tanhauser, S.M., Patterson, J.S., Nelson, K., Ewart, S.L., Marteniuk, J.V., Bowman, D.D., Kaneene, J.B., 2001. Comparison of *Sarcocystis neurona* isolates derived from horse neural tissue. *Vet. Parasitol.* 95, 167–178.

1108. Mansfield, L.S., Mehler, S., Nelson, K., Elsheikha, H.M., Murphy, A.J., Knust, B., Tanhauser, S.M. et al. 2008. Brown-headed cowbirds (*Molothrus ater*) harbor *Sarcocystis neurona* and act as intermediate hosts. *Vet. Parasitol.* 153, 24–43.

1109. Mao, J.B., Zuo, Y.X., 1994. Studies on the prevalence and experimental transmission of *Sarcocystis* sp. in chickens. *Acta Vet. Zootech. Sinica* 25, 555–559.

1110. Marchant, J., Cowper, B., Liu, Y., Lai, L., Pinzan, C., Marq, J.B., Friedrich, N. et al. 2012. Galactose recognition by the apicomplexan parasite *Toxoplasma gondii*. *J. Biol. Chem.* 287, 16720–16733.

1111. Markus, M.B., 1980. Flies as natural transport host of *Sarcocystis* and other coccidia. *J. Parasitol.* 66, 361–362.

1112. Markus, M.B., Kaiser, I.A., Daly, T.J.M., 1981. *Sarcocystis* of the vervet monkey *Cercopithecus pygerythrus*. *Proc. Electron Microsc. Soc. S. Afr.* 11, 117–118.

1113. Marsh, A.E., Barr, B.C., Madigan, J., Lakritz, J., Conrad, P.A., 1996. Sequence analysis and polymerase chain reaction amplification of small subunit ribosomal DNA from *Sarcocystis neurona*. *Am. J. Vet. Res.* 57, 975–981.

1114. Marsh, A.E., Barr, B.C., Lakritz, J., Nordhausen, R., Madigan, J.E., Conrad, P.A., 1997. Experimental infection of nude mice as a model for *Sarcocystis neurona*-associated encephalitis. *Parasitol. Res.* 83, 706–711.

1115. Marsh, A.E., Barr, B.C., Tell, L., Koski, M., Greiner, E., Dame, J., Conrad, P.A., 1997. *In vitro* cultivation and experimental inoculation of *Sarcocystis falcatula* and *Sarcocystis neurona* merozoites into budgerigars (*Melopsittacus undulatus*). *J. Parasitol.* 83, 1189–1192.

1116. Marsh, A.E., Barr, B.C., Tell, L., Bowman, D.D., Conrad, P.A., Ketcherside, C., Green, T., 1999. Comparison of the internal transcribed spacer, ITS-1, from *Sarcocystis falcatula* isolates and *Sarcocystis neurona*. *J. Parasitol.* 85, 750–757.

1117. Marsh, A.E., Denver, M., Hill, F.I., McElhaney, M.R., Trupkiewicz, J.G., Stewart, J., Tell, L., 2000. Detection of *Sarcocystis neurona* in the brain of a Grant's zebra (*Equus burchelli bohmi*). *J. Zoo Wildl. Med.* 31, 82–86.

1118. Marsh, A.E., Mullins, A.L., Lakritz, J., 2001. *In vitro* quantitative analysis of ^{3}H-uracil incorporation by *Sarcocystis neurona* to determine efficacy of anti-protozoal agents. *Vet. Parasitol.* 95, 241–249.

1119. Marsh, A.E., Johnson, P.J., Ramos-Vara, J., Johnson, G.C., 2001. Characterization of a *Sarcocystis neurona* isolate from a Missouri horse with equine protozoal myeloencephalitis. *Vet. Parasitol.* 95, 143–154.

1120. Marsh, A.E., Hyun, C., Barr, B.C., Tindall, R., Lakritz, J., 2002. Characterization of monoclonal antibodies developed against *Sarcocystis neurona*. *Parasitol. Res.* 88, 501–506.

1121. Marsh, A.E., Lakritz, J., Johnson, P.J., Miller, M.A., Chiang, Y.W., Chu, H.J., 2004. Evaluation of immune responses in horses immunized using a killed *Sarcocystis neurona* vaccine. *Vet. Therapeut.* 5, 34–42.

1122. Martínez, F.A., Troiano, J.C., Añasco, L.G., Rearte, R., Jara, D., 2002. Infección por coccidios en carívoros silvestres de cautiverio de Argentina. *Parasitol. Latinoam.* 57, 146–148.

1123. Martínez-Navalón, B., Anastasio-Giner, B., Cano-Fructuoso, M., Sánchez-Martínez, P., Llopis-Morant, A., Pérez-Castarlenas, B., Goyena, E., Berriatua, E., 2012. Short communication. *Sarcocystis* infection: A major cause of carcass condemnation in adult sheep in Spain. *Span. J. Agric. Res.* 10, 388–392.

1124. Masatani, T., Matsuo, T., Tanaka, T., Terkawi, M.A., Lee, E.G., Goo, Y.K., Aboge, G.O. et al. 2013. TgGRA23, a novel *Toxoplasma gondii* dense granule protein associated with the parasitophorous vacuole membrane and intravacuolar network. *Parasitol. Int.* 62, 372–379.

1125. Mason, F.E., 1910. Sarcocysts in the camel in Egypt. *J. Comp. Pathol. Therap.* 23, 168–176.

1126. Mason, J.R., Clark, L., 1990. Sarcosporidiosis observed more frequently in hybrids of mallards and American black ducks. *Wilson Bulletin* 102, 160–162.

1127. Mason, P., Orr, M., 1993. Sarcocystosis and hydatidosis in lamoids–diseases we can do without. *Surveillance* 20, 14.

1128. Masri, M.D., López de Alda, J., Dubey, J.P., 1992. *Sarcocystis neurona*-associated ataxia in horses in Brazil. *Vet. Parasitol.* 44, 311–314.

1129. Matuschka, F.R., 1979. The African *Bitis nasicornis* (Shaw 1802) as final host of an unknown *Sarcocystis* species. *Salamandra* 15, 264–266.

1130. Matuschka, F.R., 1981. Life cycle of *Sarcocystis* between poikilothermic hosts. Lizards are intermediate hosts for *S. podarcicolubris* sp. nov, snakes function as definitive hosts. *Z. Naturforsch.* 36C, 1093–1095.

1131. Matuschka, F.R., 1983. Infectivity of *Sarcocystis* from donkey for horse via sporocysts from dogs. *Z. Parasitenkd.* 69, 299–304.

1132. Matuschka, F.R., Melhorn, H., 1984. Sarcocysts of *Sarcocystis podarcicolubris* from experimentally infected Tyrrhenian wall lizards (*Podarcis tiliguerta*), *S. gallotiae* from naturally infected Canarian lizards (*Gallotia galloti*) and *S. dugesii* from Madeirian lizards (*Lacerta dugesii*). *Protistologica* 20, 133–139.

1133. Matuschka, F.R., Häfner, U., 1984. Cyclic transmission of an African *Besnoitia* species by snakes of the genus *Bitis* to several rodents. *Z. Parasitenkd.* 70, 471–476.

1134. Matuschka, F.R., 1985. Experimental investigations on the host range of *Sarcocystis podarcicolubris*. *Int. J. Parasitol.* 15, 77–80.

1135. Matuschka, F.R., 1986. *Sarcocystis clethrionomyelaphis* n. sp. from snakes of the genus *Elaphe* and different voles of the family Arvicolidae. *J. Parasitol.* 72, 226–231.

1136. Matuschka, F.R., Schnieder, T., Daugschies, A., Rommel, M., 1986. Cyclic transmission of *Sarcocystis bertrami* Doflein, 1901 by the dog to the horse. *Protistologica* 22, 231–233.

1137. Matuschka, F.R., 1987. *Sarcocystis chalcidicolubris* n. sp.: Recognition of the life cycle in skinks of the genus *Chalcides* and snakes of the genus *Coluber*. *J. Parasitol.* 73, 1014–1018.

1138. Matuschka, F.R., 1987. Reptiles as intermediate and/or final hosts of Sarcosporidia. *Parasitol. Res.* 73, 22–32.

1139. Matuschka, F.R., Bannert, B., 1987. Cannibalism and autonomy as predator-prey relationship for monoxenous Sarcosporidia. *Parasitol. Res.* 74, 88–93.

1140. Mácha, J., Procházkova, Z., Červa, L., Gut, J., 1985. Isolation and characterization of a lectin from *Sarcocystis gigantea. Mol. Biochem. Parasitol.* 16, 243–249.

1141. Matuschka, F.R., Mehlhorn, H., Abd-Al-Aal, Z., 1987. Replacement of *Besnoitia* Matuschka and Häfner 1984 by *Sarcocystis hoarensis. Parasitol. Res.* 74, 94–96.

1142. Matuschka, F.R., Heydorn, A.O., Mehlhorn, H., Abd-Al-Aal, Z., Diesing, L., Biehler, A., 1987. Experimental transmission of *Sarcocystis muriviperae* n. sp. to laboratory mice by sporocysts from the Palestinian viper (*Vipera palestinae*): A light and electron microscope study. *Parasitol. Res.* 73, 33–40.

1143. Hornok, S., Mester, A., Takács, N., Baska, F., Majoros, G., Fok, É., Biksi, I., Német, Z., Hornyák, Á., Jánosi, S., Farkas, R., 2015. *Sarcocystis*-infection of cattle in Hungary. *Parasit. Vectors.* 8, 69.

1144. Matuschka, F.R., 1988. Studies on the life cycle of *Sarcocystis dugesii* in the Madeirian wall lizard *Podarcis* (syn. *Lacerta*) *dugesii. Parasitol. Res.* 75, 73–75.

1145. Matuschka, F.R., Bannert, B., 1989. Recognition of cyclic transmission of *Sarcocystis stehlinii* n. sp. in the Gran Canarian giant lizard. *J. Parasitol.* 75, 383–387.

1146. Mayhew, I.G., de Lahunta, A., Whitlock, R.H., Pollock, R.V.H., 1976. Equine protozoal myeloencephalitis. In: 22nd Ann. Conv. Am. Assoc.Equine Pract., Dallas, Texas, USA, pp. 107–114.

1147. Mayhew, I.G., de Lahunta, A., 1978. Neuropathology. In: Mayhew, I.G., de Lahunta, A., Whitlock, R.H., Krook, L., Tasker, J.B. (Eds.), *Spinal Cord Disease in the Horse*, The Cornell Veterinarian, Ithaca, New York, pp. 106–147.

1148. Mayhew, I.G., Greiner, E.C., 1986. Protozoal diseases. *Vet. Clin. N. Am. Equine Pract.* 2, 439–459.

1149. McAllister, C.T., Upton, S.J., Garrett, C.M., Stuart, J.N., Painter, C.W., 1993. Hemogregarines and *Sarcocystis* sp. (Apicomplexa) in a western green rat snake, *Senticolis triaspis intermedia* (Serpentes: Colubridae), from New Mexico. *J. Helminthol. Soc. Wash.* 60, 284–286.

1150. McAllister, C.T., Upton, S.J., Barker, D.G., Painter, C.W., 1996. *Sarcocystis* sp. (Apicomplexa) from the New Mexico ridgenose rattlesnake, *Crotalus willardi obscurus* (Serpentes: Viperidae) from Sonora, Mexico. *J. Helminthol. Soc. Wash.* 63, 128–130.

1151. McCarron, R.M., Kempsie, O., Spatz, M., McFarlin, E.E., 1985. Presentation of myelin basic protein by murine cerebral vascular endothelial cells. *J. Immunol.* 134, 3100–3103.

1152. McCausland, I.P., Badman, R.T., Hides, S., Slee, K.J., 1984. Multiple apparent *Sarcocystis* abortion in four bovine herds. *Cornell Vet.* 74, 146–154.

1153. McConnell, E.E., Basson, P.A., Wolstenholme, B., de Vos, V., Malherbe, H.H., 1973. Toxoplasmosis in free-ranging Chacma baboons (*Papio ursinus*) from the Kruger National Park. *Trans. R. Soc. Trop. Med. Hyg.* 67, 851–855.

1154. McErlean, B.A., 1974. Ovine paralysis associated with spinal lesions of toxoplasmosis. *Vet. Rec.* 94, 264–266.

1155. McKenna, P.B., Charleston, W.A.G., 1980. Coccidia (Protozoa: Sporozoasida) of cats and dogs. II. Experimental induction of *Sarcocystis* infections in mice. *N. Z. Vet. J.* 28, 117–119.

1156. McKenna, P.B., Charleston, W.A.G., 1990. Production of *Sarcocystis gigantea* sporocysts by experimentally infected cats. *N. Z. J. Agr. Res.* 33, 69–76.

1157. McKenna, P.B., Charleston, W.A.G., 1990. The *in vitro* excystation of *Sarcocystis gigantea* sporocysts. *Vet. Parasitol.* 37, 207–221.

1158. McKenna, P.B., Charleston, W.A.G., 1991. The *in vivo* excystation of *Sarcocystis gigantea* and *S. tenella* sporocysts. *Vet. Parasitol.* 39, 1–11.

1159. McKenna, P.B., Charleston, W.A.G., 1992. The survival of *Sarcocystis gigantea* sporocysts following exposure to various chemical and physical agents. *Vet. Parasitol.* 45, 1–16.

1160. McKenna, P.B., Charleston, W.A.G., 1994. The outdoor survival of *Sarcocystis gigantea* sporocysts. *Vet. Parasitol.* 55, 21–27.

1161. Meads, E.B., 1976. Dalmeny disease–another outbreak–probably sarcocystosis. *Can. Vet. J.* 17, 271.

1162. Medrano, G., Hung, A., Rubio, N., 2006. Detección molecular temprana de *Sarcocystis* en el animal vivo y su estudio filogenético basado en el análisis del gen SSU rRNA en alpacas en Perú. *Mosaico Cient.* 3, 5–9.

1163. Medrano, G., Hung, A., Espinoza, J., 2008. Análisis molecular y filogenético de las especies de *Sarcocystis* que afectan a las alpacas del Perú. *Rev. Investig. (Esc. Postgrado, En linea)* 4, 171–186.

1164. Mehlhorn, H., Scholtyseck, E., 1973. Elektronenmikroskopische Untersuchungen an Cystenstadien von *Sarcocystis tenella* aus der Oesophagus-Muskulatur des Schafes. *Z. Parasitenkd.* 41, 291–310.

1165. Mehlhorn, H., Scholtyseck, E., 1974. Licht- und elektronenmikroskopische Untersuchungen an Entwicklungsstadien von *Sarcocystis tenella* aus der Darmwand der Hauskatze. I. Die Oocysten und Sporocysten. *Z. Parasitenkd.* 43, 251–270.

1166. Mehlhorn, H., Scholtyseck, E., 1974. Die Parasit-Wirtsbeziehungen bei verschiedenen Gattungen der Sporozoen (*Eimeria, Toxoplasma, Sarcocystis, Frenkelia, Hepatozoon, Plasmodium* und *Babesia*) unter Anwendung spezieller Verfahren. *Microscopica Acta* 75, 429–451.

1167. Mehlhorn, H., Sénaud, J., Scholtyseck, E., 1974. Étude ultrastructurale des coccidies formant des kystes: *Toxoplasma gondii, Sarcocystis tenella, Besnoitia jellisoni* et *Frenkelia* sp.: Distribution de la phosphatase acide et des polysaccharides au niveau des ultrastructures chez le parasite et chez l'hote. *Protistologica* 10, 21–42.

1168. Saito, M., Satho, S., Tomioka, H., Nakajima, T., Watanabe, A., Itagaki, H., 1987. Experimental swine sarcocystiasis. *J. Jpn. Vet. Med. Assoc.* 40, 442-445. (In Japanese)

1169. Mehlhorn, H., Sénaud, J., 1975. Action lytique des bactéries présentes dans les kystes de *Sarcocystis tenella* (Sporozoa, Protozoa). *Arch. Microbiol.* 104, 241–244.

1170. Mehlhorn, H., Heydorn, A.O., Gestrich, R., 1975. Licht- und elektronenmikroskopische Untersuchungen an Cysten von *Sarcocystis ovicanis* Heydorn et al. (1975) in der Muskulatur von Schafen. *Z. Parasitenkd.* 48, 83–93.

1171. Mehlhorn, H., Sénaud, J., Chobotar, B., Scholtyseck, E., 1975. Electron microscope studies of cyst stages of *Sarcocystis tenella*: The origin of micronemes and rhoptries. *Z. Parasitenkd.* 45, 227–236.

1172. Mehlhorn, H., Senaud, J., Heydorn, A.O., Gestrich, R., 1975. Comparaison des ultrastructures des kystes de *Sarcocystis fusiformis* Railliet, 1897 dans la musculature du boeuf, après infection naturelle et après infection expérimentale par des sporocystes d'*Isopora hominis* et par des sporocystes des grandes formes d'Isopora bigemina du chien et du chat. *Protistologica* 11, 445–455.

1173. Mehlhorn, H., Hartley, W.J., Heydorn, A.O., 1976. A comparative ultrastructural study of the cyst wall of 13 *Sarcocystis* species. *Protistologica* 12, 451–467.

1174. Mehlhorn, H., Heydorn, A.O., 1977. Light and electron microscopic studies of *Sarcocystis suihominis*. 1. The development of cysts in experimentally infected pigs. *Zbl. Bakt. Hyg., I. Abt. Orig. A* 239, 124–139.

1175. Mehlhorn, H., Heydorn, A.O., Janitschke, K., 1977. Light and electron microscopical study on sarcocysts from muscles of the rhesus monkey (*Macaca mulatta*), baboon (*Papio cynocephalus*) and tamarin (*Saguinus* (= *Oedipomidas*) *oedipus*). *Z. Parasitenkd.* 51, 165–178.

1176. Mehlhorn, H., Heydorn, A.O., 1978. The Sarcosporidia (Protozoa, Sporozoa): Life cycle and fine structure. *Adv. Parasitol.* 16, 43–92.

1177. Mehlhorn, H., Heydorn, A.O., 1979. Electron microscopical study on gamogony of *Sarcocystis suihominis* in human tissue cultures. *Z. Parasitenkd.* 58, 97–113.

1178. Mehlhorn, H., Matuschka, F.R., 1986. Ultrastructural studies of the development of *Sarcocystis clethrionomyelaphis* within its final and intermediate hosts. *Protistologica* 22, 97–104.

1179. Mehrotra, R., Bisht, D., Singh, P.A., Gupta, S.C., Gupta, R.K., 1996. Diagnosis of human *Sarcocystis* infection from biopsies of the skeletal muscle. *Pathology* 28, 281–282.

1180. Meingassner, J.G., Burtscher, H., 1977. Doppelinfektion des Gehirns mit *Frenkelia* species und *Toxoplasma gondii* bei *Chinchilla laniger*. *Vet. Pathol.* 14, 146–153.

1181. Meloni, B.P., Thompson, R.C.A., Hopkins, R.M., Reynoldson, J.A., Gracey, M., 1993. The prevalence of *Giardia* and other intestinal parasites in children, dogs and cats from aboriginal communities in the Kimberley. *Med. J. Aust.* 158, 157–159.

1182. Memmedov, I., 2010. Prevalence of *Sarcocystis* species in some bird in Nakhchivan Autonomous Republic. *Kafkas Univ. Vet. Fak. Derg.* 16, 857–860. (In Turkish)

1183. Mencke, N., O'Donoghue, P., Lumb, R., Smith, P., Tenter, A.M., Thümmel, P., Rommel, M., 1991. Antigenic characterization of monoclonal antibodies against *Sarcocystis muris* by Western blotting and immuno-electron microscopy. *Parasitol. Res.* 77, 217–223.

1184. Mense, M.G., Dubey, J.P., Homer, B.L., 1992. Acute hepatic necrosis associated with a *Sarcocystis*-like protozoa in a sea lion (*Zalophus californianus*). *J. Vet. Diagn. Invest.* 4, 486–490.

1185. Mercado Pezzat, M., 1971. Frequencia de *Sarcocystis* spp. en corazones de bovinos. *Veterinaria (Mexico)* 11, 6–11.

1186. Mertens, C.M., Tenter, A.M., Vietmeyer, C., Ellis, J.T., Johnson, A.M., 1996. Production of a recombinant fusion protein of *Sarcocystis tenella* and evaluation of its diagnostic potential in an ELISA. *Vet. Parasitol.* 65, 185–197.

1187. Meshkov, S., 1978. Distribution of Sarcosporidia among wild animals in Bulgaria. *Vet. Med. Nauki.* 15, 72–78. (In Russian).

1188. Meshkov, S., 1980. The jackal (*Canis aureus*) as a new host of *Sarcocystis* infecting swine. *Veterinarna Sbirka* 78, 20–21. (In Russian).

1189. Meteyer, C.U., Backos, S., Barr, B.C., Shivaprasad, H., 1990. *Sarcocystis* infection in Psittacines. *Proc. West. Poult. Dis. Conf.* 39, 76–77.

1190. Metwally, A.M., Abd Ellah, M.R., Al-Hosary, A.A., Omar, M.A., 2014. Microscopical and serological studies on *Sarcocystis* infection with first report of *S. cruzi* in buffaloes (*Bubalus bubalis*) in Assiut, Egypt. *J. Parasit. Dis.* 38, 378–382.

1191. Miescher, F., 1843. Über eigenthümliche Schläuche in den Muskeln einer Hausmaus. *Bericht der Verhandlungen der Naturforschenden Gesellschaft* 5, 198–202.

1192. Migaki, G., Albert, T.F., 1980. Sarcosporidiosis in the ringed seal. *J. Am. Vet. Med. Assoc.* 177, 917–918.

1193. Miller, I., 1993. On the occurrence of cattle sarcocystosis in different regions of Estonia. *Biologija* 1, 34–35.

1194. Miller, M.A., Sverlow, K., Crosbie, P.R., Barr, B.C., Lowenstine, L.J., Gulland, F.M., Packham, A., Conrad, P.A., 2001. Isolation and characterization of two parasitic protozoa from a pacific harbor seal (*Phoca vitulina richardsi*) with meningoencephalomyelitis. *J. Parasitol.* 87, 816–822.

1195. Miller, M.A., Crosbie, P.R., Sverlow, K., Hanni, K., Barr, B.C., Kock, N., Murray, M.J., Lowenstine, L.J., Conrad, P.A., 2001. Isolation and characterization of *Sarcocystis* from brain tissue of a free-living southern sea otter (*Enhydra lutris nereis*) with fatal meningoencephalitis. *Parasitol. Res.* 87, 252–257.

1196. Miller, M.A., Barr, B.C., Nordhausen, R., James, E.R., Magargal, S.L., Murray, M., Conrad, P.A., Toy-Choutka, S., Jessup, D.A., Grigg, M.E., 2009. Ultrastructural and molecular confirmation of the development of *Sarcocystis neurona* tissue cysts in the central nervous system of southern sea otters (*Enhydra lutris nereis*). *Int. J. Parasitol.* 39, 1363–1372.

1197. Miller, M.A., Conrad, P.A., Harris, M., Hatfield, B., Langlois, G., Jessup, D.A., Magargal, S.L. et al. 2010. A protozoal-associated epizootic impacting marine wildlife: Mass-mortality of southern sea otters (*Enhydra lutris nereis*) due to *Sarcocystis neurona* infection. *Vet. Parasitol.* 172, 183–194.

1198. Miller, M.M., Sweeney, C.R., Russell, G.E., Sheetz, R.M., Morrow, J.K., 1999. Effects of blood contamination of cerebrospinal fluid on western blot analysis for detection of antibodies against *Sarcocystis neurona* and on albumin quotient and immunoglobulin G index in horses. *J. Am. Vet. Med. Assoc.* 215, 67–71.

1199. Mirandé, L.A., Howerth, E.W., Poston, R.P., 1992. Chlamydiosis in a red-tailed hawk (*Butteo jamaicensis*). *J. Wildl. Dis.* 28, 284–287.

1200. Mitchell, M.A., 1988. The prevalence of macroscopic sarcocysts in New Zealand cattle at slaughter. *N. Z. Vet. J.* 36, 35–38.

1201. Mitchell, S.M., Richardson, D.J., Cheadle, M.A., Zajac, A.M., Lindsay, D.S., 2002. Prevalence of agglutinating antibodies to *Sarcocystis neurona* in skunks (*Mephitis mephitis*), raccoons (*Procyon lotor*), and opossums (*Didelphis virginiana*) from Connecticut. *J. Parasitol.* 88, 1027–1029.

1202. Mitchell, S.M., Zajac, A.M., Davis, W.L., Kennedy, T.J., Lindsay, D.S., 2005. The effects of ponazuril on development of apicomplexans *in vitro*. *J. Eukaryot. Microbiol.* 52, 231–235.

1203. Modrý, D., Koudela, B., Šlapeta, J.R., 2000. *Sarcocystis stenodactylicolubris* n. sp., a new sarcosporidian coccidium with a snake-gecko heteroxenous life cycle. *Parasite* 7, 201–207.

1204. Modrý, D., Votýpka, J., Svobodová, M., 2004. Note on the taxonomy of *Frenkelia microti* (Findlay & Middleton, 1934) (Apicomplexa: Sarcocystidae). *Syst. Parasitol.* 58, 185–187.

1205. Mohammed, O.B., Davies, A.J., Hussein, H.S., Daszak, P., 2000. *Sarcocystis* infections in gazelles at the King Khalid Wildlife Research Centre, Saudi Arabia. *Vet. Rec.* 146, 218–221.

1206. Mohanty, B., Misra, S.C., Rao, A.T., 1995. Pathology of *Sarcocystis* infection in naturally infected cattle, buffaloes, sheep and goats. *Indian Vet. J.* 72, 569–571.

1207. Mohanty, B.N., Misra, S.C., Panda, D.N., Panda, M.R., 1995. Prevalence of *Sarcocystis* infection in ruminants in Orissa. *Indian Vet. J.* 72, 1026–1030.

1208. Molnár, K., Ostoros, G., Dunams-Morel, D., Rosenthal, B.M., 2012. *Eimeria* that infect fish are diverse and are related to, but distinct from, those that infect terrestrial vertebrates. *Infect. Genet. Evol.* 12, 1810–1815.

1209. Monteiro, R.M., Keid, L.B., Richtzenhain, L.J., Valadas, S.Y., Muller, G., Soares, R.M., 2013. Extensively variable surface antigens of *Sarcocystis* spp. infecting Brazilian marsupials in the genus *Didelphis* occur in myriad allelic combinations, suggesting sexual recombination has aided their diversification. *Vet. Parasitol.* 196, 64–70.

1210. Moore, B.R., Granstrom, D.E., Reed, S.M., 1995. Diagnosis of equine protozoal myeloencephalitis and cervical stenoic myelopathy. *Comp. Cont. Edu. Pract. Vet.* 17, 419–428.

1211. Moore, S., 1980. Two types of ovine *Sarcocystis* macrocysts distinguished by periodic acid-Schiff staining of the cyst walls. *N. Z. Vet. J.* 28, 101–102.

1212. Moorman, T.E., Baldassarre, G.A., Richard, D.M., 1991. The frequency of *Sarcocystis* spp. and its effect on winter carcass composition of mottled ducks. *J. Wildl. Dis.* 27, 491–493.

1213. Moré, G., Basso, W., Bacigalupe, D., Venturini, M.C., Venturini, L., 2008. Diagnosis of *Sarcocystis cruzi*, *Neospora caninum*, and *Toxoplasma gondii* infections in cattle. *Parasitol. Res.* 102, 671–675.

1214. Moré, G., Pardini, L., Basso, W., Marín, R., Bacigalupe, D., Auad, G., Venturini, L., Venturini, M.C., 2008. Seroprevalence of *Neospora caninum*, *Toxoplasma gondii* and *Sarcocystis* sp. in llamas (*Lama glama*) from Jujuy, Argentina. *Vet. Parasitol.* 155, 158–160.

1215. Moré, G., Bacigalupe, D., Basso, W., Rambeaud, M., Beltrame, F., Ramirez, B., Venturini, M.C., Venturini, L., 2009. Frequency of horizontal and vertical transmission for *Sarcocystis cruzi* and *Neospora caninum* in dairy cattle. *Vet. Parasitol.* 160, 51–54.

1216. Moré, G., Bacigalupe, D., Basso, W., Rambeaud, M., Venturini, M.C., Venturini, L., 2010. Serologic profiles for *Sarcocystis* sp. and *Neospora caninum* and productive performance in naturally infected beef calves. *Parasitol. Res.* 106, 689–693.

1217. Moré, G., Abrahamovich, P., Jurado, S., Bacigalupe, D., Marin, J.C., Rambeaud, M., Venturini, L., Venturini, M.C., 2011. Prevalence of *Sarcocystis* spp. in Argentinean cattle. *Vet. Parasitol.* 177, 162–165.

1218. Moré, G., Schares, S., Maksimov, A., Conraths, F.J., Venturini, M.C., Schares, G., 2013. Development of a multiplex real time PCR to differentiate *Sarcocystis* spp. affecting cattle. *Vet. Parasitol.* 197, 85–94.

1219. Moré, G., Pantchev, N., Herrmann, D.C., Vrhovec, M.G., Öfner, S., Conraths, F.J., Schares, G., 2014. Molecular identification of *Sarcocystis* spp. helped to define the origin of green pythons (*Morelia viridis*) confiscated in Germany. *Parasitology* 141, 646–651.

1220. Moré, G., Vissani, A., Pardini, L., Monina, M., Muriel, M., Howe, D., Barrandeguy, M., Venturini, M.C., 2014. Seroprevalence of *Sarcocystis neurona* and its association with neurologic disorders in Argentinean horses. *J. Equine Vet. Sci.* 34, 1051–1054.

1221. Moré, G., Pantchev, A., Skuballa, J., Langenmayer, M.C., Maksimov, P., Conraths, F.J., Venturini, M.C., Schares, G., 2014. *Sarcocystis sinensis* is the most prevalent thick-walled *Sarcocystis* in beef for consumers in Germany. *Parasitol. Res.* 113, 2223–2230.

1222. Morgan, G., Terlecki, S., Bradley, R., 1984. A suspected case of *Sarcocystis* encephalitis in sheep. *Br. Vet. J.* 140, 64–69.

1223. Mori, Y., 1985. Studies on the *Sarcocystis* in Japanese black cattle. *Bull. Azabu Univ. Vet. Med.* 6, 51–65. (In Japanese).

1224. Morley, P.S., Traub-Dargatz, J.L., Benedict, K.M., Saville, W.J.A., Voelker, L.D., Wagner, B.A., 2008. Risk factors for owner-reported occurrence of equine protozoal myeloencephalitis in the US equine population. *J. Vet. Intern. Med.* 22, 616–629.

1225. Morley, S., Traub-Dargatz, J., Saville, W., Wagner, B., Garber, L., Hillberg-Seitzinger, A., 2001. Equine protozoal myeloencephalitis. *J. Equine Vet. Sci.* 21, 262–270.

1226. Morsy, K., Saleh, A., Al-Ghamdi, A., Abdel-Ghaffar, F., Al-Rasheid, K., Bashtar, A.R., Al Quraishy, S., Mehlhorn, H., 2011. Prevalence pattern and biology of *Sarcocystis capracanis* infection in the Egyptian goats: A light and ultrastructural study. *Vet. Parasitol.* 181, 75–82.

1227. Morsy, K., Bashtar, A.R., Abdel-Ghaffar, F., Mehlhorn, H., Al Quraishy, S., Al-Ghamdi, A., Koura, E., Maher, S., 2012. *Sarcocystis acanthocolubri* sp. n. infecting three lizard species of the genus *Acanthodactylus* and the problem of host specificity. Light and electron microscopic study. *Parasitol. Res.* 110, 355–362.

1228. Morsy, T.A., Abdel Mawla, M.M., Salama, M.M.I., Hamdi, K.N., 1994. Assessment of intact *Sarcocystis* cystozoites as an ELISA antigen. *J. Egypt. Soc. Parasitol.* 24, 85–91.

1229. Motamedi, G.R., Dalimi, A., Aghaeipour, K., Nouri, A., 2010. Ultrastructural and molecular studies on fat and thin macrocysts of *Sarcocystis* spp. isolated from naturally infected goats. *Archives of Razi Institute* 65, 91–97.

1230. Motamedi, G.R., Dalimi, A., Nouri, A., Aghaeipour, K., 2011. Ultrastructural and molecular character-ization of *Sarcocystis* isolated from camel (*Camelus dromedarius*) in Iran. *Parasitol. Res.* 108, 949–954.

1231. Moule, L.T., 1888. Des sarcosporidies et de leur frequence, principalement chez les animaux de bouch-erie. *Soc Sci Arts Vitry le-Francois* 14, 3–42.

1232. Moulé, M.L., 1886. Sarcosporidies du tissu musculaire. In: *Des Sarcosporidies*. Médicin Vétérinaire, Paris, pp. 3–43.

1233. Muangyai, M., Chalermchaikit, T., 1988. *Sarcocystis* in Thailand. I. The incidence of *Sarcocystis* in cattle and buffaloes. *Thai J. Vet. Med.* 18, 319–326. (In Thai).

1234. Muangyai, M., 1989. *Sarcocystis* in Thailand. II. Incidence in pigs. *Thai J. Vet. Med.* 19, 49–56. (In Thai).

1235. Mugridge, N.B., Morrison, D.A., Heckeroth, A.R., Johnson, A.M., Tenter, A.M., 1999. Phylogenetic analysis based on full-length large subunit ribosomal RNA gene sequence comparison reveals that *Neospora caninum* is more closely related to *Hammondia heydorni* than to *Toxoplasma gondii*. *Int. J. Parasitol.* 29, 1545–1556.

1236. Mugridge, N.B., Morrison, D.A., Johnson, A.M., Luton, K., Dubey, J.P., Votypka, J., Tenter, A.M., 1999. Phylogenetic relationships of the genus *Frenkelia*: A review of its history and new knowledge gained from comparison of large subunit ribosomal ribonucleic acid gene sequences. *Int. J. Parasitol.* 29, 957–972.

1237. Mugridge, N.B., Morrison, D.A., Jäkel, T., Heckeroth, A.R., Tenter, A.M., Johnson, A.M., 2000. Effects of sequence alignment and structural domains of ribosomal DNA on phylogeny reconstruction for the protozoan family Sarcocystidae. *Mol. Biol. Evol.* 17, 1842–1853.

1238. Muhm, R.L., Barnett, D., Bryant, D.T., Cole, J.H., Kadel, W.L., 1979. Sarcocystosis: A case study. *Proc. 22nd Ann. Meeting Am. Assoc. Vet. Lab. Diagn.* 1979, 139–146.

1239. Mullaney, T., Murphy, A.J., Kiupel, M., Bell, J.A., Rossano, M.G., Mansfield, L.S., 2005. Evidence to support horses as natural intermediate hosts for *Sarcocystis neurona*. *Vet. Parasitol.* 133, 27–36.

1240. Müller, J.J., Weiss, M.S., Heinemann, U., 2011. PAN-modular structure of microneme protein SML-2 from the parasite *Sarcocystis muris* at 1.95 A resolution and its complex with 1-thio-beta-D-galactose. *Acta Crystallogr. D. Biol. Crystallogr.* 67, 936–944.

1241. Mumba, C., Pandey, G.S., Kakandelwa, C., 2012. Sarcocystosis in an adult Brahman cow. *Int. J. Livest. Res.* 2, 271–275.

1242. Munday, B.L., Mason, R.W., Cumming, R., 1973. Observations on diseases of the central nervous sys-tem of cattle in Tasmania. *Aust. Vet. J.* 49, 451–455.

1243. Munday, B.L., 1982. Effect of preparturient inoculation of pregnant ewes with *Sarcocystis ovicanis* upon the susceptibility of their progeny. *Vet. Parasitol.* 9, 273–276.

1244. Munday, B.L., 1975. The prevalence of sarcosporidiosis in Australian meat animals. *Aust. Vet. J.* 51, 478–480.

1245. Munday, B.L., Barker, I.K., Rickard, M.D., 1975. The developmental cycle of a species of *Sarcocystis* occurring in dogs and sheep, with observations on pathogenicity in the intermediate host. *Z. Parasitenkd.* 46, 111–123.

1246. Munday, B.L., Black, H., 1976. Suspected *Sarcocystis* infections of the bovine placenta and foetus. *Z. Parasitenkd.* 51, 129–132.

1247. Munday, B.L., 1977. A species of *Sarcocystis* using owls as definitive hosts. *J. Wildl. Dis.* 13, 205–207.

1248. Munday, B.L., Humphrey, J.D., Kila, V., 1977. Pathology produced by, prevalence of, and probable life-cycle of a species of *Sarcocystis* in the domestic fowl. *Avian Dis.* 21, 697–703.

1249. Munday, B.L., Mason, R.W., Hartley, W.J., Presidente, P.J.A., Obendorf, D., 1978. *Sarcocystis* and related organisms in Australian wildlife: I. Survey findings in mammals. *J. Wildl. Dis.* 14, 417–433.

1250. Munday, B.L., 1979. The effect of *Sarcocystis ovicanis* on growth rate and haematocrit in lambs. *Vet. Parasitol.* 5, 129–135.

1251. Munday, B.L., Hartley, W.J., Harrigan, K.E., Presidente, P.J., Obendorf, D.L., 1979. *Sarcocystis* and related organisms in Australian wildlife: II. Survey of findings in birds, reptiles, amphibians and fish. *J. Wildl. Dis.* 15, 57–73.

1252. Munday, B.L., Mason, R.W., 1980. *Sarcocystis* and related organisms in Australian wildlife: III. *Sarcocystis murinotechis* sp. n. life cycle in rats (*Rattus, Pseudomys* and *Mastocomys* spp.) and tiger snakes (*Notechis ater*). *J. Wildl. Dis.* 16, 83–87.

1253. Munday, B.L., Smith, D.D., Frenkel, J.K., 1980. *Sarcocystis* and related organisms in Australian Wildlife: IV. Studies on *Sarcocystis cuniculi* in European rabbits (*Oryctolagus cuniculus*). *J. Wildl. Dis.* 16, 201–204.

1254. Munday, B.L., 1981. Premature parturition in ewes inoculated with *Sarcocystis ovicanis*. *Vet. Parasitol.* 9, 17–26.

1255. Munday, B.L., Mason, R.W., 1981. *Sarcocystis* in rats from islands. *N. Z. J. Zool.* 8, 563.

1256. Munday, B.L., 1983. An isosporan parasite of masked owls producing sarcocysts in rats. *J. Wildl. Dis.* 19, 146–147.

1257. Munday, B.L., 1985. Demonstration of viable *Sarcocystis* sporocysts in the faeces of a lamb dosed orally. *Vet. Parasitol.* 17, 355–357.

1258. Munday, B.L., 1984. The effect of *Sarcocystis tenella* on wool growth in sheep. *Vet. Parasitol.* 15, 91–94.

1259. Munday, B.L., Obendorf, D.L., 1984. Development and growth of *Sarcocystis gigantea* in experimentally-infected sheep. *Vet. Parasitol.* 15, 203–211.

1260. Munday, B.L., Obendorf, D.L., 1984. Morphology of *Sarcocystis gigantea* in experimentally-infected sheep. *Vet. Parasitol.* 16, 193–199.

1261. Munday, B.L., 1986. Effects of different doses of dog-derived *Sarcocystis* sporocysts on growth rate and haematocrit in lambs. *Vet. Parasitol.* 21, 21–24.

1262. Murphy, A.J., Mansfield, L.S., 1999. Simplified technique for isolation, excystation, and culture of *Sarcocystis* species from opossums. *J. Parasitol.* 85, 979–981.

1263. Murphy, J.E., Marsh, A.E., Reed, S.M., Meadows, C., Bolten, K., Saville, W.J.A., 2006. Development and evaluation of a *Sarcocystis neurona*-specific IgM capture enzyme-linked immunosorbent assay. *J. Vet. Intern. Med.* 20, 322–328.

1264. Murrell, T.G.C., O'Donoghue, P.J., Ellis, T., 1986. A review of the sheep-multiple sclerosis connection. *Medical Hypotheses* 19, 27–39.

1265. Musaev, M., Surkova, A.M., Iskenderova, N.G., 1989. Sarcosporidian infection in stray dogs on the Apsheron Peninsula. *Izvestiya Akademii Nauk Azerbaidzhanskoi SSR, Biol.* 1, 32–38. (In Russian).

1266. Mutalib, A., Keirs, R., Maslin, W., Topper, M., Dubey, J.P., 1995. *Sarcocystis*-associated encephalitis in chickens. *Avian Dis.* 39, 436–440.

1267. Mylniczenko, N.D., Kearns, K.S., Melli, A.C., 2008. Diagnosis and treatment of *Sarcocystis neurona* in a captive harbor seal (*Phoca vitulina*). *J. Zoo Wildl. Med.* 39, 228–235.

1268. Naghibi, A.A.G., Razmi, G.R., Ghasemifard, M., 2002. Identification of *Sarcocystis cruzi* in cattle by using of experimentally infection in final and intermediate hosts. *J. Vet. Res.* 57, 67–69.

1269. Nakamura, K., Shoya, S., Nakajima, Y., Shimura, K., Ito, S., 1982. Pathology of experimental acute sarcocystosis in a cow. *Jpn. J. Vet. Sci.* 44, 675–679.

1270. Skofitsch, G., Kepka, O., 1982. Evidence of circulating antibodies against *Frenkelia glareoli* (Apicomplexa). *Z. Parasitenkd.* 66, 355–358.

1271. Ndiritu, W., Cawthorn, R.J., Kibenge, F.S.B., 1994. Use of proteinase K in the excystation of *Sarcocystis cruzi* sporocysts for *in vitro* culture and DNA extraction. *Vet. Parasitol.* 52, 57–60.

1272. Ndiritu, W., Cawthorn, R.J., Kibenge, F.S.B., Markham, R.J.F., Horney, B.S., Chan, C.B., 1996. Use of genomic DNA probes for the diagnosis of acute sarcocystosis in experimentally infected cattle. *Vet. Parasitol.* 62, 9–25.

1273. Neill, P.J.G., Smith, J.H., Box, E.D., 1989. Pathogenesis of *Sarcocystis falcatula* (Apicomplexa: Sarcocystidae) in the budgerigar (*Melopsittacus undulatus*) IV. Ultrastructure of developing, mature and degenerating sarcocysts. *J. Protozool.* 36, 430–437.

1274. Nemeséri, L., Entzeroth, R., Scholtyseck, E., 1983. A magyarorszagi gimszarvasok *Sarcocystis*-fajanak gyakorisaga es ultrastrukturaja. *Magyar Allatorvosok Lapja* 38, 758–759.

1275. Neveu-Lemaire, M., 1912. Sarcosporidies des mammifères domestiques. *O parasitologie des animaux domestiques maladies parasitaires non-bacteriennes*, Imprimerie Societé des Sciences Naturelles, Paris, pp. 300–306.

1276. Nevole, M., Malota, L., Koudela, B., 1986. catalytic activity of selected serum enzymes in experimental sarcocystosis in calves. *Acta Vet. Brno* 55, 81–84.

1277. Nigro, M., Mancianti, F., Rossetti, P., Poli, A., 1991. Ultrastructure of the cyst and life cycle of *Sarcocystis* sp. from wild sheep (*Ovis musimon*). *J. Wildl. Dis.* 27, 217–224.

1278. Nimri, L., 2014. Unusual case presentation of intestinal *Sarcocystis hominis* infection in a healthy adult. *JMM Case Reports* 1, 1–3.

1279. Njoku, C.J., Saville, W.J.A., Reed, S.M., Oglesbee, M.J., Rajala-Schultz, P.J., Stich, R.W., 2002. Reduced levels of nitric oxide metabolites in cerebrospinal fluid are associated with equine protozoal myeloencephalitis. *Clin. Diagn. Lab. Immunol.* 9, 605–610.

1280. Noh, J.W., Jang, D.H., Kang, Y.B., Jang, H., Wee, S.H., 1988. Effects of temperature on viability of sarcocysts of *Sarcocystis cruzi* in cardiac muscle of cattle. *Korean J. Vet. Publ. Health* 12, 151–155.

1281. Nonaka, N., Nakamura, S., Inoue, T., Oku, Y., Katakura, K., Matsumoto, J., Mathis, A., Chembesofu, M., Phiri, I.G.K., 2011. Coprological survey of alimentary tract parasites in dogs from Zambia and evaluation of a coproantigen assay for canine echinococcosis. *Ann. Trop. Med. Parasitol.* 105, 521–530.

1282. Nourani, H., Matin, S., Nouri, A., Azizi, H., 2010. Prevalence of thin-walled *Sarcocystis cruzi* and thick-walled *Sarcocystis hirsuta* or *Sarcocystis hominis* from cattle in Iran. *Trop. Anim. Health Prod.* 42, 1225–1227.

1283. Nourollahi Fard, S.R., Asghari, M., Nouri, F., 2009. Survey of *Sarcocystis* infection in slaughtered cattle in Kerman, Iran. *Trop. Anim. Health Prod.* 41, 1633–1636.

1284. Novak, M.D., Fedoseenko, V.M., Orazalinova, V.A., 1987. Cyst ultrastructure in *Sarcocystis* sp. in the cattle. *Izvest. Akad. Nauk. Kazakh. SSR. Ser. Biol.* 46–49. (In Russian).

1285. O'Donoghue, P., Lumb, R., Smith, P., Brooker, J., Mencke, N., 1990. Characterization of monoclonal antibodies against ovine *Sarcocystis* spp. antigens by immunoblotting and immuno-electron microscopy. *Vet. Immunol. Immunopathol.* 24, 11–25.

1286. O'Donoghue, P.J., Weyreter, H., 1983. Detection of *Sarcocystis* antigens in the sera of experimentally-infected pigs and mice by an immunoenzymatic assay. *Vet. Parasitol.* 12, 13–29.

1287. O'Donoghue, P.J., Weyreter, H., 1984. Examinations on the serodiagnosis of *Sarcocystis* infections. II. Class-specific immunoglobulin responses in mice, pigs and sheep. *Zbl. Bakt. Hyg. A* 257, 168–184.

1288. O'Donoghue, P.J., Ford, G.E., 1984. The asexual pre-cyst development of *Sarcocystis tenella* in experimentally infected specific-pathogen-free lambs. *Int. J. Parasitol.* 14, 345–355.

1289. O'Donoghue, P.J., Rommel, M., Weber, M., Weyreter, H., 1985. Attempted immunization of swine against acute sarcocystosis using cystozoite-derived vaccines. *Vet. Immunol. Immunopathol.* 8, 83–92.

1290. O'Donoghue, P.J., Ford, G.E., 1986. The prevalence and intensity of *Sarcocystis* spp. infections in sheep. *Aust. Vet. J.* 63, 273–278.

1291. O'Donoghue, P.J., Adams, M., Dixon, B.R., Ford, G.E., Baverstock, P.R., 1986. Morphological and biochemical correlates in the characterization of *Sarcocystis* spp. *J. Protozool.* 33, 114–121.

1292. O'Donoghue, P.J., Watts, C.H.S., Dixon, B.R., 1987. Ultrastructure of *Sarcocystis* spp. (Protozoa, Apicomplexa) in rodents from North Sulawesi and West Java, Indonesia. *J. Wildl. Dis.* 23, 225–232.

1293. O'Donoghue, P.J., Obendorf, D.L., O'Callaghan, M.G., Moore, E., Dixon, B.R., 1987. *Sarcocystis mucosa* (Blanchard 1885) Labbé 1889 in unadorned rock wallabies (*Petrogale assimilis*) and Bennett's wallabies (*Macropus rufogriseus*). *Parasitol. Res.* 73, 113–120.

1294. O'Donoghue, P.J., Wilkinson, R.G., 1988. Antibody development and cellular immune responses in sheep immunized and challenged with *Sarcocystis tenella* sporocysts. *Vet. Parasitol.* 27, 251–265.

1295. O'Donoghue, P.J., Rommel, M., 1997. Australian-German collaborative studies on the immunology of *Sarcocystis* infections. *Angew. Parasitol.* 33, 102–119.

1296. O'Toole, D., 1987. Experimental ovine *Sarcocystis*: Sequential untrastructural pathology in skeletal muscle. *J. Comp. Pathol.* 97, 51–60.

1297. O'Toole, D., Jeffrey, M., Challoner, D., Maybey, R., Welch, V., 1993. Ovine myeloencephalitis-leukomyelomalacia associated with a *Sarcocystis*-like protozoan. *J. Vet. Diagn. Invest.* 5, 212–225.

1298. Obendorf, D.L., Munday, B.L., 1986. Demonstration of schizogonous stages of *Sarcocystis gigantea* in experimentally infected sheep. *Vet. Parasitol.* 19, 35–38.

1299. Obendorf, D.L., Munday, B.L., 1987. Experimental infection with *Sarcocystis medusiformis* in sheep. *Vet. Parasitol.* 24, 59–65.

1300. Obijiaku, I.N., Ajogi, I., Umoh, J.U., Lawal, I.A., Atu, B.O., 2013. *Sarcocystis* infection in slaughtered cattle in Zango abattoir, Zaria, Nigeria. *Vet. World* 6, 346–349.

1301. Oborník, M., Jirku, M., Šlapeta, J.R., Modrý, D., Koudela, B., Lukeš, J., 2002. Notes on coccidian phylogeny, based on the apicoplast small subunit ribosomal DNA. *Parasitol. Res.* 88, 360–363.

1302. Odening, K., Jakob, W., 1992. Oocysten und Sporozoan vom *Sarcocystis*-typ bei Wild- und Zoovögeln. *Vehr. Ber. Erkrg. Zootiere* 34, 309–316.

1303. Odening, K., Stolte, M., Walter, G., Bockhardt, I., 1994. The European badger (Carnivora: Mustelidae) as intermediate host of further three *Sarcocystis* species (Sporozoa). *Parasite* 1, 23–30.

1304. Odening, K., Stolte, M., Walter, G., Bockhardt, I., Jakob, W., 1994. Sarcocysts (*Sarcocystis* sp.: Sporozoa) in the European badger, *Meles meles*. *Parasitology* 108, 421–424.

1305. Odening, K., Wesemeier, H.H., Pinkowski, M., Walter, G., Sedlaczek, J., Bockhardt, I., 1994. European hare and European rabbit (Lagomorpha) as intermediate hosts of *Sarcocystis* species (Sporozoa) in central Europe. *Acta Protozool.* 33, 177–189.

1306. Odening, K., Wesemeier, H.H., Walter, G., Bockhardt, I., 1995. Ultrastructure of sarcocysts from equids. *Acta Parasitologica* 40, 12–20.

1307. Odening, K., Wesemeier, H.H., Walter, G., Bockhardt, I., 1995. On the morphological diagnostics and host specificity of the *Sarcocystis* species of some domesticated and wild Bovini (cattle, banteng, and bison). *Appl. Parasitol.* 36, 161–178.

1308. Odening, K., Stolte, M., Walter, G., Bockhardt, I., 1995. Cyst wall ultrastructure of two *Sarcocystis* spp. from European mouflon (*Ovis ammon musimon*) in Germany compared with domestic sheep. *J. Wildl. Dis.* 31, 550–554.

1309. Odening, K., Stolte, M., Bockhardt, I., 1996. Sarcocysts in exotic Caprinae (Bovidae) from zoological gardens. *Acta Parasitologica* 41, 67–75.

1310. Odening, K., Wesemeier, H.H., Bockhardt, I., 1996. On the sarcocysts of two further *Sarcocystis* species being new for the European hare. *Acta Protozool.* 35, 69–72.

1311. Odening, K., Stolte, M., Bockhardt, I., 1996. On the diagnostics of *Sarcocystis* in chamois (*Rupicapra rupicapra*). *Appl. Parasitol.* 37, 153–160.

1312. Odening, K., Stolte, M., Bockhardt, I., 1996. On the diagnostics of *Sarcocystis* in cattle: Sarcocysts of a species unusual for *Bos taurus* in a dwarf zebu. *Vet. Parasitol.* 66, 19–24.

1313. Odening, K., Stolte, M., Lux, E., Bockhardt, I., 1996. The Mongolian gazelle (*Procapra gutturosa*, Bovidae) as an intermediate host of three *Sarcocystis* species in Mongolia. *Appl. Parasitol.* 37, 54–65.

1314. Odening, K., 1997. Die *Sarcocystis*-Infektion: Wechselbeziehungen zwischen freilebenden Wildtieren, Haustieren und Zootieren. *Zool. Garten N. F.* 67, 317–340.

1315. Odening, K., Frölich, K., Kirsch, B., Bockhardt, I., 1997. *Sarcocystis*: Development of sporocysts in cell culture. *J. Protozool. Res.* 7, 9–16.

1316. Odening, K., Quandt, S., Bengis, R.G., Stolte, M., Bockhardt, I., 1997. *Sarcocystis hippopotami* sp. n. and *S. africana* sp. n. (Protozoa: Sarcocystidae) from the hippopotamus in South Africa. *Acta Parasitologica* 42, 187–191.

1317. Odening, K., 1998. The present state of species-systematics in *Sarcocystis* Lankester, 1882 (Protista, Sporozoa, Coccidia). *Syst. Parasitol.* 41, 209–233.

1318. Odening, K., Aue, A., Ochs, A., Stolte, M., 1998. *Emmonsia crescens* (Ascomycotina) und *Sarcocystis ochotonae* n. sp. (Sporozoa) bei Pfeifhasen (*Ochotona*) aus China im Zoologischen Garten Berlin. *Zool. Garten* 68, 80–94.

1319. Odening, K., Wolf, P., Kellermann, F., Stolte, M., Bockhardt, I., 1998. *Sarcocystis phoeniconaii* (Sporozoa) beim Zwergflamingo—eine Art mit Einkapselung der Wirtszelle. *Zool. Garten* 68, 56–62.

1320. Odening, K., Rudolph, M., Quandt, S., Bengis, R.G., Bockhardt, I., Viertel, D., 1998. *Sarcocystis* spp. in antelopes from southern Africa. *Acta Protozool.* 37, 149–158.

1321. Odening, K., Stolte, M., Bockhardt, I., 1999. Einheimische *Sarcocystis*-Arten (Sporozoa) in exotischen Zoosäugetieren (Gayal, Cerviden, Cameliden, Ozelot). *Zoologische Garten* 69, 109–125.

1322. Ohbayashi, M., Kitamura, Y., 1959. *Sarcocystis clethrionomysi* n. sp. from *Clethrionomys rufocanus bedfordiae* Thomas. *Jap. J. Vet. Res.* 7, 115–118.

1323. Ohno, A., Shinzato, T., Sueyoshi, T., Tominaga, M., Arakaki, M., Shiroma, S., Ohshiru, K., Saito, M., Itagaki, H., 1993. Prevalence of *Sarcocystis* infection pigs in Okinawa Prefecture. *J. Jpn. Vet. Med. Assoc.* 46, 979–982. (In Japanese).

1324. Olafson, P., Monlux, W.S., 1942. *Toxoplasma* infection in animals. *Cornell Vet.* 32, 176–190.

1325. Olias, P., Gruber, A.D., Heydorn, A.O., Kohls, A., Mehlhorn, H., Hafez, H.M., Lierz, M., 2009. A novel *Sarcocystis*-associated encephalitis and myositis in racing pigeons. *Avian Pathol.* 38, 121–128.

1326. Olias, P., Olias, L., Lierz, M., Mehlhorn, H., Gruber, A.D., 2010. *Sarcocystis calchasi* is distinct to *Sarcocystis columbae* sp. nov. from the wood pigeon (*Columba palumbus*) and *Sarcocystis* sp. from the sparrowhawk (*Accipiter nisus*). *Vet. Parasitol.* 171, 7–14.

1327. Olias, P., Gruber, A.D., Hafez, H.M., Heydorn, A.O., Mehlhorn, H., Lierz, M., 2010. *Sarcocystis calchasi* sp. nov. of the domestic pigeon (*Columba livia* f. *domestica*) and the Northern goshawk (*Accipiter gentilis*): light and electron microscopical characteristics. *Parasitol. Res.* 106, 577–585.

1328. Olias, P., Gruber, A.D., Heydorn, A.O., Kohls, A., Hafez, H.M., Lierz, M., 2010. Unusual biphasic disease in domestic pigeons (*Columba livia* f. *domestica*) following experimental infection with *Sarcocystis calchasi*. *Avian Dis.* 54, 1032–1037.

1329. Olias, P., Gruber, A.D., Kohls, A., Hafez, H.M., Heydorn, A.O., Mehlhorn, H., Lierz, M., 2010. *Sarcocystis* species lethal for domestic pigeons. *Emerg. Infect. Dis.* 16, 497–499.

1330. Olias, P., Olias, L., Krücken, J., Lierz, M., Gruber, A.D., 2011. High prevalence of *Sarcocystis calchasi* sporocysts in European *Accipiter* hawks. *Vet. Parasitol.* 175, 230–236.

1331. Olias, P., Meyer, A., Klopfleisch, R., Lierz, M., Kaspers, B., Gruber, A.D., 2013. Modulation of the host Th1 immune response in pigeon protozoal encephalitis caused by *Sarcocystis calchasi*. *Vet. Res.* 44, 10.

1332. Olias, P., Maier, K., Wünschmann, A., Reed, L., Armién, A.G., Shaw, D.P., Gruber, A.D., Lierz, M., 2014. *Sarcocystis calchasi* has an expanded host range and induces neurological disease in cockatiels (*Nymphicus hollandicus*) and North American rock pigeons (*Columbia livia* f. *dom*.). *Vet. Parasitol.* 200, 59–65.

1333. Olson, E.J., Wünschmann, A., Dubey, J.P., 2007. *Sarcocystis* sp.-associated meningoencephalitis in a bald eagle (*Haliaeetus leucocephalus*). *J. Vet. Diagn. Invest.* 19, 564–568.

1334. Omata, Y., Heydorn, A.O., Heidrich, H.G., Igarashi, I., Saito, A., Toba, H., Suzuki, N., 1993. Survey of *Sarcocystis* spp. infection in slaughtered pigs in East Hokkaido, Japan. *J. Protozool. Res.* 3, 29–30.

1335. Omata, Y., Xu, S.Z., Igarashi, I., Saito, A., Toba, H., Suzuki, N., 1994. Survey of *Sarcocystis* infection in cattle in east Hokkaido, Japan. *J. Vet. Anim. Sci.* 56, 557–558.

1336. Ono, K., Horn, K., Heydorn, A.O., 1991. Purification of *in vitro* excysted *Sarcocystis* sporozoites by passage through a DE 5 2 anion-exchange column. *Parasitol. Res.* 77, 717–719.

1337. Ono, M., Ohsumi, T., 1999. Prevalence of *Sarcocystis* spp. cysts in Japanese and imported beef (Loin: *Musculus longissimus*). *Parasitol. Int.* 48, 91–94.

1338. Oryan, A., Moghaddar, N., Gaur, S.N.S., 1996. The distribution pattern of *Sarcocystis* species, their transmission and pathogenesis in sheep in Fars Province of Iran. *Vet. Res. Commun.* 20, 243–253.

1339. Oryan, A., Ahmadi, N., Mousavi, S.M.M., 2010. Prevalence, biology, and distribution pattern of *Sarcocystis* infection in water buffalo (*Bubalus bubalis*) in Iran. *Trop. Anim. Health Prod.* 42, 1513–1518.

1340. Oryan, A., Sharifiyazdi, H., Khordadmehr, M., Larki, S., 2011. Characterization of *Sarcocystis fusiformis* based on sequencing and PCR-RFLP in water buffalo (*Bubalus bubalis*) in Iran. *Parasitol. Res.* 109, 1563–1570.

1341. Özer, E., 1984. An experimental investigation on the life cycle and pathogenicity of *Sarcocystis capracanis* (Fiescher, 1979). *Ankara Üniv. Vet. Fak. Derg.* 31, 431–451.

1342. Ovsepyan, L.A., 1987. Occurrence of *Sarcocystis* in Armenian mouflons (*Ovis orientalis* gmelini Blyth, 1841 = Armeni ana Nasonov) kept in captivity. *Biologicheskii Zhurnal Armenii* 40, 780–782. (In Russian).

1343. Özer, E., 1988. Elažig mezbahasinda kesilen sigir ve mandalarda *Sarcocystis* turleri ve incidensi uzerinde arastirmalar. *Doga Tu Vet. ve Hay.* 12, 130–139.

1344. Özer, E., Saki, C.E., Dündar, B., 1995. Tükiye'de tektirnaklilarda bulunan *Sarcocystis* türleri. *Turk. J. Vet. Anim. Sci.* 19, 177–180.

1345. Özkayhan, M.A., Karaer, Z., İlkme, A.N., Atmaca, H.T., 2007. The prevalence of *Sarcocystis* species in sheep slaughtered in municipality slaughterhouse in Kirikkale. *Türkiye Parazitol. Derg.* 31, 272–276. (In Turkish).

1346. Özmen, O., Sahinduran, S., Haligür, M., Yukari, B.A., Dorrestein, G.M., 2009. Encephalitic sarcocystosis and its prophylactic treatment in sheep. *Turk. J. Vet. Anim. Sci.* 33, 151–155.

1347. Pacheco, N.D., Fayer, R., 1977. Fine structure of *Sarcocystis cruzi* schizonts. *J. Protozool.* 24, 382–388.

1348. Pacheco, N.D., Sheffield, H.G., Fayer, R., 1978. Fine structure of immature cysts of *Sarcocystis cruzi*. *J. Parasitol.* 64, 320–325.

1349. Pacheco, R.G., Botelho, G.G., Lopes, C.W.G., 1990. Coagulograma de caprinos experimentalmente infectados com *Sarcocystis capracanis* Fischer, 1979 (Apicomplexa: Sarcocystidae). *Arq. Univ. Fed. Rur. Rio de Janeiro* 13, 17–35.

1350. Pacheco, R.G., Botelho, G.G., 1991. Bilirrubina total, bilirrubina direta e bilirrubina indireta de caprinos experimentalmente infectados com *Sarcocystis capracanis* Fischer, 1979 (Apicomplexa: Sarcocystidae). *Arq. Univ. Fed. Rur. Rio de Janeiro* 14, 13–25.

1351. Pacheco, R.G., Botelho, G.G., Lopes, C.W.G., 1991. Aspectos anátomo-histopatológicos em caprinos experimentalmente infectados com *Sarcocystis capracanis* Fischer, 1979 (Apicomplexa: Sarcocystidae). *Arq. Univ. Fed. Rur. Rio de Janeiro* 14, 189–198.

1352. Pacheco, R.G., Botelho, G.G., 1992. ASAT, CK-MB e LDH de caprinos experimentalmente infectados com *Sarcocystis capracanis* Fischer, 1979 (Apicomplexa: Sarcocystidae). *Arq. Univ. Fed. Rur. Rio de Janeiro* 15, 71–78.

1353. Pacheco, R.G., Botelho, G.G., Lopes, C.W.G., 1992. Aspectos parasitológicos de *Sarcocystis capracanis* Fischer, 1979 (Apicomplexa: Sarcocystidae) em caprinos experimentalmente infectados. *Arq. Univ. Fed. Rur. Rio de Janeiro* 15, 27–37.

1354. Page, C.D., Schmidt, R.E., Hubbard, G.B., Langlinais, P.C., 1989. *Sarcocystis* myocarditis in a red lory (*Eos bornea*). *J. Zoo Wildl. Med.* 20, 461–464.

1355. Pak, L.S., 1986. Ultrastructure of the cysts of a *Sarcocystis* sp. from *Citellus fulvus*. *Izvestiya Akademii nauk Kazakhoskoi SSR, Biologicheskaya* 5, 42–46. (In Russian).

1356. Pak, S.M., Perminova, V.V., Yeshtokina, N.V., 1979. *Sarcocystis citellivulpes* sp. n. from the yellow suslik *Citellus fulvus* Lichtenstain, 1923. In: Beyer, T.V., Bezukladnikova, N.A., Galuzo, I.G., Konovalova, S.I., Pak, S.M. (Eds.), *Toksoplazmidy, Protozoologiya*. Akad. Nauk. SSR, Moscow, Russia, pp. 111–114. (In Russian).

1357. Pak, S.M., Sklyarova, O.N., Pak, L.S., 1989. *Sarcocystis alectorivulpes* new-species and *Sarcocystis alectoributeonis* new sarcosporidian species of the chukar (*Alectoris chukar*). *Izvest. Akad. Nauk. Kazakh. Ser. Biol.* 25–30. (In Russian).

1358. Pak, S.M., Pak, L.S., Skylarova, O.N., 1989. *Sarcocystis citellibuteonis* new species of Sarcosporidia from the large-toothed suslik *Citellus fulvus*. *Izvest. Akad. Nauk. Kazakh. Ser. Biol.* 30–39. (In Russian).

1359. Pak, S.M., Sklyarova, O.N., Dymkova, N.D., 1991. Sarcocysts (Sporozoa, Apicomplexa) of some wild mammals. *Izvestiya Akademii Nauk Kazakhskoi SSR Seriya Biologicheskaya* 5, 35–40. (In Russian).

1360. Pak, S.P., 1979. Occurrence of *Sarcocystis* in *Vulpes corsac*. *Nauka Kazakhskoi SSR*, 133–134. (In Russian).

1361. Pamphlett, R., O'Donoghue, P., 1990. *Sarcocystis* infection of human muscle. *Aust. N. Z. J. Med.* 20, 705–707.

1362. Pandit, B.A., Garg, S.K., Bhatia, B.B., 1993. Preparation and assessment of precipitating antigens of *Sarcocystis cruzi* and *S. hirsuta*. *J. Vet. Parasitol.* 7, 5–15.

1363. Paperna, I., Finkelman, S., 1996. Early generation merogonies of *Sarcocystis muriviperae* in liver and muscles of white mice. *Folia Parasitol. (Praha)* 43, 91–99.

1364. Paperna, I., Finkelman, S., 1996. Ultrastructural study of *Sarcocystis muriviperae* development in the intestine of its snake hosts. *Folia Parasitol. (Praha)* 43, 13–19.

1365. Paperna, I., Martelli, P., 2000. Fine structural development of microgamonts of *Sarcocystis singaporensis* in *Python reticulatus*. *Parasitol. Res.* 86, 1022–1025.

1366. Paperna, I., Martelli, P., 2000. Fine structure of the development of *Sarcocystis singaporensis* in *Python reticulatus* from macrogamont to sporulated oocyst stage. *Parasite* 7, 193–200.

1367. Paperna, I., 2002. Ultrastructural study of the development of *Sarcocystis singaporensis* sarcocysts in the muscles of its rat host. *Parasite* 9, 161–166.

1368. Paperna, I., 2002. Fine structure of *Sarcocystis singaporensis* merogony stages preceding sarcocyst formation in the rat. *Parasitol. Res.* 88, 73–79.

1369. Parairo, J.R., Manuel, M.F., Icatlo, F.C.Jr., 1988. Ultrastructural studies of the cyst wall of *Sarcocystis* spp. in Philippine carabaos (*Bubalus bubalis*). *Phil. J. Vet. Anim. Sci.* 14, 40–54.

1370. Parenzan, P., 1947. Sarcosporidiosi (psorospermosi) da nuova specie (Prot.: *Sarcocystis atractaspidis* n. sp.) in rettile (*Atractaspis*). *Boll. Soc. Natur. Napoli* 55, 117–119.

1371. Parish, S.M., Maag-Miller, L., Besser, T.E., Weidner, J.P., McElwain, T., Knowles, D.P., Leathers, C.W., 1987. Myelitis associated with protozoal infection in newborn calves. *J. Am. Vet. Med. Assoc.* 191, 1599–1600.

1372. Park, Y.J., Kim, J.S., Joeung, D.S., Sin, M.K., Kim, K.S., Kim, T.J., 1994. Survey of *Sarcocystis* infections in slaughtered cattle and identification of *Sarcocystis cruzi*. *Korean J. Vet. Publ. Health* 18, 251–259. (In Korean).

1373. Parker, R.C., 1961. *Methods of Tissue Culture*. 3rd edn., Hoeber, New York, 145pp.

1374. Pathmanathan, P., Kan, S.P., 1981. Human *Sarcocystis* infection in Malaysia. *Southeast Asian J. Trop. Med. Pub. Health* 12, 247–250.

1375. Pathmanathan, R., Kan, S.P., 1987. Two cases of human sarcocystosis in East Malaysia. *Med. J. Malaysia* 42, 212–214.

1376. Pathmanathan, R., Jayalakshmi, P., Kan, S.P., 1988. A case of human *Sarcocystis* infection in Malaysia. *J. Malay. Soc. Hlth.* 6, 45–47.

1377. Pathmanathan, R., Kan, S.P., 1992. Three cases of human *Sarcocystis* infection with a review of human muscular sarcocystosis in Malaysia. *Trop. Geogr. Med.* 44, 102–108.

1378. Patton, W.S., Hindle, E., 1926. Notes on three new parasites of the striped hamster (*Cricetulus griseus*). *Proc. R. Soc. London, Series B* 100, 387–390.

1379. Pavlíček, A., Hrdá, Š., Flegr, J., 1999. FreeTree—Freeware program for construction of phylogenetic trees on the basis of distance data and bootstrap/jackknife analysis of the tree robustness. Application in the RAPD analysis of genus *Frenkelia*. *Folia Biologica* 45, 97–99.

1380. Pecka, Z., 1988. Parasitic protozoa of the genus *Sarcocystis* (Apicomplexa: Sarcocystidae) in pheasants, hens, and some free-living birds in Czechoslovakia. *Vest. Cs. Spolec. Zool.* 52, 266–270.

1381. Pecka, Z., 1990. Muscular sarcocystosis of fowls and pheasants in Czechoslovakia. *Veterinarstvi* 40, 314–315.

1382. Pena, H.F.J., Ogassawara, S., Sinhorini, I.L., 2001. Occurrence of cattle *Sarcocystis* species in raw kibbe from Arabian food establishments in the city of São Paulo, Brazil, and experimental transmission to humans. *J. Parasitol.* 87, 1459–1465.

1383. Pereira Lorenzo, A., 1979. Incidencia y diagnostico de *Sarcocystis miescheriana* (Kühn, 1865) Lankester, 1882. *Rev. Ibér. Parasitol.* 79, 401–409.

1384. Pereira, A., Bermejo, M., 1988. Prevalence of *Sarcocystis* cysts in pigs and sheep in Spain. *Vet. Parasitol.* 27, 353–355.

1385. Pereira, A., Wirz, S., 1989. Diagnóstico serológico de la sarcosporidiosis porcina mediante técnicas ELISA y HAI. Comparación con otras técnicas de diagnóstico. *Med. Vet.* 6, 541–542. (In Spanish).

1386. Pérez-Creo, A., Panadero, R., López, C., Díaz, P., Vázquez, L., Díez-Baños, P., Morrondo, P., 2013. Prevalence and identity of *Sarcocystis* spp. in roe deer (*Capreolus capreolus*) in Spain: A morphological study. *Res. Vet. Sci.* 95, 1036–1040.

1387. Permin, A., Yelifari, L., Bloch, P., Steenhard, N., Hansen, N.P., Nansen, P., 1999. Parasites in cross-bred pigs in the upper east region of Ghana. *Vet. Parasitol.* 87, 63–71.

1388. Perovic, M., 1991. Epidemiology and diagnosis of sarcocystosis. *Southeast Asian J. Trop. Med. Pub. Health* 22, 135–137.

1389. Perrotin, C., Graber, M., Thal, J., Petit, J.P., 1978. La sarcosporidiose chez le buffle africain (*Syncerus caffer*). *Rev. Élev. Méd. vét. Pays trop.* 31, 423–426.

1390. Pescador, C.A., Corbellini, L.G., de Oliveira, E.C., Bandarra, P.M., Leal, J.S., Pedroso, P.M.O., Driemeier, D., 2007. Aborto ovino associado com infecção por *Sarcocystis* sp. *Pesq. Vet. Bras.* 27, 393–397.

1391. Pessôa, S.B., 1935. *Sarcocystis oliverioi* n. sp., parasita do "tuim" (*Forpus passerinus* L.). *Folia Clin. Biol.* 4, 162–164.

1392. Pethkar, D.K., Shah, H.L., 1982. Attempted cross transmission of *Sarcocystis capracanis* of the goat to the sheep. *Indian Vet. J.* 59, 766–768.

1393. Phillips, P.H., Ford, G.E., 1987. Clinical, haematogical and plasma biochemical changes in specified-pathogen-free (Sporozoa) lambs experimentally infected with low numbers of *Sarcocystis tenella* sporocysts. *Vet. Parasitol.* 24, 15–23.

1394. Piekarski, G., Heydorn, A.O., Aryeetey, M.E., Hartlapp, J.H., Kimmig, P., 1978. Klinische, parasitologische und serologische Untersuchungen zur Sarkosporidiose (*Sarcocystis suihominis*) des Menschen. *Immun. Infekt.* 6, 153–159.

1395. Pitel, P.H., Pronost, S., Gargala, G., Anrioud, D., Toquet, M.-P., Foucher, N., Collobert-Laugier, C., Fortier, G., Ballet, J.-J., 2002. Detection of *Sarcocystis neurona* antibodies in French horses with neurological signs. *Int. J. Parasitol.* 32, 481–485.

1396. Pitel, P.H., Lindsay, D.S., Caure, S., Romand, S., Pronost, S., Gargala, G., Mitchell, S.M., Hary, C., Thulliez, P., Fortier, G., Ballet, J.J., 2003. Reactivity against *Sarcocystis neurona* and *Neospora* by serum antibodies in healthy French horses from two farms with previous equine protozoal myeloencephalitis-like cases. *Vet. Parasitol.* 111, 1–7.

1397. Pivoto, F.L., de Macêdo Junior, A.G., da Silva, M.V., Ferreira, F.B., Silva, D.A.O., Pompermayer, E., Sangioni, L.A., Mineo, T.W.P., Vogel, F.S.F., 2014. Serological status of mares in parturition and the levels of antibodies (IgG) against protozoan family Sarcocystidae from their pre colostral foals. *Vet. Parasitol.* 199, 107–111.

1398. Plotkowiak, J., 1973. Opis pierwszego przypadku inwazji *Isospora hominis* w Polsce. *Wiad. Parazytol.* 19, 725–730.

1399. Plotkowiak, J., 1974. Inwazja *Isospora hominis* nowy problem parazytologiczny w Polsce. *Roczniki Pomorskiej Akademii Medycznej im. K. Swierczewskiego w Szczecinie, Supl.*, 10, 53–55.

1400. Kutty, M.K., Latif, B., Muslim, A., Hussaini, J., Daher, A.M., Heo, C.C., Abdullah, S., 2015. Detection of sarcocystosis in goats in Malaysia by light microscopy, histology, and PCR. *Trop Anim. Health Prod.* 47, 751–756.

1401. Plotkowiak, J., 1976. Wyniki dalszych badań nad wystepowaniem i epidemiologia inwazji *Isospora hominis* (Railliet i Lucet, 1891). *Wiad. Parazytol.* 22, 137–147.

1402. Pober, J.S., Collins, T., Gimbrone, J., Libby, P., Reiss, C.S., 1986. Inducible expression of class II major histocompatibility complex antigens and the immunogenicity of vascular endothelium. *Transplantation* 41, 141–146.

1403. Pohl, U., Dubremetz, J.F., Entzeroth, R., 1989. Characterization and immunolocalization of the protein contents of micronemes of *Sarcocystis muris* cystozoites (Protozoa: Apicomplexa). *Parasitol. Res.* 75, 199–205.

1404. Poli, A., Mancianti, F., Marconcini, A., Nigro, M., Colagreco, R., 1988. Prevalence, ultrastructure of the cyst wall and infectivity for the dog and cat of *Sarcocystis* sp. from fallow deer (*Cervus dama*). *J. Wildl. Dis.* 24, 97–104.

1405. Pomroy, W.E., Charleston, W.A.G., 1987. Prevalence of dog-derived *Sarcocystis* spp. in some New Zealand lambs. *N. Z. Vet. J.* 35, 141–142.

1406. Porchet, E., Torpier, G., 1977. Etude du germe infectieux de *Sarcocystis tenella* et *Toxoplasma gondii* par la technique du cryodécapage. *Z. Parasitenkd.* 54, 101–124.

1407. Porchet-Henneré, E., Ponchel, G., 1974. Quelques precisions sur l'ultrastructure de *Sarcocystis tenella*: l'architecture du kyste et l'aspect des endozoïtes en microscopie electronique à balayage. *C. R. Acad. Sci. Hebd. Seances Acad. Sci. D* 279, 1179–1181.

1408. Porchet-Hennere, E., 1975. Quelques précisions l' ultrastructure de *Sarcocystis tenella*. I. L'endozoïte (après coloration négative). *J. Protozool.* 22, 214–220.

1409. Porter, R.A., Ginn, P.E., Dame, J.B., Greiner, E.C., 2001. Evaluation of the shedding of *Sarcocystis falcatula* sporocysts in experimentally infected Virginia opossums (*Didelphis virginiana*). *Vet. Parasitol.* 95, 313–319.

1410. Poulsen, C.S., Stensvold, C.R., 2014. Current status of epidemiology and diagnosis of human sarcocystosis. *J. Clin. Microbiol.* 52, 3524–3530.

1411. Powell, E.C., McCarley, J.B., 1975. A murine *Sarcocystis* that causes an *Isospora*-like infection in cats. *J. Parasitol.* 61, 928–931.

1412. Powell, E.C., Pezeshkpour, G., Dubey, J.P., Fayer, R., 1986. Types of myofibers parasitized in experimentally induced infections with *Sarcocystis cruzi* and *Sarcocystis capracanis*. *Am. J. Vet. Res.* 47, 514–517.

1413. Pozov, S.A., Komarova, L.N., Letov, I.I., Kharchenko, Y.M., 1994. Study of pathology in pigs infected with *Sarcocystis* alone or associated with other infections. In: *Diagnosis, Treatment, and Prophylaxis of Diseases in Farm Animals*. Stavropol Agricultural Institute, Stavropol, Turkmenistan, Vol. 1, pp. 20–22. (In Russian).

1414. Prakas, P., 2011. Diversity and ecology of *Sarcocystis* in Lithuanian game fauna. In: *Summary of doctoral dissertation, Biomedical Sciences, Ecology and Environmental Science (03B)*. Institute of Ecology of Nature Research Centre, Vilnius University, Lithuania, pp. 1–41.

1415. Prakas, P., Kutkiené, L., Sruoga, A., Butkauskas, D., 2011. *Sarcocystis* sp. from the herring gull (*Larus argentatus*) identity to *Sarcocystis wobeseri* based on cyst morphology and DNA results. *Parasitol. Res.* 109, 1603–1608.

1416. Prakas, P., Butkauskas, D., Sruoga, A., Švažas, S., Kutkiené, L., 2011. Identification of *Sarcocystis columbae* in wood pigeons (*Columba palumbus*) in Lithuania. *Vet. Med. Zoot.* 55, 33–39.

1417. Prasad, B.N., Jorgensen, R.J., 1976. Isolation and identification of *Sarcocystis* from adult cattle of Denmark. *Phil. J. Vet. Anim. Sci.* 17, 197–202.

1418. Prakas, P., Butkauskas, D., 2012. Protozoan parasites from genus *Sarcocystis* and their investigations in Lithuania. *Ekologija* 58, 45–58.

1419. Prakas, P., Kutkiené, L., Butkauskas, D., Sruoga, A., Žalakevičius, M., 2013. Molecular and morphological investigations of *Sarcocystis corvusi* sp. nov. from the jackdaw (*Corvus monedula*). *Parasitol. Res.* 112, 1163–1167.

1420. Prakas, P., Kutkienė, L., Butkauskas, D., Sruoga, A., Žalakevičius, M., 2014. Description of *Sarcocystis lari* sp. n. (Apicomplexa: Sarcocystidae) from the great black-backed gull, *Larus marinus* (Charadriiformes: Laridae), on the basis of cyst morphology and molecular data. *Folia Parasitol. (Praha)* 61, 11–17.

1421. Prakas, P., Oksanen, A., Butkauskas, D., Sruoga, A., Kutkienė, L., Švažas, S., Isomursu, M., Liaugaudaite, S., 2014. Identification and intraspecific genetic diversity of *Sarcocystis rileyi* from ducks, *Anas* spp., in Lithuania and Finland. *J. Parasitol.* 100, 657–661.

1422. Prasse, K.W., Fayer, R., 1981. Hematology of experimental acute *Sarcocystis bovicanis* infection in calves. II. Serum biochemistry and hemostasis studies. *Vet. Pathol.* 18, 358–367.

1423. Prathap, K., 1973. Letter: *Sarcocystis* in the Malaysian long-tailed monkey, *Macaca irus*. *Trans. R. Soc. Trop. Med. Hyg.* 67, 615.

1424. Prestwood, A.K., Cahoon, R.W., McDaniel, H.T., 1980. *Sarcocystis* infections in Georgia swine. *Am. J. Vet. Res.* 41, 1879–1881.

1425. Prickett, M.D., Prestwood, A.K., Hoenig, M., 1992. Lipid-metabolism and *Sarcocystis miescheriana* infection in growing swine. *Vet. Parasitol.* 42, 41–51.

1426. Prickett, M.D., Latimer, A.M., McCusker, R.H., Hausman, G.J., Prestwood, A.K., 1992. Alterations of serum insulin-like growth factor-I (IgF-I) and IgF-binding proteins (IGFBPS) in swine infected with the protozoan parasite *Sarcocystis miescheriana*. *Dom. Anim. Endocrinol.* 9, 285–296.

1427. Pritt, B., Trainer, T., Simmons-Arnold, L., Evans, M., Dunams, D., Rosenthal, B.M., 2008. Detection of *Sarcocystis* parasites in retail beef: A regional survey combining histological and genetic detection methods. *J. Food Prot.* 71, 2144–2147.

1428. Pucak, G.J., Johnson, D.K., 1972. *Sarcocystis* in Patas monkey (*Erythrocebus patas*). *Lab. Anim. Digest.* 8, 36–39.

1429. Purcherea, A., Radu, A., Neda, M., 1981. Cercetari asupra sarcosporidiilor de la bubaline. *Lucrari stiintifice I. A. N. B. seria C* 24, 55–58.

1430. Pusterla, N., Wilson, W.D., Conrad, P.A., Barr, B.C., Ferraro, G.L., Daft, B.M., Leutenegger, C.M., 2006. Cytokine gene signatures in neural tissue of horses with equine protozoal myeloencephalitis or equine herpes type 1 myeloencephalopathy. *Vet. Rec.* 159, 341–346.

1431. Pusterla, N., Packham, A., Wilson, W.D., White, A., Bellamy, P., Renier, A.C., Conrad, P.A., 2013. Short communication: Evaluation of the kinetics of antibodies against *Sarcocystis neurona* in serum from seropositive healthy horses without neurological deficits treated with ponazuril paste. *Vet. Rec.* 173, 249.

1432. Dubey, J.P, Verma, S.K., Dunams, D., Calero-Bernal, R., Rosenthal, B.M., 2015. Molecular and development of *Sarcocystis speeri* sarcocysts in gamma interferon gene knock out mice. In press.

1433. Pusterla, N., Tamez-Trevino, E., White, A., VanGeem, J., Packham, A., Conrad, P.A., Kass, P., 2014. Comparison of prevalence factors in horses with and without seropositivity to *Neospora hughesi* and/or *Sarcocystis neurona*. *Vet. J.* 200, 332–334.

1434. Quandt, S., Bengis, R.G., Stolte, M., Odening, K., Bockhardt, I., 1997. *Sarcocystis* infection of the African buffalo (*Syncerus caffer*) in the Krüger National Park, South Africa. *Acta Parasitologica* 42, 68–73.

1435. Quiroga, D.A., Lombardero, O.J., Zorrilla, R., 1969. *Sarcocystis tilopodi* n. sp. en guanacos (*Lama guanicoe*) de la Republica Argentina. *Gac. Vet. (Bs. As.)* 31, 67–70.

1436. Rabinovich, G.A., Baum, L.G., Tinari, N., Paganelli, R., Natoli, C., Liu, F.T., Iacobelli, S., 2002. Galectins and their ligands: Amplifiers, silencers or tuners of the inflammatory response? *Trends Immunol.* 23, 313–320.

1437. Rachinel, N., Buzoni-Gatel, D., Dutta, C., Mennechet, F.J.D., Luangsay, S., Minns, L.A., Grigg, M.E., Tomavo, S., Boothroyd, J.C., Kasper, L.H., 2004. The induction of acute ileitis by a single microbial antigen of *Toxoplasma gondii*. *J. Immunol.* 173, 2725–2735.

1438. Radchenko, A.I., Gaibova, G.D., 1989. Light and electron microscopic study of the cystic stages of *Sarcocystis fusiformis* (Sporozoa, Apicomplexa) from buffalo. *Parazitologiya* 23, 75–78. (In Russian).

1439. Radchenko, A.I., 1991. Cytochemical study of polysaccharides in coccidia from the genus *Sarcocystis*. 1. Sulfated glycosaminoglycans in the sarcocysts of *Sarcocystis muris*. *Tsitologiya* 33, 95–100. (In Russian).

1440. Radchenko, A.I., 1992. Cytochemical detection of glycoproteins and glycolipids in the sarcocysts of *Sarcocystis muris*. *Tsitologiya* 34, 91–96. (In Russian).

1441. Radchenko, A.I., Beĭer, T.V., Gaibova, G.D., 1995. An electron microscopic study of the natural death of cyst cells in Sarcosporidia. I. Ultrastructural changes in the cytoplasm of the cyst cells of *Sarcocystis muris* and *S. bovifelis. Tsitologiya* 37, 525–532. (In Russian).

1442. Radchenko, A.I., Beĭer, T.V., 2006. The ultrastructural changes of *Sarcocystis ovifelis* infested sheep tongue myofibrils. *Tsitologiya* 48, 669–673. (In Russian).

1443. Rakich, P.M., Dubey, J.P., Contarino, J.K., 1992. Acute hepatic sarcocystosis in a chinchilla. *J. Vet. Diagn. Invest.* 4, 484–486.

1444. Ramirez Romero, R., Nevarez Garza, A.M., Mateos Poumian, A., 1990. Sarcocistosis muscular en un bovino. *Vet. Méx.* 21, 419–424.

1445. Ramos-Vara, J.A., Dubey, J.P., Watson, G.L., Winn-Elliot, M., Patterson, J.S., Yamini, B., 1997. Sarcocystosis in mink (*Mustela vison*). *J. Parasitol.* 83, 1198–1201.

1446. Rangarao, G.S.C., Sharma, R.L., Hemaprasanth, 1994. Parasitic infections of Indian yak *Bos (Poephagus) grunniens*-an overview. *Vet. Parasitol.* 53, 75–82.

1447. Rangarao, G.S.C., Sharma, R.L., Shah, H.L., 1997. Occurrence of *Sarcocystis* in the camel (*Camelus dromedarius*) in India. *Indian Vet. J.* 74, 426.

1448. Rao, G., Rao, P.R., 1987. A note on sarcocystosis in domestic animals of Andhra Pradesh. *Indian Vet. J.* 64, 614–615.

1449. Rassouli, M., Ahmadpanahi, J., Alvandi, A., 2014. Prevalence of *Sarcocystis* spp. and *Hammondia* spp. microcysts in esophagus tissue of sheep and cattle, emphasized on their morphological differences. *Parasitol. Res.* 113, 3801–3805.

1450. Rátz, I., 1908. Szakosztályunk ülései. *Állattani Közlemények* 7, 177–178.

1451. Rátz, I., 1909. Az izmokban élösködö véglények és a magyar faunában elöforduló fajaik. *Allattani Kozlemenyek* 8, 1–30.

1452. Reddy, M.J., Chetty, M.S., Reddy, M.S., Reddy, G.R., Kulkarni, D., 1990. Seroprevalence of sarcocystosis in buffaloes in Hyderabad and Secunderabad. *J. Vet. Anim. Sci.* 21, 32–34.

1453. Reed, S.M., Howe, D.K., Morrow, J.K., Graves, A., Yeargan, M.R., Johnson, A.L., MacKay, R.J., Furr, M., Saville, W.J., Williams, N.M., 2013. Accurate antemortem diagnosis of equine protozoal myeloencephalitis (EPM) based on detecting intrathecal antibodies against *Sarcocystis neurona* using the SnSAG2 and SnSAG4/3 ELISAs. *J. Vet. Intern. Med.* 27, 1193–1200.

1454. Rehbein, S., Visser, M., 2007. Die Endoparasiten des Sikawildes (*Cervus nippon*) in Österreich. *Wien. Klin. Wschr.* 126, S27–S41.

1455. Rehbein, S., Visser, M., Jekel, I., Silaghi, C., 2014. Endoparasites of the fallow deer (*Dama dama*) of the Antheringer Au in Salzburg, Austria. *Wien. Klin. Wochenschr.* 126(Supp. 1), S37–S41.

1456. Reid, A.J., Vermont, S.J., Cotton, J.A., Harris, D., Hill-Cawthorne, G.A., Könen-Waisman, S., Latham, S.M. et al. 2012. Comparative genomics of the apicomplexan parasites *Toxoplasma gondii* and *Neospora caninum*: Coccida differing in host range and transmission strategy. *PLoS Pathog.* 8, e1002567.

1457. Reiner, G., Eckert, J., Peischl, T., Bochert, S., Jäkel, T., Mackenstedt, U., Joachim, A., Daugschies, A., Geldermann, H., 2002. Variation in clinical and parasitological traits in Pietrain and Meishan pigs infected with *Sarcocystis miescheriana. Vet. Parasitol.* 106, 99–113.

1458. Reiner, G., Hepp, S., Hertrampf, B., Kliemt, D., Mackenstedt, U., Daugschies, A., Zahner, H., 2007. Genetic resistance to *Sarcocystis miescheriana* in pigs following experimental infection. *Vet. Parasitol.* 145, 2–10.

1459. Reiner, G., Kliemt, D., Willems, H., Berge, T., Fischer, R., Köhler, F., Hepp, S. et al. 2007. Mapping of quantitative trait loci affecting resistance/susceptibility to *Sarcocystis miescheriana* in swine. *Genomics* 89, 638–646.

1460. Reiten, A.C., Jensen, R., Griner, L.A., 1966. Eosinophilic myositis (sarcosporidiosis; sarco) in beef cattle. *Am. J. Vet. Res.* 27, 903–906.

1461. Reiter, I., Weiland, G., Roscher, B., Meyer, J., Frahm, K., 1981. Versuche zum serologischen Nachweis der Sarkosporidiose an experimentell mit Sarkosporidien infizierten Rindern und Schafen. *Berl. Münch. Tierärztl. Wschr.* 94, 425–430.

1462. Reiter, I., Mareis, A., 1986. Zur Differenzierung von *Sarcocystis muris-* und *S.dispersa*-Infektionen der Maus mittels isoelektrischer Fokussierung und Immunoassays. *Dtsch. Tierärztl. Wschr.* 93, 433–437.

1463. Rejmanek, D., VanWormer, E., Miller, M.A., Mazet, J.A.K., Nichelason, A.E., Melli, A.C., Packham, A.E., Jessup, D.A., Conrad, P.A., 2009. Prevalence and risk factors associated with *Sarcocystis neurona* infections in opossums (*Didelphis virginiana*) from central California. *Vet. Parasitol.* 166, 8–14.

1464. Rejmanek, D., Miller, M.A., Grigg, M.E., Crosbie, P.R., Conrad, P.A., 2010. Molecular characterization of *Sarcocystis neurona* strains from opossums (*Didelphis virginiana*) and intermediate hosts from Central California. *Vet. Parasitol.* 170, 20–29.

1465. Resendes, A.R., Juan-Sallés, C., Almería, S., Majó, N., Domingo, M., Dubey, J.P., 2002. Hepatic sarcocystosis in a striped dolphin (*Stenella coeruleoalba*) from the Spanish Mediterranean coast. *J. Parasitol.* 88, 206–209.

1466. Rezakhani, A., Cheema, A.H., Edjtehadi, M., 1977. Second degree atrioventricular block and sarcosporidioses in sheep. *Zbl. Vet. Med. A* 24, 258–262.

1467. Rickard, L.G., Black, S.S., Rashmir-Raven, A., Hurst, G., Dubey, J.P., 2001. Risk factors associated with the presence of *Sarcocystis neurona* sporocysts in opossums (*Didelphis virginiana*). *Vet. Parasitol.* 102, 179–184.

1468. Rickard, M.D., Munday, B.L., 1976. Host specificity of *Sarcocystis* spp. in sheep and cattle. *Aust. Vet. J.* 52, 48.

1469. Rifaat, M.A., Salem, S.A., Khalil, H.M., Azab, M.E., Abdel Baki, M.H., Abdel Ghaffar, F.M., 1976. The epidemiology of *Sarcocystis muris* in rodents from Egypt. *J. Egypt. Pub. Health Assoc.* 51, 321–329.

1470. Rifaat, M.A., Salem, S.A., Khalil, H.M., Azab, M.E., Abdel Ghaffar, F.M., Abdel Baki, M.H., 1978. A comparative morphological study of *Sarcocystis muris* in *Mus musculus* and experimentally infected mice. *J. Egypt. Soc. Parasitol.* 8, 353–361.

1471. Rifaat, M.A., Salem, S.A., Khalil, H.M., Azab, M.E., Abdel Baki, M.H., Abdel Ghaffar, F.M., 1978. A study on the life cycle of *Sarcocystis muris*. *J. Egypt. Pub. Health Assoc.* 53, 341–357.

1472. Rimaila-Pärnänen, E., Nikander, S., 1980. Generalized eosinophilic myositis with sarcosporidiosis in a Finnish cow. *Nord. Vet. Med.* 32, 96–99.

1473. Rimoldi, G., Speer, B., Wellehan, J.F.X., Jr., Bradway, D.S., Wright, L., Reavill, D., Barr, B.C., Childress, A., Shivaprasad, H.L., Chin, R.P., 2013. An outbreak of *Sarcocystis calchasi* encephalitis in multiple psittacine species within an enclosed zoological aviary. *J. Vet. Diagn. Invest.* 25, 775–781.

1474. Roberts, J.F., Wellehan, J.F.X., Weisman, J.L., Rush, M., Childress, A.L., Lindsay, D.S., 2014. Massive muscular infection by a *Sarcocystis* species in a South American rattlesnake (*Crotalus durissus terrificus*). *J. Parasitol.* 101, 386–389.

1475. Rodrigues, J.S., Meireles, G.S., Carvalho Filho, P.R., Ribeiro, C.T., Flausino, W., Lopes, C.W.G., 2008. *Sarcocystis cruzi* (Apicomplexa: Sarcocystidae) no cachorro-do-mato (*Cerdocyon thous*). *Pesq. Vet. Bras.* 28, 561–564.

1476. Rohini, K., Hafeez, M., 2005. Serodiagnosis of sarcocystosis in buffaloes. *Indian Vet. J.* 82, 330–331.

1477. Romero J.J., Chávez V.A., Casas A.E., Maturano H.L., Puray C.N., Chileno M.M., 2012. Respuesta inmune en conejos a dos tamaños de quistes de *Sarcocystis aucheniae*. *Rev. Inv. Vet. Perú* 23, 220–227.

1478. Rommel, M., Heydorn, A.O., 1972. Beiträge zum Lebenszyklus der Sarkosporidien. III. *Isospora hominis* (Railliet und Lucet, 1891) Wenyon, 1923, eine Dauerform der Sarkosporidien des Rindes und des Schweins. *Berl. Münch. Tierärztl. Wschr.* 85, 143–145.

1479. Rommel, M., Heydorn, A.O., Gruber, F., 1972. Beiträge zum Lebenszyklus der Sarkosporidien. I. Die Sporozyste von *S. tenella* in den Fäzes der Katze. *Berl. Münch. Tierärztl. Wschr.* 85, 101–105.

1480. Rommel, M., Heydorn, A.O., Fischle, B., Gestrich, R., 1974. Beiträge zum Lebenszyklus der Sarkosporidien. V. Weitere Endwirte der Sarkosporidien von Rind, Schaf und Schwein und die Bedeutung des Zwischenwirtes für die Verbreitung dieser Parasitose. *Berl. Münch. Tierärztl. Wschr.* 87, 392–396.

1481. Rommel, M., Geisel, O., 1975. Untersuchungen über die Verbreitung und de Lebenszyklus einer Sarkosporidienart des Pferdes (*Sarcocystis equicanis* n. spec.). *Berl. Münch. Tierärztl. Wschr.* 88, 468–471.

1482. Rommel, M., Krampitz, H.E., Geisel, O., 1977. Beiträge zum Lebenszyklus der Frenkelien. III. Die sexuelle Entwicklung von *F. clethrionmyobuteonis* im Mäusebussard. *Z. Parasitenkd.* 51, 139–146.

1483. Rommel, M., Krampitz, H.E., 1978. Weitere Untersuchungen über das Zwischenwirtsspektrum und de Entwicklungszklus von *Frenkelia microti* aus der Erdmaus. *Zbl. Vet. Med. B* 25, 273–281.

1484. Rommel, M., 1979. Das Frettchen (*Putorius putorius furo*) ein zusätzlicher Endwirt für *Sarcocystis muris*. *Z. Parasitenkd.* 58, 187–188.

1485. Rommel, M., Heydorn, A.O., Erber, M., 1979. Die Sarkosporidiose der Haustiere und des Menschen. *Berl. Münch. Tierärztl. Wochenschr.* 92, 457–464.

1486. Rommel, M., Schwerdtfeger, A., Blewaska, S., 1981. The *Sarcocystis muris*-infection as a model for research on the chemotherapy of acute sarcocystosis of domestic animals. *Zbl. Bakt. Hyg.,I. Abt. Orig. A* 250, 268–276.

1487. Rommel, M., 1983. Integrated control of protozoan diseases of livestock. In: Dunsmore, J.D. (Ed.), *Tropical Parasitoses and Parasitic Zoonoses*. Proceedings of the 10th Conference of the World Association for the Advancement of Veterinary Parasitology, Perth, Australia, pp. 9–30.

1488. Rommel, M., Schnieder, T., 1985. Epizootiologie und Bekämpfung der Sarkozystose der Schweine. *Angew. Parasitol.* 26, 39–42.

1489. Ronen, N., 1992. Putative equine protozoal myeloencephalitis in an imported Arabian filly. *J. S. Afr. Vet. Assoc.* 63, 78–79.

1490. Rooney, A.L., Limon, G., Vides, H., Cortez, A., Guitian, J., 2014. *Sarcocystis* spp. in llamas (*Lama glama*) in Southern Bolivia: A cross sectional study of the prevalence, risk factors and loss in income caused by carcass downgrades. *Prev. Vet. Med.* 116, 296–304.

1491. Rooney, J.R., Prickett, M.E., Delaney, F.M., Crowe, M.W., 1970. Focal myelitis-encephalitis in horses. *Cornell Vet.* 50, 494–501.

1492. Rosenthal, B.M., Lindsay, D.S., Dubey, J.P., 2001. Relationships among *Sarcocystis* species transmitted by New World opossums (*Didelphis* spp.). *Vet. Parasitol.* 95, 133–142.

1493. Rosenthal, B.M., Dunams, D.B., Pritt, B., 2008. Restricted genetic diversity in the ubiquitous cattle parasite, *Sarcocystis cruzi*. *Infect. Genet. Evol.* 8, 588–592.

1494. Rosonke, B.J., Brown, S.R., Tornquist, S.J., Snyder, S.P., Garner, M.M., Blythe, L.L., 1999. Encephalomyelitis associated with a *Sarcocystis neurona*-like organism in a sea otter. *J. Am. Vet. Med. Assoc.* 215, 1839–1842.

1495. Rossano, M.G., Mansfield, L.S., Kaneene, J.B., Murphy, A.J., Brown, C.M., Schott, H.C., Fox, J.C., 2000. Improvement of western blot test specificity for detecting equine serum antibodies to *Sarcocystis neurona*. *J. Vet. Diagn. Invest.* 12, 28–32.

1496. Rossano, M.G., Kaneene, J.B., Marteniuk, J.V., Banks, B.D., Schott, H.C., Mansfield, L.S., 2001. The seroprevalence of antibodies to *Sarcocystis neurona* in Michigan equids. *Prev. Vet. Med.* 48, 113–128.

1497. Rossano, M.G., Murphy, A.J., Vrable, R.A., Vanzo, N.E., Lewis, S.K., Sheline, K.D., Kaneene, J.B., Mansfield, L.S., 2002. Cross-sectional study of serum antibodies against *Sarcocystis neurona* in cats tested for antibodies against *Toxoplasma gondii*. *J. Am. Vet. Med. Assoc.* 221, 511–514.

1498. Rossano, M.G., Kaneene, J.B., Schott, H.C., Sheline, K.D., Mansfield, L.S., 2003. Assessing the agreement of Western blot test results for paired serum and cerebrospinal fluid samples from horses tested for antibodies to *Sarcocystis neurona*. *Vet. Parasitol.* 115, 233–238.

1499. Rossano, M.G., Kaneene, J.B., Marteniuk, J.V., Banks, B.D., Schott, H.C., Mansfield, L.S., 2003. A herd-level analysis of risk factors for antibodies to *Sarcocystis neurona* in Michigan equids. *Prev. Vet. Med.* 57, 7–13.

1500. Rossano, M.G., Schott, H.C., Murphy, A.J., Kaneene, J.B., Sellon, D.C., Hines, M.T., Hochstatter, T., Bell, J.A., Mansfield, L.S., 2005. Parasitemia in an immunocompetent horse experimentally challenged with *Sarcocystis neurona* sporocysts. *Vet. Parasitol.* 127, 3–8.

1501. Rossano, M.G., Schott, H.C., Kaneene, J.B., Murphy, A.J., Kruttlin, E.A., Hines, M.T., Sellon, D.C., Patterson, J.S., Elsheikha, H.M., Dubey, J.P., Mansfield, L.S., 2005. Effect of daily administration of pyrantel tartrate in preventing infection in horses experimentally challenged with *Sarcocystis neurona*. *Am. J. Vet. Res.* 66, 846–852.

1502. Rosypal, A.C., Lindsay, D.S., Duncan, R., Ahmed, S.A., Zajac, A.M., Dubey, J.P., 2002. Mice lacking the gene for inducible or endothelial nitric oxide are resistant to sporocyst induced *Sarcocystis neurona* infections. *Vet. Parasitol.* 103, 315–321.

1503. Ruiz, A., Frenkel, J.K., 1976. Recognition of cyclic transmission of *Sarcocystis muris* by cats. *J. Infect. Dis.* 133, 409–418.

1504. Rzepczyk, C., Scholtyseck, E., 1976. Light and electron microscope studies on the *Sarcocystis* of *Rattus fuscipes*, an Australian rat. *Z. Parasitenkd.* 50, 137–150.

1505. Sáez, T., Ramos, J.J., García-De Jalon, J.A., Unzueta, A., Loste, A., 2003. Laryngeal hemiplegia in a ram associated with *Sarcocystis* species infection. *Vet. Rec.* 153, 27–28.

1506. Sager, H., Steiner-Moret, C., Müller, N., Staubli, D., Esposito, M., Schares, G., Hässig, M., Stärk, K., Gottstein, B., 2006. Incidence of *Neospora caninum* and other intestinal protozoan parasites in populations of Swiss dogs. *Vet. Parasitol.* 139, 84–92.

1507. Saha, A.K., Ghosh, D., 1992. Prevalence of *Sarcocystis* in black Bengal goats in Tripura. *Indian Vet. J.* 69, 82–83.

1508. Sahai, B.N., Singh, S.P., Sahay, M.N., Srivastava, P.S., Juyal, P.D., 1982. Note on the incidence and epidemiology of *Sarcocystis* infection in cattle, buffaloes and pigs in Bihar. *Indian J. Anim. Sci.* 52, 1005–1006.

1509. Sahasrabudhe, V.K., Shah, H.L., 1966. The occurrence of *Sarcocystis* sp. in the dog. *J. Protozool.* 13, 531.

1510. Saito, M., Nakajima, T., Watanabe, A., Itagaki, H., 1986. *Sarcocystis miescheriana* infection and its frequency in pigs in Japan. *Jpn. J. Vet. Sci.* 48, 1083–1090. (In Japanese).

1511. Saito, M., Mizusawa, K., Itagaki, H., 1993. Chronic *Sarcocystis* infections in slaughtered cattle. *J. Vet. Med. Sci.* 55, 757–761.

1512. Saito, M., Itagaki, H., 1994. Experimental infection of raccoon dogs with *Sarcocystis cruzi* and S. *miescheriana. J. Vet. Anim. Sci.* 56, 671–674.

1513. Saito, M., Ohuchi, Y., Kobayashi, M., Haritani, M., Itagaki, H., 1994. Preparation and applicability of *Sarcocystis cruzi* antigens and their anti-S. *cruzi* rabbit sera for serodiagnosis of bovine sarcocystosis. *J. Vet. Anim. Sci.* 56, 589–591.

1514. Saito, M., Shibata, Y., Itagaki, H., 1995. *Sarcocystis capracanis* and S. *hircicanis* from goats in Japan. *Jpn. J. Parasitol.* 44, 391–395.

1515. Saito, M., Itagaki, T., Shibata, Y., Itagaki, H., 1995. Morphology and experimental definitive hosts of *Sarcocystis* sp. from sika deer, *Cervus nippon centralis*, in Japan. *Jpn. J. Parasitol.* 44, 218–221.

1516. Saito, M., Taguchi, K., Shibata, Y., Kobayashi, T., Shimura, K., Itagaki, H., 1995. Toxicity and properties of the extract from *Sarcocystis cruzi* cysts. *J. Vet. Med. Sci.* 57, 1049–1051.

1517. Saito, M., Shibata, Y., Itagaki, H., 1996. *Sarcocystis arieticanis* of sheep in Japan (Protozoa; Apicomplexa). *Jpn. J. Parasitol.* 45, 290–294.

1518. Saito, M., Shibata, Y., Kobayashi, T., Kobayashi, M., Kubo, M., Itagaki, H., 1996. Ultrastructure of the cyst wall of *Sarcocystis* species with canine final host in Japan. *J. Vet. Med. Sci.* 58, 861–867.

1519. Saito, M., Shibata, Y., Kubo, M., Itagaki, H., 1997. *Sarcocystis mihoensis* n. sp. from sheep in Japan. *J. Vet. Med. Sci.* 59, 103–106.

1520. Saito, M., Shibata, Y., Ohno, A., Kubo, M., Shimura, K., Itagaki, H., 1998. *Sarcocystis suihominis* detected for the first time from pigs in Japan. *J. Vet. Med. Sci.* 60, 307–309.

1521. Saito, M., Shibata, Y., Kubo, M., Sakakibara, I., Yamada, A., Itagaki, H., 1999. First isolation of *Sarcocystis hominis* from cattle in Japan. *J. Vet. Med. Sci.* 61, 307–309.

1522. Saito, M., Kubo, M., Itagaki, H., 2000. *Sarcocystis* sp. from cattle slaughtered in Japan. *J. Vet. Med. Sci.* 62, 1209–1211.

1523. Salas, Y., Bello, A., Alvarado, A., Márquez, A., Rivero, J., Coronado, A., 2010. *Sarcocystis* spp. y su transmisión vertical en bovinos sacrificados en un matadero del centroccidente de Venezuela. *Gaceta de Ciencias Veterinarias* 15, 28–33.

1524. Salehi, M., Adinezade, A., Mosazade, A., Besharati, R., Bahari, P., 2014. The prevalence of *Sarcocystis* infection in sheep and cattle in Bojnurd, North Khorasan Province, Iran. *J. Zoonoses* 1, 60–63.

1525. Saleque, A., Bhatia, B.B., Juyal, P.D., 1990. *Sarcocystis* infection in goats (*Capra hircus*) in the Tarai region of Uttar Pradesh. *Indian Vet. Med. J.* 14, 276–277.

1526. Saleque, A., Bhatia, B.B., 1991. Prevalence of *Sarcocystis* in domestic pigs in India. *Vet. Parasitol.* 40, 151–153.

1527. Saleque, A., Bhatia, B.B., Juyal, P.D., Rahman, H., 1991. Toxicity of cyst extract of *Sarcocystis fusiformis* from buffalo in rabbits and mice. *Vet. Parasitol.* 38, 61–65.

1528. Saleque, A., Bhatia, B.B., Shanker, D., 1992. Prevalence of two species of *Sarcocystis* in sheep in Uttar Pradesh. *Indian Vet. J.* 69, 841–842.

1529. Samad, M.A., Bari, A.S.M., Ghimiri, N.P., 1987. Concurrent infection of babesiosis and sarcocystosis in a heifer. *Bangladesh Veterinarian* 4, 34–37.

1530. Sánchez Acedo, C., Lucientes Curdi, J., Gutierrez Galindo, J., Castillo Hernández, C., Estrada Peña, A., García Pérez, A., 1983. Incidencia de la sarcosporidiosis en animales de abasto del matadero de Zaragoza. *Rev. Ibérica Parasitol.* 43, 341–346.

1531. Sánchez-Andrade, R., Panadero, R., López, C., Lago, P., Paz, A., Morrondo, P., 2002. Parasitic forms in faeces and aegagropiles diurnal and nocturnal birds of prey in Galicia. *Rev. Ibérica Parasitol.* 62, 89–92.

1532. Sansom, F.M., Robson, S.C., Hartland, E.L., 2008. Possible effects of microbial ecto-nucleoside triphosphate diphosphohydrolases on host-pathogen interactions. *Microbiol. Mol. Biol. Rev.* 72, 765–781.

1533. Santini, S., Mancianti, F., Nigro, M., Poli, A., 1997. Ultrastructure of the cyst wall of *Sarcocystis* sp. in roe deer. *J. Wildl. Dis.* 33, 853–859.

1534. Sargison, N.D., Schock, A., Maclean, I.A., Rae, A., Heckeroth, A.R., Clark, A.M., 2000. Unusual outbreak of sporozoan encephalomyelitis in Bluefaced Leicester ram lambs. *Vet. Rec.* 146, 225–226.

1535. Saville, W.J., Reed, S.M., Granstrom, D.E., Hinchcliff, K.W., Kohn, C.W., Wittum, T.E., Stamper, S., 1997. Seroprevalence of antibodies to *Sarcocystis neurona* in horses residing in Ohio. *J. Am. Vet. Med. Assoc.* 210, 519–524.

1536. Saville, W.J., Morley, P.S., Reed, S.M., Granstrom, D.E., Kohn, C.W., Hinchcliff, K.W., Wittum, T.E., 2000. Evaluation of risk factors associated with clinical improvement and survival of horses with equine protozoal myeloencephalitis. *J. Am. Vet. Med. Assoc.* 217, 1181–1185.

1537. Saville, W.J., Reed, S.M., Morley, P.S., Granstrom, D.E., Kohn, C.W., Hinchcliff, K.W., Wittum, T.E., 2000. Analysis of risk factors for the development of equine protozoal myeloencephalitis in horses. *J. Am. Vet. Med. Assoc.* 217, 1174–1180.

1538. Saville, W.J.A., Stich, R.W., Reed, S.M., Njoku, C.J., Oglesbee, M.J., Wünschmann, A., Grover, D.L., Larew-Naugle, A.L., Stanek, J.F., Granstrom, D.E., Dubey, J.P., 2001. Utilization of stress in the development of an equine model for equine protozoal myeloencephalitis. *Vet. Parasitol.* 95, 211–222.

1539. Saville, W.J.A., Dubey, J.P., Oglesbee, M.J., Sofaly, C.D., Marsh, A.E., Elitsur, E., Vianna, M.C., Lindsay, D.S., Reed, S.M., 2004. Experimental infection of ponies with *Sarcocystis fayeri* and differentiation from *Sarcocystis neurona* infections in horses. *J. Parasitol.* 90, 1487–1491.

1540. Saville, W.J.A., Sofaly, C.D., Reed, S.M., Dubey, J.P., Oglesbee, M.J., Lacombe, V.A., Keene, R.O. et al. 2004. An equine protozoal myeloencephalitis challenge model testing a second transport after inoculation with *Sarcocystis neurona* sporocysts. *J. Parasitol.* 90, 1406–1410.

1541. Savini, G., Dunsmore, J.D., Robertson, I.D., Seneviratna, P., 1992. *Sarcocystis* spp. in Western Australian cattle. *Aust. Vet. J.* 69, 201–202.

1542. Savini, G., Dunsmore, J.D., Robertson, I.D., Seneviratna, P., 1992. The epidemiology of *Sarcocystis* spp. in cattle of Western Australia. *Epidemiol. Infect.* 108, 107–113.

1543. Savini, G., Dunsmore, J.D., Robertson, I.D., 1993. A survey of Western Australian dogs for *Sarcocystis* spp. and other intestinal parasites. *Aust. Vet. J.* 70, 275–276.

1544. Savini, G., Dunsmore, J.D., Robertson, I.D., Seneviratna, P., 1993. *Sarcocystis* spp in Western Australian sheep. *Aust. Vet. J.* 70, 152–154.

1545. Savini, G., Dunsmore, J.D., Robertson, I.D., 1994. Evaluation of a serological test system for the diagnosis of *Sarcocystis cruzi* infection in cattle using *S. cruzi* merozoite antigen. *Vet. Parasitol.* 51, 181–189.

1546. Savini, G., Robertson, I.D., Dunsmore, J.D., 1994. Risk factors associated with the occurrence of sarcocystosis in Western Australia: Results of a postal survey. *Prev. Vet. Med.* 19, 137–144.

1547. Savini, G., Robertson, I.D., Dunsmore, J.D., 1996. Viability of the sporocysts of *Sarcocystis cruzi* after exposure to different temperatures and relative humidities. *Vet. Parasitol.* 67, 153–160.

1548. Savini, G., Dunsmore, J.D., Robertson, I.D., 1996. Studies on pathogenesis, tissue infection and congenital transmission in cows experimentally infected with *Sarcocystis cruzi* by various routes. *Vet. Parasitol.* 64, 319–327.

1549. Savini, G., Robertson, I.D., Dunsmore, J.D., 1997. Class-specific antibody responses in cattle following experimental challenge with sporocysts or merozoites of *Sarcocystis cruzi*. *Vet. Parasitol.* 72, 121–127.

1550. Savini, G., Robertson, I.D., Dunsmore, J.D., 1997. Excystation rates and infectivity of sporocysts of *Sarcocystis cruzi* exposed to different treatments and storages. *Vet. Parasitol.* 73, 17–25.

1551. Sayed, F.G., Shaheen, M.S.I., Arafa, M.I., Koraa, H.M., 2008. *Sarcocystis* infection in cattle at Assiut abattoir: Microscopical and serological studies. *Ass. Univ. Bull. Environ. Res.* 11, 47–58.

1552. Scala, A., 1991. Indagini ultrastrutturali sulle pareti cistiche primarie dei Sarcosporidi in ovini della Sardegna. *Atti della Federazione Mediterranea Sanitá e Produzione Ruminati* 1, 203–208.

1553. Scanziani, E., Lavazza, A., Salvi, S., 1988. Incidienza della sarcosporidiosi in bovini macellati a Milano. *Summa* 3, 197–200.

1554. Scarratt, W.K., Wallace, M.A., 1998. Diagnosis and management of equine protozoal myeloencephalitis and concurrent infectious disease in two horses. *Equine Practice* 20, 23–24.

1555. Scarratt, W.K., Buechner-Maxwell, V.A., Karzenski, S., Wallace, M.A., Robertson, J.L., 1999. Urinary incontinence and incoordination in three horses associated with equine protozoal myeloencephalitis. *J. Equine Vet. Sci.* 19, 642–645.

1556. Schebitz, H., Hartwigk, H., 1950. Myositis sarcosporidica beim Pferd. *Tierärztl. Umschau* 5, 351–353.

1557. Schmidtová, D., 1992. Some characteristics of *Sarcocystis* spp. found in the muscles of sheep. *Folia Parasitol. (Praha)* 39, 83–84.

1558. Schmitz, J.A., Wolf, W.W., 1977. Spontaneous fatal sarcocystosis in a calf. *Vet. Pathol.* 14, 527–531.

1559. Schnieder, T., Rommel, M., 1983. Ausbildung und Dauer der Immunität gegen *Sarcocystis miescheriana* im Schwein bei kontinuierlicher Verabreichung kleiner Mengen von Sporozysten. *Berl. Münch. Tierärztl. Wschr.* 96, 167–170.

1560. Schnieder, T., Trautwein, G., Rommel, M., 1984. Untersuchungen zur Persistenz der Zysten von *Sarcocystis miercheriana* in der Muskulatur des Schweines nach ein- und mehrmaliger Infektion. *Berl. Münch. Tierärztl. Wschr.* 97, 356–359.

1561. Schnieder, T., Kaup, F.J., Drommer, W., Thiel, W., Rommel, M., 1984. Zur Feinstruktur und Entwicklung von *Sarcocystis aucheniae* beim Lama. *Z. Parasitenkd.* 70, 451–458.

1562. Schnieder, T., Zimmermann, U., Matuschka, F.R., Bürger, H.J., Rommel, M., 1985. Zur Klinik, Enzymaktivität und Antikörperbildung bei experimentell mit Sarkosporidien infizierten Pferden. *Zbl. Vet. Med. B* 32, 29–39.

1563. Schock, A., French, H., Chianini, F., Bartley, P., Katzer, F., Otter, A., 2012. Respiratory disease due to acute *Sarcocystis tenella* infection in sheep. *Vet. Rec.* 170, 571.

1564. Scholtyseck, E., Mehlhorn, H., 1970. Ultrastructural study of characteristic organelles (paired organelles, micronemes, micropores) of sporozoa and related organisms. *Z. Parasitenkd.* 34, 97–127.

1565. Scholtyseck, E., Mehlhorn, H., Müller, B.E.G., 1974. Feinstruktur der Cyste und Cystenwand von *Sarcocystis tenella, Besnoitia jellisoni, Frenkelia* sp. und *Toxoplasma gondii. J. Protozool.* 21, 284–294.

1566. Scholtyseck, E., Hilali, M., 1978. Ultrastructural study of sexual stages of *Sarcocystis fusiformis* (Railliet, 1987) in domestic cats. *Z. Parasitenkd.* 56, 205–209.

1567. Scholtyseck, E., Entzeroth, R., Chobotar, B., 1982. Light and electron microscopy of *Sarcocystis* sp. in the skeletal muscle of an opossum (*Didelphis virginiana*). *Protistologica* 18, 527–532.

1568. Schramlová, J., Blažek, K., 1978. Ultrastruktur der Cystenwand der Sarkosporidien des Rehes (*Capreolus capreolus* L.). *Z. Parasitenkd.* 55, 43–48.

1569. Schulze, K., Zimmermann, T., 1982. Sarkosporidienbefall beim Rehwild mit lebensmittel-bzw fleischhygienischer Bedeutung. *Fleischwirtschaft* 62, 1–2.

1570. Schurer, J., Davenport, L., Wagner, B., Jenkins, E., 2014. Effects of sub-zero storage temperatures on endoparasites in canine and equine feces. *Vet. Parasitol.* 204, 310–315.

1571. Scorza, J.V., Torrealba, J.F., Dagert, C., 1957. *Klossiella tejerai* nov. sp. y *Sarcocystis didelphidis* nov. sp. parasitos de un *Didelphis marsupialis* de Venezuela. *Acta Biol. Venezuelica* 2, 97–108.

1572. Scott, J.W., 1943. Life history of Sarcosporidia, with particular reference to *Sarcocystis tenella. Univ. Wyoming Agr. Exp. Stat. Bull.* 259, 5–63.

1573. Scott, J.W., 1943. Economic importance of Sarcosporidia, with especial reference to *Sarcocystis tenella. Univ. Wyoming Agr. Exp. Stat. Bull.* 262, 5–55.

1574. Scott, P., Witonsky, S., Robertson, J., Daft, B., 2005. Increased presence of T lymphocytes in central nervous system of EPM affected horses. *J. Parasitol.* 91, 1499–1502.

1575. Scott, P.R., Sargison, N.D., 2001. Extensive ascites associated with vegetative endocarditis and *Sarcocystis* myositis in a shearling ram. *Vet. Rec.* 149, 240–241.

1576. Sedlaczek, J., Zipper, J., 1986. *Sarcocystis alceslatrans* (Apicomplexa) bei einem paläarktischen Elch (Ruminantia). *Angew. Parasitol.* 27, 137–144.

1577. Sedlaczek, J., Wesemeier, H.H., 1995. On the diagnostics and nomenclature of *Sarcocystis* species (Sporozoa) in roe deer (*Capreolus capreolus*). *Appl. Parasitol.* 36, 73–82.

1578. Sedrish, S.A., Ramirez, S., 1996. What is your neurologic diagnosis? (Protozoal myeloencephalitis in a horse). *J. Am. Vet. Med. Assoc.* 209, 903–905.

1579. Sellon, D.C., Knowles, D.P., Greiner, E.C., Long, M.T., Hines, M.T., Hochstatter, T., Tibary, A., Dame, J.B., 2004. Infection of immunodeficient horses with *Sarcocystis neurona* does not result in neurologic disease. *Clin. Diagn. Lab. Immunol.* 11, 1134–1139.

1580. Sellon, D.C., Knowles, D.P., Greiner, E.C., Long, M.T., Hines, M.T., Hochstatter, T., Hasel, K.M., Ueti, M., Gillis, K., Dame, J.B., 2004. Depletion of natural killer cells does not result in neurologic disease due to *Sarcocystis neurona* in mice with severe combined imunodeficiency. *J. Parasitol.* 90, 782–788.

1581. Sellon, D.C., Dubey, J.P., 2007. Equine protozoal myeloencephalitis. In: Sellon, D.C., Long, M.T. (Eds.), *Equine Infectious Diseases.* Elsevier Inc., St. Louis, pp. 453–464.

1582. Sen, M.R., 1950. Sarcosporidiosis in cattle in Bengal. *Indian Vet.* J. 27, 261–264.

1583. Sénaud, J., 1967. Contribution a l'étude des sarcosporidies et des Toxoplasmes (*Toxoplasmea*). *Protistologica* 3, 167–232.

1584. Sénaud, J., Chobotar, B., Scholtyseck, E., 1976. Role of the micropore in nutrition of the sporozoa. Ultrastructural study of *Plasmodium cathemerium, Eimeria ferrisi, E. stiedai, Besnoitia jellisoni,* and *Frenkelia* sp. Tropenmed. *Parasitol.* 27, 145–159.

1585. Sénaud, J., Černá, Z., 1978. Le cycle de développement asexué de *Sarcocystis dispersa* (Cernà, Kolarová, et Sulc, 1977) chez la souris: étude au microscope électronique. *Protistologica* 14, 155–176.

1586. Seneviratna, P., Edward, A.G., DeGiusti, D.L., 1975. Frequency of *Sarcocystis* spp. in Detroit, metropolitan area, Michigan. *Am. J. Vet. Res.* 36, 337–339.

1587. Shah, H.L., 1983. Epidemiology of *Sarcocystis* in domestic animals. *Haryana Vet.* 22, 59–73.

1588. Shah, H.L., 1990. Human factors in dissemination and transmission of *Sarcocystis* infection. *Proceedings of First Asian Conference of Parasitology,* Patna, India, pp. 1–8.

1589. Shankar, D., Bhatia, B.B., 1993. *Sarcocystis* infection in goats in Uttar Pradesh. *Indian J. Anim. Sci.* 63, 284–287.

1590. Shapiro, K., Miller, M., Mazet, J., 2012. Temporal assocation between land-based runoff events and California sea otter (*Enhydra lutris nereis*) protozoal mortalities. *J. Wildl. Dis.* 48, 394–404.

1591. Sharma, R.K., Shah, H.L., 1990. Pariah kite (*Milvus migrans*) as a definitive host of an unknown *Sarcocystis* species. *Indian J. Anim. Sci.* 60, 804–805.

1592. Sharma, R.K., Shah, H.L., 1992. Occurrence of *Sarcocystis hircicanis* in the goat. *Indian J. Anim. Sci.* 62, 561–563.

1593. Shastri, U.V., 1988. *Sarcocystis* infection in goats in Maharashtra. *J. Vet. Parasitol.* 2, 117–119.

1594. Shastri, U.V., 1989. Prevalence of *Sarcocystis* and other coccidial infections in stray dogs in and around Parbhani Town (Maharashtra). *Indian Vet. J.* 66, 593–596.

1595. Shastri, U.V., 1990. *Sarcocystis* infection in goats in Maharashtra. *Indian Vet. J.* 67, 70–71.

1596. Shaw, J.J., Lainson, R., 1969. *Sarcocystis* of rodents and marsupials in Brazil. *Parasitology* 59, 233–244.

1597. Sheffield, H.G., Frenkel, J.K., Ruiz, A., 1977. Ultrastructure of the cyst of *Sarcocystis muris. J. Parasitol.* 63, 629–641.

1598. Sheffield, H.G., Fayer, R., 1980. Fertilization in the coccidia: Fusion of *Sarcocystis bovicanis* gametes. *Proc. Helminthol. Soc. Wash.* 47, 118–121.

1599. Shekarforoush, S.S., Razavi, S.M., Dehghan, S.A., Sarihi, K., 2005. Prevalence of *Sarcocystis* species in slaughtered goats in Shiraz, Iran. *Vet. Rec.* 156, 418–420.

1600. Shekarforoush, S.S., Shakerian, A., Hasanpoor, M.M., 2006. Prevalence of *Sarcocystis* in slaughtered one-humped camels (*Camelus dromedarius*) in Iran. *Trop. Anim. Health Prod.* 38, 301–303.

1601. Shekarforoush, S.S., Razavi, S.M., Abbasvalli, M., 2013. First detection of *Sarcocystis hirsuta* from cattle in Iran. *Iranian J. Vet. Res.* 14, 155–157.

1602. Shekhar, K.C., Pathmanathan, R., Krishnan, R., 1998. Human muscular sarcocystosis associated with neoplasms? Case report. *Trop. Biomed.* 15, 61–64.

1603. Shen, Y.L., Shi, B.K., Yu, G.W., Li, Z.X., Zong, X.J., 1987. Experimental study on the life cycle of *Sarcocystis fusiformis* in Chinese buffaloes. *J. Nanjing Agric. Univ.* 4, 111–115.

1604. Shimura, K., Ito, S., Tsunoda, K., 1981. Sporocysts of *Sarcocystis cruzi* in mesenteric lymph nodes of dogs. *Nat. Inst. Anim. Health Quart. (Jpn.)* 21, 186–187.

1605. Shrivastav, A.B., Sharma, R.K., Chaudhry, R.K., Malik, P., 1999. Sarcocystosis in a Barasingha deer (*Cervus duvauceli branderi*). *J. Zoo Wildl. Med.* 30, 454–455.

1606. Shrivastava, G., Jain, P.C., 2001. Sporocyst output in dogs fed with sarcocysts of *Sarcocystis arieticanis* and *S. tenella* of the sheep. *Indian Vet. J.* 78, 533–535.

1607. Šibalič, D., Bordjoški, A., Conić, V., Djurković-Djaković, O., 1983. A study of *Toxoplasma gondii* and *Sarcocystis* sp. infections in humans suspected of acquired toxoplasmosis. *Acta Veterinaria* 33, 39–48.

1608. Simón Vicente, F., Ramajo-Martín, V., 1984. Sarcocystosis natural en ovinos y caprinos. *Rev. Ibérica Parasitol.* 44, 367–377.

1609. Simpson, C.F., 1966. Electron microscopy of *Sarcocystis fusiformis. J. Parasitol.* 52, 607–613.

1610. Simpson, C.F., Forrester, D.J., 1973. Electron microscopy of *Sarcocystis* sp.: Cyst wall, micropores, rhoptries, and an unidentified body. *Int. J. Parasitol.* 3, 467–470.

1611. Simpson, C.F., Mayhew, I.G., 1980. Evidence for *Sarcocystis* as the etiologic agent of equine protozoal myeloencephalitis. *J. Protozool.* 27, 288–292.

1612. Singh, K.P., Agrawal, M.C., Shah, H.L., 1987. Prevalence of *Sarcocystis* sporocysts in stray dogs. *Indian J. Anim. Sci.* 57, 1101–1102.

1613. Singh, K.P., Shah, H.L., 1990. Offals from slaughtered goats as a source of infection of *Sarcocystis capracanis* for dogs. *Indian J. Anim. Sci.* 60, 1315.

1614. Singh, K.P., Shah, H.L., 1990. Viability and infectivity of sarcocysts of *Sarcocystis capracanis* of the goat after maintaining them at different temperatures. *Indian J. Anim. Sci.* 60, 429–430.

1615. Singh, K.P., Agrawal, M.C., Shah, H.L., 1990. Prevalence of sarcocysts of *Sarcocystis capracanis* in oesophagus and tail muscles of naturally infected goats. *Vet. Parasitol.* 36, 153–155.

1616. Singh, K.P., Parihar, N.S., 1996. Spontaneous sarcocystosis in brain of hill bullocks. *Indian J. Vet. Pathol.* 20, 153–154.

1617. Singh, L.A.L., Raisinghani, P.M., Kumar, D., Pathak, K.M.L., Manohar, G.S., 1992. Sporocyst output in dogs fed with sarcocysts of *Sarcocystis capracanis*. *J. Vet. Parasitol.* 6, 35–36.

1618. Singh, L.A.L., Raisinghani, P.M., Pathak, K.M.L., Kumar, D., Manohar, G.S., Bhan, A.K., Arora, J.K., Swarankar, C.P., 1992. Epidemiology of *Sarcocystis capracanis* in goats at Bikaner, Rajasthan, India. *Indian J. Anim. Sci.* 62, 1044–1045.

1619. Singh, L.A.L., Raisinghani, P.M., Kumar, D., Swarankar, C.P., 1993. Clinical and haematobiochemical changes in dogs experimentally infected with *Sarcocystis capracanis*. *Indian J. Anim. Sci.* 63, 1055–1057.

1620. Šlapeta, J.R., Modrý, D., Koudela, B., 1999. *Sarcocystis atheridis* sp. nov., a new sarcosporidian coccidium from Nitsche's bush viper, *Atheris nitschei* Tornier, 1902, from Uganda. *Parasitol. Res.* 85, 758–764.

1621. Šlapeta, J.R., Modrý, D., Votýpka, J., Jirků, M., Koudela, B., Lukeš, J., 2001. Multiple origin of the dihomoxenous life cycle in sarcosporidia. *Int. J. Parasitol.* 31, 413–417.

1622. Šlapeta, J.R., Kyselová, I., Richardson, A.O., Modrý, D., Lukeš, J., 2002. Phylogeny and sequence variability of the *Sarcocystis singaporensis* Zaman and Colley, (1975) 1976 ssrDNA. *Parasitol. Res.* 88, 810–815.

1623. Šlapeta, J.R., Modrý, D., Votýpka, J., Jirků, M., Lukeš, J., Koudela, B., 2003. Evolutionary relationships among cyst-forming coccidia *Sarcocystis* spp. (Alveolata: Apicomplexa: Coccidea) in endemic African tree vipers and perspective for evolution of heteroxenous life cycle. *Mol. Phylogenet. Evol.* 27, 464–475.

1624. Slesak, G., Tappe, D., Keller, C., Cramer, J., Güthoff, W., Zanger, P., Frank, M. et al., 2014. Muskuläre Sarkozystose nach Malaysiareise: eine Fallserie aus Deutschland. *Dtsch. Med. Wschr.* 139, 990–995.

1625. Smith, D.D., Frenkel, J.K., 1978. Cockroaches as vectors of *Sarcocystis muris* and of other coccidia in the laboratory. *J. Parasitol.* 64, 315–319.

1626. Smith, J.H., Meier, J.L., Neill, P.J.G., Box, E.D., 1987. Pathogenesis of *Sarcocystis falcatula* in the budgerigar. I. Early pulmonary schizogony. *Laboratory Investigation* 56, 60–71.

1627. Smith, J.H., Meier, J.L., Neill, P.J.G., Box, E.D., 1987. Pathogenesis of *Sarcocystis falcatula* in the budgerigar. II. Pulmonary pathology. *Laboratory Investigation* 56, 72–84.

1628. Smith, T., 1905. Further observations on the transmission of *Sarcocystis muris* by feeding. *J. Med. Res.* 13, 429–430.

1629. Smith, J.H., Neill, P.J.G., Box, E.D., 1989. Pathogenesis of *Sarcocystis falcatula* (Apicomplexa: Sarcocystidae) in the budgerigar (*Melopsittacus undulatus*) III. Pathologic and quantitative parasitologic analysis of extrapulmonary disease. *J. Parasitol.* 75, 270–287.

1630. Smith, J.H., Neill, P.J.G., Dillard III, E.A., Box, E.D., 1990. Pathology of experimental *Sarcocystis falcatula* infections of canaries (*Serinus canarius*) and pigeons (*Columba livia*). *J. Parasitol.* 76, 59–68.

1631. Smith, J.H., Craig, T.M., Dillard III, E.A., Neill, P.J.G., Jones, L.P., 1990. Naturally occurring apicomplexan acute interstitial pneumonitis in thick-billed parrots (*Rhynchopsitta pachyrhyncha*). *J. Parasitol.* 76, 285–288.

1632. Smith, T.S., Herbert, I.V., 1986. Experimental microcyst *Sarcocystis* infection in lambs: Serology and immunohistochemistry. *Vet. Rec.* 119, 547–550.

1633. Snyder, D.E., Sanderson, G.C., Toivio-Kinnucan, M., Blagburn, B.L., 1990. *Sarcocystis kirkpatricki* n. sp. (Apicomplexa: Sarcocystidae) in muscles of raccoons (*Procyon lotor*) from Illinois. *J. Parasitol.* 76, 495–500.

1634. Sobel, R.A., Colvin, R.B., 1986. Responder strain specific enhancement of endothelial and mononuclear cell Ia in delayed hypersensitivity reactions in (strain 2 x strain 13) F_1 guinea pigs. *J. Immunol.* 137, 2132–2138.

1635. Sobel, R.A., van der Veen, R.C., Lees, M.B., 1986. The immunopathology of chronic experimental allergic encephalomyelitis induced in rabbits with bovine proteolipid protein. *J. Immunol.* 136, 157–163.

1636. Sofaly, C.D., Reed, S.M., Gordon, J.C., Dubey, J.P., Oglesbee, M.J., Njoku, C.J., Grover, D.L., Saville, W.J.A., 2002. Experimental induction of equine protozoan myeloencephalitis (EPM) in the horse: Effect of *Sarcocystis neurona* sporocyst inoculation dose on the development of clinical neurologic disease. *J. Parasitol.* 88, 1164–1170.

1637. Solanki, P.K., Shrivastava, H.O.P., Shah, H.L., 1990. Endogenous life-cycle of *Sarcocystis miescheriana* in experimentally infected pigs. *Indian J. Anim. Sci.* 60, 1274–1278.

1638. Solanki, P.K., Shrivastava, H.O.P., Shah, H.L., 1991. Prevalence of *Sarcocystis* in naturally infected pigs in Madhya-Pradesh with an epidemiological explanation for the higher prevalence of *Sarcocystis suihominis. Indian J. Anim. Sci.* 61, 820–821.

1639. Solanki, P.K., Shrivastava, H.O.P., Shah, H.L., 1991. Morphology of sarcocysts of *Sarcocystis miescheriana* (Khun, 1865) Labbe, 1899 and *Sarcocystis suihominis* (Tadros and Laarman, 1976) Heydorn, 1977 from naturally infected domestic pigs (*Sus scrofa domestica*) in India. *Indian J. Anim. Sci.* 61, 1030–1033.

1640. Solanki, P.K., Shrivastava, H.O.P., Shah, H.L., 1991. Gametogonous and sporogonous life-cycle of *Sarcocystis miescheriana* of the pig in experimentally infected dogs. *Indian J. Anim. Sci.* 61, 374–378.

1641. Solanki, P.K., Shrivastava, H.O., Shah, H.L., Awadhiya, R.P., 1992. Pathology of *Sarcocystis miescheriana* in experimentally infected pigs. *Indian J. Anim. Sci.* 62(9), 797–801.

1642. Sommer, I., Mehlhorn, H., Rüger, W., 1991. Biochemical and immunological characterization of major surface antigens of *Sarcocystis muris* and *S. suicanis* cyst merozoites. *Parasitol. Res.* 77, 204–211.

1643. Sommer, I., Horn, K., Heydorn, A.O., Mehlhorn, H., Rüger, W., 1992. A comparison of sporozoite and cyst merozoite surface-proteins of *Sarcocystis. Parasitol. Res.* 78, 398–403.

1644. Somvanshi, R., Koul, G.L., Biswas, J.C., 1987. *Sarcocystis* in a leopard (*Panthera pardus*). *Indian Vet. Med. J.* 11, 174–175.

1645. Soriano, S.V., Barbieri, L.M., Pierángeli, N.B., Giayetto, A.L., Manacorda, A.M., Castronovo, E., Pezzani, B.C., Minvielle, M., Basualdo, J.A., 2001. Intestinal parasites and the environment: Frequency of intestinal parasites in children of Neuquén, Patagonia, Argentina. *Rev. Latinoam. Microbiol.* 43, 96–101.

1646. Spalding, M.G., Atkinson, C.T., Carleton, R.E., 1994. *Sarcocystis* sp in wading birds (Ciconiiformes) from Florida. *J. Wildl. Dis.* 30, 29–35.

1647. Spalding, M.G., Yowell, C.A., Lindsay, D.S., Greiner, E.C., Dame, J.B., 2002. *Sarcocystis* meningoencephalitis in a northern gannet (*Morus bassanus*). *J. Wildl. Dis.* 38, 432–437.

1648. Speer, C.A., Pond, D.B., Ernst, J.V., 1980. Development of *Sarcocystis hemionilatrantis* Hudkins and Kistner, 1977 in the small intestine of coyotes. *Proc. Helminthol. Soc. Wash.* 47, 106–113.

1649. Speer, C.A., Dubey, J.P., 1981. An ultrastructural study of first- and second-generation merogony in the coccidian *Sarcocystis tenella. J. Protozool.* 28, 424–431.

1650. Speer, C.A., Dubey, J.P., 1981. Ultrastructure of *in vivo* lysis of *Sarcocystis cruzi* merozoites. *J. Parasitol.* 67, 961–963.

1651. Speer, C.A., Dubey, J.P., 1982. *Sarcocystis wapiti* sp. nov. from the American wapiti (*Cervus elaphus*). *Can. J. Zool.* 60, 881–888.

1652. Speer, C.A., Dubey, J.P., 1982. Scanning and transmission electron microscopy of ovine mesenteric arteries infected with first-generation meronts of a *Sarcocystis tenella. Can. J. Zool.* 60, 203–209.

1653. Speer, C.A., Reduker, D.W., Burgess, D.E., Whitmire, W.M., Splitter, G.A., 1985. Lymphokine-induced inhibition of growth of *Eimeria bovis* and *Eimeria papillata* (Apicomplexa) in cultured bovine monocytes. *Infect. Immun.* 50, 566–571.

1654. Speer, C.A., Dubey, J.P., 1986. An unusual structure in the primary cyst wall of *Sarcocystis hemionilatrantis. J. Protozool.* 33, 130–132.

1655. Speer, C.A., Dubey, J.P., 1986. Vascular phase of *Sarcocystis cruzi* cultured *in vitro. Can. J. Zool.* 64, 209–211.

1656. Speer, C.A., Cawthorn, R.J., Dubey, J.P., 1986. *In vitro* cultivation of the vascular phase of *Sarcocystis capracanis* and *Sarcocystis tenella. J. Protozool.* 33, 486–490.

1657. Speer, C.A., Whitmire, W.M., Reduker, D.W., Dubey, J.P., 1986. *In vitro* cultivation of meronts of *Sarcocystis cruzi. J. Parasitol.* 72, 677–683.

1658. Speer, C.A., Burgess, D.E., 1987. *In vitro* cultivation of *Sarcocystis* merozoites. *Parasitol. Today* 3, 2–3.

1659. Speer, C.A., Burgess, D.E., 1988. *In vitro* development and antigen analysis of *Sarcocystis. Parasitol. Today* 4, 46–49.

1660. Speer, C.A., Dubey, J.P., 1999. Ultrastructure of schizonts and merozoites of *Sarcocystis falcatula* in the lungs of budgerigars *Melopsittacus undulatus. J. Parasitol.* 85, 630–637.

1661. Speer, C.A., Dubey, J.P., 2001. Ultrastructure of schizonts and merozoites of *Sarcocystis neurona. Vet. Parasitol.* 95, 263–271.

1662. Spencer, J.A., Ellison, S.E., Guarino, A.J., Blagburn, B.L., 2004. Cell-mediated immune responses in horses with equine protozoal myeloencephalitis. *J. Parasitol.* 90, 428–430.

1663. Spencer, J.A., Deinnocentes, P., Moyana, E.M., Guarino, A.J., Ellison, S.E., Bird, R.C., Blagburn, B.L., 2005. Cytokine gene expression in response to SnSAG1 in horses with equine protozoal myeloencephalitis. *Clin. Diagn. Lab Immunol.* 12, 644–646.

1664. Spickschen, C., Pohlmeyer, K., 2002. Untersuchung zum Vorkommen von Sarkosporidien bei Reh-, Rot- und Muffelwild in zwei unterschiedlichen Naturräumen des Bundeslandes Niedersachsen. *Z. Jagdwiss.* 48, 35–48.

1665. Srivastava, P.S., Saha, A.K., Sinha, S.R.P., 1985. Spontaneous *Sarcocystis* in indigenous goats in Bihar, India. *Acta Protozool.* 24, 339–345.

1666. Srivastava, P.S., Saha, A.K., Sinha, S.R.P., 1986. Effects of heating and freezing on the viability of sarcocysts of *Sarcocystis levinei* from cardiac tissues of buffaloes. *Vet. Parasitol.* 19, 329–332.

1667. Srivastava, P.S., Juyal, P.D., Sinha, S.R.P., Singh, S.P., Sahai, B.N., 1988. Attempt to infect monkeys with *Sarcocystis cruzi* sporocysts. *J. Vet. Parasitol.* 2, 63–65.

1668. Srivastava, P.S., Kumar, A., Sinha, S.R.P., Sinha, A.K., 1991. Morphological-differentiation of caprine *Sarcocystis* species-evidence of occurrence of *Sarcocystis hircicanis* in India. *Acta Protozool.* 30, 61–62.

1669. Stabenow, C.S., de Oliveira, F.C.N., Albuquerque, G.R., Lopes, C.W., 2008. *Sarcocystis lindsayi*-like (Apicomplexa: Sarcocystinae) of the opossum (*Didelphis aurita*) from Southeastern Brazil. *Rev. Bras. Parasitol. Vet.* 17(Suppl 1), 342–344.

1670. Stabenow, C.S., Ederli, N.B., Lopes, C.W.G., de Oliveira, F.C.R., 2012. *Didelphis aurita* (Marsupialia: Didelphidae): A new host for *Sarcocystis lindsayi* (Apicomplexa: Sarcocystidae). *J. Parasitol.* 98, 1262–1265.

1671. Stackhouse, L.L., Cawthorn, R.J., Brooks, R.J., 1987. Pathogenesis of infection with *Sarcocystis rauschorum* (Apicomplexa) in experimentally infected varying lemmings (*Dicrostonyx richardsoni*). *J. Wildl. Dis.* 23, 566–571.

1672. Stalheim, O.H., Proctor, S.J., Fayer, R., Lunde, M., 1976. Death and abortion in cows experimentally infected with *Sarcocystis* from dogs. In: 19th Annual Proceedings of the American Association of Veterinary Laboratory Diagnosticians, Madison, Wisconsin, pp. 317–327.

1673. Stanek, J.F., Dubey, J.P., Oglesbee, M.J., Reed, S.M., Lindsay, D.S., Capitini, L.A., Njoku, C.J., Vittitow, K.L., Saville, W.J.A., 2002. Life cycle of *Sarcocystis neurona* in its natural intermediate host, the racoon, *Procyon lotor. J. Parasitol.* 88, 1151–1158.

1674. Stanek, J.F., Stich, R.W., Dubey, J.P., Reed, S.M., Njoku, C.J., Lindsay, D.S., Schmall, L.M., Johnson, G.K., LaFave, B.M., Saville, W.J.A., 2003. Epidemiology of *Sarcocystis neurona* infections in domestic cats (*Felis domesticus*) and its association with equine protozoal myeloencephalitis (EPM) case farms and feral cats from a mobile spay and neuter clinic. *Vet. Parasitol.* 117, 239–249.

1675. Stenlund, S., Björkman, C., Holmdahl, O.J.M., Kindahl, H., Uggla, A., 1997. Characterisation of a Swedish bovine isolate of *Neospora caninum. Parasitol. Res.* 83, 214–219.

1676. Stephan, R., Loretz, M., Eggenberger, E., Grest, P., Basso, W., Grimm, F., 2012. Erster Nachweis von *Sarcocystis hjorti* bei der Fleischunteruchung von Rothirschen in der Schweiz. *Schweiz. Arch. Tierheilkd.* 154, 539–542.

1677. Stoffregen, D.A., Dubey, J.P., 1991. A *Sarcocystis* sp.-like protozoan and concurrent canine distemper virus infection associated with encephalitis in a raccoon (*Procyon lotor*). *J. Wildl. Dis.* 27, 688–692.

1678. Stojecki, K., Karamon, J., Sroka, J., Cencek, T., 2012. Molecular diagnostics of *Sarcocystis* spp. infections. *Pol. J. Vet. Sci.* 15, 589–596.

1679. Stolte, M., Odening, K., Bockhardt, I., 1996. Antelopes (Bovidae) kept in European zoological gardens as intermediate hosts of *Sarcocystis* species. *Parassitologia* 38, 565–570.

1680. Stolte, M., Bockhardt, I., Odening, K., 1996. A comparative scanning electron microscopic study of the cyst wall in 11 *Sarcocystis* species of mammals. *J. Zool. Lond.* 239, 821–832.

1681. Stolte, M., Odening, K., Walter, G., Bockhardt, I., 1996. The raccoon as intermediate host of three *Sarcocystis* species in Europe. *J. Helminthol. Soc. Wash.* 63, 145–149.

1682. Stolte, M., Bockhardt, I., Odening, K., 1997. First report of *Sarcocystis rangiferi* and a second *Sarcocystis* species with parasite-induced encapsulation in cervids from Central Europe. *Acta Protozool.* 36, 131–135.

1683. Stolte, M., Bockhardt, I., Odening, K., 1998. Scanning electron microscopic identification of *Sarcocystis gracilis* from roe deer and cattle. *J. Zool. Lond.* 244, 265–268.

1684. Stolte, M., Odening, K., Quandt, S., Bengis, R.G., Bockhardt, I., 1998. *Sarcocystis dubeyella* n. sp. and *Sarcocystis phacochoeri* n. sp. (Protozoa: Sarcocystidae) from the warthog (*Phacochoerus aethiopicus*) in South Africa. *J. Eukaryot. Microbiol.* 45, 101–104.

1685. Straka, S., Škračiková, J., Konvit, I., Szilágyiová, M., Michal, L., 1991. *Sarcocystis* species in Vietnamese apprentices. *Cesk. Epidemiol. Mikrobiol. Imunol.* 40, 204–208. (In Czech).

1686. Streitel, R.H., Dubey, J.P., 1976. Prevalence of *Sarcocystis* infection and other intestinal parasitisms in dogs from a humane shelter in Ohio. *J. Am. Vet. Med. Assoc.* 168, 423–424.

1687. Strobel, J.G., Delplace, P., Dubremetz, J.F., Entzeroth, R., 1992. *Sarcocystis muris* (Apicomplexa): A thiol protease from the dense granules. *Exp. Parasitol.* 74, 100–105.

1688. Strohlein, D.A., Prestwood, A.K., 1986. *In vitro* excystation and structure of *Sarcocystis suicanis* Erber, 1977, sporocysts. *J. Parasitol.* 72, 711–715.

1689. Stubbings, D.P., Jeffrey, M., 1985. Presumptive protozoan (sarcocystis) encephalomyelitis with paresis in lambs. *Vet. Rec.* 116, 373–374.

1690. Su, C., Evans, D., Cole, R.H., Kissinger, J.C., Ajioka, J.W., Sibley, L.D., 2003. Recent expansion of *Toxoplasma* through enhanced oral transmission. *Science* 299, 414–416.

1691. Suedmeyer, W.K., Bermudez, A.J., Barr, B.C., Marsh, A.E., 2001. Acute pulmonary *Sarcocystis falcatula*-like infection in three Victoria crowned pigeons (*Goura victoria*) housed indoors. *J. Zoo Wildl. Med.* 32, 252–256.

1692. Sugár, L., Entzeroth, R., Chobotar, B., 1990. Ultrastructure of *Sarcocystis sibirica* (Matchulski, 1947) from the Siberian roe deer, *Capreolus pygargus*. *Parasitol. Hung.* 23, 13–17.

1693. Sundar, N., Asmundsson, I.M., Thomas, N.J., Samuel, M.D., Dubey, J.P., Rosenthal, B.M., 2008. Modest genetic differentiation among North American populations of *Sarcocystis neurona* may reflect expansion in its geographical range. *Vet. Parasitol.* 152, 8–15.

1694. Suteu, E., Mircean, V., 1996. Semnalarea infestatiei cu *Sarcocystis porcifelis* la mistret (*Sus scrofa ferus*). *Rev. Rom. Med. Vet.* 6, 165–167.

1695. Suzuki, T., Miura, H., Narita, K., Suzuki, H., Miki, H., Sado, T., 1978. Quarantine and health control of the squirrel monkey (*Saimiri sciureus*). *Jikken Dobutsu* 27, 161–166.

1696. Svobodová, M., 1996. A *Sarcocystis* species from goshawk (*Accipiter gentilis*) with great tit (*Parus major*) as intermediate host. *Acta Protozool.* 35, 223–226.

1697. Svobodová, M., Vořišek, P., Votýpka, J., Weidinger, K., 2004. Heteroxenous coccidia (Apicomplexa: Sarcocystidae) in the populations of their final and intermediate hosts: European buzzard and small mammals. *Acta Protozool.* 43, 251–260.

1698. Svobodová, V., Nevole, M., 1990. Use of the muscle digestion method and indirect immunofluorescence reaction in the diagnosis of sarcocystosis in sheep. *Acta Vet. Brno* 59, 157–170.

1699. Svobodová, V., 1991. Sarkocystōza jehňat. *Vet. Med. (Praha)* 36, 235–243.

1700. Svobodová, V., Nevole, M., 1992. Diagnostika sarkocystózy ovcí metodami NFR a ELISA. *Veter. Med. (Praha)* 37, 109–112.

1701. Sykes, J.E., Dubey, J.P., Lindsay, L.L., Prato, P., Lappin, M.R., Guo, L.T., Mizisin, A.P., Shelton, G.D., 2011. Severe myositis associated with *Sarcocystis* spp. infection in 2 dogs. *J. Vet. Intern. Med.* 25, 1277–1283.

1702. Tadmor, A., Nobel, T.A., Mindel, J.B., 1966. Myositis in a mule possibly due to infestation with Sarcosporidia. *Refuah Veterinarith* 23, 106–109.

1703. Tadros, W., Laarman, J.J., 1975. The weasel, *Mustela nivalis* as the final host of a *Sarcocystis* of the common European vole, *Microtus arvalis*. *Proc. Kon. Ned. Akad. Wet. C* 78, 325–326.

1704. Tadros, W., 1976. Contribution to the understanding of the life-cycle of *Sarcocystis* of the short-tailed vole *Microtus agrestis*. *Folia Parasitol. (Praha)* 23, 193–199.

1705. Tadros, W., Laarman, J.J., 1976. *Sarcocystis* and related coccidian parasites: A brief general review, together with a discussion on some biological aspects of their life cycles and a new proposal for their classification. *Acta Leidensia* 44, 1–107.

1706. Tadros, W., Laarman, J.J., 1977. The cat *Felis catus* as the final host of *Sarcocystis cuniculi* Brumpt, 1913 of the rabbit *Oryctolagus cuniculus. Proc. Kon. Ned. Akad. Wet. C* 80, 351–352.

1707. Tadros, W., Laarman, J.J., 1977. Studies on sarcosporidiosis and *Sarcocystis*-induced coccidiosis in man, monkeys and other animals. In: *Abstracts of papers read at the 5th International Congress on Protozoology*, New York City, June 26–July 2, 1977. p. 137.

1708. Tadros, W., Laarman, J.J., 1978. A comparative study of the light and electron microscopic structure of the walls of the muscle cysts of several species of Sarcocystid Eimeriid coccidia (I). *Proc. Kon. Ned. Akad. Wet. C* 81, 469–491.

1709. Tadros, W., Laarman, J.J., 1978. Apparent congenital transmission of *Frenkelia* (Coccidia: Eimeriidae): First recorded incidence. *Z. Parasitenkd.* 58, 41–46.

1710. Tadros, W., Laarman, J.J., 1978. Note on the specific designation of *Sarcocystis putorii* (Railliet and Lucet, 1891) comb. nov. of the common European vole, *Microtus arvalis. Proc. Kon. Ned. Akad. Wet. C* 81, 466–468.

1711. Tadros, W., Laarman, J.J., 1979. Muscular sarcosporidiosis in the common European Weasel, *Mustela nivalis. Z. Parasitenkd.* 58, 195–200.

1712. Tadros, W., Laarman, J.J., 1979. Successful rodent to rodent transmission of *Sarcocystis sebeki* by inoculation of precystic schizogonic stages. *Trans. R. Soc. Trop. Med. Hyg.* 73, 350–351.

1713. Tadros, W., Laarman, J.J., 1979. Some observations on the gametogonic development of *Sarcocystis cuniculi* of the common European rabbit in a feline fibroblast cell line. *Acta Leidensia* 47, 45–52.

1714. Tadros, W., Hazelhoff, W., Laarman, J.J., 1979. The detection of circulating antibodies against *Sarcocystis* in human and bovine sera by the enzyme-linked immunosorbent assay (ELISA) technique. *Acta Leidensia* 47, 53–63.

1715. Tadros, W., 1981. Studies on the Sarcosporidia of rodents with birds of prey as definitive hosts. In: Canning, E.V. (Ed.), *Parasitological Topics, A Presentation Volume to P. C. C. Garnham, F. R. S. on the Occasion of his 80th Birthday*, Vol. 1, Allen Press, Lawrence, Kansas , pp. 248–259.

1716. Tadros, W., Laarman, J.J., 1982. Current concepts on the biology, evolution and taxonomy of tissue cyst-forming eimeriid coccidia. *Adv. Parasitol.* 20, 293–496.

1717. Tadros, W.A., Bird, R.G., Ellis, D.S., 1972. The fine structure of cysts of *Frenkelia* (the M-organism). *Folia Parasitol. (Praha)* 19, 203–209.

1718. Takla, M., 1984. Acute sporidiosis in cattle. A case of clinically natural meningoencephalitis caused by *Sarcocystis cruzi* in a bull. *Tierarztl. Prax.* 12, 167–172. (In German).

1719. Talevich, E., Kannan, N., 2014. Structural and evolutionary adaptation of rhoptry kinases and pseudo-kinases, a family of coccidian virulence factors. *BMC Evol. Biol.* 13, 117.

1720. Tan, Y.X., Li, J.H., Lin, Z., Du, J.F., 2005. Investigation of the *Sarcocystis suihominis* infection status in the Zhuang ethnic population in part of Guangxi Province. *Guangxi Medicine* 11, 295–296. (In Chinese).

1721. Tanabe, M., Okinami, M., 1940. On the parasitic protozoa of the ground squirrel, *Eutamias asiaticus* Uthensis, with special reference to *Sarcocystis eutamias* sp. nov. *Keizyo J. Med.* 10, 126–134.

1722. Tanhauser, S.M., Yowell, C.A., Cutler, T.J., Greiner, E.C., MacKay, R.J., Dame, J.B., 1999. Multiple DNA markers differentiate *Sarcocystis neurona* and *Sarcocystis falcatula. J. Parasitol.* 85, 221–228.

1723. Tanhauser, S.M., Cheadle, M.A., Massey, E.T., Mayer, B.A., Schroedter, D.E., Dame, J.B., Greiner, E.C., MacKay, R.J., 2001. The nine-banded armadillo (*Dasypus novemcinctus*) is naturally infected with *Sarcocystis neurona. Int. J. Parasitol.* 31, 325–329.

1724. Tappe, D., Abdullah, S., Heo, C.C., Kannan Kutty, M., Latif, B., 2013. Human and animal invasive muscular sarcocystosis in Malaysia—recent cases, review and hypotheses. *Trop. Biomed.* 30, 355–366.

1725. Tappe, D., Ernestus, K., Rauthe, S., Schoen, C., Frosch, M., Müller, A., Stich, A., 2013. Initial patient cluster and first positive biopsy findings in an outbreak of acute muscular *Sarcocystis*-like infection in travelers returning from Tioman island, peninsular Malaysia, in 2011. *J. Clin. Microbiol.* 51, 725–726.

1726. Tappe, D., Stich, A., Langeheinecke, A., von Sonnenburg, F., Muntau, B., Schäfer, J., Slesak, G., 2014. Suspected new wave of muscular sarcocystosis in travellers returning from Tioman Island, Malaysia, May 2014. *Euro Surveill.* 19, pii: 20816.

1727. Taylor, M.A., Boes, J., Boireau, P., Boué, F., Claes, M., Cook, A.J.C., Dorny, P. et al. 2010. Development of harmonised schemes for the monitoring and reporting of *Sarcocystis* in animals and foodstuffs in the European Union. Scientific Report submitted to EFSA. Supporting publications. EFSA-Q-2009-01074. http://www.efsa.europa.eu/en/supporting/pub/33e.htm.

1728. Teglas, M.B., Little, S.E., Latimer, K.S., Dubey, J.P., 1998. *Sarcocystis*-associated encephalitis and myocarditis in a wild turkey (*Meleagris gallopavo*). *J. Parasitol.* 84, 661–663.

1729. Tenter, A.M., 1987. Comparison of enzyme-linked immunosorbent assay and indirect fluorescent antibody test for the detection of IgG antibodies to *Sarcocystis muris*. *Zbl. Bakt. Mikrobiol. Hyg. A* 267, 259–271.

1730. Tenter, A.M., 1988. Comparison of dot-ELISA, ELISA, and IFAT for the detection of IgG antibodies to *Sarcocystis muris* in experimentally infected and immunized mice. *Vet. Parasitol.* 29, 89–104.

1731. Tenter, A.M., Johnson, M.R., Zimmerman, G.L., 1989. Differentiation of *Sarcocystis* species in European sheep by isoelectric focusing. *Parasitol. Res.* 76, 107–114.

1732. Tenter, A.M., Johnson, M.R., Johnson, A.M., 1991. Effect of tryptic or peptic digestion or mechanical isolation on the extraction of proteins, antigens, and ribonucleic acids from *Sarcocystis muris* bradyzoites. *J. Parasitol.* 77, 194–199.

1733. Tenter, A.M., Vietmeyer, C., Thümmel, P., Rommel, M., 1991. Detection of species-specific and cross-reactive epitopes in *Sarcocystis* cystozoites by monoclonal antibodies. *Parasitol. Res.* 77, 212–216.

1734. Tenter, A.M., Johnson, A.M., 1992. Comparison of *in vitro* translation products of *Sarcocystis gigantea* and *Sarcocystis tenella*. *Int. J. Parasitol.* 22, 153–164.

1735. Tenter, A.M., Baverstock, P.R., Johnson, A.M., 1992. Phylogenetic relationships of *Sarcocystis* species from sheep, goats, cattle and mice based on ribosomal-RNA sequences. *Int. J. Parasitol.* 22, 503–513.

1736. Tenter, A.M., Luton, K., Johnson, A.M., 1994. Species-specific identification of *Sarcocystis* and *Toxoplasma* by PCR amplification of small subunit ribosomal RNA gene fragments. *Appl. Parasitol.* 35, 173–188.

1737. Terrell, T.G., Stookey, J.L., 1972. Chronic eosinophilic myositis in a rhesus monkey infected with sarcosporidiosis. *Vet. Pathol.* 9, 266–271.

1738. Theis, J.H., Ikeda, R.M., Ruddell, C.R., Tay, S., 1978. Apparent absence of *Sarcocystis* and low prevalence of *Trichinella* in artificially digested diaphragm muscle removed during post-mortem examination at a Sacramento (California) medical center. *Am. J. Trop. Med. Hyg.* 27, 837–839.

1739. Thomas, N.J., Dubey, J.P., Lindsay, D.S., Cole, R.A., Meteyer, C.U., 2007. Protozoal meningoencephalitis in sea otters (*Enhydra lutris*): A histopathological and immunohistochemical study of naturally occurring cases. *J. Comp. Pathol.* 137, 102–121.

1740. Thomas, V., Dissanaike, A.S., 1978. Antibodies to *Sarcocystis* in Malaysians. *Trans. R. Soc. Trop. Med. Hyg.* 72, 303–306.

1741. Thorton, R., 1987. Protozoal abortion in cattle. *Surveillance* 14, 15–16.

1742. Thulin, J.D., Granstrom, D.E., Gelberg, H.B., Morton, D.G., French, R.A., Giles, R.C., 1992. Concurrent protozoal encephalitis and canine distemper virus infection in a raccoon (*Procyon lotor*). *Vet. Rec.* 130, 162–164.

1743. Tian, M., Chen, Y., Wu, L., Rosenthal, B.M., Liu, X., He, Y., Dunams, D.B., Cui, L., Yang, Z., 2012. Phylogenetic analysis of *Sarcocystis nesbitti* (Coccidia: Sarcocystidae) suggests a snake as its probable definitive host. *Vet. Parasitol.* 183, 373–376.

1744. Tiegs, O.W., 1931. Note on the occurence of *Sarcocystis* in muscle of python. *Parasitology* 33, 412–414.

1745. Tietz, H.J., Montag, T., Brose, E., Hiepe, T., Mann, W., Hiepe, F., Halle, H., 1986. Extracts from *Sarcocystis gigantea* macrocysts are mitogenic for human blood lymphocytes. *Angew. Parasitol.* 27, 201–206.

1746. Tietz, H.J., Montag, T., Brose, E., Widera, P., Sokolowska-Köhler, W., Mann, W., Hiepe, T., 1989. Interactions between *Sarcocystis gigantea* lectin and toxin-containing fractions in human lymphocyte cultures. *Parasitol. Res.* 76, 32–35.

1747. Tietz, H.J., Montag, T., Brose, E., Widera, P., Kiessig, S.T., Mann, W., Hiepe, T., 1990. *Sarcocystis gigantea* lectin—mitogen and polyclonal B-cell activator. *Parasitol. Res.* 76, 332–335.

1748. Tietz, H.J., Montag, T., Volk, H.D., Brose, E., Gantenberg, R., Weichold, F.F., Hiepe, T., 1991. Activation of human CD4+ and CD8+ cells by *Sarcocystis-gigantea* lectin. *Parasitol. Res.* 77, 577–580.

1749. Tillmann, T., Kamino, K., Mohr, U., 1999. *Sarcocystis muris*—a rare case in laboratory mice. *Lab. Animals* 33, 390–392.

1750. Tillotson, K., McCue, P.M., Granstrom, D.E., Dargatz, D.A., Smith, M.O., Traub-Dargatz, J.L., 1999. Seroprevalence of antibodies to *Sarcocystis neurona* in horses residing in northern Colorado. *J. Equine Vet. Sci.* 19, 122–126.

1751. Timchuk, V., Dan'Shina, M.S., Abramyan, E.N., 1988. Effect of sarcocystin on the biochemical parameters of muscle tissue in rabbits. *Izv Akad Nauk Mold SSR Ser Biol Khim Nauk* 4, 29–32. (In Russian).

1752. Tinling, S.P., Cardinet, G.H., Blythe, L.L., Cohen, M., Vonderfecht, S.L., 1980. A light and electron microscopic study of sarcocysts in a horse. *J. Parasitol.* 66, 458–465.

1753. Titilincu, A., Mircean, V., Blaga, R., Bratu, C.N., Cozma, V., 2008. Epidemiology and etiology in sheep sarcocystosis. *Bulletin UASVM, Veterinary Medicine* 65, 49–54.

1754. Tomova, C., Geerts, W.J.C., Müller-Reichert, T., Entzeroth, R., Humbel, B.M., 2006. New comprehension of the apicoplast of *Sarcocystis* by transmission electron tomography. *Biol. Cell* 98, 535–545.

1755. Toribio, R.E., Bain, F.T., Mrad, D.R., Messer, N.T., Sellers, R.S., Hinchcliff, K.W., 1998. Congenital defects in newborn foals of mares treated for equine protozoal myeloencephalitis during pregnancy. *J. Am. Vet. Med. Assoc.* 212, 697–701.

1756. Tornquist, S.J., Boeder, L.J., Mattson, D.E., Cebra, C.K., Bildfell, R.J., Hamir, A.N., 2001. Lymphocyte responses and immunophenotypes in horses with *Sarcocystis neurona* infection. *Equine Vet. J.* 33, 726–729.

1757. Torres, P., Navarrete, N., Martin, R., Contreras, A., 1996. *Sarcocystis* sp. en el diafragma de un gato doméstico Valdivia, Chile. *Bol. Chil. Parasitol.* 51, 30–32.

1758. Tran, J.Q., de Leon, J.C., Li, C., Huynh, M.H., Beatty, W., Morrissette, N.S., 2010. RNG1 is a late marker of the apical polar ring in *Toxoplasma gondii*. *Cytoskeleton (Hoboken.)* 67, 586–598.

1759. Trasti, S.L., Dubey, J.P., Webb, D.M., Blanchard, T.W., Britt, J., Fritz, D., Lewis, R.M., 1999. Fatal visceral and neural sarcocystosis in dogs. *J. Comp. Pathol.* 121, 179–184.

1760. Traub-Dargatz, J.L., Schlipf, J.W. Jr., Granstrom, D.E., Ingram, J.T., Shelton, G.D., Getzy, D.M., Lappin, M.R., Baker, D.C., 1994. Multifocal myositis associated with *Sarcocystis* sp. in a horse. *J. Am. Vet. Med. Assoc.* 205, 1574–1576.

1761. Traver, D.S., Coffman, J.R., Moore, J.N., Nelson, S.L., 1978. Protozoal myeloencephalitis in sibling horses. *J. Equine Med. Surg.* 2, 425–428.

1762. Troedsen, C., Pamphlett, R., Collins, H., 1992. Is sarcocystosis common in Sydney? *Med. J. Aust.* 156, 136.

1763. Tropilo, J., Katkiewicz, M.T., Wisniewski, J., 2001. *Sarcocystis* spp. infection in free-living animals: Wild boar/*Sus scrofa* L./, deer/*Cervus elaphus* L./, roe deer/*Capreolus capreolus* L. *Pol. J. Vet. Sci.* 4, 15–18.

1764. Tungtrongchitr, A., Chiworaporn, C., Praewanich, R., Radomyos, P., Boitano, J.J., 2007. The potential usefulness of the modified Kato thick smear technique in the detection of intestinal sarcocystosis during field surveys. *Southeast Asian J. Trop. Med. Pub. Health* 38, 232–238.

1765. Turay, H.O., Barr, B.C., Caldwell, A., Branson, K.R., Cockrell, M.K.R., Marsh, A.E., 2002. *Sarcocystis neurona* reacting antibodies in Missouri feral domestic cats (*Felis domesticus*) and their role as intermediate host. *Parasitol. Res.* 88, 38–43.

1766. Uhrin, V., 1987. Occurence of Sarcocystosis in some animal species. *Veterinarstvi* 37, 547. (In Russian).

1767. Umbetaliev, S.S., 1979. Sarcosporidia of marmota. In: X Vsesoyuznaya Konferentsiya po prirodnoi ochagovosti boleznei. Dushanbe Te zaisy dobkladov 1, 148–149. (In Russian).

1768. Unterholzner, J., 1983. Zur entwicklung einer neuen vom hund übertragenen sarkosporidienart [*Sarcocystis* sp. (Z)] der ziege. Dissertation Universität Berlin, Germany, pp. 1–64.

1769. Uzuriaga S.M., Sam T.R., Manchego S.A., Alvarado S.A., 2008. Desarrollo de estadios asexuales de *Sarcocystis aucheniae* en cultivo de células. *Rev. Inv. Vet. Perú* 19, 49–53.

1770. Vaishnava, S., Morrison, D.P., Gaji, R.Y., Murray, J.M., Entzeroth, R., Howe, D.K., Striepen, B., 2005. Plastid segregation and cell division in the apicomplexan parasite *Sarcocystis neurona*. *J. Cell Sci.* 118, 3397–3407.

1771. Valadas, S., Gennari, S.M., Yai, L.E.O., Rosypal, A.C., Lindsay, D.S., 2010. Prevalence of antibodies to *Trypanosoma cruzi*, *Leishmania infantum*, *Encephalitozoon cuniculi*, *Sarcocystis neurona*, and *Neospora caninum* in capybara, *Hydrochoerus hydrochaeris*, from São Paulo State, Brazil. *J. Parasitol.* 96, 521–524.

1772. Valinezhad, A., Oryan, A., Ahmadi, N., 2008. *Sarcocystis* and its complications in camels (*Camelus dromedarius*) of eastern provinces of Iran. *Korean J. Parasitol.* 46, 229–234.

1773. Van den Enden, E., Praet, M., Joos, R., Van Gompel, A., Gigasse, P., 1995. Eosinophilic myositis resulting from sarcocystosis. *J. Trop. Med. Hyg.* 98, 273–276.

1774. van der Lugt, J.J., Markus, M.B., Kitching, J.P., Daly, T.J.M., 1994. Necrotic encephalitis as a manifestation of acute sarcocystosis in cattle. *J. S. Afr. Vet. Assoc.* 65, 119–121.

1775. van Knapen, F., Bouwman, D., Greve, E., 1987. Onderzoek naar het voorkomen van *Sarcocystis* spp. bij Nederlandse runderen met verschillende methoden. *Tijdschr. Diergeneeskd.* 112, 1095–1100.

1776. Vangeel, L., Houf, K., Chiers, K., Vercruysse, J., D'Herde, K., Ducatelle, R., 2007. Molecular-based identification of *Sarcocystis hominis* in Belgian minced beef. *J. Food Prot.* 70, 1523–1526.

1777. Vangeel, L., Houf, K., Geldhof, P., Nollet, H., Vercruysse, J., Ducatelle, R., Chiers, K., 2012. Intramuscular inoculation of cattle with *Sarcocystis* antigen results in focal eosinophilic myositis. *Vet. Parasitol.* 183, 224–230.

1778. Vangeel, L., Houf, K., Geldhof, P., De Preter, K., Vercruysse, J., Ducatelle, R., Chiers, K., 2013. Different *Sarcocystis* spp. are present in bovine eosinophilic myositis. *Vet. Parasitol.* 197, 543–548.

1779. Vardeleon, D., Marsh, A.E., Thorne, J.G., Loch, W., Young, R., Johnson, P.J., 2001. Prevalence of *Neospora hughesi* and *Sarcocystis neurona* antibodies in horses from various geographical locations. *Vet. Parasitol.* 95, 273–282.

1780. Vashisht, K., Lichtensteiger, C.A., Miller, L.A., Gondim, L.F.P., McAllister, M.M., 2005. Naturally occurring *Sarcocystis neurona*-like infection in a dog with myositis. *Vet. Parasitol.* 133, 19–25.

1781. Vasudevan, A., 1927. A case of sarcosporidial infection in man. *Indian J. Med. Res.* 15, 141–142.

1782. Velásquez, J.N., Di Risio, C., Etchart, C.B., Chertcoff, A.V., Mendez, N., Cabrera, M.G., Labbé, J.H., Carnevale, S., 2008. Systemic sarcocystosis in a patient with acquired immune deficiency syndrome. *Human Pathol.* 39, 1263–1267.

1783. Venu, R., Hafeez, M., 1999. Prepatent periods in dogs experimentally infected with *Sarcocystis* spp. *Indian Vet. J.* 76, 574–576.

1784. Venu, R., Hafeez, M., 2000. Prevalence of sarcocystic infections in slaughtered domestic ruminants in Tirupati (AP). *Indian Vet. J.* 77, 165–166.

1785. Venu, R., Hafeez, M., Ramakrishna Reddy, P., 2000. Morphometry of sporocysts of *Sarcocystis* spp. developing in the dog. *Indian Vet. J.* 77, 354–356.

1786. Vercruysse, J., van Marck, E., 1981. Les Sarcosporidies des petits ruminants au Sénégal. *Rev. Élev. Méd. Vét. Pays Trop.* 34, 377–382.

1787. Vercruysse, J., Fransen, J., van Goubergen, M., 1989. The prevalence and identity of *Sarcocystis* cysts in cattle in Belgium. *J. Vet. Med. B* 36, 148–153.

1788. Verma, S.K., Calero-Bernal, R., Lovallo, M.J., Sweeny, A., Grigg, M.E., Dubey, J.P., 2015. Detection of *Sarcocystis* spp. Infections in bobcats (*Lynx rufus*). *Vet. Parasitol.* In press.

1789. Vershinin, I.I., 1973. Developmental cycle of *Sarcocystis tenella*. *Veterinariia* 10, 75–78. (In Russian).

1790. Vetterling, J.M., Pacheco, N.D., Fayer, R., 1973. Fine structure of gametogony and oocyst formation in *Sarcocystis* sp. in cell culture. *J. Protozool.* 20, 613–621.

1791. Vickers, M.C., Brooks, H.V., 1983. Suspected *Sarcocystis* infection in an aborted bovine foetus. *N. Z. Vet. J.* 31, 166.

1792. Vilca, L.M., Durán, O.J., Ramos, D.D., Lucas, L.J., 2013. Saneamiento y eliminación de la toxicidad de carne de alpaca (*Vicugna pacos*) con sarcocistiosis mediante ahumado y curado. *Rev. Inv. Vet. Perú* 24, 537–543.

1793. Viles, J.M., Powell, E.C., 1976. The ultrastructure of the cyst wall of a murine *Sarcocystis*. *Z. Parasitenkd.* 49, 127–132.

1794. Viles, J.M., Powell, E.C., 1981. Myofiber damage accompanying intramuscular parasitism by *Sarcocystis muris*. *Tissue and Cell* 13, 45–60.

1795. Villar, D., Kramer, M., Howard, L., Hammond, E., Cray, C., Latimer, K., 2008. Clinical presentation and pathology of sarcocystosis in psittaciform birds: 11 cases. *Avian Dis.* 52, 187–194.

1796. Vlemmas, I., Kanakoudis, G., Tsangaris, Th., Theodorides, I., Kaldrymidou, E., 1989. Ultrastructure of *Sarcocystis tenella* (*Sarcocystis ovicanis*). *Vet. Parasitol.* 33, 207–217.

1797. Voigt, W.P., Heydorn, A.O., 1981. Chemotherapy of sarcosporidiosis and theileriosis in domestic animals. *Zbl. Bakt. Hyg. I. Abt. Orig. A* 250, 256–259.

1798. Volf, J., Modry, D., Koudela, B., Slapeta, J.R., 1999. Discovery of the life cycle of *Sarcocystis lacertae* Babudieri, 1932 (Apicomplexa: Sarcocystidae), with a species redescription. *Folia Parasitol. (Praha)* 46, 257–262.

1799. Votýpka, J., Hypsa, V., Jirku, M., Flegr, J., Vavra, J., Lukes, J., 1998. Molecular phylogenetic relatedness of *Frenkelia* spp. (Protozoa, Apicomplexa) to *Sarcocystis falcatula* Stiles 1893: Is the genus *Sarcocystis* paraphyletic? *J. Eukaryot. Microbiol.* 45, 137–141.

1800. Waap, H., Gomes, J., Nunes, T., 2013. Parasite communities in stray cat populations from Lisbon, Portugal. *J. Helminthol.* 88, 389–395.

1801. Wadakar, S.V., 1990. Studies on sarcocystosis and some aspects of other cyst forming coccidial infection in goats (*Capra hircus*) from Marathwada Region (Maharashtra). *J. Vet. Parasitol.* 4, 73–74.

1802. Wadajkar, S.V., Shastri, U.V., Narladkar, B.W., 1995. Caprine sarcocystosis: Clinical signs, gross and microscopic pathology. *Indian Vet. J.* 72, 224–228.

1803. Wahlström, K., Nikkilä, T., Uggla, A., 1999. *Sarcocystis* species in skeletal muscle of otter (*Lutra lutra*). *Parasitology* 118, 59–62.

1804. Wallace, G.D., 1973. *Sarcocystis* in mice inoculated with *Toxoplasma*-like oocysts from cat feces. *Science* 180, 1375–1377.

1805. Wallace, G.D., 1975. Observations on a feline coccidium with some characteristics of *Toxoplasma* and *Sarcocystis*. *Z. Parasitenkd.* 46, 167–178.

1806. Walzer, K.A., Adomako-Ankomah, Y., Dam, R.A., Herrmann, D.C., Schares, G., Dubey, J.P., Boyle, J.P., 2013. *Hammondia hammondi*, an avirulent relative of *Toxoplasma gondii*, has functional orthologs of known *T. gondii* virulence genes. *Proc. Natl. Acad. Sci. USA* 110, 7446–7451.

1807. Wang, G., Wei, T., Wang, X., Li, W., Zhang, P., Dong, M., Xiao, H., 1988. The morphology and life cycle of *Sarcocystis microps* n. sp. in sheep of Qinghai in China. *China Vet. Technol.* 6, 9–11. (In Chinese).

1808. Wang, M., Liu, H.H., Lin, Q.W., Xiao, B.N., Zhang, C.G., Gong, Z.F., 1989. A comparative ultrastructural study on the two types of sarcocysts in Chinese buffaloes. *Acta Vet. Zootech. Sinica* 20, 356–362.

1809. Wang, M., Xiao, B.N., Lin, Q.W., Zhang, C.G., Liu, H.H., 1991. Ultrastructural study on the endogeneous development of *Sarcocystis cruzi* in Chinese buffalo infected experimentally. *Acta Zoologica Sinica* 37, 1–5.

1810. Wang, M., Lin, Q.W., Liu, H.H., Xiao, B.N., Zhang, C.G., Gong, Z.F., 1991. Utrastructure of *Sarcocystis levinei* from Chinese buffalo. *Acta Agr. Univ. Pekin.* 17, 117–120.

1811. Wang, M., Lin, Q.W., Liu, H.H., Xiao, B.N., Zhang, C.G., Gong, Z.F., 1992. Comparative ultrastructural studies on the cysts of *Sarcocystis cruzi* in cattle and buffalo after cross-infections. *Acta Vet. Zootech. Sinica* 23, 347–353.

1812. Wasmuth, J.D., Pszenny, V., Haile, S., Jansen, E.M., Gast, A.T., Sher, A., Boyle, J.P., Boulanger, M.J., Parkinson, J., Grigg, M.E., 2012. Integrated bioinformatic and targeted deletion analyses of the SRS gene superfamily identify SRS29C as a negative regulator of *Toxoplasma* virulence. *mBio* 3, e00321–12.

1813. Weber, M., Weyreter, H., O'Donoghue, P.J., Rommel, M., Trautwein, G., 1983. Persistence of acquired immunity to *Sarcocystis miescheriana* infection in growing pigs. *Vet. Parasitol.* 13, 287–297.

1814. Wee, S.H., Shin, S.S., 2001. Experimental induction of the two-host life cycle of *Sarcocystis cruzi* between dogs and Korean native calves. *Korean J. Parasitol.* 39, 227–232.

1815. Wei, T., Chang, P.Z., Dong, M.X., Wang, X.Y., Xia, A.Q., 1985. Description of two new species of *Sarcocystis* from the yak (*Poephagus grunniens*). *Scientia Agricultura Sinica* 4, 80–85. (In Chinese).

1816. Wei, T., Wang, X.Y., Zhang, P.C., Dong, M.X., Li, W.Y., 1989. Identification of *Sarcocystis* sp. in yak fetus tissues. *Chin. J. Vet. Sci. Tech.* 11, 28–29. (In Chinese).

1817. Wei, T., Zhang, P.C., Dong, M.X., Wang, X.Y., 1990. Host spectrum of two *Sarcocystis* species from yak (*Poephagus grunniens*). *Journal of Chinese Traditional Veterinary Medicine* 5, 8–10, 29. (In Chinese).

1818. Welsch, T., Burek-Huntington, K., Savage, K., Rosenthal, B., Dubey, J.P., 2014. *Sarcocystis canis* associated hepatitis in a Steller sea lion (*Eumetopias jubatus*) from Alaska. *J. Wildl. Dis.* 50, 405–408.

1819. Welsh, T.H., Bryan, T.M., Johnson, L., Brinsko, S.P., Rigby, S.L., Love, C.C., Varner, D.D., Ing, N.H., Forrest, D.W., Blanchard, T.L., 2002. Characterization of sperm and androgen production by testes from control and ponazuril-treated stallions. *Theriogenology* 58, 389–392.

1820. Wendte, J.M., Miller, M.A., Lambourn, D.M., Magargal, S.L., Jessup, D.A., Grigg, M.E., 2010. Self-mating in the definitive host potentiates clonal outbreaks of the apicomplexan parasites *Sarcocystis neurona* and *Toxoplasma gondii*. *PLoS Genet.* 6, e1001261.

1821. Wendte, J.M., Miller, M.A., Nandra, A.K., Peat, S.M., Crosbie, P.R., Conrad, P.A., Grigg, M.E., 2010. Limited genetic diversity among *Sarcocystis neurona* strains infecting southern sea otters precludes distinction between marine and terrestrial isolates. *Vet. Parasitol.* 169, 37–44.

1822. Wenzel, R., Erber, M., Boch, J., Schellner, H.P., 1982. Sarkosporidien-Infektion bei Haushuhn, Fasan und Bleßhuhn. *Berl. Münch. Tierärztl. Wochenschr.* 95, 188–193.

1823. Wesemeier, H.H., Sedlaczek, J., 1995. One known *Sarcocystis* species and one found for the first time in fallow deer (*Dama dama*). *Appl. Parasitol.* 36, 299–302.

1824. Wesemeier, H.H., Sedlaczek, J., 1995. One known *Sarcocystis* species and two found for the first time in red deer and wapiti (*Cervus elaphus*) in Europe. *Appl. Parasitol.* 36, 245–251.

1825. Wesemeier, H.H., Odening, K., Walter, G., Bockhardt, I., 1995. The black-backed jackal (Carnivora: Canidae) in Namibia as intermediate host of two *Sarcocystis* species (Protozoa: Sarcocystidae). *Parasite* 2, 391–394.

1826. Weyreter, H., O'Donoghue, P.J., 1982. Untersuchungen zur Immunoserodiagnose der *Sarcocystis*-Infektionen. I. Antikörperbildung bei Maus und Schwein. *Zbl. Bakt. Hyg. I. Abt. Orig. A* 253, 407–416.

1827. Weyreter, H., O'Donoghue, P.J., Weber, M., Rommel, M., 1984. Class-specific antibody responses in pigs following immunization and challenge with sporocysts of *Sarcocystis miescheriana*. *Vet. Parasitol.* 16, 201–205.

1828. Wilairatana, P., Radomyos, P., Radomyos, B., Phraevanich, R., Plooksawasdi, W., Chanthavanich, P., Viravan, C., Looareesuwan, S., 1996. Intestinal sarcocystosis in Thai laborers. *Southeast Asian J. Trop. Med. Pub. Health* 27, 43–46.

1829. Wilson, A.P., Thelen, J.J., Lakritz, J., Brown, C.R., Marsh, A.E., 2004. The identification of a sequence related to apicomplexan enolase from *Sarcocystis neurona*. *Parasitol. Res.* 94, 354–360.

1830. Witonsky, S.G., Gogal, R.M., Duncan, R.B., Lindsay, D.S., 2003. Protective immune response to experimental infection with *Sarcocystis neurona* in C57BL/6 mice. *J. Parasitol.* 89, 924–931.

1831. Witonsky, S.G., Gogal, R.M., Duncan, R.B., Lindsay, D.S., 2003. Immunopathologic effects associated with *Sarcocystis neurona*-infected interferon-gamma knockout mice. *J. Parasitol.* 89, 932–940.

1832. Witonsky, S.G., Gogal, R.M., Duncan, R.B., Norton, H., Ward, D., Yang, J., Lindsay, D.S., 2005. Humoral immunity is not critical for protection against experimental infection with *Sarcocystis neurona* in B-cell-deficient mice. *J. Parasitol.* 91, 830–837.

1833. Witonsky, S.G., Ellison, S., Yang, J., Gogal, R.M., Lawler, H., Suzuki, Y., Sriranganathan, N., Andrews, F., Ward, D., Lindsay, D.S., 2008. Horses experimentally infected with *Sarcocystis neurona* develop altered immune responses *in vitro*. *J. Parasitol.* 94, 1047–1054.

1834. Wobeser, G., Leighton, F.A., Cawthorn, R.J., 1981. Occurrence of *Sarcocystis* Lankester, 1882, in wild geese in Saskatchewan. *Can. J. Zool.* 59, 1621–1624.

1835. Woldemeskel, M., Gebreab, F., 1996. Prevalence of sarcocysts in livestock of northwest Ethiopia. *Zentralbl. Veterinarmed. [B]* 43, 55–58.

1836. Woldemeskel, M., Gumi, B., 2001. Prevalence of sarcocysts in one-humped camel (*Camelus dromedarius*) from southern Ethiopia. *J. Vet. Med. B* 48, 223–226.

1837. Wolfe, A., Hogan, S., Maguire, D., Fitzpatrick, C., Vaughan, L., Wall, D., Hayden, T.J., Mulcahy, G., 2001. Red foxes (*Vulpes vulpes*) in Ireland as hosts for parasites of potential zoonotic and veterinary significance. *Vet. Rec.* 149, 759–763.

1838. Wong, K.T., Pathmanathan, R., 1992. High prevalence of human skeletal muscle sarcocystosis in southeast Asia. *Trans. R. Soc. Trop. Med. Hyg.* 86, 631–632.

1839. Woodmansee, D.B., Powell, E.C., 1984. Cross-transmission and *in vitro* excystation experiments with *Sarcocystis muris*. *J. Parasitol.* 70, 182–183.

1840. Wouda, W., Snoep, J.J., Dubey, J.P., 2006. Eosinophilic myositis due to *Sarcocystis hominis* in a beef cow. *J. Comp. Pathol.* 135, 249–253.

1841. Wünschmann, A., Rejmanek, D., Cruz-Martínez, L., Barr, B.C., 2009. *Sarcocystis falcatula*-associated encephalitis in a free-ranging great horned owl (*Bubo virginianus*). *J. Vet. Diagn. Invest.* 21, 283–287.

1842. Wünschmann, A., Rejmanek, D., Conrad, P.A., Hall, N., Cruz-Martinez, L., Vaughn, S.B., Barr, B.C., 2010. Natural fatal *Sarcocystis falcatula* infections in free-ranging eagles in North America. *J. Vet. Diagn. Invest.* 22, 282–289.

1843. Wünschmann, A., Armien, A.G., Reed, L., Gruber, A.D., Olias, P., 2011. *Sarcocystis calchasi*-associated neurologic disease in a domestic pigeon in North America. *Transbound. Emerg. Dis.* 58, 526–530.

1844. Xiang, Z., Chen, X., Yang, L., He, Y., Jiang, R., Rosenthal, B.M., Luan, P., Attwood, S.W., Zuo, Y., Zhang, Y.P., Yang, Z., 2009. Non-invasive methods for identifying oocysts of *Sarcocystis* spp. from definitive hosts. *Parasitol. Int.* 58, 293–296.

1845. Xiang, Z., Rosenthal, B.M., He, Y., Wang, W., Wang, H., Song, J., Shen, P.Q., Li, M.L., Yang, Z., 2010. *Sarcocystis tupaia*, sp. nov., a new parasite species employing treeshrews (Tupaiidae, *Tupaia belangeri chinensis*) as natural intermediate hosts. *Parasitol. Int.* 59, 128–132.

1846. Xiang, Z., He, Y., Zhao, H., Rosenthal, B.M., Dunams, D.B., Li, X., Zuo, Y., Feng, G., Cui, L., Yang, Z., 2011. *Sarcocystis cruzi*: Comparative studies confirm natural infections of buffaloes. *Exp. Parasitol.* 127, 460–466.

1847. Xiao, B., Zhang, C., Gong, Z., Wang, M., 1991. Studies on cross infection of *Sarcocystis cruzi* between water buffalo and cattle. *Chin. J. Zool.* 26, 1–3. (In Chinese).

1848. Xiao, B.N., Gong, Z.F., Li, Y., Zhang, C.G., Zeng, D.N., Wang, M., 1992. Study on the development of *Sarcocystis cruzi* in dogs. *Chin. J. Zool.* 27, 1–3. (In Chinese).

1849. Xiao, B., Zeng, D., Zhang, C., Wang, M., Li, Y., Gong, Z., 1993. Development of *Sarcocystis cruzi* in buffalo (*Bubalus bubalis*) and cattle (*Bos taurus*). *Acta Vet. Zootech. Sinica* 24, 185–192. (In Chinese).

1850. Yabsley, M.J., Jordan, C.N., Mitchell, S.M., Norton, T.M., Lindsay, D.S., 2007. Seroprevalence of *Toxoplasma gondii*, *Sarcocystis neurona*, and *Encephalitozoon cuniculi* in three species of lemurs from St. Catherines Island, GA, USA. *Vet. Parasitol.* 144, 28–32.

1851. Yabsley, M.J., Ellis, A.E., Stallknecht, D.E., Howerth, E.W., 2009. Characterization of *Sarcocystis* from four species of hawks from Georgia, USA. *J. Parasitol.* 95, 256–259.

1852. Yamada, M., Yukawa, M., Mochizuki, K., Sekikawa, H., Kenmotsu, M., 1990. *Sarcocystis* in Murray Grey stock cattle introduced from Australia. *Nippon Juigaku Zasshi* 52, 883–885.

1853. Yamada, M., Yukawa, M., Sekikawa, H., Kenmotsu, M., Mochizuki, K., 1993. Studies on the morphology of sarcocysts in Thoroughbred horses in Japan. *J. Protozool. Res.* 3, 14–19.

1854. Yan, W., Qian, W., Li, X., Wang, T., Ding, K., Huang, T., 2013. Morphological and molecular characterization of *Sarcocystis miescheriana* from pigs in the central region of China. *Parasitol. Res.* 112, 975–980.

1855. Yang, J., Ellison, S., Gogal, R., Norton, H., Lindsay, D.S., Andrews, F., Ward, D., Witonsky, S., 2006. Immune response to *Sarcocystis neurona* infection in naturally infected horses with equine protozoal myeloencephalitis. *Vet. Parasitol.* 138, 200–210.

1856. Yang, Z.Q., Zuo, Y.X., Yao, Y.G., Chen, X.W., Yang, G.C., Zhang, Y.P., 2001. Analysis of the 18S rRNA genes of *Sarcocystis* species suggests that the morphologically similar organisms from cattle and water buffalo should be considered the same species. *Mol. Biochem. Parasitol.* 115, 283–288.

1857. Yang, Z.Q., Zuo, Y.X., Ding, B., Chen, X.W., Luo, J., Zhang, Y.P., 2001. Identification of *Sarcocystis hominis*-like (Protozoa: Sarcocystidae) cyst in water buffalo (*Bubalus bubalis*) based on 18S rRNA gene sequences. *J. Parasitol.* 87, 934–937.

1858. Yang, Z.Q., Li, Q.Q., Zuo, Y.X., Chen, X.W., Chen, Y.J., Nie, L., Wei, C.G., Zen, J.S., Attwood, S.W., Zhang, X.Z., Zhang, Y.P., 2002. Characterization of *Sarcocystis* species in domestic animals using a PCR-RFLP analysis of variation in the 18S rRNA gene: A cost-effective and simple technique for routine species identification. *Exp. Parasitol.* 102, 212–217.

1859. Yang, Z.Q., Wei, C.G., Zen, J.S., Song, J.L., Zuo, Y.X., He, Y.S., Zhang, H.F. et al. 2005. A taxonomic re-appraisal of *Sarcocystis nesbitti* (Protozoa: Sarcocystidae) from the monkey *Macaca fascicularis* in Yunnan, PR China. *Parasitol. Int.* 54, 75–81.

1860. Yantis, D., Moeller, R., Braun, R., Gardiner, C.H., Aguirre, A., Dubey, J.P., 2003. Hepatitis associated with a *Sarcocystis canis*-like protozoan in a Hawaiian monk seal (*Monachus schauinslandi*). *J. Parasitol.* 89, 1258–1260.

1861. Yaziroglu, O., Beyazit, A., 2005. Encephalomyelitis associated with a *Sarcocystis*-like protozoan in a ten-month-old ewe lamb. *Turk. J. Vet. Anim. Sci.* 29, 1209–1212.

1862. Yeargan, M.R., Howe, D.K., 2011. Improved detection of equine antibodies against *Sarcocystis neurona* using polyvalent ELISAs based on the parasite SnSAG surface antigens. *Vet. Parasitol.* 176, 16–22.

1863. Yeargan, M.R., Alvarado-Esquivel, C., Dubey, J.P., Howe, D.K., 2013. Prevalence of antibodies to *Sarcocystis neurona* and *Neospora hughesi* in horses from Mexico. *Parasite* 20, 29.

1864. Young, A.R., Meeusen, E.N., 2004. Galectins in parasite infection and allergic inflammation. *Glycoconjugate J.* 19, 601–606.

1865. Yu, S., 1991. Field survey of sarcocystis infection in the Tibet autonomous region. *Zhongguo Yi. Xue. Ke. Xue. Yuan Xue. Bao.* 13, 29–32. (In Chinese).

1866. Zacarías, S.F., Sam, T.R., Ramos, D.D., Lucas, A.O., Lucas, L.J., 2013. Técnicas de aislamiento y purificación de ooquistes de *Sarcocystis aucheniae* a partir de intestino del gado de perros experimentalmente infectados. *Rev. Inv. Vet. Perú* 24, 396–403.

1867. Zaman, V., Colley, F.C., 1972. Fine structure of *Sarcocystis fusiformis* from the Indian water buffalo (*Bubalus bubalis*) in Singapore. *Southeast Asian J. Trop. Med. Pub. Health* 3, 489–495.

1868. Zaman, V., Colley, F.C., 1975. Light and electron microscopic observations of the life cycle of *Sarcocystis orientalis* sp. n. in the rat (*Rattus norvegicus*) and the Malaysian reticulated python (*Python reticulatus*). *Z. Parasitenkd.* 47, 169–185.

1869. Zaman, V., 1976. Host range of *Sarcocystis orientalis*. *Southeast Asian J. Trop. Med. Pub. Health* 7, 112.

1870. Zaman, V., Colley, F.C., 1976. Replacement of *Sarcocystis orientalis* Zaman and Colley, 1975, by *Sarcocystis singaporensis* sp. n. *Z. Parasitenkd.* 51, 137.

1871. Zaman, V., Robertson, T.A., Papadimitriou, J.M., 1980. Scanning electron microscopy of *Sarcocystis fusiformis* from the water buffalo (*Bubalus bubalis*). *Southeast Asian J. Trop. Med. Pub. Health* 11, 205–211.

1872. Žąsitytė, R., Grikienienė, J., 2002. Some data on endoparasites of common mole in Lithuania. *Acta Zool. Lituanica* 12, 403–409.

1873. Zayed, A.A., El-Ghaysh, A., 1998. Pig, donkey and buffalo meat as a source of some coccidian parasites infecting dogs. *Vet. Parasitol.* 78, 161–168.

1874. Zeman, D.H., Dubey, J.P., Robison, D., 1993. Fatal hepatic sarcocystosis in an American black bear. *J. Vet. Diagn. Invest.* 5, 480–483.

1875. Zeve, V.H., Price, D.L., Herman, C.M., 1966. Electron microscope study of *Sarcocystis* sp. *Exp. Parasitol.* 18, 338–346.

1876. Zhang, D., Gaji, R.Y., Howe, D.K., 2006. Identification of a dithiol-dependent nucleoside triphosphate hydrolase in *Sarcocystis neurona*. *Int. J. Parasitol.* 36, 1197–1204.

1877. Zhang, D., Howe, D.K., 2008. Investigation of SnSPR1, a novel and abundant surface protein of *Sarcocystis neurona* merozoites. *Vet. Parasitol.* 152, 210–219.

1878. Zhang, Y.H., Zuo, Y.X., 1992. Ultrastructures of cysts of *Sarcocystis* in water buffalo and cattle. *Acta Zoologica Sinica* 38, 440–441. (In Chinese).

1879. Zhu, B.Y., Hartigan, A., Reppas, G., Higgins, D.P., Canfield, P.J., Slapeta, J., 2009. Looks can deceive: Molecular identity of an intraerythrocytic apicomplexan parasite in Australian gliders. *Vet. Parasitol.* 159, 105–111.

1880. Zielasko, B., Petrich, J., Trautwein, G., Rommel, M., 1981. Untersuchungen über pathologisch-anatomische Veränderungen und die Entwicklung der Immunität bei der *Sarcocystis suicanis*-Infektion. *Berl. Münch. Tierärztl. Wochenschr.* 94, 223–228.

1881. Zimmermann, U., Schnieder, T., Rommel, M., 1984. Untersuchungen über die Dynamik der Antikörperentwicklung bei Schweinen nach mehrfacher Immunisierung mit Sporozysten von *Sarcocystis miescheriana* und einmaliger Belastungsinfektion. *Berl. Münch. Tierärztl. Wochenschr.* 97, 408–411.

1882. Zlotnik, I., 1959. *Toxoplasma* in sheep. *Lancet* 2, 295.

1883. Zobba, R., Alberti, A., Manunta, M.L., Evangelisti, M.A., Parpaglia, M.L.P., 2014. What is your diagnosis? Cerebrospinal fluid from a sheep. *Vet. Clin. Pathol.* 2014, 1–2.

1884. Zuo, Y.X., Chen, F.Q., Li, W.Y., 1982. Two patients with *Sarcocystis* infection. In: Jiang, J.B., Arnold, K., Chang, K.P. (Eds.), Malaria and Other Protozoal Infections. *Proceedings of Chinese Soc. Protozoologists and Zhongshan (Sun Yatsen) University*, Guangzhou, China, pp. 52–53.

1885. Zuo, Y.X., Chen, F.Q., Li, W.T., 1983. *Sarcocystis* in human—report of two cases of human *Sarcocystis* intestinal infection. *Yunnan Medicine* 4, 354–358. (In Chinese).

1886. Zuo, Y.X., Chen, F.Q., Chen, X.W., Tang, D.H., 1988. Studies on *Sarcocystis* species of water buffalo with description of *Sarcocystis* sp. *Journal of Yunnan University* 10, 91–92. (In Chinese).

1887. Zuo, Y.X., Zhang, Y.H., Yie, B., 1990. A new *Sarcocystis* species from water buffalo. *Sarcocystis sinensis* sp. nov. In: *Fifth Symposium of the Chinese Society of Protozoology*. Chongqing, pp. 82–83. (In Chinese).

1888. Zuo, Y.X., Zhou, Z.B., Li, W.T., Lian, Z.Q., Xu, C.E., 1991. The prevalence of *Sarcocystis suihominis* in the natives of Yunnan Province, China. *Chin. Sci. Bull.* 36, 965–967. (In Chinese).

1889. Zuo, Y.X., 1992. *Coccidians in Livestock and Birds and Human Coccidiosis*. Science and Technology Publishing Company of Tianjin, Tianjin. (In Chinese).

1890. Zuo, Y.X., Chen, X.W., Li, Y.J., Ma, T.C., Tang, D.H., Fan, L.X., Zhao, M.L., 1995. Studies on *Sarcocystis* species of cattle and water buffalo with description of a new species of *Sarcocystis*. In: *Proceedings of the Tenth Anniversary of the Founding of China Parasitological Society*, Chinese Science and Technology Press, Beijing, PR China, pp. 20–24. (In Chinese).

1891. Anonymous, 2012. Notes from the field: Acute muscular sarcocystosis among returning travelers—Tioman Island, Malaysia, 2011. Morbidity and Mortality Weekly Report 61, 37–38.

1892. Abel, K., 1963. Untersuchungen über das Vorkomen und das saisonbedingte Auftreten von *Sarcocystis miescheriana* bei Schlachtschweinen am Schlachthof Leipzig. *Mh. Vet. Med.* 18, 621–624.

1893. Achuthan, H.N., 1983. *Sarcocystis* and sarcocystosis in buffalo (*Bubalus bubalis*) calves. *Indian Vet. J.* 60, 344–346.

1894. Adams, J.H., Levine, N.D., Todd, K.S.J., 1981. *Eimeria* and *Sarcocystis* in raccoons in Illinois. *J. Protozool.* 28, 221–222.

1895. Afshar, A., Naghshineh, R., Neshat, H., 1974. Incidence of sarcosporidiosis in sheep in Iran. *Trop. Anim. Health Prod.* 6, 192.

1896. Agarwal, M.C., Shah, H.L., Sharma, R.K., Singh, K.P., 1987. A tissue mince method for detection of intact cysts of *Sarcocystis* in musculature. *J. Vet. Parasitol.* 1, 67–68.

1897. Agholi, M., Heidarian, H.R., Moghadami, M., Hatam, G.R., 2014. First detection of acalculous cholecystitis associated with *Sarcocystis* infection in a patient with AIDS. *Acta Parasitol.* 59, 310–315.

1898. Agrawal, R.D., Chauhan, P.P.S., Ahluwalia, S.S., 1982. Occurrence of *Sarcocystis* sp. (Protozoa: Eimeriidae) in the oesophagus of a goral. *Naemorhedus goral. Indian J. Parasitol.* 6, 115–116.

1899. Akün, R., Holz, J., 1955. Histochemische Untersuchungen an Sarkozysten. *Monatshefte für Tierheilkunde* 7, 49–52.

1900. Altfeld, E., 1988. Auswirkung en der akuten Sarkozystiose auf einige gerinnungsphysiologische Parameter beim Schwein. Inaugural Dissertation, Institute of Parasitology, Hannover, Germany, pp. 1–68.

1901. Andrews, C.L., Davidson, W.R., 1980. Endoparasites of selected populations of cottontail rabbits (*Sylvilagus floridanus*) in the southeastern United States. *J. Wildl. Dis.* 16, 395–401.

1902. Apostoloff, E., Hiepe, F., 1980. Immunfluoreszenzmikroskopischer Nachweis von ds- DNS in Sarkosporidien. *Monatsh. Veterinärmed.* 35, 615–617.

1903. Arambulo, P.V., Tongson, M.S., Sarmiento, R.V., 1972. Sarcosporidiosis in Philippine carabaos. *Phil. J. Vet. Med.* 11, 53–59.

1904. Arcay-De-Peraza, L., 1966. The use of *Sarcocystis-tenella* "spores" in a new agglutination test for sarcosporidiosis. *Trans. R. Soc. Trop. Med. Hyg.* 60, 761–765.

1905. Arru, E., Cosseddu, A.M., Nieddu, A.M., 1976. La sarcosporidiosi nei carnivore della Sardegna. *Att. della Società delle Scienze Veterinarie* 29, 609–610.

1906. Arru, E., Cosseddu, A.M., Tarantini, S., 1978. L'immunofluorescenza nella diagnosi della sarcosporidiosi ovina e suina. *Clinica Veterinaria* 101, 195–200.

1907. Arther, R.G., Post, G., 1977. Coccidia of coyotes in eastern Colorado. *J. Wildl. Dis.* 13, 97–100.

1908. Aryeetey, M.E., Piekarski, G., 1978. Diaplazentarer Übergang von *Sarcocystis*-Antikörpern bei Mensch und Ratten. *Z. Parasitenkd.* 56, 211–218.

1909. Aryeetey, M.E., 1979. Morphologische und serologische Untersuchungen an Sarkosporidien (Protozoa, Sporozoa). Inaugural Dissertation, Universität Düsseldorf, Germany, pp. 1–82.

1910. Ashford, R.W., 1977. The fox, *Vulpes vulpes*, as a final host for *Sarcocystis* of sheep. *Ann. Trop. Med. Parasitol.* 71, 29–34.

1911. Avapal, R.S., Sharma, J.K., Juyal, P.D., 2003. Comparison of techniques to detect *Sarcocystis* infection in slaughtered pigs. *J. Vet. Parasitol.* 17, 65–66.

1912. Awad, F.I., 1973. The transmission of *Sarcocystis tenella* in sheep. *Z. Parasitenkd.* 42, 43–48.

1913. Balmer, T.V., Evans, E., Herbert, I.V., 1982. Prevalence of *Sarcocystis* species and other parasites in hunting dogs in Gwynedd, North Wales. *Vet. Rec.* 110, 331–332.

1914. Barutzki, D., 1980. Untersuchungen über die Wirksamkeit handelsüblicher Desinfektionsmittel auf Kokzidien-Oozysten bzw. Sporozysten (*Eimeria, Cystoisospora, Toxoplasma* und *Sarcocystis*) sowie auf Spulwurmeier (*Ascaris, Toxocara*) im Suspensionsversuch. Inaugural Dissertation. Ludwig-Maximilians-Universität München, Germany, pp. 1–44.

1915. Basson, P.A., McCully, R.M., Kruger, S.P., van Niekerk, J.W., Young, E., de Vos, V., 1970. Parasitic and other diseases of the African buffalo in the Kruger National Park. *Onderstepoort J. Vet. Res.* 37, 11–28.

1916. Batistic, B., 1965. Sarkosporidioza nekih vrsta životinja i ljundi na području Bosne i Hercegovine. Rasirenost, morfologija sarkocista i patomorfološke promjene. *Veterinaria, Sarajevo* 14, 45–64.

1917. Beaudette, F.R., 1941. Sarcosporidiosis in a black duck. *J. Am. Vet. Med. Assoc.* 99, 52–53.

1918. Bergler, K.-G., 1979. Untersuchungen zur Exzystierung und Überlebensfähigkeit von Sporozysten und Oozysten verschiedener Kokzidienarten (*Sarcocystis, Toxoplasma, Hammondia, Eimeria*). Inaugural Dissertation. Ludwig-Maximilians-Universität München, Germany, pp. 1–53.

1919. Bierschenck, A., 1979. Das Vorkommen von Sarkospoidien bei Schlachtschafen in Süddeutschland. Inaugural Dissertation. Ludwig-Maximilians-Universität München, Germany, pp. 1–41.

1920. Blandino, T., Gómez, E., Merino, N., Alonso, M., 1983. Presencia de *Sarcocystis cruzi* (Protozoa: Sarcosporidia) en perros diarreicos en Cuba. *Rev. Salud Anim.* 5, 529–534.

1921. Blewaska, S., 1981. Versuche zur Steigerung der Befallsextensität und -intensität der *Sarcocystis muris*-infektion. Inaugural Dissertation. Institute of Parasitology, Hannover, Germany, pp. 1–56.

1922. Boch, J., Böhm, A., Weiland, G., 1979. Die Kokzidien-Infektionen (*Isospora, Sarcocystis, Hammondia, Toxoplasma*) des Hundes. *Berl. Münch. Tierärztl. Wschr.* 92, 240–243.

1923. Boch, J., Walter, D., 1979. Vier verschiedene Kokzidienarten bei Katzen in Süddeutschland. *Tierärztl. Umschau* 34, 749–752.

1924. Boch, J., Mannl, A., Weiland, G., Erber, M., 1980. Die Sarkosporidiose des Hundes—Diagnose und Therapie. *Praktische Tierarzt.* 61, 636–640.

1925. Bode, H.-C., 1988. Einflüsse einer *Sarcocystis miescheriana*-Infektion auf den Stickstoff- und Mineralstoffhaushalt von Mastschweinen. Inaugural Dissertation. Institute of Parasitology, Hannover, Germany, pp. 7–89.

1926. Bordjochki, A., Conitch, V., Petrovitch, Z., Savin, Z., 1972. Diagnostic de la sarcosporidiose par l'immunofluorescence indirecte. *Rec. Méd. Vét.* 148, 217–224.

1927. Bordjochki, A., Savin, Z., Sovitsch, M., 1977. Modifications des propriétés antigéniques de *Sarcocystis tenella* exposé à températures différentes. *Bull. Acad. Vét. France* 50, 71–74.

1928. Bordjochki, A., Conitch, V., Savin, Z., Khanfar, H.M., Katitch, S., 1978. Valeur de la fixation du complément et de l'immunofluorescence indirecte dans la détection des anticorps spécifiques de plusieurs espèces de sarcosporidies. *Bull. Acad. Vét. France* 51, 189–195.

1929. Borst, G.H.A., Zwart, P., 1973. Sarcosporidiosis in psittaciformes. *Z. Parasitenkd.* 42, 293–298.

1930. Bosco, A., Rosmini, R., 1984. Incidenza delle sarcosporidiosi negli animali da macello. *Att. Soc. Ital. Sci. Vet.* 38, 618–621.

1931. Böttner, A., 1984. Vorkommen sowie licht- und elektronen- mikroskopische Differenzierung von Sarkosporidienarten bei Schlachtrindern auf der Nordinsel Neuseelands. Inaugural Dissertation. Institute of Parasitology, Hannover, Germany and Massey University Palmerston North, New Zealand, pp. 1–71.

1932. Box, E.D., Duszynski, D.W., 1980. *Sarcocystis* of passerine birds: Sexual stages in the opossum (*Didelphis virginiana*). *J. Wildl. Dis.* 16, 209–215.

1933. Bratberg, B., Landsverk, T., 1980. *Sarcocystis* infection and myocardial pathological changes in cattle from south-eastern Norway. *Acta Vet. Scand.* 21, 395–401.

1934. Bratberg, B., Helle, O., Hilali, M., 1982. *Sarcocystis* infection in sheep from south-western Norway. *Acta Vet. Scand.* 23, 221–234.

1935. Brindani, F., 1980. La sarcosporidiosi negli animali domestici e nell'uomo. *Nuovo Progresso Veterinario* 35, 571–572.

1936. Britt, D.P., Baker, J.R., 1983. Cysts of *Sarcocystis tenella* in North Ronaldsay sheep. *Vet. Rec.* 113, 516.

1937. Brown, R.J., Carney, W.P., Sudomo, M., Simandjuntak, G., 1975. *Sarcocystis* sp. in the myocardium of a water buffalo from Sulawesi (Celebes), Indonesia. *Southeast Asian J. Trop. Med. Pub. Health* 6, 284.

1938. Brugère-Picoux, J., Lacombre, B., 1987. La sarcosporidiose chez les ruminants et les suides domestiques ou sauvages. *Bull. Soc. Vét. Prat. Fr.* 71, 509–527.

1939. Bryan, H.M., Darimont, C.T., Hill, J.E., Paquet, P.C., Thompson, R.C.A., Wagner, B., Smits, J.E.G., 2012. Seasonal and biogeographical patterns of gastrointestinal parasites in large carnivores: Wolves in a coastal archipelago. *Parasitology* 139, 781–790.

1940. Bwangamoi, O., Rottcher, D., Wekesa, C., 1990. Rabies, microbesnoitiosis and sarcocystosis in a lion. *Vet. Rec.* 127, 411.

1941. Carreno, R.A., Barta, J.R., 1999. An eimeriid origin of isosporoid coccidia with Stieda bodies as shown by phylogenetic analysis of small subunit ribosomal RNA gene sequences. *J. Parasitol.* 85, 77–83.

1942. Casagrande, R.A., Cesar, M.O., Pena, H.F.J., Zwarg, T., Teixeira, R.H.F., Nunes, A.L.V., Neves, D.V.D.A., Gomes, M., Quagglia Neto, F., Milanello, L., Fontenelle, J.H., Matushima, E.R., 2009. Occurrence of *Sarcocystis* spp. in opossums (*Didelphis aurita* and *Didelphis albiventris*) in regions of the state of São Paulo, Brazil. *Braz. J. Vet. Res. Anim. Sci.* 46, 101–106.

1943. Ceretto, F., Julini, M., Cravero, G., Coscia, G., 1987. Reperti ispettivi della sarcosporidiosi in bovini da macello. *Obiettivi e Documenti Veterinari* 8, 67–71.

1944. Černá, Z., Merhautová, V., 1981. Sarcocystosis in cattle and sheep at Prague abattoir. *Folia Parasitol. (Praha)* 28, 125–129.

1945. Červa, L., Macha, J., Gut, J., Procházková, Z., 1982. *Sarcocystis gigantea*: A lectin in muscle cysts. *Z. Parasitenkd.* 67, 349–350.

1946. Challis, J.R.G., Mitchell, B.F., 1983. Endocrinology of pregnancy and parturition. In: Washaw, J.B. (Ed.), *The Biological Basis of Reproductive and Developmental Medicine*. Elsevier Biochemical, New York, pp. 105–139.

1947. Chambers, P.G., 1987. Carcass and offal condemnations at meat inspection in Zimbabwe. *Zimbabwe Vet. J.* 18, 11–18.

1948. Chaudhry, R.K., Kushawh, H.S., Shah, H.L., 1985. Biochemistry of the sarcocyst of *Sarcocystis fusiformis* of buffalo *Bubalus bubalis*. *Vet. Parasitol.* 17, 295–298.

1949. Chaudhry, R.K., Kushwah, H.S., Shah, H.L., 1986. Biochemical and histochemical studies of the sarcocyst of *Sarcocystis fusiformis* of buffalo *Bubalus bubalis*. *Vet. Parasitol.* 17, 271–273.

1950. Chaudhry, R.K., Shah, H.L., 1987. Histochemical observations on sarcocysts of *Sarcocystis cruzi* of the cattle. *Indian J. Anim. Sci.* 57, 690–691.

1951. Chauhan, P.P.S., Bhatia, B.B., Agrawal, R.D., Katara, R.P., Ahluwalia, S.S., 1977. On the gametogonic development of bubaline *Sarcocystis fusiformis* in pups-an experimental study. *Indian J. Exp. Biol.* 15, 492–494.

1952. Chauhan, P.P.S., Agrawal, R.D., Arora, G.S., 1978. Incidence of *Sarcocystis fusiformis* in Indian buffaloes. *Indian J. Parasitol.* 2, 123–124.

1953. Chávez, J.A., Volcán, G.S., 1986. Frecuencia de *Sarcocystis* sp. en el Ganado suino de Ciudad Bolívar. In: Volcán, G.S. (Eds.), *Cuadernos de Geografía Médica de Guayana*, Estado Bolívar, Venezuela, pp. 16–24.

1954. Chen, L.Y., Zhou, B.J., Li, C.Y., Yang, L.J., Yang, Z.Q., 2008. A review of *Sarcocystis* epidemiological studies. *Chin. J. Zoonoses* 24, 675–678. (In Chinese).

1955. Chhabra, M.B., Mahajan, R.C., 1978. *Sarcocystis* sp from the goat in India. *Vet. Rec.* 103, 562–563.

1956. Christie, E., Dubey, J.P., Pappas, P.W., 1976. Prevalence of *Sarcocystis* infection and of other intestinal parasitisms in cats from a humane shelter in Ohio. *J. Am. Vet. Med. Assoc.* 168, 421–422.

1957. Clark, G.M., 1958. *Sarcocystis* in certain birds. *J. Parasitol.* 44 (Suppl), 41.

1958. Collins, G.H., Charleston, W.A.G., 1979. Studies on *Sarcocystis* species. I. Feral cats as definitive hosts for sporozoa. *N. Z. Vet. J.* 27, 80–84.

1959. Collins, G.H., 1980. Host reaction to *Sarcocystis* in goats. *N. Z. Vet. J.* 28, 244.

1960. Collins, G.H., Sutton, R.H., Charleston, W.A.G., 1980. Studies in *Sarcocystis* species V. A species infecting dogs and goats; observation on the pathology and serology of experimental *Sarcocystis* in goats. *N. Z. Vet. J.* 28, 156–158.

1961. Collins, G.H., 1981. Studies in *Sarcocystis* species VIII: *Sarcocystis* and *Toxoplasma* in red deer (*Cervus elaphus*). *N. Z. Vet. J.* 29, 126–127.

1962. Collins, G.H., Emslie, D.R., Farrow, B.R.H., Watson, A.D.J., 1983. Sporozoa in dogs and cats. *Aust. Vet. J.* 60, 289–290.

1963. Cravero, G.C., 1976. L'infestione cerebrale da Sarocystidae in *Ovis aries*. *Ann. Turin. Univ. Fac. Med. Vet.* 23, 158–164.

1964. Cruz, A., Larios, F., Osuna, G., Reyes, E., 1982. Identificacion de *Sarcocystis* spp. en el cerebro de un bovino. *Tec. Pecu. Mex.* 43, 83–86.

1965. Daffron, P.N., Beasley, J.N., Briggs, D.J., 1985. Seroprevalence of bovine *Sarcocystis* in Arkansas. *Arkansas Farm Research* 34, 2.

1966. Daugschies, A., Hintz, J., Henning, M., Rommel, M., 2000. Growth performance, meat quality and activities of glycolytic enzymes in the blood and muscle tissue of calves infected with *Sarcocystis cruzi*. *Vet. Parasitol.* 88, 7–16.

1967. De Kruijf, J.M., van Logtestijn, J.G., Franken, P., Herder, K.A.M., 1974. Sarcosporidiosis bij runderen in varkens. *Tijdschr. Diergeneeskd.* 99, 303–308.

1968. De Kruijf, J.M., Bibo, T.M., 1976. Sarcosporidiosis bij runderen in schapen. *Tijdschr. Diergeneeskd.* 101, 1093–1095.

1969. Derhalli, F.S., Higashi, G.I., Hilali, M.A., 1982. Detection of Toxoplasma related antigens in *Sarcocystis fusiformis* (Railliet, 1897) buffaloes, Egypt. *J. Egypt. Vet. Med. Assoc.* 42, 47–52.

1970. Deschiens, R., Levaditi, J.-C., Lamy, L., 1957. Sur quelques aspects morphologiques de sarcosporidies de divers mammifères. *Bull. Soc. Pathol. Exot.* 50, 225–228.

1971. Deshpande, A.V., Shastri, U.V., Deshpande, M.S., 1982. Experimental infection of pups by feeding cysts of *Sarcocystis levinei* Dissanaike and Kan, 1978. *Indian J. Parasitol.* 6, 331–332.

1972. Deshpandey, A.V., Shastri, U.V., Deshpande, M.S., 1983. Ineffective treatment of *Sarcocystis levinei* infection in dogs. *Trop. Vet. Anim. Sci. Res.* 1, 92.

1973. Deshpandey, A.V., Shastri, U.V., Deshpande, M.S., 1983. Prevalence of sarcocysts in cattle and buffaloes in Marathwada (India). *Trop. Vet. Anim. Sci. Res.* 1, 92.

1974. Dessouky, M.I., Mohamed, A.H., Nassar, A.M., Hilali, M., 1984. Haematological and biochemical changes in buffalo calves inoculated with *Sarcocystis fusiformis* from cats. *Vet. Parasitol.* 14, 1–6.

1975. Destombes, P., 1957. Les sarcosporidioses au Vietnam. *Bull. Soc. Pathol. Exot.* 50, 221–225.

1976. Díez-Baños, P., 1978. Sobre la prevalencia de la sarcosporidiosis ovina en la provincia de León, con un estudio comparativo de diversos metodos diagnosticos. *An. Fac. Vet. León.* 24, 195–199.

1977. Díez-Baños, P., Cordero del Campillo, D.M., Rojo-Vázquez, F.A., 1978. Diagnóstico de la sarcosporidiosis ovina con la immunofluorescencia indirecta. *An. Fac. Vet. León.* 24, 47–55.

1978. Din Afaf, I., Shommein, A.M., 1974. Experimental transmission of sarcocysts from cattle to mice. *Trop. Anim. Health Prod.* 6, 58–59.

1979. Dissanaike, A.S., Kan, S.P., Retnasabapathy, A., Baskaran, G., 1977. Demonstration of the sexual phases of *Sarcocystis fusiformis* (Railliet, 1897) and *Sarcocystis* sp. of the water buffalo (*Bubalus bubalis*) in the small intestines of cats and dogs. *Trans. R. Soc. Trop. Med. Hyg.* 71, 271.

1980. Dollinger, P., 1974. Beitrag zur Kenntnis des Endoparasitenspektrums des Gemswildes in der Schweiz. *Jagdwiss* 20, 115–118.

1981. Douglass, E.M., Hansen, B., 1979. Sarcosporidiosis in blue and yellow tanagers. *Vet. Med. Small Anim. Clin.* 74, 1534–1535.

1982. Drost, S., 1982. Sarkosporidienfunde im Hackfleisch des Handels. *Angew. Parasitol.* 23, 185–189.

1983. Dubey, J.P., 1981. Early developmental stages of *Sarcocystis cruzi* in calf fed sporocysts from coyote feces. *J. Protozool.* 28, 431–433.

1984. Dubey, J.P., Blagburn, B.L., 1983. Failure to transmit *Sarcocystis* species of ox, sheep, goats, moose, elk, and mule deer to raccoons. *Am. J. Vet. Res.* 44, 1079–1080.

1985. Dubey, J.P., Livingston, C.W., 1986. *Sarcocystis capracanis* and *Toxoplasma gondii* infections in range goats from Texas. *Am. J. Vet. Res.* 47, 523–524.

1986. Dubey, J.P., Sheffield, H.G., 1988. *Sarcocystis sigmodontis* n. sp. from the cotton rat (*Sigmodon hispidus*). *J. Parasitol.* 74, 889–891.

1987. Dubey, J.P., Slife, L.N., 1990. Fatal encephalitis in a dog associated with an unidentified coccidian parasite. *J. Vet. Diagn. Invest.* 2, 233–236.

1988. Dubey, J.P., Duncan, D.E., Speer, C.A., Brown, C., 1992. Congenital sarcocystosis in a two-day-old dog. *J. Vet. Diagn. Invest.* 4, 89–93.

1989. Dubey, J.P., Morales, J.A., 2006. Morphologic characterization of *Sarcocystis* sp. sarcocysts from the Buffon's macaw (*Ara ambigua*). *Acta Parasitol.* 51, 231–237.

1990. Dubná, S., Langrová, I., Nápravnik, J., Jankovská, I., Vadlejch, J., Pekár, S., Fechtner, J., 2007. The prevalence of intestinal parasites in dogs from Prague, rural areas, and shelters of the Czech Republic. *Vet. Parasitol.* 145, 120–128.

1991. Dubremetz, J.F., Torpier, G., 1978. Freeze fracture study of the pellicle of an eimerian sporozoite (protozoa, coccidia). *J. Ultrastruct. Res.* 62, 94–109.

1992. Dunigan, C.E., Oglesbee, M.J., Podell, M., Mitten, L.A., Reed, S.M., 1995. Seizure activity associated with equine protozoal myeloencephalitis. *Progr. Vet. Neurol.* 6, 50–54.

1993. Ellison, S.P., Greiner, E., Dame, J.B., 2001. *In vitro* culture and synchronous release of *Sarcocystis neurona* merozoites from host cells. *Vet. Parasitol.* 95, 251–261.

1994. Elsheikha, H.M., Mansfield, L.S., 2004. Assessment of *Sarcocystis neurona* sporocyst viability and differentiation between viable and nonviable sporocysts using propidium iodide stain. *J. Parasitol.* 90, 872–875.

1995. Emnett, C.W., Hugghins, E.J., 1982. *Sarcocystis* of deer in South Dakota. *J. Wildl. Dis.* 18, 187–193.

1996. Emnett, C.W., 1986. Prevalence of *Sarcocystis* in wolves and white-tailed deer in northeastern Minnesota. *J. Wildl. Dis.* 22, 193–195.

1997. Entzeroth, R., Neméséri, L., Scholtyseck, E., 1983. Prevalence and ultrastucture of *Sarcocystis* sp. from the red deer (*Cervus elaphus L.*) in Hungary. *Parasitol. Hung.* 16, 47–52.

1998. Entzeroth, R., Goerlich, R., 1987. Monoclonal antibodies against cystozoites of *Sarcocystis muris* (Protozoa, Apicomplexa). *Parasitol. Res.* 73, 568–570.

1999. Erickson, A.B., 1940. *Sarcocystis* in birds. *Auk* 57, 514–519.

2000. Espino, M.G.R., Morales, J.F.A., 1987. Hallazgo histopatologico de un quiste de *Sarcocystis* spp en el cerebro de un ovino. Technica Pecuaria en Mexico, pp. 110–114.

2001. Ezhova, T.A., Jezova, T., 1997. Development of sarcosporidia of pigs in a cell culture infected with cystozoites. *Lietovos tsr Mokslu Akademijos Darbai, C (Biologijos Mokslai)*, pp. 81–86. (In Russian).

2002. Farmer, J.N., Herbert, I.V., Partridge, M., Edwards, G.T., 1978. The prevalence of *Sarcocystis* spp in dogs and red foxes. *Vet. Rec.* 102, 78–80.

2003. Fassi-Fehri, N., Cabaret, J., Amaqdouf, A., Dardar, R., 1978. La sarcosporidiose des ruminants au Maroc etude epidemiologique par deux techniques histologiques. *Ann. Rech. Vét.* 9, 409–417.

2004. Fayer, R., Johnson, A.J., 1974. *Sarcocystis fusiformis*: Development of cysts in calves infected with sporocysts from dogs. *Proc. Helminthol. Soc. Wash.* 41, 105–108.

2005. Fayer, R., Thompson, D.E., 1975. Cytochemical and cytological observations on *Sarcocystis* sp. propagated in cell culture. *J. Parasitol.* 61, 466–475.

2006. Fayer, R., 1977. The first asexual generation in the life cycle of *Sarcocystis bovicanis*. *Proc. Helminthol. Soc. Wash.* 44, 206–209.

2007. Fayer, R., Lynch, G.P., 1979. Pathophysiological changes in urine and blood from calves experimentally infected with *Sarcocystis cruzi*. *Parasitology* 79, 325–336.

2008. Fayer, R., Leek, R.G., 1979. *Sarcocystis* transmitted by blood transfusion. *J. Parasitol.* 65, 890–893.

2009. Fayer, R., Andrews, C., Dubey, J.P., 1988. Lysates of *Sarcocystis cruzi* bradyzoites stimulate RAW 264.7 macrophages to produce tumor necrosis factor (cachectin). *J. Parasitol.* 74, 660–664.

2010. Fedoseenko, V.M., 1989. The initial phase of cyst formation in *Sarcocystis dispersa* sarcosporidians. *Tsitologiya* 31, 247–250. (In Russian).

2011. Fernández García, J., Mirón Pérez, I.J., Álvarez Porras, J.Á., 2007. Detección en matadero de sarcosporidiosis generalizada en carne de cerdo. *ASIS Veterinaria* 42, 32–35.

2012. Filho, M.T.J., Miraglia, T., 1977. Histochemical observations on the *Sarcocystis fusiformis* cysts in ox hearts. *Acta Histochem.* 59, 160–167.

2013. Fischer, G., 1979. Die Entwicklung von Sarcocystis capracanis n. spec in der Ziege. Inaugural Dissertation Universität Berlin, Germany, pp. 1–45.

2014. Foggin, C.M., 1980. *Sarcocystis* infection and granulomatous myositis in cattle in Zimbabwe. *Zimbabwe Vet. J.* 11, 8–13.

2015. Ford, G.E., 1975. Transmission of sarcosporidiosis from dogs to sheep maintained specific pathogen free. *Aust. Vet. J.* 51, 408.

2016. Ford, G.E., 1986. Biochemical characterization for identification of ovine sarcosporidia. *Aust. J. Biol. Sci.* 39, 31–36.

2017. Framstad, K., Tharaldsen, J., Nøstvold, S., 1982. Parasitter av betydning for kjøttkontrollen. *Nor. Veterinaertidskr.* 94, 8–20.

2018. Frank, W., Häfner, U., 1981. Host range and host specificity of *Sarcocystis*. *Zentralbl. Bakteriol. Mikrobiol. Hyg. A* 250, 355–360.

2019. Frenkel, J.K., 1956. Pathogenesis of toxoplasmosis and of infections with organisms resembling *Toxoplasma*. *Ann. NY. Acad. Sci.* 64, 215–251.

2020. Frenkel, J.K., Heydorn, A.O., Mehlhorn, H., Rommel, M., 1979. Sarcocystinae: *Nomina dubia* and available names. *Z. Parasitenkd.* 58, 115–139.

2021. Furmanski, K., 1987. Häufigkeit von Infektionen mit *Sarcocystis miescheriana* und *Sarcocystis suihominis* bei Schlachtschweinen in bezug auf verschiedene Betriebssysteme. Inaugural Dissertation, Institute of Parasitology, Hannover, Germany, pp. 1–77.

2022. Garhy, A.M., Ahmed, A.M., Soliman, S.A., 2012. Survey of abnormalities on bovine carcasses in Ismailia abattoir. *SCVMJ* 17, 123–131.

2023. Gestrich, R., Schmitt, M., Heydorn, A.O., 1974. Pathogenität von *Sarcocystis tenella*-Sporozysten aus den Fäzes von Hunden für Lämmer. *Berl. Münch. Tierärztl. Wschr.* 87, 362–363.

2024. Ghosal, S.B., Joshi, S.C., Shah, H.L., 1986. A note on the natural occurrence of *Sarcocystis* in buffaloes (*Bubalus bubalis*) in Jabalpur region, M.P. *Indian Vet. J.* 63, 165–166.

2025. Ghosal, S.B., Joshi, S.C., Shah, H.L., 1987. Morphological studies of the sarcocyst of *Sarcocystis levinei* of the naturally infected water buffaloes (*Bubalus bubalis*). *Indian Vet. J.* 64, 915–917.

2026. Ghosal, S.B., Joshi, S.C., Shah, H.L., 1987. Sporocyst output in cats fed sarcocysts of *Sarcocystis fusiformis* of the buffalo. *Indian J. Anim. Sci.* 57, 1100–1101.

2027. Ghosal, S.B., Joshi, S.C., Shah, H.L., 1987. Development of sarcocysts of *Sarcocystis levinei* in water buffalo infected with sporocysts from dogs. *Vet. Parasitol.* 26, 165–167.

2028. Ghosal, S.B., Joshi, S.C., Shah, H.L., 1987. Development of sarcocysts of *Sarcocystis fusiformis* in the water buffalo (*Bubalus bubalis*) experimentally infected with sporocysts from cats. *Indian J. Anim. Sci.* 57, 413–415.

2029. Ghosal, S.B., Joshi, S.C., Shah, H.L., 1988. Studies on the morphology of sarcocysts of *Sarcocystis fusiformis* obtained from naturally infected water buffaloes (*Bubalus bubalis*). *Indian Vet. J.* 196–199.

2030. Ghosal, S.B., Joshi, S.C., Shah, H.L., 1988. Sporocyst output in dogs fed sarcocysts of *Sarcocystis levinei* of the buffalo (*Bubalus bubalis*). *Vet. Parasitol.* 28, 173–174.

2031. Gill, H.S., Singh, A., Vadehra, D.V., Sethi, S.K., 1978. Shedding of unsporulated isosporan oocysts in feces by dogs fed diaphragm muscle from water buffalo (*Bubalus bubalis*) naturally infected with *Sarcocystis*. *J. Parasitol.* 64, 551–552.

2032. Göbel, E., Rommel, M., Krampitz, H.E., 1978. Ultrastrukturelle Untersuchungen zur ungeschlectlichen Vermehrung von *Frenkelia* in der Leber der Rötelmaus. *Z. Parasitenkd.* 55, 29–42.

2033. Godoy, G.A., Volcán, G., Guevara, R., 1977. *Sarcocystis fusiformis* en bovinos del estado Bolivar, Venezuela. *Rev. Inst. Med. Trop. São Paulo* 19, 68–72.

2034. Godoy, G.A., Volcan, G.S., Medrano, P.C.E., 1979. Información adicional sobre *Sarcocystis bovicanis*, Heydorn, Gestrich, Mehlhorn, y Rommel, 1975 (*S. fusiformis*, Railliet, 1897), en el Estado Bolivar, Venezuela. *Rev. Inst. Med. Trop. São Paulo* 21, 207–215.

2035. Golemansky, V., 1975. Observations des oocystes et des spores libres de *Sarcocystis* sp. (Protozoa: Coccidia) dans le gros intestin du renard commun (*Vulpes vulpes* L.) en Bulgarie. *Acta Protozool.* 14, 291–296.

2036. Gomes, A.G., Lima, J.D., 1982. *Sarcocystis* (Lankester, 1882) em bovinos de Minas Gerais; ocorrência e métodos de diagnóstico. *Arq. Esc. Vet. U. F. M. G.* 34, 83–92.

2037. González-Castro, J., Pérez-Garro, C., 1970. Frecuencia del *Sarcocystis miescheriana* en el ganado suino de Granada. *Rev. Ibér. Parasitol.* 30, 443–447.

2038. Gorman, T.R., Lorca, M., Thiermann, E., Valenzuela, M., 1984. Diagnostico de la sarcosporidiosis porcina mediante la immunofluorescencia indirecta. *Arch. Med. Vet.* 16, 75–81.

2039. Grikieniené, J.S., 1989. Experimental investigation of the development of *Sarcocystis* from the oesophagus of sheep in the final host. *Acta Parasitol. Lituanica* 23, 59–66.

2040. Gupta, P.P., Singh, S.P., 1988. Sarcocystosis in the brain of a rabid cow. *Indian Vet. J.* 65, 1130.

2041. Gut, J., 1982. Infection of mice immunized with formalized cystozoites of *Sarcocystis dispersa* Cerná, Kolárová et Šulc, 1978. *Folia Parasitol. (Praha)* 29, 285–288.

2042. Hajimohammadi, B., Dehghani, A., Moghaddam-Ahmadi, M., Eslami, G., Oryan, A., Yasini-Ardakani, S.A., Zohourtabar, A., Mirzaei, F., 2014. Isolation of *Sarcocystis hirsuta* from traditional hamburger of Iran. *Journal of Isfahan Medical School* 32, 79–85.

2043. Hajimohammadi, B., Dehghani, A., Ahmadi, M.M., Eslami, G., Oryan, A., Khamesipour, A., 2014. Prevalence and species identification of *Sarcocystis* in raw hamburgers distributed in Yazd, Iran using PCR-RFLP. *Journal of Food Quality and Hazards Control* 1, 15–20.

2044. Hamir, A.N., Moser, G., Galligan, D.T., Davis, S.W., Granstrom, D.E., Dubey, J.P., 1993. Immunohistochemical study to demonstrate *Sarcocystis neurona* in equine protozoal myeloencephalitis. *J. Vet. Diagn. Invest.* 5, 418–422.

2045. Hamir, A.N., Dubey, J.P., Rupprecht, C.E., 1999. Prevalence of *Sarcocystis kirkpatricki* sarcocysts in the central nervous system and striated muscle of raccoons from eastern United States. *J. Parasitol.* 85, 748–750.

2046. Hasche, H.-O., 1988. Enzymaktivitäten im Blutplasma und in der Muskulatur im Verlauf der *Sarcocystis miescheriana*-Infektion der Schweine. Inaugural Dissertation, Institute of Parasitology, Hannover, Germany, pp. 1–81.

2047. Heydorn, A.O., Rommel, M., 1972. Beiträge zum Lebenszyklus der Sarkosporidien IV. Entwicklungsstadien von *Sarcocystis fusiformis* in der Dünndarmschleimhaut der Katze. *Berl. Münch. Tierärztl. Wschr.* 85, 333–336.

2048. Heydorn, A.O., Ipczynski, V., Muhs, E.O., Gestrich, R., 1974. Zystenstadien aus der Muskulatur von Isopora *hominis*-infizierten Kalbern. *Berl. Münch. Tierärztl. Wschr.* 87, 278.

2049. Heydorn, A.O., Mehlhorn, H., Gestrich, R., 1975. Licht- und elektronenmikroskopische Untersuchungen an Cysten von *Sarcocystis fusiformis* in der Muskulatur von Kälbern nach experimenteller Infektion mit Oocysten und Sporocysten der großen Form von Isopora bigemina des Hundes. 2. Die Feinstruktur der Cystenstadien. *Zbl. Bakt. Hyg., I. Abt. Orig. A* 232, 123–137.

2050. Heydorn, A.O., Gestrich, R., Ipczynski, V., 1975. Zum Lebenszyklus der kleinen Form von *Isospora bigemia* des Hundes. II. Entwicklungsstadien im Darm des Hundes. *Berl. Münch. Tierärztl. Wschr.* 88, 449–453.

2051. Levine, N.D., 1979. What is "*Sarcocystis*" *mucosa*? *Ann. Trop. Med. Parasitol.* 73, 91–92.

2052. Heydorn, A.O., Weniger, J.H., 1988. Einfluß einer akuten *Sarcocystis suihominis*-Infektion auf die Mastleistung von Absatzferkeln. *Berl. Münch. Tierärztl. Wschr.* 101, 307–310.

2053. Hiepe, F., Litzke, L.F., Scheibner, G., Jungmann, R., Hiepe, T., Montag, T., 1981. Untersuchungen zur toxischen Wirkung von Extrakten aus *Sarcocystis ovifelis*-Makrozysten auf Kaninchen. *Monatsh. Veterinärmed.* 36, 908–910.

2054. Hiepe, T., Nickel, S., Jungmann, R., Hansel, U., Unger, C., 1980. Untersuchungen zur Ausscheidung von Sporozoen-Fäkalformen bei Jagdhunden, Rotfüchsen und streunenden Hauskatzen sowie zum vorkommen von Muskelsarkosporidien bei Wildtieren. *Monatsh. Veterinärmed.* 35, 335–338.

2055. Hilgenfeld, M., Punke, G., 1973. zur Sarkosporidieninfektion des Zentralnervensystems des Schafes, ein Beitrag zur Differentialdiagnose von Protozoeninfektionen. *Archiv fur Experimentelle Veterinarmedizin* 28, 621–626.

2056. Hinaidy, H.K., 1980. Vereinfachte Homogenatmethode zum Nachweis von Sarkosporidien (Miescherschen Schläuchen) bei Schlachtrindern). *Wien. Tierärztl. Mschr.* 67, 54–55.

2057. Hintz, J., 1990. Untersuchungen zum glykolytischen Enzymmuster im Blutplasma und in der Muskulature *Sarcocystis cruzi*-infizierter Kälber. Inaugural Dissertation, Institute of Parasitology, Hannover, Germany, pp. 1–74.

2058. Hofer, J., 1981. Untersuchung über die Entwicklung von *Sarcocystis muris* und *Sarcocystis dispersa* in der Maus und über Reaktionen des Zwischen- writes nach ein- und mehrmaligen experimentellen Infektionen. Inaugural Dissertation, Ludwig-Maximilians-Universität München, Germany, pp. 1–38.

2059. Huong, L.T.T., 1999. *Sarcocystis* infections of the water buffalo in Vietnam. PhD Dissertation, Swedish University of Agricultural Sciences, Uppsala, Sweden, pp. 1–53.

2060. Jain, P.C., Joshi, S.C., Kamalapur, S.K., 1971. A concurrent infection of *Sarcocystis* Lankester, 1882 and *Gongylonema pulchrum* Molin, 1850 in a goat in Madhya Pradesh. *Food Farming and Agriculture* 3, 23–24.

2061. Jain, P.C., Shah, H.L., 1985. Prevalence and seasonal variations of *Sarcocystis* of cattle in Madhya Pradesh. *Indian J. Anim. Sci.* 55, 29–31.

2062. Jain, P.C., Shah, H.L., 1986. Gametogony and sporogony of *Sarcocystis fusiformis* of buffaloes in the small intestine of experimentally infected cats. *Vet. Parasitol.* 21, 205–209.

2063. Jain, P.C., Shah, H.L., 1986. Experimental study on gametogonic development of *Sarcocystis levinei* in the small intestine of dogs. *Indian J. Anim. Sci.* 56, 314–318.

2064. Jain, P.C., Shah, H.L., 1987. *Sarcocystis hominis* in cattle in Madhya Pradesh and its public health importance. *Indian Vet. J.* 64, 650–654.

2065. Jain, P.C., Shah, H.L., 1987. Comparative morphology of oocysts and sporocysts of bovine and bubaline *Sarcocystis* in Madhya Pradesh. *Indian J. Anim. Sci.* 57, 849–852.

2066. Jeffrey, H.C., 1974. Sarcosporidiosis in man. *Trans. R. Soc. Trop. Med. Hyg.* 68, 17–29.

2067. Jewell, D.E., Jones, D.D., Martin, R.J., Prestwood, A., Hausman, G.J., 1988. Sera from pigs infected with *Sarcocystis suicanis* and cachectin decrease preadipocyte differentiation in primary cell culture. *J. Anim. Sci.* 66, 2992–2999.

2068. Johnson, C.K., Tinker, M.T., Estes, J.A., Conrad, P.A., Staedler, M., Miller, M.A., 2012. Prey choice and habitat use drive sea otter pathogen exposure in a resource-limited coastal system. *Proc. Natl. Acad. Sci. USA* 106, 2242–2247.

2069. Jungmann, R., Knoch, W., 1980. Möglichkeiten des Sarkosporidiennachweises unter besonderer Berücksichtigung der Artendifferenzierung. Monatsh. *Veterinärmed.* 35, 947–949. 2070.   J u y a l , P.D., Sahai, B.N., Srivastava, P.S., Sinha, S.R.P., 1982. Heavy sarcocystosis in the ocular musculature of cattle and buffaloes. *Vet. Res. Commun.* 5, 337–342.

2071. Juyal, P.D., Bhatia, B.B., 1987. A note on the occurrence of *Sarcocystis* infection in goats (*Capra hircus*) and buffaloes (*Bubalus bubalis*) in Tarai region of Uttar Pradesh. *Indian Vet. Med. J.* 11, 234–235.

2072. Juyal, P.D., Ruprah, N.S., Chhabra, M.B., 1989. Rapid isolation of intact micro-*Sarcocystis* (Protozoa-Apicomplexa) cysts from muscular tissues. *Indian J. Anim. Health* June, 69–70.

2073. Kaiser, I.A., Markus, M.B., 1981. *Sarcocystis* in the avian intermediate host. *Proc. Electron Microsc. Soc. S. Afr.* 11, 115.

2074. Kaliner, G., Sachs, R., Fay, L.D., Schiemann, B., 1971. Untersuchungen über das Vorkommen von Sarcosporidien bei ostrafikanischen Wildtieren. *Zeit. Tropenmedizin* 22, 156–164.

2075. Kallab, K., 1966. Über das Vorkommen von Sarkosporidien. *Wien. Tierärztl. Mschr.* 53, 34–39.

2076. Kaltungo, B.Y., Musa, I.W., 2013. A review of some protozoan parasites causing infertility in farm animals. ISRN Tropical Medicine 2013, article ID 782609, 1–6.

2077. Kan, S.P., 1985. A review of sarcocystosis with special reference to human infection in Malaysia. *Trop. Biomed.* 2, 167–175.

2078. Kan, S.P., Pathmanathan, R., 1991. Review of sarcocystosis in Malaysia. Proceedings of the 33rd SEAMEO-TROPMED Regional Seminar. Emerging problems in food-borne parasitic zoonoses: Impact on agriculture and public health. South East Asian Journal of Tropical Medicine and Hygiene. *Southeast Asian Journal of Tropical Medicine and Public Health.* Chiang Mai, Thailand, 22, pp. 129–134.

2079. Karstad, L., Trainer, D.O., 1969. *Sarcocystis* in white-tailed deer. *Bull. Wildl. Dis. Assoc.* 5, 25–26.

2080. Kelly, A.L., Penner, L.R., Pickard, R.J., 1950. *Sarcocystis* in the moose. *J. Mammal* 31, 462–463.

2081. Kepka, O., Österreicher, H.D., 1979. Zur Häufigkeit von Sarkosporidien in Rindern der Steiermark. *Wien. Tierärztl. Mschr.* 66, 184–185.

2082. Kepper, A., 1981. Untersuchungen über den Verlauf der Antikörpertiter bei ein-und mehrmals mit *Sarcocystis muris* inokulierten NMRI- Mäusen mit Hilfe des indirekten Immunfluoreszenzantikörpertests (IFAT). Inaugural Dissertation, Institute of Parasitology, Hannover, Germany, pp. 4–60.

2083. Khan, R.A., Fong, D., 1991. *Sarcocystis* in caribou (*Rangifer tarandus terraenorae*) in Newfoundland. In: Cross, J.H. (Ed.), *Southeast Asian Journal of Tropical Medicine and Public Health*, Vol. 22, Chiang Mai, Thailand, pp. 142–143.

2084. Kislyakova, Z.I., 1969. The experimental infection of piglets with *Sarcocystis. Dokl. Vses. Akad. Sel. -Khoz. Nauk.* 2, 36–38. (In Russian).

2085. Kisthardt, K., Lindsay, D.S., 1997. Equine protozoal myeloencephalitis. *Equine Practice. Neurology* 19, 8–13.

2086. Kistner, 1978. Why are fawns of mule deer dying? *Oregon Agricultural Progress* 24, 8–19.

2087. Klima, U., 1979. Versuche zum Nachweis von Sarkosporidien-Infektionen bei Haustrieren mit dem Enzyme-Linked-Immunosorbent-Assay (ELISA). Inaugural Dissertation, Universität München, Germany, pp. 1–37.

2088. Kobelt, H., 1985. Untersuchungen zum Vorkommen und zur Fleischhygienischen Bedeutung von Sarkosporidien beim Haarwild unter Besonderer Berücksichtigung des rehwildes in hessen. Inaugural Dissertation, Universität zu Giessen, Germany, pp. 1–165.

2089. Kolářová, L., 1986. Mouse (*Mus musculus*) as intermediate host of *Sarcocystis* sp. from the goshawk (*Accipiter gentilis*). *Folia Parasitol. (Praha)* 33, 15–19.

2090. Koudela, B., 1985. Purification of cystozoites of *Sarcocystis* sp. by the chromatographic gel spheron. *Folia Parasitol. (Praha)* 32, 295–302.

2091. Kozakiewicz, B., 1982. The possible roles of the dog in the epizootiology of pig sarcosporidiosis. In: Necoechea, R.R., Pijoan, C., Casarín, A., Guzmán, M. (Eds.), *Proceedings of the Int. Pig. Vet. Soc. Congress*, Mexico D.F., Mexico, pp. 26–31.

2092. Kuraev, G.T., 1981. *Sarcocystis* infection in dromedaries and bactrian camels in Kazakhstan. *Veterinariya*, 7, 41–42. (In Russian).

2093. Kutkiené, L., Grikieniené, J., 2000. The coprophagy and transplacental transmission as possible ways of distribution of some Sarcosporidian (Protista: Coccidia) species in rodents. In: *Abstracts Int. Symp. Ecological Parasitol. on the turn of the millenium*, St. Petersburg, Russia, p. 95.

2094. Laakkonen, J., Henttonen, H., 2000. Ultrastructure of *Frenkelia* sp. from a Norwegian lemming in Finland. *J. Wildl. Dis.* 36, 362–366.

2095. Laarman, J.J., 1964. Isosporiasis in man. *Trop. Geogr. Med.* 16, 265–269.

2096. Laarman, J.J., Tadros, W., 1983. Sarcosporidiosis in the Netherlands. In: Euzeby, J. (Ed.), *Some Important Parasitic Infections in Bovines Considered from Economic and Social (zoonoses) Points of View.* Commission of the European Communities, Luxembourg, Belgium, pp. 65–71.

2097. Lam, K.K.H., Watkins, K.L., Chan, C.W., 1999. First report of equine protozoal myeloencephalitis in Hong Kong. *Equine Vet. Educ.* 11, 54–56.

2098. Landsverk, T., Gamlem, H., Svenkerud, R., 1978. A *Sarcocystis*-like protozoan in a sheep with lymph-adenopathy and myocarditis. *Vet. Pathol.* 15, 186–195.

2099. Lapointe, J.M., Duignan, P.J., Barr, B.C., Petrich, A.K., MacPherson, D.W., Gulland, F.M., Dubey, J.P., 2003. Meningoencephalitis associated with an unidentified apicomplexan protozoan in a Pacific harbor seal. *J. Parasitol.* 89, 859–862.

2100. László, N., Entzeroth, R., Scholtyseck, E., 1983. A magyarországi gimszarvasok *Sarcocystis*-fajának gyakorisága és ultrastructurája. *Magyar Allatorvosok Lapja* 38, 758–759.

2101. Laupheimer, K.E., 1978. Das Vorkommen von *Sarcocystis bovicanis*, *S. bovifelis* und *S. bovihominis* bei Schlachtrindern in Suddeutschland. Inaugural Dissertation, Universität München, Germany, pp. 1–46.

2102. Leguía, G., Herbert, I.V., 1979. The prevalence of *Sarcocystis* spp in dogs, foxes and sheep and *Toxoplasma gondii* in sheep and the use of the indirect haemagglutination reaction in serodiagnosis. *Res. Vet. Sci.* 27, 390–391.

2103. Leier, H., 1982. Versuche zur immunisierung von mäusen gegen *Toxoplasma gondii*- und *Sarcocystis muris*-infektionen. Inaugural Dissertation, Ludwig-Maximilians-Universität, München, Germany, pp. 1–50.

2104. Lele, V.R., Dhopavkar, P.V., Kher, A., 1986. *Sarcocystis* infection in man (a case report). *Indian J. Pathol. Microbiol.* 29, 87–90.

2105. Levchenko, N.G., 1962. Infection by sarcosporidia (genus *Sarcocystis*) in farm animals of Southeast Kazakhstan. *Izdat Akad Nauk Kazakh SSR, Alma-Ata*, pp. 56–62. (In Russian).

2106. Levchenko, N.G., 1962. Dynamics of an infection by *Sarcocystis tenella* sarcosporidia in sheep in southeast Kazakhstan. In: Boev, S.N. et al. (Eds.), *Parasites of Farm Animals in Kazakhstan*. Izdat. Akad. Nauk Kazakh. SSR, Alma-Ata. pp. 63–68. (In Russian).

2107. Levchenko, N.G., 1963. O rasprotranenii srkosporidii ovets (*Sarcocystis tenella* Railliet) na yugo-vostoke Kazakhstana. *Parazity Sel'skokhozyaistvennykh Zhivotnykh Kazakhstana* 2, 154–157.

2108. Levchenko, N.G., 1963. Study of the harmfulness of *Sarcocystis tenella*. Parazity Sel'skokhozyaistvennykh Zhivotnykh Kazakhstana. *Akad. Nauk. Kazakhskoi SSR.* 2, 158–162. (In Russian).

2109. Levine, N.D., Cechner, P.E., Meyer, R.C., 1974. The relationship of certain coccidia to *Sarcocystis* of swine. *Indian Vet. J.* 51, 57–59.

2110. Levine, N.D., 1977. Nomenclature of *Sarcocystis* in the ox and sheep and of fecal coccidia of the dog and cat. *J. Parasitol.* 63, 36–51.

2111. Levine, N.D., 1988. *The Protozoan Phylum Apicomplexa*. CRC Press, Boca Raton, Florida, Vol. 1, 203pp; Vol. 2, 154pp.

2112. Lindsay, D.S., Upton, S.J., Toivio-Kinnucan, M., McKown, R.D., Blagburn, B.L., 1992. Ultrastructure of *Frenkelia microti* in prairie voles inoculated with sporocysts from red-tailed hawks. *J. Helminthol. Soc. Wash.* 59, 170–176.

2113. Lindsay, D.S., McKown, R.D., Dubey, J.P., 2000. *Sarcocystis campestris* from naturally infected 13-lined ground squirrels, *Spermphilus tridecemlineatus tridecemlineatus*, from Nebraska. *J. Parasitol.* 86, 1159–1161.

2114. Linstow, V., 1903. Parasiten, meistens Helminthen, aus Siam. *Archiv fuer Mikroscopishe Anatomil und Entwicklungsmeckanik* 62, 108–121.

2115. Lombardo de Barros, C.S., Barros, S.S., Santos, M.N., Silva, C.A.M., Waihrich, F., 1986. Mieloencefalite eqüina por protozoário. *Pesq. Vet. Bras.* 6, 45–49.

2116. Lopes, C.W.G., Araújo, J.L.B., Pereira, M.J.S., 1982. *Sarcocystis levinei* (Apicomplexa: Sarcocystidae) in the water buffalo (*Bubalus bubalis*) in Brazil. *Arq. Univ. Fed. Rur. Rio de Janeiro* 5, 21–24.

2117. López-Rodríguez, R., Hernández, S., Navarrete, I., Martínez-Gómez, F., 1986. Sarcocystosis experimental en la cabra (*Capra hircus*). II. Signos clínicos e índices de eritrocitos. *Rev. Ibér. Parasitol.* 46, 115–122.

2118. Lowe, K.J., Williams, W.F., Fayer, R., Gross, T.S., 1982. #P277NE—Reproductive hormone and prostaglandin metabolite profiles of pregnant cows infected with the abortifacient *Sarcocystis cruzi*. *Proceedings of the 77th Ann. Meeting Am. Diary Sci. Assoc.* University Park, USA, p. 798.

2119. Luengo, M., Arata, N., Luengo, J., 1974. Hallazgo de sarcosporidia en cerebelo de bovino. *Bol. Chil. Parasitol.* 29, 39–41.

2120. Machul'skii, S.N., Fomovska, A., 1980. Epizootiology of *Sarcocystis* infection in swine. *Veterinariya* 0, 46–47. (In Russian).

2121. MacKay, R.J., 1997. Serum antibodies to *Sarcocystis neurona*—half the horses in the United States have them! *J. Am. Vet. Med. Assoc.* 210, 482–483.

2122. MacKay, R.J., Granstrom, D.E., Saville, W.J., Reed, S.M., 2000. Equine protozoal myeloencephalitis. *Vet. Clin. N. Am. Equine Pract.* 16, 405–425.

2123. Mayhew, I.G., de Lahunta, A., Whitlock, R.H., 1978. History and clinical evaluation. In: Mayhew, I.G., de Lahunta, A., Whitlock, R.H., Krook, L., Tasker, J.B. (Eds.), *Spinal Cord Disease in the Horse*, The Cornell Veterinarian, Ithaca, New York, pp. 24–35.

2124. Madrzak, B., 1966. Wahania ekstensywności inwazji cew Mieschera u świś rzeźńych. Doniesienie wstepne. *Medycyna Wet.* 22, 278.

2125. Madsen, S.C., Hugghins, E.J., 1979. Studies on *Sarcocystis* of wild ungulates in South Dakota. *Proc. S. D. Acad. Sci.*, 58, 169.

2126. Mahrt, J.L., 1973. *Sarcocystis* in dogs and its probable transmission from cattle. *J. Parasitol.* 59, 588–589.

2127. Wobeser, G., Cawthorn, R.J., Gajadhar, A.A., 1983. Pathology of *Sarcocystis campestris* infection in Richardson's ground squirrels (*Spermophilus richardsonii*). *Can. J. Comp. Med.* 47, 198–202.

2128. Mahrt, J.L., Colwell, D.D., 1980. *Sarcocystis* in wild ungulates in Alberta. *J. Wildl. Dis.* 16, 571–576.

2129. Mannewitz, U., 1978. Das Vorkommen von *Sarcocystis suihominis* und *Sarcocystis suicanis* bei Schlachtschweinen in Süddeutschland. Inaugural Dissertation, Universität München, Germany, pp. 1–34.

2130. Manuel, M.F., Misa, G.A., Yoda, T., 1983. Histomorphological studies of bubaline *Sarcocystis* in the Philippines. *Phil. J. Vet. Med.* 22, 24–36.

2131. Mareis, A., 1985. Speziesspezifische Serodiagnose von *Sarcocystis muris* und *S. dispersa*-infektionen der Maus mittels Isoelektrischer Fokussierung un Immunoblotting. Inaugural Dissertation, Ludwig-Maximilians-Universität München, Germany, pp. 1–47.

2132. Margolin, J.H., Jolley, W.R., 1979. Experimental infection of dogs with *Sarcocystis* from wapiti. *J. Wildl. Dis.* 15, 259–262.

2133. Mácha, J., Procházková, Z., Červa, L., Gut, J., 1985. Isolation and characterization of a lectin from *Sarcocystis gigantea*. *Mol. Biochem. Parasitol.* 16, 243–249.

2134. Markus, M.B., 1974. Serology of human sarcosporidiosis. *Trans. R. Soc. Trop. Med. Hyg.* 68, 415–416.

2135. Markus, M.B., Draper, C.C., Hutchison, W.M., Killick-Kendrick, R., Garnham, P.C.C., 1974. Attempted infection of chimpanzees and cats with *Sarcocystis* of cattle. *Trans. R. Soc. Trop. Med. Hyg.* 68, 3.

2136. Markus, M.B., 1977. *Sarcocystis* and other organisms related to *Toxoplasma*. In: Gear, J.H.S. (Ed.), *Medicine in a Tropical Environment. Proceedings Int. Symp.* South Africa, 1976. South Africa Medical Research Council, Capetown, South Africa, pp. 632–635.

2137. Markus, M.B., 1978. Terminology for invasive stages of protozoa of the subphylum Apicomplexa (Sporozoa). *S. Afr. J. Sci.* 74, 105–106.

2138. Markus, M.B., 1978. *Sarcocystis* and sarcocystosis in domestic animals and man. *Adv. Vet. Sci. Comp. Med.* 22, 159–193.

2139. Markus, M.B., 1978. Diagnosis of sarcocystosis in cats and dogs. *J. Am. Vet. Med. Assoc.* 173, 927.

2140. Markus, M.B., 1979. Antibodies to *Sarcocystis* in human sera. *Trans. R. Soc. Trop. Med. Hyg.* 73, 346–347.

2141. Markus, M.B., Mundy, P.J., 1979. Intestinal *Sarcocystis* infection in vultures and its significance. *Parasitology* 79, 39.

2142. Markus, M.B., Daly, T.J.M., 1980. Specificity of *Sarcocystis* for the intermediate host. In: *3rd European Multicolloquium of Parasitology.* Cambridge, UK, p. 141.

2143. Markus, M.B., 1981. Sarcocystosis in domestic animals. *J. S. Afr. Vet. Assoc.*, 52, 350.

2144. Markus, M.B., 1981. Admissible hapantotypical *Sarcocystis* material. *S. Afr. J. Sci.* 77, 575.

2145. Markus, M.B., Daly, T.J.M., Biggs, H.C., 1983. Domestic dog as a final host of *Sarcocystis* of the mountain zebra *Equus zebra hartmannae*. *S. Afr. J. Sci.* 79, 471.

2146. Markus, M.B., Daly, T.J.M., Mundy, P.J., 1984. Host specificity of *Sarcocystis* species of wild African ungulates. *Parasitology* 89, 43–44.

2147. Markus, M.B., Mundy, P.J., Daly, T.J.M., 1985. Vultures *Gyps* spp. as final hosts of *Sarcocystis* of the impala *Aepyceros melampus*. *S. Afr. J. Sci.* 81, 43.

2148. Markus, M.B., van der Lugt, J.J., Dubey, J.P., 2004. Sarcocystosis. In: Coetzer, J.A.W., Thomson, G.R., Tustin, R.C., Kriek, N.P.J. (Eds.), *Infectious Diseases of Livestock with Special Reference to Southern Africa*. Oxford University Press Southern Africa, Ni City, South Africa, pp. 360–375.

2149. Marot, A., 1982. Mogućnost dijagnosticiranja sarkosporidija u misićnom tkivu iako nisu vidljive mišerove mešine. *Veterinaski Glasnik* 36, 39–49.

2150. Mathey, W.J., 1966. *Isospora buteonis* Henry 1932 in an American kestrel (*Falco sparverius*) and a golden eagle (*Aguila chrysaetos*). *Bull. Wildl. Dis. Assoc.* 2, 20–22.

2151. Mathieu, A.M., Mboyo, O., 1986. Note sur la fréquence des sarcosporidies chez les bovins au Shaba (Zaïre). *Rev. Élev. Méd. vét. Pays trop.* 39, 297–299.

2152. Matuschka, F.R., 1982. Zur Biologie, Morphologie und Epidemiologie der beim Schwein parasitierenden Sporozoen *Isospora suis, Sarcocystis suicanis* und *S. suihominis. Sber. Ges. Naturf. Freunde* 22, 49–54.

2153. Mayhew, I.G., Dellers, R.W., Timoney, J.F., Kemen, M.J., Fayer, R., Lunde, M.N., 1978. Microbiology and serology. In: Mayhew, I.G., de Lahunta, A., Whitlock, R.H., Krook, L., Tasker, J.B. (Eds.), *Spinal Cord Disease in the Horse.* Cornell University, Ithaca, NY, pp. 148–160.

2154. Mayhew, I.G., 1996. Equine protozoal myeloencephalitis (EPM). *Equine Vet. Educ.* 8, 37–39.

2155. McClure, S.R., Palma, K.G., 1999. Treatment of equine protozoal myeloencephalitis with nitazoxanide. *J. Equine Vet. Sci.* 19, 639–641.

2156. McKenna, P.B., Charleston, W.A.G., 1980. Coccidia (Protozoa: Sporozoasida) of cats and dogs. I. Identity and prevalence in cats. *N. Z. Vet. J.* 28, 86–88.

2157. McKenna, P.B., Charleston, W.A.G., 1980. Coccidia (Protozoa: Sporozoasida) of cats and dogs. IV. Identity and prevalence in dogs. *N. Z. Vet. J.* 28, 128–130.

2158. McKenna, P.B., Charleston, W.A.G., 1983. *Sarcocystis* spp. infections in naturally infected cats and dogs: Levels of sporocyst production and the influence of host, environmental and seasonal factors on the prevalence of infection. *N. Z. Vet. J.* 31, 49–52.

2159. McKenna, P.B., Charleston, W.A.G., 1988. Evaluation of a concentration method for counting *Sarcocystis gigantea* sporocysts in cat faeces. *Vet. Parasitol.* 26, 207–214.

2160. McKenna, P.B., Charleston, W.A.G., 1988. Recovery of *Sarcocystis gigantea* sporocyst from cat faeces. *Vet. Parasitol.* 26, 215–227.

2161. McLeod, R., Hirabayashi, R.N., Rothman, W., Remington, J.S., 1980. Necrotizing vasculitis and *Sarcocystis*: A cause-and-effect relationship? *Southern Med. J.* 73, 1380–1383.

2162. Mehlhorn, H., Heydorn, A.O., Gestrich, R., 1975. Die Geheimnisse der Kokzidiose weiter gelüftet. *UMSCHAU* 75, 701–702.

2163. Mehlhorn, H., Heydorn, O., Gestrich, R., 1975. Licht- und elektronenmikroskopische Untersuchungen an Cysten von *Sarcocystis fusiformis* in der Muskulatur von Kälbern nach experimenteller Infektion mit Oocysten und Sporocysten der großen Form von Isopora bigemina des Hundes. 1. Zur entstehung der Cyste und der "Cystenwand". *Zbl. Bakt. Hyg. I. Abt. Orig. A* 232, 392–409.

2164. Mehlhorn, H., Heydorn, O., Gestrich, R., 1975. Licht- und elektronemikroskopische Untersuchungen an Cysten von *Sarcocystis fusiformis* in der Muskulatur von Kälbern nach experimenteller Infektion mit Oocysten und Sporocysten von *Isopora hominis* Ralliet et Lucet, 1891. 1. Zur Entstehung der Cyste und der Cystenwand. *Zbl. Bakt. Hyg., I. Abt. Orig. A* 231, 301–322.

2165. Mehlhorn, H., 1975. Elekronenmikroskopischer Nachweis von alkalischer Phosphatase und ATP-ase in Cystenstadien von *Sarcocystis tenella* (Sporozoa, Coccidia) aus der Schlundmuskulatur von Schafen. *Z. Parasitenkd.* 46, 95–109.

2166. Mehlhorn, H., Heydorn, A.O., Senaud, J., Schein, E., 1979. Les modalités de la transmission des protozaires parasites des genres *Sarcocystis* et *Theileria*, agents de graves maladies. *Ann. Biol.* 18, 97–120.

2167. Mehlhorn, H., Heydorn, A.O., Frenkel, J.K., Göbel, E., 1985. Announcement of the establishment of neohepantotypes for some important *Sarcocystis* species. *Z. Parasitenkd.* 71, 689–692.

2168. Meijer, T., Mattsson, R., Angerbjörn, A., Osterman-Lind, E., Fernández-Aguilar, X., Gavier-Widén, D., 2011. Endoparasites in the endangered Fennoscandian population of arctic foxes (*Vulpes lagopus*). *Eur. J. Wildl. Res.* 57, 923–927.

2169. Melville, R.V., 1980. *Nomina dubia* and available names. *Z. Parasitenkd.* 62, 105–109.

2170. Melville, R.V., 1984. Reply to Frenkel, Mehlhorn, and Heydorn on protozoan *nomina dubia. J. Parasitol.* 70, 815.

2171. Metsis, A.L., 1987. *Sarcocystis*-bovicanis Heydorn et al. 1975 (Sporozoa Apicomplexa) cytochemical study of tissue cysts from bovine heart. I. Nucleic acids polysaccharides lipids proteins. *Tsitologiya* 29, 955–962. (In Russian).

2172. Metsis, A.L., Rozanov, Y.M., 1988. A flow cytometrical study into heterogenecity of cyst stages of *Sarcocystis bovicanis* Sporozoa, Apicomplexa. *Tsitologiya* 30, 62–67. (In Russian).

2173. Michael, S.A., El Refaii, A.H.H., Morsy, T.A., 1979. Preparation of *Sarcocystis* zoites antigen for a slide agglutination test. *J. Egypt. Soc. Parasitol.* 9, 299–304.

2174. Michalik, R., 1982. Versuche zur chemotherapeutischen Beeinflußbarkeit der Merozoitenphase sowie unreifer und reifer Zysten von *Sarcocystis muris*. Inaugural Dissertation, Institute of Parasitology, Hannover, Germany, pp. 1–52.

2175. Modrý, D., Koudela, B., 1997. New species of coccidia genus *Sarcocystis* (Apicomplexa: Sarcocystidae) with heteroxenous snake-gecko life cycle. *J. Eukaryot. Microbiol.* 44, 25A.

2176. Montag, T., Tietz, H.J., Brose, E., Liebenthal, C., Mann, W., Hiepe, T., Hiepe, F., Coupek, J., 1987. The mitogenicity of extracts from *Sarcocystis gigantea* macrocysts is due to lectin(s). *Parasitol. Res.* 74, 112–115.

2177. Moore, L.A., Johnson, P.J., Messer, N.T., Kline, K.L., Crump, L.M., Knibb, J.R., 1997. Management of headshaking in three horses by treatment for protozoal myeloencephalitis. *Vet. Rec.* 141, 264–267.

2178. Morley, P.S., Saville, W.J.A., 1997. Equine protozoal myeloencephalitis: What does a positive test mean? *Proc. Am. Assoc. Equine Pract.* 43, 1–5.

2179. Moulé, M., 1886. Psorospermies du tissu musculaire du mouton. *Bull. Sleim. Soc. Cent. Med. Vet.* 40, 125–129.

2180. Mugera, G.M., 1968. Sarcosporidiosis in gazelles in Kenya. *East African Wildl. J.* 6, 139–140.

2181. Mukherjea, A.K., Krassner, S.M., 1965. A new species of coccidia (Protozoa: Sporozoa) of the genus *Isospora* Schneider, 1881, from the jackal, *Canis aureus* Linnaeus. *Proc. Zool. Soc. (Calcutta)* 18, 35–40.

2182. Wicht, R.J., 1981. Transmission of *Sarcocystis rileyi* to the striped skunk (*Mephitis mephitis*). *J. Wildl. Dis.* 17, 387–388.

2183. Munday, B.L., Corbould, A., 1974. The possible role of the dog in the epidemiology of ovine s sarcosporidiosis. *Br. Vet. J.* 130, 9–11.

2184. Munday, B.L., Rickard, M.D., 1974. Is *Sarcocystis tenella* two species? *Aust. Vet. J.* 50, 558–559.

2185. Munday, B.L., 1978. Cats as definitive hosts for *Sarcocystis* of sheep. *N. Z. Vet. J.* 26, 166.

2186. Munday, B.L., 1980. The significance of the toxoplasmoses in domestic animals. *Aust. Adv. Vet. Sci.* 3, 61.

2187. Murata, K., 1986. A case report of *Sarcocystis* infection in a lesser flamingo. *Jpn. J. Parasitol.* 35, 555–557.

2188. Musayev, M.A., Yolchiyev, Y.Y., Bagir-Zade, S.S., 1985. Deoxyribonuclease II activity in the cysts of *Sarcocystis fusiformis* and *S. gigantea*. *Veterinariya* 3, 193–195. (In Russian).

2189. Neumayer, F., 1982. Versuche zum nachweis der sarkosporidiose des schweines mit dem IHA, IFAT und dem ELISA. Inaugural Dissertation, Ludwig-Maximilians-Universität München, Germany, pp. 1–41.

2190. Nevole, M., Lukešová, D., 1981. Metody přímé detekce sarkocyst a jejich diagnostická spolehlivost. *Veterinární Medicína* 26, 581–584.

2191. Niederhausern, D.V., 1873. Psorospermien bei der Ziege. *Z. Vet. Wissensch. Bern* 1, 79–86.

2192. O'Toole, D., Duffell, S.J., Upcott, D.H., Frewin, D., 1986. Experimental microcyst *Sarcocystis* infection in lambs: Pathology. *Vet. Rec.* 119, 525–531.

2193. Odening, K., 1983. Sarkozysten in einer antarktischen Robbe. *Angew. Parasitol.* 24, 197–200.

2194. Odening, K., 1984. Oozysten neben Sarkozysten in der Muskulatur einer antarktischen Robbe. *Angew. Parasitol.* 25, 214–216.

2195. Odening, K., Zipper, J., 1986. Zur Ultrastruktur von *Sarcocystis hydrurgae* n. sp. (Apicomplexa: Sporozoea) aus *Hydrurga leptonyx* (Carnivora: Phocidae). *Arch. Protistenkd.* 131, 27–32.

2196. Odening, K., 1986. Tissue cyst-forming Coccidia in antarctic vertebrates. In: *Advances in Protozoological Research/Symposia Biologica Hungarica* (Budapest) 33, 351–355.

2197. Odening, K., Wesemeier, H.H., Walter, G., Bockhardt, I., 1993. First record of bovine *Sarcocystis* species found in European bison (Wisent). In: *Abstracts IX Int. Congress. Protozool.*, Berlin, Germany, p. 93.

2198. Odening, K., Wesemeier, H.H., Walter, G., Bockhardt, I., 1994. The wisent (*Bison bonasus*, Bovidae) as an intermediate host of three *Sarcocystis* species (Apicomplexa: Sarcocystidae) of cattle. *Folia Parasitol. (Praha)* 41, 115–121.

2199. Odening, K., 1996. Wild and domestic mammals in Europe as reservoir for the infection of ungulates with sarcocysts in zoos. *Parassitologia* 38, 18.

2200. Ogassawara, S., Larsson, C.E., Larsson, M.H.M.A., Hagiwara, M.K., Gouveia, G., 1977. Ocorrência de esporocistos de *Sarcocystis* sp. em cães na cidade de São Paulo. *Rev. Microbiol. (Sao Paulo)* 8, 21–23.

2201. Ogassawara, S., Benassi, S., Larson, C.E., Hagiwara, M.K., 1980. *Sarcocystis* sp.: Ocorrência de esporocistos em gatos na cidade de São Paulo. *Arq. Inst. Biol. (São Paulo)* 47, 23–26.

2202. Ogassawara, S., Benassi, S., 1980. Infecção experimental de gatos com coração de bovino parasitiado por *Sarcocystis* sp. *Arq. Inst. Biol. (São Paulo)* 47, 27–32.

2203. Opitz, H.M., Jakob, H.J., Wiensenhuetter, E., Vasandra Devi, V., 1982. A myopathy associated with protozoan schizonts in chickens in commercial farms in Peninsular Malaysia. *Avian Pathol.* 11, 527–534.

2204. Orazalinova, V.A., Fedoseenko, V.M., Sklarova, O.N., 1986. The ultrastructure of cysts of *Sarcocystis* sp. from *Alectoris chukar*. *Izvestiya Akademii Nauk SSSR, Biologicheskaya* 5, 43–48. (In Russian).

2205. Orr, M.B., Collins, G.H., Charleston, W.A.G., 1984. *Sarcocystis capracanis*: Experimental infection of goats. II: Pathology. *Int. Goat Sheep Res.* 3, 202–211.

2206. Osińska, B., Piusiński, W., 1997. Sarcocystosis miesnia sercowego zubrow (*Bison bonasus*) z puszczy Bialowieskiej. *Wiad. Parazytol.* 43, 393–398.

2207. Jog, M.M., Watve, M.G., 2005. Sarcocystosis of chital-dhole: Conditions for evolutionary stability of a predator parasite mutualism. *BMC Ecol.* 5, 3.

2208. Owens, C.G., Kakulas, B.A., 1967. Sarcosporidiosis in the sperm whale. *Aust. J. of Sci.* 31, 46–47.

2209. Page, C.D., Schmidt, R.E., English, J.H., Gardiner, C.H., Hubbard, G.B., Smith, G.C., 1992. Antemortem diagnosis and treatment of sarcocystosis in two species of psittacines. *J. Zoo Wildl. Med.* 23, 77–85.

2210. Panasyuk, D.I., Mintyugov, V.N., Pyatov, M.V., Zyablov, A.A., Golovin, V.G., 1971. Toward a species preference in Sarcosporidia. *Veterinariya* 4, 65–67.

2211. Parihar, N.S., 1987. *Sarcocystis* in sheep brain. *Indian J. Anim. Sci.* 57, 1291–1293.

2212. Partenheimer-Hannemann, C., 1991. Untersuchung zum Vorkommen von Sarkosporidien bei Reh- und Rotwild im Raum Bitburg-Prüm (Rheinland-Pfalz). Inaugural Dissertation, Institute of Parasitology, Hannover, Germany, pp. 1–114.

2213. Pas'ko, S.G., Pas'ko, V.I., Sipko, A.A., 1983. Pathology of ovine sarcocystosis. In: Nikitin, V.Y. (Ed.), Diagnostika, lechenie, profilaktika infeksionnykh i parazitarnykh zabolevanii sel' skokhozyaistvennykh zhivotnykh Sbornik Nauchnykh Trudov, pp. 65–68. (In Russian).

2214. Pereira, M.J.S., Lopes, C.W.G., 1982. The crab-eating fox (*Cerdocyon thous*) as a final host for *Sarcocystis capracanis* (Apicomplexa: Sarcocystidae). *Arq. Univ. Fed. Rur. Rio de J.* 5, 233–235.

2215. Pérez-Garro, M., Rodríguez Osorio, M., García, V.G., Castro, J.G., 1971. Contribución al estudio la sarcosporidiosis: su frequencia en el ganado vacuno de la provincia de Granada. *Rev. Ibér. Parasitol.* 31, 315–318.

2216. Perfumo, C.J., Idiart, J.R., Pigazzi, E.A., 1979. Lesiones en musculos de cerdo producida por sarcocystes. *Rev. Med. Vet. (Bs. As.)* 60, 164–167.

2217. Perrotin, C., Graber, M., 1977. Note de synthèse sur le cycle évolutif des sarcosporidies affectant les animaux domestiques. *Rev. Élev. Méd. vét. Pays trop.* 30, 377–382.

2218. Pethkar, D.K., Shah, H.L., 1982. Prevalence of *Sarcocystis* in goats in Madhya Pradesh. *Indian Vet. J.* 59, 110–114.

2219. Pethkar, D.K., Shah, H.L., 1988. Sporocyst output in dogs experimentally fed sarcocysts of *Sarcocystis capracanis* of the goat. *Indian J. Anim. Sci.* 58, 588–589.

2220. Pfeiffer, L., 1890. Ueber einige neue Formen von Miescher'schen Schläuchen mit Mikro-, Myxo- und Sarcosporidieninhalt. *Virchows Archiv Pathol. Anat. Physiol. Klinsch. Med.* 52, 552–573.

2221. Phillips, P.H., 1980. The clinical pathology of the cyst-forming sporozoa in farm animals. *Aust. Adv. Vet. Sci.* 3, 63–65.

2222. Plotkowiak, J., 1974. Badania ogniskowe nad wystepowaniem inwazji *Isospora hominis* (Railliet i Lucet, 1891) wśród ludnosci wiejskiej województwa koszalinskiego. *Wiad. Parazytol.* 20, 865–872.

2223. Polidori, G., Mura, A., Mughetti, L., Pasquai, P., Principato, M., 1994. Experimental study on ovine *Sarcocystis*. In: Özcel, M.A., Alkan, M.Z. (Eds.), *Abstracts 8th Int. Congress*. Parasitol., Izmir, Turkey, p. 2.

2224. Pond, D.B., Speer, C.A., 1979. *Sarcocystis* in free-ranging herbivores on the National Bison Range. *J. Wildl. Dis.* 15, 51–53.

2225. Pond, D.B., 1982. The pathology and cross-infectivity of *Sarcocystis* spp. in mule deer, white-tailed deer, and elk. Dissertation, University of Montana, USA, pp. 1–39.

2226. Ponse Alcocer, J., 1973. Incidencia de *Sarcocystis* spp. en bovinos nonatos. *Vet. Méx.* 4, 127–130.

2227. Porchet-Henneré, E., Ponchel, G., 1974. Quelques précisions sur l'ultrastructure de *Sarcocystis tenella*: l'architecture du kyste et l'aspect des endozoïtes en microscopie électronique à balayage. *C. R. Acad. Sc. Paris* 279, 1179–1181.

2228. Pötters, U., 1978. Untersuchungen über die Häufigkeit von Kokzidien-Oozysten und- Sporozysten (Eimeriidae, Toxoplasmidae, Sarcocystidae) in den Fäzes von Karnivoren. Inaugural Dissertation, Institute of Parasitology, Hannover, Germany, pp. 1–86.

2229. Pozov, S.A., 1983. Study of pathology and immunity in ovine sarcocystosis. *Diagnostika* 1, 59–62. (In Russian).

2230. Pozov, S.A., 1983. The agglutination reaction in the diagnosis of sarcocystosis in sheep. *Diagnostika* 1, 62–65. (In Russian).

2231. Wikerhauser, T., Džakula, N., Rapić, D., Majurdžic, D., 1981. Istraživanje sarkocistoze goveda i svinja. *Veterinarski Arhiv* 51, 275–282.

2232. Prestwood, A.K., 1982. Sarcosporidiosis in swine. *Vet. Prof. Top. Swine. Ill. Univ. Coop. Ext. Serv.* 8, 5–6.

2233. Proctor, S.J., Barnett, D., Stalheim, O.H.V., Fayer, R., 1976. Pathology of *Sarcocystis fusiformis* in cattle. *10th Ann. Proceedings Am. Assoc. Vet. Lab. Diagnosticians*, pp. 329–336.

2234. Purohit, S.K., D'Souza, B.A., 1973. An investigation into the mode of transmission of sarcosporidiosis. *Br. Vet. J.* 129, 230–235.

2235. Pyziel, A.M., Demiaszkiewicz, A.W., 2009. *Sarcocystis cruzi* (Protozoa: Apicomplexa: Sarcocystiidae) żubra (*Bison bonasus*) w Puszczy Bialowieskiej. *Wiad. Parazytol.* 55, 31–34. (In Polish).

2236. Quinn, S.C., Brooks, R.J., Cawthorn, R.J., 1987. Effects of the protozoan parasite *Sarcocystis rauschorum* on open- field behaviour of its intermediate vertebrate host, *Dicrostonyx richardsoni*. *J. Parasitol.* 73, 265–271.

2237. Quortrup, E.R., Shillinger, J.E., 1941. 3000 wild bird autopsies on Western Lake areas. *J. Am. Vet. Med. Assoc.* 99, 382–387.

2238. Quortrup, E.R., Sudheimer, R.L., 1944. Sarcosporidiosis in swallows. *J. Am. Vet. Med. Assoc.* 104, 29.

2239. Radchenko, A.I., 1986. Light and electron microscopic studies of *Sarcocystis muris* (Sporozoa, Apicomplexa). *Tsitologiya* 28, 1165. (In Russian).

2240. Radchenko, A., Gaibova, G.D., 1993. Possible routes of transport of substances in the muscle cysts of two species of Sarcosporidia (*Sarcocystis*, Apicomplexa, Sporozoa). *Tsitologiya* 35, 134–138. (In Russian).

2241. Rahbari, S., Bazargani, T.T., Rak, H., 1981. Sarcocystosis in the camel in Iran. *J. Vet. Fac. Univ. Tehran* 37, 1–10. (In Arabic).

2242. Raju, N.R., Munro, R., 1978. *Sarcocystis* infection in Fiji. *Aust. Vet. J.* 54, 599.

2243. Ramanujachari, G., Alwar, V.S., 1950. Some observations on sarcosporidiosis of domestic animals in Madras. *Indian Vet. J.* 27, 264–266.

2244. Rao, A.T., Sahoo, S.K., Nayak, B.C., Patnaik, P.K., 1982. A note on a case of generalised sarcosporidiosis in a duckling. *Indian J. Poult. Sci.* 17, 179–180.

2245. Rauscher, L., 1987. Versuche zum Einfluß einer immunsuppressiven Behandlung auf den Verlauf experimenteller *Sarcocystis muris*-Infektionen sowie zur Antikörperbildung in immunisierten BALB/c-Mäusen. Inaugural Dissertation, Institute of Parasitology, Hannover, Germany, pp. 1–74.

2246. Reed, S.M., Saville, W.J.A., 1996. Equine protozoal encephalomyelitis. *Proc. Am. Assoc. Equine Pract.* 42, 75–79.

2247. Reilly, F.K., Chester, W., 2000. Thoughts on the transmission of *Sarcocystis neurona*. *J. Am. Vet. Med. Assoc.* 216, 329.

2248. Retzlaff, N., Weise, E., 1969. Sarkosporidien beim Wasserbüffel (*Bubalus bubalis*) in der Türkei. *Berl. Münch. Tierärztl. Wschr.* 15, 283–286.

2249. Rice, D.A., Calderón, J.E., 1979. Sarcosporidiosis in cattle in El Salvador. *Trop. Anim. Health Prod.* 11, 85–86.

2250. Rickard, M.D., 1973. Current knowledge concerning the life-cycles of the sarcosporidia. *Victorian Veterinary Proceedings (1973–1974)*, 22–23.

2251. Riley, W.A., 1931. Sarcosporidiosis in ducks. *Parasitology* 23, 282–285.

2252. Rioseco, H., Cubillos, V., González, H., Díaz, L., 1976. Sarcosporidiosis en pudúes (*Pudu pudu*, Molina, 1782) primera comunicación en Chile. *Arch. Med. Vet.* 8, 122–123.

2253. Rivera, M.A., Urriola, L., 1987. Sarcocystosis en bovinos. *Rev. Fac. Ciens. Vets. UCV* 34, 1–4.

2254. Rivolta, S., 1878. Della Gregarinosi dei polli e dell'ordinamento delle gregarine e dei psorosperni degli animali domestici. *Fisiol. Patol Animali* 10, 220–235.

2255. Rodríguez-Osorio, M., Gómez-García, V., Tomas-Safont, M.J., Campos-Bueno, M., Mañas Almendros, I., 1978. Estudio comparativo de las tecnicas de digestion pepsica, muscular, inmunodifusion, e inmunofluorescencia indirecta en el diagnóstico de la sarcosporidiasis caprina. *Rev. Ibér. Parasitol.* 38, 793–804.

2256. Rommel, M., 1975. Neue Erkenntnisse zur Biologie der Kokzidien, Toxoplasmen, Sarkosporidien und Besnoitien. *Berl. Münch. Tierärztl. Wschr.* 88, 112–117.

2257. Rommel, M., Krampitz, H.E., 1975. Beiträge zum Lebenszyklus der Frenkelien. I. Die Identität von *Isospora buteonis* aus dem Mäusebussard mit einer Frenkelienart (*F. clethrionomyobuteonis* spec. n.) aus der Rötelmaus. *Berl. Münch. Tierärztl. Wschr.* 88, 338–340.

2258. Rommel, M., 1978. Vergleichende Darstellung der Entwicklungsbiologie der Gattungen *Sarcocystis, Frenkelia, Isospora, Crystoisospora, Hammondia, Toxoplasma* und *Besnoitia*. *Z. Parasitenkd.* 57, 269–283.

2259. Rommel, M., Tiemann, G., Pötters, U., Weller, W., 1982. Untersuchungen zur Epizootiologie von Infektionen mit zystenbildenden Kokzidien (Toxoplasmidae, Sarcocystidae) in Katzen, Schweinen, Rindern und wildlebenden Nagern. *Dtsch. Tierärztl. Wschr.* 89, 57–62.

2260. Rommel, M., 1985. Sarcocystosis of domestic animals and humans. *In Practice* 7, 158–160.

2261. Rommel, M., Schneider, T., 1986. Die Bedeutung des Hofhundes als Überträger gefährlicher Parasiten. *Die Milchpraxis* 24, 103–105.

2262. Wilson, W.D., Fio, L., 1996. Equine protozoal myelitis—a disease in the news. *J. Equine Vet. Sci.* 16, 1–5.

2263. Roscher, B., 1980. Die erarbeitung seerologischer verfahren (IFAT, IHA) zum nachweis einer *Sarcocystis*-infektion beim schaf. Inaugural Dissertation, Universität München, Germany, pp. 1–34.

2264. Rosenberg, B., 1892. Ein Defund von Psorospermien (Sarcosporidien) in Herzmuskel des Menschen. *Zeitschrift fur Hygiene und Infektionkrankeiten* 11, 435–440.

2265. Rosin, C., 1963. Sulla presenza di sarcosporidi nel tessuto specifico dell'apparato di conduzione del cuore. *Zooprofilassi* 11, 325–330.

2266. Rossi, L., Lanfranchi, P., Meneguz, P.G., De Meneghi, D., Guarda, F., 1988. Infezione sperimentale della capra e della pecora con sarcosporidi del muflone e del camoscio. *Parassitologia* 30(Suppl 1), 164–165.

2267. Witonsky, S., Morrow, J.K., Leger, C., Dascanio, J., Buechner-Maxwell, V., Palmer, W., Kline, K., Cook, A., 2004. *Sarcocystis neurona*-specific immunoglobulin G in the serum and cerebrospinal fluid of horses administered *S. neurona* vaccine. *J. Vet. Intern. Med.* 18, 98–103.

2268. Rüedi, D., Hörning, B., 1983. Sarkosporidiennachweis als Zufallsbefund in einem Rehwildbestand im Aargau. *Schweiz. Arch. Tierheilk.* 125, 155–158.

2269. Rynaltovskii, O.V., Dudkina, A.V., Rubina, A.P., 1973. Life cycle of *Sarcocystis* (experimental infection of kittens). *Veterinariya* 11, 71. (In Russian).

2270. Rzepczyk, C.M., 1974. Evidence of rat-snake life cycle for *Sarcocystis*. *Int. J. Parasitol.* 4, 447–449.

2271. Saha, A.K., Srivastava, P.S., Sinha, S.R.P., 1985. Toxic effects of the extracts of *Sarcocystis fusiformis* to laboratory mice. *Indian J. Anim. Sci.* 55, 656–658.

2272. Saha, A.K., Srivastava, P.S., Sinha, S.R.P., Sahai, B.N., 1986. Morphological characteristics of the developmental stages of *Sarcocystis levinei* and *S. fusiformis* in the canine, feline, and bubaline hosts. *Rivista di Parassitologia*. 3, 315–321.

2273. Sahai, B.N., Singh, S.P., Sahay, M.N., Srivastava, P.S., Juyal, P.D., 1983. Role of dogs and cats in the epidemiology of bovine sarcosporidiasis. *Indian J. Anim. Sci.* 53, 84–85.

2274. Saito, M., Yasui, Y., Itagaki, H., 1988. Therapeutic effect of sulfa drugs on canine and swine *Sarcocystis* infections and effect of thermal treatment on cysts and sporocysts. *J. Jpn. Vet. Med. Assoc.* 41, 867–869.

2275. Saito, M., Hachisu, K.I., Itagaki, H., 1988. Effects of *Sarcocystis miescheriana* infections on body weight gains in pigs. *J. Jpn. Vet. Med. Assoc.* 41, 720–722.

2276. Saito, M., 1989. *Sarcocystis* and sarcocystosis. *J. Jpn. Vet. Med. Assoc.* 42, 383–388.

2277. Saleque, A., Juyal, P.D., Bhatia, B.B., 1990. Effect of temperature on the infectivity of *Sarcocystis miescheriana* cysts in pork. *Vet. Parasitol.* 36, 343–346.

2278. Saleque, A., 1990. Studies on some aspects of epidemiology of *Sarcocystis* infection in river buffalo (*Bubalus bubalis*) and domestic pig (*Sus scrofa domestica*) with a reference to eimerian infection in buffalo. *J. Vet. Parasitol.* 4, 75.

2279. Salt, W.R., 1958. *Sarcocystis rileyi* in sage grouse. *J. Parasitol.* 44, 511.

2280. Samaraweera, H.P., Kulasiri, C.S., 1969. Sarcosporidiosis in goats in Ceylon. *Ceylon J. Med. Sci.* 18, 47–50.

2281. Samuel, W.M., Gray, D.R., 1974. Parasitic infection in muskoxen. *J. Wildl. Manage.* 38, 775–782.

2282. Sanft, S., 1990. *In vitro*-Exzystation und Lebendkonservierung von Sarkosporidien-Sporozoiten: ein Versuch zur Erhaltung Definierter *Sarcocystis*-Isolate. Inaugural Dissertation, Freien Universität Berlin, Germany, p. 69.

2283. Saville, W.J.A., Reed, S.M., Granstrom, D.E., Morley, P.S., 1997. Some epidemiologic aspects of equine protozoal myeloencephalitis. Am. Assoc. Equine Prac. *Proc. 15th Am. Coll. Vet. Int. Med. Forum*, Lake Buena Vista, USA, 43, 6–7.

2284. Saville, W.J.A., Reed, S.M., Morley, P.S., 1997. Investigations on the epidemiology of equine protozoal myeloencephalitis (EPM). Am. Assoc. Equine Prac. *Proc. 15th Am. Coll. Vet. Int. Med. Forum*, Lake Buena Vista, USA, 43, 577–480.

2285. Saville, W.J.A., Reed, S.M., Granstrom, D.E., Andrews, F.M., Morley, P.S., 1997. Response of horses exposed to *Sarcocystis neurona* when monitored biweekly. Am. Assoc. Equine Prac. *Proc. 15th Am. Coll. Vet. Int. Med. Forum*, Lake Buena Vista, USA, 43, 8–9.

2286. Schöller, F., 1982. Mikromorphologische Untersuchungen deeer Schizogonie, Endodyogenie, Zystenwandentwicklung und Zystendegeneration bei *Sarcocystis dispersa* in der Maus. Inaugural Dissertation, Ludwig-Maximilians-Universität, Müchen, Germany, pp. 1–77.

2287. Scholtyseck, E., Kepka, O., Piekarski, G., 1970. Die Feinstruktur der Zoiten aus reifen Cysten des sog. M-organismus (= *Frenkelia* spec.). *Z. Parasitenkd.* 33, 252–261.

2288. Scholtyseck, E., Mehlhorn, H., Müller, B.E.G., 1973. Identifikation von Merozoiten der vier cysten-bildenden Coccidien (*Sarcocystis, Toxoplasma, Besnoitia, Frenkelia*) auf Grund feinstruktureller Kriterien. *Z. Parasitenkd.* 42, 185–206.

2289. Scholtyseck, E., Chobotar, B., 1977. Round Table 3—Taxonomy, Cytology and evolution of the sporozoa *sensu lato*. *Proc. Fifth International Congress of Protozoology*, New York, USA, 29–41.

2290. Schulze, K., Zimmermann, T., 1981. Sarkosporidienzysten im Hackfleisch. *Fleischwirtschaft* 61, 1–7.

2291. Schulze, K., 1988. Erkrankungen nach dem Verzehr von massiv mit Sarkosporidien befallenem Rehfleisch. *Fleischwirtschaft* 68, 1139–1140.

2292. Schumacher, J., Corrier, D.E., Craig, T.M., Scrutchfield, W.L., 1981. A possible *Sarcocystis* induced diarrhea in two horses. *Southwestern Veterinarian* 34, 123–124.

2293. Schwerdtfeger, A., 1980. Untersuchungen zur Chemotherapie der *Sarcocystis muris*-infektion. Inaugural Dissertation, Institute of Parasitology, Hannover, Germany, pp. 1–79.

2294. Šebek, Z., 1975. Parasitische Gewebeprotozoen der wildlebenden Kleinsäuger in der Tschechoslowakei. *Folia Parasitol. (Praha)* 22, 111–124.

2295. Sela-Pérez, S., Cruz, M., 1979. Ciclo biológico de *Sarcocystis gigantea* (Railliet 1886) Ashford 1977. *Rev. Ibér. Parasitol.* 39, 601–610.

2296. Sela-Pérez, M.C., Martínez-Fernández, A.R., Arias-Fernández, M.C., Ares-Mazas, M.E., 1982. Ultraestructura de *Sarcocystis muris* (Blanchard, 1885), Labbé, 1889. *Rev. Ibér. Parasitol.* 42, 9–19.

2297. Sela-Pérez, M.C., Martínez-Fernández, A.R., Arias-Fernandez, M.C., Arez-Mazas, M.E., 1982. Estudio de la biologia de los Sarcosporidios encontrados en cerdos y vacas. *Rev. Ibér. Parasitol.* 42, 73–84.

2298. Sela-Pérez, M.C., MartinezFernandez, A.R., Arias Fernandez, M.C., AresMazas, M.E., 1982. Ciclo biológico de *Sarcocystis muris* (Blanchard, 1885) Labbé, 1889. *Rev. Ibér. Parasitol.* 42, 85–93.

2299. Selander, R.K., 1955. The occurence of the parasite *Sarcocystis* in mexican birds. *The American Midland Naturalist* 54, 252–253.

2300. Senaud, J., 1963. Les modalitiés de la multiplication des éléments cellulaires dans les kystes de la Sarcosporidie du mouton (*Sarcocystis tenella* Railliet, 1886). *C. R. Acad. Sci.* 256, 1009–1011.

2301. Senaud, J., Mehlhorn, H., 1974. Etude ultrastructurale des coccidies formant des kystes: *Toxoplasma gondii, Sarcocystis tenella, Besnoitia jellisoni* et *Frankelia* sp. (Sporozoa). II. Mise en évidence de l'ADN et de l'ARN au niveau des ultrastructures. *Annales de la Station Biologique de Besse-en-Chandesse* 9, 111–156.

2302. Seneviratna, P., Atureliya, D., Vijayakumar, R., 1975. The incidence of *Sarcocystis* spp. in cattle and goats in Sri Lanka. *Ceylon Vet. J.* 23, 11–13.

2303. Witonsky, S.G., Gogal Jr., R.M., Duncan Jr., R.B., Norton, H., Ward, D., Lindsay, D.S., 2005. Prevention of meningo/encephalomyelitis due to *Sarcocystis neurona* infection in mice is mediated by CD8 cells. *Int. J. Parasitol.* 35, 113–123.

2304. Shah, H.L., 1995. Sarcocystosis as a zoonosis with special reference to India. *J. Vet. Parasitol.* 9, 57–61.

2305. Sharma, R.K., 1991. Studies on the life-cycle, pathogenesis and treatment of *Sarcocystis capracanis* in goats. *J. Vet. Parasitol.* 5, 61–62.

2306. Sherkov, S.H., Leitch, B., El Rabie, Y., 1976. A survey of sarcosporidia in domestic animals in Jordan. *Egypt J. Vet. Sci.* 13, 45–51.

2307. Shi, L.Z., Zhao, H.Y., 1987. Evaluation of an enzyme immunoassay for the detection of antibodies against *Sarcocystis* spp. in naturally infected cattle in China. *Vet. Parasitol.* 24, 185–194.

2308. Shukla, D.C., Victor, D.A., 1976. The complement fixation test in the diagnosis of sarcosporidiosis in bovines. *Indian Vet. J.* 53, 852–854.

2309. Šibalić, D., 1975. Apparent isolation of *Sarcocystis* sp. from human blood (A preliminary note). *Trans. Roy. Soc. Trop. Med. Hyg.* 69, 148–152.

2310. Šibalić, S., Tomanović, B., Šibalić, D., 1977. Demonstration of sporocysts of *Sarcocystis* spp. in the feces of dogs. *Acta Parasitol. Iugoslav.* 8, 49–54. (In Serbo-Croatian).

2311. Sindilaru, E., 1974. Observations of mixed muscular parasitic infections in pigs. Lucrăi Stiintifice. II. *Zootechnie-Medicina Veterinară, Iasi, Romania*, Institut Agronomic Ion Ionescu de la Brad, pp. 215–220. (In Romanian).

2312. Singh, R.P., Dey, S., Gupta, S.L., 1992. *Sarcocystis* infection in sheep and goats—A review. *Livestock Adviser* 17, 19–23.

2313. Škárková, V., 1986. Histopathological changes in the liver tissue of house mouse and common vole during sarcocystosis. *Folia Parasitol. (Praha)* 33, 115–122.

2314. Skeels, M.R., Nims, L.J., Mann, J.M., 1982. Intestinal parasitosis among Southeast Asian immigrants in New Mexico. *Am. J. Public Health* 72, 57–58.

2315. Skibsted, S., 1945. Om forekomsten af sarkosporidier. *Dyrlaeger* 57, 27–34.

2316. Wobeser, G., Cawthorn, R.J., 1982. Granulomatous myositis in association with *Sarcocystis* sp. infection in wild ducks. *Avian Dis.* 26, 412–418.

2317. Sleiman, E.S., Sweeney, C.R., Habecker, P., 1997. Correlation of antemortem *Sarcocystis neurona* testing with postmortem findings. *Proc. 15th Ann. Vet. Med. Forum*, Lake Buenavista, USA, p. 652.

2318. Smith, T., 1905. The production of sarcosporidiosis in the mouse by feeding infected muscular tissue. *J. Exp. Med.* 6, 1–21.

2319. Smith, T.G., Walliker, D., Ranford-Cartwright, R., 2002. Sexual differentiation and sex determination in the Apicomplexa. *Trends Parasitol.* 18, 315–323.

2320. Solaymani-Mohammadi, S., Petri, W.A., 2006. Zoonotic implications of the swine-transmitted protozoal infections. *Vet. Parasitol.* 140, 189–203.

2321. Speer, C.A., Dubey, J.P., Mattson, D.E., 2000. Comparative development and merozoite production of two isolates of *Sarcocystis neurona* and *Sarcocystis falcatula* in cultured cells. *J. Parasitol.* 86, 25–32.

2322. Spickschen, C., 1990. Untersuchung zum Vorkommen von Sarkosporidien bei Reh-, Rot- und Muffelwild im raum Niedersachsen. Inaugural Dissertation, Institute of Parasitology, Hannover, Germany, pp. 139. (In German).

2323. Spindler, L.A., Zimmerman, H.E., Jaquette, D.S., 1946. Transmission of *Sarcocystis* to swine. *Proc. Helminthol. Soc. Was.* 13, 1–11.

2324. Wong, K.T., Leggett, P.F., Heatley, M., 1993. Apparent absence of *Sarcocystis* infection in human tongue and diaphragm in Northern Ireland. *Trans. R. Soc. Trop. Med. Hyg.* 87, 496.

2325. Srivastava, C.P., Sinha, B.K., Sahai, B.N., 1977. Observations on sarcosporidiasis in cattle and pig. *Indian J. Anim. Health* 14, 105–106.

2326. Srivastava, P.S., Sahai, B.N., Sinha, S.R.P., Saha, A.K., 1985. Some differential features of the developmental cycle of bubaline *Sarcocystis* sp. in canine and feline definitive hosts. *Protistologica* 21, 385–390.

2327. Srivastava, P.S., Sinha, S.R.P., Juyal, P.D., Saha, A.K., 1987. Host resistance and faecal sporocyst excretion in dogs exposed to repeated infection with *Sarcocystis levinei*. *Vet. Res. Commun.* 11, 185–190.

2328. Stiles, C.W., 1893. On the presence of sarcosporidia in birds. *USDA Bureau Anim. Ind. Bull.* 3, 79–89.

2329. Suteu, E., Coman, S., 1973. Nouvelles observations sur le cycle biologiqu de *Sarcocystis fusiformis*. *Bull. Soc. Sci. Vet. Med. Comp. Lyon* 75, 363–367.

2330. Suteu, E., 1981. Sarcosporidioza experimentala la ovine observatii ahatomoclinice si date hematologics. *Simpozionul "Ameliorarea"* 2, 233–240.

2331. Svobodova, M., Votypka, J., Vorisek, P., 1995. *Frenkelia* spp. in the populations of their final and intermediate hosts. *European J. Protistol.* 31, 467.

2332. Tadros, W., Laarman, J.J., van den Eijk, A.A., 1974. The demonstration of antibodies to *Sarcocystis fusiformis* antigen in sera of *Isospora hominis* carriers, using the indirect fluorescence technique. *Z. Parasitenkd.* 43, 221–224.

2333. Takos, M.J., 1957. Notes on sarcosporidia of birds in Panama. *J. Parasitol.* 43, 183–185.

2334. Wong, K.T., Pathmanathan, R., 1994. Ultrastructure of the human skeletal muscle sarcocyst. *J. Parasitol.* 80, 327–330.

2335. Tenter, A.M., Mencke, N., Pein, C., O'Donoghue, P.J., Rommel, M., 1989. Charakterisierung von monoklonalen Antikörpern gegen *Sarcocystis* spp. *Dtsch. Tierärztl. Wschr.* 96, 45–84.

2336. Tenter, A.M., Zimmerman, G.L., Johnson, A.M., 1991. Separation of antigens from *Sarcocystis* species using chromatofocusing. *J. Parasitol.* 77, 727–736.

2337. Tenter, A.M., 1995. Current research on *Sarcocystis* species of domestic animals. *Int. J. Parasitol.* 25, 1311–1330.

2338. Thils, E., Deom, J., Fagard, P., 1960. Considérations sur la sarcosporidiose au Katanga (Congo Belge). *Bulletin de la Societe de Pathologie Exotique* 53, 106–110.

2339. Thornton, H., 1972. Sarcosporidiosis-A review. *Trop. Anim. Health Prod.* 4, 54–57.

2340. Jog, M.M., Marathe, R.R., Goel, S.S., Ranade, S.P., Kunte, K.K., Watve, M.G., 2003. *Sarcocystis* infection in chital (*Axis axis*) and dhole (*Cuon alpinus*) in two Indian protected areas. *Zoos' Print Journal.* 18, 1220–1222.

2341. Todd, K.S., Gallina, A.M., Nelson, W.B., 1975. *Sarcocystis* species in psittaciform birds. *J. Zoo Anim. Med.* 6, 21–24.

2342. Tongson, M.S., Pablo, L.S.M., 1979. Preliminary screening of the possible definitive hosts of *Sarcocystis* sp. found in Philippine buffaloes (*Bubalus bubalis*). *Phil. J. Vet. Med.* 18, 42–54.

2343. Tongson, M.S., Molina, R.M., 1979. Light and electron microscope studies on *Sarcocystis* sp. of the Philippine buffaloes (*Bubalus bubalis*). *Phil. J. Vet. Med.* 18, 16–31.

2344. Tongson, M.S., Calingasan, N.Y., 1980. Demonstration of the developmental stages of *Sarcocystis* sp. of Philippine buffaloes (*Bubalus bubalis*) in the small intestine of dogs. *Phil. J. Vet. Med.* 19, 52–66.

2345. Tongson, M.S., Manuel, M.F., Eduardo, S.L., 1981. Parasitic fauna of goats in the Philippines. *Phil. J. Vet. Anim. Sci.* 20, 1–37.

2346. Torp, C., 1979. Untersuchungen über den einfluβ der *Sarcocystis muris*-Infektion auf die Trächtigkeit und das Aufzuchtsergebnis von NMRI-Mäusen. Inaugural Dissertation, Institute of Parasitology, Hannover, Germany, 1–71.

2347. Tuggle, B.N., Schmeling, S.K., 1982. Parasites of the bald eagle (*Haliaeetus leucocephalus*) of North America. *J. Wildl. Dis.* 18, 501–506.

2348. Wong, K.T., Clarke, G., Pathmanathan, R., Hamilton, P.W., 1994. Light microscopic and three-dimensional morphology of the human muscular sarcocyst. *Parasitol. Res.* 80, 138–140.

2349. Yakimoff, W.L., Sokoloff, I.I., 1934. Die Sarkozysten des Renntieres und des Maral (*Sarcocystis grüneri* n. sp.). *Berl. Münch. Tierärztl. Wschr.* 50, 772–774.

2350. Uggla, A., Hilali, M., Lövgren, K., 1987. Serological responses in *Sarcocystis cruzi* infected calves challenged with *Toxoplasma gondii*. *Res. Vet. Sci.* 43, 127–129.

2351. Uggla, A., Blewett, D.A., 1991. Sarcocystosis (sarcosporidiosis). In: Martin, W.B., Aitken, I.D. (Eds.), *Diseases of Sheep*. Blackwell Scientific Publications, Oxford, UK, pp. 345–348.

2352. Unterholzner, J., Heydorn, A.O., 1982. On the development of dog-transmitted *Sarcocystis* species of sheep and goat. *Zbl. Bakt. Hyg., I. Abt.* 104, 277.

2353. Upton, S.J., McKown, R.D., 1992. The red-tailed hawk, *Buteo jamaicensis*, a native definitive host of *Frenkelia microti* (Apicomplexa) in North America. *J. Wildl. Dis.* 28, 85–90.

2354. Van Hoof, J., Vandenbrande, G., Dedeken, L., 1972. Sarcosporidiose bij slachtrunderen. *Vlaams Diergeneesk. Tijdsch.* 41, 501–514.

2355. Van Thiel, P.H., Van den Berg, C., 1964. Het vóorkomen van kysten bij chronische toxoplasmose en de differentiële diagnostiek tussen *Toxoplasma gondii* en *Sarcocystis lindemanni*. *Ned. Tijdschr. Geneekd.* 108, 696–700.

2356. Vershinin, I.I., 1974. Studies on the cycle of development of *Sarcocystis hirsuta*. *Veterinariya Mosk.* 51, 77–80. (In Russian).

2357. Vershinin, I.I., 1975. *Sarcocystis* in cattle. *Doklady Veesoiuznoi Akademii Sel'skokhoziaistvennykh Nauk*, 30–32. (In Russian).

2358. Vershinin, I.I., 1982. Aberrant forms of intestinal sarcosporidia and *Isospora* from carnivores. *Kishechnye Prosteishie* 1, 35–39.

2359. Vershinin, I.I., Petrenko, V.I., kundryukova, L.I., Vasil'Eva, G.V., 1982. Pathological changes during acute experimental sarcocystosis in calves. *Profilaktika* 4, 64–70. (In Russian).

2360. Vershinin, I.I., 1982. Sarcosporidian species in cattle and sheep in central Ural. *Profilaktika I Lechenie Boleznei Sel'skokoznyaist vennykh Zhivotnykh* 4, 55–63. (In Russian).

2361. Vershinin, I.I., 1983. Sarcosporidia (*Sarcocystis*) and sarcocystosis. Teoreticheskie I Prakticheskie Voprosy Veterinarii, pp. 83–90. (In Russian).

2362. Vietmeyer, C., 1989. Charakterisierung monoklonaler Antikörper gegen *Sarcocystis muris*-Zystozoiten durch IFAT, ELISA, Dot-ELISA und IgG-Subklassenbestimmung. Inaugural Dissertation, Institute of Parasitology, Hannover, Germany, pp. 100.

2363. Vivier, E., 1979. Données nouvelles sur les sporozoaires. Cytologie—cycles—systématique. *Bull. Soc. Zool. France* 104, 345–381.

2364. Vogelsang, E.G., 1938. Contribucion al estudio de la parasitologia animal en Venezuela VIII.-*Sarcocystis iturbei* sp. n. del bovino (*Bos taurus* L.). *Sociedad Venzolana de Ciencias Naturales* 4, 278–280.

2365. Vogelsang, E.G., 2001. Beiträge zur Kenntnis der Parasitenfauna Uruguays. Sarkosporidien bei Vögeln. *Zbl. Bakt. Parasitenkd. Infekionskr. Hyg, I Abt.* 113, 206–208.

2366. Zasukhin, D.N., Gadaev, A., 1978. Life cycle of bovine sarcosporidia. *Parazitologiya* 12, 97–100. (In Russian).

2367. Walter, D., 1979. Untersuchungen über das Vorkommen von Kokzidien (*Sarcocystis, Cystoisospora, Toxoplasma, Hammondia*) bei Katzen in Süddeutschland. Inaugural Dissertation, Universität München, Germany, pp. 1–38.

2368. Warnecke, W., 1983. Histpathologische untersuchungen über den Abbau von Sarkosporidienzysten nach Chemotherapie sowie über die Regeneration und reparation der Muskulatur. Inaugural Dissertation, Institute of Parasitology, Hannover, Germany, pp. 1–97.

2369. Weber, M., 1982. Die verbesserung des Modells *Sarcocystis muris*- Maus zur Prüfung von Chemotherapeutika gegen die Akute Sarkozystose. Inaugural Dissertation, Institute of Parasitology, Hannover, Germany, pp. 1–217.

2370. Weiland, G., Reiter, I., Boch, J., 1982. Möglichkeiten und Grenzen des serologischen Nachweises von Sarkosporidieninfektionen. *Berl. Münch. Tierärztl. Wschr.* 95, 387–392.

2371. Welsch, B.B., 1991. Treatment of equine protozoal myeloencephalitis. *Comp. Cont. Edu. Pract. Vet.* 13, 1599–1602.

2372. Wenzel, R., 1981. Untersuchungen über Sarkosporidien-Infektionen bei Haus- und Wildgeflügel. Inaugural Dissertation, Universität München, Germany, pp. 1–41.

2373. Wettimuny, S.G.d.S., Abeysena, F.A., 1966. Sarcosporidiosis in slaughtered neat cattle and buffaloes in Ceylon. *Ceylon vet. J.* XIV, 2–6.

2374. Weyreter, H., O'Donoghue, P.J., 1982. Untersuchungen zur Immunoserodiagnose der *Sarcocystis*-Infektionen. I. Antikörperbildung bei Maus und Schwein. *Zbl. Bakt. Hyg. I. Abt. Orig. A* 253, 407–416.

2375. White, C.L., Schuler, K.L., Thomas, N.J., Webb, J.L., Saliki, J.T., Ip, H.S., Dubey, J.P., Frame, E.R., 2013. Pathogen exposure and blood chemistry in the Washington, USA population of northern sea otters (*Enhydra lutris kenyoni*). *J. Wildl. Dis.* 49, 887–899.

2376. Whiting, R.H., 1972. Incidence of cysticercosis and sarcosporidiosis in sheep in south eastern Australia. *Aust. Vet. J.* 48, 449–451.

2377. Bentz, B.G., Carter, W.G., Tobin, T., 1999. Diagnosing equine protozoal myeloencephalitis: Complicating factors. *Comp. Cont. Edu. Pract. Vet.* 21, 975–981.

2378. Bowman, D.D., 1991. Equine protozoal myeloencephalitis: History and recent developments. *Equine Practice* 13, 28–33.

2379. Brickson, A.W., Sondhof, A., 1998. Equine protozoal myeloencephalitis. *Iowa St. Univ. Vet.* Fall issue, 83–86.

2380. Collobert, C., Collobert, J.F., 1990. Equine encephalomyelitis caused by a protozoan: A review. *Prat. Vét. Equine* 22, 37–39.

2381. Dame, J.B., Cutler, T.J., Tanhauser, S., Ellison, S., Greiner, E.C., MacKay, R.J., 2000. Equine protozoal myeloencephalitis: Mystery wrapped in enigma. *Parasitol. Res.* 86, 940–943.

2382. Fenger, C.K., 1996. EPM-equine protozoal myeloencephalitis: Early detection means more successful treatment. *Large Animal Veterinarian* 2, 14–20.

2383. Fenger, C.K., 1997. Equine protozoal myeloencephalitis. *Comp. Cont. Edu. Pract. Vet.* 19, 513–523.

2384. Fenger, C.K., 1998. Treatment of equine protozoal myeloencephalitis. *Comp. Cont. Edu. Pract. Vet.* 20, 1154–1157.

2385. Freeman, K.P., Brewer, B., Slusher, S.H., 1989. Membrane filter preparations of cerebrospinal fluid from normal horses and horses with selected neurologic diseases. *Comp. Cont. Edu. Pract. Vet.* 11, 1100–1109.

2386. Goehring, L.S., Sloet van Oldruitenborgh-Oosterbaan, M.M., 2001. Equine protozoal myeloencephalitis (EPM) in the Netherlands? *Tijdschr. Diergeneeskd.* 126, 346–351.

2387. Granstrom, D.E., Reed, S.M., 1994. Equine protozoal myeloencephalitis. *Equine Practice* 16, 23–26.

2388. Granstrom, D.E., McCrillis, S., Wulff-Strobel, C., Baker, C.B., Carter, W., Harkins, J.D., Tobin, T., Saville, W.J., 1997. Diclazuril and equine protozoal myeloencephalitis. *Am. Assoc. Equine Prac. Proc.* 43, 13–14.

2389. Bayani, M., Kalantari, N., Sharbatdaran, M., Abedian, Z., Ghaffari, S., 2014. Demonstration of *Sarcocystis*-like parasites found in peripheral blood. *Int. J. Mol. Cell Med.* 3, 203–206.

2390. Dryburgh, E.L., Marsh, A.E., Dubey, J.P., Howe, D.K., Reed, S.M., Bolten, K.E., Pei, W., Saville, W.J.A., 2015. Sarcocyst development in raccooons (*Procyon lotor*) inoculated with different strains of *Sarcocystis neurona* culture-derived merozoites. *J. Parasitol.* 101. In press.

2391. Dubey, J.P., Hilali, M., van Wilpe, E., Calero-Bernal, R., Verma, S.K., Abbas, I., 2015. A review of sarcocystosis in camels and redescription of *Sarcocystis cameli* and *Sarcocystis ippeni* sarcocysts from the one-humped camel (*Camelus dromedarius*). *Parasitology*. In press.

2392. Dubey, J.P., Moré, G., van Wilpe, E., Calero-Bernal, R., Verma, S.K., Schares, G., 2015. *Sarcocystis rommeli*, n. sp. (Apicomplexa: Sarcocystidae) from cattle (*Bos taurus*) and its differentiation from *Sarcocystis hominis*. *J. Eukaryot. Microbiol.* In press.

2393. Farhang-Pajuh, F., Yakhchali, M., Mardani, K., 2014. Molecular determination of abundance of infection with *Sarcocystis* species in slaughtered sheep of Urmia, Iran. *Vet. Res. Forum* 5, 181–186.

2394. Herd, H.R., Sula, M.M., Starkey, L.A., Panciera, R.J., Johnson, E.M., Snider, T.A., Holbrook, T.C., 2015. *Sarcocystis fayeri*-induced granulomatous and eosinophilic myositis in 2 related horses. *Vet. Pathol.* In press.

2395. Maier, K., Olias, P., Gruber, A.D., Lierz, M., 2015. Toltrazuril does not show an effect against pigeon protozoal encephalitis. *Parasitol. Res.* 114, 1603–1606.

2396. Zuo, Y.X., Yang, Z.Q., 2015. The validity of *Sarcocystis sinensis*. *Zoological Research* 36, 109–111.

2397. Onuma, S.S.M., Melo, A.L.T., Kantek, D.L.Z., Crawshaw-Junior, P.G., Morato, R.G., May-Júnior, J.A., Pacheco, T.A., de Aguiar, D.M., 2014. Exposure of free-living jaguars to *Toxoplasma gondii*, *Neospora caninum* and *Sarcocystis neurona* in the Brazilian Pantanal. *Braz. J. Vet. Parasitol.* 23, 547–553.

2398. Prakas, P., Liaugaudait, S., Kutkiene, L., Sruoga, A., Švažas, S., 2015. Molecular identification of *Sarcocystis rileyi* sporocysts in red foxes (*Vulpes vulpes*) and raccoon dogs (*Nyctereutes procyonoides*) in Lithuania. *Parasitol. Res.* 114, 1671–1676.

2399. Sakran, T.F.A., Adam, A.A.M., Abdel-Ghaffar, F.A., 1995. Light and electron microscopic studies of sarcocysts parasitizing the camel (*Camelus dromedarius*) as intermediate host and the dogs (*Canis familiaris*) as final host. *J. Union Arab Biol.* 4, 27–47.

2400. Salehi, M., Bahari, P., Vatanchian, M., 2014. First molecular identification of *Sarcocystis ovicanis* (Protozoa, Apicomplexa) in the brain of sheep in Iran. *Iranian J. Parasitol.* 9, 286–291.

2401. Shazly, M.A., 2000. Light and electron microscopic studies on sarcocysts infecting the dromedaries (*Camelus dromedarius*) in Saudi Arabia. *Egypt. J. Zool.* 35, 273–285.

2402. Tappe, D., Slesak, G., Pérez-Girón, J.V., Schäfer, J., Langeheinecke, A., Just-Nübling, G., Muñoz-Fontela, C., Püllmann, K., 2015. Human invasive muscular sarcocystosis induces Th2 cytokine polarization and biphasic cytokine changes—An investigation among returning travelers from Tioman Island, Malaysia. *Clin. Vaccine Immunol.* 22, 674–677.

2403. Yeargan, M., Rocha, I.A., Morrow, J., Graves, A., Reed, S.M., Howe, D.K., 2015. A new trivalent SnSAG surface antigen chimera for efficient detection of antibodies against *Sarcocystis neurona* and diagnosis of equine protozoal myeloencephalitis. *J. Vet. Diagn. Invest.* 27, 377–381.

Index

Printed and bound by CPI Group (UK) Ltd, Croydon, CR0 4YY

24/10/2024

01778285-0006